Grain Legume Crops

Grain Legume Crops

Edited by
R.J. Summerfield
and
E.H. Roberts

COLLINS
8 Grafton Street, London W1

Collins Professional and Technical Books
William Collins Sons & Co. Ltd
8 Grafton Street, London W1X 3LA

First published in Great Britain by
Collins Professional and Technical Books 1985

Distributed in the United States of America
by Sheridan House, Inc.

Copyright © R.J. Summerfield and E.H. Roberts 1985

√ISBN 0-00-383037-3

Printed and bound in Great Britain by
Mackays of Chatham, Kent

Contents

Editors' Preface

This book contains eighteen chapters which, collectively, we believe
to represent the most comprehensive and authoritative review yet
available for those grain legume crops of economic significance in
world agriculture. The first four chapters (on Taxonomy, Domestication and Evolution, Biochemical and Nutritional Attributes, and
Rhizobium, Nodulation and Nitrogen Fixation) provide a foundation
for the thirteen subsequent chapters devoted to individual species.
Then, in Chapter 18, we attempt to summarise recent trends in grain
legume research in internationally orientated programmes.

At the outset, we selected six principal objectives. First,
that contributors should be recognised internationally for their
previous achievements in legume science; second, that wherever possible, and irrespective of nationality or location, groups of authors
with mutually complementary expertise should be invited to collaborate
in writing individual chapters; third, that the book should reflect
world progress in applied legume science and, in particular, the
remarkable achievements in recent years in the warmer regions of the
earth which are due in no small part to the efforts of the international agricultural research centres (as we describe briefly in
Chapter 18); fourth, that chapters devoted to individual crops should
follow a common theme (to facilitate interspecific comparisons) but
not at the expense of information particularly pertinent to individual crops; fifth, that so far as possible authors should be allowed
adequate, if not generous, space to accommodate their difficult
tasks; and, finally, that the book should be published as rapidly
as possible - an objective not easily compatible with the facts that

most (73%) of the contributing authors have their homes and places
of work in countries different from that of the editors and that
some co-authors of specific chapters were as many as 15000 km apart!

In April 1982, we invited a total of sixty-four legume scientists working in a total of sixteen different countries to participate
in this venture, sending to each of them our suggestions for a
'skeleton outline' of the subject matter for individual contributions. By July 1982, a total of forty-nine colleagues from
fourteen countries had accepted our invitations (including eighteen
researchers working in those institutes supported by the Consultative
Group for International Agricultural Research which are involved
with grain legume crop improvement), so that within three months
we had achieved our first objective: numerous contributors, internationally recognised as innovative legume scientists, had undertaken
to write for us. One month later, i.e. in August 1982, we had agreements with seventeen legume workers that, as senior authors, they
would act as coordinators and so would collate the contributions of
their co-authors and submit manuscripts to reach the editors by
1 April 1983.

Now, readers might well imagine some of the problems likely to
delay scripts, even if all our contributors had been able to assign
first priority to their tasks. Of course, many could not: some were
soon to change their jobs (sometimes anticipated, other times not!);
and unforeseen events such as vehicle accidents, major surgery,
political instability, the closing of conventional avenues of
communication and, of course, efforts devoted to attracting funds for
research combined to delay receipts of *all* scripts until beyond our
'deadline'. The first manuscript, on peas, arrived on 15 April 1983;
the last one reached us on 10 August 1984. We have included, on
page xvi, the dates of receipt of manuscripts so that readers are
better able to estimate when, during the past two years, surveys of
the literature would have been completed by each group of contributors.

The authors, we believe, are to be congratulated on the depth,
coverage and quality of their contributions within the page limits
imposed upon them. If detail has sometimes been sacrificed for
brevity, we are encouraged by the observation of Professor C.R.W.

Spedding: 'An illustration of a motor car does not enable most
individuals to build one from its 24000 constituent parts, i.e.
given perfect knowledge of a desired outcome, and provision of all
the components required, most individuals would still be quite unable
to make any useful progress.' We do not have a perfect knowledge of
what is desired in those crops considered here and we are probably
still missing a large proportion of those components necessary to
understand them more completely. Nevertheless, desired outcomes
are coming more sharply into focus and the components needed to
achieve these goals *are* becoming better identified and quantified
- as we hope will emerge in the following pages.

R.J. Summerfield and E.H. Roberts,
University of Reading

Acknowledgements

It is a pleasure for the editors to acknowledge the help they have
received during the preparation of this book from Mrs B. Whitlock,
Miss C. Chadwick, Mrs J. Allison and Mr M. Craig, all at the Plant
Environment Laboratory, University of Reading. We are grateful to
the Overseas Development Administration of the United Kingdom
Foreign and Commonwealth Office for their generous financial support,
over many years, of our research and other activities on grain
legume crops.

From the outset, Richard Miles (Editorial Director, Collins
Professional and Technical Books) and, later, Julian Grover
(Commissioning Editor), both with support from Mrs S. Moore (Senior
Processing Editor) and, later, Mrs P. Harrison (Processing Editor)
and Ms S. Kemp (Sub-editor), have always given the editors every
encouragement and welcome advice. We also thank them, on behalf of
all authors as well as ourselves, for their understanding and
tolerance of the repeated delays in bringing this venture to fruition.

Contributors

M.W. Adams, Michigan State University, Department of Crop and Soil Sciences, East Lansing, Michigan 48824, USA.

C.S. Ahn, The Asian Vegetable Research and Development Centre (AVRDC), P.O. Box 42, Shanhua, Tainan 741, Taiwan, Republic of China (*now at:* Jeonnam National University, Department of Horticulture, Kwangju, Jeonnam 500, Republic of Korea).

D.J. Allen, Centro Internacional de Agricultura Tropical (CIAT), Apartado Aéreo 6713, Cali, Colombia, South America.

J.P. Baudoin, Faculté des Sciences Agronomiques de l'État, Phytotechnie des Régions Chaudes, B 5800 Gembloux, Belgium.

G.J. Berry, Victoria Crops Research Institute, Private Bag 260, Natimuk Road, Horsham, Victoria 3400, Australia.

W.D. Beversdorf, University of Guelph, Department of Crop Science, Guelph, Ontario N1G 2W1, Canada.

F.A. Bliss, University of Wisconsin, Department of Horticulture, 1575 Linden Drive, Madison, Wisconsin 53706, USA.

D.A. Bond, Plant Breeding Institute, Maris Lane, Trumpington, Cambridge CB2 2LQ, England.

R. Bressani, Instituto de Nutrición de Centro América y Panamá (INCAP), Carretera Roosevelt zona 11, Apartado Postal 1188, Guatemala, Central America.

A.H. Bunting, University of Reading, Q7/8, No. 4 Earley Gate, White-knights Road, Reading RG6 2AR, England.

D.E. Byth, University of Queensland, Department of Agriculture, St Lucia, Brisbane 4067, Queensland, Australia.

D.P. Coyne, University of Nebraska, Department of Horticulture, 386 Plant Sciences Building, East Campus, Lincoln, Nebraska 68583-0724, USA.

J.I. Cubero, Escuela Técnica Superior de Ingenieros Agrónomos, Departmento de Génetica, Apartado 3048, Córdoba, Spain.

D.R. Davies, John Innes Institute, Colney Lane, Norwich NR4 7UH, England.

J.H.C. Davis, Centro Internacional de Agricultura Tropical (CIAT), Apartado Aéreo 6713, Cali, Colombia, South America.

T.C.K. Dawkins, University of Nottingham, School of Agriculture, Department of Agriculture and Horticulture, Sutton Bonington, Loughborough LE12 5RD, England.

G.E. Eagleton, University of Western Australia, Institute of Agriculture, Agronomy Department, Nedlands 6009, Western Australia.

W. Erskine, International Centre for Agricultural Research in the Dry Areas (ICARDA), PO Box 5466, Aleppo, Syria.

P. Farrington, CSIRO Division of Groundwater Research, Institute of Energy and Earth Resources, Private Bag, PO, Wembley 6014, Western Australia.

C.A. Francis, University of Nebraska, Department of Agronomy, Lincoln, Nebraska 68583-0724, USA.

R.W. Gibbons, International Crops Research Institute for the Semi-Arid Tropics (ICRISAT), Patancheru PO, Andhra Pradesh 502 324, India.

P.H. Graham, University of Minnesota, Department of Soil Science, 1529 Gortner Avenue, St Paul, Minnesota 55108, USA.

G.C. Hawtin, International Centre for Agricultural Research in the Dry Areas (ICARDA), PO Box 5466, Aleppo, Syria (now at: International Development Research Centre (IDRC), 5990 Iona Drive, University of British Columbia, Vancouver, British Columbia V6T IL4, Canada).

M.C. Heath, University of Nottingham, School of Agriculture, Department of Agriculture and Horticulture, Sutton Bonington, Loughborough LE12 5RD, England.

R. Hidalgo, Centro Internacional de Agricultura Tropical (CIAT), Apartado Aéreo 6713, Cali, Colombia, South America.

D.J. Hume, University of Guelph, Department of Crop Science, Guelph, Ontario N1G 2W1, Canada.

T. Hymowitz, University of Illinois, Department of Agronomy, 1102 South Goodwin Avenue, Urbana, Illinois 61801, USA.

T.N. Khan, Department of Agriculture, Jarrah Road, South Perth 6151, Western Australia.

D.A. Lawes, Welsh Plant Breeding Station, Plas Gogerddan, Aberystwyth, Dyfed SY23 3EB, Wales.

Contributors

R.J. Lawn, CSIRO Division of Tropical Crops and Pastures, The Cunningham Laboratory, 306 Carmody Road, St Lucia, Queensland 4067, Australia.

J.M. Lyman, The Rockefeller Foundation, Agricultural Sciences, 1133 Avenue of the Americas, New York 10036, USA.

F.R. Minchin, Grassland Research Institute, Plant and Crop Physiology Department, Hurley, Maidenhead SL6 5LR, England.

F.J. Muehlbauer, USDA-ARS-Western Region, Grain Legume Genetics and Physiology Research, 215 Johnson Hall, Washington State University, Pullman, Washington 99164, USA.

G. Norton, University of Nottingham, School of Agriculture, Department of Applied Biochemistry and Food Science, Sutton Bonington, Loughborough LE12 5RD, England.

J.S. Pate, University of Western Australia, Department of Botany, Nedlands 6009, Western Australia.

R.M. Polhill, Royal Botanic Gardens, Kew, Richmond, Surrey TW9 3AB, England.

E.H. Roberts, University of Reading, Department of Agriculture and Horticulture, Earley Gate, Reading RG6 2AT, England.

M.C. Saxena, International Centre for Agricultural Research in the Dry Areas (ICARDA), PO Box 5466, Aleppo, Syria.

S. Shanmugasundaram, Asian Vegetable Research and Development Centre (AVRDC), PO Box 42, Shanhua, Tainan 741, Taiwan, Republic of China.

J. Smartt, Department of Biology, Building 44, The University, Southampton SO9 5NH, England.

J.B. Smithson, International Crops Research Institute for the Semi-Arid Tropics (ICRISAT), Patancheru PO, Andhra Pradesh 502 324, India (*now at*: 5 Byemoor Avenue, Great Ayton, Cleveland TS9 6JP, England).

J.I. Sprent, Department of Biological Sciences, The University, Dundee DD1 4HN, Scotland.

W.M. Steele, International Institute of Tropical Agriculture (IITA), Oyo Road, PMB 5320, Ibadan, Nigeria.

J.H. Stephens, International Centre for Agricultural Research in the Dry Areas (ICARDA), PO Box 5466, Aleppo, Syria.

R.J. Summerfield, University of Reading, Department of Agriculture and Horticulture, Plant Environment Laboratory, Shinfield Grange, Cutbush Lane, Shinfield, Reading RG2 9AD, England.

J.A. Thompson, International Crops Research Institute for the Semi-Arid Tropics (ICRISAT), Patancheru PO, Andhra Pradesh 502 324, India (*now at*: Horticultural Research Station, PO Box 720, Gosford, New South Wales, Australia).

L.J.G. van der Maesen, International Crops Research Institute for the Semi-Arid Tropics (ICRISAT), Patancheru PO, Andhra Pradesh 502 342, India (*now at*: Vakgroep Plantentaxonomie, Postbus 8010, 6700 ED Wageningen, The Netherlands).

E.S. Wallis, University of Queensland, Department of Agriculture, St Lucia, Brisbane 4067, Queensland, Australia.

P.C. Whiteman, University of Queensland, Department of Agriculture, St Lucia, Brisbane 4067, Queensland, Australia.

W. Williams, University of Reading, Department of Agricultural Botany, Plant Science Laboratories, Whiteknights, Reading RG6 2AS, England.

J.C. Wynne, North Carolina State University, Department of Crop Science, Box 7629, Raleigh, North Carolina 27695-7629, USA.

RECEIPT OF CONTRIBUTIONS

Subject matter	*Chapter*	*Date of receipt of manuscript by the editors*
Taxonomy	1	14 June 1983
Domestication and evolution	2	26 April 1983
Biochemistry and nutrition	3	3 February 1984
Symbiosis with *Rhizobium*	4	23 June 1983
Pea	5	15 April 1983
Faba bean	6	12 September 1983
Lentil	7	8 August 1983
Chickpea	8	1 November 1983
Soyabean	9	3 October 1983
Common bean	10	31 May 1984
Lima bean	11	27 May 1983
Cowpea	12	10 August 1984
Mung bean	13	2 June 1983
Winged bean	14	26 April 1983
Pigeonpea	15	2 Decmber 1983
Lupin	16	12 September 1983
Groundnut	17	2 July 1984
International research	18	16 June 1984

PART I: General Relations

1 Taxonomy of grain legumes

R.M. Polhill and L.J.G. van der Maesen

INTRODUCTION

The taxonomy of grain legumes is relatively uncomplicated
compared to that of cereals, brassicas and some other
groups of plants because, in general, only limited gene
pools have been available for selection and subsequent
plant breeding. Then again, intergeneric legume hybrids
are not known in nature and artificial crosses attempting
to create them are seldom, if ever, successful [64].
Indeed, the genetic barriers between species and species
groups are often substantial [86,87]. The classification
of interfertile species and infraspecific variants is
inherently more difficult and the taxonomic situation in
grain legumes is not exceptional. In some instances the
available information would now seem to justify updating
of the taxonomic framework.

TRIBES

Most grain legumes belong to two tribes, the Vicieae and
the Phaseoleae. The Vicieae and many Phaseoleae have the
unusual combination of hypogeal germination and a herbaceous habit. Hypogeal cotyledons with large food reserves
are characteristic of forest plants and the twining pro-
pensity of Phaseoleae and the tendrils of Vicieae may
represent residual traits of that ancestral association.

The interchange between epigeal and hypogeal germination involves major reorganisations of seed structure, composition and behaviour such that many families are restricted to one system or the other [102]. In legumes, both systems occur widely and, as shown in *Phaseolus* for example, the interchange is under fairly simple genetic control [30]. The most advanced tribes have tended to become predominantly herbaceous and most are obligately epigeal with small hard seeds produced in large numbers, with long viability and staggered opportunistic germination.

In classifications of the last century, Phaseoleae and Vicieae have tended to be associated simply because of their comparable seed features [6,13]. In modern systems they are considered as end points of very different lines of evolution [75]. The Phaseoleae are well distributed through the tropics and subtropics, notably heterogenous and without any very obvious indications of their origin [53]. In contrast, the Vicieae belong to a small group of tribes, all of which are centred in the Sino-Indian region and appear to have spread extensively into temperate regions, mainly of the northern hemisphere, during the late Tertiary era. *Cicer* used to be included in the Vicieae, but an overall assessment reveals significant features shared with the Trifolieae. The way these characters are combined suggests that *Cicer* is a separate offshoot and best placed in a special tribe Cicereae [49]. The contiguity between these north temperate tribes is also evident.

The most notable grain legume genus outside these tribes is *Arachis*, which is morphologically very distinct. Once included in the Hedysareae, which was an artificial assemblage of papilionoid genera with segmented fruits, it is now placed in Aeschynomeneae subtribe Stylosanthinae, but with the proviso that a separate tribe Stylosantheae may become justified in due course [83]. The tribe and subtribe have their main centre of development in Central and South America. *Arachis* is unique in its tribe because

4

of obligately geocarpic fruits and an exceptional seed structure which is functionally linked to that fruiting habit.

The Mediterranean species of *Lupinus* are a recent addition to the list of major grain legumes. *Lupinus* is traditionally included in the Genisteae, but represents a distinctive herbaceous line of evolution in that tribe and warrants at least subtribal rank [9]. Like the Vicieae, the Genisteae seem to have spread from the Sino-Indian region, but are developed from a quite different part of the subfamily. The seeds of the Mediterranean species of *Lupinus* are consistently larger than those of American species of the genus, but the range of variation is not exceptional in this group of tribes.

The main implication of these collective remarks is that the grain legume tribes have had a long and separate evolutionary history. Consequently, many functional syndromes differ significantly in detail and often quite consistently between the tribes, e.g. the structure and metabolism of root nodules (see Chapter 4), stem anatomy, chemical defence mechanisms, and pollen and seed chemistry. More than once the differences found between peas and beans have provided the starting point for taxonomic surveys that have moulded current concepts of tribal evolution in the Papilionoideae [75].

Whether to subdivide the legumes into families or subfamilies is not entirely agreed. The connections between the three main groups seem fairly well demonstrated, a view expressed by one family, the Leguminosae (or Fabaceae), with three subfamilies, viz. Caesalpinioideae, Mimosoideae and Papilionoideae (or Faboideae). The alternative of three families is expressed as Caesalpiniaceae, Mimosaceae and Papilionaceae (or Fabaceae). The optional names Fabaceae and Faboideae became fashionable in North America at the beginning of this century and are equally correct.

GENERA

The only recent generic changes affecting grain legumes have been in the Phaseoleae. This is a large tropical tribe to which many species have been added since the outlines of a generic classification were laid down in the last century. *Phaseolus* and *Vigna* have been redefined since 1970 with a rather narrowly delimited American genus *Phaseolus*, and a widespread genus *Vigna*, subdivided into a number of subgenera and sections [62,98]. The reorganisation is based on an extensive survey of morphological, chemical, cytological and palynological features; it is clearly displayed as well by taximetric methods using multivariate discriminant analysis to weight taxonomically important characters, and it seems resilient to criticism [61]. *Voandzeia*, the Bambara groundnut, is also included in *Vigna* [100].

The subtribe Cajaninae is distinctive in the Phaseoleae and has two generic groups, the Rhynchosiastrae with one- or two-seeded fruits, and the Cajanastrae with two, three or more seeds [3,4]. Traditionally, *Cajanus* has comprised only two species, *C. cajan* and *C. kerstingii* but there is good evidence that the species of *Atylosia* should now be added to this genus. *Atylosia* has a large aril (strophiole), but other supposed differences, namely the angle of grooves on pod walls between the seeds and persistence of the corolla, are no longer valid. A morphological link between the former genera is indicated by *Atylosia cajanifolia* in particular. The contiguity between the two genera is emphasised by similar chromosome complements ($2n = 22$), closely homologous seed protein profiles, viable hybrids and a continuity of morphological forms. A formal revision is in preparation in which the genera will be united [96]. There are links with *Dunbaria* and *Rhynchosia* also, but more flattened fruits and only one or two ovules, respectively, provide clear differences from *Cajanus*. The recent analysis of Australian Cajaninae

6

indicates other implications for the generic realignment [74].

SPECIES

To integrate a sensible classification of crop plants with that appropriate for their wild relatives does present problems. Part of the difficulty relates to the different type of variation that arises from artificial selection of crop characteristics and the loss of a natural population structure, and part depends on the sort of material on which judgements have to be made.

In practice, it is relatively straightforward to revise groups of wild plants from specimens accumulated in herbaria and gardens, and the morphological, geographical and ecological patterns apparent from such studies can be verified to a greater or lesser extent by revisiting original populations. To piece together the evolution and domestication of crops is altogether more difficult (see Chapter 2). Some obsolete crop plant classifications have tended to persist, either because the materials needed for reassessment are not readily available or have often been lost from agricultural or other research institutes.

The extent to which the morphological species, based on visible differences related to patterns of geography, ecology and phenology, coincides with the biological species, determined by genetic barriers, is generally uncertain. As the gene pools of crop plants are explored for breeding purposes new information becomes available for taxonomic consideration. Grain legume gene pools seem to be fairly restricted in most cases. The concept of primary, secondary and tertiary gene pools advocated by Harlan and de Wet [38] can be integrated fairly easily, therefore, into a conventional framework. The primary gene pool includes the crop plant and freely interfertile spontaneous races, which may be appropriately ranked as subspecies; the secondary pool includes related species

7

that hybridise with difficulty; and the tertiary pool
includes the limits of gene transfer by artificial means.

The morphological hallmark of differentiation within
the primary gene pool is similarity among attributes apart
from the crop characteristics. In grain legumes, crop
characteristics generally involve such features as larger
plants, erect growth, loss of seed toxins, indehiscent
fruits and larger seeds (see Chapter 2). If these predictable differences are associated with successful inter-
fertility there seems no reason to exclude the wild and
weedy forms from the crop species. Thus, *Glycine max* and
G. soja, the cultivated and wild forms of the soyabean,
seem appropriately ranked as subspecies, as suggested by
Harlan and de Wet [38], but not yet generally adopted
[43,44]. Similarly, the recent discovery of a wild chickpea, *Cicer reticulatum* [54], interfertile with *C. arietinum*, and differing by a less erect habit, less persistent
fruits and rougher seeds, is more appropriately ranked as
a subspecies [65]. Subspecific ranking on similar criteria
has been recommended recently for the lentil, where formerly *Lens orientalis* and *L. nigricans* were considered
specifically distinct from *L. culinaris* [19,104].

In these three cases the broadened specific concept
coincides with the primary gene pool, fertile hybrids not
being produced with any other species. However, the genetic situation can be more complicated, and taxonomic
guesses before the variations and breeding patterns are
well known can lead to periods of some confusion. It has
been suggested, for example, that *Vigna mungo* and *V. radiata* are independently derived from *V. radiata* var. *sublobata*, and the genetic barriers are considerable between
all three [46]. If that is the case then var. *sublobata* is
better treated as a separate species. Others have argued
that material known as '*Phaseolus sublobatus*' includes
two quite separate elements, one of which is the wild form
of the mung bean, *Vigna radiata* var. *sublobata*, the other
a wild form of the black gram, *V. mungo* var. *silvestris* [60].

8

INFRASPECIFIC VARIANTS

The main subdivisions of the species, the subspecies and
variety, are used in various senses. Sometimes they indicate just a decreasing scale of morphological differentiation, sometimes the subspecies is used for geographical,
ecological or cytological segregates, and the variety for
sporadic differences, and sometimes they are regarded as
virtually synonymous (the varietal concept was exclusively
preferred in much American literature for a considerable
part of this century). The subspecies is also used to distinguish cultigens from wild relatives in the primary gene
pool, and, formerly, the variety was used for what are now
treated as forms (e.g. colour variants) and cultivars.
Some latitude is actually desirable to allow for the range
of possible biological situations, and to provide a provisional framework before the breeding behaviour is known.

The *Code of Botanical Nomenclature* insists only on the
hierarchical sequence with respect to subspecies, variety
and form, but allows for other categories to be inter-
calated if desired [88]. The *Botanical Code of Nomenclature for Cultivated Plants* encourages the use of the cultivar for any clearly recognisable assemblage of cultivated
plants that reproduces its distinguishing features [12].
The cultivar or cultivar group may be co-extensive with
the botanical subspecies or variety and designed for practical convenience. When there are two or more previously
published epithets in Latin form, the epithet that best
preserves established usage should be chosen without
regard to the botanical category in which the epithet was
published, or to priority.

The classification of *Vicia faba* has been overburdened
by traditional nomenclature, and the cultivar groups
Major, Equina and Minor might serve best for cultivated
forms defined on seed size (continued use of var. *major*
for var. *faba* is contrary to the rules of botanical
nomenclature). The status of *V. faba* subsp. *paucijuga*

9

seems to need further elucidation [36,66], and subspecific
rank may be appropriate for the present, even if only as
an 'ignorance' category. Cubero [21] suggests that Paucijuga be added simply as a fourth cultivar or land-race
group (and see Chapter 6).

The classification of *Arachis hypogaea* is a case where
the cultivar groups have a recognisable geographical basis
and the botanical categories have a shorter and simpler
history, so that the two systems have been more easily
equated [34]. Predictably, the botanical system should
become redundant.

The classification of *Vigna unguiculata* is still being
refined from several species at the beginning of the 1970s
to five subspecies [98], then into a series of subspecies
for spontaneous plants and cultivar groups, viz. Unguiculata, Biflora and Sesquipedalis for the cowpea, catjang
and yard-long bean, respectively [62]. The name Biflora
was chosen on the grounds of botanical priority, and, as
indicated above, that is inappropriate and should be sup-
planted, presumably by cultivar group Catjang. There seems
to be continuous variation between cultivar groups Unguiculata and Catjang, whereas Sesquipedalis has a distinctive morphology, origin, use and distribution; nonetheless
it seems to have been developed from cultivar group
Unguiculata [89].

The above examples are cited to indicate the flexibility of the taxonomic framework, which can be extended as
necessary. It is not for us to specify the appropriate
system for taxa we have not studied personally, but it
seems that some sensible recent updating could be adopted
more widely without fear of nomenclatural instability.

We accept the axiom that an infraspecific classification aims to represent the geographical and ecological
variation and the history of domestication and subsequent
breeding. Any expedient to group cultivars into groups or
higher categories on the basis of artificial criteria,
such as the discreteness of alternative character-

states [42,73], is liable to be internationally unacceptable, as well as scientifically retrograde.

We conclude this chapter with a more formal synopsis
of the grain legumes based on a number of recent handbooks
and floras, together with some further data from contributors to this volume [1,25,26,35,41,47,77,85,94,103]. Here,
authorities for species and infraspecific variants are
cited as these are often required by editors of scientific
publications. That imposition is part of an emphasis on
the critical need for meaningful names and correctly
identified experimental material. The importance of vou-
cher specimens and reliable naming cannot be over-
emphasised, but the citation of authorities, except for
homonyms or other confused names, has little relevance
beyond strictly taxonomic papers. Authorities for genera
and tribes can be found readily elsewhere [75].

ARACHIS HYPOGAEA (AESCHYNOMENEAE)

Specific classification

Pending the long-awaited revision by Krapovickas and
Gregory many unpublished names are in circulation since
the wild relatives play an important role in the improvement of the groundnut (*Arachis hypogaea* L.). Up to early
1983 twenty-two species had been formally described and
another forty to fifty awaited publication [34,78,81]. All
are native to South America and some are restricted to
small areas. Tribal placement has been discussed by Rudd
[83]. *Arachis* is placed in Stylosanthinae, a subtribe of
Aeschynomeneae. Other genera of Stylosanthinae are not
considered to be close relatives at present.

Gregory and Krapovickas have proposed the following
infrageneric classification [34] but the names have yet to
be validated; some of the names are not correctly formed:

11

Sect. 1 *Arachis*   Series *Annuae, Perennes* and
                           Amphiploides
Sect. 2 *Erectoides*   Series *Trifoliolatae, Tetrafolio-*
                           latae and *Procumbensae*
Sect. 3 *Caulorhizae*
Sect. 4 *Rhizomatosae*   Series *Prorhizomatosae* and
                           Eurhizomatosae
Sect. 5 *Extranervosae*
Sect. 6 *Ambinervosae*
Sect. 7 *Triseminale*

The sections are associated with the drainage systems of
the South American river basins. Most series have a chromosome number of $2n = 20$, but series *Amphiploides* (including *Arachis hypogaea*) and *Eurhizomatosae* have $4n = 40$
chromosomes. Genetic isolation is not marked between most
species of section *Arachis*. Several possible antecedents
of *A. hypogaea* have been collected in recent years from
the East Andean region [35].

Infraspecific classification

There have been many more or less detailed, formal and
informal classifications below the species level. Earlier
systems were based on growth habit, later ones on branching patterns and positions of the fruiting branches. A
recent system [81] is summarised in Table 1.1.

Table 1.1 Infraspecific classification of *Arachis* [81]

Subspecies	Variety	Cultivar groups*
hypogaea	*hypogaea* (*A. africana, A. procumbens*)	Braziliano Virginia
	hirsuta Kohler (*A. asiatica*)	Peruano p.p.
fastigiata Waldron	*fastigiata*	Peruano p.p. Valencia
	vulgaris Harz	Spanish

*Designated 'types'.

12

CAJANUS CAJAN (PHASEOLEAE-CAJANINAE)

Specific classification

As mentioned earlier, *Cajanus* will be expanded to include
Atylosia in a forthcoming revision [96]. The combined
genus has its main distribution in South and South-East
Asia and in Australia [82,96]. The nearest relative to the
pigeonpea, *Cajanus cajan* (L.) Millsp. (*C. indicus* Spren-
gel), is the West African *C. kerstingii* Harms, the only
other species considered congeneric in recent times. Among
the species formerly included in *Atylosia*, *A. cajanifolia*
Haines from India is perhaps morphologically the most
similar and produces fairly fertile hybrids with *C. cajan*.
The pigeonpea, now cultivated pantropically, reached
Africa from its origin in India before 4000 BP and was
introduced into the Americas, mainly from Africa, with
European settlement.

Infraspecific classification

Although some varieties can be distinguished in wild
species of Cajaninae, infraspecific taxa are difficult to
discern in pigeonpea. With a large germplasm at hand the
continuum is more evident than ever. *C. bicolor* DC. and
C. flavus DC., the first species described in *Cajanus* to
cover the 'Arhar' and 'Tur' groups of North and Central to
South India, respectively, with yellow-red and plain yellow flowers, are conspecific and not distinguishable even
as varieties [90,103]. The differential characters are
governed by only a few genes [24,96]. Furthermore, the
phenotype of pigeonpea is greatly influenced by responses
to day length and temperature, which can entirely change
plant appearance, so that a classification into meaningful
cultivar groups is difficult [90]. Maturity criteria
should be adhered to and ICRISAT (see Chapter 18) maintains the maturity groups shown in Table 1.2 [96].

13

Table 1.2 Maturity groups of pigeonpea [96]

Group	Days to 50% flowering	Reference cultivars
0	<60	Pant A-3
I	61 - 70	Prabhat; Pant A-2
II	71 - 80	UPAS-120; Baigani
III	81 - 90	Pusa Ageti; T-21
IV	91 - 100	ICP-6
V	101 - 110	No. 148; BDN-1
VI	111 - 130	ICP-1; 6997; ST-1; C-11
VII	131 - 140	HY-3C; ICP-7035
VIII	141 - 160	ICP-7065; 7086
IX	>160	NP(WR)-15; Gwalior-3; NP-69

Elsewhere the same groups can be discerned, albeit with modified flowering dates. The two major usage groups are pigeonpeas for dry seed (pulse) and 'dhal' preparation, and cultivars as green vegetables [77].

CICER ARIETINUM (CICEREAE)

Specific classification

A recent monograph of *Cicer* presents descriptions and keys for most species [92]. A treatment for *Flora Iranica* is also available [93]. Until recently *Cicer* was placed in the Vicieae but, as noted earlier, now seems to warrant a monotypic tribe [15,45,48,49,50,58,59]. It comprises nine species of annual herbs and thirty-five species of small perennial shrubs [92,94], including four newly described ones [16,23,54,95]. It is an Old World genus with many species in two major regions: Iran and Turkey; Central Asia and Afghanistan.

The closest relatives of the chickpea, *C. arietinum* L., are the annual species of section *Monocicer* M. Pop.: *C. bijugum* K.H. Rech., *C. echinospermum* P.H. Davis, *C. reticulatum* Ladiz. (morphologically the closest [54] and now regarded as a subspecies [65]), *C. judaicum* Boiss., *C. pinnatifidum* Jaub. & Spach, *C. cuneatum* Hochst. ex A. Rich. and *C. yamashitae* Kitam.. Section *Chamaecicer* M. Pop.

also contains an annual species, *C. chorassanicum* (Bunge)
M. Pop.. Viable and fertile hybrids involving *C. arietinum*
have so far been obtained only with *C. reticulatum*, indicating a possible role as an ancestor, and ranking as a
subspecies of *C. arietinum* has been proposed [65]. When
crossed with *C. echinospermum* the hybrids were viable but
infertile.

In contrast to most annual species, the perennial
species (sections *Polycicer* M. Pop. and *Acanthocicer* M.
Pop.[92]) are extremely difficult to cultivate and study
outside their natural habitats. Characters such as
drought resistance, high-altitude adaptation and multi-
seeded fruits are present so that introgression into
chickpea cultivars could be valuable. Disease resistance
has been established in several species (see Chapter 8)
and incorporation of blight resistance proved possible
from a source in subsp. *reticulatum*.

Further taxonomic studies are needed in *Cicer* to elu-
cidate the position of rare species such as *C. aphyllum*
Boiss., collected only once, *C. paucijugum* Nevski, *C. bal-
caricum* Galushko and several other Central Asian species
[92], *C. heterophyllum* Contandr. *et al.* [16] and *C. multi-
jugum* van der Maesen. An urgent but difficult task is to
obtain all species in a live collection for hybridisation
studies and seed distribution, as many species are physically and/or politically inaccessible [94].

Most *Cicer* species have chromosome complements of
$2n = 16$, but *C. pungens* Boiss., *C. microphyllum* Benth. and
possibly *C. anatolicum* Alef. have $2n = 14$, while some
cells of *C. montbretii* Jaub. & Spach were triploid (24
chromosomes) [50,92]. Detailed cytological studies have
revealed degrees of homology between *C. arietinum*, *C.
reticulatum* and *C. echinospermum* whereas *C. bijugum* was
more distant [50].

15

Infraspecific classification

After earlier varietal classifications by Jaubert and
Spach, and Alefeld, the Russian botanists working with
Vavilov's collections then produced detailed formal infraspecific classifications. The systems of Prosorova, and
Popova and Pavlova, inconsistent for Afghan and Turkish
chickpeas, were followed in modified form by Popova's work
in the *Flora of Cultivated Crops of the USSR* [92]. In
India, eighty-four 'Pusa' types were distinguished by Shaw
and Khan [92]. These classifications tend to attract few
followers, except for Koinov who even extended Popova's
work. Probably the taxa described by Popova need to be
reconsidered, regarding subspecies as races, the proles as
subraces, and the varieties as cultivar groups [38,92].
Infraspecific taxonomy is in need of further elucidation,
if not simplification.

Eight characters were evaluated by numerical taxonomy
at two locations in India for 5477 chickpea lines [67].
The analysis showed six clusters, each covering material
from one or more arbitrary geographical groups. This
exercise usefully pointed out, at least for the characters
studied, the strong natural and human selective pressures
that exist in conjunction with geographical isolation.

Moreno and Cubero [65] reduced the wild *C. reticulatum*
to *C. arietinum* subsp. *reticulatum* (Lad.) Cubero & Moreno,
and proposed two races, *macrosperma* and *microsperma*, in
the cultivated subsp. *arietinum*.

GLYCINE MAX (PHASEOLEAE-GLYCININAE)

Specific classification

Glycine now comprises two subgenera, subgen. *Glycine* with
seven perennial species in Australia and the Pacific, and
subgen. *Soja* (Moench) F.J. Herm., with the soyabean, *G.
max* (L.) Merrill, and its wild and weedy forms in East

Asia [40,43,44,69,97]. The breeding barriers between subgenus *Soja* and subgenus *Glycine* are strong and hybrids
have been obtained between *G. max* and *G. tomentella* Hayata
only by the use of *in vitro* ovule culture [70,105]. The
species of subgen. *Glycine*, which, like the soyabean, are
self-compatible, are rarely interfertile and sterility
barriers exist between isolated populations of certain
species [71]. *G. wightii* (R. Grah. ex Wight & Arn.) Verdc.
(*G. javanica* auctt.) is now referred to *Neonotonia* as *N.
wightii* (R. Grah. ex Wight & Arn.) Lackey [52].

   G. max differs from the wild *G. soja* Sieb. & Zucc. (*G.
ussuriensis* Regel & Maack) only by features expected of a
domesticate and there are few if any cytogenetic barriers
to hybridisation [43]. *G. soja* seems best treated as a
subspecies of the crop plant, *G. max* subsp. *soja* (Sieb. &
Zucc.) Ohashi [38,51,72]. The weedy *G. gracilis* may have
evolved as a consequence of outcrossing between *G. max* and
G. soja since it is found wherever the other two overlap
in their distribution [43]; if it seems sufficiently distinct to warrant a name, then subsp. × *gracilis* could be
validated.

Infraspecific classification

There are many soyabean cultivars and thirteen maturity
groups are now recognised in the USA [8,39]. Adaptation
ranges from very early, compact cultivars for the short
summers and long days of extreme latitudes in areas such
as North China, southern Canada and the northern United
States (Groups 000-I), to the late and tall cultivars of
the southern United States (Groups VIII-X).

LENS CULINARIS (VICIEAE)

Specific classification

Lens is a small and principally Mediterranean genus

comprising until recently only five species [14,50]. Barulina [2] wrote a classical monograph of the genus in 1930.
A recent overview is given by Cubero [19] who also dis-
cusses the close relation to *Vicia*. *Lens culinaris* Medic.,
the cultivated lentil, has a primary gene pool that
includes what were formerly regarded as two closely-
related wild species, namely *L. orientalis* (Boiss.)
Handel-Mazzetti and *L. nigricans* (M. Bieb.) Godron [19,
55]. Principal component analysis and interspecific hybridisation support the more recent arrangement of *L. orientalis* as a subspecies of *L. culinaris* [104], namely subsp.
orientalis (Boiss.) Ponert. Hybrids between *L. culinaris*
and *L. nigricans* are also normal but meiosis is more
irregular. Despite differences in karyotype and pollen
exine morphology, *L. nigricans* is now also considered to
be a subspecies of *L. culinaris*, as subsp. *nigricans* (M.
Bieb.) Thell. *L. ervoides* (Brign.) Grande, which is native
in tropical Africa as well as around the Mediterranean,
does not produce viable hybrids with *L. culinaris* and
retains its specific status [19]. *L. cyanea* (Boiss. &
Hohen.) Alef. is considered a separate species in *Flora
Iranica* [14] and as a synonym of *L. orientalis* elsewhere.

   L. montbretii (Fisch. & Mey.) P.H. Davis & Plitm., a
rare species from Turkey and Iran, has been transferred to
Vicia recently [56].

   Lens generally has a chromosome complement of $2n = 14$,
although some counts of $2n = 12$ exist [19,50].

Infraspecific classification

Barulina [2] classified *Lens culinaris* into two sub-species,
macrosperma and *microsperma*, which can be considered as
races. In *microsperma* six groups (greges) were distinguished, now reconcilable as subraces. The many formal
varieties described within the grex are at the most cultivar groups [19].

LUPINUS (GENISTEAE-LUPININAE)

Specific classification

Lupinus is a taxonomically isolated genus currently
referred to a special subtribe of Genisteae [9]. The genus
comprises about 200 species in two major groups; a cluster
of twelve large-seeded species around the Mediterranean
south to East Africa, and the rest principally in the
Andes and Rockies the length of the Americas and extending
thinly into eastern South America. Gladstones has revised
the Mediterranean species and clarified earlier miscon-
ceptions [31].

The main commercial species [32] belong to the
Mediterranean group, viz. *L. albus* L., *L. luteus* L. and
L. angustifolius L. The South American species *L. mutabi-
lis* Sweet is a small-scale subsistence crop in the Andes.
The Mediterranean species *L. cosentinii* Guss. is now being
developed as a crop in Australia.

Like other genera in the Genisteae, *Lupinus* has a complicated aneuploid and polyploid series, with different
chromosome numbers reported for all the cultivated species
[9,32]. The Mediterranean species are separated by strong
genetic barriers whereas the American species tend to
introgress readily in nature.

Infraspecific classification

L. albus has a wild variant in Greece, Turkey and the
Aegean, which Gladstones [32] prefers to call var. *graecus*
(Boiss. & Spruner) Gladst., though subsp. *graecus* (Boiss.
& Spruner) Franco & Silva is also available [25,41]. Wild
forms are detectable too in the other cultivated species
but do not warrant taxonomic status. Commercial breeding
dates from only about fifty years ago, and extensive cultivation began only within the last thirty years [32].

PHASEOLUS LUNATUS (PHASEOLEAE-PHASEOLINAE)

Specific classification

Phaseolus has been reduced in recent years to about fifty species in the New World [53,63,98]. The principal grain legume species, *P. coccineus* L., *P. vulgaris* L. and the Lima bean, *P. lunatus* L. belong to section *Phaseolus*. *P. lunatus* is rather distantly related to the other two species [62]. Morphological comparisons and crossing experiments [5,57] suggest that *P. ritensis* Jones, *P. polystachyus* (L.) Britt., Sterns & Pogg. and *P. pedicellatus* Benth. are closer to *P. lunatus* than *P. acutifolius* A. Gray and *P. filiformis* Benth.. *P. pachyrrhizoides* Harms from Peru is also very similar to *P. lunatus* var. *silvester* Baudet [62] and might be ancestral to at least part of the Lima bean complex.

Infraspecific classification

The infraspecific classification has been reviewed by Baudet, who suggests [3] the following arrangement:

Var. *silvester* Baudet   Wild forms
Var. *lunatus*   Cultivated forms, with three cultivar groups: Potato, Sieva and Big Lima

Geography

Lima beans are native to tropical America and may have originated separately in Central America (Mexico, Guatemala) and in South America (Peru) [3]. Var. *silvester* occurs both in Central America, principally from southern Mexico to Guatemala, and in South America in northern Argentina.

20

PHASEOLUS VULGARIS AND *P. COCCINEUS* (PHASEOLEAE-PHASEOLINAE)

Specific classification

As indicated above, the common (haricot) bean, *P. vulgaris* L., and the scarlet runner bean, *P. coccineus* L., form part of a complex relatively distant from *P. lunatus* in section *Phaseolus* [62]. In recent years a number of supposedly related species have been reduced to infraspecific variants. Apart from the Mexican *P. glabellus* Piper, which is doubtfully distinct from *P. coccineus*, this leaves the Tepary bean, *P. acutifolius* A. Gray, which forms infertile hybrids with both *P. coccineus* and *P. vulgaris*, as the apparently nearest species, and thereafter *P. filiformis* Benth., *P. angustissimus* A. Gray and *P. wrightii* A. Gray, all from southern North American and Central America. The other species of the section are more similar to *P. lunatus*.

Hybrids can be produced relatively easily between *P. vulgaris* and *P. coccineus* when the former is the female parent. Hybrids between *P. vulgaris* and *P. acutifolius* have been obtained and facilitated by embryo culture, but the F_1 hybrids are completely sterile [28].

Infraspecific classification

The interrelations between wild and cultivated forms of *Phaseolus* are still being actively explored, and the existing classifications are provisional and somewhat controversial.

Maréchal and co-workers [62] propose the following scheme:

P. vulgaris   var. *vulgaris*
                    var. *aborigineus* (Burk.) Baudet
P. coccineus   subsp. *coccineus*

21

subsp. *obovallatus* (Schlecht.) Maréchal,
Mascherpa & Stainier
subsp. *formosus* (Kunth) Maréchal, Mascherpa & Stainier
subsp. *polyanthus* (Greenman) Maréchal,
Mascherpa & Stainier

Var. *aborigineus* (or subsp. *aborigineus* Burk.) is considered to be the wild (or at least the weedy) form of
P. vulgaris in southern South America, but it is uncertain
whether the Central American wild forms are distinguishable, and if so whether ancestral or one of two sources
for the development of the common haricot. Recent comparisons support a separation of the mesoamerican wild
form and support the supposition of two centres of origin
for the cultivated forms (Vanderborght, pers. comm.).

In *P. coccineus*, Maréchal and co-workers [62] found
subsp. *obovallatus* from Mexico and Costa Rica most similar
and freely interfertile with subsp. *coccineus*. They regard
it as the most probable ancestral form. Subsp. *formosus*,
from the same region, is similar and has also been suggested as ancestral. Subsp. *polyanthus* from the high
plateaux of Central America south to Colombia, shares a
number of features with *P. vulgaris*, but the conformation
of hybrids suggests a closer connection with *P. coccineus*.
Indeed, *P. coccineus* subsp. *darwinianus* Hernandez &
Miranda is thought to be synonymous with subsp. *polyanthus*.
Smartt [87] suggests that subsp. *polyanthus* and the Central
American wild forms, which he refers to as *P. flavescens*
Piper, may represent a distinct taxon, more readily distinguished from *P. coccineus* than *P. vulgaris*, and best
ranked as a species. Further subspecies are likely to be
described as taxonomic studies progress (Baudoin, pers.
comm.).

PISUM SATIVUM (VICIEAE)

Specific classification

There are now only two species recognised in *Pisum*, the
cultivated pea, *P. sativum* L., and the eastern Mediterranean *P. fulvum* Sibth. & Smith [50]. Other epithets are
either reduced to subspecific rank or to synonymy, even
in *Lathyrus* [94]. The perennial wild *P. formosus* (Stev.)
Alef. is now placed in a separate monotypic genus, as
Vavilovia formosa (Stev.) Fed. [50].

    P. sativum is sympatric with wild races in the eastern
Mediterranean, but spontaneous hybridisation is apparently
very rare despite the ability to produce partly fertile F_1
hybrids when crossed artificially [7].

Infraspecific classification

The authors of the *Flora Iranica* [80], *Flora of Turkey*
[25] and *Flora of Iraq* [91] agree on the infraspecific
taxa in *P. sativum* as follows:

Subsp. *sativum*
                              var. *arvense* (L.) Poir. (the
                                          Field pea)
                              var. *sativum* (the Garden pea)

Subsp. *elatius* (M. Bieb.) Aschers. & Graebn.
                              var. *elatius* (M. Bieb.) Alef.
                              var. *pumilio* Meikle (*P. humile*
                                          Boiss. & Noë)
                              var. *brevipedunculatum* Davis &
                                          Meikle

Ben-Ze'ev and Zohary [7] treat *P. sativum-humile-elatius*
as a species aggregate of wild forms with essentially
homologous chromosomes and practically interfertile. An
earlier classification is that of Govorov [33]. Westphal
[103] distinguishes the cultivar groups Abyssinicum and

Sativum, the first with var. *elatius* forma *abyssinicum* (A. Br.) Gams as a synonym.

PSOPHOCARPUS TETRAGONOLOBUS (PHASEOLEAE-PHASEOLINAE)

Specific classification

Psophocarpus appears to be a relatively isolated genus in Phaseoleae subtribe Phaseolinae [53,62]. The genus comprises nine species probably native only to Africa and Madagascar, though the winged bean, *P. tetragonolobus* (L.) DC., may have originated in Asia from an unknown progenitor. The genus has been revised recently by Verdcourt and Halliday [101].

    P. tetragonolobus is related to *P. scandens* (Endl.) Verdc., a native of Africa and Madagascar, and also to the tropical African species *P. grandiflorus* Wilczek and *P. palustris* Desv. More wild material is needed for breeding experiments, but it has not been possible to cross *P. tetragonolobus* with *P. scandens* [27].

Infraspecific classification

The commercial production of the winged bean has been developed only recently and criteria for infraspecific classification have not yet been determined [68].

VICIA FABA (VICIEAE)

Specific classification

The taxonomic position of the faba bean, *Vicia faba* L., has met with considerable debate. Usually, *V. faba* is classified in section *Faba* (Miller) Ledeb., together with the closest morphological relatives, *V. narbonensis* L., *V. johannis* Tamamschjan, *V. galilaea* Plitm. & Zoh., and possibly *V. bithynica* L. and *V. hyaeniscyamus* Mout. [25,94].

24

Some authorities have ascribed the species to a genus of its own: *Faba*, with the species *F. vulgaris* Moench. [14].

Vicia comprises 148 species. Other sections apart from *Faba* are sect. *Vicia* (which includes many of the culti-vated vetches), sect. *Ervilia* (Link) W.D.J. Koch, sect. *Cracca* Dumort, and several more [14,59]. Detailed infra-generic accounts have been given by Kupicha [48].

Information from numerical taxonomy, breeding behav-iour and chemical studies (e.g. nuclear DNA quantity) have placed the faba bean very remote from the species of the *V. narbonensis* complex [84] despite the close morphologi-cal similarity [17,18,21,36,37].

The stipules, stipular nectary, gynoecium, inflores-cence, leaflet size and habit agree with the *V. narbonen-sis* group, but *V. faba* has distinctive fruits with spongy partitions, seeds and leaflet mucro. The uniform testa configuration of several *Vicia* species has been reported in detail [58]. *V. faba* has more reduced papillae than other *Vicia* species. Like the seed coat, the pollen data are equivocal. *V. narbonensis* and *V. johannis* are minor grain legume crops in Portugal and Turkey.

No hybrids have been obtained between *V. faba* and other species of *Vicia* despite the many attempts and the desirability of introducing new characteristics. *V. faba* has $2n = 12$ whereas $2n = 14$ is the most common number in *Vicia* [49].

Infraspecific classification

Muratova [66] classified the cultivars of *V. faba* into two subspecies, subsp. *eufaba* (now *faba*) with three varieties, var. *major* Harz (now *faba*, the broad bean, with large flattened seeds), *equina* Pers. (the horse bean, with medium-sized seeds), and *minor* Beck (the tick bean, with small rounded seeds), and subsp. *paucijuga*, with less than four leaflets per leaf. The varieties were further sub-divided into subvarieties.

25

Hanelt [36] proposed the following scheme:

Subsp. *minor*	var. *minor*	subvar. *minor*
		subvar. *tenuis*
Subsp. *faba*	var. *equina*	subvar. *equina*
		subvar. *reticulata*
	var. *faba*	subvar. *faba*
		subvar. *clausa*

Subsp. *paucijuga* was treated as a race of subsp. *minor*.

All the groups are freely interfertile [20,21,22] and, as suggested earlier, current practice might be expressed more simply by recognising just four cultivar groups, viz. Major, Equina, Minor and Paucijuga [20].

VIGNA RADIATA AND *VIGNA MUNGO* (PHASEOLEAE-PHASEOLINAE)

Specific classification

Vigna has now been broadened to include about 150 species, mostly in Africa, with twenty-two species in India, sixteen in South-East Asia and a few in America and Australia. The genus comprises seven subgenera and a number of sections [62, 98, 99]. The mung bean (green gram) and black gram (urd), *V. radiata* (L.) Wilczek and *V. mungo* (L.) Hepper, respectively, belong to subgen. *Ceratotropis* (Piper) Verdc. and probably both were domesticated in India.

Subgen. *Ceratotropis* forms a discrete group of about seventeen species largely confined to Asia and the Pacific. Most interspecific hybrids are sterile, largely due to embryo abortion, but gene transfer is possible in most cases if the appropriate female parent is used, or by embryo culture [46]. A detailed account of the subgenus would be valuable.

Verdcourt [98] broadened the concept of *V. radiata* to include two varieties, var. *glabra* (Roxb.) Verdc. and var.

sublobata (Roxb.) Verdc..Var. *glabra*, an amphidiploid, has
since been raised to specific rank as *V. glabrescens*
Maréchal, Mascherpa & Stainier [62].

Infraspecific classification

V. radiata var. *sublobata* is accepted as the wild form of
the mung bean and a probable progenitor. Recent accessions
and crossing experiments have led to the suggestion that
other forms of this taxon might be ancestral also to the
black gram, *V. mungo*. On that basis Jain and Mehra [46]
conclude that var. *sublobata* should be recognised as a
separate species. Lukoki, Maréchal and Otoul [60], on the
other hand, find that wild material of the black gram is
clearly distinguishable from var. *sublobata* at specific
level on the basis of morphological, biochemical and
breeding evidence, and treat it as *V. mungo* var. *silvestris* Lukoki, Maréchal & Otoul (for further discussion
see Chapter 13). Maréchal and co-workers [62] also
recognise *V. radiata* var. *setulosa* (Dalzell) Ohwi &
Ohashi as a further eastern variant, which overlaps to
some extent the distribution and morphological features of
var. *sublobata*.

   Older classifications in the Indian subcontinent distinguished var. *typica* Prain, var. *aurea* Prain, var. *grandis* Prain and var. *bruneus* Bose in the mung bean [10],
while in the black gram the subvarieties *viridis* Bose and
niger Bose were proposed [11], mainly on seed characters.
In line with other Indian pulses, mung and black gram were
further subdivided into forty and twenty-five 'types',
respectively, based on seed and pod wall characteristics,
plant habit and maturity.

VIGNA UNGUICULATA (PHASEOLEAE-PHASEOLINAE)

Specific classification

The cowpea, *Vigna unguiculata* (L.) Walp., belongs to *Vigna* subgen. *Catjang* (DC.) Verdc. [98]. The species has been broadened to include the catjang and yard-long bean, first as subspecies [98] and then as cultivar groups [62]. Verdcourt [98] provisionally accepted several other species in the section, but subsequently [62] most of these have been reduced to variants of *V. unguiculata*, apart from *V. nervosa* Markotter from southern Africa and the poorly known *V. brachycalyx* Bak. from Madagascar.

Attempts to cross the cowpea with other species of *Vigna* have so far failed [89], and *V. nervosa* may be the only species likely to hybridise successfully.

Infraspecific variation

The following classification has been proposed by Maréchal and co-workers [62]:

Subsp. *unguiculata*   cultivar groups Unguiculata
                                        Biflora
                                        Sesquipedalis
Subsp. *dekindtiana* (Harms) Verdc. var. *dekindtiana* (Harms)
                                        Verdc.
                                 var. *mensensis*
                                        (Schweinf.) Maré-
                                        chal, Mascherpa &
                                        Stainier
                                 var. *pubescens* (Wilczek)
                                        Maréchal, Mascherpa
                                        & Stainier
                                 var. *prostrata* (E. Mey.)
                                        Verdc.
Subsp. *tenuis* (E. Mey.) Maréchal, Mascherpa & Stainier

28

Subsp. *stenophylla* (Harv.) Maréchal, Mascherpa
& Stainier

As indicated earlier, Biflora is an inappropriate name for
the Catjang as a cultivar group. Steele and Mehra [89]
have no difficulty in accepting Catjang as a cultivar
group of the cowpea, but feel there may be a case for
maintaining subsp. *sesquipedalis* (L.) Verdc. for the yard-
long bean because it is distinct in morphology, origin,
use and distribution.

It has been suggested [89] that the cowpea was domesticated in Africa in Neolithic times from the wild/weed
complex of subspp.*mensensis* and *dekindtiana*. Cultivar
group Catjang and, less certainly, cultivar group Sesquipedalis were probably developed from cultivar group
Unguiculata in India [29,79,89].

The diversity of cultivated forms is reflected in the
IITA (see Chapter 18) *Cowpea Germplasm Catalogue* [76],
including various agronomic and botanical characters.

VERNACULAR NAMES

There are numerous common (vernacular, trivial) names of
crop plants. Widely distributed plants have names in many
languages, and even several in any one language. Grain
legumes are no exception [47] and any further scanning of
the literature will add more. For instance, the pigeonpea
has about 335 common names [86] although some are mere
orthographic variants. The difference of spelling between
British and American English needs no explanation, but it
tends to cause some confusion. The English 'soyabean'
versus the American 'soybean' is one example, and there
are many more.

Caution should be exercised in the use of vernacular
names; indeed, the worldwide acceptance of Latin names
stems from the confusion that would otherwise result.
Nonetheless, conventions may be changed by mutual

agreement to solve part of the problem. Perhaps a list of preferred common names could be devised by an appropriate international body? Suggestions for such a list are given in Table 1.3; the choice is somewhat arbitrary but with the aim of avoiding confusion.

A few examples will illustrate some of the problems. Even the words 'pea' and 'bean' are equivocal, despite the common knowledge of the shapes of peas and beans. Chickpeas are called garbanzo beans in America, which is quite a misnomer apart from garbanzo, the Spanish vernacular. Cowpeas are often bean-shaped. *Vicia faba* has as English vernaculars (among others) field bean and broad bean (*V. faba* var. *faba*) and is internationally confused. By recent agreement (see Chapter 18) the crop is now called faba bean The Portuguese name 'Frijoles de Costa' (beans of the coast) does not apply to beans, but to cowpeas (in Brazil). Some vernaculars of the cowpea are also used for *Phaseolus vulgaris* (common bean), *P. lunatus* (Lima bean), *Lathyrus sativus* (grass pea), *Vigna unguiculata* cultivar group Sesquipedalis (asparagus bean) and *Vigna subterranea* (Bambara groundnut). Many more of these examples are given elsewhere [47]. In India, chickpeas are called gram or Bengal gram, names more frequently used in the past than the name chickpea, which has become favoured only as a result of international usage.

Most scientific journals rightly insist on the mention of Latin names, and if regional or local vernaculars are included, the use of the commonest vernacular should also become mandatory in addition to the Latin.

SHORT GLOSSARY

genotype: (a) The genetic make-up of an organism (all dominant and recessive genes).
             (b) A group of organisms with the same genetic make-up.

accession: Sample in a gene bank; its number is unique

Table 1.3
A suggested list of preferred vernacular names for grain legumes

Latin	English	French	Spanish	German
Arachis hypogaea	groundnut	arachide	mani	gemeine erdnuss
Cajanus cajan	pigeonpea	pois d'Angole	guisante de paloma	kichererbse
Cicer arietinum	chickpea	pois chiche	garbanzo	hornkraut
Glycine max	soyabean	soja	soja	sojabohne
Lens culinaris	lentil	lentille	lenteja	linse
Lupinus spp.	lupins	lupines	lupino	lupines
Phaseolus lunatus	Lima bean	haricot de Lima	haba Lima	Limabohme
Phaseolus vulgaris	common bean	haricot commun	frijol	fisole
Pisum sativum	pea	pois	guisante	erbse
Psophocarpus tetragonolobus	winged bean	pois allé	sesquidilla	goabohne
Vicia faba	faba bean	fève	haba comun	ackerbohne
Vigna mungo	black gram	ambérique	judia de urd	urdbohne
Vigna radiata	mung bean	haricot doré	judia de mungo	mungobohne
Vigna unguiculata	cowpea	pois vache	chicaro de vaca	kuhbohne

31

and not re-used in cases where the sample is
lost.

landrace: A traditional cultivar not subject to scientific selection; often a population or a mixture of closely related genotypes.

variety: Botanical variety; taxonomic level below the
rank of subspecies, above the level of cultivar (see previous comments on infraspecific
variants).

cultivar: Cultivated variety; an assemblage of cultivated plants clearly recognisable from other
cultivars of the same species by structural
features and performance. Either a clone, a
self- or open-pollinated cultivar, a synthetic or a hybrid.

LITERATURE CITED

1. Allen, D.J. (1983), *The Pathology of Tropical Food Legumes: Disease Resistance in Crop Improvement.* London: John Wiley.
2. Barulina, H. (1930), [Lentils of the USSR and of other countries]. *Bulletin of Applied Botany, Genetics and Plant-Breeding* Suppl. 40, 265-304.
3. Baudet, J.C. (1977), *Bulletin de la Société Royale de Botanique de Belgique* 110, 65-76.
4. Baudet, J.C. (1978), *Bulletin du Jardin Botanique Nationale de Belgique* 48, 183-220.
5. Baudoin, J.P. (1981), *Bulletin Recherches Agronomiques de Gembloux* 16, 273-86.
6. Bentham, G. (1865), in Bentham, G. and Hooker, J.D. (*eds*) *Genera Plantarum 1.* London: Reeve, 434-600.
7. Ben-Ze'ev, N. and Zohary, D. (1973), *Israel Journal of Botany* 22, 73-91.
8. Bernard, R.L. and Nelson, R.L. (1982), *Soybean Genetics Newsletter* 9, 15-16.
9. Bisby, F.A. (1981), in Polhill, R.M. and Raven, P.H. (*eds*) *Advances in Legume Systematics.* London: Kew, 409-25.
10. Bose, R.D. (1932), *Indian Journal of Agricultural Science 2,* 607-24.
11. Bose, R.D. (1932), *Indian Journal of Agricultural Science 2,* 625-37.
12. Brickell, C.D. (*ed.*) (1980), 'International Code of Nomenclature for Cultivated Plants'. *Regnum Vegetabile* 104, 1-32.
13. Candolle, A.P. de (1825), *Mémoires sur la Famille de Légumineuses.* Paris: Belin.
14. Chrtkova-Zertova, A. (1979), *'Vicia, Faba, Lens'. Flora Iranica*

140, 57-61. Austria: Graz.
15. Clarke, G.C.S. and Kupicha, F.K. (1976), *Botanical Journal of the Linnean Society* 72, 35-44.
16. Contandriopoulos, J., Pamuksuoglu, A. and Quézel, P. (1972), *Biologia Gallo-Hellenica* 4, 3.
17. Cubero, J.I. (1973), *Theoretical and Applied Genetics* 43, 59-65.
18. Cubero, J.I. (1974), *Theoretical and Applied Genetics* 45, 47-51.
19. Cubero, J.I. (1981), in Webb, C. and Hawtin, G.C. (*eds*) *Lentils*. Farnham Royal: Commonwealth Agricultural Bureau, 15-38.
20. Cubero, J.I. (1981), in Hawtin, G. and Webb, C. (*eds*) *Faba Bean Improvement*. The Hague: Martinus Nijhoff, 91-108.
21. Cubero, J.I. (in press), in Whitcombe, J. (*ed.*) *Food Legume Germplasm Training Manual*. Aleppo: ICARDA.
22. Cubero, J.I. and Suso, M.-J. (1981), *Kulturpflanze* 29, 137-45.
23. Czerepanov, S.K. (1981), *Plantae Vasculares URSS*. Leningrad: Nauka.
24. Dahiya, B.S. (1980), *An Annotated Bibliography of Pigeonpea 1900-1977*. Patancheru: International Crops Research Institute for the Semi-Arid Tropics.
25. Davis, P.H. (1970), *Flora of Turkey 3*. Edinburgh University Press.
26. Duke, J.A. (1981), *Handbook of Legumes of World Economic Importance*. New York and London: Plenum Press.
27. Erskine, W. (1978), in *The Winged Bean*. Manila: Philippine Council for Agriculture and Resources Research, 29-35.
28. Evans, A.M. (1980), in Summerfield, R.J. and Bunting, A.H. (*eds*) *Advances in Legume Science*. London: Kew, 337-47.
29. Faris, D.G. (1965), *Canadian Journal of Genetics and Cytology* 7, 433-52.
30. Gates, R.R. (1952), *Botanical Gazette* 113, 151-7.
31. Gladstones, J.S. (1974), Lupins of the Mediterranean Region and Africa, *Western Australian Department of Agriculture Technical Bulletin* 26.
32. Gladstones, J.S. (1980), in Summerfield, R.J. and Bunting, A.H. (*eds*) *Advances in Legume Science*. London: Kew, 603-11.
33. Govorov, L.I. (1928), [The peas of Afghanistan]. *Bulletin of Applied Botany, Genetics and Plant Breeding* 19, 497-522.
34. Gregory, W.C., Krapovickas, A. and Gregory, M.P. (1980), in Summerfield, R.J. and Bunting, A.H. (*eds*) *Advances in Legume Science*. London: Kew, 469-81.
35. Halevy, A.H. (in press), *A Handbook of Flowering*. Florida: CRC Press.
36. Hanelt, P. (1972), *Kulturpflanz* 20, 75-128.
37. Hanelt, P., Schafer, H., Schultze-Motel, J. (1972), *Kulturpflanze* 20, 263-75.
38. Harlan, J.R. and de Wet, J.M.J. (1971), *Taxon* 20, 509-17.
39. Hartwig, E.E. (1973), in Caldwell, B.E. (*ed.*) *Soybeans: Improvement, Production and Uses*. Wisconsin, Madison: American Society of Agronomy, 187-210.
40. Hermann, F.J. (1962), 'A revision of the genus *Glycine* and its immediate allies'. *USDA Agricultural Research Service Technical Bulletin* 1268, 82 pp.
41. Heywood, V.H. and Ball, P.W. (1968), 'Leguminosae'. In *Flora Europaea 2*. England: Cambridge University Press.
42. Higgins, J., Evans, J.L. and Reed, P.J. (1981), *Journal of the*

National Institute of Agricultural Botany 15: 480-87.
43. Hymowitz, T. and Newell, C.A. (1980), in Summerfield, R.J. and
    Bunting, A.H. (*eds*) *Advances in Legume Science*. London: Kew,
    251-64.
44. Hymowitz, T. and Newell, C.A. (1981), *Economic Botany* 35, 272-88.
45. Ingham, J.L. (1981), in Polhill, R.M. and Raven, P.H. (*eds*)
    Advances in Legume Systematics. London: Kew, 599-626.
46. Jain, H.K. and Mehra, K.L. (1980), in Summerfield, R.J. and
    Bunting, A.H. (*eds*) *Advances in Legume Science*. London: Kew,
    459-68.
47. Kay, D.E. (1979), *Food Legumes*.Tropical Products Institute, Crop
    and Product Digest No. 3, London: Tropical Products Institute.
48. Kupicha, F.K. (1976), *Notes from the Royal Botanic Garden,
    Edinburgh* 34, 287-326.
49. Kupicha, F.K. (1977), *Botanical Journal of the Linnean Society*
    74, 131-62.
50. Kupicha, F.K. (1981), in Polhill, R.M. and Raven, P.H. (*eds*)
    Advances in Legume Systematics. London: Kew, 377-82.
51. Lackey, J.A. (1977), *Botanical Journal of the Linnean Society*
    74, 163-78.
52. Lackey, J.A. (1977), *Phytologia* 37, 209-12.
53. Lackey, J.A. (1981), in Polhill, R.M. and Raven, P.H. (*eds*)
    Advances in Legume Systematics. London: Kew, 301-27.
54. Ladizinsky, G. (1975), *Notes from the Royal Botanic Gardens,
    Edinburgh* 34, 201-2.
55. Ladizinsky, G. (1979), *Lens* 6, 24.
56. Ladizinsky, G. and Sakar, D. (1982), *Botanical Journal of the
    Linnean Society* 85, 209-12.
57. Le Marchand, G., Maréchal, R. and Baudet, J.C. (1976), *Bulletin
    des Recherches Agronomiques de Gembloux* 11, 183-200.
58. Lersten, N.R. and Gunn, C.R. (1981), *Systematic Botany* 6, 223-
    30.
59. Lersten, N.R. and Gunn, C.R. (1982), 'Testa characteristics in
    tribe Vicieae, with notes about tribes Abreae, Cicereae and
    Trifolieae (Fabaceae)'. *USDA Agricultural Research Service
    Technical Bulletin* 1667, 1-40.
60. Lukoki, L., Maréchal, R. and Otoul, E. (1980), *Bulletin du
    Jardin Botanique National de Belgique* 50, 385-91.
61. Maréchal, R. (1982), *Taxon* 31, 280-83.
62. Maréchal, R., Mascherpa, J.M. and Stainier, F. (1978), *Boissiera*
    28, 1-273.
63. Maréchal, R., Mascherpa, J.M. and Stainier, F. (1978), *Taxon* 27,
    199-202.
64. McComb, J.A. (1975), *Euphytica* 24: 497-502.
65. Moreno, M.-T. and Cubero, J.I. (1978), *Euphytica* 27, 465-85.
66. Muratova, V. (1931), 'Common beans (*Vicia faba*)'. *Bulletin of
    Applied Botany, Genetics and Plant Breeding*, Suppl. 50, 1-285.
67. Narayan, R.K.J. and Macefield, A.J. (1976), *Theoretical and
    Applied Genetics* 47, 179-87.
68. National Academy of Sciences (1981), *The Winged Bean. A High
    Protein Crop for the Tropics*. Washington DC: National Academy
    of Sciences.
69. Newell, C.A. and Hymowitz, T. (1980), *Brittonia* 32, 63-9.
70. Newell, C.A. and Hymowitz, T. (1982), *Crop Science* 22, 1062-5.
71. Newell, C.A. and Hymowitz, T. (1983), *American Journal of*

Botany 70, 334-48.
72. Ohashi, H. (1982), *Journal of Japanese Botany* 57, 29-30.
73. Parker, P.F. (1978), in Street, H.E. (*ed.*) *Essays in Plant Taxonomy*. London: Academic Press, 97-124.
74. Pedley, L. (1981), *Austrobaileya* 1, 376-9.
75. Polhill, R.M. and Raven, P.H. (*eds*) (1981), *Advances in Legume Systematics*. London: Kew.
76. Porter, W.M., Rachie, K.O., Little, T.M., Luse, R.A., Rawal, K., Singh, S.R., Wien, H.C. and Williams, R.J. (1974), *Cowpea Germplasm Catalogue*. Ibadan, Nigeria: International Institute of Tropical Agriculture.
77. Purseglove, J.W. (1968), *Tropical Crops, Dicotyledons 1*. London: Longman.
78. Ramanatha Rao, V. (in preparation), 'Botany'. In *Groundnut*. Delhi: Indian Council of Agricultural Research.
79. Rawal, K.M. (1975), *Euphytica* 24, 699-707.
80. Rechinger, K.H. (1979), *Flora Iranica* 140, 83-6.
81. Resslar, P.M. (1980), *Euphytica* 29, 813-17.
82. Reynolds, S.T. and Pedley, L. (1981), *Austrobaileya* 1, 420-28.
83. Rudd, V.E. (1981), in Polhill, R.M. and Raven, P.H. (*eds*) *Advances in Legume Systematics*. London: Kew, 347-54.
84. Schafer, H.I. (1973), *Kulturpflanze* 21, 211-73.
85. Smartt, J. (1976), *Tropical Pulses*. London: Longman.
86. Smartt, J. (1979), *Economic Botany* 33, 329-37.
87. Smartt, J. (1981), *Euphytica* 30, 445-9.
88. Voss, E.G. (*ed.*) (1983), International Code of Botanical Nomenclature. *Regnum Vegetabile*, III, 1-457.
89. Steele, W.M. and Mehra, K.L. (1980), in Summerfield, R.J. and Bunting, A.H. (*eds*) *Advances in Legume Science*. London: Kew, 393-404.
90. Summerfield, R.J. and Roberts, E.H. (in press), in Halevy, A.H. (*ed.*) *A Handbook of Flowering*. Florida: CRC Press.
91. Townsend, C.C. (1974), *Flora of Iraq* 3, 1-662.
92. van der Maesen, L.J.G. (1972), *Cicer L.: A monograph of the genus with special reference to the chickpea* (Cicer arietinum L.), its ecology and cultivation. Wageningen: Mededelingen Landbouwhogeschool, 72-10.
93. van der Maesen, L.J.G. (1979), *Flora Iranica* 140, 1-15.
94. van der Maesen, L.J.G. (1979), in Hawtin, G.C. and Chancellor, G.J. (*eds*) *Food Legume Improvement and Development*. Ottawa: IDRC Publication 126e, 140-46.
95. van der Maesen, L.J.G. (in press), in Whitcombe, J. (*ed.*) *Food Legume Germplasm Training Manual*. Aleppo: ICARDA.
96. van der Maesen, L.J.G. (in press), *Cajanus DC. and Atylosia W. & A. A revision of all taxa closely related to pigeonpea with notes on other related genera (subtribe Cajaninae)*. Wageningen: Mededelingen Landbouwhogeschool.
97. Verdcourt, B. (1970), *Kew Bulletin* 24, 235-307.
98. Verdcourt, B. (1970), *Kew Bulletin* 24, 507-69.
99. Verdcourt, B. (1971), in *Flora of Tropical East Africa, Leguminosae*, part 4. London: Crown Agents, 617-64.
100. Verdcourt, B. (1978), *Taxon* 27, 219-22.
101. Verdcourt, B. and Halliday, P. (1978), *Kew Bulletin* 33, 191-227.
102. Vogel, E.F. de (1980), *Seedlings of Dicotyledons*. Wageningen: Centre for Agricultural Publishing and Documentation.

103. Westphal, E. (1974), *Pulses in Ethiopia, their taxonomy and agricultural significance*. Agricultural Research Report 815. Wageningen: PUDOC, Centre for Agricultural Publications and Documentation.
104. Williams, J.T., Sanchez, A.M.C. and Jackson, M.T. (1974), *SABRAO Journal* 6, 133-45.
105. Grant, J.E., Grace, J.P. and Brown, A.H.D. (1983), in *Proceedings Australian Plant Breeding Conference*. Adelaide, 305-6.

2 Domestication and evolution of grain legumes

J. Smartt and T. Hymowitz

INTRODUCTION

Investigations on the domestication and evolution of crop
plants are of considerable intrinsic interest. As Harlan
[25] has stated, 'Cultivated plants mean more to man than
mere sources of food. They are part of his culture; they
influence his outlook, his religions, his social, political
and economic structures.' There are also major practical
justifications for studying domestication and evolution of
crop plants. First, it is easier to appreciate the present
state of development of our crops and to shape their future
if one is familiar with their past. The formulation of
realistic but innovative plant breeding objectives for a
specific crop is greatly facilitated by an understanding
of evolutionary events and processes which have occurred
in its development. Second, it is important that the genetic resources available for further improvement of the
crop are assessed and evaluated; its future may depend on
them. This cannot be done effectively without a comprehensive knowledge not only of our crop plants, but also of
their close relatives and how crop plants have diverged
from their ancestral forms under domestication in ancient
times and more recently. It is useful to know in a breeding programme whether what is being attempted is the re-
instatement of a quality which has been lost during
domestication, or the development of novel combinations of

characters which have evolved in geographically diverse populations, or a radical reshaping of the crop. Without some knowledge of a crop's evolutionary history it would not be possible to appreciate the precise nature of what is being attempted.

SOURCES OF EVIDENCE IN THE STUDY OF CROP EVOLUTION

In his studies of the origins of cultivated plants, de Candolle [14] clearly appreciated that very diverse kinds of evidence could shed light on crop evolution; the only practical restriction is their availability. The principal sources of evidence are biological and concern the bio-systematics of the plants themselves and their geography. Knowledge of both may be incomplete but will undoubtedly improve with further investigation.

The origin of agriculture and its subsequent evolution-ary diversification is part of man's own cultural evolu-tion. If natural selection is considered as the creative force behind natural diversity, artificial selection is equally the creative force that has produced our exceed-ingly variable domesticated plants and animals. Man has evolved both as an organism and as the producer of increas-ingly complex cultures. Thus, the evolution of crop plants can be viewed as part of man's own cultural development. It has paralleled and interacted with other facets of his cultural evolution. Progress in agricultural developments can therefore be followed in art, the production of arte-facts and the development of language. The recovery of this kind of evidence depends on the work of historians, archaeologists, anthropologists and linguists and also its broad interpretation in both the biological context (in so far as it affects specifically the development of crops) and in the cultural context (inasmuch as it broadly relates to man's cultural evolution).

It is therefore pertinent to review the evolution of grain legumes in the light of the following considerations:

(a) The classification of grain legumes and their wild
    relatives;
(b) The geographic centres of origin and domestication of
    grain legumes;
(c) The changes in grain legumes brought about under dom-
    estication; and
(d) The cultural (archaeological, anthropological, histori-
    cal and linguistic) evidence relating to the evolution
    of grain legumes.

The classification of grain legumes and their wild relatives

The difficulties which arise in a comprehensive review of
the specific evolutionary problems in the grain legumes
have been considered elsewhere [88]. Evolutionary studies
of legumes have been carried out either singly or in small
groups of related species within genera such as *Phaseolus*
and *Vigna*. It is not surprising therefore that the naming
of conspecific wild and domesticated forms has been inconsistent. Much of this work was also carried out against a
thoroughly unsatisfactory taxonomic background. However
the labours of Verdcourt [111-116], Maréchal *et al.* [61],
Maréchal [59], Hymowitz and Newell [41,96], Gregory *et al.*
[95], van der Maesen [108] and others, have in the past
decade vastly improved the state of grain legume taxonomy.
Some problems remain regarding the status of the genera
Cajanus and *Atylosia* [13] and the controversial status of
the form *Phaseolus polyanthus* [91]. Solutions to both of
these problems should not prove difficult when appropriate
biosystematic studies have been carried out (see Chapter 1).
    Probably the greatest difficulty is caused by the
inconsistent naming of wild and cultivated forms which
clearly belong to the same biological species. The wild
forms are sometimes described and named before experimental hybridisations demonstrate conspecificity with the
previously described and named cultivated forms. Even when

39

conspecificity has been established, the different forms
may be recognised at different taxonomic levels as subspecies, botanical varieties or formae. Clearly, some consistent practice would be of considerable practical advantage. A sound taxonomic system for crop plants is of
immense importance to all involved in the collection and
evaluation of the whole range of genetic resources currently in existence and which may be useful in crop
improvement. Information collection, storage, retrieval
and usage would all benefit from a standardised system,
universal if possible, but at the very least for groups of
closely related crops. As a practical expedient a modification of an approach previously formulated [26,88] might be
useful if generally accepted. Table 2.1(a) lists those
grain legumes in which both wild and cultivated forms are
known, together with their current nomenclature; Table
2.1(b) lists those species in which clearly conspecific
wild relatives have not yet been identified. An attempt is
made in Table 2.2 to produce a consistent scheme of
nomenclature in which basically wild and cultivated races
within a single species are recognized as subspecies, not
as botanical varieties [88] since this does not seem to
give sufficient scope for taxonomic recognition of differentiated populations within conspecific wild and cultivated
taxa (as within *Vigna unguiculata*, for example).

The geographic centres of origin and domestication of grain legumes

Evidence accumulated over many years enables us to identify
with reasonable certainty the regions in which crop species
originated and where they were domesticated. These centres
frequently but do not necessarily coincide. Where an ancestral wild form has a wide distribution, domestication may
have occurred in relatively few or in many locations within
the region. Much depends on the interactions between human
population densities, relative cultural achievement, the

Table 2.1(a)   Grain legume domesticates and their conspecific wild relatives

Domesticate	Conspecific wild relatives	References
Arachis hypogaea L.	*A. monticola* Krap. & Rig.	[48]
Cicer arietinum L.	*C. reticulatum* Ladizinsky	[54]
Cyamopsis tetragonoloba L.	*C. senegalensis* Guill. & Perr.	[38]
Glycine max (L.) Merr.	*G. soja* (L.) Sieb. & Zucc.	[64,78]
Lathyrus sativus L.	*L. sativus* (wild in Asia Minor)	[102]
Lens culinaris Medic.	*L. orientalis* (Boiss.) Hand.	[125]
Macrotyloma geocarpum (Harms) Maréchal & Baudet	*M. geocarpum* var. *tisserantii* (Pellegrin) Maréchal & Baudet	[15,60]
Phaseolus acutifolius Gray	*P. acutifolius* var. *acutifolius*	[16]
Phaseolus coccineus L.	*P. formosus* HBK *P. obvallatus* Schlecht	[61,106]
Phaseolus lunatus L.	*P. lunatus* var. *silvester* Baudet	[3]
Phaseolus vulgaris L.	*P. vulgaris* (wild in Mexico) *P. aborigineus* (wild in Argentina)	[17,100] [5,6,7,8,45]
Pisum sativum L.	*P. humile* Boiss. & Noë *P. elatius* (Beib.) Alef.	[127]
Vigna aconitifolia (Jacq.) Maréchal	*V. aconitifolia* (wild in India, Pakistan, Burma)	[70,100]
Vigna angularis (Willd.) Ohwi & Ohashi	*V. angularis* var. *nipponensis* (Ohwi) Ohwi & Ohashi	[61,100]
Vigna mungo	*V. mungo* var. *silvestris* Lukoki, Maréchal & Otoul	[57,100]
Vigna radiata	*V. radiata* var. *sublobata* (Roxb.) Verdc.	[57,100]
Vigna (Voandzeia) subterranea (L.) Verdc.	*V. subterranea* var. *spontanea* (Harms) Hepper	[32,81,115, 116]
Vigna umbellata (Thunb.) Ohwi & Ohashi	*V. umbellata* var. *gracilis* (Prain) Maréchal, Mascherpa & Stainier	[61,100]
Vigna unguiculata (L.) Walp.	*V. unguiculata* subsp. *dekindtiana* (Harms) Verdc.	[61,100]

Table 2.1(b)   Closest relatives of grain legume domesticates with no known conspecific wild form

Domesticate	Wild relative	Reference
Cajanus cajan (L.) Millsp.	*Atylosia lineata* W.	[35]
Psophocarpus tetragono-lobus (L.) DC.	*P. grandiflorus* Wilczek	[117]
Vicia faba (L.)	*V. galilaea* Plitm. & Zoh.	[127]

Table 2.2   A consistent nomenclatural scheme for wild forms of grain legumes

Domesticate	Present nomenclature of wild form	Suggested nomenclature of wild form
Arachis hypogaea	*A. monticola*	*A. hypogaea* subsp. *monticola*
Cicer arietinum	*C. reticulatum*	*C. arietinum* subsp. *reticulatum*
Cyamopsis tetragono-loba	*C. senegalensis*	*C. tetragonoloba* subsp. *senegalensis*
Glycine max	*G. soja*	*G. max.* subsp. *soja*
Lens culinaris	*L. orientalis*	*L. culinaris* subsp. *orientalis*
Macrotyloma geocarpum	*M. geocarpum* var. *tisserantii*	*M. geocarpum* subsp. *tisserantii*
Phaseolus acutifolius	*P. acutifolius* var. *acutifolius*	*P. acutifolius* subsp. *acutifolius*
P. coccineus	*P. coccineus* subsp. *obvallatus*	Conforms*
P. lunatus	*P. lunatus* var. *silvester*	*P. lunatus* subsp. *silvester*
P. vulgaris	*P. vulgaris* var. *aborigineus*	*P. vulgaris* subsp. *aborigineus*
Pisum sativum	*P. humile*	*P. sativum* subsp. *humile*
Vigna angularis	*V. angularis* var. *nipponensis*	*V. angularis* subsp. *nipponensis*
V. mungo	*V. mungo* var. *silvestris*	*V. mungo* subsp. *silvestris*
V. radiata	*V. radiata* var. *sublobata*	*V. radiata* subsp. *sublobata*
V. subterranea	*V. subterranea* var. *spontanea*	*V. subterranea* subsp. *spontanea*
V. umbellata	*V. umbellata* var. *gracilis*	*V. umbellata* subsp. *gracilis*
V. unguiculata	*V. unguiculata* subsp. *dekindtiana*	Conforms*

*Present nomenclature conforms with the proposed scheme.

availability of potential domesticates and local climatic
regimes.

The chronology of domestication is probably impossible
to reconstruct but radiocarbon dating can determine approximate ages of materials found in archaeological sites.
These materials are often remarkably well preserved but it
may be impossible to decide whether particular remains are
of collected wild, weedy or cultivated domesticated material. Thus, estimates of ages of domestication are cautious;
they are almost invariably underestimates. Sometimes it is
also assumed or implied that primitive farming communities
were actually responsible for the domestication of the
crops cultivated; this may or may not be so. It is not even
certain that they themselves actually evolved the agricultural technology practised. What is much more certain is
that the agriculture engaged in provided a substantial if
not the major support for an associated civilization. It
seems probable that certain communities became focal points
in which the cultivation of crops domesticated in the region
became integrated into local agricultural systems using
available technology. It is surely not sufficient to have
cultivated just a random collection of crops. Those crops
grown must have met the nutritional needs of the population
and their livestock, and their production must have been
indefinitely sustainable. There must also have been some
selection pressures for efficiency both in agricultural
production and in meeting the population's nutritional
needs.

It seems likely, then, that crops, farm livestock and
agriculture *per se* all had a very considerable evolutionary history behind them at the period when we can first
recognise their existence. Archaeological records, like
those of palaeontology, are also very fragmentary and for
rather similar reasons, depending largely on the prevalence
and persistence of conditions favourable for the preservation of both biological materials and artefacts.

Grain legumes have clearly originated in both the Old

43

and New Worlds. Botanical evidence can be critical in
determining this, especially when it is disputed as has
happened with the groundnut [105]. Old World grain legumes can readily be referred to one (or more) of the following regions, viz. the Mediterranean, central Asia, Asia
Minor, Africa, India and the Sino-Japanese area. In the
New World the Central and South American regions are the
most important centres. These regions are vast in area and
it is possible to locate within them probable centres of
origin and domestication. However, domestication of a
single species can probably occur in more than one gene
centre, as in common beans (*Phaseolus vulgaris*) and Lima
beans (*P. lunatus*) which were both domesticated in
South and Central America. A similar pattern of domestication could have occurred in the Mediterranean, Asia Minor
and central Asia where peas (*Pisum sativum*), faba beans
(*Vicia faba*), lentils (*Lens culinaris*) and chickpeas (*Cicer
arietinum*) originated. The eastern Mediterranean area is
likely to have been one of the most critical for domestication of grain legumes, and that is where Zhukovsky [124]
draws the boundary between the Mediterranean and Asia Minor
gene centres; but these two regions cannot be considered
without reference to central Asia. This group of centres
is clearly of great importance in the evolution of the
major Old World temperate grain legumes. The tropical
grain legumes of the Old World probably originated in the
Asian and African gene centres while those of the New
World were developed in Central and South America. The
Indonesian, Australian and North American regions were
apparently only involved as secondary centres of origin.
An attempt is made in Table 2.3 to summarise current views
on centres of origin and domestication.

It is probably over-optimistic to hope to define more
closely the centres of origin and domestication of these
crops. Our knowledge of present distributions of ancestral
forms is even now still far from complete.

Table 2.3 Centres of origin and domestication of grain legumes

Gene centre and location	Domesticate	Probable centre of domestication	Reference
ASIA			
India	Cajanus cajan	India	[108]
	Vigna aconitifolia	India	[61]
	V. mungo	India	[58]
	V. radiata	India	[58]
Sino-Japanese	Glycine max	China	[78]
	Vigna angularis	Japan, Far East	[61]
	V. umbellata	Far East	[61]
Asia Minor	Cicer arietinum	Middle East	[54]
(including cen-	Lathyrus sativus	Middle East	[102]
tral Asia and	Lens culinaris	Middle East	[52]
the eastern	Pisum sativum	Near East	[127]
Mediterranean)	Vicia faba	Middle East	[98]
AFRICA			
sub-Saharan	Cyamopsis tetragonoloba	India	[38]
West Africa	Macrotyloma geocarpum	West Africa	[31]
Ethiopia	Psophocarpus tetragonolobus	Asia	[89]
West Africa	Vigna subterranea	West Africa	[31]
West Africa – Ethiopia	V. unguiculata	West Africa	[99]
AMERICAS			
Central America	Pachyrrhizus erosus	Mexico	[10]
	Phaseolus acutifolius	Mexico	[44]
	P. coccineus	Mexico	[44]
	P. lunatus	Mexico	[44]
	P. vulgaris	Mexico	[44]
South America	Arachis hypogaea	Bolivia-Argentina	[95]
	Pachyrrhizus tuberosus	Brazil	[10]
	Phaseolus lunatus	Peru	[44]
	P. vulgaris	Argentina-Peru	[8]

The changes in grain legumes brought about under domestication

Schwanitz [73] considers the nature of changes brought about by selection under cultivation; those which particularly concern grain legumes are:

(a) gigantism and consequent changes in allometric growth;
(b) reduction or loss of the means of natural dissemination;
(c) reduction or loss of toxic substances and other biochemical changes;

45

(d) loss of seed dormancy;

(e) changes in branching pattern and implications for the simultaneous or sequential ripening of fruits; and

(f) differences in life form.

Not all of these changes have affected all grain legumes as they have evolved but several of them have occurred during the evolution of most grain legumes. These are considered briefly now, and in further detail later.

(a) <u>Gigantism</u>.   This has been a virtually universal change in domesticated legumes. Those that are grown primarily for their seeds often show increases in seed size by at least one order of magnitude (e.g. *Phaseolus* spp.). Larger fruits are then required and more robust stems for their support, which tend to be fewer in number. This reduces leaf number, and to partially compensate for this, leaf size often increases. Quite dramatic changes in vegetative architecture can result in the transformation of a profusely branched, vining climber with many small leaves into a dwarf, free-standing, determinate form with few relatively large leaves. These changes are usually produced without any concomitant development of genetic isolation mechanisms. The diverse range of wild and domesticated forms can therefore produce recombinants among themselves without genetic restriction.

(b) <u>Reduction or loss of the means of natural dissemination</u>. It is a characteristic of many wild legumes, and of grain legume progenitors as a whole, that they have an explosive fruit dehiscence mechanism (with a few exceptions, such as that of the groundnut). This feature is obviously undesirable in a domesticate and artificial selection pressures have resulted in fruits which are less readily dehiscent at maturity and in some instances have culminated in fruits which are indehiscent (e.g. in *Phaseolus vulgaris, Pisum sativum* and *Vicia faba*). Even in a geocarpic form such as

the groundnut (*Arachis hypogaea*) a reduced capacity for
seed dispersal is apparent. In wild *Arachis* the fruits are
characteristically segmented and may be very elongated producing a separation of about 25cm between segments. In the
domesticate the fruits are unsegmented and, in fact, of the
more typical form (an added convenience in harvesting).
This is an interesting example of artificial selection re-
versing the effect of natural selection in a morphological
sense.

Changes in pod wall anatomy and extent of lignification
have gone considerably further in the three popular vegetable species *Pisum sativum, Phaseolus coccineus* and *P. vulgaris* than in other grain legumes. In these three species
genotypes occur in which lignification of fibres is more
or less completely suppressed and so entire fruits are
edible in peas and remain edible for longer in the French
and runner beans. Garden pea genotypes have either parchmented or mangetout pod walls, whereas three types of pod
wall are recognised in *Phaseolus*: (i) parchmented, occurring in the wild forms of most species and in *P. lunatus*
and *P. acutifolius* cultivars (which are fibrous and ined-
ible); (ii) leathery, found in cultivated *P. coccineus* and
P. vulgaris which, apart from fibrous 'strings', remain
edible until the completion of growth in fruit length; and
(iii) stringless, found in both species and which remain
edible even after seed growth is quite advanced. With fruit
maturation the stringless tend to lose succulence and so
their attraction as vegetables.

(c) <u>Reduction or loss of toxic substances and other biochemical changes</u>.   Considered as a whole the Leguminosae
produce a range of very potent toxins in their seeds (and
see Chapter 3). The exploitation of legume seeds as food or
feed has been fraught with considerable danger to human and
animal life; the toxins of *Abrus* and *Laburnum* species are
notorious in this respect. Of the wide range of toxic materials formed in legume seeds, possibly the most toxic are

the alkaloids, which act rapidly. Materials with less rapid action include toxic amino acids (which may act as metabolic antagonists of protein amino acids), haemagglutinins, protease inhibitors, cyanogenetic glycosides and saponins. There are also factors which can incite clinical disorders such as goitre (possibly due to effects on thyroxine metabolism) which have been found in seeds of soyabeans, groundnuts, peas and *Phaseolus* beans. Some constituents of faba beans induce favism (an allergic haemolysis) in genetically susceptible individuals. Neuro-lathyrism is a paralytic state induced by excessive consumption of *Lathyrus sativus* (grass pea) seeds. These contain toxic non-protein amino acids which act as antagonists in protein amino acid metabolism [65,66].

The utilisation of many legume seeds for food or feed has depended on polymorphisms for presence or absence of the toxins, as in *Lupinus* [19] (see Chapter 16), or the ability to circumvent the action of toxic materials such as haemagglutinins and protease inhibitors by fermentation (soyabeans) or cooking (common beans). These strategies have been successful in the past when the toxic materials had not themselves been identified. They have been even more successful in recent times now that the compounds have been chemically identified and can all be assayed. An early example of reducing toxicity involved the seeds of Lima bean (*Phaseolus lunatus*) which contain varying concentrations of cyanogenetic glycosides. While it may not be possible to produce entirely cyanide-free seeds, the concentration of HCN liberated on hydrolysis can range from 300mg per 100g of beans to only 10mg, which is considered safe. The larger concentrations are thought to have killed people by cyanide poisoning [63]. In more recent years, attempts have been made to reduce the concentrations of protease inhibitors and phytohaemagglutinins in soyabeans and common beans. Since polymorphisms for presence or absence of these factors have been found it seems highly probable that antimetabolite-free soyabean seeds [67,68] and common bean seeds [47] could be produced and exploited, provided that no very substantial yield penalty is involved.

Sophisticated selection procedures to alter the bio-chemical composition of legume seeds require chemical information which does not yet exist for many legume crops. For example, the grass pea crop has a bad reputation because excessive consumption has been correlated with the incidence of neurolathyrism, an often irreversible and fatal progressive paralysis. Paradoxically, the problem arises from a virtue of the crop: it is very drought tolerant and can yield when other crops fail completely. Thus, the grass pea may be the sole or principal food available and it is in these circumstances that the probable incitant of lathyrism (β-N-oxalyl-α,β diamino-propionic acid) is thought to exert its detrimental effects. An extensive survey of landraces and wild forms of *Lathyrus sativus* would be useful to establish whether incitant-free genotypes can be selected or if the concentration can be reduced to safe values.

It seems probable, then, that past selections have reduced the concentrations of toxins; with modern technology the efficiency of such selection could be greatly enhanced.

(d) <u>Loss of seed dormancy</u>.   Hard-seededness is common-place in wild legumes and is of undoubted survival advantage. The dormancy it imposes can ensure long-term persistence of seeds in a given locality. Long-term seed dormancy is a very considerable drawback in grain legumes, although a short-term dormancy may still be advantageous.

Most common bean cultivars produce seeds which at maturity do not germinate in the fruit. However, cultivar Kentucky Wonder seeds will germinate readily in the fruits under humid atmospheric conditions. A similar situation exists in the groundnut. The seeds of *fastigiata* (sequentially branching) types have no dormancy and can germinate as soon as mature. In wet conditions near harvest time this can be a serious drawback not only in the loss of seed from

germination but also in the production of volunteer plants
which can carry over pests and diseases to the succeeding
crop. On the other hand, the *hypogaea* (alternately branching) types have short-term dormancy (less than three months)
which precludes loss of crop and reduces carry-over of pests
and diseases (and see Chapter 17).

(e) <u>Changes in branching pattern and implications for
fruiting</u>.   In many grain legumes two growth forms are
common; they are frequently referred to as indeterminate
and determinate. The differences between them lie in the
behaviour of the growing apices: in the indeterminates all
reproductive branches arise laterally, while in determinate
forms they are produced terminally as well. This reduces
node number and results in a more compact growth form which
can be further enhanced by shorter internodes. Shorter
internodes *per se* can produce a dwarf or bush growth habit,
as in *Phaseolus vulgaris* and other *Phaseolus* species. Similar changes have also occurred in *Pisum sativum, Glycine
max, Vigna* spp. and *Cajanus cajan*. In groundnuts, stem
apices never produce inflorescences, but a combination of
shortened internodes and ascending growth can produce a
'bunch' or bush growth form. This also reduces the effective
duration of the fruiting season since only fruiting pegs
which arise at basal nodes can successfully penetrate the
soil and produce mature fruits. This gives an approximation
to the fruiting pattern of determinate forms although
growth is actually indeterminate. True determinate growth
results when both main stems and lateral branches produce
inflorescences quickly and almost simultaneously. The
flowering season is very concentrated and the fruiting season even more so. All the fruits a determinate plant can
mature may be set in a period of one or two weeks whereas
fruit set may extend indefinitely in an indeterminate form.
    Determinate growth and simultaneous fruit maturation
is on the whole more advantageous in mechanised modern farming systems than in the more primitive ones. It is

particularly useful in the semi-arid tropics and with
destructive harvesting systems at temperate latitudes. The
indeterminate fruiting habit is advantageous in the humid
tropics where seed storage may be difficult and sequential
ripening with harvesting as required can be a more satisfactory system of production.

(f) **Differences in life form**.   Some plants taken into
cultivation may be exposed to selection pressures in favour
of a reduced lifespan. In some crops (e.g. the brassicas
and beets) this may produce biennials, but where change in
life form occurs in legumes, annuals tend to evolve.

Of the important grain legumes only the pigeonpea (*Cajanus cajan*) has clearly not been subjected intensively to
this pressure (but see Chapter 15). The geocarpic groundnut,
due to the nature of the harvesting techniques used, has
been treated as an annual ever since it was brought into
cultivation. Many important *Phaseolus* species, such as the
Lima and runner beans, are primitively perennials, from
which both perennial and annual cultivars have been selected.

The perennial life form has potentially great advantages
in cultivation but whether it is more efficient to treat
crops as perennials or annuals (if the option is open) depends on their yield over successive seasons. Plants whose
seed yield declines rapidly after the first fruiting season
are likely to be treated as annuals. If, then, the most
productive genotypes are selected, those forms will be favoured in which assimilates are directed to the developing
seeds rather than to perennating organs. Since perennation
has then become superfluous it will be selected against.

Cultural evidence relating to evolution of grain legumes

A complex of relevant cultural factors must be noted in the
study of crop evolution. Although empirically assigned to
disciplines such as archaeology, anthropology, history and
linguistics, in practice the factors often tend to merge

somewhat with each other. Of these disciplines there can
be little doubt that archaeology has been the most illumi-
nating.

(a) Archaeological evidence.   Archaeological remains of
significance are of two types: the actual remains of plants
themselves and the representations of them in artefacts and
in murals. Plant remains, if sufficiently abundant, can
allow the dating of both themselves and their associated
artefacts by radiocarbon techniques. The importance of
archaeological finds varies enormously between different
grain legumes. The pulses which originated in the Mediterranean and Asia Minor have quite a good archaeological record dating back 6000-8000 years [24]. The *Phaseolus* beans
of New World origin have been found in Mexican and Peruvian
sites of comparable age. No significant remains have yet
been discovered of African or Far Eastern grain legumes.
Indian sites have provided remains of a wide range of pulses
but not apparently of the cowpea [9,36,120]. Table 2.4
attempts to collate information on the occurrence of present-day grain legume crops in known archaeological contexts.
     Representations of legume crops in ancient art and arte-
facts are rare but quite the most remarkable representations
of any legume are to be found on Peruvian ceramics. The
depictions of groundnut fruits on an excavated Mochia vase
are extraordinarily clear and very closely resemble those
of landraces found in Peru at the present time [109,110].
Other representations have excited little comment as compared with wall paintings of cereal cultivation in Egyptian
tombs.

(b) Anthropological evidence.   Direct evidence related to
crop evolutionary processes can sometimes be obtained from
observations of the utilisation of wild forms of species
which have been domesticated or their close relatives. Burkart and Brücher [8] note that *Phaseolus vulgaris* subsp.
aborigineus is actually collected from the wild by local

Table 2.4 Ancient datings of cultivated grain legume remains (years before present, B.P.)*

Domesticate	Nature of remains	Location	Estimated age (years)	Reference
Arachis hypogaea	fruits	Huaca Prieta, Peru	2850	[21,105]
Cajanus cajan	seeds	XIIth Dynasty tombs, Egypt	4000	[74]
Cicer arietinum	seeds	Turkey	7000	[79]
Lathyrus sativus	carbonised seeds	Jarmo, Iraq	8000	[30]
Lens culinaris	seeds	Jarmo, Iraq	9000–8500	[27,28]
Phaseolus acutifolius	unspecified	Tehuacan, Mexico	7000	[44]
P. coccineus	unspecified	Tehuacan, Mexico	2200	[44]
P. lunatus	fruits and seeds	Chilca, Peru	7300	[44]
P. vulgaris	fruits	Tehuacan, Mexico	7000–5500	[45]
	seeds	Huaylas, Peru	7680	[45]
Pisum sativum	carbonised seeds	Catal Huyuk, Turkey	7850–7600	[29]
Vicia faba	seeds	Jericho	8000	[104]
Vigna mungo	carbonised seeds	Navdatoli-Maheshwar, India	3660–3440	[118,119]
V. radiata	carbonised seeds	Navdatoli-Maheshwar, India	3660–3440	[118,119]

*Note: There are no significant archaeological records for some of the major grain legumes (e.g. *Glycine max* and *Vigna unguiculata*) or others of regional importance (e.g. *Lupinus* spp. and *Psophocarpus tetragono-lobus*).

53

Table 2.5 Historical studies on grain legume crops

Domesticate	Nature of coverage	Reference
Arachis hypogaea	general dispersion	[21]
Cajanus cajan	general dispersion	[108]
Cicer arietinum	general dispersion; Near Eastern distribution; occurrence in India	[54,79, 84,103, 107,127]
Cyamopsis tetragonoloba	dispersion in Asia, trans-domestication	[38]
Glycine max	general dispersion	[37,78]
Lathyrus sativus	occurrence in India	[103,120]
Lens culinaris	comprehensive; Near Eastern and European distribution	[122,127]
Lupinus spp.	general dispersion	[19,20]
Macrotyloma uniflorum	occurrence in India	[103]
Pachyrrhizus erosus	some historical allusions	[10]
Pachyrrhizus tuberosus	some historical allusions	[10]
Phaseolus acutifolius	general dispersion	[77,93]
P. coccineus	general dispersion	[77,93]
P. lunatus	general dispersion	[77,93]
P. vulgaris	general dispersion	[77,93]
Pisum sativum	occurrence in India and Near East	[103,127]
Psophocarpus tetra-gonolobus	incidental historical allusions	[39,42]
Vicia faba	general and Near East	[75,127]
Vigna mungo	occurrence in India	[97,103]
V. radiata	occurrence in India	[97]
V. unguiculata	general dispersion	[82,89]

people in northern Argentina. In general, however, anthropological interpretation has had little impact on current understanding of the evolution of grain legumes but might become more useful in the future.

(c) <u>Historical evidence</u>.   By and large the historical records we have of grain legume crops are rather fragmentary and few comprehensive accounts have actually been

written. However, an exceptionally comprehensive account
of the groundnut's history has been produced by Hammons
[21] and is an excellent model for any would-be historians
of grain legume crops. The literature he reviewed goes back
to the early sixteenth century [69], a remarkably short
time after the discovery of the Americas and their early
exploration.

In the Old World, the written record dates back to the
time of the classical writers (e.g. Virgil's *Georgics*) and
even earlier in Sanskrit, cuneiform and Egyptian records.
For many crops, historical references are very few and comparatively recent, but an attempt has been made in Table
2.5 to indicate the nature and extent of such historical
accounts as have been written.

(d) <u>Linguistic evidence</u>. Some linguistic studies, largely
comparative in nature, of the vernacular names of a number
of grain legumes have been carried out (Table 2.6). However,
as Hymowitz and Boyd [39] point out, 'Attempts to relate
the origin of a domesticate to the number of vernacular
names in a country or region can be misleading'. A similar

Table 2.6 Linguistic studies on grain legume crops

Domesticate	Nature of study	Reference
Arachis hypogaea	S. American, N. American W. African vernacular names	[1,43,105]
Cajanus cajan	Indian Sanskrit names	[35]
Cicer arietinum	etymological evidence	[108]
Cyamopsis tetragonoloba	Indian Sanskrit names	[38]
Glycine max	analysis of pictograph 'shu'	[37]
Psophocarpus tetragonolobus	vernacular names S.E. Asia	[39]

caution is expressed by Krapovickas [105]: 'I do not necessarily conclude that the Arawaks are responsible for the
spread of the groundnut from its centre of origin as far
as the Caribbean islands. Such a hypothesis is however

55

consistent with the evidence available at the present time.'

Linguistic evidence tends to be brief, fragmentary, sometimes equivocal and not always very informative. Its value is, however, illustrated by the following example.

The groundnut's West African and Portuguese names provide some evidence to support the hypothesis which postulates its migration from South to North America via West Africa. A common name for the groundnut in Brazil is 'manduvi' [105]; the name 'manduli' has been recorded from West Africa [1]. Another West African name for this crop is 'guba' or 'nguba' the resemblance of which to the name 'goober' current in the south-east United States is probably not coincidence [43].

THE EVOLUTION OF INDIVIDUAL GRAIN LEGUME CROPS

It is impracticable in the current review to present more than an outline summary of evolution in each of the thirteen individual species included in this book and so reference will be made to fuller and more detailed treatments elsewhere (and see Chapter 1).

Arachis hypogaea ($2n = 4x = 40$; the groundnut, peanut, monkey nut, goober pea). The groundnut is generally thought to have originated in the Eastern Andean region of north-west Argentina and south Bolivia. It is an allotetraploid derived from the wild tetraploid *A. monticola* Krap. & Rig. which, in turn, is thought to have arisen from two cytologically differentiated species within section Arachis of the genus, possibly *A. batizocoi* Krap. & Greg. and *A. cardenasii* nom. nud. [95].

In pre-Columbian times groundnuts became very widely distributed in the Americas from Argentina to Mexico and the Antilles. The crop had apparently not reached areas to the north of Mexico. In post-Columbian times groundnuts were transported to Africa from Brazil by the Portuguese and to Asia via the Philippines. In Africa, a secondary

centre of diversity developed [18] and the distinctly
African types of groundnut cultivated in the south-eastern
United States were apparently transported there with West
African slaves. Current distribution of the crop is widespread in the Old World and the New. It is best suited to
warm temperate, sub-tropical and semi-arid tropical conditions and is not cultivated on an extensive scale in the
humid tropics (and see Chapter 17).

Cajanus cajan ($2n = 2x$; pigeonpea, arhar, tur).   It is
now generally accepted that the pigeonpea is of Indian
origin [108]. The question of origin was complicated by
the ease with which feral populations of this domesticate
could be established. The weight of botanical evidence
clearly favours India as the area in which domestication
occurred [13,35]. Although no clearly conspecific wild
counterpart has been identified it is obviously closely
related to species of *Atylosia*, especially *A. cajanifolia*
with which hybridisation is possible and results in fairly
fertile hybrids.
    Van der Maesen [108] using information in papers by
De]35] and Royes [80] has plotted the time-scale of pigeonpea dispersal. Movement to the New World from Africa apparently occurred in the sixteenth century. Acceptance of the
crop in Africa has been greater in the Sudan zone than
farther south; it has not been cultivated extensively by
the Bantu in southern Africa (and see Chapter 15).

Cicer arietinum ($2n = 2x = 16$; chickpea, desi, kabuli,
gram, garbanzo).   The probable progenitor type of the
cultivated chickpea was only collected quite recently
[49,54,55,56] and subsequent to the publication of van der
Maesen's monograph [107] of the genus. The species in question, *Cicer reticulatum*, was collected in Turkey and suggests that the chickpea is one of the Old World group of
pulses which appear to have been domesticated in the eastern
Mediterranean and the Fertile Crescent. The production of

fertile hybrids between the domesticate and *C. reticulatum* indicates that they together constitute a single biological species. Two major seed types have evolved, the large-seeded Mediterranean and the smaller-seeded Asiatic forms (kabuli and desi types, respectively). In pre-Columbian times chick-peas had reached the Indian subcontinent where they probably have achieved their greatest economic importance. The crop was introduced into the New World by the Spanish and Portu-guese where it is now widely cultivated in suitable climatic zones. It has also extended its range into East and Southern Africa. It is a crop well suited to Mediterranean-type cli-mates and to cultivation on residual soil moisture. The crop does not tolerate sustained heavy rainfall well and it must mature in warm dry conditions (and see Chapter 8).

Glycine max ($2n = 4x = 40$; soyabean, soybean, Japan pea). The current view [96], that soyabeans originated in the eastern half of North China at least 3000 years ago, is sup-ported by linguistic, geographical and historical evidence. Two taxa, *G. soja* and *G. gracilis*, have very close affini-ties with *G. max* and are conspecific with it. *G. soja* is truly wild, *G. gracilis* appears to be an intermediate intro-gressed weedy form that Hermann [33] included in *G. max*.

From its centre of origin in China the soyabean moved west and southwards to India, Nepal, Burma, Thailand, Indo-China, Korea, Japan, Malaysia, Indonesia and the Philip-pines. This probably occurred during or before the first millennium AD. The crop came to Europe early in the eight-eenth century and was planted for the first time in North America in 1765 [40]. Its establishment there was slow and only between the two World Wars (in the 1920s and 1930s) did production expand significantly; North America now leads world production. Considerable production potential exists in Africa and southern Europe and dramatic expansion has taken place in Brazil and could do so elsewhere in South America (and see Chapter 9).

Lens culinaris ($2n = 2x = 14$; lentil). There is general agreement that the wild species *L. orientalis* is the proto-type of the cultivated lentil [52,53,83,122,123]. The criti-cal area for domestication is the Fertile Crescent but be-cause *L. orientalis* has a wide natural distribution it is not possible to pinpoint primary domestication more exactly. Two distinct seed types, the *macrosperma* and *microsperma*, have evolved; the former is considered the more advanced.

Early spread of lentil cultivation into Greece and Bul-garia was in Neolithic times and subsequently throughout the Mediterranean and central Europe. Eastward the crop reached India and Afghanistan, and to the south, Ethiopia. In more recent times lentils have become established in the Americas but nowhere, except the Pacific northwest of the United States and in Canada, is cultivation very extensive even though it is a very acceptable pulse crop (and see Chapter 7).

Lupinus spp. (the lupins). The lupins stand apart from the other grain legume crops in that they are members of the tribe *Genisteae* whereas most others belong to either the *Vicieae* or *Phaseoleae* (with the important exception of the groundnut). Lupins have been in cultivation for 3000-4000 years in the Mediterranean region where a procedure was developed for removing toxic alkaloids from their seeds [19]. Outside the Mediterranean area lupins have found most extensive use in soil reclamation (since they grow well on infertile soils), at least prior to the isolation and pro-pagation of 'sweet', alkaloid-free seed types. Since this occurred some sixty years ago, cultivation of sweet lupins has developed in eastern Europe and in Australia.

The genus is apparently ancient and species are genet-ically well isolated from each other. Improvement of the species *Lupinus albus* ($2n = 50$), *L. luteus* ($2n = 52$), *L. angustifolius* ($2n = 40$), *L. cosentinii* ($2n = 32$) and *L. muta-bilis* ($2n = 48$) is a matter of selection within the gene pools of individual species. These all differ in chromosome

number and all except *L. mutabilis* (domesticated in the
Andes) are of Mediterranean or north African origin. Ploidy
levels are difficult to establish but if the smallest
basic numbers in the tribe are taken (i.e. 7 and 8) then
most species are either tetraploid or hexaploid (or approximately so).

Extensive breeding and selection work is now under way
in several countries, most notably in northern and eastern
Europe and more recently in Australia and the Antipodes
where a particularly concentrated effort is being mounted
[19,20,94]. These programmes have several breeding objectives including better yields, non-shattering fruits, depen-
dably small alkaloid concentrations, improved pest and disease resistance together with more suitable agronomic characters. The selection of efficient genotypes could very
significantly improve protein production for livestock feeds
and eventually for direct use as food (and see Chapter 16).

Phaseolus vulgaris ($2n = 2x = 22$; common bean, haricot,
navy bean, French bean). The American origin of this crop
has never credibly been questioned but some controversy has
arisen regarding alternative possibilities of domestication
in Mexico or South America [5,17,62]. Others accept the
possibility of domestication in both regions where wild
Phaseolus vulgaris occurs [44,45,46,106]. There is no difficulty in identifying the wild prototype. Ancient cultivation has been established in Mexico since about 7000 BP and
in Peru around 7680 BP. Where seeds have been found they
have been of essentially modern type.

Considerable dissemination occurred in pre-Columbian
times in the Americas and subsequently to Europe, Africa,
Asia and Australasia. Common beans can be grown as a summer
crop in cool temperature regions, a cool-season crop under
irrigation in the semi-arid tropics, and all year round at
high elevations in equatorial regions. It is one of the
most popular pulse crops, while immature fruits are also a
very important green vegetable. Considerable diversification

has occurred outside the Americas and significant germplasm resources have evolved, in southern Europe for example [93], away from its centre of origin and primary centres of diversity (and see Chapter 10).

Phaseolus lunatus ($2n = 2x = 22$; Lima, sieva, butter bean).

The evolutionary history of the Lima bean shows considerable parallels with that of common beans. The wild prototypes of both beans have wide distributions, and separate domestications of Lima beans have probably occurred in Mexico and Peru [44]. Peruvian domesticates are larger seeded than those from Mexico. Archaeological material dated ± 1800 BP has been obtained from Mexican sites; considerably older material (± 8000 BP) has been excavated in Peru.

Distribution of the Lima bean is less extensive geographically than that of common beans. Its climatic optimum is decidedly tropical and its penetration of cooler latitudes and altitudes is limited. It has, however, been carried to tropical and subtropical areas of Asia and Africa where it grows well. The crop withstands the combination of of hot temperature and humid air relatively well and is also towards maturity quite remarkably tolerant of drought (and see Chapter 11).

Pisum sativum ($2n = 2x = 14$; garden pea, field pea, grey pea, carlin). Peas, like chickpeas, lentils and faba beans, are a crop the evolution of which is associated with the rise of civilisations in the eastern Mediterranean and the Fertile Crescent. Peas are known to have been cultivated 8000 years ago in the Near East [76,127]. The area in which these ancient cultivated peas have been found coincides remarkably with the distribution of the wild peas *Pisum pumilio* (*humile*) and *P. elatius* which Ben-Ze'ev and Zohary [4] showed to be conspecific with cultivated *P. sativum*. These two wild forms are of particular interest in that they closely resemble, respectively, the dwarf and climbing forms of the domesticate in their growth habits. *Pisum pumilio*

61

is unusual if not unique in being a wild form but having a dwarf growth habit.

Selection has established diverse forms of peas; garden and field forms and tall or dwarf growth habits. Perhaps the greatest divergence has been produced in garden peas in which cotyledonary food reserves may be starchy or sugary (sugar peas), pod walls may be fibrous or non-fibrous (mangetout) and, recently, leafless non-lodging genotypes have been selected (and see Chapter 5).

It appears [71] that there was a progressive spread of pea cultivation westward in prehistoric times but only in historic times did it arrive in Britain and Holland. Peas have been found in the Indian Neolithic in Bihar [36]. With the European colonisation of the Americas, Africa and Australia, the pea has achieved a world-wide distribution. This popular vegetable and pulse crop can be cultivated wherever essentially cool temperate conditions prevail, as a summer crop in northern Europe or a winter crop in the south, as a cool-season crop in the semi-arid tropics, or a year-round crop at high elevations in the tropics.

Psophocarpus tetragonolobus ($2n = 2x = 18$; winged bean, Goa bean). The geographic area of origin of this crop has not been established. All wild members of the genus are African while the domesticate itself has been established in Asia and only recently introduced to Africa [117]. This peculiar distribution led Smartt [89] to suggest that the winged bean, like guar (*Cyamopsis tetragonoloba*), is a trans-domesticate [38]. In the absence of any indigenous *Psophocarpus* species in Asia it is possible that a wild African species, most probably *P. grandiflorus*, was taken from the Horn of Africa to Asia along the ancient Sabaean trade route and domesticated there. This would be an almost exact parallel of what appears to have happened to guar. The winged bean has achieved a wide distribution without being intensively cultivated anywhere except in Burma, the Philippines and New Guinea (and more recently in Thailand). It appears

to succeed best at moderate elevations in the tropics and
may disappoint the expectations of those who hope to introduce this crop into lowland humid equatorial regions. Man-
kind has in the past been very resourceful in exploiting
crop plants; if these are 'under-exploited' there are usually good reasons but unfortunately we do not always know
what they are (and see Chapter 14).

Vicia faba (2n = 2x = 12; faba, broad, horse, tick, field
beans).   The faba bean is exceptional among the grain
legumes in that no wild prototype has yet been found; it
may even be extinct although it is possible that it still
remains to be discovered [50,51,75,126]. Complete genetic
isolation appears to exist between the domesticate and
other known *Vicia* species.

While we remain for the present ignorant of its ancestry, it has been possible to reconstruct the broad dispersal
pattern of the faba bean from its presumed centre of origin
in the Near East to central Asia, the Horn of Africa, north
Africa and virtually the whole of Europe in prehistoric
times [22,23,72]. Movement to the Americas occurred in the
sixteenth century and subsequently, where the crop has
become important at high elevations in relatively low latitudes, as in Peru [11,12,75]. Morphological diversification
has occurred most notably in the seeds. The largest-seeded
types - broad beans - are used predominantly for food. The
smaller-seeded tick beans and horse beans are also used
extensively in livestock feeds. Faba beans are notorious
for the production of an allergen which precipitates a
haemolytic anaemia, favism, in susceptible subjects (and
see Chapter 6).

Vigna radiata (green gram, mungbean) and *Vigna mungo* (black
gram, urd); both 2n = 2x = 22.   These two species are
morphologically very similar but genetically are well, but
not completely, isolated [57,121]. They were both related at
one time to the wild *Vigna sublobata* which was thought to

be ancestral to each domesticate. However, Arora *et al.*
[2] have shown that what was regarded as a single species
V. sublobata actually included two distinct forms which
were not freely cross-compatible with each other but which
could be crossed easily, one with *V. radiata* the other with
V. mungo. Lukoki *et al.* [58] propose that these wild forms
be named *V. radiata* var. *sublobata* and *V. mungo* var. *silvestris*, respectively.

Although black gram is commonly grown in India (where
domestication most probably occurred) it is less popular
elsewhere than green gram (another Indian domesticate). The
latter is a very acceptable grain legume and finds considerable use as bean sprouts. Its yield potential is apparently small which has precluded wider-scale cultivation.
Since the time of Columbus it has been introduced into the
New World and is grown to supply basic ingredients for
Asiatic cuisines (and see Chapter 13).

Vigna unguiculata ($2n = 2x = 22$; the cowpea, black-eyed
pea, yard-long bean, asparagus bean). The cowpea can be
regarded very much as the Old World counterpart of the
Phaseolus beans. Its products, seeds and fruits, can be
utilised in precisely similar ways. The similarity probably
explains its ready acceptance in the New World and that of
Phaseolus beans in the Old.

It is now generally accepted that the cowpea has an African
origin [101]. A number of different wild cowpeas are found in
sub-Saharan Africa which can be crossed with the domesticate;
these are now recognised as subspecies of *Vigna unguiculata* [61].
The type most clearly related to the ancestry of the cowpea
is subsp. *dekindtiana* within which two varieties have been
recognised, var. *dekindtiana* (the truly wild cowpea) and
var. *mensensis* (the weedy form).

Steele [92] originally favoured an Ethiopian origin for
the cowpea from which location it could have been dispersed
across the Sudan zone to West Africa and in a southerly and
south-westerly direction to east, central and southern
Africa. But subsequently Steele and Mehra [99] have proposed

64

a West African origin. However, it is possible that domestication occurred in both areas and perhaps others as well.
This African domestication produced the typical cowpea selected as a pulse. This material was apparently transported
to India by the Sabaean trade route where it not only became
established but also produced two new and distinct forms
'cylindrica', an erect growing, forage type and 'sesquipedalis', a long-podded type selected for the use of immature fruits as a green vegetable.

Steele [82] believed that the cowpea was established in
West Africa some time after 3000 BC, that it reached India
between approximately 1500 and 1000 BC, the Far East
during the first millenium BC and southern Europe prior to
300 BC. The Spanish (and probably also the Portuguese)
introduced the cowpea to the New World where it became
widespread. It has flourished particularly as a crop in the
south-eastern United States where breeding and selection
have established productive, erect plant types (and see
Chapter 12).

General conclusions on grain legume evolution

Hutchinson [34] was the first to outline broad evolutionary
trends in the pulses. This review was broadened and expanded
to include the legume oilseeds by Smartt [85-88,90,91].
Comparative studies such as these are of value in that they
have demonstrated that certain common evolutionary responses
and trends have emerged under artificial selection pressures. These trends are particularly strong in herbaceous
species, less marked in shrubby forms such as *Cajanus cajan*
and *Lupinus* spp., and are well shown in the present consideration of changes established under domestication [73].

Although the responses evoked under artificial selection
provide an excellent set of examples illustrating Vavilov's
principle of homologous series, this parallelism has evolved
from markedly different biosystematic bases. This can in a
broad sense be related to the apparent evolutionary ages of

65

the different taxa involved. It is clear on the one hand
that the *Lupinus* species which have been domesticated
represent ancient evolutionary lineages [20] (since species
are genetically well isolated from each other), while the
section Arachis of the genus *Arachis* to which the cultivated
groundnut belongs is of comparatively recent origin (because
of incomplete genetical isolation between species). It is
useful to bear these considerations in mind when considering
future possibilities of the 'man-directed' evolution of our
grain legume crops.

The major practical return from the study of crop evolution is that it should improve the foundation upon which
we derive breeding strategies for crop improvement. Perhaps
most important of all it should enable us to evaluate objec-
tively the genetic resources we have at our disposal and
enable us to use these effectively. It is an inescapable
fact that the extent of readily exploitable germplasm in
the individual domesticate and its wild relatives varies
enormously (see Table 2.7). Germplasm resources are best
considered in relation to the hierarchical system of gene-
pools proposed by Harlan and de Wet [26]. This basically
categorises gene-pools as 'primary', including domesticate
and conspecific wild forms, equatable with the biological
species; 'secondary', which includes those biological species which can exchange genes with the domesticate, that is
belonging to the same coenospecies; and 'tertiary', which
includes all those coenospecies which can hybridise with
members of that coenospecies to which the domesticate
belongs, but between which gene exchange does not apparently
occur. The tertiary gene-pool may be equated with the
experimental taxonomic entity the comparium. The genetic
resources it represents will require the development of
sophisticated new techniques to transfer genetic information
from this gene-pool to the domesticate. This would probably
be difficult to achieve and the degree of expression of the
transferred character may be less in the domesticate than
in the donor species and represent a poor return for the

Genus	No. of species	Domesticate	1A	1B	2	3A	3B
					Gene-pools (GP)*		
Arachis	±60	A. hypogaea	+	+	+		+
		A. villosulicarpa	+	+	+	+	+
Cajanus (Atylosia)	2 + 35 sp. of Atylosia	C. cajan	+	+	+		+
Cicer	40	C. arietinum	+	+	+	+	
Cyamopsis	4	C. tetragonoloba	+	+			
Glycine	9	G. max	+	+		+	
Lablab	1	L. niger	+	+	?+ or	?+	+
Lathyrus	150	L. sativus	undifferentiated	+			
Lens	5	L. culinaris	+	+	+		
Macrotyloma	24	M. geocarpum	+	+			
		M. uniflorum	+		?+ or	?+	
Pachyrrhizus	6	P. ahipa	+	+			
		P. erosus	undifferentiated	+			
Phaseolus	50	P. tuberosus	undifferentiated	+			
		P. acutifolius	+	+			+
		P. coccineus	+	+	+	+	+
		P. lunatus	+	+		+	+
		P. vulgaris	+	+	+	+	+
Pisum	2	P. sativum	+	+			+
Psophocarpus	9	P. tetragonolobus	+	+	+		+
Vicia	140	V. faba	+				+
Vigna	150	V. aconitifolia	undifferentiated	+		+	
		V. angularis	+	+	+	+	
		V. mungo	+	+	+	+	
		V. radiata	+	+			
		V. subterranea	+	+			
		V. umbellata	+	+	+	+	
		V. unguiculata	+	+			

*Lupinus spp., apparently, are genetically closed gene-pools [19], and no primary (GP1) or secondary (GP2) gene-pools probably exist. GP1A and GP1B represent the domesticated and wild components of the primary gene-pool [26]; GP2 is the secondary gene-pool; GP3A and GP3B are divisions of the tertiary gene-pool referring to species which form viable and inviable hybrids with domesticates, respectively.

effort expended.

Ease of accessibility of germplasm for breeding is
obviously desirable and the most genetically accessible
germplasm resources must be collected, evaluated, conserved
and protected from erosion. The most accessible resources
comprise extant cultivars, landraces and breeding lines of
the domesticate (GP1A) together with the wild forms which
belong to the same biological species (GP1B). Because exploitation of the latter in breeding programmes inevitably
entails stringent selection against wild characters, if the
alleles selected against are closely linked to those selected for, then difficulties may arise. For this reason it
is useful, at times, to distinguish between domesticated
and wild segments of the primary gene-pool (i.e. GP1A and
GP1B, respectively). Exploitation of secondary gene-pools
implies similar but more extreme difficulties. At present
there are no good reasons for proposing any division of
the secondary gene-pool, but a case can be made for dividing the tertiary gene-pool including in one group all species forming viable but sterile hybrids with the domesticate
(GP3A) and in another group (GP3B) all species producing
inviable hybrids. This distinction recognises that the two
segments of the tertiary gene-pool could differ fundament-
ally in their developmental compatibility with their related
domesticates. Recognition of this distinction might become
more important in those domesticates in which primary and
secondary gene-pools were extremely limited and in which
the means of exploiting tertiary gene pools had been discovered.

Clearly, with limited financial resources, top priority
must be given to collection and conservation of the primary
gene-pool (GP1) with less priority to the secondary (GP2)
and least to the tertiary one (GP3). However, emergency
rescue collections of wild materials may have to be made
before their status as germplasm resources is properly
understood, as in areas where rapid agricultural development
is being undertaken. Such development must obviously proceed

but without irretrievable loss of crop genetic resources if this can possibly be avoided.

Studies on grain legume evolution have over the past three decades enabled us to arrive at a much better assessment of the value of the crops we have, the potential for their improvement and the nature and extent of the resources available for this task. They also enable us to construct more realistic blue-prints for the future man-directed evolution of these crops than would otherwise be possible.

LITERATURE CITED

1. Archer, T.C. (1853), *Popular Economic Botany*. London: Reeve.
2. Arora, R.K., Chandel, K.P.S. and Joshi, B.S. (1973), *Current Science* 42, 359-61.
3. Baudet, J.C. (1977), *Bulletin de la Société Royale de Botanique de Belgique* 110, 65-76.
4. Ben-Ze'ev, N. and Zohary, D. (1973), *Israel Journal of Botany* 22, 73-91.
5. Bergland-Brücher, O. and Brücher, H. (1976), *Economic Botany* 30, 257-72.
6. Brücher, H. (1978), *Angewandte Botanik* 42, 119-28.
7. Brücher, H. (1974), *Umschau* 74, 672-7.
8. Burkart, A. and Brücher, H. (1953), *Züchter* 23, 65-72.
9. Chowdhury, K.A., Saraswat, K.S., Hason, S.N. and Gaur, R.C. (1971), *Science and Culture* 37, 531-2.
10. Clausen, R.T. (1944), *Memoir 264*. Ithaca, New York: Cornell University Agricultural Experiment Station.
11. Cubero, J.I. (1973), *Theoretical and Applied Genetics* 43, 59-65.
12. Cubero, J.I. (1974), *Theoretical and Applied Genetics* 45, 47-51.
13. De, D.N. (1976), *Indian Journal of Genetics and Plant Breeding* 36, 141-2.
14. De Candolle, A.P. (1886), *Origin of Cultivated Plants* (2nd edn), reprinted 1959. New York: Hafner.
15. Duke, J.A., Okigbo, B.N. and Reed, C.F. (1977), *Tropical Grain Legume Bulletin* 10, 12-13.
16. Freeman, G.F. (1912), *University of Arizona Experiment Station Bulletin 68*. Tucson, Arizona: University of Arizona.
17. Gentry, H.S. (1969), *Economic Botany* 23, 55-69.
18. Gibbons, R.W., Bunting, A.H. and Smartt, J. (1972), *Euphytica* 21, 78-85.
19. Gladstones, J.S. (1970), *Field Crop Abstracts* 23, 123-48.
20. Gladstones, J.S. (1974), *Technical Bulletin No. 26*. South Perth, Western Australia: Western Australian Department of Agriculture.
21. Hammons, R.O. (1973), in *Peanuts - Culture and Uses*. Stillwater, Oklahoma: American Peanut Research and Education Association.
22. Hanelt, P. (1972), *Kulturpflanze* 20, 75-128.
23. Hanelt, P. (1972), *Kulturpflanze* 20, 209-23.
24. Hansen, J.M. (1978), *Berichte der Deutschen Botanischen Gesell-*

schaft 91, 39-46.

25. Harlan, J.R. (1977), Mimeograph Bulletin from the Crop Evolution Laboratory, University of Illinois, Urbana.
26. Harlan, J.R. and de Wet, J.M.J. (1971), *Taxon* 20, 509-17.
27. Helbaek, H. (1959), *Science* 153, 1074-80.
28. Helbaek, H. (1963), *Acta Institi Atheniensis Regni Sueciae* (Series 4) 8, 171-86.
29. Helbaek, H. (1964), *Anatolian Studies* 14, 121f.
30. Helbaek, H. (1965), *Sumer* 19, 27-35.
31. Hepper, F.N. (1963), *Kew Bulletin* 16, 395-407.
32. Hepper, F.N. (1970), *Field Crop Abstracts* 23, 1-6.
33. Hermann, F.J. (1962), *Technical Bulletin* no. 1268. Washington, DC: USDA.
34. Hutchinson, J.B. (1969), *Proceedings of the Nutrition Society* 29, 49-55.
35-36. Hutchinson, J.B. (*ed.*) (1974), *Evolutionary Studies in World Crops*. Cambridge: Cambridge University Press.
35. De, D.M., 'Pigeon pea'. 9-87.
36. Vishnu-Mittre, 'Palaeobotanical evidence in India'. 3-30.
37. Hymowitz, T. (1970), *Economic Botany* 24, 408-21.
38. Hymowitz, T. (1972), *Economic Botany* 26, 49-60.
39. Hymowitz, T. and Boyd, J. (1977), *Economic Botany* 31, 180-88.
40. Hymowitz, T. and Harlan, J.R. (1983), *Economic Botany* 37, 371-99.
41. Hymowitz, T. and Newell, C.A. (1981), *Economic Botany* 35, 272-88.
42. International Institute of Tropical Agriculture (1978), *Winged Beans - Abstracts of World Literature 1900 - 1977*. Ibadan: International Grain Legume Information Centre.
43. Johnson, F.R. (1964), *The Peanut Story*. Murfreesboro, North Carolina: Johnson Publishing Company.
44. Kaplan, L. (1965), *Economic Botany* 19, 358-68.
45. Kaplan, L. (1981), *Economic Botany* 35, 240-54.
46. Kaplan, L., Lynch, T.F. and Smith, C.E.J. (1973), *Science* 79, 76-7.
47. Klozová, E. and Turková, V. (1978), *Biologia Plantarum* 20, 373-6.
48. Krapovickas, A. and Rigoni, V.A. (1967), *Darwiniana* 11, 431-55.
49. Ladizinsky, G. (1975), *Notes from the Royal Botanic Garden, Edinburgh* 34, 201-02.
50. Ladizinsky, G. (1975), *Israel Journal of Botany* 24, 80-88.
51. Ladizinsky, G. (1975), *Euphytica* 24, 785-8.
52. Ladizinsky, G. (1979), *Euphytica* 28, 179-87.
53. Ladizinsky, G. (1979), *Botanical Gazette* 140, 449-51.
54. Ladizinsky, G. and Adler, A. (1976), *Euphytica* 25, 211-17.
55. Ladizinsky, G. and Adler, A. (1976), *Theoretical and Applied Genetics* 48, 197-203.
56. Ladizinsky, G. and Adler, A. (1976), *Israel Journal of Botany* 24, 183-9.
57. Lukoki, L. (1925), *Bulletin de Recherches Agronomiques de Gembloux* 10, 372-3.
58. Lukoki, L., Maréchal, R. and Otoul, E. (1980), *Bulletin du Jardin Botanique de Belgique* 50, 385-91.
59. Maréchal, R. (1982), *Taxon* 31, 280-83.
60. Maréchal, R. and Baudet, J.C. (1977), *Bulletin du Jardin Botanique National de Belgique* 47, 49-52.
61. Maréchal, R., Mascherpa, J.M. and Stainier, F. (1978), *Boissiera* (Genève) 28, 1-273.

62. Miranda, C.S. (1968), *Agronomia Tropical* 18, 191-205.
63. Montgomery, R.D. (1964), *West Indian Medical Journal* 13, 1-11.
64. Newell, C.A. and Hymowitz, T. (1982), *Crop Science* 22, 1062-5.
65-66. Norton, G. (*ed.*) (1978), *Plant Proteins*. London: Duckworth.
65. Fowden, L., 'Non-protein nitrogen compounds'. 109-15.
66. Liener, I.E., 'Protease inhibitor and other toxic factors in seeds'. 117-40.
67. Orf, J.H. and Hymowitz, T. (1979), *Journal of the American Oil Chemists' Society* 56, 722-6.
68. Orf, J.H., Hymowitz, T., Pull, S.P. and Pueppke, S.G. (1978), *Crop Science* 18, 899-900.
69. Oviedo y Valdés, G.F. de (1527), *Sumario Historia de las Indias*. Toledo.
70. Purseglove, J.W. (1968), *Tropical Crops - Dicotyledons*. London: Longman.
71. Renfrew, J.M. (1973), *Palaeoethnobotany*. London: Methuen.
72. Schultz-Motel, J. (1972), *Kulturpflanze* 19, 321-58.
73. Schwanitz, F. (1966), *The Origins of Cultivated Plants*. Cambridge, Massachusetts: Harvard University Press.
74. Schweinfurth, G. (1884), *Nature* 29, 312-15.
75-83. Simmonds, N.W. (*ed.*) (1976), *Evolution of Crop Plants*. London: Longman.
75. Bond, D.A., 'Field Bean'. 179-82.
76. Davies, D.R., 'Peas'. 172-4.
77. Evans, A.M., 'Beans'. 168-172.
78. Hymowitz, T., 'Soybeans'. 159-62.
79. Ramanujam, S., 'Chickpea'. 157-8.
80. Royes, W.V., 'Pigeon Pea'. 154-6.
81. Smith, P.M., 'Minor Crops'. 301-24.
82. Steele, W.M., 'Cowpeas'. 183-5.
83. Zohary, D., 'Lentils'. 163-4.
84. Singh, K.B. and van der Maesen, L.J.G. (1977), *Chickpea Bibliography 1930 - 1974*. Hyderabad, India: International Crops Research Institute for the Semi-Arid Tropics.
85. Smartt, J. (1976), *Tropical Pulses*. London: Longman.
86. Smartt, J. (1976), *Euphytica* 25, 139-43.
87. Smartt, J. (1978), *Economic Botany* 32, 185-98.
88. Smartt, J. (1980), *Economic Botany* 34, 219-35.
89. Smartt, J. (1980), *Euphytica* 29, 121-3.
90. Smartt, J. (1981), *Euphytica* 30, 415-18.
91. Smartt, J. (1981), *Euphytica* 30, 445-9.
92-100. Summerfield, R.J. and Bunting, A.H. (*eds*) (1980), *Advances in Legume Science*. Kew: Royal Botanic Gardens.
92. Davies, D.R., 'Crop structure and yield in *Pisum*'. 637-41.
93. Evans, A.M., 'Structure, variation, evolution and classification in *Phaseolus*'. 337-47.
94. Gladstones, J.S., 'Recent developments in the understanding, improvement and use of *Lupinus*'. 603-11.
95. Gregory, W.C., Krapovickas, A. and Gregory, M.P., 'Structure, variation, evolution and classification in *Arachis*'. 469-81.
96. Hymowitz, T. and Newell, C.A. 'Taxonomy, speciation, domestication, dissemination, germplasm resources and variation in the genus *Glycine*'. 251-64.
97. Jain, H.K. and Mehra, K.L., 'Evolutionary adaptation, relationships and the uses of the species of *Vigna* cultivated in India'. 459-68.

98. Lawes, D.A., 'Recent developments in understanding, improvement and use of *Vicia faba*'. 625-36.
99. Steele, W.M. and Mehra, K.L., 'Structure, evolution and adaptation to farming systems and environments in *Vigna*'. 393-404.
100. Verdcourt, B., 'The classification of *Dolichos* L. emend Verdc., *Lablab* Adans., *Phaseolus* L., *Vigna* Savi and their allies'. 45-8.
101. Summerfield, R.J., Huxley, P.A. and Steele, W. (1974), *Field Crop Abstracts* 27, 301-12.
102. Townsend, C.C. and Guest, E. (1974), *Flora of Iraq, vol. 3, Leguminales*, Baghdad: Ministry of Agriculture and Agrarian Reform.
103-106. Ucko, P.J. and Dimbleby, G.W. (*eds*) (1969), *The Domestication and Exploitation of Plants and Animals*. London: Duckworth.
103. Allchin, F.R., 'Early cultivated plants in India and Pakistan'. 323-9.
104. Hopf, M., 'Plant remains and early farming in Jericho', 355-6.
105. Krapovickas, A. 'The origin, variability and spread of the groundnut (*Arachis hypogaea*)', 427-41.
106. Smartt, J. 'Evolution of American *Phaseolus* beans under domestication', 451-62.
107. Van der Maesen, L.J.G. (1972), Cicer L.: *A monograph on the genus with special reference to the chickpea* (Cicer arietinum) *its ecology and cultivation*. Wageningen: Mededelingen Landbouwhogeschool.
108. Van der Maesen, L.J.G. (1980), *Miscellaneous Papers, Landbouwhogeschool, Wageningen* 19, 257-62.
109. Vargas, C. (1962), *Economic Botany* 16, 106-15.
110. Vargas, F.C. (1981), *Botanical Journal of the Linnean Society* 82, 313-25.
111. Verdcourt, B. (1970), *Kew Bulletin* 24, 235-307.
112. Verdcourt, B. (1970), *Kew Bulletin* 24, 379-447.
113. Verdcourt, B. (1970), *Kew Bulletin* 24, 507-69.
114. Verdcourt, B. (1971), *Kew Bulletin* 25, 65-169.
115. Verdcourt, B. (1978), *Taxon* 27, 219-22.
116. Verdcourt, B. (1981), *Kew Bulletin* 35, 474.
117. Verdcourt, B. and Halliday, P. (1978), *Kew Bulletin* 33, 191-227.
118. Vishnu-Mittre. (1962), *Technical Report on Archaeological Remains*. Poona: Deccan College.
119. Vishnu-Mittre. (1978), *Transactions Bose Research Institute Calcutta* no. 31.
120. Vishnu-Mittre. (1974), *Palaeobotanist* 21, 18-22.
121. Watt, E.E. and Maréchal, R. (1972), *Tropical Grain Legume Bulletin* 7, 31-3.
122. Webb, C. and Hawtin, G.C. (*eds*.)(1981), *Lentils*. Farnham Royal: Commonwealth Agricultural Bureau, 15-38.
123. Williams, J.T., Sanchez, A.M.C. and Jackson, M.T. (1974), *SABRAO Journal* 6, 133-45.
124. Zhukovsky, P.M. (1925) *World Gene Pool of Plants for Breeding*. Leningrad: USSR Academy of Sciences.
125. Zohary, D. (1972), *Economic Botany* 26, 326-32.
126. Zohary, D. (1977), *Israel Journal of Botany* 26, 39-40.
127. Zohary, D. and Hopf, M. (1973), *Science* 182, 887-94.

3 Biochemical and nutritional attributes of grain legumes

G. Norton, F.A. Bliss and R. Bressani

INTRODUCTION

Pulse legumes (i.e. those species harvested traditionally
for their mature seeds) are a major source of dietary proteins and calories in food and feed products throughout the
world. They are especially important as food in those regions where animal proteins are scarce or where poverty,
religious or ethnic preferences preclude the consumption of
meat. Oilseed legumes are the major contributor to the
world supply of edible oil, as well as furnishing protein-
rich residues. Unfortunately, the proteins in grain legume
seeds are regarded as dietetically inferior to animal protein because of their low sulphur-amino acid concentrations.
This limitation in quality, coupled with the relatively
small and notoriously variable yields of many grain legume
crops, represent two major areas for improvement in the
future. In the following discussion, aspects of the chemistry, biochemistry and nutritional properties of grain
legume seeds will be considered as a prelude to the formulation of breeding strategies aimed at improving these
crops.

CHEMICAL COMPOSITION OF GRAIN LEGUME SEEDS

Characteristically, grain legume seeds have large protein
contents which may account for as much as 40% of their dry

matter (see Table 3.1) [106]. The pulse legumes are also
rich in digestible carbohydrate, mainly starch, and concentrations of 50% or greater are common; they also contain
appreciable amounts of dietary fibre. In general, legume
seeds rich in carbohydrate contain relatively small amounts
of lipid (1-2%) mainly as phospholipid. Oilseeds such as
soyabeans (*Glycine max*) and groundnuts (*Arachis hypogaea*)
are grown primarily for their oil but the residues after
oil extraction are valuable sources of protein for feed
and food.

Storage materials in legume seeds are located mainly
in the cotyledonary tissues. Embryonic axes, although rich
in protein, which is largely functional, contribute little
to the total seed protein content because they are so small.
Similarly, the testas, because they contain small amounts
of protein and only represent small proportions of the seed
mass, make only a small contribution to total seed protein
concentration.

Proteins

The major storage proteins in legume seeds are the globulins
which usually account for about 70% of the total protein
[17]. Glutelins (10-20%) and albumins (10-20%) make up the
remainder. The principal storage globulins in most legumes
are those which sediment at 11S (legumin) and 7S (vicilin)
[51,94]. Minor amounts of other globulins are also present.
While legumin is the major storage globulin in peas (*Pisum
sativum*) and faba beans (*Vicia faba*), vicilin predominates
in common beans (*Phaseolus vulgaris*) and in mung beans
(*Vigna radiata*).

Details of the subunit structures and amino acid compositions along with other physical properties of storage
proteins have been collated and in many respects legumins
from different sources resemble each other closely [51,94].
The majority of legumins (M_r*300 000-400 000) contain large

*M_r denotes molecular mass.

Table 3.1 Composition of legume seeds (g 100g seed^{-1} or, for limiting amino acids, mg g N^{-1})*.

| Crops | Water | Protein (N × 6.25) | Lipid | Carbohydrate | | | Dietary fibre | Limiting amino acids | | | | |
				Sugar	Starch	Total		Met	Cys	Thr	Try	Val
PULSES												
Peas	13.3	21.6	1.3	2.4	47.6	50.0	16.7	60	70	250	60	290
Faba beans	13.8	25.0	1.2	ND	ND	56.9	ND	44	50	210	ND	280
Lentils	12.2	23.8	1.0	2.4	50.8	53.2	11.7	50	60	250	60	310
Chickpeas	9.9	20.6	5.7	10.0	40.0	50.0	15.0	80	90	240	50	280
Common beans	11.0	22.1	1.7	3.0	42.0	45.0	25.0	70	50	250	60	290
Lima beans	11.6	19.1	1.1	3.6	46.2	49.8	21.6	90	90	260	60	320
Cowpeas	11.5	22.7	1.6	ND	ND	61.0	ND	75	70	225	70	280
Mung beans (*V. radiata*)	12.0	22.0	1.0	1.2	34.4	35.6	22.0	30	40	210	ND	260
(*V. mungo*)	10.6	21.0	1.6	ND	ND	63.4	19.5	90	60	230	ND	370
Winged beans	9.7	32.8	17.0	ND	ND	36.5	ND	75	100	270	60	300
Pigeonpeas	10.0	20.0	2.0	9.0	45.0	54.0	15.0	80	70	180	30	230
Lupins	8.0	44.3	16.5	ND	ND	28.2	ND	50	90	225	60	250
OILSEEDS												
Soyabeans	7.0	36.8**	23.5	11.2	12.3	23.5	11.9	80	100	240	80	300
Groundnuts	4.5	24.3+	49.0	2.1	3.8	5.9	8.1	70	80	160	70	260

ND: Not determined

*Compiled from: Aykroyd, W.R. (1982), in Aykroyd, W.R., Doughty, J. and Walker, A. (eds), *Legumes in Human Nutrition* (2nd edn). FAO Nutrition Paper no. 20. Rome: FAO; Anon. (1975), *The Winged Bean: A High Protein Crop for the Tropics*. Washington, DC: National Academy of Sciences; and Paul, A.A. and Southgate, D.A.T. (1978), *McCance and Widdowson's The Composition of Foods* (4th edn). Medical Research Council Special Report no. 297. London: HMSO. It is important to note that reliable information on composition is difficult to find and that values are often contradictory. The distinction between 'pulse' and 'oilseed' species is somewhat arbitrary for winged beans and lupins.

** N × 5.71   + N × 5.41

amounts of acidic amino acids and their amides, as well as arginine. In general, the amounts of sulphur-amino acids, half-cystine and methionine, are relatively small and carbohydrate is absent. Legumins from different sources seem to have the same subunit structures and similar amino acid compositions. Two types of subunits, one acidic (α-subunit, M_r about 40000) and rich in glutamic acid and glutamine, and the other basic (β-subunit, M_r about 20000) with larger amounts of alanine, valine and leucine, appear to be present in equimolar amounts. Pairs of acidic and basic subunits are generally held together by disulphide bonds to form a mono-mer (M_r about 60000) [51,94,108]. Legumin consists of six monomers arranged in a specific manner so that the subunits are packed in two identical hexagons arranged one on top of the other with each acidic subunit opposite three basic subunits and vice versa. Legumins from different sources have been found to be related immunologically [51].

Vicilins are usually rich in acidic amino acids, mainly present as amides, but contain little or no half-cystine, methionine or tryptophan. The 7S proteins from faba beans, soyabeans, mung beans and common beans are glycosylated. Although the 7S globulins from different sources resemble each other chemically, they are a heterogeneous group with complex subunit structure. Few vicilins have been investi-gated in detail. The major vicilin-like globulin in soya-beans, β-conglycinin, was shown to exist in six isomeric forms, B_1-B_6 conglycinins (M_r about 150 000-175 000) [136]. Each conglycinin isomer contained three subunits per mole-cule and three major subunits were identified, viz. α and α^1, with identical amino acid composition (M_r 54000), and β (M_r 42000). The subunit composition of these isotrimers was identified as B_1, $\alpha^1 \beta_2$; $B_2, \alpha\beta_2$; $B_3, \alpha\alpha^1 \beta$; $B_4 \alpha_2\beta$; $B_5\alpha_2\alpha$; and $B_6\alpha_3$; thus, only B_6 was a homotrimer. Many of the similarities and differences in properties between the isomeric conglycinins are consistent with their sugar, N-terminal and amino acid compositions. A cyclic structure for the subunits in the 7S protein was proposed. These dissociated at extreme pH but associated to form a dimer of two identical assemblies facing each other. All the subunits were acidic, lacked both cysteine and methionine and contained 3.5-5.0% carbohydrate (mannose and glucosamine). Glycoprotein II (Phaseolin or G1 globulin) from common beans was found to consist of four major subunits (M_r 53000, 50000, 47000 and 43000) and named α, β, γ and δ, respectively [117]. These subunits, but mainly α, β and δ, were associated into heterotrimers (M_r 142 000) which were converted into stable tetramers at extreme pH (4 trimers or 12 subunits; M_r 560 000). Glycoprotein II was microheterogeneous and a number of fractions were obtained in which the proportions of the major subunits varied from 1:1:1 to 3:3:1. These ratios changed during seed development.

Three phaseolin types, viz. T (Tendergreen), S (Sanilac) and C (Contender) have been identified following a survey of more than one hundred bean cultivars [12,31]. T and S phaseolins possessed specific polypeptide patterns while C types had intermediate patterns. A total of fourteen constituent polypeptides (M_r 45000-51000 and pI* 5.6-5.8) were found in these phaseolins. Because of the similarities in molecular weight and the homology observed from peptide mapping it was concluded that the polypeptides were similar proteins. Vicilins from other species including peas, faba beans [60] and mung beans [79] may also have a trimeric structure. From current information it seems that vicilins are heterogeneous glycoproteins devoid of disulphide linkages and which usually consist of three subunits not covalently linked.

Another major storage protein has been isolated from peas and faba beans [46]. This protein was named convicilin because it was normally associated with the vicilin fraction and because of its cross-reactivity with the antisera of such fractions. Convicilin (M_r 290 000) is readily separated from vicilin by means of non-dissociating techniques. It

---

*pI denotes the isoelectric pH, i.e. that pH at which the net charge on the molecule is zero.

consists of four subunits (M_r 71000) and is devoid of carbohydrate. Each subunit contains a single half-cystine and
methionine residue per molecule. Convicilin is antigenic-
ally dissimilar to legumin.

Legume seeds also contain appreciable amounts of albumins. These proteins are water-soluble and mainly functional. They include the enzymes necessary for cellular
functions at the onset of germination and the proteinase
inhibitors which, in some seeds, can account for a large
proportion of the albumins. In terms of composition the
albumins are richer in the sulphur-amino acids and other
essential amino acids than the globulins [17].

Several heat-labile antinutritional factors, including
proteinase and carbohydrase inhibitors and haemagglutinins,
are also found in legume seeds. Proteinase inhibitors (PI)
are small proteins devoid of carbohydrate with M_r values
ranging from 4000 to 20000 [120]. In general, they consist
of a single peptide which may exist in several forms in iso-
inhibitors. Although the majority of PI inhibit trypsin,
many possess a broad specificity and will also inhibit
chymotrypsin and other serine proteinases. They are usually
all inactive or only weakly active towards the sulphydryl-,
metallo- and acidic-proteinases. A few PI are extremely
specific for either trypsin or chymotrypsin. Some, with a
broad specificity, possess a single reactive site whereas
others are capable of inhibiting two enzymes (trypsin and
chymotrypsin) simultaneously and are termed double-headed
or polyvalent. The amino acid sequence of a number of PI
is known including the Kunitz and Bowman-Birk inhibitors
from soyabeans and those from Lima beans (*Phaseolus lunatus*)
and common beans [119]. Apart from the Kunitz inhibitor,
many PI are rich in half-cystine which may account for as
much as 15% of the amino acid residues in some cases. The
function of these inhibitors remains a matter of conjecture.
They may be involved in defence mechanisms against insects
and micro-organisms, the regulation of endogenous proteinases or may simply serve as protein reserves rich in

sulphur-amino acids. This last alternative may be especially important in seeds with small concentrations of these acids. For example, in common beans between 30% and 40% of the total seed cystein has been calculated to be present in these inhibitors [112]. The location of PI in cells is uncertain: they are present mainly in the protein bodies in common beans [115] but are in the cytoplasm of cells of mung beans [40].

An inhibitor specific for mammalian α-amylase has been isolated from common beans and purified to homogeneity [90]. This inhibitor was identified as a glycoprotein (M_r 49000) and formed a stable complex with pancreatic α-amylase in a 1:1 molar ratio. Information on the amino acid composition, cellular location and the nutritional significance of this inhibitor is not yet available.

Lectins are widely distributed in legume seeds. They are glycoproteins except for concanavalin A from *Concanavalia ensiformis* and the lectins from groundnuts. In some species, lectins can achieve large concentrations; in certain common beans, for example, they can amount to 10% of the total protein. Such large concentrations are five or six times greater than those found in *Concanavalia ensiformis* and soyabeans. Lectins from soyabeans, common beans and Lima beans have been characterised [94]. All have a tetramic structure (M_r about 120 000) but different subunit compositions.

The lectins of common beans have been investigated in some detail. Bean cultivars were classified into four types (A, B, C and D) according to the specific blood cell agglutinating properties and mitogenic activity of their lectins [73]. Pusztai *et al*. [114], however, identified three lectin types (1, 2, and 3) and described their subunit composition after one-dimensional sodium dodecyl-sulphate-polyacrylamide gel electrophoresis (SDS-PAGE). Following a survey of the lectin-containing globulin-2/albumin proteins of more than one hundred cultivars by two-dimensional electrophoresis, Brown *et al*. [34] identified eight types based on polypeptide patterns. All types, with the exception

of Pi_{G2} (which was devoid of globulin-2/albumins), contained
polypeptides with M_r 30000-41000 and pI 5.0-5.5. Each type
was classified according to agglutination pattern using
the criteria of Jaffé *et al.* [73]. Thus, A types (T_{G2}, B_{G2},
M_{G2}) predominated; B types (S_{G2}, V_{G2}, Pr_{G2}) were less numerous and C (P_{G2}) and D (Pi_{G2}) types were rare. The specific
lectin polypeptides involved in the agglutination of erythrocytes were identified. Lectins of common beans seem to
be restricted to the protein bodies and their membranes.

Lectins from soyabeans and common beans contain 4.5%
neutral sugar (mannose) and 1.5% N-acetylglucosamine, and
6.0-6.5% neutral sugar and 1.6% glucosamine, respectively.
The carbohydrate-peptide linkage in lectins from soyabeans
and Lima beans is N-glycosidic involving N-acetyl glucosamine and asparagine [126].

Lipids

Soyabeans and groundnuts are the two most important oilseeds cultivated. Winged beans (*Psophocarpus tetragonolobus*) and lupins (*Lupinus* spp.) are also being investigated
as potential oilseed crops. Approximately 25% and 50% of
the dry matter of soyabeans and groundnuts, respectively,
is lipid, largely in the form of oil or triacyglycerol
(TAG; see Table 3.1). As well as neutral lipid, these
seeds also contain functional phospho- and glycolipids but
these, together, account for less than 10% of the total
lipid. About 60% of the fatty acids in soyabean oil are
the polyunsaturated, linoleic (18:2) and linolenic (18:3)
acids. Groundnut oil contains 50% and 30% of its constituent fatty acids as oleic acid (18:1) and (18:2), respectively, and as a consequence is of poorer quality than soyabean oil. Pulse legume seeds (Table 3.1) contain small
amounts of lipid mainly in the form of phospholipid. This
lipid is rich in (18:1) and (18:2). Functional lipids and
TAG are discussed in greater detail below.

80

Carbohydrates

Commercially important pulse legume seeds contain starch
as the major reserve polysaccharide in amounts which vary
from 15% to 65% of seed dry weight. Starch is also present
in mature soyabean seeds but in much smaller amounts [152].
The proportions of amylose and amylopectin in legume star-
ches vary considerably both within and between species. In
round and wrinkled pea seeds, for example, the starch contained 37% and 69% amylose, respectively [76]. Variable
amounts of small molecular weight carbohydrates, such as
sucrose, as well as sucrosyl oligosaccharides including
raffinose, stachyose and verbascose, are found in most
legume seeds [39,86,152]. In some instances these alcohol-
soluble sugars may represent 10% of the seed dry weight
[39]. Sucrose, stachyose and verbascose usually predominate
but there is considerable variation between species. Since
all of these oligosaccharides contain galactopyranose residues in α-(1→6) linkages, they are indigestible and are
partly responsible for the flatulence which can result from
the ingestion of legume seeds.

Evidence has recently been presented [93] that the
galactose-rich polysaccharides (galactans) responsible for
the heavy thickening of the cell walls of *Lupinus* spp., and
sometimes accounting for as much as 30% of seed weight, act
as a reserve carbohydrate which is mobilised during germination. Many legume seeds have been found to contain 'amy-
loids' or xyloglucans in either cell walls or endosperm.
As with the cell wall galactans, these xyloglucans may be
reserve carbohydrates which are utilised during germination
[93].

SEED DEVELOPMENT

Embryo and seed development

These topics have been reviewed fully in recent years

81

[53,97] and only broad outlines will be considered here. Legume seed embryogenesis can be divided into three stages. Following fertilisation of the ovule, rapid growth of the pod wall, testa and seed endosperm ensues. Cell division in the embryo is rapid and is completed by the end of the first stage. At this time the embryo is still small and the seed coat accounts for the major part of the seed. During the second stage, growth of the embryo is rapid to accommodate the intracellular deposition of storage materials (proteins, carbohydrates or oil and phytates) in cotyledonary cells. RNA and nuclear DNA continue to be synthesised in the cotyledonary cells and levels of the latter may achieve 32-64C* or greater. Cellular contents of DNA and RNA, mainly ribosomal RNA (rRNA) and transfer RNA (tRNA), reach constant maxima well before the end of this second stage and while globulin synthesis is still active. Messenger RNA (mRNA) or polyadenylated $^{+}$RNA (poly A^{+}RNA), accounting for a constant 1-2% of the total RNA, turns over rapidly. By the end of this stage the predominant tissues in the seed are cotyledons whose cells contain the storage proteins, carbohydrates and/or oil. The embryo proper is small and consists of a radicle, plumule and embryonic axis Desiccation of the seed occurs throughout stage three. The durations of these stages vary between species and are influenced by external factors such as temperature and water stress.

Protein synthesis and deposition

The process of protein biosynthesis has been thoroughly reviewed only recently [97,149] and will not be discussed here.

Sites and products of protein synthesis. Early work indicated that rough endoplasmic reticulum (RER) was an important site for the synthesis of storage protein which

---

*2C represents the diploid nuclear DNA content.

subsequently accumulated (after a period of 20-30 minutes) in protein bodies [5]. Specific probes for detecting storage protein (ferritin coupled antibodies) identified their location in the ER cisternae [8]. That the RER is the site of storage protein synthesis has recently been confirmed [41]. *In vivo* labelling experiments with developing pea cotyledons followed by fractionation of the tissue extracts on sucrose density gradients revealed that newly synthesised protein was sequestered by the RER prior to transportation to the protein bodies. Legumin was present in the ER as a family of peptide precursors (M_r 60000-65000) while vicilin-like peptides appeared only as molecules with M_r 40000, 50000, 70000 and 75000. The subunits present in mature legumin (M_r 40000 and 20000) and those of vicilin (M_r 49000) were absent. Glycosylation of the polypeptides M_r 50000 and 70000 of vicilin took place in the RER. After synthesis, the storage protein precursors were transported to the protein bodies within 20-30 minutes in vesicles derived from the ER. Prior to transportation, the legumin and vicilin precursor peptides were assembled in the ER to form 8S (three polypeptides M_r about 60000) and 7S (three polypeptides M_r 49000) oligomers, respectively [42]. In the protein bodies, proteolytic processing of the polypeptides while still associated in oligomers occurred. Each legumin monomer (M_r about 60000) was cleaved to produce the subunits M_r 40000 and 20000, characteristic of mature legumin. This process, and the assembly of the subunits to form the 12S hexamer, occurred within two or three hours. The appearance of the smaller vicilin polypeptides (M_r 12000-34000) was much slower (6-20 hours) and these were cleaved from the polypeptides M_r 49000-50000. Only the M_r 14000 polypeptide contained carbohydrate.

In peas, true vicilin, as distinct from convicilin, synthesised *in vitro* and *in vivo* consisted of polypeptides M_r 50000 (including polypeptides M_r 47000) [60]. *In vivo* these were assembled into trimers (M_r 170 000) and in this association only the polypeptides M_r 47000 were processed

83

(within about four hours) to produce the smaller peptides [61].

Translation products of polysomes and poly A-containing RNA from developing pea, faba bean and soyabean seeds in *in vitro* synthesising systems must undergo considerable co- and post-translational modification to produce the subunit polypeptides characteristic of the mature storage proteins. Vicilin, convicilin and legumin are formed as precursor polypeptides [44,45,60,70], probably in accordance with the 'signal' hypothesis [60]. Legumin was synthesised *in vitro* in pea and broad bean systems as a precursor polypeptide (M_r 60000) which *in vivo* would be cleaved to yield the classical legumin subunits (M_r 40000 and 20000). Vicilin precursors are synthesised *in vitro* as polypeptides (M_r 50000) which would be glycosylated and cleaved to yield the spectrum of subunit polypeptides found in the vicilins of mature seeds.

Timing of synthesis of individual globulins.   The major storage globulins are not synthesised in the same proportions throughout seed development. Although legumin is synthesised during very early pea seed development [52], vicilin accumulates faster than legumin initially but the latter continues to be synthesised for much longer. Convicilin also accumulates towards the end of the storage protein deposition stage in peas and faba beans [46]. Since very little turnover of storage protein occurs, the changes in capacity for the synthesis of specific proteins must be a consequence of changes in the amounts of the respective mRNA. Increasing proportions of legumin-specific mRNA in the total poly A^+RNA from faba bean seeds have been demonstrated from *in vitro* protein synthesis studies [97]. Regulation of the synthesis of legumin and other storage proteins is therefore mediated at the transcription level.

Storage protein localisation.   Storage proteins are located entirely in the protein bodies of various cells in mature legume seeds. These bodies are spherical organelles normally 1-3μm in diameter but may be as large as 10-20μm, and bounded by a single unit membrane [107]. Protein bodies contain 70-80% storage protein, 10% salts of phytic acid, as well as certain enzymes and inorganic constituents. Globoid inclusions are now regarded as a feature of all protein bodies [87]. These globoids are large and numerous in the protein bodies of groundnuts, *Clianthus formosus* and *Cassia artemisoides* but small and rare in those of faba beans, soyabeans, common beans, peas and *Acacia conferta*. Large concentrations of orthophosphate, K^+, Mg^{2+} and occasionally Ca^{2+} have been found in globoids which is consistent with them being rich in phytin. The protein matrix of some protein bodies also contains phytin and other ions including Mn^{2+}, Cl^- and Na^+. Acid phosphatase (phytase), but not protein-hydrolysing enzymes, can also be present.

Protein bodies often contain other proteins apart from the major storage ones. Lectins, for example, may account for almost 10% of the protein bodies of common beans and could be classified as storage proteins [114]. Protein bodies can also be heterogeneous with regard to their constituent storage proteins. Both legumin and vicilin are generally present in the same protein body but some contain only vicilin.

The membrane fraction from the protein bodies of common beans consists of five polypeptides in addition to small molecular weight glycolipid-type material [116]. Two of these polypeptides (M_r 25000 and 78000) were regarded as integral components of the membrane structure and the smaller component had an amino acid composition similar to that of other plant membranes. A polypeptide (M_r 30000) was identified as an agglutinating globulin (isolectin) but the role of this component in the membranes was unknown. The origin of these membranes, i.e. whether derived from the tonoplast or ER, was not investigated.

85

Lipid synthesis and accumulation

Lipid changes in developing legume oilseeds.

Studies on the lipid changes which occur during seed development have been concentrated on soyabeans. Little or no TAG are synthesised prior to the completion of embryo cell division. Seeds shortly after anthesis have small lipid contents which are mainly functional phospholipid and glycolipid (PL and GL) [118,121,144]. Extensive synthesis of TAG occurs during the phase of rapid dry matter accumulation (oil and protein) until, at maturity, oil accounts for 90% of the total lipid which, in turn, is approximately 20% of final seed dry weight. About 94% of the mature soyabean seed weight is in cotyledonary tissue.

Shortly after anthesis almost 30% of the total seed lipid was glycolipid (diacylgalactosyl - and diacyldigalactosyl-glycerol) which must have been located in the plastids [118]. The proportion of GL in the total lipid decreased with time and by maturity it had declined to 1.6%. Phospholipids were the major functional lipids of soyabean seeds throughout development. Shortly after flowering approximately 50% of the total lipid was phosphatidyl choline (PC) but this declined to less than 10% in mature seeds. Although the proportion of the total lipid in PL and GL declined with time, there was considerable synthesis of these lipids especially in the cotyledonary tissue. Conflicting reports have described the composition of the PL fraction [118,131, 144]. Provided suitable precautions are taken to inhibit phospholipase D during lipid extraction, the major PL in immature soyabean seeds were PC (45-58%), phosphatidyl ethanolamine (PE; 24-28%), phosphatidyl inositol (15-18%) and small amounts of phosphatidyl glycerol (PG; 4-6%) [109, 124]. Phosphatidic acid (PA) was only a very minor component throughout.

Changes in the content and composition of lipids throughout soyabean seed development were reflected in fatty acid (FA) composition [118,145]. Shortly after anthesis,

seed lipids were rich in 16:0 (about 40%) and 18:0 (about
25%) but contained smaller amounts of 16:1, 18:1, 18:2 and
18:3. Within a few days the content of 16:0 and 18:0
decreased (to 28% and 10%, respectively) while 18:2 and
18:3 increased proportionally (to about 20% each). With the
onset of TAG synthesis, 18:2 and 18:1 increased while 18:3
and 16:0 decreased until about midway through seed development when the FA composition resembled that of mature seeds.
In such seeds the lipid, largely TAG, had the following
composition: 16:0, 12%; 18:0, 4%; 18:1, 25%; 18:2, 50% and
18:3, 8% [121, 145]. The FA composition changes which follow
the onset of oil formation are due to changes in the spectrum of TAG molecular species with time. Thirteen TAG species were identified shortly after oil deposition commenced
with D_3, M_3, T_3, S_2D, S_2M, SM_2, SD_2 and M_2D* predominating
(33.0, 17.0, 8.5, 8.3, 7.8, 7.5, 6.6 and 5.0 mol %, respec-
tively) [145]. At maturity, D_2T, D_3, SD_2, M_2D, SMD and MD_2
combined accounted for 80% of the total TAG and T_3, DT_2 and
MDT were absent. Although such changes in TAG composition
have been attributed to turnover, particularly during early
seed development, the evidence from this assertion seems to
be equivocal.

TAG biosynthesis and FA desaturation.   Two pathways of
TAG biosynthesis have been recognised in oil seeds [123].
The Kennedy or Glycerol Phosphate Pathway is important in
the synthesis of TAG containing large proportions of mono-
enoic and saturated FA, as outlined below:

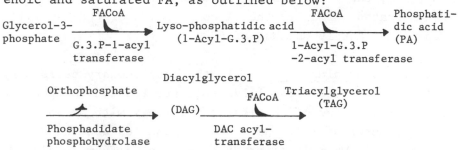

* S-saturated FA (16:0 and 18:0); M-monoenoic acid (18:1); D-dienoic
  acid (18:2); T-trienoic acid (18:3).

An alternative route for the synthesis of TAG containing a large proportion of polyunsaturated FA has been proposed for soyabeans and other seeds [130]. Labelled precursors, ^{14}C-acetate and ^{3}H-glycerol, were incorporated rapidly into PL, predominantly PC but not PA, then DAG and more slowly TAG, which indicates that PC and not PA was the precursor of DAG. Initially, ^{14}C-acetate was incorporated into 18:1 but, with time, 18:2 became labelled at an increasing rate. In pulse-chase experiments, both types of label were conserved in glycerolipid. Label incorporated into 18:1 was chased into 18:2, first in PC, then DAG and much later into TAG. Similarly, glycerol was located first in PC and DAG containing 18:1 and later in TAG containing 18:2. It was proposed for soyabeans that PC acted as substrate for the desaturation of 18:1 and, as appropriate, of 18:2, and a donor of DAG containing polyunsaturated FA from TAG biosynthesis. These processes were thought to occur on the ER [123]. A simplified scheme for TAG biosynthesis in soyabeans which is based mainly on *in vivo* labelling studies is presented below:

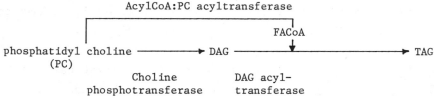

Apart from acylCoA:PC acyltransferase, which catalyses the ATP-independent acyl exchange between oleoyl CoA and PC [132], the other enzymes have been implicated but not demonstrated. Clearly, the FA composition of the TAG will depend ultimately on that of PC.

Oil-bodies or oleosomes.   In the cotyledonary cells of all oilseeds, the triacylglycerols are located in spherical structures called oil-bodies or oleosomes with an approximate diameter of 1-1.5 μm. These organelles are found in the cytosol. When isolated, they appear uniform in size and are bounded by a half unit membrane. The lipophilic surface

of this membrane is oriented inwards to the TAG located
within and the hydrophilic surface faces outwards [150].
The membrane confers specific properties on the oleosomes
and prevents them fusing in cells and during isolation
[129]. Carefully prepared oleosomes from mature groundnut
cotyledons contained 99.5% TAG, 0.1% PL and 0.2% protein
[151], whereas those from germinating seeds, with only 45%
of the original lipid remaining, contained 97.3% TAG, 0.8%
PL and 1.3% protein [72]. The composition of the protein
was similar to that of membrane proteins from other organelles and the conclusion was that both protein and PL were
localised in the bounding membrane of the oleosome [150].
This view has been supported by other workers and it must
be concluded that proteins of the oleosome are not involved
either in the synthesis or in the degradation of the TAG
and that oleosomes are metabolically inert [123]. As yet,
no conclusive information is available on the origin and
development of oleosomes in legume oilseeds.

NUTRITIONAL CONSIDERATIONS

Food legumes not only increase the protein contents of diets
otherwise composed largely of carbohydrates from cereals or
starchy root crops but also considerably enhance the protein
quality of cereal-based diets [19,137]. In diets based on
starchy foods, grain legumes can only provide protein of a
quality in the mixture equal to that as tested with starch
[25]. In such circumstances, supplementation with methionine
will improve the protein quality of the diet [24].
   Most food legumes, except those which accumulate oil
rather than carbohydrate, provide only small quantities of
calories in the diet. Most food legumes contain about 2%
ether-extractables, composing highly unsaturated fatty
acids. On the other hand, the lipid contents of lupins,
winged beans, soyabeans and groundnuts vary from 16% to
44%, providing good quality protein in adequate quantities,
as well as significant amounts of energy. Oil-free flours,

particularly those from a soyabean, are the basis of several protein sources with a wide range of applications as human food.

One aspect which complicates the study of food legumes is that, with few exceptions, they must be processed in order to inactivate antiphysiological substances such as trypsin inhibitors and haemagglutinin compounds. Appropriately controlled processing is therefore required in order to study their nutritional values. Although some interesting positive attributes such as hypocholesterolaemic effects and diabetic management have recently been associated with food legume consumption, they still cause undesirable side-effects in many situations.

Legume seeds as sources of dietary protein: factors affecting quality

Amino acid composition of storage protein.   Comparisons of the essential amino acid contents to reference patterns as well as many animal-feeding trials indicate that the sulphur-containing amino acids are deficient in all food legumes and that tryptophan is deficient in only a few [75]. However, because of differences in the sulphur-amino acid requirements between rats and humans, legume proteins may contain sufficient amounts of sulphur-amino acids to satisfy adult human needs [125]. Some nitrogen balance studies using soyabean proteins with and without methionine supplementation have concluded that soyabean proteins are not deficient in these amino acids for humans provided they are fed at large intake levels. However, other results (Table 3.2) suggest that methionine addition improves the quality of soyabean protein as measured by nitrogen balance in human subjects [54,59,63,80,81,100,105,125,155]. Similar studies for other food legumes in human subjects are available only for common beans [64]. Here, prolonged feeding of methionine-enriched beans at 8-10% protein calories supported satisfactory growth. Even if it is accepted that sulphur-amino acids are not deficient in soyabean proteins, when fed in appropriate proportions of protein intake, the same conclusion cannot be applied to other food legumes because these have smaller sulphur-amino acid contents and, for some species, protein digestibility is less than that of soyabean.

Methods of improving sulphur-amino acid content.   The sulphur-amino acid limitation to protein quality in legume seeds is especially significant when food legumes are fed alone, either to experimental animals or human subjects. Its significance is reduced when food legumes are components of cereal-based diets or those rich in starchy foods such as cassava or plantains [24,25]. Improvement of the sulphur-amino acid status of diets based on grain legumes would not be appropriate through incorporation of the crystalline amino acid or sulphur compounds which may partially meet methionine needs. Rather, improvements in such diets could best be attained by stimulating the consumption of those foods whose proteins are rich sources of sulphur-amino acids [2].

An alternative approach is to take advantage of a complementary effect between cereal and food legume proteins. Evidence shows that when the legume provides more than 50% of dietary protein in a food legume diet, sulphur-amino acids are still the most deficient. If, however, cereal protein predominates, such a deficiency ceases to be significant [18,19,137].

The methionine content of food legume proteins, as well as their protein concentrations, are particularly important for populations subsisting on cassava and plantains, foods very poor in protein and in sulphur-amino acid content. In such circumstances, increasing food legume concentration or intake improves animal performance; nevertheless, methionine supplementation induces more dramatic effects [24]. These situations, which are common around the world, could be improved through the selection of food legumes with greater

Table 3.2 Effects of methionine supplementation on the protein quality of soyabean protein in human subjects

Subject age	Soya product	Intake	Results of methionine supplementation	Reference
1- 3 years	Soya milk	2.0 g N kg^{-1} d^{-1}	Positive	[54]
8- 9 years	Whole soya flour	1.2 g protein kg^{-1} d^{-1}	Positive	[105]
12-16 years	TVP*	4.0 g N d^{-1}	Positive	[81]
Adults	TVP*	4.0 g N kg^{-1} d^{-1}	Positive	[80]
Adults	Soya protein	8.0 g N d^{-1}	No effect	[80]
Adults	Soya protein	3.0 g N d^{-1}	Positive	[155]
Adults	Soya protein	45.0 g N d^{-1}	Positive	[155]
Adults	Soya protein	6.0 g N d^{-1}	Positive	[155]
Adults	Soya protein	0.51 g protein kg^{-1} d^{-1}	Positive	[125]
Adults	Soya protein	5.6 g N d^{-1}	Positive	[125]
Adults	Soya protein	9.5 g N d^{-1}	Small	[125]
Adults	Soya protein	0.6 g protein kg^{-1} d^{-1}	Positive	[100]
Infants	Soya milk	$194-242$ mg N kg^{-1} d^{-1}	Positive	[64]
Infants	Soya milk	$317-320$ mg N kg^{-1} d^{-1}	No effect	[64]

*Textured vegetable protein.

sulphur-amino acid content.

The sulphur-amino acid limitation is further complicated by the poor digestibility of most food legume proteins. Improved digestibility without more sulphur-amino acids does not improve efficiency of utilisation, but an increase in sulphur-amino acid concentrations without an increase in digestibility would be of significant value.

Digestibility. Food legumes are known to contain trypsin inhibitors, lectins, and flatulence factors which can variously interfere with the biological utilisation of the nutrients they contain. The first two groups of compounds are inactivated by appropriate thermal processing. However, as important as these compounds are, the full utilisation of protein is restricted by the small concentrations of sulphur-containing amino acids and, even more important, but not sufficiently emphasised, by the poor digestibility of these proteins, well-demonstrated long ago in experimental animals [19,137]. Research with isolated proteins from raw food legumes has shown their inhibitory effects *in vitro*, and the inactivation of these inhibitory effects by thermal treatment. These findings have prompted suggestions that the poor digestibility of food legume proteins is due to residual proteinase inhibitors not affected by heat, or to reactions between the proteins and compounds such as polyphenols, or to the tertiary structure of protein bodies which prevents proteolytic enzymes from acting effectively [137].

Data on protein digestibility in humans comes from studies with soyabeans and, more recently, with common beans [25]. Soyabean protein digestibility, although slightly less than that of animal protein foods, is significantly better than that of common beans. Apparent protein digestibility values with beans of different colour are quite small (51-62%) as compared to animal protein digestibilities. Recent studies in humans and animals also reveal that the water-soluble fraction of cooked beans is poorly digestible [22,25].

93

Importance of tannins.   There is sufficient evidence now
available to indicate that many food legumes contain polyphenolic compounds [20] which, both *in vitro* and *in vivo*
[21,67], interfere with protein digestibility. After cooking
of common beans, it was not possible to recover all of the
polyphenolic compounds measured in the raw product [20].
Furthermore, significant amounts have been found in the
liquors which separate from the cooked endosperm. Therefore,
in cooked materials there may be interactions between protein and the bound polyphenolics, or the free polyphenolics
in the cooked cotyledons, or those bound or free polyphenols in
the cooking liquors. Bound polyphenols could affect protein
digestibility by decreasing enzyme efficiency and the free ones
could inhibit digestive enzymes directly, thus reducing nitrogen availability. These compounds also directly or indirectly
affect protein quality. Results with humans indicate that polyphenolic intake was positively associated with faecal nitrogen
output and negatively correlated with protein digestibility
[20,21]. Polyphenolic intake decreased protein digestibility by about 10%.

Assay methods for protein quality

Sample preparation for protein quality assays.   Food legumes must be cooked in some way to inactivate the antiphysiological substances before subjecting them to *in vivo*
or *in vitro* protein quality assays. Processing must be carried out under well-controlled and standardised conditions
otherwise protein quality will diminish through a decrease
in the availability of lysine [1]. In general, recently-
harvested seeds require shorter cooking times than those
stored for long periods under poor conditions. Cooking seeds
previously soaked may lead to greater losses in protein
quality than when seeds are processed without pre-treatment.
Protein quality in common beans will also depend on whether
or not the cooking liquor is included in the sample [56].
Since the cooked samples must be dehydrated to be included
in the basal diet for assay, the dehydration process must

also be controlled, otherwise it will introduce additional problems. Clearly, it is essential to use well-standardised sample preparation conditions, particularly to be able to detect differences between cultivars.

Protein digestibility.   In principle, the bioassay for protein digestibility depends on the relation involving the difference between the intake of nitrogen and loss of faecal nitrogen expressed as a proportion (%) of the intake value. If true digestibility estimates are required, endogenous faecal N losses must be included from a group of animals fed N-free diets (NFD). Although the method is simple and straightforward it must be standardised. In studies on this topic, dietary protein values of 8-10% have been recommended for rats 28-35 days old [55]. Rat assays tend to give larger values than human assays for the same samples but rankings are in the same relative order.

Protein quality.   Numerous methods are available for estimating protein quality and it is difficult to recommend a single method as being most appropriate. All methods have disadvantages; however, there are two which are generally useful. The first, Protein Efficiency Ratio (PER), has the advantage that the assay period is twenty-eight days, per-mitting a better interaction between the animal and the protein consumed. The second method, Net Protein Ratio (NPR), involves an assay time of ten days and the use of endogenous losses of nitrogen from animals fed an NFD. Other proced-ures, such as Net Protein Utilisation (NPU) and Relative Protein Value (RPV) have been used but these are expensive and it is doubtful that they can provide additional useful information in comparison with the NPR or PER [139]. Indeed, good correlations between methods have been demonstrated.

Data extrapolation to humans.   While it is accepted that equal numerical values cannot be obtained, data from animal trials are relatively well correlated with those for humans.

95

From studies with common beans, for example, it is clear that white-coloured beans are rated better than black and red-coloured ones in both subjects [20].

In vitro protein digestibility.   *In vitro* assays for food legumes are subject to the same problems which affect *in vivo* assays. Thus, food legumes must be thermally treated for the enzyme to be able to act. Native glycoprotein 11 (phaseolin or globulin 1) in common beans was markedly resistant to pepsin, trypsin or chymotrypsin but heat treatment induced a significant increase in digestion by each enzyme. *In vivo* digestibility increased from 57.0% to 92.5% [85].

Different *in vitro* enzymatic assays, developed for foods in general, have been used for protein digestibility of food legumes. The use of multiple enzymes seems to reflect better the differences between cultivars or the effects of processing. A multiple enzyme system [146] tested on seventeen cultivars of common beans gave a very significant correlation ($r = -0.762$) between the *in vivo* and *in vitro* assay. The method is sensitive to polyphenolic compounds, is able to discriminate between cultivars, and is becoming useful as a rapid procedure for screening programmes. Protein digestibility estimates can also be obtained from *Tetrahymena* bioassays [50] (and see below).

In vitro assay for protein quality.   The Computed-PER (C-PER) [147] assay used to evaluate *in vitro* protein quality of a relatively large number of samples did not predict accurately the quality of nine food legumes included in the study. In contrast, the *Tetrahymena* Relative Nutritive Value (T-RNV) may prove to be a good system for detecting differences between cultivars of the same species and those between species [50,142]. Although antiphysiological factors may cause interference [91], this assay is sensitive enough to detect differences between processed food legumes [50]. Other *in vitro* digestibility and protein quality assays are only variations of basically the same methodology tested on a relatively large variety of foods, but with few if any samples from any of the food legumes. Any new method should be tested for its validity with food legumes before being adopted.

Microbiological assays for amino acids.   These assays can be useful for only one amino acid as shown [77] in the preliminary evaluation of large numbers of genotypes of common beans for methionine concentration. In this and other studies, *Streptococcus zymogenes* bioassays have been used [77, 98]. *Tetrahymena pyriformis* has also been used for the assay of available lysine and methionine [156].

Chemical analyses.   Various workers have used several chemical methods which, besides being precise, enable a relatively large number of samples to be analysed each working day. A colorimetric method for protein concentration could be used for both whole seeds and cotyledons of pigeonpeas (*Cajanus cajan*) and for dhal samples [128]. The Acid Orange-12 dye-binding procedure was better for excised cotyledons since the seed coats absorbed some of the dye. Although the Kjeldahl method is more accurate it is slower than these two methods and so less appropriate when examining large numbers of samples. Other methods, as well as the advantages and disadvantages of those mentioned above, are discussed elsewhere [139]. Efforts have also been made to use sulphur analysis as an indicator of total sulphur-amino acid content [58,74], and lysine may be determined by colorimetric procedures [38].

Other compounds of interest.   Trypsin inhibitor concentration may be assayed by the official method of the American Association of Cereal Chemists [3]. The method most commonly used for polyphenolic content is the vanillin assay, which expresses polyphenols as catechin equivalents [23].

97

Other considerations concerning the consumption of legume seed proteins

Gastrointestinal hypersensitivity and immunological responses to legume protein.

The supplementary value of food legume protein to cereal grains has been an argument for using these foods for infants and small children. However, the results of surveys carried out in various countries with relatively large common bean intakes among the adult populations, show that mothers do not feed food legumes to small children, much less to infants. The reasons given are that they are not well-tolerated and cause diarrhoea and vomiting. Haemagglutinin activity has been shown to persist in cooked samples collected from field studies [10,82].

Only recently it has been found that when rats are fed pure lectins isolated from common beans they develop circulating antibodies of poor avidity to dietary lectins [113]. In several serum samples from rats fed raw beans, small amounts of a protein reactive with rabbit antilectin antibodies were detected. Based on this evidence the authors proposed that bean lectins induce toxicity not only by interfering with normal intestinal digestion and/or absorption of nitrogen through damaging enterocytes, but also, in rats, by a systemic immune response to the internalised lectin which increases nitrogen output in the urine.

Favism.

This disease, of particular importance in the Middle East, is induced by the consumption of faba beans by individuals with the genetic deficiency of erythrocyte glucose-6-phosphate dehydrogenase [9]. The organic compounds suggested as being responsible for favism include vicine, convicine and dihydroxyphenylalanine (L-DOPA) and its glycoside. The first two compounds, together with iso-uramil, can oxidise reduced glutathione (GSH) in glucose-6-phosphate dehydrogenase-deficient erythrocytes but not in normal cells. Immature and fresh beans, either raw or cooked, are reported to be more toxic in young humans than dried or cooked beans in adult subjects. The concentrations of these substances are large in immature seeds but decrease towards maturity and then remain constant. Proteins isolated by wet processes are free or virtually free of the glucosides. An alternative possibility to processing is to decrease these concentrations genetically. Both genotype and environment have very significant effects on the glucoside concentrations of faba bean samples, convicine showing greater variability among cultivars than vicine [10].

Hypocholesterolaemic effects of food legumes.   There are a relatively large number of studies which indicate that consumption of food legumes often decreases serum cholesterol concentrations in experimental animals and humans. Such effects have been observed following consumption of common beans, Lima beans, peas, soyabean protein, faba beans or other legume foods. Some authors have concluded that the effect is due to the carbohydrate component in food legumes but it has also been observed with protein concentrates prepared from food legumes [141]. Other factors suggested include saponins, the lysine to arginine ratio and poor protein digestibility.

Legume seeds as sources of energy

Utilisation and digestibility of carbohydrates.   Apart from the oil-containing species (see Table 3.1), other food legumes commonly eaten by humans are not considered as dietary energy sources. Depending on their intake per day, however, they can make a significant contribution to the daily dietary energy intake.

Some reports have described the digestible and metabolisable energy of mixed diets containing common beans. In one study using American males fed a rural Guatemalan diet of beans, maize, rice, bread and small amounts of vegetables and food animal products, an energy digestibility of 89%

99

was reported, as compared to 97% from the control diet based on eggs [37]. With pre-school children fed 1.73 g protein kg^{-1} d^{-1} from a diet based on common beans and maize [138], energy digestibilities of 91%, 80% and 85% were observed when the crude energy intake was 89, 91, and 81 kcal kg^{-1} d^{-1}, respectively.

When adults were fed rice, beans and other vegetable foods [140] their average energy digestibility was 93.9%, and when the diet was supplemented with milk or additional energy, the metabolisable energy values varied from 92.1% to 93.2%. *In vivo* experiments have shown that when children are fed food legumes containing starch with long-chain amylose units then digestion is difficult. The *in vivo* digestibility of the starch from common beans was superior to that from pigeonpeas or chickpeas (*Cicer arietinum*) and boiling and pressure cooking improved digestibility more than roasting [62]. Starch granule size is also an important factor in digestibility and those carbohydrates that escape digestion undergo fermentation in the caecum, leading to flatulence.

Other considerations of carbohydrate in legume seeds
Flatulence factors. The oligosaccharides of the raffinose family are always found in mature food legumes, and many studies have shown that these compounds might be responsible for flatulence, characterised by significantly large amounts of methane and hydrogen in the intestinal gas [43]. The three oligosaccharides which have been identified are raffinose, stachyose and verbascose, all of which are characterised by galactose-glucose and galactose-galactose bonds. The concentrations of these sugars vary among food legume species; verbascose is most concentrated, followed by stachyose and raffinose. Flatulence can be reduced by processing, using methods such as alcohol extraction, germination, soaking prior to cooking and enzymatic techniques. Since there is genetic variability in the relative abundance of these compounds it may be possible to eliminate,

or at least minimise, their effects by breeding [43].

Diabetes management.   Food legume consumption has an
hypocholesterolaemic effect on patients with large cholesterol concentrations and this fact has prompted other observations on glucose tolerance in both normal and diabetic
subjects fed bean-rich diets. Although further research is
needed, these observations are probably related to the
carbohydrate and dietary fibre components of food legumes
and their reactions at the gastrointestinal level [84].

BREEDING

Objectives and priorities

Larger and more stable yields are primary objectives of
most grain legume improvement programmes. When available
land is limited and competition from other crops is intense,
increased productivity is essential to make legumes as
economically attractive as the productive cereals and to
obtain the seeds necessary for food and feed.

Soyabeans and groundnuts are grown primarily for their
edible oil and protein-rich meal. Production of protein and
oil per unit area can be maximised by improving seed yield,
or percentage protein or percentage oil while other trait(s)
remain constant, or by increasing two or more traits simultaneously.

Priorities for the improvement of pulses grown for
direct consumption are often different. In these crops
(Table 3.1), oil contents are small and the production of
total protein per hectare is usually less important than in
the oilseed legumes. It is more important to produce calories and proteins to supplement other foods. Thus, protein
concentration assumes increased importance, since only
limited amounts of seeds can be ingested comfortably, particularly by small children. Increasing the protein concentrations of the pulse seeds to values similar to those of

soyabean, whilst maintaining seed yields, would reduce the protein imbalance often found in diets where legumes are a major component [19].

The poor nutritional value of grain legumes is usually attributed to insufficient sulphur-amino acids, the presence of antinutritional compounds, and to poor protein digestibility. To increase the amounts of methionine and cystine to nutritionally-adequate levels requires an approximate doubling of the current concentrations. Lectins and proteinase inhibitors are usually present in small quantities and are inactivated with proper heat treatment. However, prevailing cooking patterns may be inadequate and genetic reductions would be useful. Polyphenols present in the seed coats appear to reduce the nutritional value of legume seed proteins, but their potential beneficial effects related to pest resistance are not fully known.

Reliable production resulting from timely maturity, better symbiosis with *Rhizobium* [66,92], resistances to environmental stress, pests and diseases, and stable yields may aid subsistence farmers more than total yield, particularly if over-abundant seed production can be neither stored reliably nor sold profitably.

Scope for and constraints to improvement

Plant genetic resources.   The primary, secondary and tertiary gene-pools (see Chapter 2) contain the principal genetic resources available for improvement of biochemical and nutritional attributes. As additional knowledge of the molecular properties of legume seed proteins becomes available, they appear to be prime candidates for *in vivo* manipulation and improvement using recombinant DNA techniques [11,32].

The germplasm used most widely for breeding grain legumes has been domesticated material. Hybridisation among closely related taxa within the primary and secondary gene-pools offers the greatest opportunity for short-term

improvement. The transfer of genes between less closely
related crops (e.g. soyabeans and groundnuts or common
beans) may be of limited value since nearly all grain legumes have similar biochemical and nutritional deficiencies.

The potential value of the available germplasm is difficult to estimate fully since the nature of the traits
that confer superior biochemical and nutritional properties
are complex. Furthermore, traits such as protein concentration are usually quantitatively inherited and influenced
substantially by non-genetic factors, making it difficult
to evaluate plant materials accurately without extensive
testing.

Variation and heritable factors.   Substantial genetic
variation for total seed protein concentration (percentage
protein) exists in the gene-pools of the grain legumes [4,
13,47,68,71,153,154]. Although environmental effects are
usually large, genotype × environment interactions are often
small, which indicates that the relative differences between
genotypes should be similar in several environments. Heritability estimates for percentage protein range from 0.25 to
0.60, with occasional larger values [13,27,48,57,83,104].

Although grain legumes in general are considered to be
deficient in sulphur-amino acids [103], actual amounts vary
among species. There are substantial differences in amino
acid composition of individual protein fractions [6,7,12,16],
and synthesis and accumulation of different fractions and
the polypeptide composition of the fractions are under separate genetic control [96]. Major genes which affect the
amounts of seed storage proteins have been reported in peas
[49] and in common beans [122,133].

The factors responsible for the antinutritional properties of raw legume seeds show both quantitative and qualitative variations. Quantitative variation in trypsin inhibitor activity has been reported in faba beans [89]. In soyabeans, some accessions lack detectable inhibitor activity
[101] and at least three different forms of the Kunitz

trypsin inhibitor have been identified [95].

Seed lectins are widely distributed in many legume species. However, about 10% of the accessions of common beans surveyed so far have been found to be lectin-free [33,35], and there is extensive variation for type of lectin and ability to agglutinate different cell types [34]. Occurrence of soyabean seed lectin (SBL) varies greatly within the genus *Glycine*, being confined to members of the subgenus *Soja*, and most common in *G. max* [111]. The presence of SBL in *G. max* is controlled by a single dominant gene, with some recessive strains being free of SBL [102].

The polyphenols found in the testas of coloured legume seeds have been associated with reduced nutritional value. However, considerable variability exists, with heritability estimates in common beans ranging from 0.48 to 0.97 [88].

Oil concentration (%) is quantitatively inherited and determined largely by the maternal parent [30]. In soyabeans and groundnuts, heritability estimates for percentage oil can be large [36,135].

Relations between traits: bioenergetic considerations.
The potential biomass yield of seeds is related to the photosynthate production and distribution of the parent plant and the chemical composition of the seeds produced. Depending on the relative and actual amounts of proteins, carbohydrates, lipid and ash in seeds, the nitrogen demand during seed maturation may also limit seed yield. Among nine grain legumes considered in one study [127], soyabeans and groundnuts had the smallest potential biomass production per gram of photosynthate. Soyabeans are also unique in that they have the largest nitrogen requirements per gram of photosynthate used. With the possible exception of groundnuts, all the grain legumes have a large demand for nitrogen during seed maturation when uptake may be inadequate to meet these demands. Since any nitrogen shortfall must be met by nitrogen metabolised and translocated from vegetative tissues, these sources may become depleted, with vegetative

tissues losing their physiological activity. Rates and durations of seed fill, and hence yield, are intimately related to rates of nitrogen uptake which, in turn, are influenced by seed chemical composition.

From a genetic analysis of energy production in soya-beans, Hanson *et al.* [69] concluded that 'one would expect the genetic correlations between percent protein and yield to be small and positive'. Since approximately 0.75g, 1.14g and 0.40g of sugar carbon are required to produce one gram of each of protein, oil and residual, respectively, it should prove possible to produce large yields of seeds with small oil concentration and, if nitrogen is not limiting and other physiological restrictions are not serious, large yields of protein-rich seeds.

Correlations between traits.   Genotypic and phenotypic correlations between yield and percentage protein in soy-bean seeds ranged from -0.42 to 0.22 and -0.58 to 0.35, respectively [27], while in groundnut populations, correlations between seed yield and protein concentration were positive. Similar correlations between seed yield and per-centage protein, with values ranging from -0.12 to 0.64 for common beans, -0.57 to 0.22 for peas, -0.38 to -0.14 for cowpeas (*Vigna unguiculata*) and -0.29 to 0.23 for faba beans, have been cited elsewhere [57]. Correlations between seed yield and specific amino acid contents have seldom been significant.

The occurrence of negative correlations between total protein concentration and sulphur-amino acid content is due largely to the differential accumulation of storage proteins with different proportions of amino acids [16]. This differ-ential accumulation depends both on genetic control of pro-tein synthesis and on environmental effects that influence synthesis and deposition. Correlations between percentage protein and methionine content in common beans have ranged from small and negative to small and positive [77,98].

The correlations between protein and oil concentrations

in soyabeans have all been negative, and usually large.
Such relations would be expected, particularly in the
absence of variation in the residual fraction (e.g. starch)
[69]. In groundnuts, both negative and positive values have
been recorded [135].

Correlated selection response.   Correlated responses
(e.g. changes in population means) of other traits not subjected to direct selection are indicative of the strengths
of the relations between traits. In soyabeans, recurrent
selection for increased yield did not lead to correlated
effects on seed size, oil or protein concentration [78].
Selection for percentage protein (the primary trait) had
no adverse effects on yield, but oil concentration was
reduced, whereas selection for increased oil concentration
led to seeds with smaller percentage protein [28,36]. Pedigree and modified pedigree selection for increased percentage protein in common beans lines was successful, but the
protein-rich genotypes gave poorer seed yields [13].

Improvement strategies

Improvement strategies for the oilseed legumes will differ
from those for the pulses because oil content must be considered in addition to other important agronomic and culinary traits. Breeding procedures that emphasize larger
yields whilst considering protein and/or oil concentration
concurrently or secondarily should be utilised [14]. Since
these traits are usually quantitatively inherited, methods
that allow replicated testing of early generation lines
[15] in many target environments are preferred. When insect
and disease resistance or simply-inherited, qualitative
traits are to be incorporated, selection should begin in
the early segregating generations on either a single plant
or line basis prior to or concurrent with selection for
quantitative traits. Choice of the target area for future
production will determine the extent of testing to ensure

appropriate adaptability of new cultivars.

In pulses, particularly where seed traits related to consumer acceptability are important, pedigree selection, modified pedigree selection and production of inbred back-cross lines can be used effectively [26,134]. Recurrent selection, where the selection criteria are based on indices, can be used where acceptability traits are less important.

The genetic alteration of specific protein fractions provides a means for increasing total protein content, raising limiting essential amino acid concentrations by differentially regulating fractions with different amino acid composition, and reducing the concentrations of undesirable antinutritional compounds. The major gene mutants which affect quantity and type of seed protein fractions that have been identified so far in peas and common beans can be incorporated into adapted cultivars using standard procedures.

Prospects

Prospects for improving the productivity and nutritional properties of grain legumes will depend heavily on research efforts and funding. There is adequate natural variation to allow substantial improvement of most, if not all, the important traits. While yields of the grain legumes grown primarily by subsistence farmers have often remained stagnant or even declined slightly [65] in the recent past, soyabean yields in the United States, where there has been recent concerted breeding activity, have increased (albeit by far less relatively than the area sown to the crop). These increases are attributable to both improved cultivars and better production practices. Better yields of soyabean, without noticeable reductions in either oil or protein concentration, suggest that improvements at least equally as large should be possible in other grain legumes, since soyabeans have the largest energy demand on photosynthate

production [127]. It is questionable whether protein concentration of soyabean can be increased much more whilst maintaining large yield and/or suitable oil content.

Advances in research methodology contribute to the
favourable prospects for grain legume improvement. Percentage protein can now be estimated accurately and rapidly
using near-infrared reflectance [143] rather than more
laborious procedures. Oil content can be estimated using
the same near-infrared reflectance instruments or by nuclear magnetic resonance [29]. Advanced techniques for isolating and purifying proteins allow the production of specific antibodies that can be used to identify rapidly and
accurately small quantitative and qualitative differences
in protein fractions [99,148].

Finally, the seed proteins of legumes appear to be
particularly well suited for manipulation using recombinant
DNA techniques. They are generally tissue-specific, synthesised rapidly during a short time-period, well characterised and represent gene products close to the site of gene
action. Substantial progress has already been made toward
cloning the genes that encode for seed proteins in common
beans, peas and soybeans [11,32].

LITERATURE CITED

1. Almas, K. and Bender, A.E. (1980), *Journal of the Science of Food and Agriculture* 31, 448-52.
2. Al-Nouri, F.F., Siddigi, A.M. and Markakis, P. (1980), *Food Chemistry* 5, 309-13.
3. American Association of Cereal Chemists (1976), *Approved Methods of the AACC*. St Paul, Minnesota: AACC.
4. Amirshahi, M.C. and Tavakoli, M. (1970), in *Improving Plant Proteins by Nuclear Techniques*. Vienna: International Atomic Energy Agency, 331-5.
5. Bailey, C.J., Cobb, A. and Boulter, D. (1970), *Planta* 95, 103-18.
6. Bajaj, S., Mickelson, O., Lillevik, H.A., Baker, L.R., Bergen, W.G. and Gill, J.L. (1971), *Crop Science* 11, 813-15.
7. Basha, S.M.M. and Cherry, J.P. (1976), *Journal of Agriculture and Food Chemistry* 24, 359-65.
8. Baumgartner, B., Tokuyasu, K.T. and Chrispeels, M.J. (1980), *Planta* 150, 419-25.
9. Belsey, M.A. (1973), *Bulletin of the World Health Organisation* 48, 1-13.

10. Bender, A.E. and Reaidi, G.B. (1982), *Journal of Plant Foods* 4, 15-22.
11. Bliss, F.A. (1984), *Hortscience* 19, 43-8.
12. Bliss, F.A. and Brown, J.W.S. (1982), *Qualitas Plantarum Plant Foods for Human Nutrition* 31, 269-79.
13. Bliss, F.A. and Brown, J.W.S. (1983), in Janick, J. (*ed.*) *Plant Breeding Reviews.* Westport, Connecticut: Avi Publishing, 59-102.
14. Boerma, H.R. and Cooper, R.L. (1975), *Crop Science* 15, 225-9.
15. Boerma, H.R. and Cooper, R.L. (1975), *Crop Science* 15, 313-15.
16. Boulter, D. (1977), in Muhammed, A., Aksel, R. and von Borstel, R.C. (*eds*) *Genetic Diversity in Plants.* New York: Plenum Press, 387-96.
17. Boulter, D. and Derbyshire, E. (1978), in Norton, G. (*ed.*) *Plant Proteins.* London and Boston: Butterworth, 3-24.
18. Bressani, R. (1974), in White, P.L. and Fletcher, D.C. (*eds*) *Nutrients in Processed Foods.* Acton, Massachusetts: Publishing Sciences Group.
19. Bressani, R. (1975), in Milner, M. (*ed.*) *Nutritional Improvement of Food Legumes by Breeding.* New York: Wiley, 15-42.
20. Bressani, R. and Elias, L.G. (1979), in Hulse, J.H. (*ed.*) *Polyphenols in Cereals and Legumes.* St Louis, Missouri: Institute of Food Technology.
21. Bressani, R., Elias, L.G. and Braham, J.E. (1982), *Journal of Plant Foods* 4, 43-55.
22. Bressani, R., Elias, L.G. and Molina, M.R. (1977), *Archivos Latin American Nutrition* 27, 215-31.
23. Bressani, R., Elias, L.G., Wolzak, A., Hagerman, A.E. and Butler, L.G. (1983), *Journal of Food Science* 48, 1000-01.
24. Bressani, R., Navarrete, D.A. and Elias, L.G. (in press), *Qualitas Plantarum Plant Foods for Human Nutrition.*
25. Bressani, R., Navarrete, D.A., Vargas, E. and Gutierrez, O.M. (1981), in Torun, B., Young, V.R. and Rand, W.M. (*eds*) *Protein-Energy Requirements of Developing Countries. Evaluation of New Data.* UNU World Hunger Programme Food and Nutr. Bulletin Suppl. 5, 108.
26. Brim, C.A. (1966), *Crop Science* 6, 220.
27. Brim, C.A. (1973), in Caldwell, B.E. (*ed.*) *Soybeans: Improvement, Production and Uses.* Madison, Wisconsin: American Society of Agronomy, 115-58.
28. Brim, C.A. and Burton, J.W. (1979), *Crop Science* 19, 494-98.
29. Brim, C.A., Schutz, W.M. and Collins, F.I. (1967), *Crop Science* 7, 220-22.
30. Brim, C.A., Schutz, W.M. and Collins, F.I. (1968), *Crop Science* 8, 517-18.
31. Brown, J.W.S., Ma, Y., Bliss, F.A. and Hall, T.C. (1981), *Theoretical and Applied Genetics* 59, 83-8.
32. Brown, J.W.S., Ersland, D.R. and Hall, T.C. (1982), in Khan, A.A. (*ed.*) *The Physiology and Biochemistry of Seed Development and Germination.* New York: Elsevier-North Holland Biomedical Press, 3-42.
33. Brown, J.W.S., Osborn, T.C., Bliss, F.A. and Hall, T.C. (1981), *Theoretical and Applied Genetics* 60, 245-50.
34. Brown, J.W.S., Osborn, T.C., Bliss, F.A. and Hall, T.C. (1982), *Theoretical and Applied Genetics* 62, 263-71.
35. Brucher, O. (1968), *Proceedings of the Tropical Region of America*

Society of Horticultural Science 12, 68-85.

36. Burton, J.W. and Brim, C.A. (1981), *Crop Science* 21, 31-4.
37. Calloway, D.H. and Kretsch, M.J. (1978), *American Journal of Clinical Nutrition* 31, 1118-26.
38. Carpenter, K.J. (1981), in Bodwell, C.E., Adkins, J.S. and Hopkins, D.T. (eds) *Protein Quality in Humans: Assessment and in vitro Estimation*. Westport, Connecticut: Avi Publishing, 234-60.
39. Cerning-Beroard, J. and Filiatre, A. (1976), *Cereal Chemistry* 53, 968-78.
40. Chrispeels, M.J. and Baumgartner, B. (1978), *Plant Physiology, Lancaster* 61, 617-23.
41. Chrispeels, M.J., Higgins, T.J.V., Craig, S. and Spencer, D. (1982), *Journal of Cell Biology* 93, 5-14.
42. Chrispeels, M.J., Higgins, T.J.V. and Spencer, D. (1982), *Journal of Cell Biology* 93, 306-13.
43. Cristoforo, E., Motta, E. and Wahrmann, J.J. (1973), *Nestlé Research News*, Lausanne, Switzerland.
44. Croy, R.R.D., Gatehouse, J.A., Evans, M.I. and Boulter, D. (1980), *Planta* 148, 49-56.
45. Croy, R.R.D., Gatehouse, J.A., Evans, M.I. and Boulter, D. (1980), *Planta* 148, 57-63.
46. Croy, R.R.D., Gatehouse, J.A., Tyler, M. and Boulter, D. (1980), *Biochemical Journal* 191, 509-16.
47. Dahiya, B.S. and Brar, J.S. (1976), *Tropical Grain Legume Bulletin* 3, 18-20.
48. Dahiya, B.S. and Brar, J.S. (1977), *Experimental Agriculture* 13, 193-200.
49. Davies, D.R. (1980), *Biochemical Genetics* 18, 1207-19.
50. Davis, K.R. (1981), *Cereal Chemistry* 58, 454-60.
51. Derbyshire, E., Wright, D.J. and Boulter, D. (1976), *Phytochemistry* 15, 3-24.
52. Domoney, C., Davies, D.R. and Casey, R. (1980), *Planta* 149, 454-60.
53. Dure, L.S. (1975), *Annual Review of Plant Physiology* 26, 259-78.
54. Dutra de Oliveira, J.E. (1975), in *Memorias de la Conferencia Latinoamericana sobre la Proteina de Soya*. Mexico: Assoc. Amer. Soya, 65.
55. Elias, L.G. and Bressani, R. (1982), *Efecto del tiempo de coccion y del nivel de proteina sobre la digestibilidad apparente del Frijol comun (Phaseolus vulgaris)*. Annual Report 1982. Guatemala: INCAP.
56. Elias, L.G., Cristales, F.R., Bressani, R. and Miranda, H. (1976), *Turrialba* 26, 375-80.
57. Evans, A.M. and Gridley, H.E. (1979), *Current Advances in Plant Science* 32, 1-17.
58. Evans, M.I. and Boulter, D. (1974), *Journal of the Science of Food and Agriculture* 25, 311-22.
59. Fomon, S.J. and Ziegler, E.E. (1979), in Wilcke, H.L., Hopkins, D.T. and Waggle, D.G. (eds) *Soy Protein and Human Nutrition*. New York: Academic Press, 79-99.
60. Gatehouse, J.A., Croy, R.R.D., Morton, H., Tyler, M. and Boulter, D. (1981), *European Journal of Biochemistry* 118, 627-33.
61. Gatehouse, J.A., Lycett, G.W., Croy, R.R.D. and Boulter, D. (1982), *Biochemical Journal* 207, 629-32.
62. Geervani, P. and Theophilus, F. (1981), *Journal of the Science of Food and Agriculture* 32, 71-8.

63. Graham, G.G. (1969), in Scrimshaw, N.S. and Altschul, A.M. (eds) *Amino Acid Fortification of Protein Foods*. Cambridge, Massachusetts: MIT Press.

64. Graham, G.G., Morales, E., Placko, R.P. and Maclean, W.C. (1979), *American Journal of Clinical Nutrition* 32, 2362-66.

65. Graham, P.H. (1978), *Field Crops Research* 1, 295-317.

66. Graham, P.H. (1981), *Field Crops Research* 4, 93-112.

67. Griffiths, D.W. (1979), *Journal of the Science of Food and Agriculture* 30, 458-62.

68. Grohlich, W.G., Pollmer, W.G. and Christ, W. (1974), *Zeitschrift für Pflanzenzüchtung* 72, 160-65.

69. Hanson, W.D., Leffel, R.C. and Howell, R.W. (1961), *Crop Science* 1, 121-6.

70. Higgins, T.J.V. and Spencer, D. (1981), *Plant Physiology, Lancaster* 67, 205-11.

71. Hymowitz, T., Palmer, R.G. and Hadley, H.H. (1972), *Tropical Agriculture, Trinidad* 49, 245-50.

72. Jacks, T.J., Yatsu, L.Y. and Altschul, A.M. (1967), *Plant Physiology, Lancaster* 42, 585-97.

73. Jaffé, W.G., Levy, A. and Gonzalez, D.J. (1974), *Phytochemistry* 13, 2685-93.

74. Jambunathan, R. and Singh, U. (1981), *Qualitas Plantarum Plant Foods for Human Nutrition* 31, 109-17.

75. Jansen, G.R. (1973), in Jaffé, W.G. (ed.) *Nutritional Aspects of Common Beans and Other Legume Seeds as Animal and Human Foods*. Caracas, Venezuela: Archivos Latinoamericanos de Nutricion, 217-32.

76. Jenner, C.F. (1982), in Loewus, F.A. and Tanner, W. (eds) *Encyclopedia of Plant Physiology* N.S.13A. Berlin: Springer-Verlag, 700-47.

77. Kelly, J.D. and Bliss, F.A. (1975), *Crop Science* 15, 753-57.

78. Kenworthy, W.J. and Brim, C.A. (1979), *Crop Science* 19, 315-18.

79. Khan, M.R.I., Gatehouse, J.A. and Boulter, D. (1980), *Journal of Experimental Botany* 31, 1599-611.

80. Kies, C.V. and Fox, H.M. (1973), *Journal of Food Science* 38, 637-38.

81. Korte, R. (1972), *Ecology of Food & Nutrition* 1, 303-07.

82. Kushawa, J.S. and Tawar, M. (1973), *Indian Journal of Agricultural Science* 43, 1049-54.

84. Leeds, A.R. (1982), *Journal of Plant Foods* 4, 23-7.

85. Liener, I.E. and Thompson, R.M. (1980), *Qualitas Plantarum Plant Foods for Human Nutrition* 30, 13-25.

86. Longe, O.G. (1980), *Food Chemistry* 6, 153-61.

87. Lott, J.N.A. and Buttrose, M.S. (1978), *Australian Journal of Plant Physiology* 5, 89-111.

88. Ma, Y. and Bliss, F.A. (1978), *Crop Science* 18, 201-04.

89. Marquardt, R.R., McKirdy, J.A., Ward, T. and Campbell, L.D. (1975), *Canadian Journal of Animal Science* 55, 421-9.

90. Marshall, J.J. and Lauda, C.M. (1975), *Journal of Biological Chemistry* 250, 8030-37.

91. McCurdy, S.M., Scheier, G.E. and Jacobson, M. (1978), *Journal of Food Science* 43, 694-7.

92. McFerson, J.R. and Bliss, F.A. (1982), *Hortscience* 17, 476.

93. Meier, H. and Reid, J.S.G. (1982), in Loewus, F.A. and Tanner, W. (eds) *Encyclopedia of Plant Physiology*, N.S.13A. Berlin: Springer-Verlag, 418-71.

94. Miège, M.N. (1982) in Boulter, D. and Parthier, B. (eds) *Encyclopedia of Plant Physiology* N.S.14A. Berlin: Springer-Verlag, 291-35.

111

95. Mies, D.W. and Hymowitz, T. (1973), *Botanical Gazette* 134, 121-255.
96. Millerd, A. (1975), *Annual Review of Plant Physiology* 26, 53-72.
97. Muntz, K. (1982), in Boulter, D. and Parthier, B. (*eds*) *Encyclopedia of Plant Physiology* N.S.14A. Berlin: Springer-Verlag, 505-58.
98. Mutschler, M.A. and Bliss, F.A. (1981), *Crop Science* 21, 289-94.
99. Mutschler, M.A., Bliss, F.A. and Hall, T.C. (1980), *Plant Physiology, Lancaster* 65, 627-30.
100. Navarrete, D.A., Elias, L.G., Braham, J.E. and Bressani, R. (1979), *Archivos Latin American Nutrition* 29, 386-401.
101. Orf, J.H. and Hymowitz, T. (1979), *Crop Science* 19, 107-09.
102. Orf, J.H., Hymowitz, T., Pull, S.P. and Pueppke, S.G. (1978), *Crop Science* 18, 899-900.
103. Orr, M.L. and Watt, B.K. (1957), in *Home Economics Research. Report No. 4*, Washington, DC: USDA, 16-54.
104. Panday, S. and Gritton, E.T. (1975), *Journal of the American Society of Horticultural Science* 100, 87-90.
105. Parthasarathy, H.N., Doraiswamy, T.R., Panemangalore, M., Narayana Rao, M., Chandrasekhar, B.S., Swaminathan, M., Srinivasan, A. and Subrahmanyan, V. (1964), *Canadian Journal of Biochemistry and Physiology* 42, 377-84.
106. Paul, A.A. and Southgate, D.A.T. (1978), *McCance and Widdowson's The Composition of Foods*. London: HMSO.
107. Pernollet, J.C. (1978), *Phytochemistry* 17, 1473-80.
108. Pernollet, J.C. and Mossé, J. (1983), in Daussant, J., Mossé, J. and Vaughan, J. (*eds*) *Seed Proteins*. Ann. Proc. Phytochemical Soc. Eur. No. 20, London: Academic Press, 155-91.
109. Phillips, F.C. and Privett, O.S. (1979), *Lipids* 14, 949-52.
110. Pitz, W.J., Sosulski, F.W. and Rowland, G.G. (1981), *Journal of the Science of Food and Agriculture* 32, 1-8.
111. Pueppke, S.G. and Hymowitz, T. (1982), *Crop Science* 22, 558-60.
112. Pusztai, A. (1972), *Planta* 107, 121-9.
113. Pusztai, A., Clarke, E.M.W., Grant, G. and King, T.P. (1981), *Journal of the Science of Food and Agriculture* 32, 1037-46.
114. Pusztai, A., Croy, R.R.D., Grant, G. and Stewart, J.C. (1983), in Daussant, J., Mossé, J. and Vaughan, J. (*eds*) *Seed Proteins*. Ann. Proc. Phytochemical Soc. Eur. No. 20. London: Academic Press, 53-82.
115. Pusztai, A., Croy, R.R.D., Grant, G. and Watt, W.B. (1977), *New Phytologist* 79, 61-71.
116. Pusztai, A., Croy, R.R.D., Stewart, J. and Watt, W.B. (1979), *New Phytologist* 83, 371-8.
117. Pusztai, A. and Stewart, J.C. (1980), *Biochemica et Biophysica Acta* 623, 418-28.
118. Privett, O.S., Dougherty, K.A., Erdahl, W.L. and Stolyhwo, A. (1973), *Journal of the American Oil Chemistry Society* 50, 516-20.
119. Ramshaw, J.A.M. (1982), in Boulter, D. and Parthier, B. (*eds*) *Encyclopedia of Plant Physiology* N.S. 14A. Berlin: Springer-Verlag, 229-90.
120. Richardson, M. (1977), *Phytochemistry* 16, 159-69.
121. Roehm, J.N. and Privett, O.S. (1970), *Lipids* 5, 353-8.
122. Romero-Andreas, J. and Bliss, F.A. (1982), *Hortscience* 17, 504.
123. Roughan, P.G. and Slack, C.R. (1982), *Annual Review of Plant Physiology* 33, 97-132.
124. Roughan, P.G., Slack, C.R. and Holland, R. (1978), *Lipids* 13, 497-503.
125. Scrimshaw, N.S. and Young, V.R. (1979), in Wilcke, H.L., Hopkins,

D.T. and Waggle, D.H. (eds) *Soy Protein and Human Nutrition*. New York: Academic Press, 121-48.
126. Sharon, N. and Lis, H. (1979), *Biochemical Society Transactions* 7, 783-99.
127. Sinclair, T.R. and de Wit, C.T. (1975), *Science* 189, 565-7.
128. Singh, U. and Jambunathan, R. (1981), *Journal of the Science of Food and Agriculture* 32, 705-10.
129. Slack, C.R., Bertaud, W.S., Shaw, B.D., Holland, R., Browse, J. and Wright, H. (1980), *Biochemical Journal* 190, 551-61.
130. Slack, C.R., Roughan, P.G. and Balasingham, N. (1978), *Biochemical Journal* 170, 421-33.
131. Stearns, E.M. and Morton, W.T. (1977), *Lipids* 12, 451-4.
132. Stymne, S. and Glad, G. (1981), *Lipids* 16, 298-305.
133. Sullivan, J.G. (1981), Ph.D. Thesis, University of Wisconsin, Madison, USA.
134. Sullivan, J.G. and Bliss, F.A. (1983), *Journal of the American Society of Horticultural Science* 108, 787-91.
135. Tai, Y.P. and Young, C.T. (1975) *Journal of the American Oil Chemistry Society* 52, 377-85.
136. Thanh, V.H. and Shibasaki, K. (1978), *Journal of Agriculture and Food Chemistry* 26, 692-5.
137. Tobin, G. and Carpenter, K.J. (1978), *Nutrition Abstracts and Reviews* 48, 919-36.
138. Torun, B. and Viteri, F.E. (1981), in Torun, B., Young, V.R. and Rand, W.M. (eds) *Protein-Energy Requirements of Developing Countries: Evaluation of New Data*. UN University Food and Nutrition Bulletin Suppl. No. 5.
139. United Nations University (1980), in Pellet, P.L. and Young, V.R. (eds) *Nutritional Evaluation of Protein Foods*. Food and Nutrition Bulletin Suppl. No. 4.
140. Vargas, E., Bressani, R., Navarrete, D.A., Braham, J.E. and Elias, L.G. (in press). *Archivos Latin American Nutrition*.
141. Walker, A.F. (1982), *Journal of Plant Foods* 4, 5-14.
142. Wang, Y.Y.D., Miller, J. and Beuchat, L. (1979), *Journal of Food Science* 44, 540-44.
143. Williams, P.C. (1975), *Cereal Chemistry* 55, 561-76.
144. Wilson, R.E. and Rinne, R.W. (1974), *Plant Physiology, Lancaster* 54, 744-7.
145. Wilson, R.E. and Rinne, R.W. (1978), *Plant Physiology, Lancaster* 61, 830-33.
146. Wolzak, A., Bressani, R. and Gomez Brenes, R. (1981), *Qualitas Plantarum Plant Foods for Human Nutrition* 31, 31-43.
147. Wolzak, A., Elias, L.G. and Bressani, R. (1981), *Journal of Agriculture and Food Chemistry* 29, 1063-8.
148. Yabuchi, S., Lister, R.M., Axelrod, B., Wilcox, J.R. and Nielsen, N.C. (1982), *Crop Science* 22, 333-7.
149. Yarwood, A. (1978), in Norton, G. (ed.) *Plant Proteins*. London and Boston: Butterworth, 41-5.
150. Yatsu, L.Y. and Jacks, T.J. (1972), *Plant Physiology, Lancaster* 49, 937-43.
151. Yatsu, L.Y., Jacks, T.J. and Hensanling, T.P. (1971), *Plant Physiology, Lancaster* 48, 673-82.
152. Yazdi-Samadi, B., Rinne, R.W. and Seif, R.D. (1977), *Agronomy Journal* 69, 481-6.
153. Yohe, J.M. and Poehlman, J.M. (1972), *Crop Science* 12, 461-4.

154. Young, C.T. and Hammons, R.O. (1973), *Oléagineux* 28, 293-7.
155. Zezulka, A.Y. and Calloway, D.H. (1976), *Journal of Nutrition* 106, 212-21.
156. Ford, J.E. (1981) in Bodwell, C.E., Adkins, J.S. and Hopkins, D.T. (*eds*) *Protein Quality in Humans: Assessment and* in vitro *Estimates*. Westport, Connecticut: Avi Publishing, 278-305.

4 *Rhizobium*, nodulation and nitrogen fixation

J.I. Sprent and F.R. Minchin

INTRODUCTION: LEGUME NODULE TYPES AND THEIR FORMATION

This chapter considers the formation and functioning of the
symbioses between legumes and rhizobia and the role of
nitrogen* fixation in grain legume productivity. It concentrates on the differences amongst species and highlights
areas where there is insufficient knowledge, but does not
attempt to cover the biology of rhizobia in soil or details
of inoculant production and use. The volume edited by Vincent [150] serves as an excellent introduction to these
aspects of *Rhizobium* biology. Similarly, for a general
account of the biochemistry of nitrogen fixation, readers
are referred to Postgate [100].

All the grain legumes discussed in this book are classi
fied taxonomically within the Leguminosae subfamily Papilio
noideae (see Chapter 1); most of them occur in either the
tribe Phaseoleae (largely tropical or sub-tropical) or in
the Vicieae (temperate). These taxonomic divisions are correlated with groups of features associated with root nodules (see also Sprent [125]) and symbioses with *Rhizobium*.
The major features currently known are listed in Table 4.1.

---

* Nitrogen gas is correctly termed 'dinitrogen' but since the colloquialism 'nitrogen' is used more or less universally for the gaseous
  element N_2, especially in the literature pertaining to the fixation
  of N_2 by legumes, it is used in that context throughout this book.

Table 4.1 Some major characteristics of nodules of grain legumes

Tribe of legume host		Rhizobium		Mode of infection	Nodule growth	Principal product exported
Genus	Species	Species	Growth*			
PHASEOLEAE						
Cajanus	cajan	promiscuous	slow	root hair	determinate when young	ureides
Glycine	max	japonicum	slow/fast**	root hair	determinate	ureides
Phaseolus	lunatus	cowpea type	slow	root hair	determinate	ureides
Phaseolus	vulgaris	phaseoli	fast	root hair	determinate	ureides
Psophocarpus	tetragonolobus	slow growing	slow	root hair	determinate when young	ureides
Vigna	mungo	cowpea type	slow	root hair	determinate	ureides
Vigna	radiata	cowpea type	slow	root hair	determinate	ureides
Vigna	unguiculata	cowpea type	slow	root hair	determinate	ureides
VICIEAE						
Lens	culinaris	leguminosarum	fast	root hair	indeterminate	amides
Pisum	sativum	leguminosarum	fast	root hair	indeterminate	amides
Vicia	faba	leguminosarum	fast	root hair	indeterminate	amides
CICEREAE						
Cicer	arietinum	sp.	fast	root hair	indeterminate	amides, ureides
AESCHYNOMENEAE						
Arachis	hypogaea	various	fast/slow	lateral root junctions	indeterminate	ureides, amides
GENISTEAE						
Lupinus	spp.	lupini	fast/slow	root hair	indeterminate, collar-shaped	amides

*On defined media; **fast-growing strains have recently been isolated in China [162].

116

Infection and nodulation

Phaseoleae.   There are interesting variations amongst the
rhizobia which nodulate the grain legumes in the Phaseoleae.
Those which infect soyabeans (*Glycine max*) have generally
been assigned species status as *R. japonicum*. They are
typical of the slow-growing rhizobia and differences between *R. japonicum* and the fast-growing species may be sufficient to warrant their separation into distinct genera
[57]. *Bradyrhizobium* is the suggested name for the genus
infecting soyabeans [57]. The so-called 'cowpea miscellany'
of rhizobia which infect *Vigna* and other genera (including
Glycine) has so far defeated taxonomists: the organisms
are promiscuous in their host range and, arguably, primitive [144]. The genus *Phaseolus* is infected by both fast-
and slow-growing rhizobia, but not indiscriminately so.
Common beans (*P. vulgaris*) and runner beans (*P. coccineus*)
are nodulated by the fast-growing *R. phaseoli*, which shares
many features with rhizobia nodulating temperate legumes
(see [126] and references therein; also [8]). *P. coccineus*
is probably the most cold-tolerant of the *Phaseolus* species
although some cultivars of *P. vulgaris* are also relatively
cold-tolerant. Lima beans (*P. lunatus*) on the other hand,
are not cold-tolerant and are nodulated by slow-growing
rhizobia of the cowpea type. Further details of these
relations are considered by Shantaram and Wong [115].

Root infection and subsequent nodule development appear
to be similar, respectively, in *Glycine, Phaseolus* and
Vigna, but there is not yet sufficient information for
Psophocarpus or *Cajanus* to determine whether or not they
correspond to the *Glycine* pattern. The nodules of these last
two genera often remain active for more than one season,
as befits their short-lived perennial (rather than annual)
life-span. Whilst young, the nodules closely resemble those
described below, but after several months they may become
lobed and so their growth is not strictly determinate.

Infection in all genera studied can occur via root

hairs, but may also be via root epidermises (see Bal and Wong [2]). The general sequence of events has been described by Sprent [123] and Newcomb [86]. Rhizobia can live saprophytically in soil, using various forms of combined nitrogen for growth. They are attracted (possibly chemotactically) to the roots of a suitable host legume and multiply within the rhizosphere, probably stimulated by energy-rich nutrients released from or sloughed off the roots. Root hairs remain susceptible to infection only transiently [11], and perhaps only whilst they are growing actively. All models of infection invoke modification of hair growth, but they vary in detail (compare Hubbell [54] with Bauer [4]). A general prerequisite for infection is curling of root hairs and most, but not all, infections occur in the crooks of such curls. Attachment of rhizobia to root hairs is polar and associated with a specific recognition system whereby only compatible rhizobia penetrate a root hair. The recognition process involves a sequence of events, including lectin binding and formation of cellulose microfibrils, but not all stages are equally well defined. Entry through the host wall involves extremely localised digestion of wall components, followed by secretion of new wall material around the invading bacteria. The latter forms the outer layer of the infection thread which grows down the hair (preceded by the host cell nucleus) and then into the outer cortex of the root. Invading bacteria divide inside the thread and secrete around themselves an extracellular matrix.

Before the infection thread reaches the cells just below the root epidermis, the latter are stimulated to divide. Some of the resultant cells are penetrated by infection threads, others are not. In those which are infected, bacteria are released into small packets enclosed by membranes of plasmalemma origin. The knot of uninfected and infected cells (together with the enclosed bacteria) continues to divide and eventually ruptures the epidermis of the root to become a visible nodule. It is towards the end of this period of division that rapid cell enlargement occurs and several unique proteins are produced. Two of these, nitrogenase and leghaemoglobin, are responsible for the reduction of nitrogen to ammonia and the transport of oxygen, respectively. The functions of the other proteins are generally unknown [160], although one has recently been identified in soyabean nodules as uricase, necessary for the synthesis of ureides [159]. The induction of nitrogenase within the bacteria is probably coupled to the repression of the ammonia-assimilating machinery, so that when nitrogen is reduced to ammonia, the product passes to the host cell for incorporation into organic molecules. Leghaemoglobin is formed outside the bacteroids, probably on both sides of the peribacteroid membrane [7], and serves to maintain a rapid flux of oxygen (necessary for ATP production) at a low concentration (necessary to avoid oxygen inactivation of nitrogenase).

Another feature which is essential to gaseous exchange in nodules is an intercellular space system. Although opinions differ (see discussion in Bergersen [7]), the consensus is that an airspace continuum exists within the infected region and inner nodule cortex, but that a shallow layer without airspaces occurs somewhere in the middle or outer cortex. Communication with the soil atmosphere is via lenticels. Gaseous transport in soyabean nodules has been discussed elsewhere [121,142].

Branches of the root stele pass into the nodule cortex, encircle the infected tissue, fuse at the distal end and thus generally terminate nodule growth. Individual nodules usually remain active in nitrogen fixation for several weeks, but then senesce rapidly from the centre outwards. The onset of senescence is usually indicated by changes in pigmentation in the centre of the nodules (from pink-red to greenish-brown). Recent evidence suggests that this colour change and the loss of nitrogenase activity which accompanies it are to some extent reversible [97]. As the first-formed nodules begin to senesce, nitrogen fixation is continued by successive generations of nodules, usually formed on secondary and tertiary roots.

Vicieae. The rhizobia which nodulate the Vicieae are always fast-growing. The initial processes of root hair infection appear to be broadly similar to those described above. Differences in nodule development become apparent at about the time when the infection thread reaches the base of the root hair. Again, divisions are stimulated in the root cortex, but in this case in the inner cells which lie just outside the endodermis. The planes of division are more ordered than in the nodules of the Phaseoleae and a meristem and differentiation region can soon be distinguished. The infection thread passes through and between the cortical cells until it arrives at the inner region of dividing cells, i.e. those which are soon to differentiate. Many of these cells are penetrated individually by branches of the infection thread. In such newly infected cells, bacteria are released as before into membrane-bound packets, but the infected cells do not divide. Instead, considerable cell enlargement occurs to produce vacuolate cells with a broad layer of peripheral cytoplasm which contains differentiated bacteroids and leghaemoglobin. Bacteroids are much larger and contain more DNA than their free-living counterparts. They often adopt various modified shapes, the so-called 'T', 'X' and 'Y' forms being the most common. Modifications to the cell wall/membrane system accompany this differentiation and lead to increased permeability, together with reduced viability.

The nodule meristem continues to divide and many of the newly formed cells eventually become infected by other branches of the original infection thread. The young nodule ruptures the root epidermis, continues to grow for many weeks and often branches. Outgrowths from the vascular system of the subtending root continue to grow in the cortex of the distal ends of the nodules. After a few weeks of active nitrogen fixation, infected cells senesce and so infected regions adjacent to the subtending root become progressively less active and usually turn greenish-brown.

In this type of nodule there are intercellular spaces too, but whether or not they form a continuum in the active region has been questioned by Dixon *et al.* [29] (see also Bergersen [7]). Gaseous exchange with the soil atmosphere occurs through intercellular spaces scattered over the entire nodule surface, rather than being confined to lenticels, as in soyabeans.

Table 4.2   Some differences between nodules of the tribes Phaseoleae and Vicieae

Attribute	Phaseoleae	Vicieae
Infected cells	non-vacuolate	vacuolate
Infection threads	non-persistent	persistent
Newly infected cells	divide frequently	do not divide
Bacteroids compared with free-living forms	little increase in size or DNA content; some viability	large increase in size and DNA; little viability

What governs these differences (Table 4.2) in nodule development? A major factor is likely to be the balance of hormones produced by host and endophyte. Hormones have been implicated ever since the original observations that infection threads stimulate divisions in cells which they have not penetrated, nor even reached. Since then, evidence has accumulated that hormones are produced by both partners in the symbiosis, but no clear picture of the control of cell division and enlargement has yet emerged. The fact that one strain of *Rhizobium* produces determinate nodules on one host and indeterminate ones on another [58], and that strains of *Rhizobium* of different origin can produce nodules of different structure on the same host [3], indicate that both symbionts play an active role. Furthermore, the nutritional status of the endophyte is critical. The nutrient requirements of fast- and slow-growing rhizobia show differences in both the free-living and symbiotic states

[31]. An interesting possibility is that nodule host cells may become partially anaerobic, producing succinate in peas (*Pisum sativum*) [151] and ethanol and acetaldehyde in soyabeans [139]. Succinate is a most effective substrate to support nitrogen fixation in *R. leguminosarum*, and it may further assist in the differentiation of the bacteroid form [148].

Differences between host tribes with respect to ammonia assimilation have been discussed elsewhere [124].

Cicereae. This monogeneric tribe was, until recently, included in the Vicieae (see Chapter 1). Chickpeas (*Cicer arietinum*) grow well in warmer regions and their nodules probably export a mixture of amides and ureides [103]. Otherwise, such knowledge as we have suggests that nodulation conforms to the pattern described for the Vicieae.

Aeschynomeneae. Groundnuts (*Arachis hypogaea*) are the only economically important grain legume from this tribe. Symbiosis in this species (as well as the geocarpic fruiting habit) is rather unusual. It was the first case described in which infection occurred other than through root hairs [19]. The related genus *Stylosanthes* [20] and the unrelated one *Neptunia* [109] are now known not to use root hairs for infection, so the apparent exception of groundnuts may yet prove to be common in certain tribes. Full details of infection can be found in Chandler [19]. Briefly, rhizobia penetrate the epidermis at the bases of multicellular hairs which arise at the junctions of the lateral roots. Thus, hairs are associated with, if not a route for, infection and mutants without hairs do not nodulate [83]. Bacteria penetrate the root via the middle lamellae and enter cells of the cortex of the emerging lateral root where the cell wall is structurally modified. Such cells and the bacteria they contain divide repeatedly to form a more or less spherical nodule with a broad attachment to the subtending root. Bacteroids, which do not differentiate until host cell

122

division is complete, are spherical, with little or no cell wall. This last point is particularly interesting as *Arachis* may nodulate with cowpea-type rhizobia which form enlarged, rod-shaped bacteroids in *Vigna* spp. Thus, the host cell clearly affects the form of the active endophyte. Furthermore, rates of nitrogen fixation per unit bacteroid tissue in groundnut are six times greater than those in cowpea (values averaged over four strains of *Rhizobium*) [114], which suggests that the spheroplast form may allow rapid exchange of metabolites between symbionts. Unlike nodules of the Phaseoleae and Vicieae, infected cells in groundnut nodules are not interspersed with uninfected ones. In the former tribes, these 'interstitial' cells have large vacuoles and frequent plasmodesmatal connections with their infected neighbours. In *Glycine*, *Phaseolus* and *Vigna* they contain prominent microbodies (peroxisomes) which are thought to contain urate oxidase, an enzyme necessary for ureide biosynthesis [40,87]. If *Arachis* is confirmed regularly to export a significant amount of its combined nitrogen as ureides, the question is posed as to where these compounds are synthesised.

Genisteae. *Lupinus* is considered to be an atypical member of this tribe for a number of taxonomic reasons (and see Chapter 1): its nodules are also rather anomalous in that they have one or more meristematic regions and so their growth is indeterminate. However, these meristems frequently grow laterally so that the nodules encircle the roots in a collar-like structure. Bacteroids are similar in appearance to those of soyabeans and the major export products of nodules are amides. Alone amongst nodules outside the Vicieae and Trifolieae their vascular strands include transfer cells.

PHYSIOLOGY AND BIOCHEMISTRY

Fixation and assimilation of nitrogen by root nodules

In outline, nitrogenase catalyses the following reactions:

$$N_2 + 6H^+ + 6e^- \rightarrow 2NH_3 \qquad\qquad (1)$$

$$2H^+ + 2e^- \rightarrow H_2 \qquad\qquad (2)$$

These two reactions occur simultaneously and at least one molecule of hydrogen is evolved for every nitrogen molecule reduced. About 15 ATP are thought to be needed for each N_2 reduced or H_2 evolved. Hydrogen evolution can be regarded as a waste of energy both in terms of ATP and in the potential energy of the reducing power used in its formation. Some rhizobia possess an uptake hydrogenase which oxidises hydrogen to water with the production of ATP. This partly offsets the energy lost and may also act to scavenge excess oxygen.

Ammonia produced by nitrogenase action passes from the bacteroids to the host where it is assimilated, initially into glutamine, at the expense of another ATP. Further reactions may occur to produce asparagine, ureides (allantoin and allantoic acid) and other amino compounds which are exported to the rest of the plant.

Further details of these processes and their implications for carbon (C) balance and pH regulation are discussed in Sprent [127]. The nitrogenase enzyme contains both iron and molybdenum. The latter is also required for nitrate reductase and for xanthine dehydrogenase, a key enzyme in ureide biosynthesis. Nitrogenase shows considerable substrate versatility; as well as reducing N_2 and H^+, in the presence of acetylene it will divert all its incoming electrons to the production of ethylene. This reaction forms the basis of the much used (and, as we discuss below, abused) acetylene reduction assay for nitrogenase activity.

As would be expected of such a complex process, nodule functioning is greatly affected by the soil and aerial environments (see the recent review by Sprent *et al.* [129]). One question which has seldom been posed is whether nitrogen-fixing plants show sensitivities to environmental factors (such as drought or extremes of temperature) different from those of the same genotype grown on combined nitrogen (N). Furthermore, many of the recommendations for non-N fertilizers for grain legumes are based on trials involving crops grown on combined nitrogen. Are these requirements necessarily the same as for nodulated plants? Before best use can be made agronomically of both legumes and fertilisers, these questions must be answered. Only then can the obviously desirable aim of improving nitrogen fixation by grain legumes be achieved. Progress so far has been mainly based on glasshouse and controlled environment studies which have yet to be translated into results applicable in the field. We now consider the reasons for this failure.

Errors in measurement techniques

Fundamental to any nitrogen fixation improvement programme is an ability to measure accurately N_2 fixation by whole plants, in both controlled environments and the field. Several techniques have been developed which purport to accomplish this task, and they have been widely employed with differing degrees of enthusiasm. The two most commonly used are the acetylene reduction assay and the ^{15}N dilution technique [6].

The acetylene reduction assay. This technique is based on the reaction:

$$C_2H_2 + 2e^- + 2H^+ \rightarrow C_2H_4 \tag{3}$$

It has been recognised for several years that the major difficulty with this assay is in the conversion from moles ethylene produced to moles N_2 which would have been fixed in the absence of acetylene [120,147]. The theoretical molar ratio is 3:1 (based on two electrons being used for the reduction of acetylene to ethylene and six for the reduction of N_2 to $2NH_3$; see equations (1) and (3)) and this was the value originally proposed as the conversion factor [41,42,46]. However, a wide range of values based on experimental data was being reported even at that time [42] and the reduction of protons to H_2 was suggested as a possible source of error [18]. Indeed, Burris [18] concluded that there was no reliable standard conversion ratio and that the assay should be calibrated for each experimental condition, preferably using $^{15}N_2$. This caution has proved remarkable, both for its foresight and the extent to which it has subsequently been ignored.

The role of H_2 production became more apparent with the acceptance that it is a common phenomenon in legume nodules [112] and may well be an inescapable consequence of nitrogen fixation by nitrogenase [111,131]. If one H_2 is produced for every N_2 reduced, then 25% of the electron flow to nitrogenase is utilised in H_2 production and the $C_2H_2:N_2$ conversion factor should be increased to four. However, as noted by Turner and Gibson [147], if the ratio of N_2 consumed to H_2 produced is greater than 1:1, the conversion factor will be greater than four. Furthermore, some nodules contain an uptake hydrogenase [112] which renders it impossible to measure total H_2 production without techniques such as the use of carbon monoxide to inhibit that enzyme [12].

The conversion ratio has also been reported to vary seasonally [36,128] and to alter in response to environmental stress [33,36]. This could reflect seasonal or environmentally induced changes in H_2 production, which are known to occur in some cases [9], but the possibility of a more fundamental error in the acetylene reduction assay as it is frequently performed should be considered as well.

The usual way in which the acetylene reduction assay is carried out is for nodulated roots to be enclosed in an air-tight vessel containing an atmosphere supplemented with 10% acetylene. The amount of ethylene which has accumulated after a given time (usually 30-60min) is then measured. Using a flow-through system, Cooper *et al.* [22] and Minchin *et al.* [80,81] have now demonstrated for several legume species that rates of acetylene reduction decline markedly during the 30-minute period following the introduction of acetylene. This decline is not inevitable [74,143]. However, linearity of ethylene production should always be checked, preferably with a flow-through system, since small changes in rate are difficult to detect by the cumulative method.

The implications of a slowly declining rate of acetylene reduction are: (a) underestimation of N_2 fixation; (b) obscuring or introduction of small variations in N_2 fixation due to environmental manipulations and diurnal or seasonal changes; and (c) errors in measurement of the relative efficiency (RE) of H_2 use, normally calculated as:

$$RE = 1 - \frac{\text{hydrogen evolved in air}}{\text{acetylene reduced}} \qquad (4)$$

Clearly, it is prudent to use a flow-through system and to calibrate it at the outset, preferably by $^{15}N_2$.

From these implications it is also clear that the acetylene reduction assay should be confined to the detection of the presence of an active nitrogenase system and as a crude assay for major changes in N_2 fixation activity. It has limited use as a reliable field assay for rates of N_2 fixation (since flow-through systems are difficult to set up and replicate under field conditions) and care should be exercised in using it as a screening technique to detect symbiotic systems with superior N_2 fixation (e.g. see [50]). The enthusiasm which this technique attracted a few years ago should now be tempered.

The ^{15}N dilution technique.   In essence, this technique requires the soil N pool to be labelled with the stable isotope ^{15}N which is then taken up by the legume and diluted with $^{14}N_2$ fixed from the atmosphere. The degree of dilution compared with the original ^{15}N soil labelling allows calculations of the amount of $^{14}N_2$ fixed by the plant. Unfortunately, it is not possible to measure the ^{15}N labelling of the soil N around the roots without destroying the crop. Instead, the ^{15}N labelling of a non-fixing, companion crop is measured, on the assumption that this will reflect the ^{15}N available to the legume.

The need to select an appropriate control crop, with a rooting pattern similar to the legume, is well recognised [105,106,152], as are other problems such as non-uniformity of ^{15}N distribution in the labelled plant [49,106]. However, a previously neglected source of error is the change in the $^{15}N:^{14}N$ ratio of the soil pool, as ^{15}N is taken up by the crop, or lost through leaching, or ^{14}N is released through mineralisation [155,156]. This latter process can be particularly significant if the application of large amounts of fertiliser N to the non-fixing companion crop has a 'priming' effect, i.e. it increases mineralisation [26,48,82].

The change in the ^{15}N labelling pattern of the soil pool requires that companion crops be selected not only for their similarity of rooting pattern, but also for similarity for soil N uptake profiles, so that the companion and legume crops 'see' soil N pools with the same $^{15}N:^{14}N$ ratio. Failure to match the N uptake profiles can lead to errors as large as 50% in the estimated N fixed [156]. In practice, the matching of companion crops in terms of both parameters will be very difficult, if not impossible to achieve.

An alternative approach is to label the soil organic pool with ^{15}N by either incorporating ^{15}N labelled organic matter [47] or adding ^{15}N fertiliser together with a carbon source [65]. In these circumstances ^{15}N will be released into the soil N pool by mineralisation, thus maintaining the $^{15}N:^{14}N$ ratio. However, such an approach requires a protracted incubation period (of at least one year) and the application of organic matter may alter the activity of the microflora and physical properties of the soil [16,106]. The use of slow-release ^{15}N fertilisers (e.g. by gypsum pelleting [156]) is an approach which needs to be actively investigated.

Other ^{15}N methods.   Two other potentially useful field assay techniques are: (a) to utilise the difference in natural abundance of ^{15}N (δ^{15}N) between the atmosphere and soils, and (b) to expose the legume roots to ^{15}N$_2$ gas.

The natural abundance technique is an expensive assay (it requires a mass spectrometer, rather than the comparatively cheap micro-mass spectrometers used in ^{15}N dilution or ^{15}N$_2$ measurements), and suffers from errors due to isotopic fractionation in the soil [28], in the legume tested [60] and even in grazing animals [130]. Furthermore, there may be differences in the distribution patterns of ^{15}N within legumes [116,117]. Consequently, some authors have concluded that this approach is only useful as a qualitative technique [14,61,130] whilst others are more optimistic about its quantitative potential [105,146].

The use of ^{15}N$_2$ is not only limited by the cost of the gas, but also by the need to enclose the legume root system, which will isolate it from the rest of the root environment and require careful environmental control [49,59]. Furthermore, it is necessary to maintain the pN$_2$ at 78%, as replacement of most of the N$_2$ with inert gases such as argon or helium will cause a decline in nitrogenase activity similar to that effected by acetylene [80]. This necessitates the replacement of the existing N$_2$ atmosphere in the root enclosure with one containing ^{15}N$_2$, a process which can be very difficult and may physically damage the enclosed root system [59].

Sap analysis techniques.   Despite reports of the presence of ureides in soyabeans [61a] and common beans [30], the possible role of these compounds in nodule metabolism was largely ignored for many years [75,76]. Furthermore, with the exception of Pate [90], xylem sap collected from detached roots or nodules was analysed only for amino acids and amides [39,132,158] on the mistaken assumption that all legumes showed the same pattern of export products as found in the Vicieae and Trifolieae [94]. Largely as the result of several publications from Japan [35,71,72,73], it is now recognised that the ureides allantoin and allantoic acid are the major export compounds of many tropical legumes [124]. The fact that the occurrence of ureides in xylem sap is largely restricted to nodulated plants has prompted optimism that sap analyses of ureide-producing species can provide a quantitative assay for nitrogen fixation [25,63,66,67,93]. However, xylem sap collection under field conditions is often very difficult, especially from unwatered plots [51], and plant genotype may affect the ureide:amide ratio; an alternative technique based on the ureide contents of plant parts has now been proposed [51, 52]. Although these techniques have potential for field measurements, they are of course limited to species (or even symbiotic combinations) which produce ureides. Then again, significant ureide production may occur even in non-nodulated plants (see [63,140]).

Problems in the interpretation of data

Comparative respiratory costs of nitrogen and nitrate reduction and implications for seed yield.   Numerous studies on the respiratory costs of nitrate reduction in legumes have found the process to be either equal to or, more frequently, less than the cost of fixing an equivalent amount of N_2 [78,92]. This diversity is sometimes portrayed as contradictory [17,55,98]. However, both types of result can be accommodated in a unifying hypothesis. Thus, the costs in terms of carbohydrate utilisation will be similar if both N sources require reductant to be produced via respiration [44,113], but nitrate-fed plants can 'escape' a proportion of these costs by transporting nitrate to the leaves and utilising photosynthetically-produced reductant [55,110, 113] (the actual savings may not be very great and may vary with the necessary acid produced for pH regulation [161]).

If this hypothesis is correct then 'simple' comparisons between nitrate-fed and N_2-fixing plants in terms of respiratory and N accumulation activities of roots are not very meaningful [1,62,77,79,95]. Instead, comparisons should be made in terms of C/N economies of whole plants [69,79,108].

The corollary of these studies is that where nitrate-fed plants 'escape' the carbohydrate utilisation costs, their growth rates will be greater than N_2-fixing plants accumulating a comparable amount of N. However, this assumption is not valid if the nitrate supply is sub- or supra-optimal [17,95,138] or if the photophosphorylation energy is all required for carbon assimilation. Nevertheless, it has been suggested [64] that seed yields of legume crops could be greater for nitrate-fed plants because of their beneficial carbon costs. However, most evidence suggests that the effects of nitrogen on growth rates are not directly related to ultimate seed yield [107,137]. Indeed, such a link should not be expected, given the numerous differences between plants relying on nitrate and those relying on N_2 [91].

Two of these differences may be particularly relevant. Firstly, organic N products of fixation are transported from the nodules via the xylem and certain compounds, allantoin and amides for example, can by-pass leaves by xylem-to-phloem exchange in stems or petioles [91]. Conversely 'cost escaping'nitrate-fed plants must produce most of their organic N compounds in the leaves. If leaves and fruits are considered to compete for N, then nodule-based production allows this competition to be directed in favour of developing reproductive structures. Secondly, it is well documented that many legumes have less capacity for nitrate uptake and reduction during the late reproductive period whilst N_2 fixation can still be comparatively active [107]. This also gives N_2 fixation a distinct advantage in terms of supplying

131

N to seeds. It is, therefore, not surprising that for many of the legume crops included in this book reports can be found where applications of large amounts of fertiliser N have resulted in little or no improvement in seed yield when compared with actively-fixing plants.

Nitrate effects on N_2 fixation.   There are two distinct hypotheses for the inhibitory effects of nitrate on nodular nitrogen fixation: (a) a direct effect through production of nitrite in the nodules and (b) an indirect effect through limitation of carbohydrate supply to the nodules [131]. Much of the evidence for the second hypothesis relates to observations that less carbon flows to nodules after plants are given nitrate (e.g. Ursino *et al.* [149]). This evidence continues to gain credence despite the fact that Gibson and Pagan [37] have pointed out that a direct effect on nodule activity would reduce sink strength and thus carbohydrate movement to nodules. However, Streeter [133] has demonstrated that a nitrate-induced decline in N_2 fixation activity of soyabean nodules was not accompanied by any significant changes in the concentrations of carbohydrate pools within the nodules. Such changes would be inevitable if the carbohydrate deprivation hypothesis were correct.

Then again, the nitrite production hypothesis was apparently disproved by the observation that nitrate still inhibited the activity of nodules formed by nitrate reductase-deficient mutants of *Rhizobium* [37,70]. Bacteroids were the putative site of nitrate reductase in nodules, but Streeter [134] has shown that significant amounts of nitrate reductase activity are associated with the cytosol, which confirms a previously isolated report [21]. This fact is not altogether surprising as the cytosol of host cells of nodules is derived from root cells which are known to be capable of nitrate reductase activity. Previous failures to detect cytosol activity should perhaps have aroused suspicions about the reliability of measurement techniques.

Although controversy over the effects of nitrate will,

no doubt, continue (see [88]), it is apparent that the question can only be resolved satisfactorily by a detailed examination of all potentially important parameters and that the answer may not be the same for all species. Low concentrations of nitrate, for example, may never reach the nodules of plants with a strong nitrate-reducing system in their roots.

Nitrate effects on leaf growth.   One of the few examples where nitrate improves both growth and nitrogen fixation in some legume species is when it is applied at an early stage of vegetative growth, the so-called 'starter' nitrogen effect. The exact mechanisms leading to such benefits are not clear, but for those legumes where nitrate reduction is largely a leaf-located process, it may be via enhanced leaf expansion. For example, Radin and Boyer [101] showed that expansion of sunflower leaves was directly related to the flux of nitrate to them. They concluded that nitrate increased the hydraulic conductivity which, in turn, allowed leaves to expand during the daytime under conditions when they would normally be limited by mild water stress. Similar data have now been obtained for soyabeans [102], and areas of first trifoliolate leaves of common beans are much greater when plants are grown on nitrate than on either ammonium or fixed nitrogen (S.M. Allen, pers. comm.). This species reduces much of its nitrate in the leaves, and often benefits from starter nitrogen. Conversely, those species, such as faba beans (*Vicia faba*), which actively reduce applied nitrate in their roots, show little response to starter nitrogen. Effects of nitrate also vary with other environmental factors, such as temperature [56].

Carbon and oxygen supplies as limiting factors in N_2 fixation.   The evidence that carbohydrate supply to nodules is the major limiting factor for N_2 fixation is well documented [45], although the universality of this hypothesis has been questioned [99,135] and in many instances carbo-

133

hydrate supply may exert its effect indirectly via growth of the whole plant [78,96]. Nevertheless, improved carbohydrate supply to nodules is regarded as a major objective for legume breeders [63,136].

On the other hand, researchers who study the mechanisms and regulation of O_2 supply to bacteroids have often concluded that this is a major limiting factor [5,7,154]. The apparent contradiction between these two hypotheses has rarely been noted (but see Hardy et al. [43]).

Recent work [119,157] has demonstrated that nodules can be classified into two groups: (a) those that can adapt very rapidly (i.e. within ten minutes) to changes in pO_2, and (b) those that can adapt only slowly, if at all. Other data suggest two types of response to increased rates of host shoot photosynthesis: (a) a more or less linear increase in fixation activity (e.g. in lucerne [118]) and (b) a saturation at comparatively low rates of photosynthesis (e.g. in soyabeans [118,153]). In combination, these observations produce a unifying hypothesis, whereby nodules which can rapidly alter their resistance to O_2 diffusion can adapt to short-term changes in carbohydrate supply and can be limited by this supply, but nodules which cannot rapidly influence their O_2 diffusion are unable to utilise increased carbohydrate supplies because they are O_2-limited.

Hardy et al. [43] dismissed the idea of O_2-limitation because nodules of their soyabean plants could adapt to changes in pO_2 within twenty-four hours [23]. However, this response would be far too slow to take advantage of ephemeral increases in carbohydrate supply and would probably render the nodules O_2-limited. Furthermore, the observation [24] that nitrogenase activity of attached soyabean nodules did not increase significantly between atmospheric pO_2 and the optimum of 33% O_2 does not demonstrate that O_2 was non-limiting, merely that those particular nodules were unable to utilise concentrations of O_2 greater than ambient (but see Pankhurst and Sprent [89]).

Clearly, the possibility that O_2-limitation is a major

factor in N_2 fixation has important implications for breeding programmes aimed at increasing carbohydrate supply to nodules.

Limitations in plant and crop physiology research

Single environment studies.   Many experimental data have been obtained for plants grown in single artificial environments (which may not allow for normal growth or development; see Summerfield [137]) but then conclusions have often been presented as if they apply to that species in all symbiosis/ environment combinations, or even to all grain legumes. However, unless a particular physiological attribute is studied under several environmental conditions which represent the agricultural habitat of that species it is not possible to appreciate the variability which can be encountered in crop situations. Therefore, single-environment studies may be of academic importance but are of limited practical value to breeders and agronomists.

Studies on environmental effects.   Plant physiological studies on N_2 fixation are of special interest to breeders and agronomists when they deal with environmental effects and differences between genotypes in responsiveness to them. However, much of this research has been descriptive and has added little to our understanding of how the environmental effects operate. Consequently, for most of the major environmental factors it is not possible to conclude whether the effects are directly on nodules, when they may be amenable to *Rhizobium* strain selection, or indirectly via the host plant, when alleviation requires manipulation of both symbionts, agronomically, genetically or both [129].

Neglected areas of research.   In any thorough review of the physiology of nitrogen-fixing legumes it is possible to discern two areas which have received very little attention: hormonal interactions between the symbiotic partners, and

the biochemical pathways of carbohydrate utilisation within
the nodules. There is evidence for hormonal control of nodulation and N_2 fixation [129] and examples of strain of *Rhizobium* influences on growth, flowering and fruit production
are increasingly common [32,34,85,145]. Then again, carbohydrate metabolism has been extensively studied *in vitro*
for both vegetative rhizobia and isolated bacteroids, but
in vivo nodule studies are few [7,104]. The reluctance of
researchers to become involved in these areas is understand-
able in terms of the costs of instrumentation and materials,
and the time required before useful findings are likely.
However, studies of this sort are essential to the full
realisation of the potential benefits of genetic manipulation (see [15,63]).

GENETICS

The genetics of individual host species is considered in
Chapters 5-17 and the genetics of *Rhizobium* will be considered only briefly here. Since the early 1970s the whole
emphasis in *Rhizobium* genetics has changed from one in which
selection of improved strains was the standard procedure to
one where the technology for *producing* (by genetic manipulation) new strains can be utilised. Important advances have
been achieved through the world-wide cooperation of microbial geneticists. Fundamental to this change was the discovery that the genes for nitrogenase production, uptake
hydrogenase production and the induction of nodules on host
plants are located, not in the bacterial chromosome, but in
accessory units of DNA known as plasmids. By suitable techniques these can now be transferred between rhizobia and
also sections of DNA cloned in other bacteria such as *Esch-
erichia coli*. Whole series of mutants are being produced
which should be of great value in studies of host/rhizobial
interactions, as well as mapping the genetic material.

The technology of microbial genetics has generated an
extensive and unfamiliar vocabulary, and it is very easy for

more applied scientists to be discouraged by the 'jargon'. Beringer *et al.* [8] have outlined the processes and terminology very clearly. If workers involved in higher plant studies could define the properties they would like to see in strains of *Rhizobium* for particular regions or crops, it is likely that the 'manufacture' of such rhizobia will soon be possible. However, to define these properties at the required biochemical/physiological level is likely to be the more difficult task, and to take longer than the optimistic forecast of five years suggested in 1979 [13].

One area in which there has been progress is in the transfer of Hup^+ (the gene for uptake hydrogenase) into strains of *Rhizobium* which lack it and there is evidence that the new strains are able to fix more nitrogen than their parents [27]. In this work, it is tacitly assumed that the oxidation of H_2 yields ATP. Although this is often the case, not all Hup^+ strains couple H_2 oxidation to ATP synthesis [84]. This may be why some workers have been unable to demonstrate increased N_2 fixation when H_2 is being oxidised [38]. Furthermore, the benefits obtained from saving energy in coupled Hup^+ systems will depend upon how the particular symbiosis behaves under different environmental conditions. Since there is no obvious disadvantage of inserting Hup^+ DNA into rhizobia, it may be a wise general practice in case favourable conditions for its expression obtain.

Attempts to select rhizobia which will fix nitrogen in the presence of nitrate have proved less successful and there may be better prospects for obtaining nitrate-resistant nodules by host rather than rhizobial selection [68].

CONCLUSIONS

It is clear that there is considerable variation in each of the diverse aspects of the symbioses of different grain legumes. For N_2 fixation improvement in crops, rhizobiologists, plant breeders and agronomists all require a knowledge of the physiological and biochemical functions which they are attempting to influence. Hitherto, consistent improvements in grain legume yields have been small compared with those for cereal crops [53]. This in part stems from the complex nature of the legume plant which has many levels of inter-action between the host and the rhizobial symbiont. These may be further confounded in tripartite symbioses involving the legume, *Rhizobium* and vesicular-arbuscular mycorrhizas. The latter may enhance [122] or lessen [10] nitrogen fixa-tion, depending on species and mineral nutrient supply.

Central to the problem is a need for greater knowledge of (and consequently financial investment in research on) plant and crop physiology if the benefits of N_2 fixation to grain legumes are to be realised. In particular, reliable techniques for field measurements must be developed, and the mechanisms of environmental responses must be under-stood. Furthermore, more research must be encouraged into both the overall biochemistry of nodules and rhizobial/plant interactions in relation to hormonal and nutritional effects.

ACKNOWLEDGEMENT

The Grassland Research Institute is funded by the UK Agri-culture and Food Research Council (AFRC). Both authors would like to thank the AFRC for financial assistance.

LITERATURE CITED

1. Atkins, C.A., Pate, J.S., Griffiths, G.J. and White, S.T. (1980), *Plant Physiology* 66, 978-83.
2. Bal, A.K. and Wong, P.P. (1982), *Canadian Journal of Microbiology* 28, 890-96.
3. Bal, A.K., Shantaram, S. and Wong, P.P. (1982), *Applied Environ-mental Microbiology* 44, 965-71.
4. Bauer, W.D. (1981), *Annual Review of Plant Physiology* 32, 407-49.
5. Bergersen, F.J. (1977), in Hardy, R.W.F. and Silver, W.S. (*eds*) *A Treatise on Dinitrogen Fixation. Section III Biology*. New York: Wiley Interscience, 519-55.
6. Bergersen, F.J. (*ed.*) (1980), *Methods for Evaluating Nitrogen Fixation*. Chichester: Wiley Interscience.
7. Bergersen, F.J. (1982), *Root Nodules of Legumes: Structure and Functions*. Chichester: Research Studies Press (Wiley).
8. Beringer, J.E., Brewin, N.J. and Johnstone, A.W.B. (1982), in

Broughton, W.J. (ed.) *Nitrogen Fixation Vol. 2*. Oxford: The Clarendon Press, 167-82.

9. Bethlenfalvay, G.J. and Phillips, D.A. (1979), *Plant Physiology* 63, 816-20.

10. Bethlenfalvay, G.J., Pacovsky, R.J., Bayne, H.G. and Stafford, A.E. (1982), *Plant Physiology* 70, 446-50.

11. Bhuvaneswari, T.V., Bhagwat, A.A. and Bauer, W.D. (1981), *Plant Physiology* 68, 1144-9.

12. Bothe, H., Tennigkeit, J., Eisbrenner, G. and Yates, M.G. (1977), *Planta* 133, 237-42.

13. Boulter, D. (1979), in Hewitt, E.J. and Cutting, C.V. (eds) *Nitrogen Assimilation in Plants*. London: Academic Press, 659-67.

14. Bremner, J.M. (1977), in Ayanaba, A. and Dart, P.J. (eds) *Biological Nitrogen Fixation in Farming Systems of the Tropics*. Chichester: John Wiley, 335-53.

15. Brewin, N.J., Johnstone, A.W.B. and Beringer, J.E. (1980), in Stewart, W.D.P. and Gallon, J.R. (eds) *Nitrogen Fixation*. London: Academic Press, 365-76.

16. Broadbent, F.E., Nakashima, T. and Chang, G.Y. (1982), *Agronomy Journal* 74, 625-8.

17. Broughton, W.J. (1979), in Clysters, H., Marcello, R. and Paucke, M. (eds) *Photosynthesis and Plant Development*. The Hague: Martinus Nijhoff/Dr W. Junk, 285-99.

18. Burris, R.M. (1974), in Quispel, A. (ed.) *The Biology of Nitrogen Fixation*. Amsterdam: North Holland, 9-33.

19. Chandler, M.R. (1978), *Journal of Experimental Botany* 29, 749-55.

20. Chandler, M.R., Date, R.A. and Roughley, R.J. (1982), *Journal of Experimental Botany* 33, 47-57.

21. Chen, P.-C. and Phillips, D.A. (1977), *Plant Physiology* 59, 440-42.

22. Cooper, R.E., Sheehy, J.E. and Minchin, F.R. (1982), *Annual Report of the Grassland Research Institute*, Hurley: AFRC, 119-22.

23. Criswell, J.G., Havelka, U.D., Quebedaux, B. and Hardy, R.W.F. (1976), *Plant Physiology* 58, 622-5.

24. Criswell, J.G., Havelka, U.D., Quebedaux, B. and Hardy, R.W.F. (1977), *Crop Science* 17, 39-44.

25. Day, J.M. and Roughley, R.J. (1978), *Annual Report of Rothamsted Experimental Station*, Harpenden: AFRC, 241-2.

26. Deibert, E.J., Bijeriego, M. and Olson, R.A. (1979), *Agronomy Journal* 71, 717-23.

27. DeJong, T.M., Brewin, N.J., Johnstone, A.W.B. and Phillips, D.A. (1982), *Journal of General Microbiology* 128, 1829-38.

28. Delwiche, C.C. and Steyn, P.L. (1970), *Environmental Science and Technology* 4, 929-35.

29. Dixon, R.O.D., Blunden, E.A.G. and Searl, J.W. (1981), *Plant Science Letters* 23, 109-16.

30. Drift, C.V.D. and Vogels, G.D. (1966), *Acta Botanica Neerlandica* 15, 209-14.

31. Elkan, G.H. and Kuykendal, L.D. (1982), in Broughton, W.J. (ed.) *Nitrogen Fixation Vol. 2* Oxford: Clarendon Press, 147-66.

32. El-Sherbeeny, M.H., Mytton, L.R. and Lawes, D.A. (1977), *Euphytica* 26, 149-56.

33. Engin, M.E. and Sprent, J.I. (1973), *New Phytologist* 72, 117-26.

34. Evensen, H.B. and Blevins, D.G. (1981), *Plant Physiology* 68, 195-8.

139

35. Fujihara, S., Yamamoto, K. and Yamaguchi, M. (1977), *Plant and Soil* 48, 233-42.
36. Gibson, A.H. (1981), in Gibson, A.H. and Newton, W.E. (*eds*) *Current Perspectives in Nitrogen Fixation*. Canberra: Australian Academy of Science, 334.
37. Gibson, A.H. and Pagan, J.D. (1977), *Planta* 134, 17-22.
38. Gibson, A.H., Dreyfus, B.L., Lawn, R.J., Sprent, J.I. and Turner, G.L. (1981), in Gibson, A.H. and Newton, W.E. (*eds*) *Current Perspectives in Nitrogen Fixation*. Canberra: Australian Academy of Science, 373.
39. Halliday, J. (1976), 'An interpretation of seasonal and short-term fluctuations in nitrogen fixation'. PhD. thesis, University of Western Australia.
40. Hanks, J.F., Tolbert, N.E. and Schubert, K.R. (1981), *Plant Physiology* 68, 65-9.
41. Hardy, R.W.F., Burns, R.C., Hebert, R.R., Holsten, R.D. and Jackson, E.K. (1971), *Plant and Soil* Special Volume, 561-90.
42. Hardy, R.W.F., Burns, R.C. and Holsten, R.D. (1973), *Soil Biology and Biochemistry* 5, 47-81.
43. Hardy, R.W.F., Criswell, J.G. and Havelka, U.D. (1977), in Newton, W.E., Postgate, J.R. and Rodriguez-Barrueco, C. (*eds*) *Recent Developments in Nitrogen Fixation*. London: Academic Press, 451-67.
44. Hardy, R.W.F. and Havelka, U.D. (1975), *Science* 188, 633-43.
45. Hardy, R.W.F., Havelka, U.D. and Heytler, P.G. (1980), in Corbin, F.T. (*ed.*) *World Soybean Research Conference II: Proceedings*. London: Granada, 57-72.
46. Hardy, R.W.F., Holsten, R.D., Jackson, E.K. and Burns, R.C. (1968), *Plant Physiology* 43, 1185-207.
47. Hauck, R.D. (1973), *Journal of Environmental Quality* 2, 317-27.
48. Hauck, R.D. and Bremner, J.M. (1976), *Advances in Agronomy* 28, 219-66.
49. Haystead, A. (1981), in Hodgson, J., Baker, R.D., Davies, A., Laidlow, A.S. and Leaver, J.D. (*eds*) *Sward Measurement Handbook*. Hurley: British Grassland Society, 229-42.
50. Heichel, G.H. (1982), *Iowa State Journal of Research* 55, 255-80.
51. Herridge, D.F. (1982), *Plant Physiology* 70, 1-6.
52. Herridge, D.F. (1982), *Plant Physiology* 70, 7-11.
53. Hood, A.E.M. (1982), *Philosophical Transactions of the Royal Society* B296, 315-28.
54. Hubbell, D.M. (1981), *BioScience* 31, 832-7.
55. Hunter, W.J., Fahring, C.J., Olsen, S.R. and Porter, L.K. (1982), *Crop Science* 22, 944-8.
56. Jones, R.S., Patterson, R.P. and Raper, C.D. (1981), *Plant and Soil* 63, 333-44.
57. Jordan, D.C. (1982), *International Journal of Systematic Bacteriology* 32, 136-9.
58. Kidby, D.K. and Goodchild, D.J. (1966), *Journal of General Microbiology* 45, 147-52.
59. Knowles, R. (1981), in Gibson, A.H. and Newton, W.E., (*eds*) *Current Perspectives in Nitrogen Fixation*. Canberra: Australian Academy of Science, 327-33.
60. Kohl, D.H. and Shearer, G. (1980), *Plant Physiology* 66, 51-6.
61. Kohl, D.H., Shearer, G. and Harper, J.E. (1980), *Plant Physiology* 66, 61-5.
61a Kushizaki, M., Ishizuka, J. and Akamatsu, F. (1964), *Journal of*

Soil Science and Manure Japan <u>35</u>, 323-7.

62. Lambers, H., Layzell, D.B. and Pate, J.S. (1980), *Physiologia Plantarum* <u>50</u>, 319-25.

63. LaRue, T.A. (1980), in Stewart, W.D.P. and Gallon, J.R. (*eds*) *Nitrogen Fixation*. London: Academic Press, 355-64.

64. LaRue, T.A. and Patterson, T.G. (1981), *Advances in Agronomy* <u>34</u>, 15-38.

65. Legg, J.O. and Sloger, C. (1975), in Klein, J. (*ed.*) *Proceedings of the Second International Conference on Stable Isotopes*. Illinois: Oak Brook, 661-6.

66. McClure, P.R. and Israel, D.W. (1979), *Plant Physiology* <u>64</u>, 411-16.

67. McClure, P.R., Israel, D.W. and Volk, R.J. (1980), *Plant Physiology* <u>66</u>, 720-25.

68. McNeil, D.L. (1982), *Applied Environmental Microbiology*, <u>44</u>, 647-52.

69. Mahon, J.D. and Child, J.J. (1979), *Canadian Journal of Botany* <u>57</u>, 1687-93.

70. Manhart, J.R. and Wong, P.P. (1980), *Plant Physiology* <u>65</u>, 502-05.

71. Matsumoto, T., Yatazawa, M. and Yamamoto, Y. (1977), *Plant and Cell Physiology* <u>18</u>, 353-9.

72. Matsumoto, T., Yatazawa, M. and Yamamoto, Y. (1977), *Plant and Cell Physiology* <u>18</u>, 459-62.

73. Matsumoto, T., Yatazawa, M. and Yamamoto, Y. (1977), *Plant and Cell Physiology* <u>18</u>, 613-24.

74. Mederski, K.J. and Streeter, J.G. (1977), *Plant Physiology* <u>59</u>, 1076-81.

75. Meeks, J.C., Wolk, C.P., Schilliong, N., Shaffer, P.W., Avissor, T. and Chien, W-S. (1978), *Plant Physiology* <u>61</u>, 980-83.

76. Miflin, B.J. and Lea, P.J. (1976), *Phytochemistry* <u>15</u>, 873-85.

77. Minchin, F.R. and Pate, J.S. (1973), *Journal of Experimental Botany* <u>25</u>, 295-308.

78. Minchin, F.R., Summerfield, R.J., Hadley, P., Roberts, E.H. and Rawsthorne, S. (1981), *Plant, Cell and Environment* <u>4</u>, 5-26.

79. Minchin, F.R., Summerfield, R.J. and Neves, M.C.P. (1980), *Journal of Experimental Botany* <u>31</u>, 1327-45.

80. Minchin, F.R., Witty, J.F., Sheehy, J.E. and Muller, M. (1983), *Journal of Experimental Botany* <u>34</u>, 641-9.

81. Minchin, F.R., Witty, J.F. and Sheehy, J.E. (1983), in Davies, R.D. and Jones, D.G. (*eds*) *The Physiology, Genetics and Nodulation of Temperate Legumes*. London: Pitmans, 201-17.

82. Moore, A.W. and Cranwell, E.T. (1976), *Communications in Soil Science and Plant Analysis* <u>7</u>, 335-44.

83. Nambiar, P.T.C., Nigam, S.N. and Dart, P.J. (1983), *Journal of Experimental Botany* <u>34</u>, 484-8.

84. Nelson, L.M. and Salminen, S.O. (1982), *Journal of Bacteriology* <u>151</u>, 989-95.

85. Neves, M.C.P., Summerfield, R.J., Minchin, F.R., Hadley, P. and Roberts, E.H. (1982), *Plant and Soil* <u>68</u>, 249-60.

86. Newcomb, W. (1981), *International Review of Cytology Supplement* <u>13</u>, 247-98.

87. Newcomb, E.H. and Tandon, S.R. (1981), *Science* <u>212</u>, 1194-6.

88. Noel, H.D., Carneal, M. and Brill, W.J. (1982), *Plant Physiology* <u>70</u>, 1236-41.

89. Pankhurst, C.E. and Sprent, J.I. (1975), *Journal of Experimental*

Botany <u>26</u>, 287-304.

90. Pate, J.S. (1973), *Soil Biology and Biochemistry* <u>5</u>, 109-19.
91. Pate, J.S. (1980), *Annual Review of Plant Physiology* <u>31</u>, 313-40.
92. Pate, J.S., Atkins, C.A. and Rainbird, R.M. (1981), in Gibson, A.H. and Newton, W.E. (*eds*) *Current Perspectives in Nitrogen Fixation*. Canberra: Australian Academy of Science, 105-16.
93. Pate, J.S., Atkins, C.A., White, S.T., Rainbird, R.M. and Woo, K.C. (1980), *Plant Physiology* <u>65</u>, 961-5.
94. Pate, J.S., Gunning, B.E. and Briarty, L.G. (1969), *Planta* <u>85</u>, 11-34.
95. Pate, J.S., Layzell, D.B. and Atkins, C.A. (1979), *Plant Physiology* <u>64</u>, 1083-8.
96. Pate, J.S. and Minchin, F.R. (1980), in Summerfield, R.J. and Bunting, A.H. (*eds*) *Advances in Legume Science*. London: HMSO, 105-14.
97. Pfeiffer, N.E., Malik, N.S.A. and Wagner, F.W. (1983), *Plant Physiology* <u>71</u>, 393-9.
98. Phillips, D.A. (1980), *Annual Review of Plant Physiology* <u>31</u>, 29-49.
99. Phillips, D.A., DeJong, T.M. and Williams, L.E. (1981), in Gibson, A.H. and Newton, W.E. (*eds*) *Current Perspectives in Nitrogen Fixation*. Canberra: Australian Academy of Science, 117-20.
100. Postgate, J.R. (1982), *The Fundamentals of Nitrogen Fixation*. Cambridge: Cambridge University Press.
101. Radin, J.W. and Boyer, J.S. (1982), *Plant Physiology* <u>69</u>, 771-5.
102. Radin, J.W. (1983), *Plant, Cell and Environment* <u>6</u>, 65-8.
103. Rainbird, R.M. (1982), 'Elements in the Cost of Nitrogen Fixation with Special Reference to The Legume Cowpea *Vigna unguiculata* (L.) Walp. cv. Caloona'. PhD thesis, University of Western Australia.
104. Rawsthorne, S., Minchin, F.R., Summerfield, R.J., Cookson, C. and Coombs, J. (1980), *Phytochemistry* <u>19</u>, 341-55.
105. Rennie, R.J. (1982), *Canadian Journal of Botany* <u>60</u>, 856-61.
106. Rennie, R.J., Rennie, D.A. and Fried, M. (1978), *Isotopes in Biological Dinitrogen Fixation*. Vienna: IAEA, 107-33.
107. Rigaud, J. (1981), in Bewley, J.D. (*ed.*) *Nitrogen and Carbon Metabolism*. The Hague: Martinus Nijhoff/Dr W.Junk, 17-48.
108. Ryle, G.J.A., Powell, C.E. and Gordon, A.J. (1978), *Annals of Botany* <u>42</u>, 637-48.
109. Schaede, R. (1940). *Planta* <u>31</u>, 1-21.
110. Schrader, L.E. and Thomas, R.J. (1981), in Bewley, J.D. (*ed.*) *Nitrogen and Carbon Metabolism*. The Hague: Martinus Nijhoff/Dr. W. Junk, 49-93.
111. Schrauzer, E.N. (1977), in Newton, W.E., Postgate, J.R. and Rodriguez-Barrueco, C. (*eds*) *Recent Developments in Nitrogen Fixation*. London: Academic Press, 109-18.
112. Schubert, K.R. and Evans, H.J. (1976), *Proceedings of the National Academy of Sciences* <u>73</u>, 1207-11.
113. Schubert, K.R. and Ryle, G.J.A. (1980), in Summerfield, R.J. and Bunting, A.H. (*eds*) *Advances in Legume Science*. London: HMSO, 85-96.
114. Sen, D. and Weaver, R.W. (1981), *Plant and Soil* <u>60</u>, 317-19.
115. Shantaram, S. and Wong, P.P. (1982), *Applied Environmental Microbiology* <u>43</u>, 677-85.
116. Shearer, G., Feldman, L., Bryan, B.A. Skeeters, J.L., Kohl, D.H.,

Amarger, N., Marrioti, F. and Marrioti, A. (1982), *Plant Physiology* <u>70</u>, 465-8.

117. Shearer, G., Kohl, D.H. and Harper, J.E. (1980), *Plant Physiology* <u>66</u>, 57-60.

118. Sheehy, J.E., Fishbeck, H.A., DeJong, T.M., Williams, L.E. and Phillips, D.A. (1980), *Plant Physiology* <u>66</u>, 101-4.

119. Sheehy, J.E., Minchin, F.R. and Witty, J.F. (1983), *Annals of Botany* <u>52</u>, 565-71.

120. Silvester, W.B. (1981), in Gibson, A.H. and Newton, W.E. (*eds*) *Current Perspectives in Nitrogen Fixation*. Canberra: Australian Academy of Science, 334.

121. Sinclair, T.R. and Goudriaan, J. (1981), *Plant Physiology* <u>67</u>, 143-5.

122. Smith, S.E. and Daft, M.J. (1978), *Australian Journal of Plant Physiology* <u>4</u>, 403-13.

123. Sprent, J.I. (1979), *The Biology of Nitrogen Fixing Organisms*. London: McGraw-Hill.

124. Sprent, J.I. (1980), *Plant, Cell and Environment* <u>3</u>, 35-43.

125. Sprent, J.I. (1981), in Polhill, R.M. and Raven, P.H. (*eds*) *Advances in Legume Systematics*. London: HMSO, 671-6.

126. Sprent, J.I. (1982), *Philosophical Transactions of the Royal Society* B <u>296</u>, 387-95.

127. Sprent, J.I. (1984), in Wilkins, M.B. (*ed.*) *Advanced Plant Physiology*. London: Pitmans, 249-76.

128. Sprent, J.I. and Bradford, A.M. (1977), *The Journal of Agricultural Science (Cambridge)* <u>88</u>, 303-310.

129. Sprent, J.I., Minchin, F.R. and Thomas, R.J. (1983), in Davies, D.R. and Jones, D.G. (*eds*) *The Physiology, Genetics and Nodulation of Temperate Legumes*. London: Pitmans, 269-318.

130. Steele, K.W. (1981), in Gibson, A.H. and Newton, W.E. (*eds*) *Current Perspectives in Nitrogen Fixation*. Canberra: Australian Academy of Science, 366-7.

131. Stiefel, E.I. (1977), in Newton, W.E., Postgate, J.R. and Rodriguez-Barrueco, C. (*eds*) *Recent Developments in Nitrogen Fixation*. London: Academic Press, 69-108.

132. Streeter, J.G. (1972), *Agronomy Journal* <u>64</u>, 311-14.

133. Streeter, J.G. (1981), *Plant Physiology* <u>68</u>, 840-44.

134. Streeter, J.G. (1982), *Plant Physiology* <u>69</u>, 1429-34.

135. Streeter, J.G., Mederski, H.J. and Ahmad, R.A. (1980), in Corbin, F.T. (*ed.*) *World Soybean Research Conference II Proceedings*. London: Granada, 129-37.

136. Subba Rao, N.S. (1980), in Subba Rao, N.S. (*ed.*) *Recent Advances in Biological Nitrogen Fixation*. London: Edward Arnold, 1-7.

137. Summerfield, R.J. (1980), in Hurd, R.G., Biscoe, P.V. and Dennis, C. (*eds*) *Opportunities for Increasing Crop Yields*. London: Pitmans, 51-69.

138. Summerfield, R.J., Dart, P.J., Huxley, P.A., Eaglesham, A.R.J., Minchin, F.R. and Day, J.M. (1977), *Experimental Agriculture* <u>13</u>, 129-42.

139. Tajima, S. and LaRue, T.A. (1982), *Plant Physiology* <u>70</u>, 388-92.

140. Thomas, R.J., Feller, U. and Erismann, K.H. (1980), *Journal of Experimental Botany* <u>31</u>, 409-17.

141. Thomas, R.J., McFerson, J.R., Schrader, L.E. and Bliss, F.A. (1984), *Plant and Soil* <u>79</u>, 77-88.

142. Tjepkema, J.D. (1981), in Gibson, A.H. and Newton, W.E. (*eds*)

Current Perspectives in Nitrogen Fixation. Canberra: Australian Academy of Science, 368.

143. Tjepkema, J.D. and Winship, L.J. (1980), *Science* 209, 279-81.
144. Trinick, M.J. (1982), in Broughton, W.J. (*ed.*) *Nitrogen Fixation Vol.2*. Oxford: Clarendon Press, 76-146.
145. Triplett, E.W., Heitholt, J.J., Evensen, K.B. and Blevins, D.G. (1981), *Plant Physiology* 67, 1-4.
146. Turner, G.L. (1981), in Gibson, A.H. and Newton, W.E. (*eds*) *Current Perspectives in Nitrogen Fixation*. Canberra: Australian Academy of Sciences, 335.
147. Turner, G.L. and Gibson, A.H. (1980), in Bergersen, F.J. (*ed.*) *Methods for Evaluating Biological Nitrogen Fixation*. Chichester: John Wiley, 111-38.
148. Urban, J.E. and Dazzo, F.B. (1982), *Applied Environmental Microbiology* 44, 216-26.
149. Ursino, D.J., Hunter, D.M., Laing, R.D. and Highley, J.L.S. (1982), *Canadian Journal of Botany* 60, 2665-70.
150. Vincent, J.M. (*ed.*) (1982), *Nitrogen Fixation in Legumes*. Sydney: Academic Press.
151. Vries, G.E. de, Veld, T. and Kijne, J.W. (1980), *Plant Science Letters* 20, 115-23.
152. Wagner, G.H. and Zapata, F. (1982), *Agronomy Journal* 74, 607-12.
153. Williams, L.E., DeJong, T.M. and Phillips, D.A. (1982), *Plant Physiology* 69, 432-6.
154. Wittenberg, J.B. (1980), in Newton, W.E. and Orme-Johnson, W.H. (*eds*) *Nitrogen Fixation II*. Baltimore: University Park Press, 53-67.
155. Witty, J.F. (1983), in Davies, D.R. and Jones, D.G. (*eds*) *The Physiology, Genetics and Nodulation of Temperate Legumes*. London: Pitmans, 253-67.
156. Witty, J.F. (1983), *Soil Biology and Biochemistry* 15, 631-9.
157. Witty, J.F., Minchin, F.R., Sheehy, J.E. and Minguez, M.I. (1984), *Annals of Botany* 53, 13-20.
158. Wong, P.P. and Evans, H.J. (1971), *Plant Physiology* 47, 750-55.
159. Bergmann, H., Preddie, E. and Verma, D.P.S. (1983), *EMBO Journal* 2, 2333-9.
160. Bisseling, T., Been, C., Klugkist, van Kammen, A. and Nadler, K. (1983), *EMBO Journal* 2, 961-6.
161. Raven, J.A. (1984), *Annual Proceedings of Phytochemical Society Europe* 24, 89-98. Oxford: University Press.
162. Keyser, H.H., Bohlool, B.B., Hu, T-S and Weber, D.F. (1982), *Science* 215, 1631-2.

PART II: Individual Species

5 Pea (*Pisum sativum* L.)

D.R. Davies, G.J. Berry, M.C. Heath and T.C.K. Dawkins

INTRODUCTION

HISTORICAL TRENDS, CURRENT STATUS AND FUTURE PROJECTIONS
FOR WORLD-WIDE PRODUCTION

Peas *(Pisum sativum)* have been a staple diet of man and
livestock since prehistoric times. In certain regions,
people have always relied on peas and other pulses to provide protein to complement the cereals in their diets.
During the middle ages, dried peas were widely grown in
English cottage gardens to mix with lentils and make pottage
[94] - a food still prepared in many warmer countries [265].
    The origin, history and domestication of peas have been
considered elsewhere [57,262]. Four possible centres of
origin are recognised [57,245,262]: Abyssinian (Ethiopia),
Mediterranean (Turkey, Greece, Yugoslavia, Lebanon), Near
Eastern (Iran, Iraq, Caucasia) and Central Asian (north-
west India, Pakistan, USSR, Afghanistan). Archaeological
evidence indicates that peas were cultivated in Neolithic
times but, although among the first crops to be exploited
by early man, it was not until Tudor times that the garden
pea was first cultivated for use as a fresh vegetable [94].
Present commercial pea varieties are the result of hybridisations begun in England in 1787 by Thomas Knight [94,262].
    Annual total world production of green peas is estimated
close to 4 million tonnes: the USA and UK are the largest

producers, followed by France, India (where the crop is often harvested for immediate use as a fresh vegetable), the USSR and China (see Table 5.1). In comparison, dried pea crops are far more extensive and constitute one of the world's four most important grain legumes [57,265]. The total area for global dried pea production, currently estimated at almost 8 million hectares, is approximately ten times greater than that for green pea production (Table 5.1), and annually realises 8 to 10 million tonnes. Russia and China jointly produce almost 80% of the world's dried pea supply (Table 5.1 and Figs, 5.1 and 5.2), but the USA is the single largest exporter [248]. The UK dried pea area declined in the 1950s following the expansion of the vining pea industry [251], but demands for an alternative 'break' crop and for home-produced protein are stimulating a gradually increasing production of dried peas within the EEC.

Dried pea yields, in common with those of other grain legumes, are unstable but they are one of the few food legumes for which the world average yield has improved over the last quarter of a century [245]. Figure 5.3 shows that this is largely due to more productive crops in Europe and, to a lesser extent, in North and Central America.

The relatively short growing season of peas usually varies between 80 and 100 days [265]. Growing seasons of 95 to 100 days are typical of the semi-arid regions of Canada [248] and the Middle East [173], but in humid, temperate areas such as the UK the season extends to 150 days [213]. Seed maturity (from anthesis to harvest) requires 40 to 45 days in Australia, Canada and England, but only 30 to 35 days in India [245]. Average yields for western Canada and for the UK are about 1.6 and 3 t ha^{-1}, respectively [248, 254], but when productivity is calculated on the basis of yield of dry seed per day of the growing season it is not vastly different between the two countries (16 and 20 kg ha^{-1} d^{-1}, respectively). Both regions are at similar latitudes, hence crops experience similar daylengths - an important factor in the interpretation of such calculations [245].

Pea (*Pisum sativum* L.)

Table 5.1 Average areas (1000 ha), yields (t ha^{-1}) and production
(1000 t) of peas for the two 3-year periods 1969-71 (a) and
1979-81 (b) in different regions [78]

Region/Country	Area		Yield		Production	
	(a)	(b)	(a)	(b)	(a)	(b)
Green peas						
Africa	19	21	4.65	5.96	89	126
North and Central America	192	171	6.52	7.57	1254	1300
Canada	20	19	2.89	3.67	57	69
Mexico	13	16	2.23	3.19	29	54
USA	160	136	7.32	8.64	1168	1177
South America	45	42	2.48	2.93	111	122
Asia	132	151	3.96	4.00	523	603
China	33	42	5.47	5.38	181	225
India	75	88	2.87	2.88	215	253
Europe	280	295	6.88	6.96	1926	2054
France	48	60	8.36	7.50	400	477
Hungary	30	35	5.03	6.09	152	215
Romania	16	28	2.66	2.56	41	71
UK	52	56	10.56	12.03	546	679
Oceania	30	23	4.96	6.83	147	154
Australia	21	14	5.58	7.83	117	111
USSR	57	67	3.04	3.73	174	251
World	755	770	5.59	5.99	4224	4610
Dry Peas						
Africa	416	433	0.66	0.77	275	335
Burundi	38	31	0.80	1.19	30	37
Ethiopia	105	131	0.65	0.98	68	128
Morocco	61	49	0.64	0.48	39	25
Rwanda	76	60	0.85	0.78	64	47
Zaire	61	73	0.63	0.63	39	47
North and Central America	160	131	1.64	1.95	262	255
Canada	32	47	1.35	1.76	44	84
USA	118	70	1.80	2.29	212	162
South America	138	146	0.73	0.71	101	103
Colombia	44	56	0.68	0.59	30	33
Asia	3138	2160	0.97	1.25	3055	2692
China	2100	1533	1.03	1.50	2167	2300
India	982	571	0.85	0.57	835	330
Iran	21	25	1.23	1.26	26	31
Europe	404	273	1.57	2.15	636	590
Czechoslovakia	14	31	1.62	2.11	23	64
France	12	26	3.31	4.13	38	107
Hungary	80	53	1.40	2.03	111	104
Poland	50	42	1.24	1.55	62	65
UK	25	36	2.96	2.96	75	107
Oceania	51	71	1.57	1.73	81	122
Australia	29	47	0.97	1.17	28	55
New Zealand	22	24	2.35	2.82	53	67
USSR	3316	4181	1.51	1.05	4989	4339
World Total	7624	7395	1.23	1.15	9399	8434

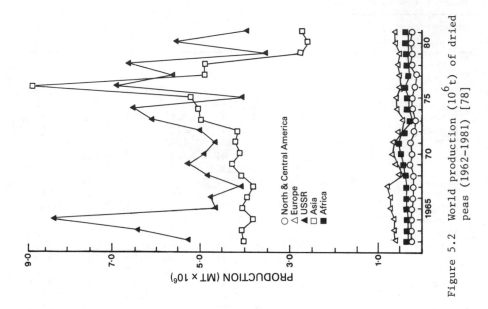

Figure 5.2 World production (10^6t) of dried peas (1962–1981) [78]

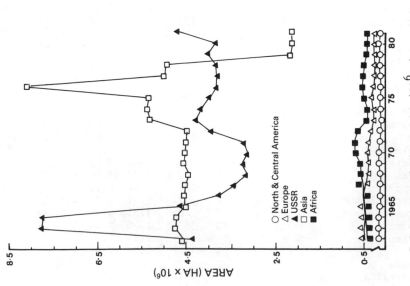

Figure 5.1 Areas of the world (10^6ha) used for growing dried peas (1962–1981) [78]

Pea (*Pisum sativum* L.)

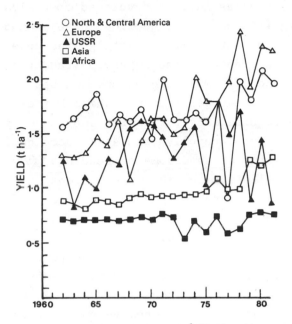

Figure 5.3 Average yields (t ha^{-1}) of dried peas in different regions of the world (1962–1981) [78]

PRINCIPAL USES OF THE CROP

While pea crops are grown mainly for their seeds, the vegetative parts can be useful too, e.g. the residuum left after harvesting either immature or mature seeds, or the whole crop, may be used for silage. Seeds are harvested when immature (vining, green, processing and freezing peas) or mature (dry, field, combining and threshed peas) and are used in various ways for food as well as for feed. Freedom from toxic components means that peas are eminently suited for consumption with or without cooking.

The seed crop

Harvested when immature.   A substantial area of pea crops has always been grown for the production of fresh peas for

151

immediate consumption, and while this is still true of many
countries, the areas in most developed countries have now
become insignificant. These declines have been due to the
labour-intensive nature of harvesting, the development of
the frozen pea industry, and improved standards of living
creating demands for more convenient and better quality
products. The vining pea industry which has expanded to
satisfy the requirements for frozen and canned products is
sophisticated and carefully regulated: seed quality is
assessed in terms of tenderometer readings (95-105 for
freezing; 115-120 for canning), sugar content, texture and
colour, and care is taken to cool the seeds before degrada-
tive processes occur. Vining peas are usually wrinkled-
seeded genotypes; their yields in northern Europe vary
according to the stage harvested but average between 4 and
5 t ha^{-1}. Crops are harvested by mobile viners which either
cut all the foliage and recover the seeds in a single field
operation, or alternatively, remove only the fruits before
isolating the seeds from them.

Accelerated freeze-drying is also used to process
immature seeds, but the market is small. Demand for edible
podded or 'sugar' peas, which lack the parchment layer normally present on the inner surface of pod walls, is also minor.

Harvested when mature.   Dried peas can be used directly
as a cooked vegetable, or resoaked and canned; alternatively
they can be ground and incorporated in soups or convenience
foods, or used as animal feed. Traditional and regional
preferences define the size, texture, testa thickness and
colour of the varieties used for these different purposes
but, almost invariably, they are round-seeded. For milling,
and in convenience foods, white or cream seeded varieties
which can be dehulled easily are used. For animal feeds,
seeds rejected because of physical or pest damage as
unsuitable for human consumption are often used. Their
value for feed depends on the protein, carbohydrate and
mineral contents as well as the amino acid composition of

the seed, and will vary according to the nutritional
requirements of the particular animal fed. The protein concentration of current varieties of round, dried peas is
approximately 22% (see below). Peas have smaller concentrations of trypsin inhibitor than soyabeans, very small urease
concentrations and very good protein digestibility [125].
Air classification of meal provides a concentrate the
protein concentration of which is enriched by a factor of
between two and three [12]. This can be used for extrusion
as a meat analogue, or as a meat extender in processed
foods. If pea flour is incorporated into pasta it allows
hard wheat to be used as the basic raw material instead of
the more expensive durum wheat [201]. Pea seed coats are
used as a fibre additive in bread or health foods and the
protein has been incorporated into bread to increase both
the amount and quality of its protein [126]. Another minor
use is for sugar-coated peas as a confectionery.

The yields of dry peas are small but within a range
comparable with other grain legumes; for example, the world
average yield for 1974 was 1.23 t ha^{-1}, which compares with
1.1 t ha^{-1} for faba beans and 1.26 t ha^{-1} for soyabeans
[162]. In Britain and northern Europe the current average
yield is about 3 t ha^{-1}, and it has been suggested [251]
that this could be improved to 5 t ha^{-1}.

The vegetative parts

Peas are grown for silage either alone or as a component
of a mixture which can include a cereal such as oats, and
sometimes another legume such as a vetch [218,255]. The
yields obtained from pure stands can be 6 t ha^{-1} of dry
matter with a crude protein concentration of about 14%.
The haulm which remains after the harvest of vining peas
is sometimes used for silage: about 2.5 t ha^{-1} of dry
matter can be obtained with a crude protein concentration
of 16.7% [13]. The straw left after harvesting dry seed
yields about 3 t ha^{-1} of dry matter which contains about

10% crude protein; in *in vitro* tests 50-60% of the dry
matter is digestible. This provides a good source of energy
and protein for ruminants and is equivalent to a hay crop
from a ryegrass/clover mixture.

PRINCIPAL FARMING SYSTEMS IN WHICH PEAS ARE A COMPONENT

Peas are a cool-season crop that generally grow best between
10°C and 30°C and are therefore largely confined to the
cooler temperate zones between the Tropic of Cancer and
50°N [245]. Peas are particularly sensititive to high
temperature and drought stress, especially during flowering.
However, the crop can be grown in the tropics at high elevations [265] or during cooler parts of the year in those
regions where summers are hot [173]. At latitudes beyond
50°N, peas are generally confined to the lowlands of maritime areas [245]. Time of sowing and place in the rotation
inevitably depend on regional climate [173]. All Canadian
pea crops and the majority of those in northern and western
Europe and in the USA are spring-sown. Winter-hardy
varieties are autumn-sown in southern Europe while others
succeed as winter annuals in relatively hot countries such
as India and Australia [265].

Dry peas are primarily grown as a 'break' or 'catch'
crop in cereal rotations. Estimates of the residual soil
nitrogen legacy provided by pea crops vary; in India, peas
reduced the fertiliser nitrogen requirement for maize by
20 to 32 kg ha^{-1} compared with wheat or fallow, respectively [2]. UK growers have estimated that fertiliser rates
for cereals could be reduced by 19-75 kg N ha^{-1} and by
25-30 kg ha^{-1} of both P and K, after a crop of combining
peas [44]. In France, estimates suggest that about 50 kg N
ha^{-1} are returned to the soil by peas [217]. The nitrogen
concentration of the pea haulm is double that of cereal
straw; thus, the burial and subsequent decomposition of pea
crop residues does not immobilise soil nitrogen to any
great extent [217]. Nitrate released from pea residues can

Pea (*Pisum sativum* L.)

be leached between harvest and spring [44] and so autumn-sown rather than spring-sown cereals are better able to exploit any increased soil nitrogen-reserve. Peas are arguably the ideal entry crop for winter wheat.

Peas are an attractive alternative to pasture in the present pasture-cereal rotations of winter-rainfall areas in Australia [156]. In western Europe, combining peas provide an alternative to oilseed rape as a cereal 'break' crop, especially following the introduction of EEC subsidies which are designed to stimulate home-grown protein production. The introduction of new 'leafless' varieties may enable peas to be grown as a 'break' crop in the wetter northern and western regions of the UK. Growing combining peas provides an agronomically sound way of extending and improving crop rotations in western Canada [248]. In general, pea crops must be grown in long rotations, preferably of at least five years, to prevent the accumulation of persistent soil pests and diseases [214,248]

Vining peas are generally considered a principal crop in their own right wherever they are intensively grown in developed countries. They provide one of the most mechanised, closely regulated and technologically advanced forms of arable crop production. A combination of scheduled sowing dates and the use of early, mid-season, and late-maturing varieties ensures that the processing factories are continuously supplied with optimal quality fresh peas during the six to seven week harvest period. Sowing dates can be determined in several ways, the most common of which is the Accumulated Heat Unit System [90].

Peas are attractive as a 'break' crop because they do not require specialist equipment to grow them. They also help to maintain cereal yields by reducing the incidence of cereal pests and diseases and allow cheaper weed control. The residual soil nitrogen lessens fertiliser inputs, and as peas mature early in the year they can spread the labour requirements on the farm at harvest time and allow a more timely entry for the following winter cereal.

BOTANICAL FEATURES OF THE CROP

Morphology

Pisum has often been considered to be a monospecific genus
[37,153,262,284] but two species are now recognised taxonomically (see Chapter 1). Wild forms are designated as
races or ecotypes, since all can be intercrossed and progeny
are at least partially fertile. *Pisum sativum* is the specific
name commonly accepted for the cultivated pea [60,262]
although *P. arvense* is sometimes used [153]. The *Pisum
Genetics Association* has proposed that the standard genotype
of Blixt [40] be adopted for genetic studies.

Leaves.   In the standard genotype the leaves are compound.
From the junction with the stem, where there are two (pseudo-)
stipules, a petiole carrying one or more pairs of leaflets
terminates in a simple or compound tendril. Leaf form
changes during ontogeny from one to two or three pairs of
leaflets. Leaf size usually increases up to the node of
first flower, and decreases thereafter [37].

Thirty-three genes are known to modify the size or form
of leaves [39]. The most important of these are *af* and *st*
(Table 5.2). The gene *af* (replacement of leaflets with
tendrils [144]) produces the 'semi-leafless' pea (Plate 5.1)
which has advantages of better standing ability through
interplant support, more uniform ripening, and reduced
susceptibility to pathogen attack [58,250].

The 'leafless' pea (Plate 5.1) results from the combined
effects of *af* and *st* (reduced stipule size [211]). This
has formed the basis of new plant ideotypes (see below) and
its potential is also being evaluated in drier areas because
of its smaller water requirement compared with standard-
leafed types [27,104]. The background genotype influences
the size of the vestigial stipules *(st)* and the vigour and
profusion of tendrils of *af* plants [251]. Leaflessness may
cause the plants to be poor competitors with an inherently

small yield per plant [108], although other evidence indicates that this need not necessarily be so, and may relate to the background genotype of the plant [6].

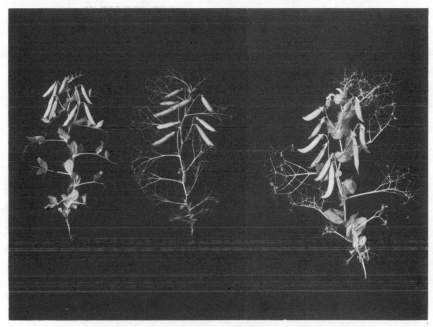

Plate 5.1 Conventional, leafless *(afafstst)* and semi-leafless
(afafStSt) forms in *P. sativum* (left, centre
and right, respectively)
Photograph kindly provided by B. Snoad.

Stems.   Within the stems there are four cortical vascular bundles, two of which consist only of phloem fibres and connect to the petioles, while the other two consist of xylem and phloem and connect to the stipules [267]. There are twenty-seven mutant genes known to affect stems [39]. Genotypes may vary from single-stemmed to profusely branched. Branches may be produced at the base of the main stem (e.g.*fr* and *fru*) or from nodes immediately below that of the first flower (e.g.*ram*). Other important genes are those which influence internode length (*le*, *la* and *cry*), fasciation (*fa* and *fas*) and stem strength (*rms*) (Table 5.2).

157

Table 5.2 Agriculturally important genes controlling morphological
traits in peas

Gene(s)	Phenotypic expression	PGA Type*	Reference
af	Leaflets converted to tendrils	1692	[144]
st	Stipules reduced in size; lanceolate	232	[211]
fr,fru	Basal branching; >7 in spaced plants	851	[150]
ram	Aerial branching; nodes immediately below node of first flower	2077	[191]
le,la,cry	Control internode length	102	[100,186,222]
fa,fas	Fasciation of stem, usually only of reproductive nodes	6	[151,186]
rms	Many aerial branches; stems more lignified	5237	[38]
fn,fna	Number of flowers per inflorescence; *Fn Fna,*[1] *Fn fna* or *fn Fna,*[2] *fn fna,*[3] or more	1308	[148,273]
dt,pr,pre	Shorten peduncle length	578	[149]
Br, Bra	Bracts subtending flowers on peduncle	2	[152,154]
p, v	Edible-podded; no parenchyma	754	[186]

*Line number of the *Pisum* Genetics Association-Weibullsholm
Collection

Roots and nodules.   Root systems usually consist of a
primary root and first and second order laterals [268].
The primary root axis may extend to a depth of 1.2 m or
more [106]. Laterals are regularly spaced along the primary
root but their frequency varies between cultivars [27].
Nodule morphology and distribution on roots are extremely
variable and difficult to characterise precisely [146].

Inflorescences.   For the standard genotype the inflorescences are axillary, each with one or two flowers on a
peduncle. The flowers have five united sepals, five petals,
and ten anthers. The gynoecium consists of one carpel (the
pod), which contains a number of ovules.

Thirteen genes influence the form of the inflorescence and another twenty-two affect its colour [39]. The inflorescence may have one to many flowers (*fn, fna*), the peduncle may vary in length (*dt, pr, pre*), and leafy bracts (*Br, Bra*) may subtend the flowers (Table 5.2).

Fruits.   The pod walls of the standard genotype consist of an exocarp (a single layer of thick-walled epidermal cells), a mesocarp (large parenchymatous cells with thin walls) and a fibrous endocarp [69,83]. The vasculature consists of two adjacent longitudinal veins which supply traces alternately to the seeds, and another vein on the fruit's lower surface. All three converge at both ends of the fruit and are interconnected with a network of minor veins which traverse the pod walls [266].

Twenty-six genes are known to affect fruit form and colour [39]. In primitive landraces the fruits readily dehisce as they reach maturity. Even in modern cultivars, shattering is still a problem. The genes *p* and *v* may be useful in overcoming it by altering the form of the pod wall so that the parenchymatous cells are not produced (Table 5.2); the *p v* pods, after initially inflating, shrink around the seeds at maturity.

Seeds.   Two cotyledons, which can nourish the embryonic plant for about two weeks, dominate the seed [215]. They are enclosed in a testa of maternal tissue. Almost fifty genes are known to influence the shape, size and colour of the seeds; most of them control the patterns and colours of the testa [39].

Phenology

Flowering is the most important phenological event with regard to crop productivity. In many areas early flowering can lead to poor yields because of frost damage, while late flowering can also reduce yields due to moisture and high temperature stresses.

Depending on genotype and environment, the first flowers may appear as soon as three weeks after sowing or not until forty weeks [28]. Field peas are sown usually in late autumn or in spring, depending on the region, and take between about eight and sixteen weeks to flower. The period of flowering for indeterminate cultivars varies from four to eight weeks or longer [227].

Effects of genotype on flowering.   The genetics of flowering in peas has been analysed by quantitative methods [229,252,253,283] and also in terms of major genes [193-198, 264]. The actions and interactions of the major genes have been reviewed in detail by Murfet [264]. In long-photoperiod, cool-temperature environments (e.g. >12 h and <20°C) the effects of individual genes are generally less than the underlying environmental variation, thereby necessitating quantitative analyses. In short-photoperiod, warm-temperature conditions (e.g. 8 h and 25°C) the gene effects are magnified so that clear discontinuities occur in the frequency distributions of segregating populations.

The gene Sn confers sensitivity to both vernalisation and photoperiod through the production of a flowering inhibitor at warm temperatures and in short days. Further sensitivity to vernalisation and photoperiod is imparted by the genes Lf and Dne, respectively. The gene Hr enhances the effects of Sn, and e interacts with the other genes so that floral bud abortion occurs in some genotypes (Table 5.3).

Effects of environment on flowering.   Temperature and photoperiod are the major determinants of flowering [28, 29], although irradiance [174], light quality [224], soil nutrient status [25,256] and soil moisture [256] have some influence. The temperature required for maximum rate of reproductive development depends on photoperiod for Sn Dne cultivars. For example, for cultivar Dun, 18°C was the optimum temperature in an 8 h photoperiod, compared with

Table 5.3 Major genes controlling flowering in peas

Gene	Phenotypic expression	PGA	Type*	Reference
lf	Minimum flowering node decreases in order: $Lf^d > Lf > lf > lf^a$. Lf^d and Lf confer sensitivity to vernalisation, and are epistatic to E.	*lf* *lf* *Lf* *Lf*	102 1792 2687 1771	[264,273]
e	Low flowering node; responsible for flower bud abortion in short days. E is epistatic to Sn and hypostatic to Lf.	*E* *e*	1793 1792	[193]
sn	Day neutral; Sn confers sensitivity to photoperiod and vernalisation; increases reproductive nodes.	*Sn* *sn*	2698 1972	[274]
Hr	With Sn, flowering time and node very responsive to photoperiod; *hr* causes rapid apical senescence.	*Hr* *hr*	2686 1792	[195]

*Line number of the *Pisum* Genetics Association–Weibullsholm Collection.

24°C in a 12 h photoperiod, and 24°C in longer photoperiods [28]. Changes in temperature account for nearly all the variation in development rate seen in day-neutral genotypes (e.g. cultivar Alaska); the optimum here is warmer than 27°C. Rate processes such as leaf appearance, node production and floral bud development can be explained by linear or logarithmic functions of temperature, depending on the range considered [25].

Photoperiod can influence plants during the period from sowing to floral induction [209] but the period from induction to initiation is independent of photoperiod, although short days can cause flower buds to abort in some genotypes [73,264].

Agronomy

Seedling Establishment.   Fungal pathogens (e.g. *Pythium* spp. and *Fusarium* spp.) and the vigour of seeds affect emergence [263]. Muehlbauer and Kraft [192] examined the effects of seed genotype on pre-emergence damping-off and resistance to *Fusarium* and *Pythium* root rots. Lines with Gene *A* (anthocyanin pigmentation) had significantly better emergence than those with *a* (no anthocyanin), presumably due to the fungistatic effects of phenols associated with anthocyanin pigmentation. In addition, the *M* (marbled testa) gene appeared to confer better emergence than *m*, independently of *a*. The dominant alleles *F* (purple-spotted testa), *I* (yellow cotyledons) and *R* (smooth seeds) with *M* had a cumulative effect in reducing susceptibility to pre-emergence damping-off and root rots.

Harvesting.   In wetter pea-growing areas such as the UK, inclement weather at harvest can cause significant yield losses [58,251], while in drier areas such as Canada and Australia, yield losses due to fruit shattering are more important. Rapid increases in temperature at the end of the growing season promote shattering; plant breeders in the UK [251] and in Australia [26] are incorporating the genes *p* and *v* into new cultivars to reduce losses from this cause.

Reproductive biology

Peas are self-pollinated, with usually less than 1% outcrossing [80]. Pollen is liberated from the anthers approximately twenty-four to thirty-six hours before the flowers open. One of the male nuclei from the pollen tube fertilises the egg, whilst the other unites with the two polar nuclei giving rise to a triploid endosperm, which disappears by the time the seed matures [51,223].

All ovules within a fruit are usually fertilised but, depending upon genotype and environment, a number may abort [161]. In one trial, the proportion of ovules which developed into mature seeds varied from 36% to 42% for large-seeded cultivars (e.g. Mansholts and Zelka), from 55% to 60% for cultivars such as Alaska and Dun, and from 72%

162

to 80% for primitive land races [25]. The percentage seed set was strongly and negatively correlated with seed size.

The growth of the seed is typically sigmoidal [87,241] or may exhibit a biphasic pattern [84,189]. Seeds may be capable of germination as early as eighteen days after pollination, although twenty-four to thirty-six days are generally necessary to ensure a good percentage germination [80]. There is no dormancy or after-ripening in peas, but many wild types have hard seed coats.

Components of yield

Reproductive nodes.   For indeterminate cultivars, the number of reproductive nodes depends on the duration of the flowering period and plant density, and usually varies from five to twenty or more [25,77]. The gene *hr* may accelerate apical senescence once flowering has commenced. More determinate genotypes than those with *hr* are known, indicating that other genes are also involved [142]. The number of reproductive nodes which produce fruits is affected by stress (e.g. high temperatures, hot winds and dry soil) so that flowers at the last few reproductive nodes often fail to set fruits.

Flowers per node.   Two duplicate recessive genes, *fn* and *fna*, control the number of flowers on each inflorescence. Their expression depends to a large extent on environment and background genotype [113,114,148]. Variation from one (*Fn Fna*) to seventeen (*fn fna*) flowers has been reported [114]. The number was positively correlated with temperature and irradiance for *fn fna* genotypes. On the other hand, *fn Fna* and *Fn fna* genotypes had negative relations with these two environmental factors. Most cultivars have one or two flowers at each node (e.g. genotypes *fn Fna, Fn fna* or *Fn Fna*). 'Multipodded' lines usually set fewer seeds per fruit, and similar numbers of seeds per node can be produced by 'single podded' types with a high ovule fertility [25,251].

Ovules per fruit.   Ovule numbers range from four to fifteen, but are usually in the range of six to ten in cultivars. The *bt* gene (fruit apex shape) appears to have a
pleiotropic effect on ovule number [25,175,176]. Pointed
fruits *(bt)* have more ovules than blunt ones *(Bt)*.

Seed size.   Seed size varies from about 30 mg to more
than 480 mg, although most cultivars are within the range
180-300 mg. Three genes *(par, sq^{-1} and sq^{-2})* which control
seed size are known [39].

Yield component interactions.   Yield per plant is more
strongly correlated with seed number per plant than with seed
size [25,143,206]. Within the range 100 mg to 300 mg, seed
number and seed size are largely independent, both within
progenies of crosses as well as among inbred lines [25].
Numbers of reproductive nodes, fruits per node and seeds
per fruit are not strongly correlated, which suggests that
yields could be increased by selecting for any one component. However, other studies have reported significant
interactions between components [206]. No general conclusion
can be reached since the significance or otherwise of a
particular interaction is dependent on the environment and
on the range of component values within the genotypes
investigated.

SYMBIOTIC NITROGEN FIXATION

It is generally accepted that peas grown in soil in which
fully effective strains of *Rhizobium* are present do not
benefit from nitrogen fertiliser (see below). Applications
of larger amounts of fertiliser nitrogen reduce both nodulation [208,230] and nitrogen fixation - the severity of
the reductions depending on the time of application of the
fertiliser [204,275]. However, small amounts of combined
nitrogen can benefit the early stages of seedling development prior to nodule establishment [30,171,208] which takes

place between fourteen and thirty days after germination
[103]. Once established, the nodules continue to function
throughout most of the plant's life [289] and it is difficult to show any benefit from added combined nitrogen.

The relative contributions to total plant nitrogen of
symbiotically fixed and inorganic nitrogen remain a matter
of discussion. To a large extent this is due to uncertain-
ties regarding the reliability of the methods available
for measuring nitrogen fixation; those which have been
used with peas are reviewed by Sims *et al*. [130] (and see
Chapter 4). Virtanen and von Hausen [282] calculated that
symbiosis with *Rhizobium* contributed 67%, and the soil 33%,
of the nitrogen required by the plant. In comparison,
Anderson *et al*. [275] calculated that without fertiliser
nitrogen, 81% of the total nitrogen in the pea crop was
derived by fixation. Then, on the basis of experiments in
which the roots of growing plants were exposed to ^{15}N in a
sealed lysimeter, Sims *et al*. [130] estimated that 92-95%
was from fixation. Field studies in the western USA [170],
using the acetylene reduction technique, gave estimates of
only 22% to 33% for the contributions from fixation, a range
at variance with the much larger proportions quoted by other
workers.

Estimates of the absolute amounts of nitrogen fixed
per annum by peas also vary: 17-69kg ha^{-1} according to site
and aspect [170], 52kg ha^{-1} [163], and about 100kg ha^{-1}
[282]. Of this last amount, approximately 30% was available
for the following cereal crop, but only approximately 5%
for a second cereal crop. Phillips [216] has suggested that
the 52kg ha^{-1} estimate should be doubled because the cal-
culation was based on the growth of a legume every other
year rather than every year. Two other estimates [67,85]
are 71kg ha^{-1} and 119kg ha^{-1}.

There are many estimates of the cost of nitrogen fixation in terms of carbon utilisation. These values, as well
as the problems associated with the methods of estimation,
have been summarised elsewhere [129,216]; they range from

1.5-8.7g C per g N fixed. The full energetic implications
of the fixation process itself, as well as those of nodule
formation and maintenance, remain to be fully evaluated
(and see Chapter 4).

Several environmental factors affect fixation [265],
but most of them cannot be regulated precisely under field
conditions. Improvements may therefore depend largely on
manipulations of the genotype either of the host or of the
micro-symbiont. The most clearly established genetic variation in the host is that for resistance or susceptibility
to infection. One group of peas is not nodulated by European
strains of *Rhizobium leguminosarum* but is nodulated by
rhizobia from the Middle East; another group is nodulated
by both classes of bacteria [159,160,290]. The presence of
a particular plasmid (pRL5JI) in the bacteria confers the
ability to develop a successful symbiosis with both groups
of peas [290]. Resistance to the European strains of
Rhizobium is due to a single gene [115]. The main significance of these observations is that they demonstrate that
pea genotypes resistant to particular strains do exist. If
improved strains of *Rhizobium* become available, then some
specificity of infection will be required to ensure that
peas are selectively infected by these strains, rather than
by the ubiquitous unimproved forms. Yields of pea straw
and of seed in the USSR and Czechoslovakia have been
improved by 8-23% and 8-15%, respectively, after artificial
inoculation [190]. More recent work has demonstrated genetic
variation for fixation rates (measured by acetylene reduction techniques [111]), both among host genotypes and also
among strains of *Rhizobium* [64,112] as well as host-symbiont
genotype interactions [112].

While improved strains of *R. leguminosarum* may be
selected on the basis of the improved productivity of the
peas which they infect, future advances are also likely to
result from a manipulation of the genetic system of the
bacteria. An enhancement of the fixation process *per se*
may be possible, but the first trait to be modified is

likely to be the hydrogen uptake (*Hup*) system. Not all strains of *R. leguminosarum* are *Hup*[+] but there is evidence that those which are can fix more nitrogen [64], presumably as a result of the energy conserved (and see Chapter 4). Furthermore, transfer of the plasmid bearing the determinant(s) of *Hup* activity to strains which lack it enhances fixation and improves biomass production [64].

PRINCIPAL LIMITATIONS TO YIELD

Soil physical conditions

Peas are able to tolerate a wide range of soil types but extremes need to be avoided; they grow best on slightly acid soils (pH 6.5) [249]. Many poor vining pea crops were previously attributed to poor seed vigour exacerbated by early sowing into cold wet soil [263], but the advent of seed vigour tests has helped to overcome this problem [182]. However, poor soil physical conditions can still lead to poor establishment. The effects of soil compaction on pea yields have been intensively investigated [61,62,63,107]. The yields of dried and vining peas were reduced by up to 50% and 70%, respectively, as a result of topsoil compaction by tractor wheelings, and losses of up to 22% occurred in vining peas as a result of subsoil compaction. Yield losses were most severe on compacted areas when the seed bed was dry at emergence [61] and ethylene may be implicated in the effect [61,95]. If this is so, then early applications of substances such as 3,5 diiodo-4-hydroxy benzoic acid (DIHB) might counter the effects of compaction. Experiments in which it has been incorporated into compacted soils have shown that DIHB produces significant increases in root length, whereas it has no effect in non-compacted soils [155,288].

Thorough subsoil loosening failed to improve yields but deep loosening may improve them where subsoil compaction restricts rooting, or under dry conditions where deeper rooting can allow greater extraction of stored water at depth [61].

In drier regions, salinity can limit the productivity of the crop [42,47,92,172,279].

Stand establishment

Pulse crop yields, measured in terms of seed weight, demonstrate the classical flat-topped parabolic yield/population relation [116,117]. Most population and agronomic studies have involved vining peas but combining peas demonstrate the same relation [14,76,260]. The optimum plant population, in terms of profitability, is largely influenced by pea seed costs (Fig. 5.4) which in turn can be the single largest contributor to variable costs [134]. Combining peas have slightly smaller optimum populations than vining peas; the latter are harvested when immature, and before their full yield potential is realised [14]. Optimum planting densities for combining peas vary with seed (or embryonic axis) size [108] and, under UK conditions, range from 65 plants m^{-2} for large-seeded marrowfat types to 95 plants m^{-2} for small-seeded white and small-blue types [214]. Optimum populations for vining peas range between 80 and 100 plants m^{-2} [139], being larger under irrigated conditions [260] and for leafless (*afafstst* mutant) genotypes [108]. Yield per unit area can be limited by relatively sparse plant populations but yields are not seriously reduced at suboptimal densities in freely branching varieties capable of producing compensatory increases in the number of fruiting nodes per plant [8,76,184,225]. Sparse plant populations are undesirable in vining pea production because they lead to wide variations in date of maturity.

Correct planting patterns are essential for maximising yields. Before introduction of efficient herbicides, combining peas were grown in 61 cm rows to allow inter-row cultivations. For constant plant populations, yield increases of 4% and 35% could be obtained by reducing row width to

to ten days after full bloom [131,147]. Analysis of the yields
of vining peas and of the corresponding weather data from
sixty-two location-years in Oregon [221] showed that yields
were negatively correlated with thermal sum above a base
temperature of 25.6°C during blooming and fruit filling. It
was predicted that each degree-day above 25.6°C during these
growth stages reduced yield by 13 kg ha^{-1}. High temperature
reduced the number of floral structures per node [205] and
so yield reductions due to this stress are primarily the
result of fewer fruits per plant [41,202,257]. Temperature
stressed plants are likely to suffer also from moisture stress.

Water relations

Water deficit.   The effects of water deficit on the
growth and yield of peas have been the subject of much
research [14,24,41,86,104,110,183,187,219,235,236,239,244].
The critical period appears to be at anthesis and, to a
lesser extent, during seed growth, notwithstanding losses
due to lack of water at germination. Drought stress during
the vegetative phase has little effect on the date of
anthesis but the duration of the reproductive phase is
reduced, resulting in fewer flowers and fruits per square
metre; flower and fruit abortion can also occur [219,244].
An interesting corollary is that there may be a requirement
for short periods of water stress prior to and immediately
after floral inception [86]. Without such stress vining
and dried peas tended to assume an indeterminate habit of
growth [61,244].

Yield losses due to late sowings may in part be attributed to drought stress. As sowing is delayed, rooting
depth is restricted and so less water is available to the
crop [244]. Furthermore, late sowing increases the chance
of the moisture-sensitive stages coinciding with dry periods.
Winter sowing lessens sensitivity to drought stress because
the moisture-sensitive stages are completed earlier in the
season when water is more freely available [244].

Conventional-leafed peas are more sensitive to drought stress than leafless lines [104,244] and so the latter could be more productive in arid or semi-arid regions.

Irrigation.   Although many researchers have investigated the yield responses of vining peas to irrigation at different crop growth stages and in diverse soil conditions, few [14, 260,261,285] have investigated the response of dried peas. Well-defined moisture-sensitive periods of growth occur during flowering and seed-swelling [238]. Irrigation during the vegetative stage does not increase yields [235,236]. Irrigation at the start of flowering increases seed set, hence seed number per fruit, whereas irrigation during active seed growth increases mean pea weight and volume [236]. Both treatments increased the number of fruits per plant and weight of peas per fruit. Irrigation at the stage between completion of seed-set and the start of fruit swelling did not increase yield. A combination of irrigations at the start of flowering and during fruit swelling was best [187,236]. Under Oregon field conditions irrigation at fruit swelling also gave the largest increases in yield [220]. Little root growth occurs after flowering begins, which may partly explain why peas become sensitive then to reduced soil moisture [237]. Optimal plant density may [260] or may not [14,239] be increased under irrigated conditions. The optimum plant density was greater under irrigated conditions for vining peas harvested at the green pea but not the seed pea stage [286]. In New Zealand, irrigation during fruit swelling was most beneficial when applied at a 15% soil moisture value on a shallow silt-loam, and at one of 20% on a deep loam [259,260]. In the UK, application of 25 mm of water at a soil moisture deficit of >25 mm is recommended at both the early flowering and fruit swelling stages [165].

Waterlogging.   Peas are sensitive to ephemeral exposures to anaerobic soil conditions [68], being most sensitive to waterlogging immediately before flowering and then during

172

fruit filling [45] (Table 5.4). Root growth and photosynthesis were both reduced and recovered little after draining the profile, except where waterlogging occurred in the vegetative phase. Damage from waterlogging in legumes is generally more severe at warmer temperatures [109]. Jackson [123] reported that waterlogging for as short a period as 24h decreased vegetative growth. Filby, a leafless cultivar, was found to be less sensitive to waterlogging than a conventional-leafed type [123,124]. Attempts to reduce the damage to pea plants as a result of waterlogging by applications of growth regulators have not been effective [122].

Table 5.4 The effect of waterlogging on peas at different stages of development [45] (dry weight of seeds given as % of control)

Duration of waterlogging	Stage of waterlogging			
	Vegetative	Pre-flowering	Flowering	Pod filling
2 days	72	62	71	73
5 days	44	42	60	60

Crop protection

Birds, especially pigeons, and field mice attack pea crops, particularly at seedling emergence, and can cause serious yield losses.

Peas, because of their relatively small initial growth rates, compete poorly with weeds [245]. Weed competition reduced vining pea yields in Scotland by 8-10% over a range of densities [157]. Weeds also delay drying, impede combine harvesting, and weed seeds increase seed cleaning costs and can even lead to crop rejection [166]. Weeds were traditionally controlled by inter-row cultivations [135] and although the entire UK pea crop is now treated with one or more herbicide applications [137], crops continue to suffer from varying degrees of weed competition [157]. Broad-leaved weeds are controlled best by pre-emergent residual herbicides, the choice of which will depend on climate, soil

173

type and weed spectrum. Prometryne, cynazine, and the mixtures of terbutryne with terbuthylazine or trietazine with simazine are commonly used in the UK [166,214], whereas trifluralin is used extensively in North America, and linuron in India. Spring-sown peas have a relatively short growing season and so the duration of the required weed control is also limited. Use of persistent residual herbicides is restricted by the possible effects of phytotoxic residues on subsequent winter cereal crops.

Dinoseb-amine is used widely for post-emergent broad-leaved weed control but its mammalian toxicity hazard has led to the use in the UK of mixtures of bentazone with 4-(4 chloro-2-methylphenoxy) butyric acid (MCPB) or cyanazine with MCPB [102]. Wild oats *(Avena fatua)*, which can reduce yields by 60% [93], are controlled best by soil-incorporated triallate [15,16,137]. A range of chemicals including barban, diclofop-methyl, alloxydim-sodium and trichloroacetic acid (TCA) are used for post-emergent control of wild oats and other grass weeds [166,169].

Fungal pathogens, insect pests, and virus diseases of peas are listed in Table 5.5. Downy mildew *(Peronospora viciae)*, leaf- and pod-spot *(Ascochyta pisi)* and grey mould *(Botrytis cinerea)* are the most important foliar diseases in Europe, yet none is controlled satisfactorily by field applications of fungicides [167]. Effective control of seed-borne leaf- and pod-spot can be achieved however with seed treatments containing thiabendazole [31,33,167]. Seed treatments containing systemically-active metalaxyl or fosetyl-aluminium have given good control of primary and secondary downy mildew infections in the UK [33,89,138,212]. Genes for disease resistance exist for both of these pathogens; and those which confer resistance to downy mildew have been incorporated into combining pea varieties. Powdery mildew *(Erysiphe polygoni/pisi)* is a major problem in dry regions such as southern Europe and India. Varieties selected for resistance to powdery mildew in France and the USA carry the resistant genes *er1* and/or *er2* [180].

174

Pea (*Pisum sativum* L.)

Table 5.5 Diseases and pests of peas

A - FUNGAL DISEASES

Peronospora viciae or *P. pisi*	– DOWNY MILDEW
Pythium ultimum or other *Pythium* spp.	– PRE-EMERGENT DAMPING OFF
Erysiphe polygoni or *E. pisi*	– POWDERY MILDEW
Botrytis cinerea	– GREY MOULD
Foot-rot fungi complex consisting of:	
Fusarium oxysporum f. sp. *pisi* race 1	– PEA WILT
Fusarium oxysporum f. sp. *pisi* race 2	– NEAR WILT
Fusarium solani f. sp. *pisi*	– FUSARIUM FOOT-ROT
Ascochyta pisi	– ASCOCHYTA LEAF AND POD SPOT
Ascochyta pinodella (Phoma medicaginis var. *pinodella)*	– ASCOCHYTA ROOT-ROT
Mycosphaerella pinodes	– ASCOCHYTA BLIGHT
Aphanomyces euteiches	– COMMON ROOT-ROT
Thielaviopsis basicola	– BLACK ROOT-ROT
Sclerotinia sclerotiorum	– SCLEROTINIA

B - VIRUS DISEASES

Pea Early Browning Virus (PEBV)	– Transmitted by free-living nematodes *(Trichodorus* spp.*)*
Pea Enation Mosaic Virus (PEMV)	– Aphid-transmitted, e.g. Pea aphid *(Acyrthosiphon pisum)*
Pea Mosaic Virus (PMV)	– Aphid-transmitted
Pea Top Yellows (PTYV)	– Aphid-transmitted
Pea Seed-borne Mosaic Virus (PSbMV)	
Pea Streak Virus (PSV)	

C - BACTERIAL DISEASES

Pseudomonas pisi	– BACTERIAL BLIGHT

D - PESTS

Sitona lineatus	– PEA AND BEAN WEEVIL
Cydia nigricana	– PEA MOTH
Acyrthosiphon pisum	– PEA APHID
Heterodera gottingiana	– PEA CYST NEMATODE
Contarinia pisi	– PEA MIDGE
Kakothrips pisivorus	– PEA THRIPS
Cnephasia interjectania	– TORTRIX MOTH
Thrips angusticeps	– CABBAGE/FIELD THRIPS

Soil-borne pathogens which affect root development cause large yield losses. Although combining pea varieties resistant to *Fusarium* wilt (*Fusarium oxysporum* f.sp. *pisi*) have been developed, the only effective means of controlling a combination of associated soil-borne pathogens (the so-called foot-rot complex, see Table 5.5) is by crop rotation. Incorporation of organic matter is another possible cultural control method [185,207]. Yield loss due to severe *Fusarium* foot-rot (*Fusarium solani* f.sp. *pisi*) and drought was estimated at 22.7% in commercial pea crops in eastern Canada [21]. It is also estimated that approximately 10% of the total pea crop in the USA is lost annually because of foot-rot infestations [233]. *Aphanomyces euteiches* is the most destructive member of the foot-rot complex in the major pea-growing areas of the USA and although it is responsible for approximately 80% of all foot-rot damage [233], control by soil-applied fungicides is not economically effective [207]. Dinitroaniline herbicides (e.g. trifluralin) increase the yields of processing peas as a result of root-rot suppression [98,105,272].

Pea moth (*Cydia nigricana*) is the most important insect pest in the UK and the only one to justify treatment with insecticide [32,167,168]. Although yield losses (by weight) only amount to about 1%, average losses from seed damage vary from less than 3% to more than 18% [20]. Heavy infestations of pea aphid (*Acyrthosiphon pisum*) and pea and bean weevil (*Sitona lineatus*) can reduce yield [164,167]. Aphid-borne virus diseases rarely affect yield but the nematode-transmitted Pea Early Browning Virus can be serious locally on light soils [251].

Plant growth regulators

Chlormequat reduces the height of peas [23], and small increases in yield followed the application of cycocel to peas in the UK [214]. Treatments with gibberellic acid (GA3) have produced a number of effects ranging from more rapid development to improved yields [65,91,120]. Applications of 2,4,D or 2,4,5,T have increased fresh weight, length and number of fruits [145]. Reliable chemical control of lodging could lead to larger yields.

Standing ability

The quality of harvested dried peas influences marketable
yield. Typically, 4-18% of the yield (depending on variety)
is classed as 'waste and stain' (i.e. seeds stained or
chitted in the lodged crop or damaged during harvesting)
under UK conditions [213]. The proportions of 'waste and
stain' losses are variable and can be large in unfavourable
weather [58,219]. Yield instability and difficulties associated with harvesting lodged combining pea crops are
responsible for the reluctance of many growers to undertake
pea growing in the UK. Leafless peas have better standing
ability and are easier to combine-harvest than conventional
peas [108]. 'Waste and stain' figures for the leafless pea
variety Filby are smaller than those of conventional-leafed
peas [212,251] (but there are no values available yet
regarding the areas of leafless or semi-leafless pea crops).

Summary

To a large extent the losses attributable to delayed
sowings, sub-optimal plant populations, poor soil physical
conditions, severe weed and soil-borne disease infestations
and infections can be minimised by good husbandry. Farmers
in less well-developed regions, however, are often unaware
of, or unwilling to adopt, the currently recommended package
of agronomic practices [200].

NUTRITIONAL REQUIREMENTS

Nitrogen (and see SYMBIOTIC NITROGEN FIXATION)

The general consensus, based on extensive trials in many
parts of the world, is that neither dried nor vining peas
respond to applied combined nitrogen [10,11,56,61,81,88,

217,226,265,275,287].

In contrast, some benefits have been evident from applications of up to 45 kg N ha^{-1} in a wide range of experiments, but the background nitrogen status, form of nitrogen (although this seems to be of little significance [217]) and scale of application have varied widely. Improved leaf growth and dry-matter production, and increases in plant height were observed, but quality, yield and yield components were rarely affected [1,3,171,199,222,228,230,231,232]. Some experiments have demonstrated deleterious effects of applied nitrogen on the growth and yield of pea crops [232].

Where peas are sown into poor seed beds or when rainfall at sowing is intense, some nitrogen may aid crop establishment [56]. Where nitrogen was applied as a top dressing to vining pea crops in a compacted seed bed there was no yield advantage [61]. The typical accumulation of nitrogen in the total above-ground components during the period June to July is shown in Fig. 5.5 and that of the seed in Table 5.6.

Table 5.6 The composition of dried and fresh peas [210]

	Water	Carbohydrate	Fat	Protein	Nitrogen
		(g 100 g^{-1}; %)			
Dried peas	13.3	50.0	1.3	21.6	3.45
Fresh peas	78.5	10.6	0.4	5.8	0.9

	Na	K	Ca	Mg	P	Fe	Cu	Zn	S	Cl
				(mg 100 g^{-1})						
Dried peas	38	990	61	116	300	47	0.5	3.5	130	60
Fresh peas	1.0	340	15	30	100	1.9	0.2	0.7	2.0	3.9

Phosphate

Response to phosphate fertilisers depends on the residual concentration in the soil which, in turn, is governed at least in part by previous cropping history [56,226]. Many

178

workers have shown some response to phosphate in terms of
improved dry-matter production and yield under varying
conditions but have often failed to indicate background
concentrations [3,11,74,121,199,243,269,278]. In the UK,
Crowther *et al.* [56] have demonstrated equal numbers of
positive and negative responses to phosphate fertilisers
but even where they occurred the responses were negligible.
Staud [258] found no effect of phosphate but interactions
between phosphate applications and irrigation have been
reported in terms of improved nodulation and root growth
[75,158]. Plants grown from seed produced from phosphate-
deficient parents yielded 25% less than those in which the

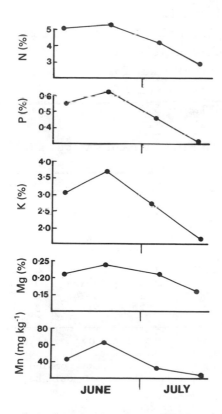

Figure 5.5   Seasonal changes in the concentrations (%) of N,
P, K, and Mg and in the amounts of Mn (mg kg^{-1}) in total
above-ground dry-matter of vining peas sown on
20 April in the UK

phosphate supply to the parent plant had been adequate [17]. Phosphate had little effect on the quality of canning peas [1]. Typical seasonal variations in the phosphorus concentrations in total above-ground dry matter and that present in seeds are shown in Fig. 5.5 and Table 5.6, respectively. Current phosphate recommendations for the UK are summarised in Table 5.7.

Potash

Potash is one of the most important major nutrients for peas [56]. A broadcast dressing of 150 kg K_2O ha^{-1} at sowing resulted in twenty instances of increased yield over the control and five decreases; although the average response was small, gains were much larger where the soil was poor in readily available potash [226]. Potash did not affect the quality of peas for canning [1]. Current potash recommendations for the UK are summarised in Table 5.7 (and see Fig. 5.5 and Table 5.6).

Table 5.7 Current UK manurial recommendations for peas [167]

| Soil index (N,P or K) | Soil analysis values (mg l^{-1}) | | | Application rate (kg ha^{-1}) | | | | |
| | | | | Broadcast | | | Combine-drilled* | |
	N(Nitrate)	P	K	N	P	K	P	K
0 very low	0–25	0–9	0–60	0	50	150	50	150**
1 low	26–50	10–15	61–120	0	25	40	25	50
2 medium	51–100	16–25	121–240	0	0	0	0	40
3 high	100	25	240	0	0	0	0	0

*Peas are more responsive to combine-drilled fertiliser and so optimum rates are larger than those broadcast.
**Not more than 50 kg K_2O ha^{-1} should be combine-drilled otherwise germination will be affected. The rest should be broadcast.

Trace element nutrition

The role of trace (minor) elements in legumes has been

extensively reviewed [101,280]. Peas require several trace
elements including boron, zinc, magnesium, sulphur, calcium,
iron, copper, molybdenum, cobalt, chlorine and manganese.
Manganese deficiency causes 'Marsh spot' [240]; a disorder
easily recognised by a brown spot in the centre of the
seed. The problem is exacerbated by alkalinity on organic
soils combined with any conditions which result in stress
[52], but can be alleviated by sprays of manganese sulphate
[214]. Seasonal variations in above-ground manganese and
magnesium concentrations are illustrated in Fig. 5.5
trace elements in seeds are shown in Table 5.6.

Method of fertiliser application

Peas do not develop an extensive root system during their
short growing season [61,244]. Fertilisers therefore need
to be applied so that they are within the limited rooting
zone in order to maximise utilisation by the crop. The most
profitable method of application for wide-drilled crops is
band placement 5 cm to the side of the seed row and 2.5 cm
below it, subject to residual status [43,53]. However, with
the introduction of narrower rows, fertiliser can either
be combine-drilled or broadcast but amounts may have to be
adjusted according to the method used (Table 5.7). Fertiliser which is broadcast should be incorporated by ploughing
or cultivation prior to sowing [167].

YIELD AND QUALITY

The definition of pea seed quality differs according to
use; in the wrinkled $(r_a r_a)$ forms used for vining, quality
is a function of tenderometer value, texture, sweetness
and size, with the criteria varying according to regional
and national preferences. For vining peas there is a general
inverse relation between quality and yield once the seeds
have developed beyond a certain stage.

In dry peas grown for food, where round $(R_a R_a)$ forms

181

are usually used, quality is again a function of texture, size and colour with different traditions and uses defining the exact criteria. Where peas are an important component of the diet, and as such are an important source of protein, different criteria are used. Peas satisfy human requirements for essential amino acids apart from the sulphur-amino acids methionine and cysteine [79].

Round varieties have 43-56% of their dry-matter present as starch and about 6% as sugars, while wrinkled genotypes have 27-37% starch and about 10% sugars [50,133,140]. Two genes influence carbohydrate composition [140] (and see Chapter 3).

The amount of protein also differs with genotype; crude protein concentrations between 16% and 33% of the dry seed weight have been reported [40,132,178]. However, there is evidence [9,181,246,276] that much of this variation is due to environmentally induced rather than genetically regulated differences. Where protein concentration is estimated by multiplying that for nitrogen by a factor of 6.5 this may overestimate the true value. Holt and Sosulski [118] and Colonna *et al*. [49] have suggested that a factor of 5.5 or 5.8, respectively, may be more appropriate for peas. Either no relation [206,276] or a negative correlation [247] has been reported between the percentage protein in seeds and economic yield. An inconsistent relation has been found between percentage protein and mean seed weight by some [97,247,276], whereas others [206] have found a negative correlation. The r_a locus may affect the protein as well as the carbohydrate content of the seed. Wrinkled seeds may have as much as 3% more protein than round ones [242,247,276], but since the average weight of the wrinkled seed is less, the absolute amounts of protein per seed are similar. In contrast, others [177,179] could find no difference between round and wrinkled forms.

The two main storage proteins of peas are legumin and vicilin; since the latter has a very small sulphur-amino acid content, improvement in quality should be attained by increasing the proportion of the former in the seed (and see Chapter 3). Genetic variation exists for both the proportion of these two proteins as well as for the amino acid composition of legumin. It has been calculated [46] that if a genotype were selected which had 50% of its storage protein present as legumin, and this legumin had the best sulphur-amino acid composition of those studied, then the mean sulphur-amino acid concentration of the seed could be improved from about 1.9 g% to about 2.2 g% - a value still smaller than the FAO 'Standard Protein' of 3.5 g% [79]. As round seeds have more legumin than wrinkled ones [59] the former genotypes are usually a better source of protein.

Colonna *et al.* [48] showed that a wrinkled $(r_a r_a R_b R_b)$ variety had twice as much lipid as a round one $(R_a R_a R_b R_b)$, and Coxon and Davies [55] demonstrated that genotypes $R_a R_a R_b R_b$, $r_a r_a R_b R_b$, $R_a R_a r_b r_b$ and $r_a r_a r_b r_b$ had, respectively, 2.4%, 4.2%, 4.7% and 5.6% crude lipid; the last genotype had 300% more neutral lipid than the first. The implication of this variation in lipid concentration for the use of peas as animal feed has yet to be evaluated.

The absence or presence of only very small amounts of antinutritional or toxic factors, and the good digestibility of the protein of peas makes them well suited for food and feed. Varieties with dark testas contain tannins which can lessen digestibility and the availability of the proteins [277]. If pea protein is to be exploited on a large scale as a meat analogue then new criteria of quality may need to be defined which relate both to the physical properties of the protein, and also to the palatability and flavour of the derived product.

GERMPLASM RESOURCES AND CROP IMPROVEMENT

All pea improvement programmes are nationally based; some are state-sponsored and others are fostered by industry. While the latter often have multinational interests, no international improvement programme exists. Even within national programmes, the resources devoted to peas are often meagre. This may in part account for the fact that while pea yields have increased over the last three decades, they have not matched the increases achieved in some of the temperate cereals. In many countries the different requirements for peas harvested when immature (for fresh consumption or for processing) and those harvested when mature and dry have further diluted the effort and resources devoted to the crop.

The directory of germplasm collections of food legumes published in 1980 by the International Board for Plant Genetic Resources (IBPGR; see Chapter 18) [18] lists twenty-four centres which have collections of peas; this is probably an incomplete list since the earlier (1973) *World List of Grain Legume Collections* [234] mentions eight countries not included in the IBPGR list. Table 5.8 summarises the information available for the most important of these centres (in terms of numbers of accessions).

Computer-based information storage and retrieval systems are becoming an integral part of the efficient use of these collections. One of the earliest systems for a pea collection was developed in the UK [128]. Extensive cytogenetic data on the Swedish collection of peas are now available [40] and in a form which allows ready transfer to other data storage systems [177]. Data on the collections held by the USDA and by the Plant Experiment Station at Wiatrowo, Poland, respectively, have been recently updated as readily available publications [66,271].

BREEDING STRATEGIES

Selection within land races or established cultivars is the oldest method of producing new cultivars. A recent Australian example is provided by cultivar Dundale, released in 1979 [203]; it is an early flowering selection out of cultivar Dun. The two cultivars are isogenic except for one allele influencing flowering, possibly at the *Lf* locus.

Pea (*Pisum sativum* L.)

Table 5.8   Centres having large collections of germplasm of *Pisum sativum*

Country	Institute and Curator (in parentheses)	Number of accessions	Reference
Bulgaria	Institute of Genetics and Plant Breeding, Partschewitsch-Str. 22-A, Sofia.	2000	[234]
Czechoslovakia	Plant Breeding Research Institute of Technical Crops and Legumes, 787 12 Sumperk, Tumenice (J. Lahola).	800+	[18]
Ethiopia	Ethiopian Plant Genetic Resource Centre, PO Box 30726, Addis Ababa (M. Worede).	1000	[177]
German Democratic Republic	Zentralinstitut für Genetik und Kulturpflanzenforschung, Correnstrasse 3, 4325 Gatersleben (C.O. Lehmann).	2000	[177]
Hungary	National Institute of Agrobotany, Tapioszele (A. Szucs).	720	[234]
India	National Bureau of Plant Genetic Resources, IARI Campus, New Delhi 110012 (K.L. Mehra).	1400	[18]
Italy	Istituto del Germoplasma del CNR, Via G.Amendola 165-A, 70126, Bari (E. Porceddu).	5000	[18]
Netherlands	Foundation for Agricultural Plant Breeding (SVP), Institute de Haaf, PO Box 117, 6700 AC Wageningen (H. Lamberts).	800	[18]
Poland	Plant Experiment Station, Wiatrowo, Laboratory of Pea Breeding & Collection, 62-100 Wagrowiec (W. Swiecicki).	1400	[177]
Sweden	Weibullsholm Plant Breeding Institute, Landskrona (S. Blixt).	5000	[18]
Turkey	Regional Agricultural Research Institute (K. Temiz).	2000	[177]
UK	John Innes Institute, Colney Lane Norwich (P. Matthews).	2000	[177]
USA	USDA-SEA National Seed Storage Laboratory, Colorado State University, Fort Collins, CO.80523 (L.N. Bass).	1150	[18]
USA	USDA-SEA, NY State Agricultural Experiment Station, Northeastern Regional Plant Introduction Station, Geneva, NY 14456 (D.D. Dolan).	1825	[18]
USSR	N.I. Vavilov All-Union Institute of Plant Industry, 44 Herzen St, Leningrad (V.F. Dorofeev).	5550	[18]

Hybridisation followed by single-seed descent [127,251], pedigree [27,127,251,270], backcross [27] or bulk [8] methods of selection are used in field pea breeding programmes. The method used depends upon the type of cross, the mode of inheritance of the characters concerned and the facilities available. Wide crosses have resulted in the production of cultivars (e.g. Buckley, from ecotype *sativum* x *abyssinicum* [5]), and induced mutation has given useful breeding lines (e.g. fasciated [96] and improved stem strength [38] mutants).

Where single-seed descent methods are used, three generations can be grown each year [251]. In the field, early generation yield testing (F_3 to F_5) is usually carried out in small plots ($1m^2$). Later generations are usually tested at a number of locations in larger plots ($>10m^2$).

Mandates and objectives of breeding programmes

Larger and more stable yields of marketable seeds are the common objectives of all breeding programmes. To this end specific objectives are concerned with disease resistance, plant architecture and ability to cope with stresses.

Resistance to foliar pathogens of the *Ascochyta* spp. complex is an important objective of breeding programmes in Canada [8], Chile [141], the UK [251] and in Australia [7]. Other diseases for which genetic resistance is being sought or bred into new cultivars include bacterial blight *(Pseudomonas* spp.*)* [27], powdery mildew *(Erysiphe polygoni)* [8,141], root-rots *(Fusarium* spp., *Pythium* spp. and *Aphanomyces eutiches)* [127,270] and viruses (BYMV, PEMV and PSbMV) [127,270] (and see Table 5.5).

Breeding programmes in Australia, Canada, New Zealand, Poland and the UK are making use of semi-leafless *(af St)* and leafless *(af st)* genotypes to modify plant architecture. Table 5.9 summarises ideotypes (ideal plant types) which have been proposed for field pea. Seed quality is an important breeding objective, and increased protein concentration may also be sought.

Pea (*Pisum sativum* L.)

Table 5.9 Suggested field pea ideotypes

Author	Vegetative traits		Reproductive traits	
	Leaf	Stem	Flowering	Yield
Bandel and Gottschalk [19] (Germany)		dwarf, fasciated	early, indeterminate	many seeds per fruit; large seeds
Berry [26] (Australia)	semi-leafless	dwarf, fasciated, branched	determinate, short peduncles	many seeds per fruit; reduced fruit shattering
Hraska [119] (USSR)	leafless	dwarf, branched		
Hedley and Ambrose [108] (UK)	leafless	non-branched	early, indeterminate	many seeds per fruit; one fruit per node; poor competitor; small seeds
Snoad [251] (UK)	leafless, semi-leafless	branched	indeterminate	many seeds per fruit; reduced fruit shattering

AVENUES OF COMMUNICATION

The need for avenues of communication between researchers
is unquestioned; what is less certain is whether there is
a need for communications which deal solely with a single
crop. Agronomists might well feel that many of their
problems are region, rather than crop, specific, and those
interested in nitrogen fixation would seek information
from those working on other legumes and not feel the need
for crop-specific avenues of communication. In these disciplines, therefore, a thorough knowledge of the standard
scientific literature may be adequate. However, an area
where there is common ground amongst researchers throughout
the world is plant breeding. While the precise objectives
and the attributes sought differ between regions, the
genetic resources available to be exploited and the means
of their exploitation are common to all plant breeders. In

order to help cope with these needs, the *Pisum Genetics Association* was formed in 1969 to foster the collection, maintenance, documentation and exchange of genetic material. An annual newsletter, the *Pisum Newsletter*, edited by Dr G.A. Marx and an international board of editors, is published by the Department of Seed and Vegetable Sciences, New York State Agricultural Experiment Station, Geneva, New York. The annual price of membership for 1983 was $3 (and see Chapter 18).

PROSPECTS FOR LARGER, MORE STABLE YIELDS

Although the field pea is not a new crop plant, it has received far less attention in agricultural research than the temperate cereals which it complements in many farming systems. Ironically, the field pea is one of the best known crop species in genetical terms, with some 346 genes recorded [40].

In recent times, plant breeders have begun to exploit the genetic variation of the species more fully. The most exciting development in this area is that of using morphological and physiological traits to design new plant types better adapted to modern farming systems. New semi-leafless and leafless cultivars have proved their ability to yield as well as the traditional leafed cultivars, with the advantage of being easier and quicker to harvest. They should also have greater stability of yield, being less subject to attack by diseases, due to their more open canopy.

A number of agronomic aspects may need to be revised for the new plant types. Already, there is evidence that for some leafless genotypes, at least, larger seeding rates are required [108]. Responses to fertiliser, nitrogen fixation potential, competition with weeds and susceptibility to stresses may also be influenced by the modified plant type.

Further improvements in yield are likely to come from refinements of the semi-leafless and leafless types. In drier areas, overcoming fruit shattering will be important, while in wetter areas disease resistance may do more to increase yields. Selection for tolerance to stress factors such as drought, temperature extremes [70-72] and water-logging should improve the stability of yield.

LITERATURE CITED

1. Adam, W.B. (1952), *Agriculture* 59, 38-42.
2. Ahlawat, I.P.S., Singh, A. and Saraf, C.S. (1981), *Experimental Agriculture* 17, 57-62.
3. Ahmad, S. and Shafi, M. (1975), *Journal of Agricultural Research Pakistan* 13, 417-32.
4. Aitken, Y. (1978), *Australian Journal of Agricultural Research* 29, 983-1001.
5. Aitken, Y., Patton, C.T. and Mann, A.P. (1964), *Journal of Agriculture of Victoria* 62, 545-6.
6. Ali, S.M. (1980), *Australian Plant Breeding & Genetics Newsletter* 30, 69-70.
7. Ali, S.M., Nitschke, L.F., Dube, A.J., Krause, M.R. and Cameron, B. (1978), *Australian Journal of Agricultural Research* 29, 84-9.
8. Ali-Khan, T. Personal Communication.
9. Ali-Khan, S.T. and Young, C.G. (1973), *Cereal News* 15, 11-12.
10. Ambrus, P. (1976), *Vedecke Prace Vyzkumneho Ustavu Rastlinnej Vyroby v Piestanoch* 13, 37-45.
11. Amma, A.T. (1971), *Revista de Investigaciones Agropecurias* 2, (8) 67-79.
12. Anon. (1974), *PFPS Bulletin 1.* Canada: University of Saskatchewan, Saskatoon.
13. Anon. (1976), *Protein Feeds for Farm Livestock in the UK.* London: Agricultural Research Council.
14. Anderson, J.A.D. and White, J.G.H. (1974), *New Zealand Journal of Experimental Agriculture* 2, 165-71.
15. Armsby, W.A. and Gane, A.J. (1962), *Proceedings of 6th British Weed Cont. Conference.* 329-43.
16. Armsby, W.A. and Gane, A.J. (1964), *Proceedings of 7th British Weed Cont. Conference.* 742-8.
17. Austin, R.B. (1966), *Plant and Soil* 24, 359-68.
18. Ayad, G. and Anishetty, N.M. (1980), *Directory of Germplasm Collections I. Food Legumes.* Rome: IBPGR.
19. Bandel, G. and Gottschalk, W. (1978), *Zeitschrift für Pflanzenzüchtung* 81, 60-76.
20. Bardner, R. (1978), *ADAS Quarterly Review* 31, 159-72.
21. Basu, P.K., Jackson, H.R. and Wallen, V.R. (1978), *Canadian Journal of Plant Science* 58, 159-64.
22. Batch, J.J. (1981), *Outlook on Agriculture* 10, 371-8.
23. Baz, A., Omar, F.A. and Safwat, M.S.A. (1980), *Research Bulletin of the Faculty of Agriculture, Ain Shams University.* No. 1230 21pp.
24. Behl, N.K., Sawhney, J.S. and Moolani, M.K. (1968), *Indian*

Journal of Agricultural Science 38, 623-6.
25. Berry, G.J. (1981), PhD. thesis, University of Melbourne.
26. Berry, G.J. (1982) *Australian Plant Breeding & Genetics Newsletter* 32, 113-4.
27. Berry, G.J. Unpublished data.
28. Berry, G.J. and Aitken, Y. (1979), *Australian Journal of Plant Physiology* 6, 573-87.
29. Berry, G.J., Halloran, G.M. and Aitken, Y. (in press), *Australian Journal of Agricultural Research* 35.
30. Bethlenfalvay, G.J., Abu-Shakra, S.S. and Phillips, D.A. (1978), *Plant Physiology* 62, 127-30.
31. Biddle, A.J. (1980), *Annual Report of the Processors and Growers Research Organisation.* Peterborough, England. 17.
32. Biddle, A.J. (1981), *Information sheet no. 118. Processors and Growers Research Organisation.* Peterborough, England.
33. Biddle, A.J. (1982), *Information sheet no. 137. Processors and Growers Research Organisation.* Peterborough, England.
34. Bleasdale, J.K.A. and Thompson, R. (1962), *Annual Report of the National Vegetable Research Station, Wellesbourne (1961)* 12, 36-7.
35. Bleasdale, J.K.A. and Thompson, R. (1963), *Annual Report of the National Vegetable Research Station, Wellesbourne (1962)* 13, 35-9.
36. Bleasdale, J.K.A. and Thompson, R. (1964), *Annual Report of the National Vegetable Research Station, Wellesbourne (1963)* 14, 39-42.
37. Blixt, S. (1972), *Agri Hortique Genetica* 30, 1-293.
38. Blixt, S. (1976), *Agri Hortique Genetica* 34, 83-7.
39. Blixt, S. (1977), *Pisum Newsletter,* Suppl. 9, 1-59.
40. Blixt, S. (1978), *Agri Hortique Genetica* 36, 56-87.
41. Boswell, V.R. (1926), *Proceedings of the American Society of Horticultural Science* 23, 162-8.
42. Bresler, E., McNeal, B.L. and Carter, D.L. (1982), *Saline and Sodic Soils. Principles, Dynamics, Modeling. Advanced Series in Agricutlural Sciences No. 10.* Berlin: Springer-Verlag.
43. Bullen, E.R., Dadd, C.V. and Cooke, G.W. (1954), *Agriculture* 61, 19-22.
44. Burns, S.M. (1980), *Agricultural Enterprise Studies in England and Wales. Economic Report No. 73.* Department of Agricultural Economics and Management, University of Reading, England.
45. Cannell, R.G., Suhail, B.A. and Snaydon, R.W. (1976), *Letcombe Laboratory Annual Report ARC (1975),* 37-8.
46. Casey, R. and Short, M.N. (1981), *Phytochemistry* 20, 21-3.
47. Cerda, A., Caro, M., Fernandez, F.G. and Guillen, M.G. (1979), *Anales de Edafologia y Agrobiologia* 38, 1827-37.
48. Colonna, P., Galland, D. and Mercier, C. (1980), *Journal of Food Science* 45, 1629-36.
49. Colonna, P., Gueguen, J. and Mercier, C. (1981), *Sciences des Aliments* 1, 415-26.
50. Colonna, P. and Mercier, C. (1979), *Lebensm-Wiss.u.Technol.* 12, 1-12.
51. Cooper, D.C. (1938), *Botanical Gazette* 100, 123-32.
52. Cooke, G.W. (1975), *Fertilizing for Maximum Yield.* London: Granada.
53. Cooke, G.W. and Dadd, C.V., (1953), *Agriculture* 60, 34-8.
54. Cousin, R. (1976), *Annales de l'Amélioration des Plantes* 26, 235-63.

Pea (*Pisum sativum* L.)

55. Coxon, D.T. and Davies, D.R. (1982), *Theoretical and Applied Genetics* 64, 47-50.
56. Crowther, E.M., Reynolds, J.D. and Shorrock, R.W. (1952), *Agriculture* 58, 584-8.
57. Davies, D.R. (1976), in N.W. Simmonds (ed.) *Evolution of Crop Plants*. Longman, London: 172-4.
58. Davies, D.R. (1977), *Science Progress (Oxford)* 64, 201-14.
59. Davies, D.R. (1980), *Biochemical Genetics* 18, 1207-19.
60. Davis, P.H. (1970), *Flora of Turkey, Vol. III*. Edinburgh: Edinburgh University Press.
61. Dawkins, T.C.K. (1982), PhD. Thesis, University of Nottingham, England.
62. Dawkins, T.C.K., Hebblethwaite, P.D. and McGowan, M. (1980), *Arable Farming*, April 1980, 24.
63. Dawkins, T.C.K., Hebblethwaite, P.D., McGowan, M. and King, J. (1981), *Journal of Soil and Water Management* 9, 19-21.
64. DeJong, T.M. and Phillips, D.A. (1981), *Plant Physiology* 68, 309-13.
65. Doijode, S.D. (1977), *Mysore Journal of Agricultural Science* 11, 114.
66. Dolan, D.D. (1982), *Pisum Seed Available and Descriptive Notes*, Publication 22C, Geneva, New York: Northeast Regional Plant Introduction Station.
67. Duggar, J.F. (1898), *Alabama Agricultural Experimental Station Bulletin* 96, 183-208.
68. Erickson, A.E. and Van Doren, D.M. (1960), *Transactions of 7th International Congress on Soil Science* 3, 428.
69. Esau, K. (1965), *Plant Anatomy*. New York: Wiley.
70. Eteve, G. Personal communication.
71. Eteve, G. and Derieux, M. (1982), *Agronomie* 2, 813-17.
72. Eteve, G., Hiroux, G. and Catoir, J.M. (1979), *Annales de l'Amélioration des Plantes* 29, 557-62.
73. Evans, L.T. (ed.) (1969) *The Induction of Flowering: Some Case Histories*. Melbourne: Macmillan. Haupt, W. 393-408.
74. Fageria, N.K. (1977), *Agrochimica* 21, 75-8.
75. Fageria, N.K. and Bajpai, M.R. (1971), *Indian Journal of Agricultural Research* 5, 233-8.
76. Falloon, P.G. and White, J.G.H. (1978), *Proceedings, Agronomy Society of New Zealand* 8, 27-30.
77. Falloon, P.G. and White, J.G.H. (1980), *New Zealand Journal of Agricultural Research* 23, 243-8.
78. FAO (1981) *FAO Production Yearbook* Vol. 35; and previous volumes dating back to 1973, (Vol. 27).
79. FAO/WHO Ad Hoc Expert Committee (1973), *Energy and Protein Requirements*. Technical Reports Series No. 522, Rome, Italy: WHO.
80. Fehr, W.R. and Hadley, H.H. (eds) (1980), *Hybridization of Crop Plants*. Madison: American Society of Agronomy. Gritton, E.T., 347-56.
81. Fiuczek, M. (1976), *Hodowla Roslin, Aklimatyzacja i Nassiennictwo* 20, 315-20.
82. Fletcher, H.F., Ormrod, D.P., Maurer, A.R. and Stanfield, B. (1966), *Canadian Journal of Plant Science* 46, 77-85.
83. Flinn, A.M. (1974), *Physiologia Plantarum* 31, 275-8.
84. Flinn, A.M. and Pate, J.S. (1968), *Annals of Botany* 32, 479-95.
85. Fred, E.W., Baldwin, I.L. and McCoy, E. (1932), *Root Nodule*

Bacteria and Leguminous Plants. Madison: University of Wisconsin Press.

86. Frohlich, M. and Henkel, A. (1961), *Archiv für Gartenbau* 9, 405-28.
87. Frydman, V.M., Gaskin, P. and MacMillan, J. (1974), *Planta* 115, 11-15.
88. Gane, A.J. (1963), *Field Crop Abstracts* 16, 67-70.
89. Gane, A.J. (1982), *Information sheet No.133. Processors and Growers Research Organisation*, Peterborough, England.
90. Gane, A.J., King, J.M. and Gent, G.P. (1971), *Pea and Bean Growing Handbook. Vol. I-Peas*. Peterborough, England: Processors and Growers Research Organisation.
91. Garcia-Martinez, J.L. and Carbonell, J. (1980), *Planta* 147, 451-6.
92. Garg, B.K. and Garg, O.P. (1980), *Proceedings of the Indian Natural Science Academy Part B* 46, 694-8.
93. Gargouri, T. and Seeley, C.I. (1972), *Research Progress Report Western Society of Weed Science*. 102-3.
94. Genders, R. (1972), *The Complete Book of Vegetables and Herbs*. London: Ward Lock Ltd.
95. Goeschl, J.D., Rappaport, L. and Pratt, H.K. (1966), *Plant Physiology* 44, 877-84.
96. Gottschalk, W. (1970) *Improving Plant Protein by Nuclear Techniques*. Vienna: IAEA, 201-15.
97. Gottschalk, W. and Wolff, G. (1974), *Pisum Newsletter* 6, 18.
98. Grau, C.R. and Reiling, T.P. (1977), *Phytopathology* 67, 273-6.
99. Gritton, E.T. and Eastin, J.A. (1968), *Agronomy Journal* 60, 482-5.
100. Haan, H. de (1927), *Genetica* 9, 481-98.
101. Hallsworth, E.G. *(ed.)* (1958), *Nutrition of the Legumes. Proceedings of the University of Nottingham Fifth Easter School in Agricultural Science*. London: Butterworth. Hewitt, E.J., 15-42.
102. Handley, R.P. and King, J.M. (1976), *Proceedings of 1976 British Crop Protection Conference - Weeds*, 425-31.
103. Hardy, R.W., Burns, R.C., Hebert, R.R., Holsten, R.O. and Jackson, E.K. (1971), *Plant and Soil* (Special Volume), 561-90.
104. Harvey, D.M. (1980), *Annals of Botany* 45, 673-80.
105. Harvey, R.G., Hagedorn, D.J. and DeLoughery, R.L. (1975), *Crop Science* 15, 67-71.
106. Hayward, H.E. (1938), *The Structure of Economic Plants*. New York: Macmillan.
107. Hebblethwaite, P.D. and McGowan, M. (1980), *Journal of the Science of Food and Agriculture* 31, 1131-42.
108. Hedley, C.L. and Ambrose, M.J. (1981), *Advances in Agronomy* 34, 225-77.
109. Heinrichs, D.H. (1972), *Canadian Journal of Plant Science* 52, 985-90.
110. Hiler, E.A., Van Bavel, C.H.M., Hossain, M.M. and Jordan, W.R. (1972), *Agronomy Journal* 64, 60-64.
111. Hobbs, S.L.A. and Mahon, J.D. (1982), *Canadian Journal of Plant Science* 62, 265-76.
112. Hobbs, S.L.A. and Mahon, J.D. (1982), *Canadian Journal of Botany* 60, 2594-600.
113. Hole, C.C. (1977), *Pisum Newsletter* 9, 10.
114. Hole, C.C. and Hardwick, R.C. (1976), *Annals of Botany* 40, 707-22.

Pea (*Pisum sativum* L.)

115. Holl, F.B. (1975), *Euphytica* 24, 767-70.
116. Holliday, R. (1960), *Nature (London)* 186, 22-4.
117. Holliday, R. (1960), *Field Crop Abstracts* 13, 159-67 and 247-54.
118. Holt, N.W. and Sosulski, F.W. (1979), *Canadian Journal of Plant Science* 59, 653-60.
119. Hraska, S. (1975), *Debreceni Agraria Egyetem Tudomanyegyetem Kozlemenyei* 19, (Suppl.), 185-98.
120. Israelstam, G.F. and Davis, E. (1979), *Canadian Journal of Botany* 57, 1089-92.
121. Iswaran, V. and Sen, A. (1973), *Science and Culture* 39, 405-6.
122. Jackson, M.B. (1977), *Letcombe Laboratory Annual Report, ARC (1976)*, 69-71.
123. Jackson, M.B. (1978), *Letcombe Laboratory Annual Report, ARC (1977)*, 61-3.
124. Jackson, M.B. and Cannell, R.Q. (1979), *Letcombe Laboratory Annual Report, ARC (1978)*, 39-40.
125. Jaffe, W.G. (1950), *Proceedings of the Society for Experimental Biology and Medicine* 75, 219-22.
126. Jeffers, H.C., Rubenthaler, G.L., Finney, P.L., Anderson, P.D. and Bruinsma, B.L. (1978), *Bakers Digest* 52, 36-40.
127. Jermyn, W.A. Personal communication.
128. Johnson, M.W., Snoad, B. and Davies, D.R. (1971), *Euphytica* 20, 126-30.
129. Jones, D.G. and Davies, D.R. (*eds*) (1983), *Temperate Legumes: Physiology, Genetics and Nodulation*, London: Pitman. Minchin, F.R., Witty, J.F. and Sheehy, J.C., 201-18.
130. Jones, D.G. and Davies, D.R. (*eds*) (1983), *Temperate Legumes: Physiology, Genetics and Nodulation*, London: Pitman. Sims, A.P., Folkes, B.F., Barber, D.J. and Walls, D., 159-74.
131. Karr, E.J., Link, A.J. and Swanson, C.A. (1959), *American Journal of Botany* 46, 91-3.
132. Kaul, A.K. (1971), *IARI Research Series* 6, 7-32.
133. Kellenburger, S., Silveira, V., McCready, R.M., Owens, H.S. and Chapman, J.L. (1951), *Agronomy Journal* 43, 337-40.
134. Kerr, M.W.T. (1973), *Agricultural Enterprise Studies in England and Wales, Economic Report 15*, Department of Agriculture and Horticulture, University of Nottingham, England.
135. King, J.M. (1966), *Pea Growing Research Organisation Ltd, Misc. Publ. No.18*, Peterborough, England.
136. King, J.M. (1967), *Agriculture* 74, 167-70.
137. King, J.M. (1976), *British Crop Protection Council Monograph No.18*, 73-9.
138. King, J.M. (1980), *Annual Report of Processors and Growers Research Organisation*, Peterborough, England. 16-17.
139. King, J.M. (1981), *Information sheet No. 114, Processors and Growers Research Organisation*, Peterborough, England.
140. Kooistra, E. (1962), *Euphytica* 11, 357-73.
141. Krarup, H.A. Personal communication.
142. Krarup, H.A. (1974), *Agro Sur* 2, 28-9.
143. Krarup, H.A. and Davis, D.W. (1970), *Journal of the American Society for Horticultural Science* 95, 795-7.
144. Kujala, V. (1953), *Archivum Societatis Zoologicae Botanicae Fennicae Vanamo* 8, 44-5.
145. Kumar, K.V. and Sreekumar, V. (1981), *South Indian Horticulture* 29, 65-7.

146. Personal observations.
147. Lambert, R.G. and Linck, A.J. (1958), *Plant Physiology* 33, 347-50.
148. Lamprecht, H. (1947), *Agri Hortique Genetica* 4, 79-98.
149. Lamprecht, H. (1949), *Agri Hortique Genetica* 7, 112-33.
150. Lamprecht, H. (1950), *Agri Hortique Genetica* 8, 1-6.
151. Lamprecht, H. (1952), *Agri Hortique Genetica* 10, 158-68.
152. Lamprecht, H. (1953), *Agri Hortique Genetica* 11, 40-54.
153. Lamprecht, H. (1966), *Die Entstehung der Arten und hoheren Kategorien*. Wien: Springer-Verlag.
154. Lamprecht, H. and Mrkos, H. (1950), *Agri Hortique Genetica* 8, 153-62.
155. Larque-Saaverda, A., Wilkins, H. and Wain, R.L. (1975), *Planta* 126, 269-72.
156. Laurence, R.C.N. (1979), *Australian Journal of Experimental Agriculture and Animal Husbandry* 19, 495-503.
157. Lawson, H.M. (1982), *Weed Research* 22, 27-38.
158. Lenka, D. and Gautam, O.P. (1972), *Indian Journal of Agricultural Science* 42, 676-80.
159. Lie, T.A. (1971), *Plant and Soil* 34, 751-2.
160. Lie, T.A. (1978), *Annals of Applied Biology* 88, 462-5.
161. Linck, A.J. (1961), *Phytomorphology* 11, 79-84.
162. Lovett, J.V. (1980), *Perspectives in World Agriculture*. Farnham Royal: Commonwealth Agricultural Bureau, 91-122.
163. Lyon, T.L. and Bizzell, J.A. (1934), *Journal of the American Society of Agronomy* 26, 651-6.
164. MAFF (1978), *Leaflet HVD56*. Alnwick, Northumberland, England: MAFF Publications.
165. MAFF (1979), *Booklet 2067*. Alnwick, Northumberland, England: MAFF Publications.
166. MAFF (1981), *Booklet 2262(8l)*. Alnwick, Northumberland, England: MAFF Publications.
167. MAFF (1982), *Leaflet 801*. Alnwick, Northumberland, England: MAFF Publications.
168. MAFF (1982), *Leaflet 334*. Alnwick, Northumberland, England: MAFF Publications.
169. MAFF (1982), *Agricultural Chemicals Approval Scheme, Approved Products for Farmers and Growers*. Alnwick, Northumberland, England.
170. Mahler, R.L., Bezdicek, D.F. and Witters, R.E. (1979), *Agronomy Journal* 71, 348-51.
171. Mahon, J.D. and Child, J.J. (1979), *Canadian Journal of Botany* 57, 1687-93.
172. Malik, Y.S., Pandita, M.L. and Jaiswal, R.C. (1977), *Haryana Journal of Horticultural Science* 6, 181-5.
173. Mann, M.M. (1947), *The Empire Journal of Experimental Agriculture* 15, 249-59.
174. Marx, G.A. (1969), *Crop Science* 9, 273-5.
175. Marx, G.A. and Mishanec, W. (1962), *Proceedings of the American Society for Horticultural Science* 80, 462-7.
176. Marx, G.A. and Mishanec, W. (1967), *Crop Science* 7, 236-9.
177. Matthews, P. Personal communication.
178. Matthews, P. and Dow, P. (1974), *John Innes Annual Report*, 29-31.
179. Matthews, P. and Dow, P. (1983), Personal communication.

180. Matthews, P. and Bayer, O. (1980), *John Innes Annual Report*, 28-29.
181. Matthews, P. and Dow, P. (1975), *John Innes Annual Report*, 28-30.
182. Matthews, S. and Bradnock, W.T. (1967), *Proceedings of the International Seed Testing Association* 32, 553-63.
183. Maurer, A.R., Ormrod, D.P. and Fletcher, H.F. (1968), *Canadian Journal of Plant Science* 48, 129-37.
184. Meadley, J.T. and Milbourn, G.M. (1970), *Journal of Agricultural Science, Cambridge* 74, 273-8.
185. Mehrotra, R.S. and Garg, D.K. (1977), *Plant and Soil* 46, 691-4.
186. Mendel, G. (1866), Reprinted in (1951), *Journal of Heredity* 42, 3-47.
187. Miller, D.G., Manning, C.E. and Teare, I.D. (1977), *Journal of the American Society of Horticultural Science* 102, 349-51.
188. Milbourn, G.M. and Hardwick, R.C. (1968), *Journal of Agricultural Science, Cambridge* 70, 393-402.
189. Millerd, A. and Spencer, D. (1974), *Australian Journal of Plant Physiology* 1, 331-41.
190. Mishustin, E.N. and Shil'nikova, V.K. (1971), *Biological Fixation of Atmospheric Nitrogen*. London: Macmillan.
191. Monti, L.M. and Scarascia-Mugnozza, G.T. (1967), *Genetica Agrara* 21, 301-12.
192. Muehlbauer, F.J. and Kraft, J.M. (1978), *Crop Science* 18, 321-3.
193. Murfet, I.C. (1971), *Heredity* 26, 243-57.
194. Murfet, I.C. (1971), *Heredity* 27, 93-110.
195. Murfet, I.C. (1973), *Heredity* 31, 157-64.
196. Murfet, I.C. (1975), *Heredity* 35, 85-98.
197. Murfet, I.C. (1978), *Pisum Newsletter* 10, 48-52.
198. Murfet, I.C. and King, W. (1982), *Abstracts of the Australian Society for Plant Physiology*, annual meeting.
199. Nandpuri, K.S., Singh, H. and Kumar, J.C. (1973), *Journal of Research, Punjab Agricultural University* 10, 141-4.
200. Narayanappa, A. (1978), Thesis, Department of Agric. Extension, Hebbal, India.
201. Nielsen, M.A., Sumner, A.K. and Whalley, L.L. (1980), *Cereal Chemistry* 57, 203-6.
202. Nonnecke, I.L., Adedipe, N.O. and Ormrod, D.P. (1971), *Canadian Journal of Plant Science* 51, 479-84.
203. Nourse, H. (1977), *Fact Sheet 146/77*. Department of Agriculture & Fisheries, South Australia.
204. Oghoghorie, C.G.O. and Pate, J.S. (1971), *Plant and Soil* (Special Volume), 185-202.
205. Ormrod, D.P., Maurer, A.R., Mitchell, G. and Eaton, G.W. (1970), *Canadian Journal of Plant Science* 50, 201-2.
206. Pandey, S. and Gritton, E.T. (1975), *Crop Science* 15, 353-6.
207. Papavizas, G.C. and Ayers, W.A. (1974), *US Department of Agriculture Technical Bulletin 1485*. Washington, DC, USA.
208. Pate, J.S. and Dart, P.J. (1961), *Plant and Soil* 15, 329-46.
209. Paton, D.M. (1969), *Australian Journal of Biological Science* 22, 303-10.
210. Paul, A.A. and Southgate, D.A.T. (1978), *McCance and Widdowson's The Composition of Foods*. London: HMSO.
211. Pellew, C. and Sverdrup, A. (1923), *Genetics* 13, 125-31.
212. PGRO (1981), *Annual Report, Processors and Growers Research*

Organisation, Peterborough, England.
213. PGRO (1981), *Advisory Leaflet No. 4. Processors and Growers Research Organisation,* Peterborough, England.
214. PGRO (1982), *Notes on growing combining peas. Processors and Growers Research Organisation.* Peterborough, England.
215. Phillips, D.A. (1971), *Physiologia Plantarum* 25, 482-7.
216. Phillips, D.A. (1980), *Annual Review of Plant Physiology* 31, 29-49.
217. Plancquaert, P. (1978), *Perspectives agricoles (Institut Technique des Cereales et des Fourrages) No. 13,* March 1978, 24-35.
218. Potts, M.J. (1978), *West of Scotland Agricultural College. Technical Note no. 39,* Ayr, Scotland.
219. Procter, J.M. (1963), *Journal of Agricultural Science, Cambridge* 61, 281-9.
220. Pumphrey, F.V. and Schwanke, R.K. (1974), *Journal of the American Society of Horticultural Science* 99, 104-06.
221. Pumphrey, F.V., Ramig, R.E. and Allmaras, R.R. (1979), *Journal of the American Society of Horticultural Science* 104, 548-50.
222. Rasmusson, J. (1928), *Hereditas* 10, 1-152.
223. Reeve, R.M. (1948), *American Journal of Botany* 35, 591-602.
224. Reid, J.B. and Murfet, I.C. (1977), *Journal of Experimental Botany* 28, 1357-64.
225. Reynolds, J.D. (1950), *Agriculture (London)* 56, 527-37.
226. Reynolds, J.D. (1960), *Agriculture (London)* 66, 509-13.
227. Ridge, P. Personal communication.
228. Rodriguez, J.M. (1976), *Revista de Investigaciones Colombia Agropecurias* 11, 1-22.
229. Rowlands, D.G. (1964), *Genetica* 35, 75-94.
230. Rubes, L. (1977), *Rostlinna Vyroba* 23, 1147-58.
231. Rubes, L. and Kralova, M. (1973), *Rostlinna Vyroba* 19, 397-408.
232. Rubes, L. and Neuberg, J. (1973), *Rostlinna Vyroba* 19, 15-31.
233. Sacher, R.F., Hopen, H.J. and Jacobsen, B.J. (1978), *Weed Science* 26, 589-593.
234. Saint, S. (1973), *Plant Genetic Resources Newsletter* 29, 28-38.
235. Salter, P.J. (1962), *Journal of Horticultural Science* 37, 141-9.
236. Salter, P.J. (1963), *Journal of Horticultural Science* 38, 321-34.
237. Salter, P.J. and Drew, D.H. (1965), *Nature* 206, 1063-4.
238. Salter, P.J. and Goode, J.E. (1967) *Commonwealth Bureau of Horticulture and Plantation Crops Research Review* 2, Farnham Royal, England, 49-51.
239. Salter, P.J. and Williams, J.B. (1967), *Journal of Horticultural Science* 42, 59-66.
240. Samuel, G. and Piper, S. (1929), *Annals of Applied Biology* 16, 493-524.
241. Scharpe, A. and Parijs, R. van (1973), *Journal of Experimental Botany* 18, 65-77.
242. Shia, G. and Slinkard, A.E. (1977), *Crop Science* 17, 183-4.
243. Shukla, D.N., Samarjit Singh, and Subrahmanyam, T. (1977/1978), *Journal of Science Research, Banaras Hindu University* 28, 27-30.
244. Silim, S.N. (1982), PhD. Thesis, University of Nottingham, England.
245. Sinha, S.K. (1977), *FAO Plant Production and Protection Paper 3. AGPC MISC/36.* Rome: FAO.
246. Slinkard, A.E. (1977), *Production, Utilization and Marketing of Field Peas. Annual Report 3,* University of Saskatchewan, Saskatoon, Canada.

Pea (*Pisum sativum* L.)

247. Slinkard, A.E. (1981), *Pisum Newsletter* 13, 49.
248. Slinkard, A.E. and Drew, B.N. (1977), *Field Crops Publication No. 225,* University of Saskatchewan, Extension Division, Agricultural Science,Saskatoon, Canada.
249. Small, J. (1946), *pH and Plants.* London: Balliere, Tindall and Cox.
250. Snoad, B. (1974), *Euphytica* 23, 257-265.
251. Snoad, B. (1980), *ADAS Quarterly Review* 37, 69-86.
252. Snoad, B. and Arthur, A.E. (1973), *Euphytica* 22, 327-37.
253. Snoad, B. and Arthur, A.E. (1973), *Euphytica* 22, 510-19.
254. Sosulski, S.W., McClean, L.A. and Austenson, H.N. (1974), *Canadian Journal of Plant Science* 154, 247-51.
255. Spedding, C.R.W. and Diekmahns, E.C. *(eds)* (1972), *Grasses and Legumes in British Agriculture.* Farnham Royal: Commonwealth Agriculture Bureau.
256. Sprent, J.I. (1967), *Annals of Botany* 31, 608-18.
257. Stanfield, B., Ormrod, D.P. and Fletcher, H.F. (1966), *Canadian Journal of Plant Science* 46, 195-203.
258. Staud, J. (1974), *Rostlinna Vyroba* 20, 1193-201.
259. Stoker, R. (1973), *New Zealand Journal of Experimental Agriculture* 1, 73-6.
260. Stoker, R. (1975), *New Zealand Journal of Experimental Agriculture* 3, 333-7.
261. Stoker, R. (1977), *New Zealand Journal of Experimental Agriculture* 5, 233-6.
262. Sutcliffe, J.F. and Pate, J.S. *(eds)* (1977), *The Physiology of the Garden Pea,* London: Academic Press. Marx, G.A., 21-43.
263. Sutcliffe, J.F. and Pate, J.S. *(eds)* (1977), *The Physiology of the Garden Pea,* London: Academic Press. Matthews, S., 83-118.
264. Sutcliffe, J.F. and Pate, J.S. *(eds)* (1977), *The Physiology of the Garden Pea,* London: Academic Press. Murfet, I.C., 385-430.
265. Sutcliffe, J.F. and Pate, J.S. *(eds)* (1977), *The Physiology of the Garden Pea,* London: Academic Press. Pate, J.S., 349-83.
266. Sutcliffe, J.F. and Pate, J.S. *(eds)* (1977), *The Physiology of the Garden Pea,* London: Academic Press. Pate, J.S. and Flinn, A.M., 431-68.
267. Sutcliffe, J.F. and Pate, J.S. *(eds)* (1977), *The Physiology of the Garden Pea,* London: Academic Press. Sachs, T., 213-33.
268. Sutcliffe, J.F. and Pate, J.S. *(eds)* (1977), *The Physiology of the Garden Pea,* London: Academic Press. Torrey, J.G. and Zobel, R., 119-52.
269. Svoboda, J. (1974), *Rostlinna Vyroba* 20, 1183-91.
270. Swiecicki, W.K. Personal communication.
271. Swiecicki, W.K., Swiecicki, W. and Czerwinska, S. (1981), *The Catalogue of Pisum Lines.* Poznan.
272. Teasdale, J.R., Harvey, R.G. and Hagedorn, D.J. (1978), *Weed Science* 26, 609-13.
273. Tedin, H. (1897), *Sveriges Utsadesforenings Tidskrift* 7, 111-29.
274. Tedin, H. and Tedin, P. (1923), *Hereditas* 4, 351-62.
275. Thompson, R. and Casey, R. *(eds)* (1983), *Proceedings of Symposium on Perspectives for Peas and Lupins as Protein Crops.* The Hague: Martinus Nijhoff. Anderson, A.J., Haahr, V., Jensen, E.S. and Sandfaer, J., 205-18.
276. Thompson, R. and Casey, R. *(eds)* (1983), *Proceedings of Symposium on Perspectives for Peas and Lupins as Protein Crops.* The Hague: Martinus Nijhoff. Cousin, R., 146-64.

277. Thompson, R. and Casey, R. *(eds)* (1983), *Proceedings of Symposium on Perspectives for Peas and Lupins as Protein Crops*. The Hague: Martinus Nijhoff. Griffiths, D.W., 322-6.

278. Tyagim, D.N., Gupta, G.P. and Singh, K. (1971/1972), *Journal of Science Research Banaras Hindu University* 22, 79-84.

279. Uprety, D.C. and Sarin, M.N. (1975), *Acta Agronomica Academiae Scientarum Hungaricae* 24, 452-7.

280. Van Schreven, D.A. (1958), in Hallsworth E.C. *(ed.) Nutrition of the Legumes. Proceedings of the University of Nottingham Fifth Easter School in Agricultural Science*. London: Butterworth.

281. Vincent, C.L. (1958), *Washington Agricultural Experimental Station Bulletin 594*. Pullman: Washington, USA.

282. Virtanen, A.I. and S-von Hausen, S. (1952), *Plant and Soil* 4, 171-7.

283. Watts, L.E., Stevenson, E. and Crampton, M.J. (1970), *Euphytica* 19, 405-10.

284. Wellensiek, S.J. (1925), *Genetica* 7, 1-64.

285. White, J.G.H. and Anderson, J.A.D. (1974), *New Zealand Journal of Experimental Agriculture* 2, 159-64.

286. White, J.G.H., Sheath, G.W. and Meiher, G. (1982), *New Zealand Journal of Experimental Agriculture* 10, 155-60.

287. Widdowson, F.V. and Cooke, G.W. (1958), *Journal of Agricultural Science, Cambridge* 51, 53-61.

288. Wilkins, S.M., Wilkins, H. and Wain, R.L. (1976), *Nature (London)* 259, 392-94.

289. Young, J.P.W. (1982), *Annals of Botany* 49, 135-9.

290. Young, J.P.W., Johnston, A.W.B. and Brewin, N.J. (1982), *Heredity* 48, 197-201.

6 Faba bean (*Vicia faba* L.)

D.A. Bond, D.A. Lawes, G.C. Hawtin,
M.C. Saxena and J.H. Stephens

INTRODUCTION

Historical trends and geographical status

There are many common names for *Vicia faba* L. in the English
language but most of them refer to a particular subgroup
rather than the whole species. The smaller-seeded types,
vars *minor* and *equina*, are often called field beans in the
UK [22] and Europe (not to be confused with *Phaseolus vulgaris*, the common bean, but frequently referred to as field
beans in North America) while the large-seeded var. *major*
types are often known as broad beans. Other common names
include horse, tick, tic, longpod, Windsor, Spanish,
Egyptian, Mazagan, pigeon, winter, spring, garden, faba,
Vicia and common beans. In an attempt to standardise English
usage, *the term 'faba bean' has now been widely adopted to
denote the whole species.*

The wild progenitor and exact region of origin of faba
beans remain unknown. Cubero [39] has proposed that the crop
was first domesticated in west Asia, while Ladizinski [106]
has suggested that it originated further east in central
Asia. Unlike several other pulse crops, e.g. lentils (*Lens
culinaris*), peas (*Pisum sativum*) and chickpeas (*Cicer arietinum*), seeds of faba beans are very scarce in archaeological
remains of the early farming settlements in the Near East,
and it appears that the crop was not among the very first

to be domesticated. Shultze-Motel [170] concluded that
domestication occurred in the Neolithic period, and by the
third millenium BC the crop was widely distributed throughout the Mediterranean region, and was even known in northern Europe.

Faba beans were well known in the ancient world, being
a common food of many of the Mediterranean and Near Eastern
civilisations including the ancient Egyptians, Greeks and
Romans. Until the introduction of *Phaseolus* beans from the
New World in the post-Columbian era, *Vicia faba* was the
only commonly grown bean in Europe. It was widely used for
both food and feed.

The crop is believed to have spread along the Nile
Valley and into Ethiopia at an early stage of its domestication and also further east into Afghanistan and the Hima-
layan region of northern India. The date of introduction
of *V. faba* var. *minor* into China is thought to be about
100 BC [186], although var. *major* may not have been introduced until the silk trade became established sometime
after 1200 AD [85]. The crop was unknown in the New World
before the arrival of Columbus in 1492 and it is believed
to have been introduced into Central and South America by
the Spanish and Portuguese in the sixteenth century [19].

Today, *V. faba* ranks among the world's most important
grain legume crops: it was grown on 3.6 million hectares
in more than fifty countries in 1981 and total production
in that year exceeded 4 million tonnes (Table 6.1) [62].

By far the largest producer in 1981 was China, which
accounted for more than 60% of the total area and almost
65% of world production. Ethiopia, the second largest producer, only accounted for 9% of the total world area.

Egypt has the largest proportion of its total arable
land devoted to faba beans (4.07%), whereas of the major
producers, only in Ethiopia, Morocco, Tunisia, China, Italy
and Portugal is at least 1% of the arable land sown to the
crop.

Production throughout the world can be divided into

Faba bean (*Vicia faba* L.)

Table 6.1 Area, yield and production of faba beans in various countries in 1981 (After FAO [62]).

Region/Country	Area		Yield	Production
	(1000 ha)	(% total arable)	(kg ha^{-1})	(1000 t)
	1981	1980	1981	1981
Africa	*708*		*1012*	*716*
Algeria	46	0.67	630	29
Egypt	105	4.07	2495	262
Ethiopia	325	2.47	852	277
Morocco	130	2.15	500	65
Sudan	17	0.13	1243	22
Tunisia	77	2.32	702	54
North and Central America	*75*		*1285*	*97*
Guatemala	20	1.35	450	9
Mexico	46	0.18	1716	79
South America	*232*		*498*	*115*
Bolivia	11	0.34	909	10
Brazil	173	0.32	358	62
Ecuador	8	0.46	575	5
Paraguay	16	0.99	875	14
Peru	23	0.74	913	21
Asia	*2259*		*1236*	*2791*
China	2200	2.34	1227	2700
Cyprus	3	0.82	1231	3
Iraq	16	0.30	1049	17
Syria	8	0.13	1834	15
Turkey	30	0.12	1767	53
Europe	*380*		*1487*	*565*
Czechoslovakia	41	0.81	1705	71
France	21	0.12	3129	65
German Democratic Republic	6	0.13	1898	11
German Federal Republic	4	0.05	3266	14
Greece	6	0.21	1403	9
Italy	162	1.74	1270	206
Portugal	31	1.11	514	16
Spain	63	0.55	810	51
UK	45	0.69	2710	122
Australia	*10*	*0.02*	*1500*	*15*
Eastern Europe and USSR	*47*	*0.02*	*1728*	*82*
World Total	*3619*		*1154*	*4178*

about nine major agro-geographical regions, described below.

Northern Europe.   According to Moreno and Martinez [125]
this region includes countries north of a line through
northern Spain, northern Italy, and southern France and
includes north-west Greece and Yugoslavia. Faba beans have
been grown successfully as far north as Sweden and Finland.
The largest producers in northern Europe are the UK, Czechoslovakia and France (Table 6.1). The crop is mainly sown
in spring, and small- and medium-seeded types predominate
throughout the region. The crop reached a peak in the UK
in the late nineteenth and early twentieth century [23].
During the middle decades of the twentieth century, production declined markedly, as fewer beans were used to feed
decreasing numbers of draught horses and cheaper imported
protein feeds became available. The dramatic increase in
world prices of soyabeans in the early 1970s, however, has
stimulated interest in the crop in north-west Europe.

Mediterranean region.   The countries surrounding the
Mediterranean Basin are all important producers of faba
beans, with the largest areas being in Italy, Morocco and
Spain. The crop is planted in the autumn and harvested before the onset of the hottest summer weather. Large-seeded
(var. *major*) types predominate throughout this region, and
are used mainly as food. As in northern Europe, there has
been an overall decline in production over the past decade
or so, largely due to increasing labour costs. Mechanisation of production is expected to become increasingly important and a consequent shift to greater production of the
more readily handled *minor* and *equina* types is predicted.
Large-seeded types are likely to decrease in importance as
a dry-seed crop, but may become more common as a vegetable
crop for the fresh market and for processing.

The Nile Valley.   Faba beans are one of the more important
field crops in Egypt and are widely grown in the northern

provinces of Sudan along the Nile Valley. Since 1950, average yields have increased by nearly 40% even though the area
under cultivation has remained fairly static [97]. Production is almost exclusively of *minor* and *equina* types, which
are grown as irrigated winter crops and are harvested dry
as a pulse for food.

Ethiopia.   Faba beans occupy almost 2.5% of the total cultivated area in Ethiopia. They are the main highland pulse
crop, and are most commonly found in the 'Weyna Dega', i.e.
the temperate zone between 1800 m and 2400 m elevation
[196]. The normal growing season is from July to December.

Central Asia.   Although of relatively minor importance in
this region, faba beans can be found as a winter crop under
irrigation in central and southern Iraq and Iran, and as a
spring-planted rain-fed crop in highland areas around the
Caspian Sea and in Afghanistan. They are used almost exclusively as food.

East Asia.   China is by far the world's largest producer
of faba beans, although over the past ten years the area
sown has declined considerably and, in some provinces, production has now ceased. In Zhejiang Province, for example,
there were about 167 000 ha under faba beans in the 1950s,
compared to less than 67000 ha grown there in 1980 [186].
Major types for consumption as a green vegetable predominate
and these are grown as a winter crop in the southern regions
of the country and as a spring-sown crop in the north.

Oceania.   Faba beans are of very minor importance in the
countries of Oceania; the crop is grown on less than 10000
ha annually.

Latin America.   More than 300 000 ha are sown to faba
beans in Central and South America. Brazil, by far the
largest producer, accounted for almost 60% of the area in

203

1981. The crop is mainly grown in the cooler winter months and production is mostly of large-seeded types for food.

North America. Faba beans have never become established as a major crop in the USA. However, interest in the crop has increased in recent years in the Pacific north-western states and in Canada where, by 1990, estimates predict about 42000 ha will be devoted to the crop, largely in Manitoba, Saskatchewan and Alberta Provinces [180].

Current status of the crop

Faba beans have played an important role in the agriculture of many regions in the past. However, the crop has received a substantial set-back in many countries during this century with declining demands for feed for draught horses, increasing imports of cheaper proteins, and through competition with more profitable crops such as the productive semi-dwarf wheats introduced in the 1960s. In addition, rising labour costs have also reduced profitability in many parts of the world and have prompted a need for greater mechanisation.

More recently, there has been a renewed interest in the crop throughout the world. Rising costs of protein-rich food and feed, national desires for greater self-sufficiency in food production, the need for greater agricultural diversification and, perhaps most importantly, rapidly increasing human populations, have all contributed to this revival. Provided the main production problems can be resolved, faba beans could well have a promising future.

Crop duration and productivity

Cropping season duration varies widely from about three months (e.g. in Sudan and Canada) to eleven months (e.g. winter beans in north-west Europe). Hence, satisfactory economic yields, within the context of the local farming system, can be obtained over a wide range of daily productivity values.

Faba bean (*Vicia faba* L.)

However, a series of spring-sown trials of eight cultivars in twenty-two environments within the European Economic
Community (EEC) [44] showed that acceptable seed yields
(larger than 4.5 t ha^{-1}) were only obtained from environments where the calculated mean daily productivity of seed
during the generative phase (i.e. from the onset of flowering until maturity) was greater than 50 kg ha^{-1} day^{-1}. A
long generative phase, which in these trials was associated
with a long vegetative period, was unable to compensate for
poor daily productivity (Figure 6.1). Crop duration is more
often determined by the need to avoid limiting factors (e.g.
drought and disease) than by a minimum duration of either
vegetative or reproductive growth.

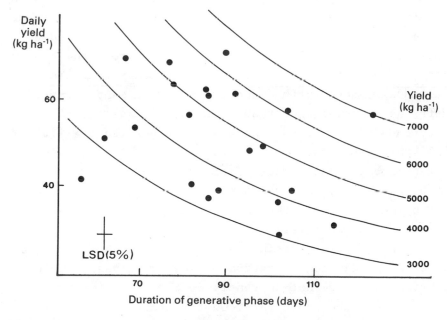

Figure 6.1 Duration of the generative phase (days) and calculated
average daily productivity (kg seed ha^{-1}d^{-1}) in twenty-two
environments within Europe (after Dantuma *et al*. [44]).

205

PRINCIPAL ECONOMIC YIELD AND USES OF CROP PRODUCTS

Economic return

In modern agriculture, faba beans are generally a relatively inexpensive crop because home-saved seeds are often used, few fertilizers are required, and pests and diseases can usually be partially controlled by husbandry methods. In many societies, faba beans are eaten in traditional dishes and may contribute a necessary part of the diet; in such circumstances faba beans are not necessarily in competition with more productive agricultural crops.

In mechanised agricultural systems it has been possible to adjust the machines developed for cereals, to the sowing, harvesting and handling of faba beans. However, the use of such machinery has led to a preference for small-seeded types. It is generally accepted that standard machines can only deal effectively with cultivars which produce seeds individually smaller than about 0.8 g, and many farmers prefer beans not larger than about 0.5 g. As some of the most productive cultivars have large seeds (each > 1.0 g) [44], this preference may restrict potential yield and thereby the economic return from the crop.

Dry seeds are the component of the crop most commonly used by man and where these are used for animal feed their value lies in their large protein concentrations compared with those of cereal grains. However, although there is a range in protein concentration between cultivars, protein-rich faba beans do not currently command a better price than less enriched cultivars. On the other hand, small-seeded types which are suitable for domestic pigeons do command a small premium in some countries. In the EEC, a farmer's return is now also improved by the receipt of a subsidy [36] which is intended to encourage the cultivation of adapted grain legumes so as to relieve dependence on imported protein-rich products for feed.

Faba bean (*Vicia faba* L.)

Uses of the crop

<u>Dry seeds for feed</u>.   Faba beans, surplus to human require-
ments, have probably been fed to animals ever since they
were domesticated, and the use of faba beans in livestock
production systems in developed countries long preceded that
of soyabeans and other imported protein feeds. In the UK,
about 30% of faba beans are fed on the farm where they are
grown, mainly to pigs and poultry. They can substitute for
soyabeans (*Glycine max*) satisfactorily [107] at inclusion
rates of up to 15% for pigs [67] or 10% for poultry [46].
Faba beans can also be fed to cattle and sheep though it is
generally less economical to do so [123] except in special
management systems where ripe faba beans are grazed by
sheep [6]. Faba bean protein is rapidly degraded in the
rumen and so if fed to dairy cows it is mainly for its
energy value and should be given in conjunction with un-
degradable protein [123]. Nevertheless, faba bean meal has
been incorporated into some compound feeds [107], but when
prices of faba beans are inflated and soyabeans are cheap,
least cost formulations do not permit the inclusion of faba
beans. Whilst it is clear that EEC subsidies help this situation, discontinuity of supply still limits faba bean use.
For pigs and poultry, deficiency in methionine has to be
rectified [67]; other factors, possibly connected with the
favism-inducing compounds vicine and convicine [33], also
depress egg-laying [46].
　　Tannin-free faba beans have improved animal performance
in feeding trials with monogastric animals [64,116] and such
genotypes now feature in some plant breeding programmes.
Autoclaved [33,55] and dehulled [54] (testas removed mechanically) beans have a better nutritive value than untreated
ones, but these processes are not yet common commercially.
The hulls (i.e. the testas) themselves also have a useful
feeding value - equivalent to that of good meadow hay [174].
Antinutritive factors (trypsin inhibitors, lectins and phytates) do not normally limit the use of faba beans in animal

diets at the commonly-used inclusion rates.

In some regions (e.g. in north-west Europe) a small proportion of the crop is fed to domestic pigeons. For this purpose small, round seeds are preferred.

Dry seeds for food.   Whereas most of the faba beans used for feed are of the small- or medium-seeded types, those used for food include all three types. In northern Europe, few faba beans are conserved dry for human consumption but in the Mediterranean area, the Middle East (particularly in Egypt), Ethiopia and China, one of the major uses is in dishes where the dried beans are soaked, cooked for long periods and seasoned [91]. A common breakfast dish through-out the Middle East, 'Foul Medames', is canned in several countries, e.g. Egypt, Jordan, Syria and Lebanon, and is even exported to that region from China. Another ancient use is to make flour from the beans [41] and a more modern one is a salty, fried, whole-bean snack [124]. In France, a small amount of faba bean flour is used in bread making. Elsewhere (e.g. in Egypt) faba beans are a regular part of the staple diet and an important source of protein for a large proportion of the population. Clearly, cooking quality and organoleptic properties are important attributes of faba beans in these situations; for example, some large-seeded cultivars (var. *major*) are more palatable and have a better sugar concentration than cultivars of vars *equina* and *minor* [2]. The only contraindication to human consumption of faba beans is among people who are susceptible to favism (see below and Chapter 3). Not all beans are consumed in the country of production; some are exported from Europe and Canada to the Middle East and Far East.

Industrial processing of dry seeds.   Attempts have been made in some industrialised countries (e.g. the UK, Canada, France and Denmark) to utilise faba beans in processed foods in a similar manner to the now commercially exploited soyabean products [103,139]. The general conclusion has been that faba beans could compete successfully with equivalent soya products [174]. However, of the three food ingredients, (a) flour from hulled beans (about 30% protein), (b) concentrate produced by air classification (about 70% protein), and (c) isolate produced by aqueous or alkaline extraction (90% protein), the concentrate seems to have the greatest potential [104]. Faba bean flour and concentrates are also more likely to be exploited in the functional protein products (e.g. in bread and pasta) than in textured vegetable protein [104]. The starch in faba beans may also come to be exploited in industrialised processes.

Immature seeds used as a vegetable.   In almost all countries where faba beans are grown, a certain proportion of the crop is eaten either as a fresh or cooked green vegetable. In a few regions, such as the Mediterranean Basin and in India, whole immature pods are eaten. Where the immature seeds are preserved, mainly in developed countries, this is achieved by canning or freezing [75]. This industry is often an extension of that engaged in preserving green peas with some modification to the vining and processing machinery.

Only white-flowered (i.e. tannin-free) cultivars are acceptable for canning in the UK since they do not discolour when cooked. However, pigmented cultivars are used for freezing and are said to be superior in flavour and texture, though they are more prone to post-harvest discoloration [75]. Some white-flowered, small-seeded cultivars are used as an ingredient of mixed vegetable packs and as small broad-beans.

Whole-plant use.   Faba beans have a long tradition as a green manure. Whole plants are ploughed in with considerable benefit to the subsequent crop. This technique suits farming systems in arid areas and where the climate permits multiple cropping. The stems (haulm) are commonly used as green manure in Chinese agriculture after the removal of the fruits for green-seed use. Empty pods (husk) are used for making compost.

Use of the whole green plants for animal fodder, either in pure stands or mixed with cereals, is found more often at higher altitudes or extreme latitudes where ripening of the seeds is slow. Biological yields (yields of above-ground dry matter) in the range 5-9 t ha^{-1} for spring cultivars [193] and 9-12 t ha^{-1} for winter cultivars [192] have been reported. Faba bean silage has given satisfactory results with dairy cows [120] and lambs [191], while dried pellets are a useful supplement to poor-protein silage [111], though artificial drying is usually expensive. Growing faba beans in mixtures with cereals reduces pest and disease infections of the beans, and harvesting whole plants gives greater flexibility in timing of harvest than exists when crops are harvested for seeds alone. Growing mixtures, however, poses problems about choice of herbicides.

Crop residues.   Straw from combine harvesters has little value and is normally burned in the field, ploughed in, or used for bedding animals. However, where the plants harvested for seeds are cut whilst partly green the straw makes useful fodder. In Egypt and Sudan most of the crop residue is used for animal feed, fuel, or in brick-making. There is a cash market for bean straw in Egypt; it has been estimated to contribute 9-14% of the revenue from faba bean crops [166].

Special uses.   A few recipes mention the green tops of faba bean plants which can be eaten as a salad or boiled like cabbage.

Pigmented faba beans are rich in 3,4-dihydroxyphenylalanine (L-DOPA) and it has been suggested that the pharmaceutical industry could extract this for use in the treatment of Parkinson's disease [133].

Faba bean (*Vicia faba* L.)

PRINCIPAL FARMING SYSTEMS IN WHICH FABA BEANS ARE A COMPONENT

Faba beans are grown in a great diversity of farming systems. That this has been possible is due to the plasticity of response of plants to a wide range of environments and it also reflects the genetic variation within the species which man has been able to exploit. There is variation in tolerance of cold temperature, photoperiodic requirement, life-cycle duration and phenology. Thus, the crop is grown from the equator almost to the Arctic circle and from sea level to high altitudes. Restrictions as to the farming systems in which faba beans can be incorporated are imposed more by lack of tolerance to very hot temperatures, arid conditions and extremely acid or saline soils.

Crop rotation

Faba beans are rarely the main crop in a farming system, and they are seldom grown continuously on the same land. The most common place for faba beans is to be rotated with cereals, especially wheat or barley. However, in warm climates they are often included in rotations with rice, cotton, maize or sorghum (e.g. in southern China, Egypt and Sudan [164]), or rapeseed (China [96], Canada and north-west Europe). In some dry regions (e.g. western Canada) a fallow prior to faba beans is necessary to allow the accumulation of sufficient soil moisture.

The inclusion of a 'break' crop in continuous cereal-growing systems is mainly to prevent the build-up of soil-borne cereal fungi [52]. Several legumes fill this role and also supply nitrogen to the subsequent crop [150] (see below). Faba beans are particularly suitable in that they also improve soil structure, their stems are erect and, provided the seeds are not too large, no special equipment is needed other than that which farmers already have for culture of cereals. Where a two-year break is relatively more beneficial to the cereals [52], faba beans can conveniently be used as one of these. Faba beans are also included as a 'break' crop in farming systems where more expensive crops such as sugar beet or potatoes are not well suited, and they are often used as a 'break' from wheat because of their suitability to heavy clay soils.

An interval between growing successive faba bean crops on the same land is usually needed because of the possibility of building up to dangerous numbers populations of soil-borne parasites such as stem nematodes (*Ditylenchus dipsaci*), diseases such as stem rot (*Sclerotinia trifoliorum*) in northern Europe, and the parasitic weed broomrape (*Orobanche crenata*) in southern Europe, west Asia and north Africa. Other possibilities are *Thielaviopsis, Rhizoctonia* and *Fusarium* spp. Then again, certain leaf pathogens which also live saprophytically on crop residues (e.g. *Ascochyta fabae* and, to a lesser extent, *Botrytis fabae*) can increase if faba beans are grown too frequently on the same land. There is also the danger of re-infection from volunteer seedlings in adjacent fields. The duration required between faba bean crops depends on the region, the kinds of parasite that are prevalent and whether other susceptible crops are included in the rotation. As an example, a period of seven years is advised where nematodes are a problem. When producing pure seed of a commercial cultivar it is general practice in northern Europe to select a field which has not grown faba beans for the previous three years.

Double and mixed cropping

In intensive farming systems, such as those in parts of China where two or even three crops are taken in one year, faba beans are double-cropped with rice, maize, cotton or sweet potato, and then rotated with a cereal or rapeseed in the following winter season [96]. In Egypt, faba beans are grown as the winter crop and maize, rice or cotton as the summer one, on the same land [164]. Clearly, fast-growing types have to be used for this purpose and where there is not time to harvest the beans before the succeeding crop must be sown, as for example in parts of China, faba beans are undersown with rice or cotton, or interplanted with maize or cotton [52]. In China, interplanting cereals (wheat or barley) with faba beans is practised, and in southern Egypt it is common to interplant faba beans with sugar cane.

Mixed cropping of faba beans and cereals, mainly for animal fodder, was popular in Europe in the early part of the nineteenth century. An advantage of this system was that the beans suffered less from pests, diseases and flower-drop because of the greater spacing between bean plants, but the practice declined due to the specificity of the herbicides that have become part of modern agriculture and the uneven seed ripening that is a feature of such mixtures. However, some mixtures are still used for silage or fodder and in China, Ethiopia and Afghanistan faba beans are sometimes grown in mixtures with peas, *Lathyrus* and/or vetches.

Subsistence and developed systems

The range of systems varies from small patches of faba beans on marginal land with the produce being consumed locally as fresh beans, to intensively mechanised farming of fields as large as 100 ha, where the dry seeds are sold for export.

In the Mediterranean region, faba beans are sometimes to be found in small areas within orchards [124], often of olives. In Egypt, as well as the usual fields, strips or even single rows of faba beans are sometimes planted around the perimeter of fields of other crops, e.g. Berseem clover [164], and along the banks of irrigation canals and ditches. In some semi-arid areas, small fields may be planted with faba beans wherever there is a chance of an economic return on the cost of the seed (which would be home-saved) since, in these subsistence systems, other costs are very small.

213

The frequency of this practice, and the intensity of faba bean culture, increase in those regions where rainfall is adequate or irrigation facilities are available and proximity to a town provides a market for the produce. In the Nile Valley [164] and on the east coast of Spain [124] most crops of faba beans are irrigated and there is a more intensive type of agriculture and also a market for dry as well as fresh beans.

Where *V. faba minor* and *equina* are grown for harvesting dry their cultivation is often integrated with a livestock system [41]. Thus, faba beans may be grown on the home farm to provide protein for pigs or poultry and the farmyard manure is then used to help production of other crops. Faba beans used for silage or grazing are also clearly linked to livestock enterprises.

In northern Europe, however, dry faba beans are more often sold to merchants for incorporation into compound feeds, and they therefore fit into a cash-cropping system. When harvested green, faba beans are usually grown either in private gardens or as a horticultural crop for immediate use, or in fields in the vicinity of factories where the crop is conserved by freezing or canning.

BOTANY, MORPHOLOGY, PHENOLOGY, AGRONOMY, REPRODUCTION AND COMPONENTS OF YIELD

Botany

Taxonomically, *Vicia faba* is a member of the section *Faba* of the genus *Vicia*. The section includes *V. narbonensis*, *V. galilaea* and *V. hyaeniscumus* [106]. Neither these nor any other *Vicia* species have yet been successfully crossed with *V. faba* though attempts are still being made. It is therefore still to be considered as an isolated species, though with considerable variation within it (and see Chapters 1 and 2).

Some authors have recognised subspecies [85,127] but

Cubero [94] argued that, because there are no fertility barriers within *V. faba*, only the four botanical varieties *faba* (= *major*), *equina, minor* and *paucijuga* should be distinguished. Classification of cultivars or other genotypes into these four botanical varieties has been based mainly on seed size (though there are some fairly strongly associated characters, e.g. pod length). However, as a result of breeding from wide crosses, there is now a tendency for the merging of botanical varieties, and for descriptive purposes a classification of cultivars has been proposed which utilises readily discernible discontinuous characters, e.g. colour of flower, testa or hilum [94]. Within botanical varieties, winter and spring types are recognised, though more distinctly within *equina* than the others; and within *major, equina* and *minor* there are both northern European and Mediterranean types.

Morphology

Faba bean plants are distinctly annual with strong, hollow, erect stems bearing usually one or more basal branches arising from leaf axils. There is a robust tap root with profusely branched secondary roots, but the rooting system is not as deep as that of wheat and where moisture is available in the surface layer of soil there is a strong development of secondary and adventitious roots. In most soils the tap root and, to a lesser extent, the secondary ones bear white or grey nodules formed by *Rhizobium leguminosarum* (see below).

Alternate pinnate leaves comprise between two and six entire, oval leaflets each up to 8 cm long. The stipules have extra-floral nectaries but, unlike other *Vicia* species, tendrils are absent or only rudimentary. Plant height of the common cultivars varies considerably with environment, but is usually within the range 50-200 cm. Some true dwarfs are, however, available and semi-dwarfs are also being developed. Growth is indeterminate though genotypes

215

vary in degree of competition between the vegetative apices
and flowers, pods or branches; determinate mutants have
also been described [63,69,130,176,183].

Phenology

European winter genotypes and those grown in the Mediterranean winters respond to vernalisation by flowering earlier
[184] and at a lower node, but unvernalised plants do
eventually produce flowers. North European spring cultivars
do not require vernalisation.

The majority of cultivars show a quantitative response
to long days though early-flowering types, mostly originating from low latitudes, are effectively day-neutral.
However, response to long photoperiods in European cultivars is accentuated by warm temperatures [184] and seasonal
differences in temperature can markedly affect dates of
flowering [143].

Rapid vegetative dry-matter production in warm temperatures enables the species to grow in short, warm seasons
(e.g. in Canada and Sudan), but it may be this same feature
which, in long seasons of fluctuating temperatures (e.g.
in north-west Europe), gives rise to the strongly competitive growth of stem apices and other vegetative organs
which sometimes results in excessive flower-drop. Supra-
optimal temperatures at flowering time can also adversely
affect fertilisation, particularly in African or Indian
genotypes [194].

Response to irrigation can be in vegetative rather
than in reproductive growth unless the water is applied
after many ovules have been fertilised and 'sinks' have
been established; then, increases in pod numbers and yield
of seeds can be substantial. There are numerous interactions between the effects of different climatic factors;
for example, plentiful soil-moisture can increase leaf
area but this may lead to more mutual shading and so
reduce carbon assimilation in the lower leaves.

216

Faba bean (*Vicia faba* L.)

Agronomy

Date of sowing is critical in many farming systems because of the need to avoid frost, drought, pests or diseases, which may occur or appear early or late on in the growing season. The relatively large seeds of *V. faba* allow deep sowing, which is necessary when certain pre-emergence weedkillers are used, and this also protects them from predators.

A wide range of crop densities is used, from 15-60 plants m^{-2} according to cultivar and region. In the small-seeded spring types there is often a positive response in yield up to 60 plants m^{-2}, but for large-seeded types densities greater than 20 plants m^{-2} are seldom economic. Winter bean densities greater than 25 plants m^{-2} increase the risk of excessive flower-drop and disease, particularly aggressive chocolate spot [99]. At constant densities with European indeterminate types, row widths within the range 15-60 cm have little effect on yield, but this result may not apply to genotypes of different architectures.

Wider plant spacing allows weed control by hand or machine but this is now common only in subsistence agriculture. In modern farming systems an increasing number of herbicides are available and harvesting is usually by machines designed for cereals. Earliness and uniformity of ripening still need to be improved but beyond about 53°N latitude, because of their erect stems and relatively open canopy, the harvesting of faba beans is more reliable than that of peas [190], the only other grain legume grown to any extent in northern Europe. Research on growth regulators [35,74] suggests that it may become possible to impose chemical control on the tall, indeterminate types, with a benefit to yield in cool, humid regions.

Reproduction

Traditional cultivars produce between two and ten typically

papilionaceous flowers on each axillary raceme starting at between the fifth and tenth node. Each flower has ten stamens, one of them free. The style, which is slender and hairy, bends at an angle to the ovary of slightly less than 90°, and is terminated by a stigma covered with papillae. Anthers and stigma are enclosed within the keel petal and are only exposed (i.e. the flowers are tripped) when a bee depresses the keel by landing on the wing petals.

The main insect pollinators in central and northern Europe are the long-tongued bumblebees *Bombus hortorum*, *B. ruderatus*, *B. pascuorum* and *B. lapidarius*; they pollinate whilst probing for and collecting nectar as well as when collecting pollen. The short-tongued bees *B. terrestris* and *B. lucorum* rob nectar from the base of the corolla, and like honeybees (*Apis mellifera*), which rob nectar through the holes pierced by short-tongued bumblebees, they assist in tripping and cross pollination only when collecting pollen. In the Mediterranean region, in addition to *Bombus* and *Apis*, some of the solitary-bee genera (*Anthophora*, *Andrena*, *Osmia*, and *Eucera*) visit and pollinate faba-bean flowers.

Tripping assists in self-pollination and the presence of bees provides the opportunity for natural cross-pollination. On average, about 30% of the plants in a population are crossbreds [70]. There is little evidence of self-incompatibility in non-inbred plants but some genotypes, particularly crossbred plants, are capable of self-fertilisation without tripping, i.e. they are auto-fertile [50] (Plate 6.1).

The proportion of ovules which are fertilised varies considerably, depending on genotype and environment; early flowers on some UK winter beans are poorly fertilised [206]. In general, however, more ovules are fertilised than develop into mature seeds [161]. Topping (i.e. decapitating) plants can induce flowers to set [73] and hence many abscised flowers must have contained some fertilised ovules [72]. Furthermore, there may be early abortion of some embryos within pods which also contain fully mature seeds.

As flowers normally open sequentially, the first ones on a raceme, and sometimes the earliest racemes, compete

Plate 6.1 Autofertile plant with good seed set at nodes where flowers were left untripped as well as alternate nodes which were tripped. (Copyright: Plant Breeding Institute, No.8340).

most strongly [51] and may in fact induce the formation of
abscission layers in flowers fertilised later. Such competitive effects might be reduced by selecting genotypes
in which flower opening is more nearly synchronous or in
which there is an independent vascular supply to each
flower [72].

Components of yield

The number of ovules per ovary varies from two to ten,
and average seed size from 0.2-3.0 g (Plate 6.2). Cultivars with more than five seeds per fruit have large seeds
(individually usually heavier than 1.0 g) but some breeding
programmes are now aimed at a large number of small seeds
per fruit [154]. The length of the fruit is loosely
related to the number of seeds in it, and the angle of the
fruit to the stem tends to be more erect when it contains
a few small seeds and more pendent when it contains a
larger number of heavy seeds.

There is also much variation, due to environment and
genotype, in the number of fruit-bearing nodes and number
of fruits per node. However, since seed size and number
of seeds per fruit have genetically-determined upper limits,
there is a restriction on the distribution of assimilates
once the number of seeds on a plant has been determined
[190]. Thompson and Taylor [190] concluded from observations on trials comparing crops grown on fertile and
nutrient-poor soils, and seeded at a range of plant
spacings, that initial sink numbers are a major factor
limiting yield. Nevertheless, the upper limit of seed
size of a given genotype is sometimes not achieved and a
positive association between large yield and large seeds
has been demonstrated in trials in a large number of
environments in France [149].

There is greater response to selection for the individual components of yield, and their inheritance is
nearer to additivity than it is for yield *per se* [17]

Faba bean (*Vicia faba* L.)

(which is much influenced by dominance or overdominance
due to heterosis). Increase in one component, however, is
likely to be compensated for by a decrease in another.
Manipulation of yield components to produce the same or a
larger seed yield from fewer fruiting nodes mav also

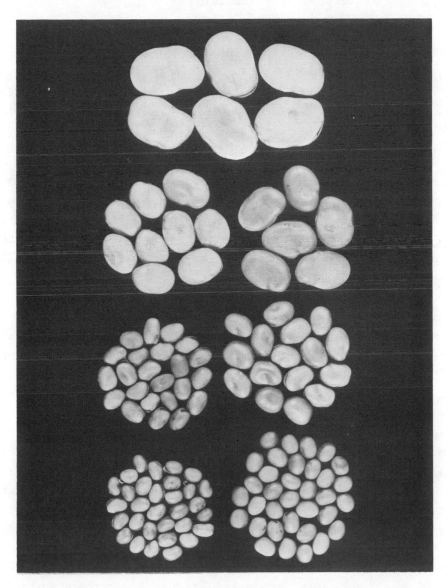

Plate 6.2   Range in seed size and shape in *V. faba*.
(Copyright: Plant Breeding Institute, No. 25177)

improve the harvest index, an important objective since total dry-matter production (biological yield) rarely limits seed yields [190]. The possibility of obtaining relatively few large sinks so as to reduce competition between them is, however, sometimes limited by the agricultural requirement for small seeds.

SYMBIOTIC NITROGEN FIXATION POTENTIAL, ITS EXPRESSION IN THE FIELD AND IMPLICATIONS FOR FARMING SYSTEMS AND SOIL FERTILITY

Rhizobium leguminosarum is common in nearly all faba bean growing areas of the world. However, where the crop is being introduced for the first time (e.g. in parts of Australasia and Canada) inoculation may be needed. In other areas where faba beans are sometimes grown (e.g. in Sudan), the amount of natural nodulation which occurs has been related to the time since previous cropping with faba beans or a similar legume [128]. In Upper Egypt, Hamdi [83] reported that inoculation gave a significant positive response in seed yield at three out of ten locations.

In most areas however, the potential for nitrogen fixation is probably sufficient to sustain large yields, though this may not always be achieved in practice. Environmental conditions such as pH, salinity or temperature, can affect the presence, density and effectiveness of *Rhizobium* populations; for example, Olivares *et al.* [138], found a greater yield increase from a combination of nitrogen and sulphur applications and *Rhizobium* inoculation than was achieved by any of the three treatments separately (and see Chapter 4).

The diversity of strains of *Rhizobium* within any given region is also normally large, with strains displaying different environmental tolerances [10] and a range of nitrogen fixing capabilities [59]. An instance of environmental tolerance and differing effectiveness was noted by Clark [37] who reported that a strain isolated in Morocco and tolerant of hot temperatures displayed greater nitrogen fixing capabilities in mid-Canada than did strains isolated from within this colder region.

Sprent and Bradford [182] calculated that nitrogen fixation by *R. leguminosarum* in association with *Vicia faba* may exceed 600 kg ha^{-1} a^{-1}. This amount would only be achieved under optimal environmental conditions and smaller values of 59 and 146 kg N ha^{-1} a^{-1} have been given by Saxena [166] and K.W. Clark (pers. comm.) respectively.

In terms of self-sufficiency for N, Saxena [166] calculated that two Syrian landraces of faba beans, ILB 1814 and ILB 1813, fixed 84-96% and 94-97%, respectively, of their total nitrogen requirements. K.W. Clark (pers. comm.) whilst reporting that cv. Herz Freya in Manitoba fixed from 68-146 kg N ha^{-1} a^{-1}, noted that 60% of this was in the plants at harvest; some 28-58 kg N ha^{-1} being returned to the soil by the ploughed-in stubble, contributing a useful addition to soil fertility.

Faba beans thus have a rotational value (see also earlier comments) and, as in UK cereal growing systems, they play an important role in disease control (particularly 'take-all'; *Gaeumannomyces graminis*) and make nitrogen available to the succeeding crop. Amounts of residual N of the order of 50 kg N ha^{-1} have been quoted by Dyke [52]. Further experimental evidence of the value of faba beans in a rotation has been given by results from Drayton Experimental Husbandry Farm, Stratford-upon-Avon, UK [187]. An experiment, over the three years 1980-82, assessed the yield response of subsequent wheat crops following beans or wheat, to varying applications of nitrogen in the spring. Response to additional nitrogen was small following beans but substantial following wheat (Figure 6.2) and there was little seasonal variation. It is not possible to separate the various factors which may have contributed to such results but certainly the nitrogen fixed by the faba beans will have been a major factor. In the warmer Mediterranean climates, regrowth after seed harvest may occur and this 'second crop' can be turned in as green manure.

An interesting agricultural development is a return to the use of legumes to supply all the nitrogen requirements in a farming system, and Patriquin *et al*. [141] report that faba beans are used as the principal component of what appears to be a successful attempt to realise nitrogen self-sufficiency on two Nova Scotia farms.

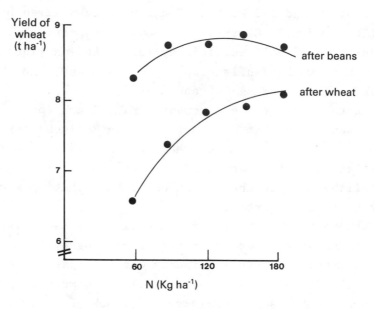

Figure 6.2 Yield of winter wheat (t ha^{-1} at 85% dry-matter) in relation to the amount of N applied to crops following faba beans or wheat (from Tas [187])

The marked influence that specific faba bean genotype x strain of *Rhizobium* interactions may have on dry-matter production, amount of nitrogen fixed and efficiency of nitrogen utilisation, has been demonstrated in sterile culture by Mytton and El-Sherbeeny [58,59,110]. However, these interactions have not yet been exploited in crop conditions because of the differential competitiveness [7,68] and the instability [32] of individual strains of *Rhizobium* in the field. Where there is not an indigenous vigorous *Rhizobium* population and seed inoculation is possible, there should be greater opportunity for taking advantage of specific associations. In this context it is relevant to note that

if specific associations were to be developed it would be
more appropriate if the cultivars used were pure lines
rather than populations.

PRINCIPAL LIMITATIONS TO PRODUCTION AND YIELD

Crop priorities and attitudes of farmers

Faba beans can be an unpredictable and unreliable crop and,
particularly in more intensive farming systems, interest in
the crop has declined during the twentieth century. In part,
at least, this has been because only average yields have
generally been achieved and the crop has proved less profitable than cereals as well as more difficult to grow.

When making generalised statements about the position
of faba beans on a world-wide basis in the 1980s, three
different situations can be recognised:

(a) those where faba beans are grown for food, often as
    part of a subsistence farming system;
(b) those where home-grown protein is required for feed and
    alternative sources perhaps imported from elsewhere are
    not available; and,
(c) those where protein is required on the farm for feed
    and where both home-grown crops and imported protein
    are available. Then, the economic advantages of the
    different sources of protein have to be taken into
    account.

In most of Europe, faba beans in the third category
have to compete with soyabeans (mainly imported) which have
larger protein concentrations and are more uniform within
each cargo, or with peas (mainly home-grown) which are earlier to harvest, have rather more predictable yields and
have a more flexible, dual-purpose role (for feed or food).
However, if improvements in faba beans could be achieved,
particularly in yield stability, protein concentration,

amino acid balance, and removal of tannin, convicine and
vicine, this would elevate the crop into a stronger competitive position. There is, in western Europe in any case,
a large shortfall between the quantity of protein feed
required and that which is home-produced; and while the
aim is increased self-sufficiency, faba beans are likely
to continue to fill part of this need as peas are less well
suited to heavy soils.

In all farming systems there is a tendency for the
greatest investment, care and interest to be devoted to cash
crops. These crops are often the principal source of a
farmer's income and with them he is looking for innovation
and improvement as means of increasing yields. Many home-
grown crops, and in particular those used for feed, do not
get this degree of attention and are given less priority in
the farming system. Similarly, because the profit margin
cannot be so easily gauged, such crops do not attract their
proportionate share of breeding, research and development
resources.

Yield-limiting factors vary considerably with geographical regions and farming systems but the major ones are discussed below.

Pests and diseases

Details of these may be found elsewhere [92] and a summary
of genetic resistances is given later, but it is important
to note here that severe yield losses can be caused by
aggressive attacks of *Botrytis fabae* on winter beans in
the UK, France and in humid areas of the Mediterranean.
Husbandry methods and fungicides give only partial control
and undoubtedly any innovation which improves control or
decreases the prevalence of this disease would improve
yield stability and so increase interest in faba bean
production.

Ascochyta is seed-transmitted and dispersed by rain-
splash. It can limit yields when uncontrolled. In the UK,

226

freedom from infection is an important feature of seed certification. *Uromyces* appears late in the growing season in
some regions but is earlier, and important, elsewhere. *Fusarium* foot-rots limit production on soils of poor fertility or
structure, and some stem infections of *Botrytis*, associated
with other species of *Fusarium*, combine to cause lodging.

There are many troublesome viruses, but yield-reduction
following infection by aphid-transmitted viruses is often
only moderate, probably due to some natural resistance. On
the other hand, weevil-transmitted viruses, which are often
also seed-transmitted, may cause substantial yield losses on
the less frequent occasions when the combined presence of
the virus and the vector in large numbers leads to heavy
infections.

Direct damage by *Aphis fabae* is common but in some regions its arrival is forecast and it can be controlled efficiently by insecticides. Nematodes can multiply to yield-
reducing population densities in certain rotations. The
effect of the weed-parasite *Orobanche crenata* can be devastating, especially in the Mediterranean region where it
limits both the area cultivated and yield. Control with glyphosate is showing promise though dose rates are critical
[122].

Chemical control of many of the pests and diseases will
not be economic unless the crop has a potential for large
yields. There are also often mechanical problems with field
spraying, the crop being liable to damage by application
equipment.

Weeds

In subsistence agriculture, and in the developed countries
before mechanisation, weed control was achieved by either
hand-hoeing or by implements pulled by draught animals. In
modern agricultural systems weed control is now usually
achieved by herbicides but these are still not universally
effective and some may be phytotoxic.

Frost damage

The most hardy winter beans tolerate about -12°C. In north-
west Europe this limits their cultivation to areas in southern UK and western France where temperatures fall below this
value not more often than one year in about fifteen; but
when this happens yields are small or crops are even destroyed. Frost damage can also reduce yields at higher altitudes and, in many areas of the world, late frosts can damage flowers with subsequent yield losses or variable maturity of fruits.

Late maturity

In extreme latitudes, where ripening is slow, and also closer to the equator, where seasons are short due to rainfall
distribution (e.g. in Sudan), the life-cycle of faba beans
may be too long to obtain good yields of dry seeds. Sowing
date and cultivar have to be chosen carefully. In a few
areas desiccants are used to impose ripeness.

Drought and waterlogging

Although faba bean plants show plasticity in response to
changes in environment such as stand density, they are
nevertheless sensitive to other environmental stresses, particularly soil moisture. Drought is an important factor
limiting yield and most faba bean crops in arid climates
give a substantial and often economic response to well-time
irrigation. However, faba beans are sensitive to too much a
well as to too little water, in that untimely intense rainfall or irrigation (e.g. before flowering in cool, humid
climates) can reduce fruit-setting and yields significantly

Hot temperature

The temperature range for seed setting is rather narrow

228

(10°-30° C) and the maximum is cool compared with some other species [26]. This rather poor tolerance of heat, together with the need to conserve moisture, are reasons why the crop is to be found more often at higher altitudes in warm climates.

Pollination

Lack of pollination may limit yields in some regions, in some seasons, but the distribution of insect pollinators and the frequency of their visits are difficult to quantify. Sometimes lack of pollination of some flowers can be compensated for by setting of others, although the distribution of fruits on the plant may be affected. Also where inadequate pollination lessens the degree of natural cross-fertilisation, there may be a reduction in vigour and yield of the next generation [201].

Prospects

There is some research in progress aimed at alleviating the effects of those factors which limit yield but, for the reasons given above, this is often small in relation to the task and compared with that directed towards other more widely grown crops. Progress is, however, being made in identifying the major limiting factors in a range of environments and also towards the further domestication of the species through improved types of cultivars. In all these efforts, international collaboration is playing a major role as national programmes are often too small and inadequately funded to tackle the problems thoroughly.

FERTILISER REQUIREMENTS AND THEIR INTERACTION WITH WATER SUPPLY

For the full yield potential of faba beans to be realised, it is necessary that adequate and well-balanced quantities of all the essential mineral nutrient elements are present in the rooting medium. A deficiency of macronutrients, such as phosphorus and potassium, in field-grown crops can affect growth and nitrogen fixation adversely [109]. An appropriate application of fertiliser is therefore necessary on soils of poor fertility to bring them up to the required status.

Provided the supply of other essential mineral nutrients is not limiting and effective strains of *Rhizobium leguminosarum* are present, the faba bean crop is able to meet most of its nitrogen requirements through symbiosis [157].

Nutrient removal

The removal of some major nutrients by seeds, haulms (stems and empty pods) and whole shoots by the cultivar Giza-2 at maturity is shown in Table 6.2 [98].

Table 6.2 Removal of some major nutrients by seeds, haulms and whole shoots of faba bean cultivar Giza-2 (recalculated from data presented by Ibrahim and Kabesh [98])

Nutrients	kg nutrients per 1000 kg dry weight of:		
	Seeds	Haulms	Total shoots
N	40.1	12.5	25.0
P	3.7	0.7	2.1
K	12.3	21.5	17.3
Ca	3.2	16.9	10.7

These estimates are similar to those reported in field trials of faba beans in Moscow Province, USSR, where during the season the crop took 12-14 kg K_2O, 2.5-3.0 kg P_2O_5, 10-13 kg CaO and 3.0-4.0 kg MgO, all per 1000 kg above-ground dry-matter [172].

Maximum uptake of N, P and K by faba beans has been observed to occur at the mid-flowering stage, before the fruits are formed, whereas maximum uptake of Ca and Mg occurs at the pre-flowering stage [172]. The pattern of nitrogen uptake in relation to dry-matter accumulation over the period of crop growth [157] is shown in Figure 6.3. It is apparent that the most rapid uptake of N and dry-matter accumulation took place between the pre-bloom and full-bloom stages and decreased after the onset of seed filling. The period of rapid uptake of nitrogen commenced before that of rapid dry-matter accumulation.

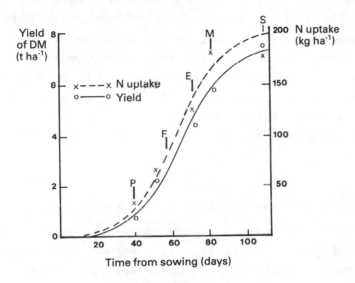

Figure 6.3 Relation between N uptake and dry-matter accumulation by nodulated faba beans in the Seven Sisters field trial at Manitoba, Canada (after Richard and Soper [157]). (P,F,E,M and S denote pre-bloom, full-bloom, early pod-fill, mid-pod-fill and senescence, respectively)

Nitrogen fertilisation

Many studies on the effects of nitrogen fertiliser on field-grown faba beans have shown that no fertiliser nitrogen is needed for the attainment of either large yields or seed protein concentrations [8,43,47,117,157,181]. These studies have all suggested that a vigorous symbiotic relation is able to provide for the nitrogen needs of the crop and that fertiliser nitrogen application affects the contribution of symbiotically fixed nitrogen adversely. The occasional positive responses to large dressings (about 300 kg N ha^{-1})

of nitrogen fertiliser which have been obtained in experiments at Rothamsted, UK, may be explained by attacks on the nodules of *Sitona* weevil larvae which reduced the nitrogen supply available from symbiosis [117].

However, it is common practice in Europe to apply small 'starter' dressings of nitrogen in the seed bed to aid establishment, and a beneficial effect from applications not exceeding 36 kg N ha^{-1} has also been obtained in the Nile Valley [84] on soils with small concentrations of inorganic nitrogen. Also, in the Canadian prairie, faba beans grown on stubbles have responded favourably [160,163] to applications of N at either 52 or 90 kg ha^{-1} although no response was obtained when the crop was grown on land which had been summer-fallowed [160]. In pot experiments in Iran, N fertilisation gave some alleviation of the adverse effects of salinity [165].

Phosphorus fertilisation

A deficiency of available phosphorus seriously limits the growth of faba beans [185]. Studies on the effects of the presence or absence of P at different periods of growth have revealed that the period from the third to ninth week after sowing is the most critical; it is during this period that plants absorb most of the P they require for normal growth [4]. Responses to phosphorus fertiliser are related to the available phosphorus status of the soil and also to the method of application. Positive and significant responses to an application of 50 kg P_2O_5 ha^{-1} have been recorded on the silty clay loam soils at Tel Hadya research farm of the International Centre for Agricultural Research in the Dry Areas (ICARDA), Aleppo, Syria (see Chapter 18) when available phosphorus concentration has been less than 5 ppm [167]. A large number of field trials conducted on clay alluvial and loam soils of moderate P status in the Nile Valley of Egypt have shown good responses of faba beans to phosphorus and an application of 72 kg P_2O_5 ha^{-1}

is therefore a standard recommendation for the crop in that region [84]. Phosphorus application has been found to reduce the incidence of *Botrytis fabae* in Egypt [93] and in the UK the disease is more severe on P-deficient soils [76]. However, responses to phosphorus applications in general have been rare in the UK and the recommendation that 50 kg P_2O_5 ha^{-1} be applied to faba beans is to compensate for the phosphorus that will be removed by the crop and to maintain soil reserves (J. McEwen, pers. comm.).

Placement of phosphorus fertiliser with or near to the seeds is necessary to ensure a good response, particularly on alkaline soils rich in calcium carbonate [3].

Mycorrhiza

Although vesicular-arbuscular mycorrhizal association with faba bean roots has been observed [202] and inoculation with *Glomus moooeae* has increased dry-matter production even in the presence of indigenous mycorrhizal spores [203], mycorrhizal effects on the availability of phosphorus to plants has been little investigated. Studies by Pang and Paul [203] have revealed that faba beans with mycorrhizal association transferred a greater proportion of assimilated carbon to below-ground parts than the non-mycorrhizal plants because of the greater respiratory rates of below-ground parts of mycorrhizal plants. The agronomic implications of this change in the partitioning of carbon in relation to the availability of phosphorus or even other mineral nutrients are not yet clear.

Potassium fertilisation

Culture solution studies on potassium nutrition have highlighted its importance in the growth and development of faba bean plants and in the promotion of an effective symbiotic nitrogen-fixing system [82]. Rates of glycolysis and the functioning of the tricarboxylic acid cycle may be reduced in roots if K supply becomes suboptimal [137].

Field responses to potassium applications have been rare in the faba bean growing areas of west Asia and north Africa [3,84] because of the good available K status of the soils. Also, no response to K application was obtained in Denmark for the same reason [8]. However, positive responses to K at up to 200 kg K ha^{-1} have been obtained in West Germany [95], the Netherlands [156], Poland [65], and particularly in the UK [197] where the exchangeable K in many soils is less than 200 mg kg soil^{-1}. Potassium application in West Germany has also been found to improve the keeping quality of faba beans in storage [95]. McEwen et al. [118], citing the work of A.E. Johnston, have indicated that faba bean yields increased with increasing soil-K up to at least 200 mg exchangeable K kg soil^{-1}. They attributed part of the variation in the maximum yields (3.4, 5.0 and 6.4 t seed ha^{-1}) in their experiments at Rothamsted between 1976 and 1978, to differences in the exchangeable K contents of the soil which, after the harvest of the crop, were 133, 181 and 280 mg K kg soil^{-1}, respectively. Farmyard manure is sometimes used to supply K to faba beans in the UK. As with P, infection of Botrytis fabae can be exacerbated by K deficiency [198].

Micronutrient fertilisation

A few reports from East Europe have suggested positive responses to seed soaking or foliar applications of some micronutrients. In a field trial on acid alluvial soil in Bulgaria [135], a foliar spray of Mo as ammonium molybdate at 33, 65 or 130 g Mo ha^{-1} increased seed yields by 13%, 32% and 38%, respectively. Smaller increases were obtained from the same rates of vanadium application. Crude protein concentration was also increased by Mo application. These studies indicate that the yield of faba beans may be increased through micronutrient fertilisation, a possibility

234

which has not yet attracted adequate attention in many
faba bean growing areas.

Fertiliser requirements and their interaction with water supply

From reviews of the water relations of faba beans [60,181]
it is clear that water shortage affects all aspects of the
symbiosis with *Rhizobium*, from survival of bacteria in the
soil, through size and number of nodules formed, to amounts
of nitrogen fixed and hence nitrogen nutrition of the crop.
The maintenance of soil moisture status near to field
capacity seems to be ideal for nitrogen fixation. On the
other hand, frequent rain, causing ephemeral flooding, resulted in increased nitrate uptake and reduced nitrogen
fixation in the Netherlands [47].

There have also been a few studies on the interrelations
between fertiliser applications and moisture supply. Oknina
[136] reported larger increases in yields from P, K and Mo
fertiliser applications when irrigation was also applied.
Some reports seem to suggest that more positive responses
to applied phosphorus, particularly on soils with only
moderate or poor available phosphorus status, are obtained
in drier soils [57,61]. Thus, phosphate fertilisation
improves water use efficiency under these conditions [57].
Response to farmyard manure has also been greater when
lack of soil moisture has generally limited growth [144],
though this is largely due to moisture conservation.

NUTRITIVE VALUE AND QUALITY OF SEED

Nutritive value

An indication of the overall nutritive value of faba beans
is given in Table 6.3.

Seed protein concentrations vary between 22% and 36%
depending on cultivar [21,80], and where winter and spring

forms are recognised winter beans have slightly smaller
concentrations than spring beans [53]. Protein concentration
is influenced by both genetic and environmental factors,
and inheritance of this trait is largely due to additive
effects, with some partial dominance [21,80,148,171].

World protein prices are based on those of soyabeans,
and faba beans have the disadvantage that protein concentration is smaller than that in extracted soyabean meal.
However, faba beans are also a useful source of carbohydrate and can partly replace cereal grains. One view is
that a lessening of protein concentration (by breeding)
might lead to improved yields, the crop then being regarded
mainly as an energy source [190].

Table 6.3   Average analyses of winter and spring beans and their
various fractions (from Simpson [173]). (dehulled = testas
removed)

| Component | Winter | | Spring | |
concentration (%)	Whole	Dehulled	Whole	Dehulled
Moisture	14.1	14.9	11.8	12.3
Ether extract	0.95	1.02	1.00	1.08
Protein (N x 6.25)	23.3	25.3	27.7	30.5
Crude fibre	7.05	2.95	6.30	2.07
Ash	3.5	3.5	3.5	3.5
N-free extract	51.1	52.3	49.7	50.6
Calcium	0.08	0.03	0.08	0.03
Average phosphorus	0.1	0.1	0.1	0.1
Salt	0.8	0.7	0.8	0.7

The protein in faba beans is relatively expensive
compared to soyabean protein, but the commodity has also
been unpopular with the feed industry because of uncertainty
of supplies and difficulties with milling [173]. The poor
reputation for processing may have been partly due to
inadequate drying and unsatisfactory storage. However,
storage techniques are being improved and propionic acid
treatment may allow storage of moist beans for feed [131].

Faba bean (*Vicia faba* L.)

Chemical analyses appear to overestimate the nutritive value of faba beans and it may be that the protein is relatively indigestible [121]. For example, faba beans have been most successful in pig feeds when they contribute not more than 20% of the diet and only half the supplementary protein [45,66]. Furthermore, the feeding value for monogastric animals may be increased if the testas are removed (dehulling), both because of the removal of a large part of the fibre and also because of the removal of the antinutritional factors which they contain [121,173].

Except for the sulphur-amino acids, particularly methionine, faba beans protein is reasonably well-balanced with respect to the essential amino acids. Table 6.4 gives the typical amino acid composition of faba beans in comparison with soyabean meal. When protein concentration is improved by breeding and selection the relative proportion of lysine decreases [18,80,126], the proportion of histidine remains roughly constant [126], but the proportion of arginine increases [80,126,177]. Unfortunately, the proportion of methionine remains small in protein-rich lines [86,142].

In contrast with many crops, seed protein concentration and seed yields in faba beans are not negatively correlated [80,178] but there is the possibility that large seed protein concentration may be associated with late maturity [148]. Breeding and selection for increased protein concentration has been carried out in several programmes and, although improvement may be sought by hybridising appropriate parents, positive selection can also be made within most faba bean populations and this has been a method of making considerable progress.

Thus, breeding offers the possibility of increasing seed protein concentration, but it has been alternatively suggested [21,31] that the greatest contribution to increasing protein production per unit area of land may be by improving and stabilising seed yields. However, these two approaches are not necessarily mutually exclusive.

Table 6.4. Amino acid composition (g 16g N^{-1}) of faba beans and
soyabean meal [56]

Amino acid	Faba beans	Soyabean meal
Aspartic acid	11.21	10.85
Threonine	3.71	3.89
Serine	5.00	5.03
Glutamic acid	17.72	18.06
Proline	4.33	5.51
Glycine	4.51	4.28
Alanine	4.21	4.19
Valine	4.81	5.22
Isoleucine	4.39	4.58
Leucine	7.53	7.48
Tyrosine	4.66	3.33
Phenylalanine	4.48	5.42
Lysine	6.26	6.08
Histidine	2.59	2.66
Arginine	10.14	7.23
Methionine	0.87	1.56
Cystine	1.52	1.49
Tryptophan	0.94	1.29

Antinutritive factors

Faba beans contain small amounts of several possible antinutritive factors. Examples are protease inhibitors (of
which trypsin inhibitors are the best known), haemagglutinins (lectins), glucosides, phytates and tannins. However,
the effects of these antinutritive compounds in faba beans
are, in general, less acute than they are in many other
grain legumes (see Chapter 3).

Protease inhibitors are present in faba beans but these
are at much smaller concentrations than found in soyabeans
[112,199] and are probably sufficiently small not to influence markedly nutritive quality [140]. They have been detected in a longpod type of broad bean [195] and a wide range
of variation in differing cultivars and populations has been

reported [14,78,115,121].

The greatest concentrations of lectins reported in faba beans have been much smaller than those contained in common beans and soyabeans. Cultivar differences in lectin concentration have been found [78,115] and significant responses to positive and negative selection have been reported [179].

The glucosides vicine, convicine and DOPA-glucoside are believed to be the main factors responsible for favism, a haemolytic anaemia which can occur in susceptible people of Mediterranean and Middle East origin, after eating faba beans. There is also evidence that these glucosides, especially convicine, are responsible for reduced biological value of faba beans when fed to animals [206], including poultry whose egg-weight is depressed [207]. A study of fifty-eight cultivars [100] showed little difference in the vicine concentrations of the testas, but there were cultivar differences in the concentrations in the cotyledons. Other genetic differences in vicine and con-vicine [15,206] and in L-DOPA concentration [158] have also been reported.

Griffiths and Jones [79] found that different cultivars of faba beans contain varying concentrations of phosphorous, of which 40-60% is phytate-phosphorous, but phytate concentration is also influenced by environmental factors.

The testas of coloured-flower cultivars contain 4-8% tannin; absence of tannin in the testas is associated with white flowers [20,49,146,162]. Only tannin-free cultivars are suitable for canning [49]. Some variation in tannin concentration within coloured-flower genotypes has also been reported [116]. There is need for more evidence of the importance of the antinutritive factors in faba beans. Some of them are heat-labile but with evidence that there is genetic variation for most of them, there is the possibility of reducing concentrations or even removing these compounds altogether.

GERMPLASM RESOURCES AND STATUS OF NATIONAL AND INTER-
NATIONAL CROP IMPROVEMENT PROGRAMMES

Germplasm resources

Faba beans have been cultivated in many parts of the world
for as long as 6000 years; many landraces exist and they
represent a wide range of variation. However, most
collections made of these geographical races [127] and of
the various spontaneous mutants [175] in the 1920s and
1930s have been lost, and trials of large numbers of named
stocks in England [77] and in France [145] in the 1950s
did not identify much useful variation in terms of consistently better-than-average yields.

Subsequently, many breeders have assembled working
collections while more comprehensive gene banks are maintained at certain specialised institutes including ICARDA
in Syria, the University of Bari in Italy, at Braunschweig,
West Germany and at Gatersleben, East Germany. However,
Witcombe [200], in a recent review of faba bean germplasm,
concluded that existing collections are incomplete and that
many items are not yet fully characterised. Germplasm from
China, India and Iran is under-represented and although
collections have been made by recent expeditions in
Cyprus [48], Egypt [1] and Afghanistan [200] many of the
samples have yet to be evaluated. A list and brief description of genetic variation, including induced mutations,
has however been published [34], and the International
Board for Plant Genetic Resources (IBPGR; see Chapter 18)
has included faba beans among those crops which should
have centres responsible for maintaining their genetic
resources. Collaboration for this purpose in the EEC is
through the Bari Germplasm Laboratory, and it is likely
that ICARDA will assume responsibility for faba bean germplasm world-wide [5].

The two main problems in maintaining faba bean
collections and germplasm are, firstly, the storage space

required for the large seeds and, secondly, intercrossing, leading to genetic impurity, that can occur between multiplication plots in the field [89]. However, seed viability can be maintained for many years provided seeds are harvested dry and stored at cool temperatures in dry air. Also, although it is difficult without insect-proof enclosures to eliminate intercrossing between plots completely, contamination decreases rapidly with distance between them and with size of plots.

Germplasm may be maintained as populations, inbred lines or trait-specific gene pools [200]. Inbreeding uncovers genetic variability but to maintain all the variation it is also necessary to preserve the populations from which it was derived. Gene pools with common geographic origins and/or morphological traits provide a means of maintaining variability without handling large numbers of lines, but they should not be subjected to selection pressure. The ICARDA germplasm bank has both a base collection, which preserves variability in the form of populations or gene pools, and a working collection, mainly in the form of inbred lines, which is used for distribution to breeders [89].

What is now required is international agreement on descriptors, both morphological and agronomic, for items in collections together with a common form of evaluation. There is also a need for dissemination of information about the germplasm available; it will then become clear where the major omissions are both in geographical origins and in evaluation, so that steps can be taken to fill them.

Status of improvement programmes

Except where faba beans are a major crop, most national programmes are small. In European countries there are, typically, one or two state-aided research stations with small programmes of research on faba beans, and some short-term projects at one or two universities. In some countries there is also a contribution from private companies where, for example, new cultivars may result from research.

In the EEC, research on faba beans, like their cultivation, has been motivated since 1973 by a policy of increased self-sufficiency in protein-rich grains. Some research stations and universities have contracts to the EEC, and regular meetings and exchanges between faba bean researchers are arranged (see below).

In several countries of the Middle East and north Africa, research on faba beans has a long history. In Egypt, for example, research was started by the Ministry of Agriculture in 1929. However, until the establishment of ICARDA in 1976, research on faba beans in most countries of the region tended to be on a small scale and fragmented.

In most other regions of the world research has started only recently, although increasingly it is being organised in a context of international collaboration. ICARDA has assumed an important role in this process.

BREEDING STRATEGIES AND OBJECTIVES

The breeding strategy is determined by the natural breeding system and the range of genetic variation available, both of which have an influence on the methods which can be usefully employed.

Natural breeding system

Bees visiting the front of faba bean flowers assist self-pollination and cause about 30% natural cross-fertilisation on average in the UK [70,108]. Some plants of certain genotypes and F_1 hybrids, however, are autofertile [50] (see earlier and Plate 6.1).

Most cultivars and landraces are populations with some degree of genetic heterogeneity; thus natural crossing within these populations provides an intermediate degree of heterozygosity. This heterozygosity is thought

242

to provide a buffer against unpredictable numbers of pollinating insects as well as being the basis of an intermediate amount of general heterosis which may be expressed both in vigour of growth and yield of seed [17]. Thus, breeders have to aim at a balance between, on the one hand, uniformity to give suitability for a particular purpose or location (and in some countries to meet statutory requirements for uniformity), and on the other, heterogeneity to provide heterosis and buffering against the unpredictable variations in the environment (particularly the prevalence of bees) which may occur over successive generations.

Sources of genetic variation

The basic chromosome number is $n = 6$. There are no sterility barriers between subspecies or between botanical varieties but crosses with other species have not yet succeeded. As a result of the natural breeding system, considerable variation exists within populations and this may be exposed by controlled inbreeding.

A considerable range of variation is now available to breeders due to plant collections [200] and induced mutations [1,129,176] which have been assembled as germ-plasms (see earlier). International collaboration in the analysis of this variation and the use of modern cyto-genetical techniques is expected to result in progressively improved chromosome maps. Translocations and gene inter-changes present opportunities for genetic manipulation and provide a further extension of the variation available to the breeder. Fertile tetraploids and aneuploids have already been induced [155] but these are likely to be exploited in chromosome mapping rather than as the basis of cultivars.

Release of variability

As with most crops, gene recombination is brought about by controlled hybridisation. A feature of faba beans, however, is that recombination can take place by natural crossing between or within cultivars which are themselves populations and, as a result, improvement can be brought about by selection within cultivars or following natural crosses between them.

Breeding methods

Breeding may be carried out under either open or controlled pollination. Mass selection and bulk-pedigree selection are widely used where facilities for controlled pollination are not available. These methods are probably best suited to situations where the degree of natural crossing is relatively infrequent, and where the characters undergoing selection have a narrow range or can be recognised easily. Hand emasculation and pollination of faba beans is shown in Plate 6.3.

Mass selection.   Plants with desired characters are chosen from populations and their bulked seed is used to form the next generation. Repeated mass selection can bring about useful changes in populations in simply-inherited and contrasting characters, such as seed size and flowering date, but improvements in yielding ability have rarely been effected this way. Better progress is made if the progenies of the selected plants are tested before a decision is made as to which of them should enter the new population.

Bulk-pedigree selection.   Variation can be introduced by controlled introgression of parental populations, then, after a number of generations of selection under open pollination, selected families are bulked and multiplied

244

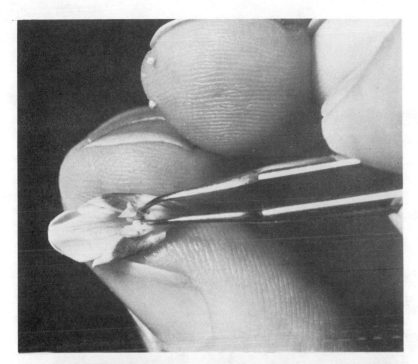

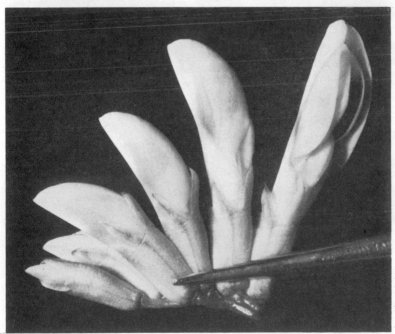

Plate 6.3 Hand emasculation and pollination of faba beans:
(a) Upper buds are at a suitable stage for emasculation (b) Anthers being removed during emasculation

Plate 6.3 (contd) (c) Flower being pollinated with forceps

Plate 6.3 (contd)

(d) Close-up of pollination (note pollen on right print of forceps)

(e) Flower being hand-tripped (note pollen on stigma and petals)

(Copyright: Plant Breeding Institute; (a) No. 10786; (b) No. 10785; (c) No. 25174; (d) No. 25175; (e) No. 25176)

in isolation. Further selection during multiplication and maintenance is usually necessary to remove types which do not conform to the limits of the desired population.

Recurrent selection.　Selfed progenies of selected plants are intercrossed to provide recombinations from which further cycles of selection, selfing and intercrossing can be continued. The method is expected to increase the frequency of favourable genes, to make use of epistasis and additive gene effects, and to break linkages [71]. However, selection pressure has to be intense and, if hybridisations are carried out manually, considerable effort is required.

Synthetic or composite varieties.　The components may be inbreds or populations but they are not necessarily tested for combining ability as in the original definition of a 'synthetic' [5]. Nevertheless, in faba bean breeding the best lines usually make the best synthetics. The system is long-term in that inbreds cannot be adequately evaluated until I_4 and synthetics do not reach a plateau of yield until syn-3. For maximum performance a large degree of natural intercrossing is necessary in the generation before commercial use [201]. A number of cultivars presently grown in Europe are synthetics [25].

Partially inbred populations.　A compromise between the homogeneity but depressed yields of most inbred lines and the variability but vigour of selected populations is to mass select in F_2 (following controlled hybridisation), self in F_3 and F_4 and multiply the bulks of selected F_4 families in isolation [88].

Inbred lines.　Most inbred lines suffer loss of fitness but there are reports of some that are large yielding [27,42,153]. It is possible that if obligate autogamy can be developed, the scale of testing of inbreds could be increased and this would improve the chances of selecting large-yielding lines.

F1 hybrids.   Much effort has been put into trying to exploit heterosis by the use of male sterility but, as yet, despite the isolation of effective restorer genes and a range of maintainer genotypes [12,16], the rate of reversion to fertility over generations has been too great to make F1 hybrid varieties practicable commercially.

Scale of breeding

In most countries the resources invested in plant breeding of a particular crop are approximately proportional to its agricultural importance, so that the scale of breeding in crops of minor importance, such as faba beans, is small and progress is relatively slow. In faba beans there is an additional need for resources, out of proportion to its cultivated area, because of the constraints of the natural breeding system and the crop's relative lack of domestication. A typical national faba bean programme may involve the annual testing of only a few hundred lines compared with the thousands or tens of thousands that are common in cereal breeding programmes. A limiting factor in faba bean improvement is the ability to produce lines rather than to test them.

As agronomy and disease control are not yet well developed in many regions, productivity *per se* is often difficult to measure because of the presence of confounding factors. Thus, in many countries, new varieties are being produced but advances in yield have been small and as yet without spectacular changes in morphology or disease resistance. Failing major changes in national policies, the way to broaden the bases of breeding programmes is likely to be by international cooperation making genetic variation available to all collaborators (see earlier). At ICARDA, the scale of the breeding programme is

proportionately larger than could be undertaken by a single
nation because of investment by a number of donor agencies.
Possibly the problem may be resolved by an international
organisation producing well-defined gene-pools which can
undergo further selection nationally or locally.

Breeding objectives

The faba bean is so widely distributed geographically and
used for such diverse purposes that few objectives are
common to all regions. However, more breeding effort is
aimed at reducing yield losses and making the crop more
reliable than in attempts to improve seed quality.

Yield and yield stability.   The main aim of most trials
of advanced selections is to measure yield but some
breeders also measure stability of yield over a range of
environments. There is evidence of variation in yield
stability among genotypes [44] but those with large mean
yields are more often responsive to large-yielding environments than tolerant of poorer ones.

Tolerance of factors limiting yield.   The principal
factors have been briefly described earlier, and there is
some breeding for tolerance or resistance to most of them,
though in many cases the degrees of resistance so far
obtained are relatively small.

Resistance to diseases and pests. Pathogens include:
Botrytis and *Ascochyta*, and some resistance is now available
to both diseases [87,101,102]; for *Uromyces*, resistance
genes have been isolated [13]; for *Fusarium*, some resistance
has been reported [105]; resistance to some viruses is
common in existing cultivars [38];for *Aphis fabae*, partial
resistance is known [29]; for stem nematode, resistance
has been reported from a Moroccan cultivar [169]; and for
Orobanche, active selection programmes have produced some

Faba bean (*Vicia faba* L.)

resistance [40,132] or tolerance [1].

Winter hardiness.   Selection for hardiness is common in
winter bean breeding because crosses of hardy x non-hardy
parents are necessary to transfer characters from spring
types in which there is wider general genetic variation.

Tolerance of drought and excess water.   Tolerance of
drought is generally rather poor, but some genotypes
yield more, relative to local control cultivars, in dry
than in wet years [151], and selection for reduced
stomata density has been suggested [134]. An approach to
prevention of excess growth in wet conditions is through
changes in plant architecture (see below).

Maturity date.   Earlier maturity is a common breeding
objective in slow-ripening climates. There is considerable
genetic variation in maturity date though often an association between early maturity and small yields.

Tolerance of salinity.   In regions where saline soils
are common, e.g. in parts of Egypt, Sudan and Iraq, the
development of cultivars which can tolerate saline
conditions is receiving attention.

Plant architecture.   Most breeders have an 'ideotype'
in mind which they seek to achieve. This may vary with
programmes but in general a plant is sought with a strong
stem, few branches, the first fruit at not less than 20 cm
from the ground, a concentrated region of fruit on the
stem, and little vegetative growth after the last flower.
Improved uniformity of ripening results from such a model
though in arid regions resistance to shattering may also
be required. Genetic variation exists for most of the
attributes of such an 'ideotype' [143], including various
mutants with a terminal inflorescence, synchronous
flowering within racemes, independent vascular supply to

251

each floret [72], and increased degrees of autofertility (see earlier and below).

Autofertility.   Breeding for the improvement of auto-fertility (Plate 6.1) is, in any case, proceeding in regions where independence of insect pollinators is required [108] and where an association with cleistogamy is sought so as to allow a fundamental shift in breeding methods. The problem of abscission of fertilised flowers may be reduced by breeding for independent vascular supply to florets and there is a hypothesis that change in plant stature may decrease internal competition for assimilates and thereby reduce flower abscission.

Symbiosis with *Rhizobium*.   This could be improved since advantageous interactions between host genotype x strain of *Rhizobium* have been demonstrated [58].

Seed quality.   This is being improved in terms of concentration of protein [148]. The antinutritive factors, lectins and trypsin inhibitors are known to respond to selection [179], while cultivars differ slightly in concentrations of phytate [81] and favism-causing glucosides [15,206]. Absence of tannin in the testas is associated with white flowers and this is controlled by a single recessive gene [147].

Breeding strategies have then the overall aim of further domesticating the faba bean. Cultivars of the future may be entirely self-fertilised, semi-dwarf, uniform and determinate, and their seeds free of tannin but with a choice of large or small protein concentrations.

AVENUES OF COMMUNICATION

During the 1970s there has been a considerable reawakening of interest internationally in the potential of faba beans.

Research and breeding programmes are now carried out in several countries and, where these are being conducted by a government-sponsored organisation and at more than one centre, there is usually some broad planning and communication of results nationally. Until the 1970s, international collaboration was mainly on an *ad hoc* basis, but more recently there has been an increasing number of scientific meetings devoted to grain legumes in general or to faba beans in particular, e.g. the seminar on 'Nutritional Quality and Breeding of Faba Beans' at Göttingen University, West Germany in 1974 [159], and a symposium on 'Production, Processing and Utilisation' at the Scottish Horticultural Research Institute in 1977 [188].

Two events in the mid-1970s, in particular, greatly increased the opportunities for communication between faba bean researchers. One was the recognition within the EEC that faba beans are one of the home-grown, protein-rich grain crops which could substitute for imported soyabean meal in animal feeds; this was accompanied by both the consequent subsidising of faba bean production and the encouragement of related research. Since that time some faba bean research projects have been under contract to the EEC, and the Protein Advisory Committee has arranged annual meetings, either as seminars (published) [24,30,148, 189] or workshops (unpublished) to review progress made under these contracts as well as other research on grain legumes which has often included work outside the EEC. Faba beans have featured prominently among the grain legumes discussed at these meetings, and the programme has included collaborative trials and exchanges of germplasms as well as visits of research workers.

The second event was the establishment, in 1976, of the International Centre for Agricultural Research in the Dry Areas (ICARDA) by the Consultative Group on International Agricultural Research (CGIAR). ICARDA (see Chapter 18) is based near Aleppo in northern Syria and primarily serves

the region extending from Morocco to Pakistan and from
Turkey to Sudan. Among other responsibilities, the Centre
works internationally on the improvement of faba beans. In
1979, the Centre started an information service known as
'FABIS' which aims to assist communications between faba
bean scientists throughout the world. It produces a regular
(twice-yearly) newsletter and by arrangement with the Commonwealth Agricultural Bureau (CAB) produces a quarterly
abstract journal *Faba Bean Abstracts*. In addition to these
regular publications, FABIS produced a *Directory of World
Faba Bean Research* and a list of *Genetic Variation in* Vicia
Faba in 1981. Further information on this service may be
obtained from the Coordinator, FABIS, ICARDA, PO Box 5466,
Aleppo, Syria.

ICARDA also hosted an international conference on faba
beans in Cairo in 1981, in collaboration with the Egyptian
Ministry of Agriculture [90].

Several books have included chapters on *Vicia faba* but
the most recent one, *Faba Beans: A Basis for Improvement*
[92], is dedicated entirely to faba beans and provides a
basic reference text.

PROSPECTS FOR LARGER AND MORE STABLE YIELDS

Mean yields of faba beans in northern Europe are about
twice as large as those in southern Europe, north Africa
and South America (see Table 6.1), and although larger
yields are obtained where the crop is regularly irrigated
(e.g. in Egypt) these are still short of the 3 t ha^{-1} yields
averaged by France, the UK and West Germany. The north-south
difference is partly due to soil type and length of growing
season but, in view of the yields obtained in industrialised
countries with short seasons (e.g. Canada and Australia),
and the fact that improved soil fertility often leads to
large total dry-matter yields but not to large yields of
seed [190], it may be concluded that the difference is also
associated with the stage of development of general

agricultural technology. As improved agriculture becomes more widespread so average yields of faba beans in the world should increase.

Over the period 1950-80, however, yields of faba beans in northern Europe have not risen to the same extent as those of cereals. In particular, stability of yield is still unsatisfactory for these areas where land is expensive and farmers hesitate to take the risk of growing unreliable crops. Where faba beans compete successfully with cereals and with the alternative crops such as oilseed rape and peas (which, in the EEC, are also subsidised), it is often only because beans can be easily handled and input costs are small. Thus, although knowledge of the agronomic require- ments of the crop is improving, the expected yields are not always achieved and, in many farming systems, the costs of the treatments have to be assessed in relation to the crop's traditional role as an inexpensive enterprise.

For example, yield increases have been demonstrated when soil-borne pathogens such as *Fusarium* and nematodes have been controlled with aldicarb [118]. Also *Sitona* larvae (controllable with permethrin or phorate) are reported to do as much damage as *Aphis fabae* [11], which is more often the target of insecticides. Broad Bean Mosaic Virus can severely depress yields in certain environments [38] but can be greatly reduced by control of its weevil vectors; the problem is that its incidence is difficult to forecast. *Botrytis* infection can be reduced with benomyl or carben- dazin [114,118] or with a seed-dressing [9]; all of these give considerable savings of crop losses but only when there are early aggressive attacks. Similarly, partial control of *Ascochyta* can be achieved with chlorothalonil [113] and of *Orobanche* with glyphosate [168]. The range of herbicides available for use in faba bean crops is increasing and a number of products have now to be assessed. Irrigation, pro- vided the water is applied at the appropriate time and in the correct quantity, can give important yield increases. There is also the possibility that growth regulators could be used to exert control over excessive vegetative growth.

In practice, few of these measures are applied universally either because of unpredictability of the losses which
might be caused by the limiting factor or doubt about the
cost-effectiveness of the treatments. The experimental
results which most urgently need translating into practice,
therefore, are those which investigate the economics of
such treatments. An example of such experiments is that
carried out at Rothamsted Experimental Station in 1981
(Table 6.5) [119]. There, irrigation of spring beans, and
an economic regime of chemical control of pests and pathogens in winter beans, gave almost as large yields as the
full comprehensive range of treatments that growers rightly
hesitate to use [119]. With this kind of information from
research the prospects for increasing yield and yield stability by improved husbandry practices must be good.

Table 6.5 Effects of irrigation and control of pests and pathogens
on seed yield of faba beans (t ha^{-1}) (*from McEwan* et al.
[119]).

| Crop/treatment | Pest and pathogen control: | | |
	Standard	Economic	Full
Spring beans			
Unirrigated	3.9	4.5	4.6
Irrigated	4.7	5.0	5.0
SED ± 0.12 (±0.10 within level of irrigation)			
Winter beans			
Unirrigated	2.9	4.5	4.8
SED ± 0.20			

There is less uncertainty for the farmer concerning the
cost of introducing new cultivars. Two distinct contributions from plant breeding are likely, one for the near
future and the other longer term. At present breeders are
searching for, and in some cases finding (see earlier),
degrees of resistance or tolerance to the major pathogens

and to the other limiting factors. It can be expected that
yield losses will be reduced as this work progresses. Some
breeders, particularly in areas or environments to which
faba beans have been recently introduced, are also achieving a fairly rapid response to selection for yield *per se*.
This may be via the components of yield or changes in
maturity date which together make the crop better suited
to its new environment. In regions where there is a much
longer tradition of growing faba beans there are usually
populations which are already adapted to the local environment, including its fluctuations, and here improvements in
mean yield, which must include yield stability, are likely
to be more gradual. However, there is evidence that, even
among well adapted populations, selection of large seeds
for sowing [28] and the selecion of large-seeded genotypes
per se can give significant increases in yield [44]. Some
variation in yield stability also exists among modern genotypes [25,44].

In summary, breeders in the longer term are likely to
make progress towards a fuller domestication of the species
and this should lead to a more efficient plant, larger and
more stable yields, and a more manageable and better quality
crop. These longer-term breeding objectives include complete self-fertilisation [108,152] (ideally without inbreeding depression [27]) and a determinate growth habit [176].
Other gross changes may be made in plant architecture,
which might include shorter racemes [190] with an independent vascular supply to each floret [72] and a much improved
harvest index. Furthermore, it is possible that flowers will
be white and testas free of tannin.

Genetic variation for most of these characters is now
known to exist [143], but it will require intensive and
collaborative breeding to bring the attributes together
effectively into one gene pool and finally into productive
cultivars. With the proposed radical change in plant architecture there will almost certainly have to be a period of
readjustment of the yield components, and a general

257

readaptation to environments, of what will be effectively
a new crop plant. In the event, the crop to emerge is expected to be shorter than most existing cultivars, to
require no insects for pollination, to be similar in uniformity to soyabeans or peas but to be erect and manageable
like the major cereals.

ACKNOWLEDGEMENTS

We thank Professor P.R. Day (Director), Dr A.J. Thomson
and Dr G. Lockwood of the Plant Breeding Institute, Cambridge, for helpful comments on the text, and Mrs Karen L.
Parr for typing the manuscript.

LITERATURE CITED

1. Abdalla, M.M.F. (1982), in Hawtin, G. and Webb, C. (eds), *Faba Bean Improvement*. The Hague: Martinus Nijhoff, 83-90.
2. Abdalla, M.M.F., Morad, M.M. and Roushdi, M. (1971), *Zeitschrift für Pflanzenzüchtung* 77, 72-79.
3. Ageeb, O.A.A. (1982), in Hawtin, G. and Webb, C. (eds) *Faba Bean Improvement*. The Hague: Martinus Nijhoff, 117-25.
4. Ahmed, M.B. (1958), *Indian Journal of Agricultural Science* 28, 43-56.
5. Allard, R.W. (1960), *Principles of Plant Breeding*. New York: John Wiley.
6. Allden, W.G. and Geytenbeek, P.E. (1980), *Proceedings of Australian Society for Animal Production* 13, 249.
7. Amarger, N. (1974), 'Competition pour la formation des nodosités sur la féverole entre souches de *Rhizobium leguminosarium* apportés par inoculation et souches indigènes'. *Comptes Rendus Hebdomadaires de Seances de l'Académie de Sciences, Série D,* 279-530.
8. Augustinussen, E. (1972), *Tidsskift für Planteavl.* 76, 6-12. (Abstract No. 5644, *Field Crop Abstracts* 26).
9. Bainbridge, A. (1982), *Rothamsted Experimental Station Report for 1981*, 33-4 and 198.
10. Bardawaj, K.K.R. (1975), *Plant and Soil* 43, 377-85.
11. Bardner, R., Fletcher, K.E., Griffiths, D.C. and Hamon, N. (1982), *FABIS* 4, 49-50.
12. Berthélem, P. and Le Guen, J. (1974), 'Rapport d'activité de Station d'Amélioration des Plantes. Le Rheu, France, 1971-74'.
13. Bernier, C.C. and Conner, R.L. (1982), in Hawtin, G. and Webb, C. (eds) *Faba Bean Improvement*. The Hague: Martinus Nijhoff, 251-7.
14. Bhatty, Y.R.S. (1974), *Canadian Journal of Plant Science* 54, 413-21.
15. Bjerg, B., Poulsen, M.H. and Sørensen, H. (1980), *FABIS* 2, 51-2.
16. Bond, D.A. (1957), PhD Thesis, University of Durham, UK.
17. Bond, D.A. (1966), *Journal of Agricultural Science, Cambridge* 67, 325-36.

18. Bond, D.A. (1970), *Proceedings of the Nutrition Society* 29, 74-9.
19. Bond, D.A. (1976), in Simmonds, N.W. (*ed.*) *Evolution of Crop Plants*. Harlow: Longman, 179-82.
20. Bond, D.A. (1976), *Journal of Agricultural Science, Cambridge* 86, 561-6.
21. Bond, D.A. (1977), in *Protein Quality from Leguminous Crops*. EEC Seminar, Dijon, EUR 5686 EN: 348-53.
22. Bond, D.A. (1979), *FABIS* 1, 15.
23. Bond, D.A. (1979), *FABIS* 1, 5-6.
24. Bond, D.A. (*ed.*) (1980), Vicia faba: *Feeding Value, Processing and Viruses*. World Crops, Vol. 3. The Hague: Martinus Nijhoff.
25. Bond, D.A. (1982), in Hawtin, G. and Webb, C. (*eds*) *Faba Bean Improvement*. The Hague: Martinus Nijhoff, 41-51.
26. Bond, D.A., Lawes, D.A. and Poulsen, M.H. (1980), in Fehr, W. and Hedley, H.H. (*eds*) *Hybridization of Crop Plants*. Madison: American Society of Agronomy, 203-13.
27. Bond, D.A., Lockwood, G., Toynbee-Clarke, G. and Pope, M. (1981), *Annual Report of Plant Breeding Institute, Cambridge, 1980*, 41.
28. Bond, D.A., Lockwood, G., Toynbee-Clarke, G. and Pope, M. (1983), *Annual Report of Plant Breeding Institute, Cambridge, 1982*, 51.
29. Bond, D.A. and Lowe, H.J.B. (1975), *Annals of Applied Biology* 81, 21-32.
30. Bond, D.A., Scarascia-Mugnozza, G.T. and Poulsen, M.H. (*eds*) (1979), *Some Current Research on* Vicia faba *in Western Europe*. EEC Seminar, Bari. EUR 6244 EN.
31. Boulter, D. (1982), *Proceedings of Nutrition Society* 41, 1-6.
32. Brockwell, J., Schwinghamer, E.A. and Gault, R.R. (1977), *Soil Biology and Biochemistry* 9, 19-24.
33. Campbell, L.D. and Marquadt, R.R. (1977), *Poultry Science* 56, 442-8.
34. Chapman, G.P. (1981), 'Genetic Variation in *Vicia faba*'. Aleppo, Syria: ICARDA.
35. Chapman, G.P. and Sadjadi, A.S. (1981), *Zeitschrift für Pflanzen-physiologie* 104, 265-73.
36. Charlton, P.M. (1980), in D.A. Bond (*ed.*) Vicia faba: *Feeding Value Processing and Viruses*. The Hague: Martinus Nijhoff, 273-80.
37. Clark, K.W. (1980), *FABIS* 2, 5-7.
38. Cockbain, A.J. (1980), in Bond, D.A. (*ed.*) Vicia faba: *Feeding Value Processing and Viruses*. The Hague: Martinus Nijhoff, 297-308.
39. Cubero, J.-I. (1974), *Theoretical and Applied Genetics* 45, 47-51.
40. Cubero, J.-I. and Martinez, A. (1980), *FABIS* 2, 19-20.
41. Cubero, J.-I. and Suso, M.-J. (1981), *Kulturpflanze* 24, 137-45.
42. Dantuma, G. (1980), Personal Communication.
43. Dantuma, G. and Klein Hulze, J.A. (1979), in Bond, D.A., Scarascia-Mugnozza, G.T. and Poulsen, M.H. (*eds*) *Some Current Research on* Vicia faba *in Western Europe*. EEC Seminar, Bari. EUR 6244 EN, 396-406.
44. Dantuma, G., von Kittlitz, E., Frauen, M. and Bond, D.A. (1983), *Zeitschrift für Pflanzenzüchtung* 90, 85-105.
45. Davidson, J. (1977), in *Protein Quality from Leguminous Crops*. EEC Seminar, Dijon, EUR 5686 EN, 243-50.
46. Davidson, J. (1977), *Proceedings of Nutrition Society* 36, 51.
47. Dekhujzen, H.M., Verkerke, D.R. and Houwers, A. (1981), in Thompson, R. (*ed.*) Vicia faba: *Physiology and Breeding*. The Hague: Martinus Nijhoff, 7-33.

259

48. Della, A. (1980), *Plant Genetic Resources Newsletter* 44, 17-19.
49. Dickinson, D., Knight, M. and Rees, D.I. (1957), *Chemistry and Industry* 16, 1503.
50. Drayner, J.M. (1959), *Journal of Agricultural Science, Cambridge* 53, 387-404.
51. Duc, G. and Picard, J. (1981), in Thompson, R. (*ed.*) Vicia faba: *Physiology and Breeding*. The Hague: Martinus Nijhoff, 283-98.
52. Dyke, G.V. (1982), *FABIS* 4, 15-16.
53. Eden, A. (1968), *Journal of Agricultural Science, Cambridge* 70, 299-301.
54. Edwards, D.G. and Duthie, I.F. (1973), *Journal of Science, Food and Agriculture* 24, 486-7.
55. Eggum, B.O. (1980), in Bond, D.A. (*ed.*) Vicia faba: *Feeding Value, Processing and Viruses*. The Hague: Martinus Nijhoff, 107-20.
56. Eggum, B.O. and Jacobsen, I. (1976), *Journal of Science, Food and Agriculture* 27, 1190.
57. El-Gibally, M.H. (1969), 'Effect of soil moisture regime, phosphorus and nitrogen application on the consumptive use of some crops in Assiut (UAR)'. *Transactions of the 7th Congress on Irrigation and Drainage*, Mexico City, 1969.
58. El-Sherbeeny, M.H., Lawes, D.A. and Mytton, L.R. (1977), *Euphytica* 26, 377-83.
59. El-Sherbeeny, M.H., Mytton, L.R. and Lawes, D.A. (1977), *Euphytica* 26, 149-56.
60. Elston, J. and Bunting, A.H. (1980), in Summerfield, R.J. and Bunting, A.H. (*eds*) *Advances in Legume Science*. Kew: Royal Botanical Gardens, 37-42.
61. Eweida, M.H.T. and Wetwally, M.A. (1975), *Agrokemia es Talajtan* 24, 46-52.
62. FAO (1981), *FAO Production Yearbook*. Rome: FAO.
63. Filippetti, A., De Pace, C. and Scarascia-Mugnozza, G.T. (1982), *FABIS* 4, 19-21.
64. Ford, J.E. and Hewitt, D. (1980), in Bond, D.A. (*ed.*) Vicia faba: *Feeding Value, Processing and Viruses*. The Hague: Martinus Nijhoff, 125-39.
65. Fordonski, G., Paprocki, S. and Rutkowski, M. (1980), *Zeszyty Naukowe Akademii Rolniczo-Technicznej W. Olsztynie, Rolnictwo (1980).* No. 30, 173-80. (Abstract No. 263, *Faba Bean Abstracts* 2 (3).)
66. Fowler, V.R. (1977), in R. Thompson (ed.) *Proceedings of Symposium on the Production, Processing and Utilisation of the Field Bean* (Vicia faba L.). Bulletin No. 15, Scottish Horticultural Research Association: 73-79.
67. Fowler, V.R. (1980), in Bond, D.A. (*ed.*) Vicia faba: *Feeding Value, Processing and Viruses*. The Hague: Martinus Nijhoff, 31-43.
68. Franco, A.A. and Vincent, J.M. (1976), *Plant and Soil* 45, 27-48.
69. Frauen, M. and Brimo, M. (1983), *Zeitschrift für Pflanzenzüchtung* 91, 261-3.
70. Fyfe, J.L. and Bailey, N.T.J. (1951), *Journal of Agricultural Science, Cambridge* 41, 371-8.
71. Gallais, A. (1977), *Annales de l'Amélioration des Plantes* 27, 281-329.
72. Gates, P., Yarwood, J.N., Harris, N., Smith, M.L. and Boulter, D. (1981), in Thompson, R. (*ed.*) Vicia faba: *Physiology and Breeding*. The Hague: Martinus Nijhoff, 299-316.

Faba bean (*Vicia faba* L.)

73. Gehriger, W. and Keller, E.R. (1979), *Revue Suisse Agriculture* 11, 215.
74. Gehriger, W., Bulluci, S. and Keller, E.R. (1979), in Bond, D.A., Scarascia-Mugnozza, G.T. and Poulsen, M.H. (*eds*) *Some Current Research on* Vicia faba *in Western Europe*. EEC Seminar, Bari. EUR 6244 EN, 421-35.
75. Gent, G.P. (1981), *FABIS* 3, 18-20.
76. Glassock, H.H., Ware, W.M. and Pizer, N.H. (1944), *Annals of Applied Botany* 31, 97-9.
77. Greenwood, H.N. (1959), *Journal of the Royal Agricultural Society of England* 119, 70-77.
78. Griffiths, D.W. (1979), *Report of the Welsh Plant Breeding Station for 1978*. 227-43.
79. Griffiths, D.W. and Jones, D.I.H. (1977), in *Protein Quality from Leguminous Crops*. EEC Seminar, Dijon. EUR 5686 EN, 105-14.
80. Griffiths, D.W. and Lawes, D.A. (1977), in *Protein Quality from Leguminous Crops*. EEC Seminar, Dijon. EUR 5686 EN, 361-8.
81. Griffiths, D.W. and Thomas, T.A. (1981), *Journal of Science, Food and Agriculture* 32, 187-92.
82. Haghparast, M.R. and Mengel, K. (1973), *Zeitschrift für Pflanzenernahrung u. Bodenkunde*. 135, 150-55.
83. Hamdi, Y.A. (1982), in Hawtin, G. and Webb, C. (*eds*) *Faba Bean Improvement*. The Hague: Martinus Nijhoff, 139-43.
84. Hamissa, M.R. (1974), in *Proceedings of the First FAO/SIDA Seminar on Improvement and Production of Field Crops for Plant Scientists from Africa and the Near East*. Rome: FAO, 410-16.
85. Hanelt, P. (1972), *Kulturpflanze* 20, 75-128.
86. Hanelt, P., Rudolph, A., Hammer, K., Jank, H.W., Muntz, K. and Scholz, F. (1978), *Kulturpflanze* 26, 183.
87. Hanounik, S.B. and Hawtin, G.C. (1982), in Hawtin, G. and Webb, C. (*eds*) *Faba Bean Improvement*. The Hague: Martinus Nijhoff, 243-50.
88. Hawtin, G.C. (1982), in Hawtin, G. and Webb, C. (*eds*) *Faba Bean Improvement*. The Hague: Martinus Nijhoff, 15-32.
89. Hawtin, G.C. and Omar, M. (1980), *FABIS* 2, 20-22.
90. Hawtin, G.C. and Webb, C. (*eds*) (1982), *Faba Bean Improvement*. Proceedings of the Faba Bean Conference, Cairo 1981. World Crops, Vol. 6. The Hague: Martinus Nijhoff.
91. Hawtin, L. (1980), *Faba Bean Cook Book*. Aleppo, Syria: ICARDA.
92. Hebblethwaite, P.D. (*ed.*) (1983), *Faba Beans: A Basis for Improvement*. London: Butterworth.
93. Hegazy, M.F.H. (1968), PhD Thesis, Faculty of Agriculture, Ain Shams University, Egypt.
94. Higgins, J., Evans, J.L. and Reed, P.J. (1981), *Journal of the National Institute of Agricultural Botany* 15, 480-87.
95. Hong, J.U. (1960), *Bayer Landwirtschaft Jahrbuch* 37, 729-34.
96. Huan-Ziam, J. (1982), *FABIS* 4, 9-10.
97. Ibrahim, A.A., Nassib, A.M. and El-Sherbeeny, M.H. (1979), in *Food Legume Improvement and Development*. Ottawa: IDRC, 39-46.
98. Ibrahim, N.E. and Kabesh, M.O. (1971), *United Arab Republic Journal of Soil Science* 11, 271-83.
99. Ingram, J. and Hebblethwaite, P.D. (1976), *Agricultural Progress* 51, 27-32.
100. Jamalian, J. (1978), *Journal of Science, Food and Agriculture* 29, 137-40.

101. Jellis, G.J., Bond, D.A. and Old, J. (1982), *FABIS* 4, 53-4.
102. Jellis, G.J., Lockwood, G. and Old, J. (1982), *FABIS* 4, 46.
103. Jonas, D.A. (1980), in Bond, D.A. (*ed.*) Vicia faba: *Feeding Value Processing and Viruses*. The Hague: Martinus Nijhoff, 217-31.
104. Jonas, D.A. (1981), *FABIS* 3, 11-12.
105. Kiselev, A. (1976), *Trudy Tul'Skoi S Kh Op Stantsii* 5, 182-6.
106. Ladizinski, G. (1975), *Israel Journal of Botany* 24, 80-88.
107. Lange, W. (1980), in Bond, D.A. (*ed.*) Vicia faba: *Feeding Value Processing and Viruses*. The Hague: Martinus Nijhoff, 205-213.
108. Lawes, D.A. (1974), *Span* 17, 21-3.
109. Lawes, D.A. (1980), in Summerfield, R.J. and Bunting, A.H. (*eds*) *Advances in Legume Science*. Kew: Royal Botanical Gardens, 625-36.
110. Lawes, D.A., Mytton, L.R., El-Sherbeeny, M.H. and Swrwli, F.K. (1978), *Annals of Applied Biology* 88, 466-8.
111. Lonsdale, C.R. and Taylor, J.C. (1969), *Journal of British Grassland Society* 24, 299-301.
112. Learmonth, E.M. (1958), *Journal of Science, Food and Agriculture* 9, 269-73.
113. Liew, R.S.S. and Gaunt, R.E. (1980), *New Zealand Journal of Experimental Agriculture* 8, 67-80.
114. MAFF (1981), ADAS Booklet No. 2373, Alnwick, UK: MAFF, 21-2.
115. Marquadt, R.R., McKirdy, J.A., Ward, T. and Campbell, L.D. (1975), *Canadian Journal of Animal Science* 55, 421-9.
116. Martin-Tanguy, J., Guillaume, J. and Kossa, A. (1977), *Journal of Science, Food and Agriculture* 28, 757-65.
117. McEwen, J. (1970), *Journal of Agricultural Science, Cambridge* 74, 61-6.
118. McEwen, J. *et al.* (1981), *Journal of Agricultural Science, Cambridge* 96, 129-50.
119. McEwen, J. *et al.* (1982), *Rothamsted Experimental Station Report for 1981*, 32-3.
120. McKnight, D.R. (1974), *Canadian Journal of Animal Science* 54, 474.
121. McNab, J.M. and Wilson, B.J. (1977), in Thompson, R. (*ed.*) *Proceedings of the Symposium on the Production, Processing and Utilisation of the Field Bean* (Vicia faba L.). Bulletin No. 15, Scottish Horticultural Research Institute Association, 63-72.
122. Mesa-Garcia, J., Garcia-Torres, L. and Moreno, M.T. (1983), *FABIS* 4, 26-7.
123. Miller, E.L. (1980), in Bond, D.A. (*ed.*) Vicia faba: *Feeding Value, Processing and Viruses*. The Hague: Martinus Nijhoff, 17-30.
124. Moreno, M.T. and Cubero, J.-I. (1982), *FABIS* 4, 10-13.
125. Moreno, M.T. and Martinez, A. (1980), *FABIS* 2, 18-19.
126. Mossé, J. and Baudet, J. (1977), in *Protein Quality from Leguminous Crops*. EEC Seminar, Dijon. EUR 5686 EN, 48-56.
127. Muratova, V.S. (1931), *50th Bulletin of Applied Botany, Genetics and Plant Breeding, Leningrad*, Supplement, 1-298.
128. Musa, M.M. (1982), in Hawtin, G. and Webb, C. (*eds*) *Faba Bean Improvement*. The Hague: Martinus Nijhoff, 139-43.
129. Nagl, K. (1978), *International Atomic Energy Agency, Vienna, Report for 1978*, 243-52.
130. Nagl, K. (1979), in Bond, D.A., Scarascia-Mugnozza, G.T. and Poulsen, M.H. (*eds*) *Some Current Research on Vicia faba in Western Europe*. EEC Seminar, Bari. EUR 6244 EN, 355-69.
131. Nash, M.J. (1977), in Thompson, R. (*ed.*) *Proceedings of the Symposium of the Production, Processing and Utilisation of the Field*

Faba bean (*Vicia faba* L.)

Bean (Vicia faba *L.*). Bulletin No. 15, Scottish Horticultural
Research Institute Association, 43-6.

132. Nassib, A.M., Ibrahim, A.A. and Khalil, S.A. (1982), in Hawtin, G.
and Webb, C. (*eds*) *Faba Bean Improvement*. The Hague: Martinus
Nijhoff, 199-206.

133. Natelson, B.H. (1969), *The Lancet*, 20 September 1969, 640-41.

134. Nerkar, Y.S., Wilson, D. and Lawes, D.A. (1981), *Euphytica* 30,
335-45.

135. Nikolov, B.A. (1969), *Godish. Sofiisk, Univ. Biol. Fac. 1967-68*
62 (2) 225-32. (Abstract No. 3713, *Field Crop Abstracts* 25.)

136. Oknina, R.M. (1969), 'Uptake of mineral nutrients and D.M. accumulation by *Vicia faba*.' *Izv. timiryazev. Sel'.-Khoz. Adad. No. 4*,
93-110. (Abstract No. 1318, *Field Crop Abstracts* 22.)

137. Okomato, S. (1968), *Soil Science and Plant Nutrition* 14 175-82.

138. Olivares, J., Martin, E. and Recalde-Martinez, L. (1983), *Journal
of Agricultural Science, Cambridge* 100, 149-52.

139. Olsen, H.S. (1980), in Bond, D.A. (*ed.*) Vicia faba: *Feeding Value,
Processing and Viruses*. The Hague: Martinus Nijhoff, 233-55.

140. Palmer, R. and Thompson, R. (1975), *Journal of Science, Food and
Agriculture* 26, 1577-83.

141. Patriquin, D.G., Burton, D. and Hill, N. (1980), in Lyons, J.M.
et al. (*eds*) *Genetic Engineering for Nitrogen Fixation and Conservation of Fixed Nitrogen*. New York: Plenum Press, 651-71.

142. Paul, C. (1977), *Zeitschrift für Pflanzenzüchtung* 78, 97-112.

143. Peat, W.E. (1982), *Outlook on Agriculture* 2, 179-83.

144. Penman, H.L. (1962), *Journal of Agricultural Science* 58, 365-70.

145. Picard, J. (1953), *Annales de l'Amélioration des Plantes* 3, 57-106.

146. Picard, J. (1963), *Annales de l'Amélioration des Plantes* 13, 97-117.

147. Picard, J. (1976), *Annales de l'Amélioration des Plantes* 26, 101-06.

148. Picard, J. (1977), in *Protein Quality from Leguminous Crops*. EEC
Seminar, Dijon. EUR 5686 EN, 339-47.

149. Picard, J. and Berthélem, P. (1980), *FABIS* 2, 20.

150. Plancquaert, Ph. (1982), *FABIS* 4, 13-14.

151. Plant Breeding Institute (1982), *Annual Report for 1982*, Cambridge:
134.

152. Poulsen, M.H. (1977), *Journal of Agricultural Science, Cambridge*
88, 253-6.

153. Poulsen, M.H. (1979), in Bond, D.A., Scarascia-Mugnozza, G.T. and
Poulsen, M.H. (*eds*) *Some Current Research on* Vicia faba *in Western
Europe*. EEC Seminar, Bari. EUR 6244 EN, 342-54.

154. Poulsen, M.H. and Knudsen, J.C.N. (1980), *FABIS* 2, 26-8.

155. Poulsen, M.H. and Martin, A. (1977), *Hereditas* 87, 123-6.

156. Prummel, J. (1979), *Bedrijfsontwikkeling* 10, 77-80.

157. Richard, J.E. and Soper, R.J. (1982), *Canadian Journal of Soil
Science* 62, 21-30.

158. Rivoira, G. (1979), *FABIS* 1, 29-30.

159. Röbbelen, G. (*ed.*) (1974), 'Ernahrungsqualität und Züchtung von
Ackerbohnen (*Vicia faba minor*)'. *Göttingen Pflanzenzüchter-
Seminar* 2, Göttingen University, DFR.

160. Rogalsky, J.R. (1972), 'Summary of faba bean fertility trials'.
Papers presented at the 16th Annual Manitoba Soil Science Meeting,
University of Manitoba, Winnipeg, 169-170.

161. Rowland, G.G. and Bond, D.A. (1983), *Journal of Agricultural
Science, Cambridge* 100, 35-41.

162. Rowlands, D.G. and Corner, J.J. (1962), in *Proceedings of 3rd*

Congress of Eucarpia. Paris: INRA, 229-34.

163. Sadler, J.M. (1975), 'Soil, fertilizer and symbiotically-fixed nitrogen as sources of nitrogen for faba beans under prairie conditions.' Papers presented at the 19th Annual Manitoba Soil Science Meeting, University of Manitoba, Winnipeg, 121-8.

164. Salkini, A.B., Halimeh, H., Mazid, A. and Nordblom, T. (1982), *FABIS* 4, 5-8.

165. Sameni, A.M. and Bassiri, A. (1982), *Iran Agricultural Research* 1, 49-60.

166. Saxena, M.C. (1982), *FABIS* 4, 24-25.

167. Saxena, M.C. and Wassimi, N. (1980), *FABIS* 2, 31-3.

168. Schmitt, U. (1982), in *Proceedings of the International Faba Bean Conference, Cairo, Egypt*. Aleppo, Syria: ICARDA, 79-82.

169. Schreiber, E.R. (1978), *Bulletin de Protection des Cultures* 4, 1-30.

170. Schultze-Motel, J. (1972), *Kulturpflanze* 19, 321-8.

171. Selim, A.K.A., Shahin, S.A. and Mamoud, S.A. (1974), *Egyptian Journal of Genetics and Cytology* 3, 229-35.

172. Shatilov, I.S. and Okinina, P.M. (1968), *Izv. timiryazev'. Sel'. Khoz. Akad. 1968* 6, 55-70. (Abstract No. 1612, *Herbage Abstracts*, 39.)

173. Simpson, A.D.F. (1977), in Thompson, R. (*ed.*) *Proceedings of Symposium on the Production, Processing and Utilisation of the Field Bean* (Vicia faba *L.*). Bulletin No. 15, Scottish Horticultural Research Institute Association, 47-56.

174. Simpson, A.D.F. (1980), in Bond, D.A. (*ed.*) Vicia faba: *Feeding Value, Processing and Viruses*. The Hague: Martinus Nijhoff, 257-72.

175. Sirks, M.J. (1931), *Genetica* 13, 209-631.

176. Sjodin, J. (1971), *Hereditas* 67, 155-80.

177. Sjodin, J. (1974), in *Proceedings of First FAO/SIDA seminar on Improvement and Production of Field Bean Crops*. Rome: FAO, 295-301.

178. Sjodin, J. and Martensson, P. (1979), in Bond, D.A., Scarascia-Mugnozza, G.J. and Poulsen, M.H. (*eds*) *Some Current Research on* Vicia faba *in Western Europe*. EEC Seminar, Bari. EUR 6244 EN, 313-23.

179. Sjodin, J., Martensson, P. and Magyarosi, T. (1981), *Zeitschrift für Pflanzenzüchtung* 86, 231-47.

180. Slinkard, A.E. and Buchan, J. (1980) 'The potential for special crop production in Western Canada.' Canadian Wheat Board, Prairie Production Symposium, Saskatoon.

181. Sprent, J.I. (1976), in Kozlowski, T.T. (*ed.*) *Water Deficits and Plant Growth*. Volume 4. New York: Academic Press, 291-315.

182. Sprent, J.I. and Bradford, A.M. (1977), *Journal of Agricultural Science, Cambridge* 88, 303-10.

183. Steuckardt, R., Dietrich, M. and Griem, H. (1982), *Archiv für Züchtungsforschung, Berlin* 12, 33-42.

184. Summerfield, R.J. and Roberts, E.H. (1985), in Halevy, A.H. (*ed.*) *A Handbook of Flowering*. Florida: CRC Press (in press).

185. Tamaki, K. and Naka, J. (1971), *Technical Bulletin of the Faculty of Agriculture, Kagawa University* 22, 73-82.

186. Tao, Z.X. (1981), *FABIS* 3, 24-5.

187. Tas, M.V. (1983), *Arable Farming*, February 1983, 71-2.

188. Thompson, R. (*ed.*) (1977), *Proceedings of Symposium on the Production, Processing and Utilisation of the Field Bean* (Vicia faba *L.*). Bulletin No. 15, Scottish Horticultural Research Institute Association.

189. Thompson, R. (*ed.*) (1981), Vicia faba: *Physiology and Breeding*. World Crops Vol. 4. The Hague: Martinus Nijhoff.
190. Thompson, R. and Taylor, H. (1982), *Outlook on Agriculture* 2, 127-33.
191. Thorlacius, S.O. and Beacom, S.E. (1981), *Canadian Journal of Animal Science* 61, 663-8.
192. Toynbee-Clarke, G. (1970), *Journal British Grassland Society* 25, 228-32.
193. Toynbee-Clarke, G. (1973), *Journal British Grassland Society* 28, 69.
194. Toynbee-Clarke, G. (1978), *Report of the Plant Breeding Institute, Cambridge for 1977*, 72-3.
195. Warsy, A.S., Norton, G. and Stein, M. (1974), *Phytochemistry* 13, 2481-6.
196. Westphal, E. (1974), *Pulses in Ethiopia, Their Taxonomy and Agricultural Significance*. Agricultural Research Report No. 815, Department of Tropical Crops, Wageningen.
197. Williams, R.J.B. and Cooke, G.W. (1970), 'Results of the Rotation I experiment at Saxmundham, 1964-69'. *Report of Rothamsted Experimental Station (1970)*, Part 2, 68-97.
198. Wilson, A.R. (1937), *Annals of Applied Biology* 24, 258-88.
199. Wilson, B.J., McNab, J.M. and Bentley, H. (1972), *Journal of Science, Food and Agriculture* 23, 679-84.
200. Witcombe, J.R. (1982), in Hawtin, G. and Webb, C. (*eds*), *Faba Bean Improvement*. The Hague: Martinus Nijhoff, 1-13.
201. Wright, A.J. (1977), *Journal of Agricultural Science, Cambridge* 89, 495-501.
202. Abdel-Wahab, A.M. (1974), *Egyptian Journal of Botany* 17, 191-3.
203. Pang, P.C. and Paul, E.A. (1980), *Canadian Journal of Soil Science* 60, 241-50.
204. Kucey, R.M.N. and Paul, E.A. (1983), *Canadian Journal of Soil Science* 63, 87-95.
205. Stoddard, F.L. and Lockwood, G. (1984), in Hebblethwaite, P.D. *et al.* (*eds*) Vicia faba: *Agronomy, Physiology and Breeding*. The Hague: Martinus Nijhoff, 247-54.
206. Bjerg, B., Eggum, B.O., Olsen, O. and Sørensen, H. (1984), in Hebblethwaite, P.D. *et al.* (*eds*) Vicia faba: *Agronomy, Physiology and Breeding*. The Hague: Martinus Nijhoff, 287-96.
207. Olaboro, G., Marquardt, R.R. and Campbell, L.D. (1981), *Journal of Science, Food and Agriculture* 32, 1074-80.

7 Lentil (*Lens culinaris* Medic.)

F.J. Muehlbauer, J.I. Cubero and R.J. Summerfield

INTRODUCTION

Lentils (*Lens culinaris* Medic.) are an Old World legume and
were probably one of the first plant species to be domesticated; they are coeval with wheat and barley. The earliest
forms of the legume so far documented (dated 9500-10000 BP)
were found at Mureybit in northern Syria, and authenticated
remains dated 8000-9000 BP are common in early Neolithic
sites in the same region. Undoubtedly, the crop had been
widely disseminated throughout the Mediterranean region,
Asia, and Europe by the Bronze Age [8,28,41,58,95,100,101].

Lentils are known by many trivial names in different
languages, e.g. adas (Arabic), masur (Hindi), mercimek
(Turkish) and heramame (Japanese) [38]. They are now cultivated mainly in central and south-west Asia, (particularly
on the Indian subcontinent) southern Europe, north Africa,
Ethiopia, and North and South America.

PRODUCTIVITY AND PRODUCTION

Recent world production (1978-80) was estimated at 1.1 million tonnes per annum which represents an increase of about
13% from that in 1969-71. However, average yields have
remained virtually static during this period (Table 7.1)
and so production has improved because of increased area
sown [15,54]. Whereas lentils are of minor importance in

266

overall world agriculture, they are a major pulse in certain areas, particularly in the Indian subcontinent, where the greatest proportion of the crop is grown (Table 7.1), and in the semi-arid parts of the Middle East, Turkey, and northern Africa.

Yields of lentils vary widely between countries and years. Unfortunately, they tend to be smallest in those countries where the crop is most important and where a large proportion of production is consumed domestically. In India, for example, where more than 40% of the world's lentils are produced, average yields are less than 500 kg ha^{-1}, partly because of untimely planting dates [70]. Production of the crop on marginal lands by subsistence farmers who seldom fertilise or irrigate them, or remove weeds, is primarily responsible for poor yields. Then again, improved cultivars are seldom available for these areas and so farmers have relied on indigenous landraces that sometimes more closely resemble wild forms than productive genotypes.

Shifts in production and importance of lentils have taken place during the past twenty years. During that time, southern Europe (Italy, France and Greece) has gradually but steadily reduced production so that current output is quite small. In Italy, for example, the area sown to lentils declined from about 340 000 ha in 1927 to only 2000 ha in 1977 [74] - a decline due partly to heavy infestations of the parasitic weed *Orobanche*. Reduced production in southern Europe also reflects increased emphasis on cereals and other more profitable crops in that region.

Declining production in Egypt, traditionally an important producer and consumer of the crop, can be linked to the completion of the Aswan High Dam in 1965 that made possible the irrigation of nearly the entire Nile Valley. This expanded irrigation capability promoted the production of profitable crops such as sugar cane, wheat and berseem, but at the expense of lentils. However, lentil production could expand again if improved cultivars and management practices became available, or if production were extended to recently

267

Table 7.1 Production (1000 t), area (1000 ha) and average seed yield (kg ha^{-1}) of lentils in the principal producing countries [15,54]*

Country	Average annual production			Area	Seed yield	Relative change (%) from 1979/71 to 1978/80	
	1961/65	1969/71	1978/80	1980	1980	Area	Yield
India	348	348	428	925	486	+ 19.6	− 7.1
Turkey	94	100	199	201	1169	+ 75.5	+13.5
Syria	64	73	72	84	952	− 15.3	+17.5
USA	25	37	59	53	1148	+ 79.3	−12.2
Bangladesh	46	55	51	85	600	+ 12.2	−17.3
Argentina	11	11	29	40	875	+ 95.0	+35.2
Ethiopia	95	46	29	58	470	− 42.0	+ 7.8
Pakistan	35	22	36	86	422	+ 44.6	+16.2
Iran	39	26	27	38	737	−*	+ 3.7
Morocco	12	17	17	37	432	+ 4.0	−13.9
Chile	13	10	26	53	507	+181.2	−10.2
Egypt	50	36	11	6	1111	− 54.5	−34.9
USSR	90	57	10	11	1818	− 74.4	−20.4
Spain	−	37	54	63	905	+ 37.3	+ 9.7
Canada	−	−	15	44	450	−	−
World	985	995	1123	1842	655	+ 13.3	− 0.3

*Dashes denote 'no information' or 'cannot be calculated from information available'.

reclaimed lands; but competition for irrigated land for other more profitable crops may prove an insurmountable barrier.

Various reasons have contributed to dramatic increases in production during the past decade in Turkey, North America (USA and Canada) and in South America (Argentina and Chile). The increased output from Turkey, now the largest exporter of the crop, was influenced by studies that demonstrated the feasibility of replacing summer fallowing in cereal production systems with annual grain legume crops such as lentils or chickpeas (*Cicer arietinum*) [65]. Large areas of Turkey receive sufficient rainfall to support a lentil crop in alternate years with wheat, and the crop is now being used more extensively in such rotations. Estimates of the area sown to lentils in Turkey in 1983 are in the region of 500,000 ha.

Increased production in North America has been prompted by strong demands in international markets and the recent development of herbicides for the control of weeds that otherwise can reduce yields dramatically [85]. In contrast, the prevalence of lentil rust (*Uromyces fabae*) and other diseases greatly reduced production in South America during the 1950s [56]. Nevertheless, recent increases in production from that region are due to strong world market demands and the development of cultivars moderately tolerant to rust [89]. There are, however, pressing needs in South America for additional cultivars resistant to rust [56,59,89] and for the development of better agronomic practices (appropriate herbicides, seed bed preparation and seeding methods are particularly important).

In those countries where production has remained relatively stagnant (i.e. in India, Bangladesh, Jordan, Syria and Iran), lentils are a dietary mainstay and there are few alternative crops which can be grown on the marginal lands which lentils now occupy. Better availability of labour for handling the crop, or development of suitable mechanisation schemes, may be key factors to sustain production in these regions [40]. In countries with modern, mechanised agricultural systems (e.g. the USA, Canada and in Argentina), production is likely at least to remain more or less constant; it may even increase in the future because farmers wish to include legumes in rotations with their cereals. They now recognise that benefits from such rotations can include better control of grassy weeds, suppression of some cereal diseases, gains from symbiotic nitrogen fixation and, in the case of sloping terrain in the Palouse region of the USA, better control of soil erosion because the legume replaces summer fallowing, a practice conducive to severe erosion [29].

Researchers at the International Centre for Agricultural Research in the Dry Areas (ICARDA) in Syria [11] and in several national programmes (e.g. the All India Coordinated Pulse Improvement Project and the USA Department of Agriculture) are attempting to reverse, or at least to arrest, declining production trends in traditional lentil producing regions. These research programmes have placed emphasis on breeding for adaptation to mechanised harvest, machinery development, reducing hazards from pests, diseases, and weeds, improved symbiotic nitrogen fixation, and improved product quality [28].

World lentil production will probably expand in the future, but perhaps especially in regions with marginal lands unsuited to other crops and so average yields are likely to remain relatively small. In these marginal areas, the urgent need is for well-adapted, short-duration cultivars capable of exploiting limited moisture resources and restricted growing seasons; and with quality standards (seed and straw) to ensure local acceptability.

Economic yields and principal uses of crop products

Economic yields in the major lentil-producing countries range from about 500 kg ha^{-1} in India to about 1200 kg ha^{-1} in Turkey (Table 7.1). Yields in Syria in 1980 averaged

952 kg ha^{-1}, but in previous years they had ranged from 500 to 679 kg ha^{-1}. Yields in the USA have averaged about 1100 kg ha^{-1} for many years and in Egypt, where the crop is entirely irrigated and a single cultivar 'Giza 9' dominates production, recent yields have also been about 1100 kg ha^{-1}

The primary product of lentil crops is the mature seed which contains relatively large amounts of protein, calories, and carbohydrates (Table 7.2) but few antinutritional factors. An average protein concentration of 24.2% makes lentils an extremely desirable dietary constituent in many countries. The fast cooking characteristic also makes lentils especially attractive in regions where supplies of fuel are limited. In these areas, lentil seeds are eaten as 'dhal' (boiled), in soups or made into flour and mixed with cereal flour for baking; in some areas they are fried and seasoned and consumed as a snack food. The seeds are often used as a source of commercial starch for the textile and printing industries [54].

Residues from threshing (straw, fruit walls, leaves and other debris) have good livestock feeding values and contain about 10.2% moisture, 1.8% fat, 4.4% protein, 50.0% carbohydrate, 21.4% fibre and 12.2% ash [38]. Indeed, when production of traditional forage is reduced in dry seasons (e.g. in Syria, Jordan and elsewhere in the Middle East) lentil residues have commanded prices equal to or better than the seeds. Because of the great value placed on lentil fodder, research on mechanical harvesting in many programmes has focussed on the collection and salvaging of as much of the straw and leaves as possible [40,69]. To this end, cultivars that retain their leaves until maturity and have large total biological yields are desirable breeding objectives for forage-deficient regions.

Principal farming systems which include lentils

Lentils are used in subsistence, mixed cropping, and modern systems of crop production, but are only rarely grown in

Table 7.2 Estimated nutrient contents of lentils (g 100g seed^{-1}, i.e. %) compared with other important grain legumes (calculated from [16])

Species	Calories	Water	Protein	Oil	Fibre	Carbohydrate (by difference)	Ash
Chickpeas	358	11.0	20.1	4.5	4.9	56.6	2.9
Soyabeans	335	11.0	36.8	17.4	4.7	25.5	4.6
Peas	346	11.0	22.5	1.8	5.5	53.7	5.5
Faba beans	348	11.0	23.4	2.0	7.8	52.4	3.4
Cowpeas	342	11.0	23.4	1.8	4.3	56.0	3.5
Lentils	346	11.0	24.2	1.8	1.8	59.0	2.2

monoculture. Such a wide range of uses attests to their adaptability to variations in duration of growing seasons, temperature, and the availability of water. Indeed, the adaptability and relatively short crop durations of many cultivars is exploited in fitting them into cropping systems during cooler and often wetter seasons between other warm-season crops. In the Mediterranean region, for example, lentils are planted soon after the onset of autumnal rains and they ripen the following spring. Similar farming practices extend across the same latitudes in Iran, Pakistan and into India. In other areas of India, Pakistan and Bangladesh, lentils are planted after the harvest of rice and are grown as a winter ('rabi') crop on residual soil moisture. The crop is spring-sown in areas of the USA, Turkey, Canada, Chile and Argentina, usually in rotation with cereal crops, where it is grown on stored soil moisture supplemented by limited rainfall during the growing season. Winter cultivars, developed in Turkey to withstand cold injury on the Anatolia Plateau, are becoming increasingly important in that country as an autumn-sown crop capable of utilising winter rainfall.

Lentil (*Lens culinaris* Medic.)

PRINCIPAL TAXONOMIC FEATURES OF THE GENUS *LENS*

The genus *Lens* is classified taxonomically in the Order
Rosales, Suborder Rosineae, Family Leguminosae and Subfamily
Papilionaceae. Within the Papilionaceae, *Lens* occupies an
intermediate position between *Vicia* and *Lathyrus*, but it is
more closely related to *Vicia*. Formerly, the genus was
reported to contain the cultivated lentil, *L. culinaris*
(synonym *L. esculenta* Moench.) and four wild species: *L.
ervoides* (Brign.) Grande, *L. nigricans* (M. Bieb.) Godron,
L. orientalis (Boiss.) Handel-Mazzetti and *L. montbrettii*
Fisch. and Mey. [12,96]. However, this classification seems
rather dubious because it is based primarily on examinations
of herbarium specimens and not on analyses of crossability
and chromosome structure within the genus. Indeed, *L. montbrettii* has $2n = 12$ chromosomes whereas the other *Lens* species have $2n = 14$. The former therefore, should not be considered a *Lens* species, but would be more correctly classified within *Vicia* [44]. Synonyms for the wild species have
appeared in the literature, but they are too numerous to
mention individually here (and see Chapter 1).

Within *Lens*, the species designated as *L. culinaris*, *L.
orientalis* and *L. nigricans* are somewhat similar, and intermediate forms between them are not infrequent both in living
material and in herbarium specimens. More recently, authors
have considered them as conspecific (*nigricans, orientalis*
and *culinaris* being subspecies of *L. culinaris*), based on
the fact that they can form fertile or, at least, partially
fertile interspecific hybrids, and on considerations of comparative morphology and distribution [8,97]. All three subspecies have aristate or awned peduncles and calyx teeth
generally greater or subequal in length to the corolla, the
plants being generally glabrous. *L. culinaris* subspecies
nigricans has both toothed stipules and aristate peduncles,
with a short rachis (≤ 20 mm) bearing less than ten leaflets. *L. culinaris* subspecies *orientalis* has entire/simple
stipules, non-aristate peduncles, usually with an awn, and

273

a short rachis (≤ 20 mm) with six to twelve leaflets.
Finally, subspecies *culinaris*, to which all the cultivated
forms belong (although, occasionally, both subspecies *nigricans* and *orientalis* have been recorded as being cultivated, either alone or in mixtures), has entire stipules
(or, if toothed, with a long rachis, ≥ 40 mm), aristate
peduncles, shorter or subequal to the rachis (the two wild
subspecies have peduncles longer or subequal to rachis), a
long rachis and peduncles (40-50 mm) with six to sixteen
leaflets. Figure 7.1 shows the main morphological distinctions in the *L. culinaris* group and Fig. 7.2 the geographical distribution of wild lentils. *Lens ervoides* lies somewhat distant from the other species and is differentiated
by its non-aristate peduncles (rarely with an awn), calyx
teeth generally much shorter than the corolla, very short
rachis (≤ 10 mm) with four to eight leaflets, and overall
pubescence.

Ladizinsky *et al.* [46] have now shown how two groups
within *L. nigricans* could be differentiated, one with horizontal stipules (oriented perpendicular to the stems) and
the other group with upright stipules (parallel to the
stems). These two *nigricans* groups are isolated from each
other by three reproductive barriers (complete cross-
incompatibility, albino F_1 plants and sterility). Further
analyses reveal that the horizontal stipule type within
nigricans can be more aptly classified within the *culinaris*
group because of similarities in chromosome structure and
the possibilities for geneflow [18]. This new subspecies
was designated *odemensis* Ladizinsky. The upright stipule
type with reproductive barriers to the *culinaris* group is
more aptly classified as subspecies *nigricans* within *L.
nigricans*, together with subspecies *ervoides*. Supporting
this revised classification is the fact that subspecies
nigricans produces partially fertile hybrids with subspecies *ervoides*. In summary, Ladizinsky *et al.* [46], using
cytological evidence and morphological distinctions, suggest the following classification within *Lens*:

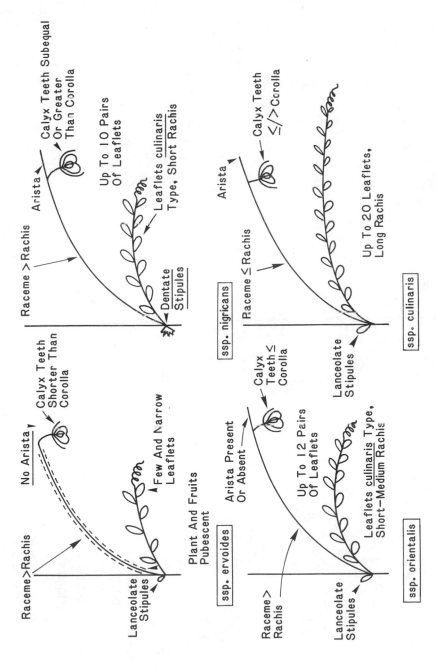

Figure 7.1 Morphological distinctions among the various subspecies of *Lens*

Figure 7.2 Distribution of wild lentils

Lentil (*Lens culinaris* Medic.)

1. Stipules lanceolate, entire if semi-hastate, situated at right angles to the stem, and calyx teeth ± as long as corolla .... *L. culinaris*
   stipules semi-hastate .... ssp. *odemensis*
   stipules lanceolate, mature fruits dehisce when dry .... ssp. *orientalis*
   mature fruits do not dehisce, or only slightly so, cultivated .... ssp. *culinaris*
2. stipules semi-hastate .... *L. nigricans*
   stipules entire or slightly dentate, calyx teeth much shorter than corolla .... ssp. *ervoides*
   stipules pointing upward and dentate, calyx teeth as long or longer than corolla .... ssp. *nigricans*

The classification of some herbarium specimens such as *nigricans* or *orientalis* has often depended on the botanist concerned. Boissier [5] described his *Ervum cyaneum* as showing a complete mixture of *ervoides, nigricans* and *orientalis* characters, and both *orientalis* and *nigricans* types have been considered as 'companion weeds' of the domesticate, the former in the eastern and the latter in the western extreme of the geographical distribution. *Lens ervoides* could also be a secondary companion weed together with *nigricans*, and perhaps even a primary one in Ethiopia [5].

In summary, the botanical features of *L. culinaris* subspecies *culinaris*, i.e. the cultivated lentil, can be described as follows: Annual, slender, almost erect plants, 15-50 cm tall, rarely up to 75 cm. Slightly pubescent, except for some very hairy specimens from Palestine, Syria and Turkey and several primitive cultivars from India. Ten to sixteen (rarely as few as six) leaflets are subtended on a long rachis (40-50 mm). Lower leaves are mucronate, upper ones have simple (rarely branched) tendrils. Leaflets are oval or linear elliptical, generally 12-15 mm by 3-4 mm, but with lengths and widths ranging from 7-20 mm and 1.5-8 mm, respectively.

Gracile, aristate peduncles bear one to three, but

rarely four, tiny flowers. The peduncles are shorter or
subequal in length to each rachis, but some primitive genotypes have peduncles longer than the rachis (a character
usually present in wild lentil of the *culinaris* group).
Flowers are 5-7 mm long, the calyx (6-8 mm) is white or
blue, with teeth longer, shorter or subequal in length to
the corolla.

Glabrous (but sometimes pubescent; see above) rhomboid
fruits 6-20 mm long by 4-12 mm wide, contain one or two but
rarely three seeds, which vary from small to large (3-9 mm
in diameter), and are flattened or subglobose, with large
qualitative and quantitative variations. The size of seeds
increases from the eastern to the western populations (and
see below).

Some researchers find it convenient and useful to adopt
the classification of Barulina [2,3] who divided the domesticate into two subspecies:

(a) *macrosperma*, which are found mainly in the Mediter-
    ranean region and the New World and have large seeds
    (6-9 mm in diameter), normally yellow cotyledons
    and little or no pigmentation in the flowers or
    vegetative structures; and
(b) *microsperma*, most common in the Indian sub-continent
    and parts of the Near East, which have smaller seeds
    (2-6 mm in diameter) with red, orange or yellow
    cotyledons. Cultivars in the *microsperma* are more
    polymorphus as a group but are generally shorter,
    more pigmented and have smaller fruits, leaves and
    leaflets than *macrosperma* types.

There are several reasons why this system is inappropriate. First, the rules of taxonomic nomenclature require that
one subspecies must retain the same name as that given to
the species; in this case, one of Barulina's subspecies
would have to be named *L. culinaris* ssp. *culinaris*, for
example, instead of *L. culinaris* ssp. *macrosperma*. It would

have to be the same with the taxa under the subspecific
rank. Secondly, and more important, according to the biological concept of species, the subspecies is seen as the
'bud' of a new species [23]. Furthermore, both *macrosperma*
and *microsperma* forms do not show reproductive barriers
between them but rather intergradation and intermediate
forms [68,74,96]. Differences between them are easily
ascribed to human selection and so they should be included
in the same subspecies in agreement with the system proposed by Harlan and de Wet [23] for cultivated species in
general. Finally, the botanical categories of variety and
subspecies share equal rank [77].

Taking all these facts into consideration, Cubero [8]
has proposed the merging of both Barulina's subspecies, as
races of ssp. *culinaris*; 'race' he interprets as a group of
forms showing morphological similarities without any further
considerations involving genetic and/or geographical features. In this scheme, Barulina's six geographical groups for
microsperma are considered as grex (i.e. 'groups'). The
latin names under the subspecific rank are retained for
practical convenience. A short description of these races
and groups of cultivated lentils is given below along with
Barulina's map of their distribution (Fig. 7.3):

(a) race *macrosperma*: large fruits (15-20 mm by 7.5-10.5
    mm), generally flat.
        Large flattened seeds (6-8 mm). Flowers large (7-8
        mm long), white, rarely bluish. Peduncles 2-3 flow-
        ered. Calyx teeth long. Large, oval leaflets.
        Height 25-75 cm.
(b) race *microsperma*: fruits small or medium (6-15 mm by
    3.5-7 mm) and convex.
        Seeds flattened-subglobose, small or medium (3-6
        mm). Flowers small (5-7 mm), white to violet. Ped-
        uncles 1-4 flowered. Leaflets small (8-15 mm by
        2-5 mm), linear or lanceolate. Height 15-35 cm.
            (i) grex *europeae*: 2-4 white flowers on each

279

peduncle. Seeds 4-5 mm in diameter.
Leaflets medium. Calyx teeth much longer
than corolla. Mid-season forms.

(ii) grex *asiaticae*: 1-3 blue or bluish flowers
on each peduncle. Seeds 3-5 mm in diameter. Leaflets small, narrow. Calyx
teeth equal to or rarely longer than
corolla. Early forms.

(iii) grex *intermediae*: 1-3 white flowers. Seeds
5-6 mm in diameter. Calyx teeth equal
to or longer than corolla. Leaflets
small or medium.

(iv) grex *subspontaneae*: Fruits dehiscent, coloured black or blackish. Seeds small
(3-3.5 mm in diameter). 1-2 flowers on
each peduncle. Calyx teeth much shorter
than corolla. Leaves with 3-5 pairs of
small leaflets. Dwarf plants. Afghanistan endemic.

(v) grex *aethiopicae*: 1-2 blue or violet flowers
on each peduncle. Calyx teeth much
shorter than corolla. Leaves with 3-7
pairs of elongated leaflets. Fruits with
a typically elongated apex. Seeds black
or brown with black spots. Ethiopian
endemic.

(vi) grex *pilosae*: 1, but rarely 2 flowers on each
peduncle. Calyx teeth much shorter than
corolla. Leaves with 3-6 pairs of leaflets and short. Sometimes rudimentary
tendrils. Seeds black or reddish with
black spots. Dwarf plants, typically
pubescent.

The origin of *Lens*

The actual origin of lentil is unclear. Barulina [3]

Lentil (*Lens culinaris* Medic.)

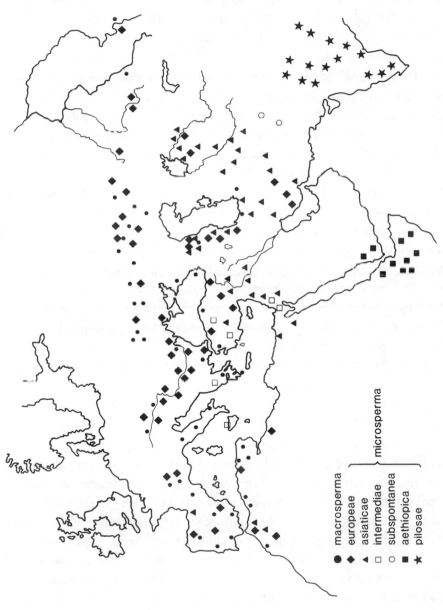

Figure 7.3 Distribution of cultivated lentils

macrosperma
europeae
asiaticae
intermediae
subspontanea
aethiopica
pilosae

} microsperma

● ◆ ◀ □ ○ ■ ★

suggested that *L. culinaris* evolved from *L. orientalis* in the region between the Hindu-Kush and the Himalayas, but Renfrew [57,58] and Zohary [100] seem to favour the Fertile Crescent. More compelling cytological evidence presented by Ladizinsky [41,46] also suggests that *L. culinaris* evolved from *L. orientalis*, but probably in Turkey. The primary determinant of domestication was indehiscent fruits, a trait controlled by a single recessive gene [42]. Whether the wild species, including *L. ervoides* Brign, were at one time domesticated but then, because of consumer preferences and other competitive advantages of the domesticate (which came to be known as *L. culinaris*), escaped to the wild, is unknown. Collection and identification of cultivated types with morphological and cytological similarities to the wild species could be proof that the wild species were at one time cultivated. Clearly, these currently non-cultivated species are most often found in areas of previous human activity but seldom in nearby undisturbed habitats. Dissemination of the wild species as 'weedy' contaminants of cultivated types remains a distinct possibility.

The wild species represent a potential gene-pool for the improvement of cultivated forms [41]. The importance of the wild gene-pool for breeding should not be overlooked, especially since the cultivated types probably originated from a single gene mutation involving a change from dehiscent to indehiscent fruits, and most likely quite recently in the evolution of the genus. Extensive collections of cultivated types might therefore all trace back to a single mutational event, a situation that would indicate an extremely narrow genetic base compared to that in wild populations. Solh and Erskine [74] correctly point out the importance of the wild species and stress that urgent priority be given to the collection, maintenance, and utilisation of such material in breeding programmes. The wild species are considered to be potentially valuable sources of germplasm for improving tolerance to environmental stresses because they evolved in dry, stony, shallow soils (i.e. in habitats

Lentil (*Lens culinaris* Medic.)

Plate 7.1   Typical habitat of wild *Lens* species

unsuited to many other species; Plate 7.1). These habitats
are similar and often adjacent to areas where lentil crops
have been grown in the past. Ladizinsky and Muehlbauer [45]
have now made extensive collections of wild species which
are available on request.

PRINCIPAL BOTANICAL FEATURES AND REPRODUCTIVE BIOLOGY

Germination of lentils is hypogeal which could mean that
seedlings are less likely to be killed by freezing, desiccating wind, insect damage, grazing, or by toxic doses of
agrochemicals than if germination were epigeal. If young
lentil shoots are damaged, new buds can be initiated from
nodes below ground.

Three types of root systems have been described and
related to soil type, branching pattern and seed size [38].

283

Small-seeded genotypes are said to produce rather shallow
root systems and to be adapted to light soils, whereas
large-seeded types have a tendency to deeper root systems
and are better adapted to heavy soils. The agronomic relevance of these observations has yet to be established [78].

Lentils are slender, semi-erect annuals usually between
30 cm and 75 cm tall (Plate 7.2) but, if flowering is
delayed and/or crop durations are extended, by, for example,

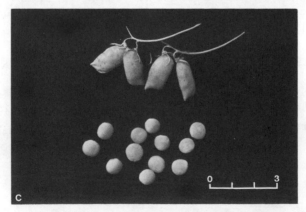

Plate 7.2 The cultivated lentil: (a) typical morphology;
(b) flowers; and (c) fruits and seeds (scales in (a) and (c) are in
cm; (b) in mm; and see [66])

cool temperatures, their indeterminate habit (i.e. they
do not initiate flowers at stem apices) can result in
excessively tall plants. The crop is remarkably plastic
and many cultivars will branch profusely - depending, of
course, on crop density; individual plants may vary from
single stems to vigorous, bushy forms in dense and sparse
stands, respectively [99]. For example [65], plants of the
same cultivar produced between 4.2 and 75.3 branches as
population densities changed from 444 to 56 plants m^{-2}.

Stems may be pigmented at the base or along their
length. The leaves, like those of chickpeas, are relatively
small compared with, for example, soyabeans (*Glycine max*)
and common beans (*Phaseolus vulgaris*); they are described
as pinnate or imparipinnate and comprise perhaps as many
as fourteen sessile, ovate or elliptic or obovate or lanceolate leaflets, each about 1 cm long. Each leaf is subtended by two small stipules and may or may not terminate
in a tendril. Leaflets can vary from pale to bluish green.
However, cool temperatures can lead to excessive anthocyanin
synthesis, when leaflets become purple. This can be confused
with symptoms of phosphorus deficiency. As in many other
legumes, the number of leaflets per leaf and their size and
shape can change from the base to the top of the plant (but
the differences are not as prevalent as in other Viciae).
The photosynthetic capacity of lentil leaves seems neither
greater nor less than other (more productive) grain legumes, is equally variable and presents the same problems
for measurement and interpretation of comparative data. It
seems unlikely that selection for 'photosynthetic rate'
would be a worthwhile objective for lentil breeders [83].

Reproductive nodes bear either single flowers or two,
three, but rarely four-flowered racemes on short peduncles,
although Hawtin [24] has recorded seven flowers per peduncle
on one selection from Afghanistan when grown in a glasshouse. The typical Papilionaceous ('butterfly-like') flowers
are small (4-8 mm long), white, pale purple or purple-blue
and usually open before 10.00 h on cloudless days, but

perhaps not until 17.00 h when the sky is overcast; the
corolla petals fade within three days and fruits appear
three to four days later [28,78]. The oblong fruits are
flattened and smooth, about 1-2 cm long, and usually contain one or two, but rarely three seeds. For a more complete description of flowering in lentils see Summerfield
et al. [85,87].

A single plant can produce 10-150 peduncles, each
2-5.5 cm long, and bearing one to four flowers. Flowering
proceeds acropetally; lower nodes may bear fruits close to
maturity while younger nodes are still initiating flowers.
The flowers are naturally, perhaps almost obligatorily, self-
pollinated. Cross-pollination, vectored by thrips or other
small insects, but not by wind or honey bees, is estimated
to be less than 0.8% [98]. In general, cold temperature
vernalisation of seeds, warm air temperatures, and long
days promote early flowering, but we still have only a cur-
sory understanding of the relative importance of these
responses [61] and their significance in the adaptation of
lentils to different agro-climates [78].

However, current research [82,83] is seeking to redress
this limitation. It seems that over wide ranges of temperature and daylength, rates of progress towards flowering in
diverse genotypes can be predicted accurately from the following equation:

$$1/f = a + b\overline{t} + cp \qquad (1)$$

where f denotes days from sowing until the first flower
opens, $\overline{t}$ is average air temperature and p is photoperiod;
a, b and c are constants, the values of which vary between
genotypes (Summerfield and Roberts, in preparation). Thus,
the relation between rate of progress towards flowering is
positive and linear with both mean temperature and photoperiod, with no interaction between these two environmental
factors. Experiments in progress are seeking to quantify
the effects of vernalisation, by appropriate modifications

286

Lentil (*Lens culinaris* Medic.)

to equation (1), and to assess the implications of such
relations for the screening of germplasm for photo-thermal
effects on development in carefully selected field locations.

Fruits are rhomboidal and compressed laterally; one or
two fruits usually mature on each peduncle. As in other
legumes, the number of fruits per plant is strongly influenced by environment and genotypic variation; mainly between *microsperma* and *macrosperma* types [48].

Seeds are typically lens-shaped and range from 2-9 mm
in diameter. The testas can be uniformly coloured, spotted,
or marbled with different colours [13]. Frequent colours
are tan, green, brown, grey and black. The weight of 100
seeds varies from about 1 g to almost 9 g. Seed yields are
strongly correlated with the number of fruits per plant,
but the number of seeds per fruit, mean seed weight and
seed shape are more strongly heritable characters. There
are also genotypic differences for seed dormancy, which can
last for as long as three or four weeks in freshly-harvested
seeds, or even for three months when seeds are produced in
a glasshouse. Dormancy can be overcome by soaking seeds in
aerated water for 4-8 h or by freezing them at about -15°C.

Components of yield

Research on the characters affecting lentil yield has been
fragmentary and has utilised a limited range of germplasm;
genotypes have not been commonly tested in different studies
and, consequently, few comparisons of data can be useful.
Then again, the work done at different latitudes is difficult to interpret because of the different genotypic responses to photoperiod and temperature which may have biased the
results (e.g. the *macrosperma* types are known to be especially sensitive to daylength).

Components of yield in lentils have been studied using
correlation coefficients, path and factor analyses [52,64,
71,90,91,92] but, not surprisingly, with inconsistent

results [79]. In two different studies, the number of secondary branches had either positive or negative effects on yield. The order of factors positively and most strongly influencing yield was different in five other studies, viz (a) the numbers of primary branches, fruits per cluster and seeds per fruit; (b) fruit length, plant height and days to flowering; (c) the numbers of secondary branches, days to maturity and fruits per plant (plant height and days to flowering in this case had very small effects); (d) the numbers of fruits and branches per plant; and (e) the numbers of days to flowering and fruits per plant (the number of secondary branches showing here a negative effect). Seed yield has also been correlated positively with plant height, days to maturity, and average seed weight (itself correlated with fruit length, number of secondary branches, plant height and days to flowering). Factorial analyses showed a complex of four characters (the numbers of primary and secondary branches, fruits per plant and days to flower) determined the largest proportion of variation in yield (66%) in the thirty-five cultivars studied. In summary, the most important characters related to yield seem to be branching pattern and number of fruits to reach maturity.

It does not always appear to have been appreciated that heritability and genotypic correlations are dependent on the range of genotypes included in the study and on the environment in which they are grown. Also, it seems that the previous emphasis on yield components as means of selecting improved lines of lentils has not been justified. The problem with this approach is that the components involved (e.g. degree of branching and fruit number) are influenced markedly by agronomy and environment (e.g. plant spacing, available moisture, time of planting, and so on) and so these components vary from year to year and location to location, even for the same genotype. Rather than commit limited resources to repetitive counting of yield components, breeders need to be aware of these limitations and to use procedures that not only recognise the importance of branching patterns

and fruit set but also avoid wasting time collecting
uninterpretable data.

Lentil seeds are more susceptible to mechanical damage
than peas (*Pisum sativum*), faba beans (*Vicia faba*), lupins
(*Lupinus* spp.) and chickpeas. Their shape influences this
susceptibility: seeds that are thick and have blunt edges
are less prone to damage. Thickness and bluntness of lentil
seeds are controlled by four and two genes, respectively [62].

Composition of lentil seeds varies with genotype and
location; it changes as the seeds mature, and is influenced
by the availability of various inorganic nutrients [78].
Differences in their mineral composition are known to affect
cooking time of seeds [194].

SYMBIOTIC NITROGEN FIXATION POTENTIAL

Our knowledge of nitrogen fixation in lentils is rather
fragmentary. In fact, only when ICARDA (see Chapter 18)
instigated its programme on the lentil/*Rhizobium* symbiosis
in 1978 did systematic and coordinated work begin.

Nodules on lentil roots are elongate, sometimes branched,
indeterminate, and with an apical meristem; they are seldom
longer than 5 mm [76]. Lentil crops may or may not respond
favourably to artificial inoculation, depending on the previous cropping history and soil type, and the concentration
of inorganic nitrogen available to the crop.

Lentils are nodulated by *Rhizobium leguminosarum* and it
is often stated that strains, which are also capable of nodulating *Pisum* spp. and *Vicia faba*, are common in soils of
the entire area where lentils are grown. However, in spite
of this assertion, positive increases in both nodulation
and yield have been reported following inoculation. With
some selected strains of *Rhizobium*, the increases in yield
were as large as 32% in loamy soils or 50-90% on sandy
loams. Interactions between strains of *Rhizobium* and lentil
cultivars have also been detected. Poorly nodulating genotypes (15-25 nodules per plant) and well nodulated ones

(30-60 nodules per plant) have now been identified [30].
These interactions of *Rhizobium* strain and host genotype
pose a problem because indigenous strains in certain areas
may not be the ones capable of fixing most nitrogen; the
alternatives are to either change the host genotype (hence,
the necessity of developing cultivars for, perhaps, rather
restricted areas) or to change the strain of *Rhizobium*. The
latter may be relatively simple to achieve in a laboratory,
but is far more difficult to achieve agronomically: persistent indigenous strains may have competitive advantage because of their population density and invasive characteristics. Furthermore, the economics of the production and distribution of inoculated seeds must be considered
especially in the subsistence farming systems in which
lentils are often grown.

From the meagre data which are available, from crops
that produced only poor yields, symbiotic nitrogen fixation
by lentils has been estimated [53] to vary from 35-77 kg N
$ha^{-1} a^{-1}$. For an average seed yield of just 600 kg ha^{-1},
which is close to the world average (Table 7.1), and with
a mean protein concentration of 25%, a lentil crop which
matures in 120-150 days from sowing would produce about
150 kg protein ha^{-1}. This represents about 24 kg N ha^{-1} in
seeds alone. Thus, if estimates of *annual* fixation are realistic, the 'average' lentil crop could deplete soil nitrogen
status, which contradicts the assumed beneficial role of
lentils in cropping rotations. Other estimates [30] have
indicated an annual average fixation of about 100 kg ha^{-1}
(range 83-114); a value similar to that of chickpeas but
inferior to cowpeas and faba beans. Although some reports
[78,86] cite values as large as 103-115 kg N ha^{-1} for symbiotic fixation by lentil crops in Egypt and the USA, they
make no mention of the time required to achieve these
amounts, or how they were calculated. In India, the fertiliser needed by a maize crop can be reduced substantially
if the preceeding crop is either lentils, chickpeas, peas,
or *Lathyrus sativus* rather than wheat or, indeed, winter

fallow [1]. Clearly, cultural and agronomic practices which optimise the seasonal fixation activity of symbiotic associations well adapted to their aerial and edaphic environments need to be identified, and studies towards this end merit some priority in lentil improvement programmes.

Nitrogen fixation rates and durations are limited by many factors (see Chapter 4). Temperature is among the most important, and experience with other legumes suggests that selection of *Rhizobium* strains capable of better fixation in hot, dry soils should be possible [30]. Some data indicate that lentils can also nodulate well in short, rather dull, days (10-12 h with an illuminance of 12000 lux), conditions said to be typical of geographical regions where lentils are autumn-sown; but with strains of *Rhizobium* collected from ten countries, from Spain to Syria, lentils nodulated poorly in natural conditions and with traditional cultural practices [31]. However, the introduction of more effective strains seems promising, especially since *Rhizobium*/lentil host co-adaptation is suggested by the existence of several strains that show better nodulation in dry soil when rainfall is limited.

Lentil crop improvement programmes should include 'symbiotic potential' among selection criteria and if breeders are to select for nodulation and nitrogen fixation, they will need to know the ecology of *Rhizobium* in their breeding plots and to restrict applications of inorganic fertilisers to them. It may also be prudent to rotate the locations of lentil breeding plots with those of cereal breeders. If these criteria and practices are neglected, the symbiotic potential of lentils could decline in proportion to the breeding effort put into the crop [79].

The importance of research on symbiotic nitrogen fixation cannot be over emphasised because lentils are grown in many places where no other legume and almost no other crop (except, perhaps, barley) can grow. Even in less extreme circumstances, lentils are usually grown on poor soils, generally by poor people who are unlikely to apply

291

inorganic N to the crop. Lentil nodules can therefore
represent the only source of nitrogen additions to the soil
in such situations. Any research producing a little more
yield and releasing a little more nitrogen to the soil
could represent significant progress.

PRINCIPAL LIMITATIONS TO ECONOMIC YIELD

Lentils can tolerate extreme environmental conditions typical of desert fringes where rainfall is minimal (about 325
mm a^{-1}) and temperatures are hot. They can be sown in the
autumn or in spring, up to altitudes of 1000 m in Mediterranean climates and 600-800 m in Mediterranean-Continental
areas. Even though early and late maturing types are available, growers often use the same or similar cultivars in
diverse agro-ecological areas. Such 'plasticity' can be considered a response to selection but, perhaps, it leads to
small yields (Table 7.1). More specifically-adapted cultivars might be more productive, but before such a breeding
objective can be successful, it will be necessary to change
either the attitudes of producers or the environmental
limitations to yield, or both.

It seems probable that farmers grow lentils in extreme
environmental conditions because they expect at least some
yield, albeit small, from the crop and know that other
pulses are more likely to fail. The reproductive potential
of many genotypes is small: plants are often tiny and delicate and so cannot support large seed weights. If erect and
tall types with better standing ability could be developed,
yields would probably be much improved.

Extreme environmental conditions and producer attitudes
are by no means the only factors which limit lentil yields;
infestations by insects, diseases, and weeds (including
parasitic plants) can combine to cause extensive crop losse

Lentil (*Lens culinaris* Medic.)

Insects

Lentil crops are attacked by several pests wherever and whenever they are cultivated; the most commonly cited are pod borers (*Etiella zinckenella*), aphids (*Aphis* spp.), weevils (*Sitona lineatus*), bruchids (*Bruchus* spp.) and cutworms (*Agrotis* spp.) [27,72]. Insect problems are not confined to the field; they can continue during storage. Indeed, lentils taken from archaeological sites in Egypt have revealed a similar degree of infestation and by a member of the same Bruchidae genus as could be expected today [75].

The bruchids, *Sitona* weevils and the pod borers merit special mention [21,63,88]. Bruchids not only destroy a large quantity of seed, they also reduce germination of seeds which are damaged but not eaten completely. Several species are troublesome; the most important are *Bruchus lentis*, *B. ervi*, *B. signaticornis* (especially in the Mediterranean region) and *Callosobruchus cinensis* (cosmopolitan). The degree of infestation can also depend on the cultivar, but the variation established so far in available germplasm is insufficient to warrant a breeding programme for resistance.

Pod borers are reported to be the major insect pest in India, but are not restricted to this region. The larvae feed upon immature seeds within the fruits and yield losses can be substantial.

Infestation by *Sitona* weevils (several species) can lead to economic losses in many regions. The larvae feed on roots and nodules and the adults damage leaves, producing a typical crenellated margin. When conditions are conducive to only small crop growth rates, *Sitona* attacks can severely damage, and even destroy, the young plants. Insecticides can provide useful control, but genotypic differences in sensitivity seem to be minor and not yet sufficient to warrant breeding towards resistance.

Soil-borne insects of occasional economic importance

293

are cutworms (*Agrotis ipsilon*), bud weevils (*Apion* spp.),
seedcorn maggots (*Delia platura*) and wireworms (*Limonius*
and *Ctenicera* spp.). Larvae of these insects destroy plants
by feeding on stem apices of seedlings - a syndrome well
known in other similar crops. Seedcorn maggots and wireworms
attack lentil seeds soon after planting and/or the stems and
roots of seedlings. Seed treatments with insecticides are
the only economic means of preventing severe stand losses.

Other insect pests known to cause economically significant damage include pea aphids (*Macrosiphum pisi*), cowpea
aphids (*Aphis craccivora*) and thrips (*Megalurothrips* spp.).
Among the aphids, *Aphis craccivora* is the most important:
severe infestations cause plants to be variously stunted,
deformed and barren (i.e. flowers and fruits are dropped
prematurely). Aphids also transmit viruses from clover,
alfalfa and other legumes growing near lentil fields. Pea
Enation Mosaic Virus, Pea Streak Virus and various mosaic
viruses are vectored in this way [85]. Thrips have mouth
parts capable of rasping buds, flowers, and leaves and
sucking the exuded juices. Typical symptoms of thrip damage
are distorted and discoloured flowers, white streaking and
mottling of leaves, and brown streaking of the fruits, but
damage is seldom sufficiently serious to warrant the use of
insecticides.

Lygus bugs have been shown recently to be more important
than suspected hitherto [86] and all species cause abscission of reproductive structures and deformations of the
seeds, which may result in loss of economic yield *per se*
or of product quality, or both. The damage to lentil seeds
known colloquially in the USA as 'Chalky Spot' is caused by
Lygus bugs feeding on immature reproductive structures.

Many of these insects, particularly *Lygus* bugs, are
common pests in forage legumes from which they migrate to lentils when the forage crop is harvested.

Lentil (*Lens culinaris* Medic.)

Diseases

The most serious and ubiquitous diseases of lentils are
caused by organisms comprising the root-rot/wilt complex
(i.e. *Pythium, Rhizoctonia, Fusarium,* and *Sclerotium* spp.
[28,36,37]). *Ascochyta* blight (*Ascochyta lentis*), rust
(*Uromyces fabae*), downy mildew (*Peronospora lentis*), collar-
rot (*Corticium rolfsii*) and several viruses can also attack
the crop [39]. Nevertheless, lentils are relatively free of
severe disease problems and so breeding for disease resistance is not a primary objective in international programmes
[27].

The most important organism in the root-rot/wilt complex is the vascular or true wilt, caused by *Fusarium oxysporum* f.sp. *lentis*. The fungus invades the vascular system,
restricts moisture uptake, and causes plants to wilt and
die, usually before blooming. Temperatures between 17°C and
31°C, sandy loam soil with about 25% moisture, and pH values
between 7.6 and 8.0, favour the disease. Thus, agronomic
adjustments to the acidity, structure, moisture content and
composition of the soil, and timely sowing, can combine to
avoid excessive damage. Because of the presence of several
fungi in the wilt complex (e.g. *F. avenaceum* and *Rhizoctonia
solani* are often associated) no resistant cultivar has yet
been developed. Selections for resistance to components of
the complex followed by recombination may provide improved
breeding material.

Collar- and root-rots have been reported to cause severe
losses [39]. Infected plants droop, become chlorotic, and
eventually die. The diseases are seed-borne, and favoured
by humid conditions and warm temperatures. Collar- and root-
rots can often be associated with vascular wilt, which is
unfortunate because there seems to be resistance to collar-
rot available in several germplasm collections. Root-rot is
caused by a complex of no fewer than fifteen pathogens,
including several species of *Fusarium, Sclerotinia*, and
Pythium. The disease is especially severe when *Thanatephorus*

cucumeris is associated with *F. solani* [39].

Rust is clearly the most important foliar disease in India, Ethiopia, and most of South America [39,59]. The causal agent, *Uromyces fabae*, is an autoecious fungus that completes its life cycle on lentils. The disease is favoured by mild temperatures (20°-22°C), cloudy weather, and humid conditions. Several resistant genotypes have now been identified.

Other depredations by fungi, nematodes and viruses have been reported, but only mildew and *Ascochyta* blight are of local and occasional importance [34,39].

Ascochyta blight (*Ascochyta lentis*), a seed-borne disease, attacks leaves, stems, fruits, and seeds. Common symptoms include purple blotching of infected organs and some seed discoloration. The disease has been reported from Canada [73] where, in areas that receive large amounts of summer rainfall, it is of economic importance.

Pea Seedborne Mosaic Virus attacks lentils causing stunting, malformation of the leaves, twisting of stems, and flower, fruit and seed abortion. Those seeds which do mature are smaller than normal, and variously misshapen. Immunity has been identified and is controlled by a single recessive gene [20].

Weeds

Lentils are notoriously poor competitors with weeds, probably because crop growth rates are so small during early vegetative growth, especially in cool weather. Inadequate weed control may reduce seed yield by as much as 75%, although the period of crop growth during which competition from weeds is most deleterious varies in different locations: 30-60 days after planting seems most critical in India, compared with 60-90 days in Syria [28]. Basler [4] has reported that at a sowing rate of 150 kg ha^{-1}, a clean-weeded plot yielded approximately 600 kg ha^{-1}, while the unweeded control gave only 90 kg ha^{-1}. Larger or smaller sowing rates had no effect because of increased intraspecific competition or greater competition from weeds respectively.

Hand-weeding, the traditional control method, is time-consuming and expensive and is often only practised on small farms with family labour. Recent research has shown that several herbicide formulations are effective with lentils [4,85], although for many subsistence farmers hand-weeding or mechanical cultivations remain the principal methods of control. Unfortunately, both methods can cause damage to lentil seedlings, increase the incidence of stem and root diseases, and expose even more weed seeds which can germinate and create additional problems later during crop life.

It seems that many lentils are more sensitive to herbicide toxicities than other pulses. One of the most important factors which influence the success of herbicides is the moisture content of the soil, and lentils are mainly grown in dry regions where absorption by roots is much more intense than under wetter conditions [4]. Other edaphic conditions, such as soil temperature and pH, combine to complicate effective weed control, especially by subsistence farmers. Research and extension programmes should provide these farmers with simple equipment and appropriate instructions to correctly apply the most effective herbicides in each production zone.

Broomrape, an obligate parasitic angiosperm, can be very serious on lentils grown in Mediterranean and Near East regions. Three different species (*Orobanche crenata, O. ramosa* and *O. aegyptiaca*) can parasitise the crop. Indeed, lentil culture has been abandoned in some areas because of broomrape (e.g. in Andalucia, Southern Spain [63]). Control methods for broomrape have been reviewed extensively [9,10]. Hand-pulling is practised occasionally but no effective herbicides have yet been found. There are two principal lines of research work: genetic resistance and development of selective herbicides. Differences in susceptibility among lentil germplasm are available but, as yet, the degree

of resistance is not sufficient to warrant improvement programmes to develop commercial cultivars. Nevertheless, useful genetic variations might be found (as in faba beans) if
germplasm collections are screened systematically. Lentil
tolerance to glyphosate (a herbicide which is recommended
in Egypt for *Orobanche* control in *Vicia faba*) is very poor
and this chemical should not be used on the crop [4]. Breeding for host plant resistance shows some promise as do the
synthetic germination stimulants reported by Johnson *et al.*
[35].

FERTILISER REQUIREMENTS AND INTERACTIONS WITH WATER SUPPLY

Researchers are unsure of the demand for nutrients by lentil
crops and especially those producing large yields. Based on
experience around the world, advisors have considered it
worthwhile economically, and a reasonable 'insurance' for
growers, to apply molybdenum as a seed-dressing (essential
for good nodulation and nitrogen fixation), sulphur (the
concentrations in seeds of the nutritionally-limiting sulphur-amino acids can be improved in other legumes by applying S fertilisers, e.g. Fox *et al.* [17]), and phosphorus
(again, essential for good symbiotic performance and overall
plant growth). In general, 44-66 kg P_2O_5 ha^{-1} are applied if
soil tests (acetate extract) reveal that available P concentrations are 4 ppm or less. Responses to P fertiliser are
more common on severely eroded soils. On sandy or severely
eroded substrates, potassium applications of about 22 kg K_2O
ha^{-1} may prove beneficial and also improve the cooking
quality of seeds [94].

Effectively nodulated lentil crops seldom respond to an
application of inorganic N fertiliser. However, if crops are
seeded early into cool, wet soil a small 'starter' dose,
applied adjacent to but not in contact with seeds, may prove
effective and circumvent the 'nitrogen hunger' phase often
experienced by grain legume crops before the advent of significant symbiotic nitrogen fixation. Inoculation with an

Lentil (*Lens culinaris* Medic.)

appropriate strain of *Rhizobium leguminosarum* is essential
when lentils are seeded into new fields for the first time
(or after a lapse of several years) and in these circumstances special care will be needed with seed-dressings of
fungicides potentially toxic to *Rhizobium* [60].

Table 7.3 Estimates [80,84] of the nutrients removed (kg) in lentil
seeds for crops producing an economic yield of 1000 kg ha^{-1}
(data for soyabeans are included for comparison [67])

Crop	Nutrient (kg 1000 kg seeds^{-1})					
	N	P	K	Ca	Mg	S
Lentil*	43	5.0	11.7	0.7	1.2	2.0
Soyabean**	71	6.1	20.3	3.0	3.0	1.7

* Mean values for cultivars Chilean '78, Tekoa and Brewer.
** USA crops.
All data have been converted to elemental equivalents to facilitate
comparisons.

Based on chemical analyses of seed lots produced in the
same year at the same location (Pullman, USA 46°46'N), a
lentil crop yielding 1000 kg ha^{-1} would remove from the
soil in the seeds alone substantial amounts of P, K, Ca,
Mg and S (Table 7.3). Depletion of soil N reserves will,
of course, depend on the relative contributions of symbiotic
nitrogen fixation and uptake of inorganic sources. It will
be interesting to compare these preliminary data with those
for other lentil crops grown in a range of edaphic conditions and locations. Do different cultivars need to assimilate into seeds more or less similar amounts of these elements, or are there appreciable differences in seed composition (either inherent differences or those due to genotype
x environment interactions, or both)? Subsequent studies
should also assess the role of lentil crop residues (hay)
in maintaining long-term soil fertility status, and as a
feed for cattle [19]. In the USA, lentil hay cannot be fed
to cattle because herbicide tolerances have not been
determined or approved.

Environmental stress

Sensitivity to hot temperatures and drought during the critical flowering and fruit-setting periods is a major factor affecting lentil yields. These stresses are often interrelated and difficult to separate because hot seasons tend to be dry, and cooler ones more humid. To be 'well adapted' to the environment, lentil genotypes need to take full advantage of time and space available during the growing season. It is therefore necessary to fit the developmental timetables of different genotypes, namely their vegetative (including juvenile), mature (time to flower), reproductive (flowering and setting of fruits) and senescent stages (including ripening of fruits), to the natural environments which prevail in different ecological zones. Notwithstanding this objective, it is difficult to ensure that any given genotype is well adapted to the duration of the growing season at a particular location, especially when the time favourable for crop production is limited, because a thorough understanding is needed not only of the genetic factors which control crop duration, but also of the environmental factors which react with them [78,81].

Susceptibility to cold temperatures and desiccating winds are severe environmental stresses affecting winter survival. Many lentils are resistant to cool temperatures, but they need to be more tolerant of cold in order to ensure survival in the severe winters typical of the Anatolia Plateau of Turkey, the USSR, the USA and Canada. Screening of the ICARDA collection identified 238 accessions with suitable degrees of cold tolerance [74]. If this degree of winter hardiness is still insufficient, crosses among the more resistant accessions from different origins (to avoid genetic similarity) followed by selection for transgressive segregants, could ensure steady improvement. Crosses between cultivated accessions and wild materials (ssp. *orientalis* and *nigricans*) could complement breeding programmes for winter hardiness; although the wild species seems to have

300

less variation for this trait than the cultivated accessions [62]. In any case, wide crosses of this type are long-term projects. The problems related to soil structure and composition (e.g. compaction) and moisture content (e.g. waterlogging) need to be studied as well. Jana and Slinkard [32] identified salt tolerance in the world collection, but the variation is not considered sufficiently broad to be usable in a breeding programme [51].

Sensitivity to photoperiod and temperature is probably a major reason limiting *macrosperma* cultivation in India (W. Erskine, pers. comm.). Germplasm collections can be useful in developing suitable large-seeded types for that area, not only by crossing typical Indian materials with *macrosperma* types, but by selecting within available accessions including, for example, germplasm from South America.

NUTRITIONAL QUALITY

Lentil seeds are rich in protein; concentrations between 22.4% and 34.6% were found by Hawtin *et al*. [26] who evaluated 1688 accessions. As in other grain legumes, the nutritional value of these proteins is limited by their small concentrations of the sulphur-containing amino acids, methionine and cystine [80]. Interestingly, improved cultivars may not only yield more seed, but they may also assimilate seed proteins of better quality [80]. Indeed, we concur with Boulter [6] that while larger supplies of grain legume protein are most likely to result from increased *and* stabilised yields of seed rather than improved seed biochemical efficiency, since the energy required is approximately the same for synthesising proteins with good or poor amino acid profiles [7], then improvements in protein quality seem distinctly possible. In passing, we should note that if the seeds are for human consumption, then a smaller concentration of protein of improved quality may be a better breeding objective than a larger concentration of protein of poorer quality. The current status and future potential of lentils as a diverse food source for various ethnic groups have been reviewed and discussed by Leung and Salunkhe [47].

Numerous data on protein concentration and quality in lentil seeds have now been assembled and will be described elsewhere (Summerfield and Muehlbauer, in preparation). However, it seems that useful genetic variation is lacking. Variations due to crop location, year, and and laboratory performing analyses seem to be equally as large, if not larger, than differences between cultivars grown at a given location for a specific period and then analysed by a single technique in the same laboratory. The large variations in amino acid spectrum observed by different researchers may also reflect considerable effects of different analytical methods rather than useful, heritable variations. For example, whether or not the sulphur-containing amino acids are protected from destruction (to a greater or lesser extent) during acid-hydrolysis by prior oxidation of methionine to methionine sulfone and of cystine to cysteic acid using performic acid, could explain a significant proportion of the respective variations in these traits which have been described in the literature [80]. And so, at present, the lentil breeder cannot expect to include 'desirable culinary aspects' such as nutritionally balanced proteins among his principal selection criteria with any surety of success. Attention should focus on increasing economic yield and on the stability of yield between sites and years.

GERMPLASM RESOURCES

The most comprehensive collection of lentil germplasm is maintained by ICARDA at Aleppo in Syria [11]. The collection contains more than 4700 accessions from forty-seven different countries [74]; it includes major national collections from India, Turkey, the USA, Ethiopia and elsewhere [14,93]. Many collection expeditions to important regions have been made, but many other places of paramount importance (e.g. the mountainous regions of Afghanistan, Iran, Kurdistan and

Turkey) have yet to be sampled intensively. Collections of the wild relatives have been assembled and evaluated and are being maintained [43,51].

The impressively large ICARDA collection is still somewhat unbalanced; some regions are very well represented while others scarcely so, or not at all. The importance of collecting material from remote lentil growing areas should be stressed, even in regions where the crop is rather recent. The concept of 'microcentres' seems relevant here [22]: a small area with a difficult geography can be richer in forms than larger but more easily sampled areas. Besides, useful genes can be found far away from the original regions of the species. Thus, cold tolerance was found in lentil accessions from Chile (among other countries) and adaptations to short days in *macrosperma* types in accessions from Argentina. In both countries, lentils were introduced by the Spaniards well after 1550 [10].

Lentils grown in many small and internationally insignificant regions can also be very important sources of variability. Morocco and Algeria are, for example, scarcely represented in the ICARDA collection, but they could be good sources of resistance to extreme conditions. Afghanistan is another pertinent example: although well represented in world collections, it is doubtful that all the primitive forms described by Barulina [2] are present. Indeed, collectors were not able to find the primitive group during visits to Afghanistan in the mid-seventies.

Of the botanical main groups already described, *subspontaneae* is the only one now lacking. Further collections are needed urgently, not only to provide breeding material but also to minimise losses of potentially useful botanical variation.

PRINCIPAL BREEDING STRATEGIES AND SCALES, MANDATES AND OBJECTIVES

Lentil breeding is a relatively recent endeavour when

compared to other major crops. Past crop improvement has
been based largely on materials introduced from one location and then selected for adaptation to particular areas.
Thus, most 'cultivars' in farmers' fields today have been
derived from selections within heterogeneous populations
and are not the outcome of hybridisation. Progress through
this approach is limited by the diversity available in these
populations at the outset; further improvement efforts will
need to include vigorous hybridisation and selection.
Indeed, hybridisation of lentils to improve yield potential
is becoming increasingly important and there are now active
breeding programmes in all principal lentil growing regions
of the world (for details on lentil hybridisation, see
Muehlbauer *et al.* [49] and Muehlbauer and Slinkard [50]).
In most of those programmes, yield and reliability of yield
are of utmost importance; adaptation to stress environments,
broad adaptation, and disease and insect resistance are also
receiving significant attention.

Most of the countries where lentils are cultivated have
also instigated research programmes on the crop. In many
cases, the research is restricted to studies on local crop
agronomy, cultural practices, and production techniques.
The most common work on breeding is the screening of new
types from collections of local or imported genetic stocks.
The ICARDA world collection now provides breeders with a
broad genetic base to use in programmes designed to produce
cultivars better adapted to local environments and with
desired agronomic traits and improved crop quality. The
general characteristics sought are larger and more stable
yields and acceptable quality.

Since the establishment of ICARDA in 1977, not only its
germplasm collections but also its programmes of research
have strongly influenced the national research efforts of
many other countries. Without doubt, ICARDA has provided
an essential link among lentil researchers. The primary
objectives of the ICARDA programme (summarised by Hawtin
[25] and see Chapter 18) are:

304

Lentil (*Lens culinaris* Medic.)

(a) Large and stable dry-seed yields as well as large total biological yields. The main ways to achieve these aims are seen to be:

  (i) an inherently large yield potential;
  (ii) tolerance to stress conditions; and
  (iii) resistance to major diseases and pests (including *Orobanche*).

(b) Wide adaptability and plasticity, mainly modifying photo-thermal sensitivity.

(c) Characteristics suited for mechanical harvesting:

  (i) erect growth habit;
  (ii) resistance to lodging and fruit dehiscence;
  (iii) first fruit position high enough off the ground to allow for cutting.

It is of course likely that as lentil cultivation expands, new problems will arise. But there must be considerable optimism that in due course, and sooner rather than later, the efforts of scientists of different countries, coordinated on a voluntary basis by ICARDA, will produce the basic genotypes needed for improved production.

PROSPECTS FOR LARGER AND MORE STABLE YIELDS

Breeding programmes in lentils which attempt to produce cultivars for a very narrow range of environments (national programmes) differ from those which seek adaptation to a wide range of environments (international programmes). Accordingly, techniques of breeding may vary. (For more information on genetics and breeding methodology, see Muehlbauer and Slinkard [50].)

Under a *narrow range of environments*, programme objectives are usually well defined and include specific agronomic traits required to improve locally grown cultivars. All standard breeding techniques are appropriate; however, there may be advantages to using pedigree selection, which is more appropriate for simply-inherited traits. There is a tendency

305

for local programmes to deal more in qualitative traits and to devise selection indices for them.

When breeding for a *wide range of environments*, there is a tendency to use bulk population methods because of their simplicity and the ability of the programme to undertake numerous crosses. This approach also provides seeds of early generation breeding populations in sufficient quantities to make testing in numerous ecogeographical locations possible. Selections within these bulks are made based on the needs of specific local environments. Also, with this approach, an evaluation of bulk populations over a wide range of environments is possible, and decisions are then made concerning those populations most likely to yield selections with a wide range of adaptability. Selections made in specific local environments are entered into multi-location trials to evaluate their range of adaptation. Natural selection may influence bulk populations at certain locations and may bring about an increased frequency of desirable genotypes, thereby improving selection potential within the affected bulk populations.

To be successful, the international programmes need to identify photo-thermal sensitivities and local needs for disease resistance, insect resistance, and possibly environmental stress resistance (or avoidance) that would allow breeding materials and selections to be grown over as wide a range of environmental conditions as possible. In order to realise these objectives, materials must be tested in carefully chosen sites in major production areas. Selection indices that give proportional weights to the different production regions will ensure that those regions receive appropriate consideration in the development of improved cultivars.

AVENUES OF COMMUNICATION BETWEEN RESEARCHERS

Regional programmes in many countries (e.g. the USA, India, Turkey, Jordan, Canada, Argentina, Hungary and Chile) now have strong commitments to lentils. If we accept that the only valid criterion for judging plant breeding work is its effect on national or regional food production [33], then this simple criterion demands collaboration and cooperation between multidisciplinary breeding teams world-wide. The complementary roles of international and national programmes for lentil crop improvement cannot be over-emphasised if any improvements in farming potential are to be exploited.

An annual newsletter, *LENS*, first published in 1974, has proved to be an excellent vehicle for improving communications among lentil researchers. This, and other coordination promoted by ICARDA, have fostered fruitful contacts among scientists and extension workers in many countries. The workshop held at ICARDA in 1979 was a very important date in lentil research. A book entitled *Lentils* (e.g. see [8]) resulted from the meeting and was the first book about the crop to be written in English. There are indications that these activities have stimulated new lines of research, and intensified others, world-wide. An international conference would now seem to be justified for a thorough review of these expanded efforts.

Researchers should be encouraged to share and exchange their experiences in temperate and more tropical locations; germplasm accessions should be made freely available to others (with appropriate quarantine controls); and breeders, researchers and commercial producers should foster any opportunity for dialogue. The wishes of consumers (e.g. the type, form, colour, texture and taste of product preferred, and the price they are willing to pay for it) must, of course, feature strongly in formulating research objectives and breeding strategies. Local preferences are unlikely to be acceptable universally and market opportunities must be identified and exploited accordingly.

LITERATURE CITED

1. Ahlawat, I.P.S., Singh, A. and Saraf, C.S. (1981), *Experimental*

Agriculture 17, 57-72.
2. Barulina, H. (1928), *Bulletin of Applied Botany, Genetics and Plant Breeding*, Leningrad. [Russian with English summary]
3. Barulina, H. (1930), *Bulletin of Applied Genetics and Plant Breeding* (Leningrad), Supplement 40, 1-319. [In Russian.]
4. Basler, F. (1981), in Webb, C. and Hawtin, G.C. (*eds*) *Lentils*. Farnham Royal: Commonwealth Agricultural Bureau, 143-154.
5. Boissier, E. (1972), *Flora Orientalis*, Vol. II. Vasilea.
6. Boulter, D. (1977), in Muhammed, A., Aksel, R., and von Borstel, R.C. (*eds*) *Genetic Diversity in Plants*. New York: Plenum Press, 387-96.
7. Boulter, D. (1980), in Summerfield, R.J. and Bunting, A.H. (*eds*) *Advances in Legume Science*. London: HMSO, 127-34.
8. Cubero, J.I. (1981), in Webb, C. and Hawtin, G.C. (*eds*) *Lentils*. Farnham Royal: Commonwealth Agricultural Bureau, 15-30.
9. Cubero, J.I. (1983), in Hebblethwaite, P.D. (*ed.*) *The Faba Bean*. London: Butterworth, 493-522.
10. Cubero, J.I. and Moreno, M.T. (1985), in Cubero, J.I. and Moreno, M.T. (*eds*) *Leguminosas de Grano*. Madrid: Mundi Prensa. (in press).
11. Darling, H.S. (1979), *Span* 22, 55-7.
12. Davis, P.E. and Plitmann, U. (1970), in Davis, P.H. (*ed.*) *Flora of Turkey* 3, 325-8.
13. Duke, J.A. (1981), *Handbook of Legumes of World Economic Importance*. New York: Plenum Press, 110-13.
14. Esquinas-Alcazar, J. (1985), in Cubero, J.I. and M.T. Moreno (*eds*) *Leguminosas de Grano*. Madrid: Mundi Prensa. (in press).
15. FAO (1980), *FAO Production Yearbook* No. 34. Rome: FAO.
16. FAO (1981), *Joint FAO/WHO Food Standards Program*. Geneva: Codex Alienentarius Commission.
17. Fox, R.L., Kang, B.T. and Nangju, D. (1977), *Agronomy Journal* 69, 201-05.
18. Goshen, D., Ladizinsky, G. and Muehlbauer, F.J. (1982), *Euphytica* 31, 795-9.
19. Gupta, P.S. and Jayal, M.M. (1965), *Agricultural Research (India)* 5, 211-12.
20. Haddad, N.I., Muehlbauer, F.J. and Hampton, R.O. (1978), *Crop Science* 18, 613-15.
21. Hariri, G. (1981), in Webb, C. and Hawtin, G.C. (*eds*) *Lentils*. Farnham Royal: Commonwealth Agricultural Bureau, 173-89.
22. Harlan, J.R. (1951), *American Naturalist* 85, 97-103.
23. Harlan, J.R. and de Wet, J.M.J. (1971), *Taxon* 20, 509-17.
24. Hawtin, G.C. (1977), *LENS* 4, 1-3.
25. Hawtin, G.C. (1979), in Hawtin, G.C. and Chancellor, G.J. (*eds*) *Food Legume Improvement and Development*. Ottawa: International Development Research Centre, Publication IDRC-126e, 147-54.
26. Hawtin, G.C., Rachie, K.O. and Green, J.M. (1977), *Breeding Strategy for the Nutritional Improvement of Plants*. Ottawa: International Development Research Centre, Publication TS7e, 43-51.
27. Hawtin, G.C. and Chancellor, G.J. (*eds*) (1979), *Food Legume Improvement and Development*. Ottawa: International Development Research Centre, Publication IDRC-126e.
28. Hawtin, G.C., Singh, K.B. and Saxena, M.C. (1980), in Summerfield, R.J. and Bunting, A.H. (*eds*) *Advances in Legume Science*. London: HMSO, 613-24.

Lentil (*Lens culinaris* Medic.)

29. Horner, G.M., Overson, M.M., Baker, G.O. and Pawson, W.W. (1960), *Bulletin* 1. US Department of Agriculture and the Idaho, Oregon and Washington Agricultural Experiment Stations, 1-25.
30. Islam, R. (1981), in Webb, C. and Hawtin, G.C. (*eds*) *Lentils*. Farnham Royal: Commonwealth Agricultural Bureau, 155-61.
31. Islam, R. (1982), *LENS* 9, 23-4.
32. Jana, M.K. and Slinkard, A.E. (1979), *LENS* 6, 25-6.
33. Jennings, P.R. (1974), *Science* (New York) 186, 1085-8.
34. Jimenez-Diaz, R. (1985), in Cubero, J.I. and Moreno, M.T. (*eds*) *Leguminosas de Grano*. Madrid: Mundi Prensa. (in press).
35. Johnson, A.W., Rosebery, G. and Parker, C. (1976), *Weed Research* 16, 223-7.
36. Kaiser, W.J. (1981), *Economic Botany* 35, 300-20.
37. Kaiser, W.J. and Horner, G.M. (1980), *Canadian Journal of Botany* 58, 2549-56.
38. Kay, D.E. (1979), *Tropical Products Institute, Crop and Product Digest No. 3*. London: HMSO, 210-23.
39. Khare, M.V. (1981), in Webb, C. and Hawtin, G.C. (*eds*) *Lentils*. Farnham Royal: Commonwealth Agricultural Bureau, 163-72.
40. Khayrallah, W.A. (1981), in Webb, C. and Hawtin, G.C. (*eds*) *Lentils*. Farnham Royal: Commonwealth Agricultural Bureau, 131-41.
41. Ladizinsky, G. (1979), *Euphytica* 28, 179-87.
42. Ladizinsky, G. (1979), *Journal of Heredity* 70, 135-7.
43. Ladizinsky, G. (1979), *LENS* 5, 24.
44. Ladizinsky, G. and Sakar, D. (1982), *Botanical Journal of the Linnean Society* 85, 209-12.
45. Ladizinsky, G. and Muehlbauer, F.J. (1983), *Collection and Assessment of the Wild Genepool of* Lens. Final Report, BARD Project, US 79-7, Pullman, Washington.
46. Ladizinsky, G., Braun, D., Goshen, D. and Muehlbauer, F.J. (1984), 'The biological species of the genus Lens L.' [submitted to *Evolution*].
47. Leung, H.K. and Salunkhe, D.K. (1985), in Graham, H. (*ed.*) *Chemistry and Technology of Tropical Food Crops*, Connecticut: Avi. (in press).
48. Muehlbauer, F.J. (1974), *Crop Science* 14, 403-06.
49. Muehlbauer, F.J., Slinkard, A.E. and Wilson, V.E. (1980), in Fehr, W.R. and Hadley, H.H. (*eds*) *Hybridization of Crop Plants*. Madison: American Society of Agronomy, 417-26.
50. Muehlbauer, F.J. and Slinkard, A.E. (1981), in Webb, C. and Hawtin, G.C. (*eds*) *Lentils*. Farnham Royal: Commonwealth Agricultural Bureau, 69-90.
51. Muehlbauer, F.J. and Slinkard, A.E. (1985), in *Proceedings of a Workshop 'Faba Beans, Kabuli Chickpeas, and Lentils in the 1980's'*. Aleppo, Syria: ICARDA. (in press).
52. Nandan, R. and Pandya, B.P. (1980), *Cytologia* 38, 195-203.
53. Nutman, P.S. (1969), *Proceedings of the Royal Society, London, Series B* 172, 417-37.
54. Nygaard, D.F. and Hawtin, G.C. (1981), in Webb, C. and Hawtin, G.C. (*eds*) *Lentils*. Farnham Royal: Commonwealth Agricultural Bureau, 7-13.
55. Papp, E. (1980), *LENS* 7, 8.
56. Paredes, O.M. and Tay, J.L. (1981), *LENS* 8, 1-3.
57. Renfrew, J.M. (1969), in Ucko, P.J. and Dimbleby, G.W. (*eds*) *The Domestication and Exploitation of Plants and Animals*. London: Duckworth, 148-72.

58. Renfrew, J.M. (1973), *Palaeoethnobotany, The Prehistoric Food Plants of the Near East and Europe*. New York: Columbia University Press.
59. Riva, E.A. (1975), *LENS* 2, 9-10.
60. Roughley, R.J. (1980), in Summerfield, R.J. and Bunting, A.H. (*eds*) *Advances in Legume Science*. London: HMSO, 97-104.
61. Saint-Clair, P.M. (1972), 'Responses of *Lens esculenta* Moench to controlled environmental factors'. Mededelingen Landbouwhogeschool Wageningen, Nederland, 72-12, 1-84.
62. Sakar, D. (1983), PhD thesis, Washington State University, Pullman, Washington.
63. Santiago, C. (1985), in Cubero, J.I. and Moreno, M.T. (*eds*) *Leguminosas de Grano*. Madrid: Mundi Prensa. (in press).
64. Sarwar, D.M., Kaul, A.K. and Puader, M. (1982), *LENS* 9, 22-3.
65. Saxena, M.C. (1981), in Webb, C. and Hawtin, G.C. (*eds*) *Lentils*. Farnham Royal: Commonwealth Agricultural Bureau, 111-29.
66. Saxena, M.C. and Hawtin, G.C. (1981), in Webb, C. and Hawtin, G.C. (*eds*) *Lentils*. Farnham Royal: Commonwealth Agricultural Bureau, 39-52.
67. Scott, W.O. and Aldrich, S.R. (1970), *Modern Soybean Production*. Illinois: S & A Publications.
68. Sharma, B. and Kant, K. (1975), *LENS* 2, 17-20.
69. Sheldrake, A.R. and Saxena, N.P. (1979), in Mussell, H. and Staples, R.C. (*eds*) *Stress Physiology in Crop Plants*. New York: Wiley-Interscience, 465-83.
70. Singh, H.P. and Saxena, M.C. (1982), *LENS* 9, 30-31.
71. Singh, K.B. and Singh, S. (1969), *Indian Journal of Agricultural Science* 39, 737-41.
72. Singh, S.R., van Emden, H.F. and Ajibola Taylor, T. (1978), *Pests of Grain Legumes, Ecology and Control*. London: Academic Press.
73. Slinkard, A.E. and Drew, B.N. (1982), *Lentil Production in Western Canada*. University of Saskatchewan, Saskatoon, Publication No. 413.
74. Solh, M. and Erskine, W. (1981), in Webb, C. and Hawtin, G.C. (*eds*) *Lentils*. Farnham Royal: Commonwealth Agricultural Bureau, 53-67.
75. Southgate, B.J. (1978), in Singh, S.R., vanEmden, H.F. and Ajibola Taylor, T. (*eds*) *Pests of Grain Legumes, Ecology and Control*. London: Academic Press, 219-29.
76. Sprent, J.I. (1980), *Plant, Cell and Environment* 3, 35-44.
77. Stebbins, G.L. (1967), *Variation and Evolution in Plants*. New York: Columbia University Press.
78. Summerfield, R.J. (1981), in Webb, C. and Hawtin, G.C. (*eds*) *Lentils*. Farnham Royal: Commonwealth Agricultural Bureau, 91-110.
79. Summerfield, R.J. (1981), in Hurd, R., Biscoe, P. and Dennis, C. (*eds*) *Opportunities for Increasing Crop Yields*. London: Pitman, 51-69.
80. Summerfield, R.J. (1981), *Preliminary Observations and Comments on the Chemical Composition of Lentil Seeds*, US Department of Agriculture, Legume Breeding and Production Internal Comm. No. 4, Pullman, Washington.
81. Summerfield, R.J., Minchin, F.R., Roberts, E.H. and Hadley, P. (1983), in Smith, W.H. and Yoshida, S. (*eds*) *Potential Productivity of Field Crops Under Different Environments*. The Philippines: IRRI, 249-80.

Lentil (*Lens culinaris* Medic.)

82. Summerfield, R.J. and Muehlbauer, F.J. (1981), *Experimental Agriculture* 17, 363-72.
83. Summerfield, R.J. and Muehlbauer, F.J. (1982), *Experimental Agriculture* 18, 3-15.
84. Summerfield, R.J. and Muehlbauer, F.J. (1982), *Communications in Soil Science and Plant Analysis* 13, no. 4, 317-33.
85. Summerfield, R.J., Muehlbauer, F.J. and Short, R.W. (1982), *Production Research Report No. 181*. Washington, DC: US Department of Agriculture.
86. Summerfield, R.J., Muehlbauer, F.J. and Short, R.W. (1982), Agricultural Research Service, ARM, Western Series No. 29. Washington, DC: US Department of Agriculture.
87. Summerfield, R.J., Muehlbauer, F.J. and Roberts, E.H. (1985) in Halevy, A.H. (ed.) *A Handbook of Flowering*. Florida: CRC Press. (in press).
88. Tahhan, O. and Hariri, G. (1982), *LENS* 9, 34-7.
89. Tay, J., Parades, M. and Kramm, V. (1981), *LENS* 8, 30.
90. Tikka, S.B.S. (1981), *LENS* 8, 19-20.
91. Tikka, S.B.S., Asawa, B.M., Goyal, S.N. and Jaimini, S.N. (1977), *LENS* 4, 17-20.
92. Tikka, S.B.S., Asawa, B.M., Goyal, S.N. and Jaimini, S.N. (1977), *LENS* 4, 20-22.
93. Van der Maesen, L.J.G. (1979) in *Publication IDRC-126e*. Ottawa: IDRC, 140-46.
94. Wassimi, N., Abu-Shakra, S., Tannous, R. and Hallab, A.H. (1978), *Canadian Journal of Plant Science* 58, 165-8.
95. Westphal, E. (1974), *Pulses in Ethiopia, Their Taxonomy and Agricultural Significance*. Agricultural Research Report 815, Wageningen: Centre for Agricultural Publishing and Documentation.
96. Williams, J.T., Sanchez, A.M.C. and Jackson, M.T. (1974), *SABRAO Journal* 6, 133-45.
97. Williams, J.T., Sanchez, A.M.C. and Carasco, J.F. (1975), *SABRAO Journal* 7, 27-36.
98. Wilson, V.E. and Law, A.G. (1972), *Journal of the American Society of Horticultural Science* 97, 142-3.
99. Wilson, V.E. and Teare, I.D. (1972), *Crop Science* 12, 507-10.
100. Zohary, D. (1972), *Economic Botany* 26, 326-32.
101. Zohary, D. and Hopf, M. (1973), *Science* (New York) 182, 887-94.

8 Chickpea (*Cicer arietinum* L.)

J.B. Smithson, J.A. Thompson and R.J. Summerfield

INTRODUCTION

Chickpea (*Cicer arietinum* L.), with a total annual production
of around 7 million tonnes of dry seed from an area of about
10 million hectares, ranks among the world's three most
important pulse crops [225]. The main producing country is
India, where the crop is the principal pulse and provides a
major source of protein in the diet of the predominantly
vegetarian population. However, in recent times in north
India, area and production have declined as chickpeas have
been displaced by the rapid expansion of irrigated areas and
the introduction of modern, productive cultivars of wheat.

   Two main categories of chickpea are distinguished, based
primarily on seed characteristics: the 'desi' types, having
relatively small, angular seeds with rough, usually yellow
to brown testas; and the 'kabuli' types, which have larger,
more rounded and cream coloured seeds [68] (Plate 8.1). The
desi types, also known as Bengal gram, constitute about 85%
of annual world production and are confined entirely to the
Indian subcontinent, Ethiopia, Mexico and Iran. The kabuli
(or garbanzo) types, comprise only a minor part of Indian
area and production, but account entirely for the crops of
Afghanistan through West Asia to northern Africa, southern
Europe and the Americas, except Mexico. Other, locally important, categories are the 'gulabi' (pea-shaped) types of
central India, and green-seeded desi types of central and

Chickpea (*Cicer arietinum* L.)

Plate 8.1. Desi (left) and kabuli type chickpea seeds

north-western India. Though distinguished principally by
seed appearance, desi and kabuli chickpeas differ also in
several other morphological and physiological features and
it will be one purpose of this chapter to compare and contrast the characters of these two major groups and to consider their implications for adaptation, utilisation and
crop improvement.

Ecologically, the crop is adapted to cool, dry conditions and grows almost exclusively on residual soil moisture
in the post-rainy seasons of subtropical winter or spring
of the northern hemisphere. Within these situations, chickpea crops encounter a diversity of physical and biological
environments, which exert profound influences on crop
growth, development and productivity.

In north India, the principal producing area, the crop
is sown from October to November and can produce dry seed
yields as large as 5000 kg ha^{-1} at the end of a growing
season lasting from 160-170 days. Elsewhere, the growing
season is restricted to 100-110 days, and potential seed
yields of rainfed crops are reduced to 1500-2000 kg ha^{-1},
southwards by the earlier onset of heat and moisture
stresses, and eastwards by the commencement of the rains.
Thus, maximum productivity declines from about 30 kg to
about 20 kg seed ha^{-1} day $^{-1}$.

Outside the Indian subcontinent, chickpeas are traditionally sown in spring, although in the Mediterranean
basin consistently large yield improvements have recently
been demonstrated from sowing in winter, provided the crop
is protected against *Ascochyta* blight [71]. The environments, growth durations and seed yields of these spring and
winter-sown crops bear striking similarities to those of the
peninsular and northern India situations and suggest at
least some degree of compatibility, which may be exploitable
in crop improvement.

While chickpea crops are attacked by fewer pests than
most other grain legume species, they suffer depredations
by several pathogens, some of which can be devastating. The

resultant instability of economic yield, and the tendency
of existing cultivars to produce excessive vegetative growth
when soil moisture is adequate or on fertile soils (which
aggravates disease problems, causes lodging and reduces
yields) are major factors in the crop's inability to compete
with wheat in areas or regions where irrigation is possible.

It is against this background that we describe the status, growth and development, and the constraints to productivity of the chickpea crop, and the prospects for yield
improvement. We shall not attempt an exhaustive literature
review but, instead, we refer readers to several reviews
published elsewhere [6,9,89,184,217,225,238] and mention
more specialised studies only where they are recent or pertinent.

HISTORY AND PRODUCTION

The genus *Cicer* and its relations

The taxonomy and distribution of the genus *Cicer*, and the
origins of the cultivated species have been recently reviewed in the contexts of floral ontogeny [225], plant diseases [6] and in earlier chapters of this book [144,218] and
so are not considered in detail here.

The genus comprises about forty herb and small shrub
species of which eight are annuals more or less closely related to the cultivated, *Cicer arietinum* L. [144,238]. The
wild species have been infrequently collected, are difficult
to maintain, and so have been inadequately studied. Nevertheless, several of them possess attributes which could be
usefully transferred to *C. arietinum*. For example, *C. judaicum* is said to be resistant to *Fusarium* wilt; some accessions of *C. bijugum*, *C. judaicum* and *C. reticulatum* are
resistant to *Ascochyta* blight [241]; and *C. pinnatifidum* is
resistant to *Botrytis* grey mould [196].

Although crosses between the wild and cultivated species
are generally unsuccessful, *C. reticulatum* can be readily

315

crossed with *C. arietinum* to form hybrids exhibiting only
slightly reduced chiasmata formation, pollen fertility and
seed set [102]. Hybrids have also been obtained from crosses
of *C. echinospermum* with *C. arietinum* and *C. reticulatum*
[102]; among *C. bijugum*, *C. judaicum* and *C. pinnatifidum*
[102,147]; and between *C. judaicum* and *C. cuneatum* [147]
and, though naturally invariably sterile, produced seeds
when hand-selfed or back-crossed to their parents.

The chromosome complements of all accessions of *C.
arietinum*, except the one designated 58 F (a white-flowered
introduction from Ethiopia), and *C. reticulatum* were structurally similar but differed from that of *C. echinospermum*
in respect of a single reciprocal translocation [102].
Based on these observations and evidence from seed coat
[100] and seed protein [101] characteristics, *C. reticulatum*
was suggested to be the progenitor of the cultivated species
[102]. However, the *C. arietinum* accession 58 F differed
from *C. echinospermum* by two paracentric inversions and
from *C. reticulatum* by a paracentric inversion and a reciprocal translocation, suggesting that the situation may be
more complex and deserving of further study.

The wild species merit increased attention. More extensive collection is needed urgently in order to assess the
extent of variability in their physiological, biochemical
and disease- and insect-resistance properties, as well as
their crossability. Also necessary are studies of the barriers to crossability and the application of techniques
such as embryo culture and protoplast fusion, which have
not yet been attempted in *Cicer*.

Domestication and dissemination

It appears extremely likely that chickpeas were domesticated
in the Fertile Crescent some 7000 years ago [6,152,225]. The
centre of diversity is in western Asia [242]; the postulated
ancestor, *C. reticulatum*, occurs in south-eastern Turkey
[101,102]; and the earliest record is from Hacilar in Turkey

[69], dated around 7450 BP, followed by findings at various early Neolithic sites throughout the Middle East.

The crop probably reached the Mediterranean area in the sixth or fifth millenium BP [152], where the kabuli type was differentiated. In India, the earliest record dates from about 4000 BP at Atranjikhera in Uttar Pradesh. The crop spread into peninsular India via overland routes or through maritime trade with the Middle East. Ethiopia probably received the crop from the Mediterranean; and in the sixteenth century, chickpeas were transported to the New World by the Spanish and Portuguese. The introduction of the kabuli type into India is probably even more recent, as is the spread of the crop into the USA, Canada and Australia, where it is now being cultivated.

Recent trends in area and production

World trends and fluctuations in chickpea production are dominated by events in the Indian subcontinent and especially in India which accounts for 77% of global production from 76% of the area devoted to the crop. In India, both area and production increased progressively until the early sixties, but between 1960 and 1978 declined at average annual rates of 10000 hectares and 25000 tonnes [157]. These declines have been attributed primarily to the expansion of irrigation and the introduction of modern cultivars of wheat in the mid-sixties (the so-called 'green revolution'), displacing chickpea from its traditional growing areas. Support for this suggestion derives from the greater relative declines in the states of Punjab, Haryana, Uttar Pradesh and Bihar, all on the alluvial soils of the Indo-Gangetic Plain, where the modern wheats have had the most impact; while in Rajasthan and Madhya Pradesh, the area and production of chickpea crops increased during this period [176]. However, although accelerated by the introduction of the productive wheats, these declines were in progress well before the mid-sixties [176], indicating the

expansion of irrigation alone to have been a factor in the
move away from chickpeas [35].

Area and production patterns are also characterised by
pronounced seasonal fluctuations (Fig. 8.1). These volatile
trends are associated with factors such as relative yields
[35] and prices [18,19], and annual rainfall receipts [18,
19,192]. Indeed, the area sown must be primarily determined
by the timely occurrence of rain during October and November
without which establishment is impossible on lighter soils,
but which on heavier soils can condition a swing to rainfed
wheat. The amount and distribution of rainfall also contribute, directly and indirectly, to seasonal fluctuations in
economic yield. For example, widespread and severe drought
in north India in 1979/80 resulted in a 35% fall in production, while the reduction in 1967/68 was due to heavier-
than-normal rainfall during the growing season, creating
conditions favourable for the development and spread of
leaf diseases, notably *Ascochyta* blight. The poor yields
reported for Pakistan in 1978/79 and in 1979/80 [112] were
also due to *Ascochyta* blight, which continued to wreak
havoc in the chickpea crops of both Pakistan and north India
during the two subsequent seasons.

On the Indian subcontinent, production instability and
overall poor productivity are responsible for the declining
area sown to chickpeas. With unit prices which range up to
two or three times larger than those of wheat, the smaller
crop yields from chickpeas can be competitive, but in terms
of stability and assurance of economic returns, chickpeas
are currently markedly inferior.

Similar influences have caused changes in area and production outside the Indian subcontinent. In Europe, since
1950, area and production (about 60% of which are accounted
for by Spain) have declined by average annual rates of
14000 hectares and 5500 tonnes to just over 130 000 hectares
and 100 000 tonnes in 1980 (Fig. 8.2), due to the inability
of the crop to compete economically with other agricultural
activities [37]. In Ethiopia, area and production increased

318

Chickpea (*Cicer arietinum* L.)

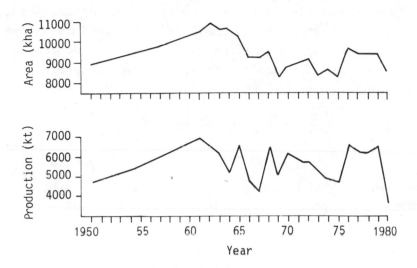

Figure 8.1   Trends in area and production of chickpeas in the
Indian subcontinent (1950-80) [48]

progressively up to 1973 but then fell abruptly as a consequence of drought in the mid-seventies and have since
failed to recover; averages are now about 200 000 hectares
and 80 000 tonnes annually, less than 50% of the former
maximum values.

North African area and production have shown few obvious
trends but have been characterised by pronounced seasonal
fluctuations due to the erratic onset and distribution of
rainfall.

The regions where area and production have increased
during this period are: the Near and Middle East (principally Turkey, Syria and Iran); North, Central and South
America (almost entirely in Mexico); and East Africa, though
the latter is due at least partly to the later inclusion of
first Tanzania and then Malawi in FAO statistics [48]. In
contrast, countries such as Australia, Canada, the USA and
Kenya are not included in FAO statistics [48] but their
areas and production are small and make little difference
to the general pattern. Everywhere, chickpeas have declined

in importance relative to other crops, pointing to a need for improvements in cultivars and cultivation practices, which present considerable challenges to breeders and agronomists alike. In the following sections, we examine current constraints to production and assess the progress being made to overcome them.

UTILISATION, COMPOSITION AND QUALITY

Uses

Chickpeas are probably used in more diverse food preparations than any other pulse. Shah Jehan, Moghul Emperor of India from 1628 -58, given a choice of food grains to accompany him in captivity, selected chickpea as capable of providing the most varied diet. A comprehensive treatment of the subject may be found in van der Maesen [238].

The crop is used predominantly as a pulse, but the manner of use varies with seed type and between regions. In the Indian subcontinent, the desi types are generally milled to remove the testas and produce a split pea composed solely of cotyledonary tissue and known as 'dhal'. Dhal is utilised either in the preparation of a thin spiced porridge of the same name, which forms an accompaniment to most Indian meals, or further ground to flour ('besan') for the preparation of fried, sweet or savoury snacks or besan curry. Whole chickpea seeds are spiced and eaten by north Indians as 'chole'. The by-products of the milling process (about 15-20% of the whole seed), composed mainly of testas, are used as cattle feed. Kabuli and green-seeded desi types are principally utilised whole in soups, curries and stews.

Outside the Indian subcontinent, the predominantly kabuli types are consumed as whole seeds in soups and stews or, increasingly, in developed countries, in salads as a 'health' food. In the Mediterranean area, cooked seeds are mixed with sesame oil and other flavouring to prepare a savoury paste ('hommos bi-tehineh') served as a side-dish

Chickpea (*Cicer arietinum* L.)

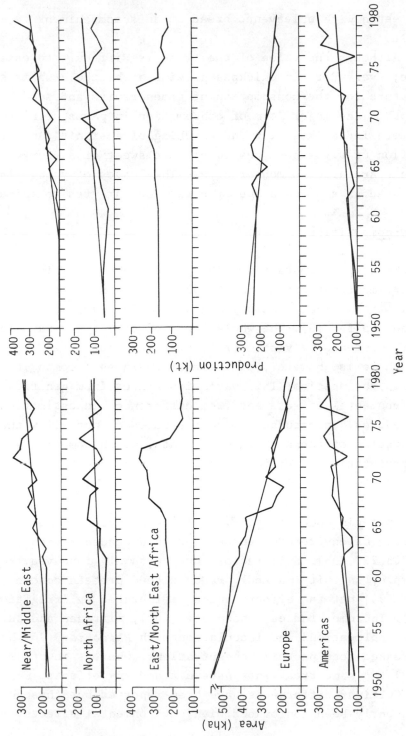

Figure 8.2 Trends in regional area (1000 ha) and production (1000 t) of chickpeas (1950-1980) [48]

and eaten with unleavened bread as an accompaniment to main
meals.

Although the value of the trade is difficult to estimate, everywhere in chickpea growing areas the crop is sold
immature for the consumption of green fruits and seeds,
probably the major form of consumption in parts of north-
eastern India. Roasting and parching of a significant proportion of all seed types to produce sweet or savoury
snacks are also common practice. The leaves of young plants
are used as a spinach and crop residues are fed to animals.

Seed composition

Chickpea seeds contain smaller protein concentrations than
other important grain legumes but, except for groundnuts
(*Arachis hypogaea*) and soyabeans (*Glycine max*), they are
richer in fat [3]. The whole seeds of desi types have large
fibre contents by virtue of their thick testas.

There is, however, wide variation in seed composition
within the species. This variation arises from genetic and
environmental effects and from differences in estimation
procedures. Unfortunately, most studies do not allow the
separation of these three components, which complicates the
interpretation of differences both within- and between-
species.

Seed protein concentration. There are clearly large
effects of environment on seed protein concentration [40,46,
92,208,227,233]. Soil factors are certainly involved. For
example, salinity markedly reduces seed protein concentration [94] but the effect can be ameliorated by irrigation
[75]. In some studies, phosphate [206], nitrogen and sulphur
[56] applications and inoculation with *Rhizobium* [76] have
increased seed protein concentrations, whereas in others
[159] the same treatments have not had any effects.

Cultivar differences have been reported but these have
frequently been confounded with environmental effects.

Nevertheless, genetic differences do exist [40,82,91,92,104, 158,178,208] and are demonstrable in suitably designed tests. Furthermore, one genotype, designated T-1-A, has consistently maintained very large protein concentrations in all tests [75,235].

Protein concentrations are often considered to be greater in kabuli than in desi type seeds, but the seed coats are much thicker in the desi (105-205 μm) than in kabuli types (37-106 μm) and so they account for a much greater proportion of the seed weight (9.7-19.4% compared with 3.7-8.8%) [82,103,104,209]. When decorticated samples are analysed [82], the differences in protein concentration between desi and kabuli seed types either disappear or are substantially reduced. Genotype × environment interactions have been reported [40,178,208,233] but, like cultivar differences, they are usually small compared with environmental effects.

Breeding and selection efforts for improved seed protein concentration have been limited. The trait has been shown to be a property of the maternal genotype [235]; broad sense heritabilities are large; and variation is both additive and non-additive [178]. The large seed protein concentration of T-1-A is being transferred to improved backgrounds [75], but appears to be associated with defective carbohydrate metabolism [210] and smaller seeds [98] and so may not be of practical value. More realistic approaches to improvement may be to ensure that the seed protein concentrations of breeding lines do not fall below those of existing cultivars and to use as parents genotypes combining moderate concentrations of seed protein with larger seeds.

Seed protein quality. Legume proteins are generally rich in lysine and arginine and most deficient in the sulphur-amino acids, cystine and methionine [24] (and see Chapter 3). In chickpeas, threonine, valine and tryptophan were thought to be deficient, but recent analyses [84,219] suggest that tryptophan is adequate and threonine and valine

323

are, at worst, marginal.

Variations both between- and within-species probably exist [3,84,91], but have been inconsistent, even between species. It was recently concluded [211] from fractionation, electrophoresis and amino acid analyses that the seed proteins of desi and kabuli types were similar. Other studies [101,235] also indicate that the seed protein profiles of chickpea cultivars of diverse origin are relatively uniform, so the prospects for improvement of protein quality by breeding may be limited.

Mineral concentration.   Whole legume seeds are rich in calcium, phosphorus, magnesium and potassium. Chickpeas tend to have larger concentrations of calcium and iron than other important grain legumes, except soyabeans.

In contrast with other seed components, cultivar differences account for a much greater proportion of the variations in mineral concentrations than environmental effects [83]. The seed coats of kabuli types are richer in calcium, zinc, copper, iron and manganese than those of desi types but, because of the greater proportion of seed coat in desi type seeds and the concentration of minerals in the seed coat, the differences between the mineral compositions of whole seed and dhal fractions of desi and kabuli types are small, except in the case of calcium.

Cooking and processing quality

Cooking and taste characteristics of chickpea seeds have been neglected by researchers and breeders and yet these could be key factors in consumer acceptance of improved cultivars. The cooking times of whole seeds are long and extremely variable, ranging from 55 minutes to longer than 200 minutes [249]. Cooking times differ considerably between genotypes [73,75,76] and are probably related to seed size [249]. They are also negatively correlated with seed phytic acid and magnesium concentrations and with meal texture as

324

measured in Instron units [75].

Cooking times (and cultivar differences) can be considerably shortened by pre-treatment of seeds. They are more than halved (to 22-50 minutes) in dhal samples [73,75, 76] or by pre-soaking in water (to 16-49 minutes) [75,248], and are further reduced by pre-soaking in salt solutions, for example, sodium chloride, carbonate and bicarbonate (to 10-28 minutes) [75,248]. The effects of environment on cooking characteristics are not known. In tests of seeds of five cultivars grown at two locations in India, cultivars were responsible for most of the difference in cooking time and other characteristics, while the effects of locations and the location × cultivar interaction were small (U. Singh, pers. comm.). Duration and method of storage and seed moisture content are likely to have important consequences, but have not yet been investigated.

Taste tests have not revealed differences in dhal preparations from desi and kabuli cultivars [76]. In kabuli types, large seeds are commonly preferred to small ones. However, seed size did not affect the preferences for 'hommos bi-tehineh' in tests in Syria [229]. Indeed, when prepared from desi type seeds, the dish was preferred to the traditional preparation, provided the testas were removed before processing [71.

Improved processing equipment has been developed [7,27] and studies indicate that milling [7] and puffing (P.P. Kurien, pers. comm.) characteristics are affected by genotype and environment, particularly storage conditions.

CROPPING SYSTEMS

Chickpeas are adversely affected by excessive moisture and hot temperatures and are therefore grown exclusively in post-rainy reasons and almost always on residual soil moisture, without irrigation. At low elevations and latitudes, they are generally sown in the autumn or early winter and they mature in spring; at higher latitudes they are sown in spring to mature in summer, but this pattern is disturbed by altitudinal differences. Distinct cropping systems and genotypes have evolved in response to the environments produced by these differing practices and situations. Their descriptions below are based on country reports presented at workshops and conferences at ICRISAT and ICARDA (see Chapter 18), reviews [238,247], and personal observations of the authors and their colleagues.

The Indian subcontinent

In the Indian subcontinent, including Burma, chickpeas are entirely autumn- or winter-sown and the desi types predominate. Several major producing zones may be distinguished, each with its own characteristic cropping patterns and constraints to production:

(a) the alluvial soils of the Indo-Gangetic Plains;
(b) the rice fallows of north-eastern India, Bangladesh and Burma;
(c) the sandy soils of southern Punjab and Haryana and northern Rajasthan states of India, and the Thal area of Pakistan;
(d) the black soils of central India; and
(e) the black soils of peninsular India.

Chickpeas are sown in October and November and mature from February to April. Crop duration ranges from about 110 days in peninsular India to 170 days in north-western areas, being curtailed by increasing temperatures and water stress. Soils are neutral to alkaline, and coarse sands to heavy clays.

Throughout the region, cultivation, sowing and weeding are accomplished by hand or by animal-drawn equipment. Fertilisers and insecticides are seldom used and crops are rarely if ever, irrigated, For harvest, the crop is cut or uprooted by hand, dried in small heaps in the field and then transported to a central area for hand or animal threshing.

Seeds are commonly sown by placing them into furrows
made by an animal-drawn plough, or by broadcasting. In the
north-eastern rice fallow areas, 'keshari' (*Lathyrus sativus*), lentils (*Lens culinaris*) and chickpeas are broadcast
into standing crops of rice ('utera' or 'pyra' cultivation).
Chickpeas also offer an alternative to blackgram (*Vigna
mungo*) for a similar situation in the rice fallows of the
Krishna/Godavari flood plains of eastern Andhra Pradesh.
In sandy soils, seeds are sown very deep (up to 25 cm) so
as to place them into a moist seed bed.

Chickpeas follow various rainy season crops. On the
alluvial soils of the Indo-Gangetic Plains, they are sown
after pearl millet, sorghum, maize, sugar-cane or guar
(*Cyamopsis tetragonoloba*) and eastwards, after sesame or
early rice. They succeed rice in the rice fallow areas. In
the sandy soils of Rajasthan and surrounding areas they
commonly follow pearl millet. On the black soils of central
and peninsular India, chickpeas are usually sown after a
period of fallow as these heavy soils are difficult to work
during the rainy season. Fallowing during the rainy season
is also common on sandy and alluvial soils; cultivation and
'planking' (i.e. dragging a flat, heavy implement across
the soil surface) are practised to conserve moisture; and
heavy rolling to raise moisture to the surface.

Chickpeas are frequently grown as a sole crop but are
also intercropped in alternate rows with sugar cane or between wide rows of rape or mustard (*Brassica* spp.); with
linseed, in the Indo-Gangetic Plains; with 'taramira'
(*Eruca sativa*) in the sandy soil areas; or with safflower,
in peninsular India. The crop is also grown mixed with
wheat, barley or linseed in the northern and central areas
of the subcontinent and with sorghum in peninsular India.

The 'ghed' area of the Gujarat state of India deserves
special mention. In coastal alluvium, inundated during the
rainy season, the entire area supports continuous post-
rainy season chickpea crops which give seed yields as large

327

as 3000 kg ha^{-1}. The 'bhal' area, also in Gujarat, with medium black soils inundated for four months of the year, supports mainly cotton in the post-rainy season but chickpeas are a very productive alternative crop.

Constraints to production also differ between areas. In north-western India and Pakistan, the main limitation is *Ascochyta* blight; further eastwards *Botrytis* grey mould becomes important. Another major cause of yield instability, and so of the failure of chickpeas to compete with wheat in irrigated areas in north India, is a tendency to produce excessive vegetative growth which causes crops to lodge, aggravates disease problems and reduces seed yields. In the north-east, the early onset of rains during the reproductive period can inhibit fruiting, delay maturity, and interfere with harvesting. Dry seed beds are an important constraint in Rajasthan and other areas with sandy soils. In peninsular India, the limited availability of water throughout crop growth is a major limiting factor and seed yields have been consistently doubled by timely irrigation. Soil salinity and alkalinity are important limitations in parts of the Indo-Gangetic Plain, in the sandy soils of Rajasthan and the bhal and ghed tracts of Gujarat.

West Asia, North Africa and Europe

Chickpeas are exclusively spring-sown (except in the Nile Valley) throughout an extensive and ecologically diverse region, extending westwards from Afghanistan through the Middle and Near East, into North Africa and southern Europe. Crops are kabuli types except in a small area in Iran where desi types are grown.

Several major producing zones may be distinguished:

(a) the mountainous areas of the Middle East (northern Afghanistan, Iran and Turkey);
(b) the lowlands of the Middle East (Iraq, Syria, Jordan, Lebanon and Israel);

(c) the Nile Valley;

(d) north Africa; and

(e) southern Europe.

In these five regions combined, pulse production is less than in the Indian subcontinent and chickpeas are superseded in importance by lentils in the Near East, by faba beans (*Vicia faba*) in the Nile Valley and north Africa, and by these species and others in southern Europe.

Chickpeas are sown between February and May, timed to coincide with rising temperatures in spring and the cessation of the winter rains. Crop durations vary from three to four months and growth is terminated by increasing temperatures and water stress in summer. In the Nile Valley, in contrast, chickpeas are sown from September to November and mature in the spring. The soils are clays to sandy clays, and neutral to alkaline.

Limited mechanisation of sowing and weeding operations has been accomplished throughout most of the area, but harvesting is still by hand and the crop is dried in small heaps in the field prior to hand, mechanical or animal threshing.

Chickpeas are sole-cropped throughout the region, although they are frequently seen as undercrops in olive groves in northern Syria. They are grown generally where annual rainfall exceeds 400 mm, in two-course (annual) or longer rotations with winter cereals (bread or durum wheats or barley), summer crops (melons) and forages. At the drier end of their range, three-course rotations include a fallow year, which is replaced by a summer crop or a forage as rainfall increases. In wetter zones, cereals and chickpeas are grown in alternate years.

The seeds are either drilled, hand-sown into furrows or broadcast followed by ploughing, to form rows, sometimes very far apart (for example, in north Africa, 2 m and wider), which are easily cultivable by animal or tractor-drawn equipment. Crops are rainfed, except in parts of the

329

Nile Valley. Phosphatic fertilisers are frequently applied and, in the Near East, the use of fungicides and insecticides to control *Ascochyta* blight and *Heliothis* larvae, respectively, are not uncommon.

Ascochyta blight is an important problem and, with rising labour costs, hand-harvesting is becoming increasingly uneconomic and has been the major factor in recent area reductions. Tall genotypes, with fruits borne well above the ground, are being evaluated as they will facilitate mechanical harvesting. Improved *Ascochyta*- and frost-resistance, weed control and (in highland areas) cold tolerance will be necessary if the practice of winter sowing is to be widely adopted as an alternative to seeding in the spring (and see Chapter 7).

Ethiopian and East African highlands

Chickpeas are cultivated at altitudes between 1400 m and 2300 m in a zone extending from Ethiopia in the north, through East Africa to Zambia and Malawi in the south. They are the chief pulse in the northern and central highlands of Ethiopia but southwards are of only recent introduction and of minor importance.

Sowing generally coincides with the end of the rainy season, from August to September in Ethiopia (harvesting January to February), and southwards from February to April (harvesting June to August). Where rainfall is bimodal, chickpeas may be sown at the ends of both rainy periods (e.g. in April and October in Kenya). The crop is usually sown on clayey soils which are neutral to alkaline.

They are almost always grown as monocrops but are also found, in Ethiopia, mixed with safflower and sorghum and, in Kenya, with maize. In Ethiopia, chickpeas occur in rotation with wheat, barley and tef (*Eragrostis tef*).

The seeds are broadcast or sown into furrows made by animal-drawn ploughs. Crops are seldom irrigated or fertilised and pesticide use is minimal. Weeding is by hand or

330

animal-drawn equipment; harvesting is by hand. The main problem appears to be the wilt/root-rot complex, which causes crop losses as large as 80%.

The Americas and Australia

The Americas and Australia constitute an extensive zone of considerable diversity in climate and cropping practices. Chickpeas are a minor crop except in Mexico, which is by far the largest producer. The crop is mainly kabuli, but desi types are grown in Australia and comprise 80% of Mexico's chickpea production for animal feed.

In South and Central America, chickpeas are sown in the autumn or winter and in rotation with maize, soyabeans, sesame, wheat or pasture. Operations are mainly by hand or animal-drawn equipment and the use of fertilisers, pesticides or irrigation is negligible.

In the USA and Australia chickpeas are of comparatively recent introduction. The main concentrations are in the states of California, Washington and Idaho in the USA, and south-eastern Queensland in Australia. Inputs are large and operations are completely mechanised from sowing to harvest, the latter being successful even though cultivars are of conventional plant habit. Current production is very small, but chickpeas offer promise as an alternative to spring cereals in areas where rainfall is marginal.

A principal constraint to production throughout the zone appears to be the combined effects of wilts and root-rotting fungi.

MORPHOLOGICAL AND PHENOLOGICAL CHARACTERISTICS, THEIR INHERITANCE AND RELATIONS WITH ADAPTATION AND PRODUCTIVITY

Morphology and development

The morphology of the cultivated chickpea and of its variants has been comprehensively reviewed [9,238] and only the main features are outlined here.

Chickpeas are herbaceous annuals. Germination is hypogeal and development commences with the elongation of a
main shoot, the first two nodes of which invariably subtend
small, scale-like structures. The imparipinnate leaves
arise alternately from the third node upwards. Mutants
with simple leaves are well known, and rare forms with an
altered, spiral phyllotaxy have been described [148]. Typically, leaflets range in number from eleven to eighteen per
leaf; they are subsessile, opposite or subopposite, and
ovate with serrate margins, but wide variations in each of
these traits occur. These involve expressions of extremes
of size, shape and number of leaflets, alternate arrangement and in one case, commonly known as tiny leaf [45], a
bipinnate form.

Plant habit varies from prostrate to erect. Those forms
which are taller than 100 cm are of interest for purposes
of mechanisation. Foliage colour ranges from pale yellow
through green to dark purple.

Branches are usually described as primary, secondary or
tertiary but all three categories can arise from the base
of the plant and, since they are often longer than the main
stem, are difficult to distinguish in the strict botanical
sense. We therefore prefer to classify branches as either
basal or upper, which is more easily applied and probably
is of greater agronomic significance. Basal branches arise
from the lower nodes of the plant, including the cotyledonary node if the main shoot is damaged, and are similar in
structure to the main shoot. Upper branches arise from the
nodes immediately beneath the first reproductive nodes of
the main stem and basal branches and their extension appears
to be initiated by the onset of reproductive activity.

Because of the association between branching and reproductive activity among existing cultivars, early flowering
types tend to have fewer basal branches and earlier and
much better upper branch development than later flowering
plants. In the latter, basal secondary branches arise on

basal primary branches and upper branch development is
delayed so that plant form resembles a basal rosette during
early and intermediate stages of growth. Moreover, late
genotypes in which flowering is accelerated by extending
daylength artificially to 24 hours assume a similar branching pattern to early flowering genotypes, with fewer basal
branches and earlier and better developed upper ones [76].
Branching characteristics have received little attention,
yet branch disposition and development must have important
influences on the timing and amount of economic yield and
so merits much more intensive study.

Flowers are borne singly on pedicels subtended by single
peduncles in the axils of the leaves. The normal condition
is one pedicel (and flower) per peduncle but double-flowered
genotypes are quite common. The proportion of double-flow-
ers which set fruit varies with genotype and environment
but, when well expressed, the 'double-podded' character
(Plate 8.2) contributes to slightly improved and more stable
yield [193]. The corolla is generally pink in desi types
and white in kabuli or gulabi types although blue, and various shades of blue and pink have also been reported [156].

Flowering commences on the main stem and basal branches
and, in southern India, proceeds acropetally at intervals
averaging 1.5 to 2 days between successive nodes (J.B.
Smithson, unpublished). Thus, the node number of the first
flower, which varies from around the fourteenth mainstem
node in early plants to the thirtieth node or above in late-
flowering plants, accounts for most of the variation in
time to flowering at Hyderabad (17°N).

Factorial experiments carried out in controlled environments on a number of each of desi and kabuli genotypes show
that this species is a quantitative long-day plant, but
that the time taken to flower can also be strongly affected
by temperature. If rates of progress towards flowering are
considered as the reciprocals of the times taken to flower,
the evidence suggests that there is a positive linear relation between both mean temperature, photoperiod and rate of

333

Plate 8.2. Branches of 'double-podded' (left) and conventional
desi chickpea plants.

progress towards flowering, with no interaction between
these two environmental factors. Thus:

$$\frac{1}{f} = a + b\overline{t} + cp \tag{1}$$

where f is the time taken to first flower (days), $\overline{t}$ is mean
air temperature (°C), p is photoperiod (h), and a, b and c
are constants the values of which vary between genotypes
(E.H. Roberts *et al.*, in preparation).

The first flowers to appear may desiccate and abort at
different stages before opening and have been described as
'pseudo-flowers' [14]. The numbers of pseudo-flowers vary
among genotypes [76] and are greater in shorter photoperiods.

cool temperatures and where moisture supply is ample [76, 223]. The agronomic significance of pseudo-flower production is not clear but it could have adaptive value in prolonging vegetative growth and thereby increasing source size in favourable conditions. It may also explain some of the anomalies we have encountered in interpreting flowering records. Flowers may also abort after opening, especially if nights are cold [76]. This phenomenon is an important feature of crop development in north India and, like pseudo-flower production, may be of adaptive significance, delaying reproductive growth until conditions become favourable for fruit development and thereby prolonging vegetative growth and increasing yield potential. Genotypes capable of fruiting in cold temperatures have been identified [76] and may have value in reducing the excessive vegetative growth which can otherwise occur in north India.

Chickpea is essentially self-pollinated. Reported degrees of out-crossing are less than 2% [51,130]. The fruit is an inflated pod containing between two and four ovules of which one or two usually develop into seeds. Genotypes having as many as six ovules per carpel have been identified but seldom do they all become seeds [184]. Fruits containing six seeds have been observed but expression of the multi-seeded character is variable and influenced strongly by environment (O. Singh, pers. comm.).

Following fertilisation, the pedicels recurve so that fruit development proceeds beneath the leaf canopy. Pedicel mutants [146], in which recurving does not occur and the fruit develops above the canopy, have been reported but the character does not appear to contribute to improved yield [186]. Various other rare abnormalities of the reproductive organs have also been described including female and/or male sterility, polycarpy and other distortions of the floral parts. Male sterility may be useful for hybrid production or for the adoption of population improvement strategies but expression of the character is unstable [74].

Average seed weight ranges from about 40-600 mg. Testa

335

colour varies from black through shades of brown to orange
and yellow to white and even green. Seed shape varies from
irregular to rounded and the testa texture may be smooth,
rough or rarely tuberculate. A productive traditional desi
plant type at reproductive maturity is shown in Plate 8.3.

Plate 8.3.   A productive desi chickpea plant
at reproductive maturity.

Inheritance of qualitative characters

Studies of the inheritance of many characters in chickpea
have already been listed [9,238]; we will indicate only the
main conclusions here and draw attention to studies of
potential interest in crop improvement.

Flower colour.   The inheritance of pink, blue and white
corolla colour has been attributed variously to segregation
at one, two or three independent loci. The most plausible
hypothesis [13] is that three factors (designated C, B and
P) are involved, of which C and B are complementary and P

is supplementary to B. The corolla is pink when all three factors are present in the homozygous dominant condition; blue when only P is absent; and white with all other constitutions.

More recent studies are not explicable in these terms, but in one case [142], the expected ratios in F_3 are wrongly calculated. In others, only F_1 and F_2 data were reported and there was no attempt to relate between crosses; in one case a single genotype (Chikodi V.V.) was allotted three different genic constitutions in separate studies [164,165,166]. However, the appearance of white flowers in crosses of pink-flowered parents [142] and the recognition of shades of pink and blue, indicate the need for further studies of the inheritance of flower colour, which could provide information of considerable evolutionary value.

Seed characters.   The inheritance of seed characters is rather more complex than corolla colour. Parents and methods of classification used and gene symbols assigned vary among studies, while seed coat colour is known to change during seed development and ageing.

It is clear that several factors are involved, that each interacts with the others, and that some have pleiotropic effects. Thirteen seed-colour categories ranging from dark brown to yellow were explained in terms of segregation of five loci, two of which (B and P) were also concerned in the inheritance of corolla colour [13]. Two loci (B and F^r) also conferred round seed shape, B reduced seed size, and F^r caused 'puckering' of the seed surface, but their effects were modified by certain combinations of the other factors. Later, the operation of two more loci was proposed [17], one of which (T_4) caused black testa colour. It is difficult to relate between studies, especially in view of different interpretations of flower colour inheritance, but other data in general indicate the operation of several interacting factors and can be explained

337

in similar terms.

A recessive allele (g^r) causes green colour of the cotyledon and seed coat and blue-green foliage. Black versus cream testa colour has been attributed to the segregation of as many as four loci [131] but this needs substantiating. One source of black testa colour is unstable [17, 132], so that black, mottled and mixed seed types can occur on the same plant. The inheritance of rough and smooth seed surface has been ascribed to the segregation of one to three loci [123]; and the presence of tiny spines on the seed coat (spinate) is controlled by two complementary genes.

The seeds of desi and kabuli types differ in several respects including size, colour, shape and testa texture and thickness. Although kabuli genotypes were included in the studies summarised above, the inheritance of the characters distinguishing them was not directly examined. However, the infrequent recovery of plants with kabuli type seeds from desi × kabuli crosses suggests that kabuli seed characters are conferred mainly by recessive alleles.

Plant habit.   Basal branching appears to be dominant over absence of basal branches; erect and semi-erect plant types are dominant over spreading forms; and various dwarf mutants have been due to a single recessive allele.

Other characters.   The inheritances of other leaf and foliage characters all appear to be conferred by single recessive alleles.

Character association.   Considering the large number of variants which have been studied genetically, remarkably few cases of character association have been reported until recently. Some of the loci affecting corolla colour have pleiotropic effects on seed characters but this has been considered a case of very close linkage [120]. Corolla, stem and pedicel colour [38,118,141,164,165], and green seed coat, cotyledon and foliage colour [119] appear to result from pleiotropic action.

Cases of linkage include: branching habit, leaflet arrangement and earliness [21]; seed coat colour loci, F^r and T^2 [20]; bunchy habit with one of the loci controlling seed coat colour [25]; corolla colour, flower number per axil, seed coat colour and seed shape [38,118,147]; tiny leaf and corolla colour [164]; and one of the spinate seed loci with a locus affecting seed coat colour [123]. In contrast, there was no evidence for linkage among seven loci controlling corolla and seed coat colour [17], nor among eight loci controlling several mutant forms [11].

Quantitative variation

Various forms of analysis of quantitative characters and their relations have been conducted with chickpea. Results have differed depending on the nature of the plant materials used, the environments in which tests have been performed, and the method of analysis; but some conclusions are possible.

Character association.   Fruit number per plant has been significantly positively correlated with seed yield per plant in all of more than sixty cases we have seen reported, with r values ranging from 0.28 to 0.96 [117]. Where seed numbers have been determined, this character has also been significantly and positively correlated with seed yield and with fruit number per plant. Path coefficient studies confirm that fruit and seed numbers have the largest direct effects on seed yield.

With few exceptions, seed yield is also significantly and positively correlated with the number of branches per plant. Both primary and secondary branches play important roles and operate principally through fruit number, which has a strong positive association with branching, and through which branching frequently has large indirect

339

effects on seed yield in path coefficient studies.

Associations between seed yield and other characters are smaller and/or less consistent, differing with the nature of the materials used and the environment in which they were tested. However, seed yield is almost always positively correlated with seed number per fruit and, although this association is significant in only about a third of the cases reported, we have not seen any published data indicating a significant negative association between the two characters.

Seed yield and seed size are also often significantly and positively correlated (about 40% of the cases reported), though negative associations have been recorded (10% of cases), and the enhancing effects of increased seed size on yield are frequently countered by its negative associations with fruit number per plant and seed number per fruit (40% and 50% of cases, respectively). Path coefficient studies also show that while seed size often has positive and direct effects on seed yield, its indirect effects through other characters are negative.

Correlations of seed yield with time to flowering and to maturity vary from positive to negative or are absent depending on the duration of the growing season and the material studied.

Genetic variation.   Estimates of other genetic parameters are more conflicting. Combining ability analyses of F_1 and F_2 populations of diallel and line × tester sets indicate genetic variation to be additive in some cases and non-additive in others, varying with parental combinations, which have always been small in number and an inadequate sample of the variability available; the generation studied and the environment [52,59]; and the character measured. Based on current evidence, and although genetic variation for some characters such as seed size and time to flowering is additive in nature, other yield components (for example, the numbers of fruits, seeds and branches per plant) and

340

seed yield are subject to predominantly non-additive gene
action [137]. The predominance of non-additive gene action
is reflected also in significant and very large heterosis
for seed yield and fruit and branch numbers (up to 163%
over the better parent for seed yield) [22] and the consequent inbreeding depression.

Similar inferences may be drawn from the relative mag-
nitudes of heritability and genetic advance [86]. While
most estimates are inordinately large (for heritability of
seed yield they range to more than 90% [58,160]), they are
broad-sense estimates calculated from the total genetic
variance among genotypes in single environments and are
markedly inflated by non-additive variation and genotype ×
environment interaction. Thus, narrow-sense estimates are
much smaller. For seed yield the largest reported is 44.1%
[12] declining to as little as -2.9% [151].

Heritability estimates are further reduced when genotype × environment interactions are taken into account. For
seed yield, genotype × environment variances are often
greater than genetic variances. Stability analyses [43]
indicate that both linear and non-linear components are
important [50,105,115,153,207] and that genotype × location
and genotype × agronomic practice interactions are evidently
larger than genotype × year effects [30,154], although the
latter are based on relatively few seasons.

Variability and intraspecific classification

There have been few attempts at detailed analysis of the
total spectrum of variability in cultivated chickpeas.
Earlier classifications distinguished four subspecies [145]
or races [238]. Some recent studies have been conflicting
because they have focussed on quantitative morphological
and physiological characteristics which although of considerable adaptive importance, are prone to marked environmental modification. Thus, D^2 and canonical analysis of a
chickpea germplasm collection of more than 5400 accessions

341

grown in New Delhi identified, on the one hand, plant type
[122] and, on the other, seed size [121] as the most important cause of genetic divergence. Also, factor analysis
showed growth attributes, including seedling establishment
and photosynthetic activity, to be the most important in
one study [121] and reproductive capacity in another [122].
Similarly, thirty-two genotypes formed different clusters
at each of the eight locations in which they were grown in
north India [81], while in other studies, the same genotypes invariably fall into different groups. There was no
evidence of association between geographic origin and genetic diversity in any of these studies.

In contrast, correlation and principal component analysis of 150 accessions, representative of a larger collection
of chickpeas grown in Spain, distinguished two major groups
- the *microsperma* and the *macrosperma*, based on fruit, seed
and leaf characters [120]. The *microsperma* are characterised by small fruits (less than 23 mm long), seeds (lighter
than 35 mg), leaves (rachis shorter than 4 cm) and leaflets
(less than 12 mm long). The seeds (one to three per fruit)
are diverse in form and colour and plant parts are pigmented. Furthermore, they are concentrated at the eastern
end of the distribution of the species and conform broadly
with the desi type. The *macrosperma* have large fruits,
seeds, leaves and leaflets. Their seeds are mainly white
or pinkish, strongly sheep-headed with rough testas, and
rarely more than one per fruit. The plants usually lack
pigmentation, so that flowers are white and vegetative parts
are yellow-green. The *macrosperma* are most frequent at the
western end of the distribution of the species and correspond broadly to the kabuli types. Desi and kabuli types
also formed distinct groups in other studies in north India
[15,81]. The kabulis exhibited greater within-group divergence than the desis, and had more primary branches, larger
seeds and were taller but they had fewer seeds per fruit
and fewer fruits and secondary branches per plant [15].
Seed characteristics thus appear to be a more consistent

basis for grouping than other characters and have a greater practical significance.

SYMBIOTIC NITROGEN FIXATION

As in other legumes (see Chapter 4) the factors which can affect the establishment (nodulation) and functioning (symbiotic nitrogen fixation) of the *Rhizobium*/chickpea symbiosis are numerous, and the interactions among them are complex. Thus, it is not surprising that nodulation, nitrogen fixation and responses to inoculation are extremely variable, or that the factors and the interactions among them are little understood in this crop.

Variation in nodulation and nitrogen fixation and responses to inoculation

Considerable variation in nodulation has been recorded in chickpeas in India [174] and in countries in west Asia and north Africa [71,79]. Nodulation is often extremely poor or fails completely. In Syria, for example, uninoculated chickpeas only nodulated at two out of seven locations whereas faba beans and lentil crops nodulated at all sites [71].

Both nodulation and nitrogen fixation activity vary between environments. In Vertisols at Hyderabad, nodules are concentrated on the taproots and within 15 cm of the soil surface; almost all nodules occur at soil depths less than 30 cm [187]. Nodules develop rapidly and nitrogenase activity is measurable within 20 days of sowing. Nodule numbers and weights and nitrogen fixation activities are largest between 50 and 70 days after sowing but then decline, so that by mid-fruiting (70 to 80 days after sowing) fixation has ceased. Further north in India, at Hissar (29°N), nodules are borne on roots at greater soil depths; their numbers, weights and nitrogen fixation activities are much greater than at Hyderabad; and nitrogen fixation activity persists for much longer (until 130-150 days after sowing) and so extends further into the seed filling period [75].

The quantities of nitrogen fixed by chickpeas are extremely variable. Estimates of these quantities range from 1-114 kg N ha^{-1} (by difference in uptake from a non-legume) in farmers' fields in Syria [71]. The earliest estimate (103 kg N ha^{-1}) is ascribed to Rizk in Egypt [133], others, from experimental stations, are: 42 kg N ha^{-1} (in a spring-sown crop) and 75 kg N ha^{-1} (winter-sown) at Tel Hadya in Syria [71], and 63 kg N ha^{-1} at New Delhi in India [221] (by ^{15}N dilution techniques); and 104 kg N ha^{-1}, in Punjab state of north India [139] (by acetylene reduction). Also in north India, chickpeas supplied 60-70 kg ha^{-1} more N than fallow to a succeeding maize crop [5]; and, in Syria, the yields of wheat following chickpeas were at least equal to those of wheat following fallow [71].

Responses to inoculation also vary between environments. In India, for example, seed yield responses to inoculation have often been marginal and seldom statistically significant [183]. In comparison, significant yield responses to inoculation were reported from ten out of seventeen [226] and six out of twelve [220] locations mainly in central and northern India, and all on experimental stations. At Hyderabad, in traditional situations, significant yield responses to inoculation have rarely been obtained. A similar situation exists in Syria, though at Tel Hadya in 1980-81, substantial improvements in seed yields were reported [70].

In very few cases have the reasons for the variations in nodulation, nitrogen fixation and in responses to inoculation been satisfactorily established but some of the factors probably involved are discussed below.

Factors affecting nodulation, nitrogen fixation and responses to inoculation

Specificity of host and rhizobia. The rhizobia which nodulate chickpeas also nodulate wild *Cicer* species (O.P. Rupela, unpublished) but do not consistently nodulate other

344

genera with the possible exception of *Sesbania* [49], thus supporting the taxonomic status of *Cicer* as a monogeneric tribe. This marked specificity has enabled dramatic responses to inoculation in those areas where chickpea or its relatives are not normally grown: for example, in Australia [36], Israel [134] and even in areas of India where the crop was sown for the first time.

Despite this marked specificity there is certainly diversity among populations of *Cicer* rhizobia. When tested against eight anti-sera, twenty-four isolates collected from ten chickpea cultivars growing on a research station in north India cross-reacted, but fell into eleven different groups [39]; four other isolates did not cross-react. Agglutination tests distinguished seven groups among other isolates collected in north India [28].

Although the term 'interaction' frequently appears in the literature, host × strain interactions in the statistical sense are comparatively rare and, even when evident for seed yields, they may have no obvious biological relation with other variables such as nodule weight [149] and competitive ability [57]. In Australia, field studies in the absence of natural rhizobia did not show interactions between chickpea cultivars and strains of *Rhizobium* [36]. Conversely, interactions between cultivars and strains of *Rhizobium* have been reported in both field [78,174] (O.P. Rupela, unpublished) and controlled environments [222] - at least for nodulation if not for seed yield. However, significant interactions for plant weights and N uptakes were recorded between ten chickpea cultivars and ten *Rhizobium* isolates from an area of India where chickpeas are traditionally grown [39]. So there is some evidence for the existence of specificity between cultivars and strains of *Rhizobium* and it is possible that this would be observed more frequently were it not obscured by interactions among the many other factors involved.

Soil populations of rhizobia.   Variations in the numbers

of rhizobia present in soils must be a major cause of differences in nodulation and responses to inoculation in this crop.

Numbers of rhizobia are large in the alluvial soils of the Indo-Gangetic Plain (generally exceeding 1000 rhizobia g soil^{-1}); they are smaller at Gwalior in central India (10-1000 g soil^{-1}); and in vertisols at Hyderabad in peninsular India [234] they range from 100 to more than 1000 g soil^{-1}. At Hyderabad, numbers are smaller in alfisols, soils in which chickpeas are seldom grown, and following rice paddy (<100g soil^{-1}) [73]. In vertisols at Hyderabad, numbers are largest at soil depths of 5-15 cm, but rhizobia are still present below 1 m [234].

There are serious deficiencies in the information available on nodulation and responses to inoculation. Poor nodulation in farmers' fields may or may not be due to lack of rhizobia but experiments to test responses are almost always conducted on research stations, and soil populations are seldom quoted. Interestingly, responses have been found even under these conditions [220,226] and so they may reasonably be expected in farmers' fields too, especially where numbers of rhizobia are small.

At Hyderabad, seed yields of chickpeas have been increased by inoculation in rice paddies [174] and alfisols [76], where soil rhizobia populations are small, but not in other situations.

Chickpea rhizobia have proved less tractable to the development of high titre anti-sera than other strains of *Rhizobium* (J.A. Thompson, unpublished) so that studies to monitor the success of infection, which are usually based on serological techniques, are uncommon. Strains from northern India used as inocula differed in competitive ability in the face of local strains [28]. This finding is common but deserves further investigation.

The soil environment.   There is considerable scope for research on the soil factors which influence *Rhizobium*

346

concentrations, and their ability to infect the host and enhance nitrogen fixation and seed yields.

The infection and fixation processes are both adversely affected by hot soil temperatures - no nodules are formed at temperatures warmer than 32°C and nitrogenase activity fails to recover after plant roots are subjected to 35°C [41]. Other studies in controlled environments [116] indicate that infection is inhibited at mean diurnal temperatures warmer than 24°C. For nodulation and nitrogen fixation, optimal mean temperatures appear to lie between 18°C and 22°C. Nodulation is also better, and more nitrogen is fixed, in shorter than in longer days [41,116].

At Hyderabad, in 1981, soil temperatures at a depth of 10 cm exceeded 30°C for 6 h each day for the first 60 days after sowing, while at Hissar they did not exceed 30°C until 160 days, so that temperature differences may be at least partly responsible for the differences observed between locations in nodulation and nitrogen fixation activities.

There is good evidence that nodulation and nitrogen fixation are affected adversely by water stress (see Chapter 4) At Hyderabad, irrigation increased the nodule weights of chickpeas ten-fold, and nitrogen fixation activity four-fold, compared with unirrigated treatments [74]. One effect of irrigation may be to enable the migration of rhizobia from the surface of inoculated seeds into the root zone [76]. Growth and nodulation responses at Hyderabad, India (O.P. Rupela, unpublished) and in Syria [80] were greater following inoculation with a soil-applied liquid suspension than with conventional seed application.

Rhizobial populations presumably adapt to the soil environments in which they occur. It has been claimed that strains of *Rhizobium* should be used (as inoculants) only in soil and climatic conditions similar to their original habitat [189], and that inoculant strains are seldom better than natural populations [183]. However, there is little direct evidence for these assertions in spite of the popularity of collecting local strains [179]. Differences among strains

347

in sensitivity to hot temperatures have been recorded in
controlled environments in the UK [41] and in India (O.P.
Rupela, unpublished); and in Sudan, chickpeas nodulated
better and produced heavier seed yields in a saline soil
when inoculated with a strain of *Rhizobium* isolated from a
saline soil in India than with another strain of equal
effectiveness [174].

The effectiveness of inoculation is also influenced by
the quality of the inoculant used. To the authors' knowledge, the only important chickpea-producing countries wher
inoculants are commercially available are Israel, where
inoculants are subject to careful control and are of good
quality, and India, where quality control is minimal and
inoculants are generally poor [232]. The latter observation
could explain some of the variation in response to inoculation in India, and also raises the problem of deterioration
of inoculants during distribution.

The host.   Nodulation and the quantities of nitrogen
fixed must be extremely dependent upon the general 'health'
and 'vigour' of the host, and differences in these complex
attributes may account for many of the interactions reportec
in the literature. For example, the differences between
Hyderabad and Hissar chickpea crops in nodulation and nitrogen fixation may arise from differences in growth - crop
durations are prolonged at Hissar and result in far greater
biological yields than at Hyderabad.

Nevertheless, large and repeatable differences among
chickpea genotypes in nodulation and quantities of nitrogen
fixed have been established [78,174] and nodulation parameters and seed yield have been well correlated [175]. The
exploitation of these differences in breeding has been ham-
pered by the lack of satisfactory screening techniques [116]
leading to large plant-to-plant variability in nodulation
and nitrogen fixation even within lines considered to be
pure.

However, some progress has been made. The good

348

correlations among nodulation parameters have been utilised to develop a rapid visual score for nodulation [175], so that plants may be repotted for seed production following assessment. Methods of rooting cuttings are also available [42,173,175]. Crosses of well- and poorly-nodulating lines have been made and generations advanced using these methods [76].

RESPONSES TO FERTILISERS

There are no reports of visible symptoms of nitrogen, phosphate or potash deficiency in chickpea crops [184] and yet yield responses to nitrogen and phosphate applications are common [182], although responses to phosphate are less consistent and frequently small. An economic analysis of 123 fertiliser experiments conducted in India indicated that 30-34 kg N ha^{-1} was profitable for rainfed chickpeas especially on alluvial soils, but responses to P were small and often uneconomic [150]. Surprisingly, neither nitrogen nor phosphorus applications appeared profitable for irrigated chickpeas, possibly because of the excessive vegetative growth and lodging which can result from irrigation [184].

Lack of response to N and P has been attributed to intense fixation of P by soils [182] and the accumulation of applied nutrients in the upper soil layers, which dry out soon after sowing and so do not release nutrients to the deeper active roots [75]. However, neither foliar nutrient application [182] nor deep placement [75] has proved any more effective than conventional broadcast applications. Also, in Syria, chickpeas failed to respond to the application of phosphate in soils where lentils and faba beans gave good seed yield responses. So, the uptake of P may have been more efficient in chickpeas than in the other two legumes and symbiotic nitrogen fixation was probably adequate to meet the chickpeas' N requirements [182].

PRINCIPAL LIMITATIONS TO YIELD

Potential chickpea yields are large. More than 3000 kg ha^{-1}
of dry seeds have been consistently harvested from large,
irrigated plots in peninsular India [73,75,76] and even in
excess of 5000 kg ha^{-1} in north India and, more recently,
at Lattakia in Syria [71].

That such yields are not attained by farmers is due at
least partially to the relegation of chickpeas to marginal
lands, without irrigation or other inputs. However, with
existing cultivars, even where inputs are given, large
yields cannot be guaranteed. Susceptibility to diseases and
pests, sensitivity to salinity and other edaphic conditions,
and tendency to excessive vegetative growth at the expense
of fruiting in favourable conditions all contribute to poor
and unstable yields.

Diseases

Fifty-two pathogens of chickpea, including eleven species
of nematodes, have been documented [125]. Their distribution and importance [6,124] are determined principally by
rainfall and temperature. Apart from *Ascochyta* blight, individual pathogens are responsible only for locally severe
crop losses. Collectively, plant diseases are probably the
principal cause of the instability which characterises the
yields of this crop. Here, discussion is confined to the
four major diseases: wilt (caused by *Fusarium oxysporum* f.
sp. *ciceris*), blight (*Ascochyta rabiei*), grey mould (*Botrytis cinerea*), and stunt (Pea Leaf Roll Virus).

Fusarium wilt.   Numerous pathogens have been isolated
from wilted plants and various causes of wilt have been
proposed. Recently, the separate agents of the so-called
'wilt complex' have been diagnosed [126] and forms of *Fusarium oxysporum, Rhizoctonia bataticola, R. solani, Sclerotium rolfsii, Sclerotinia sclerotiorum* and *Operculella*

padwickii are now established as the principal causes of this syndrome.

Predominant among the wilts and root-rots is *Fusarium* wilt, incited by *F. oxysporum* f.sp. *ciceris*. It is widespread, being reported throughout the Indian subcontinent and in Burma, the USSR, Ethiopia, Malawi, Tunisia, Spain, the USA, Mexico and Peru.

In northern India, the disease is expressed either in the seedling stage or after flowering, being suppressed at least partially in the intervening period, probably by cooler temperatures, which are known to reduce the incidence of wilt [34]. However, late-wilting genotypes have been identified [62,236] in which the time of wilting is also delayed beyond the seedling stage.

For individual plants, yield reduction is greater with earlier wilting [62], but crop losses may be smaller due to compensatory growth of healthy plants. In India, losses have been estimated between 10% and 15% [114,124,197], but can be more severe locally. For example, there are reports of up to 70% mortality in Bihar [230,231], although other pathogens may have been involved as well.

The fungus is soil-borne and can survive in organic matter for longer than four years [76]. It is systemic and has been isolated from all parts of infected plants, including seeds [129]. It also infects other grain legumes, including pigeonpeas, lentils and peas [63].

Because of its seed-borne nature and its longevity, it is not easy to prevent infestation nor to eradicate this pathogen once established. Seed dressing with Benlate-T eliminates seed transmission [66] which is less in resistant than in susceptible cultivars [129]. Crop rotation is relatively ineffective. In north India, wilt incidence was reduced by late sowing [135] and on heavier, more acid soils [33,44].

Host plant resistance offers the most effective means of control and the basic requirements for its utilisation already exist [127]. Infested field ('wilt-sick') plots

351

have been developed at several centres in India, and in
Algeria, and are extremely suited to the screening for
resistance of large numbers of germplasm and breeding
materials. They offer the additional possibility of identifying resistance to other soil-borne pathogens which may be
present. Since the first report of resistance [109] sixty
years ago, many other sources of resistance have been identified [128] and fifty-eight genotypes, all desi types,
which have shown resistance in field and pot tests at Hyderabad are listed from a germplasm collection exceeding 12000
accessions. Resistant sources have been used in breeding
programmes and have been successfully transferred to improved desi [140] and kabuli [96,97] backgrounds.

Pot and water culture techniques have also been developed and used to study the specificity and the inheritance
of resistance. Evidence has been presented for cultivar-
specific pathogenicity [32,64]. The latter report distinguishes four races among five isolates of the pathogen,
collected from different locations in India, but the distribution and importance of the races are not yet known.
Inheritance studies [61,95,236,237] indicate that resistance
to race 1 is controlled by at least two loci. Recessive
alleles at each locus separately result in late wilting and
together confer almost complete resistance. More recent studies [177] indicate the presence of third alleles at both
loci which singly confer strong resistance. Studies of both
aspects are continuing and are vital to the development of
an effective breeding strategy.

Ascochyta blight.   Among the diseases of chickpea, the
most serious cause of crop loss is unquestionable blight,
caused by *Ascochyta rabiei*. Disease development is most
rapid at temperatures close to 20°C [246] and requires continuous moisture cover of the leaf surface for at least 6
hours to facilitate release of pycnidiospores, so that disease incidence is closely correlated with rainfall [108].

The pathogen causes appreciable damage in most years,

over an area extending from north-west India, across northern Pakistan and through west Asia and the Mediterranean
region, as far as Spain and northern Morocco. In west Asia
and the Mediterranean, it is especially damaging to winter-
sown chickpeas, a practice which is still in the experimental phase but may soon be extended to commercial cultivation, although *Ascochyta* also causes losses in the traditional, spring-sown crop when conditions are favourable
for disease development. In Pakistan and north-west India,
disease incidence is greater in wetter years (when it is
also evident in areas to the south and east) as occur
periodically in north India (for example in the 1967-68 and
1981-82 seasons). In Pakistan, chickpea production was
halved by *Ascochyta* blight during the two seasons from 1979
to 1981 [136] and damage was equally severe in 1981-82.

A. *rabiei* can survive on infected plant debris in the
soil for more than two years [108]. It is also seed-borne
to the extent of 70% [110] and is capable of remaining viable in seeds for at least two years [87]. Alternate hosts
have not been recorded previously but recently *A. rabiei*
has been found to be pathogenic on field peas (*Pisum sativum*), common beans (*Phaseolus vulgaris*) and cowpeas (*Vigna
unguiculata*) [76].

The pathogen can be eradicated from seeds by treatment
with tridemorph/maneb formulations [161,163]. Cultural
measures such as destruction of infected debris, appropriate
crop rotation, roguing of infected plants and intercropping
can each achieve a measure of control where incidence is
moderate [108]. Foliar application of chlorothalonil effectively controlled disease development in Syria where incidence was very great [76]. However, combinations of these
measures are unlikely to provide effective control where
conditions are most favourable for disease development and,
in any case, chemicals are expensive to obtain and to apply,
so that host plant resistance must be a key feature of control strategy.

Sources of resistance have been known for more than

fifty years [99]. However, immunity has never been reported and the history of breeding for resistance to *A. rabiei* is one of a succession of 'resistant' genotypes which have subsequently been found to be susceptible. These events are well illustrated in recent summaries of breeding work in the Punjab [10,112,243]. The original source of resistance, cultivar F 8, was wilt-susceptible and replaced by cultivar C 1234 developed from a cross of cultivars F 8 and Pb.7. Cultivar C 1234 was found to be susceptible to infection by *A. rabiei*. In India, cultivar C 1234 was succeeded by C 235 (C 1234 × IP 58) and then G 543 (C 168 × C 235), and both of these are now known to show only moderate resistance to blight. More recently, genotypes reported resistant in Syria have subsequently been rated susceptible in other locations and seasons [202].

Based on these cyclic fluctuations in resistance and differential reactions in disease nurseries, many authors have argued that resistance to *A. rabiei* is extremely specific. However, these events are at least as well explained by differences in environment causing fluctuations in disease intensity [6,10]. Undoubtedly, the pathogen is extremely variable, in pathogenicity as well as other features but in tests of isolates against genotypes, disease reactions have been essentially quantitative [55,245] and a properly analysed and repeatable case of a specific host/pathogen interaction remains to be documented.

Resistance is said to be controlled by a single locus with resistance dominant in four parents and recessive in another [203], but in the majority of crosses recognisable ratios are not obtained. Furthermore, the spectrum of disease reactions recorded among genotypes, from extreme susceptibility to near immunity, indicates the situation to be far more complex and that polygenic resistance may be at least equally important [23].

Five genotypes, with cream-coloured kabuli or intermediate type seeds, have maintained good resistance across several locations and years [71]. In desi types there appea

354

to be an association between resistance and black testas, but resistance is less effective than that of the kabuli and intermediate types. However, in north India in 1981-82, several desi types (including derivatives of cultivar F 8) expressed a moderate resistance [243] which may be useful in the short term, especially if used in conjunction with other cultural and chemical control measures. All of these sources of resistance have been incorporated into breeding programmes in India and elsewhere [76,198] and programmes involving horizontal resistance are also in progress in Syria [198] and Morocco [143]. Meanwhile investigations into the degree of specificity of the host/pathogen relations and the nature of the inheritance of resistance continue as a basis for breeding strategy.

Botrytis grey mould.   Losses due to grey mould, incited by *Botrytis cinerea*, are a recurring feature among the chickpea crops of north-eastern India and Bangladesh; and in years of excessive rainfall the problem extends into north-west India and has been recorded as far west as Pakistan (J.B. Smithson, unpublished). The pathogen has also been reported from North and South America and from Australia.

Where conditions are favourable, *B. cinerea* can cause almost total crop loss. Incipient symptoms include flower drop, which may cause significant yield losses or contribute to excessive vegetative growth, but is difficult to detect and assess. The pathogen has a very wide host range and survives in plant debris.

Screening for resistance has been conducted in field nurseries, using both artificial and natural infections in areas where the pathogen is endemic [31], or in isolation plant propagators [65]. Immunity has not been recorded but sources of moderate resistance have been identified [65,138, 196,244] and have been used in crosses. Information on the inheritance of resistance is not yet available.

355

Stunt.   Chickpea stunt, known to be caused by Pea Leaf
Roll Virus, was first described in Iran [88] and has since
been found, albeit infrequently, throughout many chickpea
producing areas including north Africa [162], Spain [26]
and Australia (E.J. Knights, pers. comm.). Incidence can
be as severe as 90% and cause substantial crop loss. The
virus is transmitted by several species of aphid and has a
wide range of alternate hosts amongst the legumes, which
probably serve as sources of primary infection. It is not
transmissible mechanically. Field screening has led to the
identification of several genotypes which have consistently
exhibited less than 10% incidence, and have been incorporated in breeding programmes [76].

Pests

Chickpeas suffer remarkably few insect problems compared
to other grain legumes, almost certainly due to the secretion of an extremely acidic (pH about 1.0) exudate from
the profuse glandular hairs which cover all plant parts.
Twenty-two insect species damaging chickpea have been
reported [238], thirteen in the Middle East [212], but
relatively few of them are considered of economic importance
except in localised situations.

Distribution and importance.   *Heliothis armigera* (Hb.)
is undoubtedly the most damaging pest [167]. It is distributed throughout chickpea producing areas of the Old World.
In peninsular India, where the weather is not sufficiently
cool to restrict insect activity during the growing season,
the larvae can cause damage from the seedling stage through
to near maturity. In northern India and other areas of its
distribution, damage is caused only during the reproductive
stages, when temperatures are warm enough for the larvae to
be active.
    Vegetative parts can be defoliated but yield loss is
negligible as the damage is either mild or the crop is able

356

to compensate by the growth of axillary branches. During
the reproductive stages, the larvae penetrate pod walls
and damage the immature seeds which fail to mature. Surveys conducted in nine states in the principal chickpea
growing areas of India during the four seasons between 1977
and 1981 revealed pronounced seasonal fluctuations in fruit
damage, which ranged between states from 0.4-30.5% [75] but
was rarely more than 10%. However, other reports estimate
total crop losses in some situations in north India.

The larvae of *H. zea* and *H. virescens* are the major pod
borers in the Americas [107] while, around the Mediterranean, there appears to be a mixture of species, including
H. armigera, *H. viriplaca* and *H. peltigera* [212,215]. Their
relative abundance varies seasonally but *H. viriplaca* predominates in at least some years [228]. As in India, crop
losses are relatively small [212] but can range up to 40%
being largest in southern Syria and Jordan [73].

The leaf miner, *Lyriomyza cicerina*, which is distributed
throughout the Mediterranean basin, is the second most widespread pest of the crop. The larvae burrow under the leaf
epidermis causing premature defoliation but, although not
yet estimated precisely, crop losses are probably small.

Other insects of local importance are termites, cutworms (*Agrotis* spp.) [75] and the semi-looper, *Autographa
nigrisigna* [172] in northern India; aphids, notably *Aphis
craccivora*, which cause little direct damage, but are the
vectors of Pea Leaf Roll and the Mosaic Viruses; and bruchids, mainly *Callosobruchus* spp., which can cause significant losses of seed in storage.

Host plant resistance.   The search for host plant resistance has concentrated hitherto on *Heliothis armigera*.
Screening for resistance has been mainly in field nurseries
with natural infestations, supplemented infrequently by
laboratory-reared adults. Immunity has not been found and
is indeed improbable with a pest of the polyphagous nature
of *Heliothis*. Nevertheless, differences in resistance have

been reported. For example, kabuli types are generally recognised to be more susceptible than the desi types [167]. Screening of the world collection [75] has revealed several sources of resistance, notably accession ICC-506, a landrace collected in southern India, which has shown less than 5% fruit damage accompanied by consistently better seed yields relative to other, more susceptible desi types [74,76]. Accession ICC-506 has a particularly acid exudate [171] and has the largest concentration of polyphenols in the seed of any genotype yet tested [76]. Since the polyphenols are concentrated in the testas, which are generally removed before consumption, they may provide a useful mechanism of resistance. The relative ability to compensate for earlier damage is another important factor in host plant resistance [167].

Insecticide use.   In India, the use of insecticides on the commercial crop is minimal [75] and varies seasonally with degree of infestation. In numerous tests, responses to insecticide have been variable and frequently uneconomic [8]. Applications of DDT and BHC can control *Heliothis* effectively but are now being replaced by endosulfan. Incorrect timing of application may be one factor contributing to the variable yield responses reported. Another may be the ability of the crop to compensate for earlier damage in protracted growing seasons. Limited availability of water may hamper the use of water-based sprays in many situations. The use of ultra-low-volume equipment combined with the synthetic pyrethroids, which have been shown to be as equally effective as conventional insecticides, is an obvious strategy for improving the feasibility of chemical pest control on chickpea crops.

Cultural control.   Insect populations and their effects on crops and crop losses may be modified by various cultural practices. In north India, delayed sowing beyond early October results in linear increases in fruit damage due to

Heliothis and corresponding decreases in seed yield [106],
although the latter may well be associated also with the
effects of sowing date *per se*. On susceptible genotypes,
Heliothis population densities consistently increase with
increasing plant density [72,74,75,76] but their effects
on fruit damage and seed yields are variable. Neither plant
density nor insecticide application affected *Heliothis*
populations, fruit damage nor seed yield in the resistant
cultivar ICC-506 [75,76].

In Syria, fruit damage due to *Heliothis* is greater on
winter-sown than on spring-sown chickpeas but seed yields
are unaffected and crop losses due to leaf miner are likely
to be less on the more advanced, earlier-sown crop [212].

These, albeit limited, studies indicate that although
changes in cultural practice can affect pest populations,
these necessarily presage neither effects on pest damage
nor on crop loss. Thus, while a major innovation such as
winter-sowing in the Mediterranean could have profound
effects on pest complexes, which must be monitored, they
are unlikely to outweigh the yield advantages to be gained
from advancing the sowing date [212].

Natural enemies.   *Heliothis* attacks are generally associated with fairly heavy parasitism [167] but its nature and
incidence are characterised by distinct temporal and spa-
tial fluctuations. Incidence ranges from negligible to
more than 30% and is less in larvae on resistant cultivars
[214]. The hymenopterans, *Campoletis* and *Diadegna* spp., and
the dipteran, *Carcelia illota*, are most often involved
[72,74,75,76]. The possibility of using *Eucelatoria* sp., a
dipteran parasite from the USA, for the control of *Heliothis*
is being examined [73].

The bird species *Acridotheres tritis* L. and *Dicrurus
adsimilis* prey on *Heliothis* [75,167] although their effects
on pest populations have not been quantified. The reduced
numbers of large green larvae on genotypes with anthocyanin-
pigmented foliage has also been attributed to predation by
birds [213].

359

Weeds

Although chickpeas are traditionally grown on residual soil
moisture, weeds are major problems in many situations.
Potential yield losses due to weeds range between 22% [183]
and 100% [77]. Common annual weed species include *Cheno-
podium album, Asphoedelus tennifolius, Argemone mexicana,
Fumaria parviflora, Polygonum plebejum, Lathyrus* spp.,
Vicia sativa, Euphorbia dracunuculoides and *Phalaris minor*.
Common perennial species are *Cyperus rotundus, Cynodon
dactylon* and *Cirsium arvense* [183].

Hand weeding at thirty and again at sixty days after
sowing essentially eliminates the adverse effects of weed
competition [182]. In commercial practice, the cultivation
of preceding rainy-season fallows not only helps to capture
and conserve moisture but also reduces weed infestations.
On black soils, in the wetter areas of central India,
'haveli' cultivation (the practice of containing water by
bunding in the rainy season) serves similar purposes. Inter-
row cultivation by tractor or animal-drawn implements is
common, facilitated in north Africa by sowing the crop in
very wide rows.

When properly used, pre-emergent herbicides such as
prometryne, terbutryne, pronamide, cyanazine and methabenz-
thiazuron accomplish effective and economic weed control,
accompanied by chickpea seed yields similar to or only
slightly smaller than those of weed-free treatments [93,
183].

Weed growth is conspicuously heavy in chickpeas sown
in winter in the Mediterranean area. Pre-emergent herbicides are one solution. A second possibility is the mechanical removal of early weed growth by slightly delaying the
sowing of chickpea crops. An effective means of weed control is crucial to the widespread adoption of winter sowing, especially in view of the expense of labour in the
region.

Physical constraints

Inadequate soil moisture.   Dry seed beds reduce germination and establishment and are an important cause of sparse plant populations in rainfed crops [185]. Germination is reduced to about 20% at soil moisture potentials of -1.5 MPa [60]. Procedures to screen seeds for their ability to germinate and emerge from dry seed beds have been developed for use in the laboratory [74] and field [185], and cultivar differences have been reported [74].

Reports of large responses of chickpeas to irrigation are common [182,183]. Estimates of water use range between 110 mm in an unfertilised crop at Dehra Dun in north India [195] and 422 mm in early-sown cultivar ILC 482 at Jindiress in Syria [90]. In peninsular India, limited soil moisture is the principal constraint to chickpea productivity and seed yields have been consistently doubled by irrigation [73,75,76]. Heavy yielding cultivars adapted to peninsular India have greater shoot water potentials than unadapted cultivars [216] and in field studies using varying degrees of stress, short duration cultivars of heavy yield potential produced the largest seed yields when soil moisture was limiting [74]. Cultivars capable of producing consistently heavy seed yields in stress conditions have been identified [184].

Excessive soil moisture.   Paradoxically, chickpeas are also extremely sensitive to soil moisture concentrations in only slight excess of requirements. Thus, responses to irrigation in north India have been inconsistent and in some cases negative [183], accompanied by excessive vegetative growth and lodging [184]. The identification of genotypes which are not prone to excessive growth or which maintain seed yields when their sowing is delayed is receiving increased attention in that region.

Cold temperatures.   Ability to withstand cold temperatures,

desiccating winds and frost are important components of
adaptation to winter sowing at high altitudes. Chickpeas
are relatively tolerant of cold temperatures and genotypic
differences in seedling growth rates in cool temperatures,
and in frost tolerance, have been reported [223]. In a
recent screening of more than 3000 chickpea germplasm
accessions at Hymana in Turkey, during which temperatures
as cold as -26.8°C were recorded, all plants of four accessions survived and two others showed 67% survival; four of
the six accessions originated in India [201]. In field tests
at Tel Hadya in Syria, 1808 out of 10800 germplasm accessions and breeding materials remained free from cold damage
and 1806 showed substantial tolerance despite exposure to
thirty-one nights at subzero temperatures [71]. In the
laboratory, exposure of seedlings with five to seven leaves
to a temperature of -5°C for 18 hours gives comparable
results to field screening.

Heat stress.   The growth of chickpeas is normally curtailed by increasing heat and drought stress in spring and
summer. The separate effects of heat and drought stress are
difficult to distinguish and are probably interrelated.
Significant interactions in relative growth rates between
cultivars and sowing dates [187] and differential responses
of cultivars to transfer to hot temperatures in controlled
conditions [224] have been attributed to differences in
heat tolerance. More recently, additional evidence has been
obtained that heat stress *per se* during the reproductive
period curtails growth and reduces yield significantly [222].

Soil salinity and alkalinity.   Soil salinity and/or alkalinity are major agronomic problems of the Indian subcontinent. There are an estimated 7 million hectares of salt-
affected soils in India, more than one third of which are
in the Indo-Gangetic Plain [1]. Saline soils are characterised by large concentrations of soluble salts (EC >4 mmhos
cm^{-1}), exchangeable sodium concentrations smaller than 15%

362

[2], and may be sulphate or chloride dominated. The latter are more detrimental to the growth of chickpeas [113]. Alkaline (or sodic) soils (pH > 8.5) have moderate concentrations of soluble salts (EC < 4 mmhos cm^{-1}) but exchangeable sodium concentration is greater than 15%.

Earlier work on the effects of salinity and alkalinity on chickpeas has been comprehensively reviewed [238]; the crop is extremely sensitive to such conditions [29]. Seed yields were more than halved [181] by irrigation with water of EC 1.0 mmhos cm^{-1} and crops were killed [180] in a soil of EC 1.8 mmhos cm^{-1} irrigated with water with an EC of 2.6 mmhos cm^{-1}. Soil amelioration is expensive and not very effective [29]. Techniques to screen chickpeas for tolerance to salinity have been developed and although differences in sensitivity have been detected [29,184] useful degrees of tolerance have not yet been identified.

Iron deficiency.   Iron chlorosis has been reported in chickpeas grown in Syria and Lebanon [182], India [188] and Pakistan (J.B. Smithson, unpublished). The symptoms include a general yellowing of young leaves, extending in severe cases to distortion, necrosis and shedding of terminal pinnae [126]. The condition is aggravated by rainfall or irrigation during crop growth [184] but is usually transient. The incidence of symptoms is not related to iron concentration in the rooting medium [4] and is more likely to be due to interferences with absorption and translocation [111]. Cultivars exhibit marked differences in sensitivity to iron chlorosis [54,187,188] and this appears to be directly related to time to flowering [54]. The recovery of sensitive genotypes is hastened by foliar applications of solutions of iron compounds, which can increase seed yields by up to 50%. Insensitive cultivars neither show symptoms nor give yield responses to iron application.

Zinc deficiency.   There are frequent reports of zinc deficiency in chickpeas [182,184]. Symptoms are evident initially as yellowing of the middle and lower leaves followed by bronzing and necrosis; they can be corrected, and seed yields increased, by foliar or soil application of $ZnSO_4$. Chickpea genotypes differ in sensitivity to zinc deficiency. The balance of P and Zn appears critical for maximum nitrogen fixation [194].

GERMPLASM RESOURCES AND CROP IMPROVEMENT PROGRAMMES

Chickpea improvement has received only minor attention compared with that allotted to other crops. Even where work has been undertaken, it has frequently represented only a small part of a larger effort on several legume species. The only countries which have had significant national programmes on chickpeas are India and Pakistan. More recently the International Development Research Centre (IDRC) in Canada has funded the establishment of legume research programmes, of which chickpeas are a component, in Algeria, Bangladesh, Jordan and Pakistan.

International programmes have included: the Regional Pulses Improvement Project (RPIP), instigated by USAID and the Agricultural Research Service of the USDA in 1963, in collaboration with the governments of India and Iran; and the Regional Food Legume Project of the Arid Lands Agricultural Development Program (ALAD) of the Ford Foundation, started in the Near East in 1972. Recently, international research centres were established at Hyderabad, India (the International Crops Research Institute for the Semi-Arid Tropics, ICRISAT) and at Aleppo, in Syria (International Centre for Agricultural Research in the Dry Areas, ICARDA), and assigned global (ICRISAT) and regional (ICARDA) responsibility for chickpea improvement by the Consultative Group for International Agricultural Research (see Chapter 18).

The development and present status of these national and international programmes and their progress in crop improvement are described below.

364

Chickpea (*Cicer arietinum* L.)

Germplasm resources

Large-scale, organised collecting and evaluation of chickpea germplasm resources started only in the mid-sixties.
Before then the crop had been included among the collections
of Vavilov and small collections had been assembled and
studied at various centres in India and elsewhere but the
largest of these comprised no more than a few hundred accessions and there was little systematic evaluation.

The first major attempt to assemble chickpea germplasm
was undertaken by RPIP. The project was terminated in 1970
but by then had assembled a collection of 4177 accessions
from twenty-four countries [168,169] which formed the basis
of subsequent collections. Parts of the original collection
are now stored at the National Seed Storage Laboratory at
Pullman, Washington, USA. In India, the collection has expanded to more than 6000 accessions from twenty-one countries [77].

The second major collection was initiated by ALAD in
1973 and evaluated in Lebanon, Egypt and Sudan [67]. In
1977, ICARDA assumed responsibility for the collection
which, following additions from ICRISAT and several national
programmes, and reselection, now numbers more than 4400
accessions, mainly kabuli types, from thirty-two countries
[200]. At least 3300 accessions have so far been evaluated
for twenty-nine characters, including cold tolerance and
resistance to *Ascochyta* blight.

The third, and the largest, collection is maintained
at ICRISAT [239,241]. This now comprises a total of 12502
desi and kabuli accessions from thirty-nine countries [76],
and has been evaluated for several morphological, agronomic
and resistance characteristics.

The origins and characteristics of most of the accessions in these collections have been described elsewhere
[85,170,200,240].

365

National and international crop improvement programmes

India.   Agricultural research is conducted under the aus-
pices of the Indian Council for Agricultural Research (ICAR)
The Council was established in 1929 in accordance with the
recommendations of the Royal Commission on Agriculture to
promote, guide and coordinate agricultural research throughout India [155]. Following reorganisations after independence and in 1966, the activities of the central institutes,
which had developed under departments and commodity commit-
tees, were unified under the administrative and technical
control of ICAR.

Before 1966, germplasm was collected and evaluated and
improved chickpea cultivars were developed by selection
within landraces and from crossing programmes conducted at
a number of centres, notably the Imperial (now Indian)
Agricultural Research Institute (IARI) at Pusa in Bihar.
The cultivars which were developed and released during this
period are listed in Table 8.1. Where parentage is not shown
the cultivars are direct introductions or selections from
local materials.

Development of the present-day structure began with the
second reorganisation in 1966 when, coincident with the
RPIP, All India Coordinated Research Projects were established for the main groups of crops, including pulses.
Based initially at the IARI in New Delhi, with fifteen
centres throughout India, the Directorate, comprising a
Director and a cadre of Principal Investigators responsible
for eight disciplines, is now at Kanpur and coordinates the
activities of twenty-one main centres and affiliated subcentres (Fig. 8.3).

Scientists in the disciplines of breeding, pathology,
entomology, agronomy and microbiology are established at
each main centre. Centres are allocated research responsibility for specific pulses and for chickpea, four agro-
ecological zones are recognised (Fig. 8.3) [205]. Coordinated trials and cultivar releases are zonally based.

366

Chickpea (*Cicer arietinum* L.)

Table 8.1   Chickpea cultivars released in India prior to establishment
of the All India Coordinated Pulse Improvement Project in
1966

State	Cultivars*
Andhra Pradesh	No. 63-8-12-41; Warangal; Jyothi
Bihar	S.T.4; BR 17; BR 65; BR 77
Gujarat	Chafa; Dohad Yellow
Karnataka	18-12; Kadale-2; Kadale-3; Annigeri-1
Madhya Pradesh	Adhartal Type-V; D 8; No. 10; EB 28; Gwalior-2; Gwalior-3; Ujjain Pink 2^g; Ujjain 21; Ujjain 24
Maharashtra	Chafa; Early Gulab^g; N 10; N 29; N 30; N 31; N 59; N 68; N 74
New Delhi (IARI)	N.P. 2^k; N.P. 27; N.P. 25; N.P. 28; N.P. 58
Punjab (includes Haryana and Himachal Pradesh)	Pb 7; F 8; C 1234 (Pb 7 × F 8); G 24; S 26; C 104^k (Pb 7 × Rabat); C 235 (C 1234 × I.P. 58)
Rajasthan	RS 10; RS 11
Tamil Nadu	Co 1; Co 2
Uttar Pradesh	T 87; T 1 (Banda × T 87); T 2 (Banda × T 87); T 3 (No. 197 × No. 195); K 4^k; K 5^k
West Bengal	B 75; B 98; B 108; B 110

*All yellow or brown desi seed types except where indicated, i.e. k = kabuli, g = gulabi.

Recently, there have been movements to designate centres for specific operations, such as breeding for disease resistance, and screening as a support service for other centres, and to refine further agro-ecological divisions.

In addition, all states conduct their own research programmes and have subcentres at which coordinated and state trials are conducted. ICRISAT has its headquarters near Hyderabad and regional centres at Hissar in Haryana State and Gwalior in Madhya Pradesh. The Institute supports coordinated and state programmes by supplying germplasm and breeding materials, both segregating and stabilised, in the form of trials or in response to specific requests, and has considerably widened the range of diversity available to national researchers.

367

Figure 8.3 All India Coordinated Pulse Improvement Project Centres
and agro-ecological zones

Improved materials from all of these centres are contributed to coordinated cultivar trials, the numbers of
which expand annually and which provide excellent coverage
of the agro-ecological conditions where chickpeas are
grown. In 1982-83, for example, a total of 154 trials were
undertaken as part of the coordinated programme. A steady
stream of improved cultivars has been identified and
developed for zonal and state release since the establishment of this system, and their origins and important characteristics are shown in Tables 8.2 and 8.3

Pakistan.   Until recently, agricultural research has

368

Chickpea (*Cicer arietinum* L.)

Table 8.2 Origins and seed characteristics of zonally released chickpea cultivars in India

Cultivar	Origin	Year of identi- fication	Zone	Seed type	Weight of 1000 seeds (g)
Chafa	Local selection	1973	Peninsular	Desi	150
JG 62	Local selection	1973	Central plains	Desi	150
T 3	197 × 195	1973	North plains, peninsular	Desi	240
Annigeri 1	Local selection	1975	Peninsular	Desi	180
C 235	C 1234 × I.P. 58	1976	North, central plains	Desi	130
G 130	No. 708 × C 235	1976	North, central plains	Desi	135
L 550	C 104 × N.P. 12	1976	Irrigated conditions	Kabuli	210
H 208	(S 26 × G 24) × C 235	1977	North, central plains	Desi	110
H 355	G 140 × S 26	1977	North, central plains	Desi	120
BG 203	P 827 × C 235	1977	North plains	Desi	115
Pant G 114	G 130 × 154	1978	North plains	Desi	120
K 468	Germplasm	1978	North west plains	Desi	120
BG 209	P 827 × C 235	1980	North plains	Desi	122
BC 212	P 340 × G 130	1982	Central	Desi	130
GG 588	G 130 × G 549	1982	North west plains	Desi	130

Table 8.3 Origins and seed characters of chickpea cultivars released
on a state basis in India since 1968

State	Cultivar	Origin	Year	Seed type	Weight of 1000 seeds (g)
Bihar	BR 78	BR 103 × BR 192	1974	Green	182
Gujarat	ICCC 4	H 208 × T 3	1983	Desi	160
Haryana	C 214	(I.P. 58 × G 24) × G 24	1971	Desi	137
	G 130	No. 78 × C 235	1971	Desi	135
	L 144	S 26 × Rabat	1975	Kabuli	300
	H 208	S 26 × G 24	1977	Desi	115
	H 355	V 140 × S 26	1977	Desi	128
	Gaurav (H75 35)	C 235 × E 100Y	1983	Desi	215
Madhya Pradesh	JG 62	Local selection	1972	Desi	150
	JG 221	Germplasm	1979	Desi	155
	JG 5	Pink 2 × GC 654	1979	Gulabi	196
	L 550	C 104 × N.P. 12	1979	Kabuli	200
	JG 315	Germplasm (P 315)	1982	Desi	155
Maharashtra	BDN9 3	Local selection	1978	Desi	155
	Vikas (Phule G 1)	Khanpur 6 × A.F. 7-10	1982	Desi	180
	Vikram (Phule G 5)	B 110 × N 31	1983	Desi	270
Punjab	C 214	(I.P. 58 × G 24) × G 24	1971	Desi	137
	G 130	C 235 × No. 708	1971	Desi	130
	Hare Chhole 1	S 26 × Gg Bijapur	1973	Green	150
	L 550	C 104 × N.P. 12	1974	Kabuli	200
	G 543	C 168 × C 235	1977	Desi	130
	GL 769	H 223 × L 168	1981	Desi	—
Uttar Pradesh	Radhey	Local selection	1968	Desi	180
	K 468	Germplasm	1977	Desi	120
	K850	Banda local × Etah bold	1979	Desi	325
	Avarodhi	K 315 × T 3	1981	Desi	190
West Bengal	B 115	N 31 × B 75	1977	Desi	195
	B 124	N 31 × B 75	1977	Desi	179

Chickpea (*Cicer arietinum* L.)

remained primarily with the provincial governments and
agricultural universities and colleges. Earlier, without
research establishments and a role limited to financial
support and some coordination of federal research activities, the Pakistan Agricultural Research Council was
charged in 1978 with major responsibility for planning,
coordination and evaluation of national agricultural research and for strengthening provincial research capabilities. The first coordinated research programme for pulses
was initiated in 1980-81.

Early efforts with chickpeas were concentrated on the
identification of and breeding for resistance to *Ascochyta*
blight. Three resistant sources, all introductions from
France, were identified at Campbellpur [108], and one of
them, cultivar F 8, which also combined large yield with
satisfactory seed characteristics, was distributed for cultivation. Subsequent breeding work led to the development
of the cultivars C 1234 (Pb.7 × F 8), C 612 (F 8 × C 144),
and C 727 for blight-affected areas [112]. Cultivar C 235
(I.P. 58 × C 1234), developed in Punjab State of India, is
also being widely grown. The resistance of all these cultivars has proved to be inadequate in epiphytotic conditions.

More recently, a black-seeded multiline (AUG 480)
developed at Faisalabad Agricultural University, and the
mutants CM 68 and CM 72, from the Nuclear Institute for
Agriculture and Biology in Faisalabad, have shown better
resistance [112]. These are being evaluated in national
trials, together with cultivars contributed by the Ayub
Agricultural Research Institute, also in Faisalabad, for
suitability for release.

Bangladesh.   The collection, introduction and evaluation
of chickpeas commenced in 1949 at the East Pakistan Agricultural Research Institute in Dacca (now Dhaka) with subcentres at Joydebpur and Rajshahi, first by the Cereals
and, from 1957 onwards, by the Pulses and Oilseeds Divisions. Agricultural research has recently been consolidated

under the Bangladesh Agricultural Research Institute at Joydebpur near Dhaka, where a coordinated pulse improvement programme was instigated in 1979.

The cultivars Sabour 4 and Bhangura 45 were released in 1955-56 and Faridpur 1 and Pabna Local in the early sixties [53], all being selections from local materials. The most recent release has been Hyprosola, produced by Co^{60} treatment of Faridpur 1, at the Institute of Nuclear Agriculture in Mymensingh [190]. A wide range of local and introduced germplasm is currently under evaluation [53].

Mexico.   Centres responsible for chickpea improvement have been established at Culiacan in Sinaloa State, concentrating on irrigated kabuli chickpeas, and at Celaya, in Guanajuato State, for desi types for irrigated and rainfed situations. A small germplasm collection has been assembled and evaluated and active breeding programmes are in progress. A number of improved cultivars have been released: the kabuli types Union, Culiacancito 860, Macerena and Surutato 77 and the desi types Cal Grande, Grande 12 and, more recently, Carreta 145. Interestingly, Union, Macerena and Surutato 77 have a single, incised leaf blade (simple leaf) in contrast to the normal composite leaf.

Other programmes.   In other countries, programmes were earlier confined to the evaluation of small collections of local germplasm and a little basic agronomy, without active breeding work. The regional programmes, RPIP and ALAD and, more recently, the international centre ICARDA, have provided a major stimulus to chickpea research in the Middle East and north Africa by making available a much wider range of germplasm and breeding materials than hitherto existed, by training, and by encouragement to establish local chickpea improvement programmes. Coordinated improvement activities are now being conducted in Algeria by L'Institut de Développement des Grandes Cultures; in Jordan, by the University of Jordan; and in Ethiopia by the National

Crop Improvement Committee. A large, active breeding programme is now being conducted in Syria by ICARDA, which
has recently also initiated work in Tunis, in collaboration
with the government of Tunisia (see Chapter 18.)

Despite these efforts, cultivation is still based for
the most part on local landraces. Known exceptions are:
Giza 1 and Family 2, in Egypt; Pyrouze, Kaka, Jam and Kou-
rosh, in Iran; and Dubie, DZ-10-2, -4 and -11, in Ethiopia.
The cultivar ILC-482, a selection made at ICARDA in a germplasm accession from Turkey, has been released in Syria;
and a sister accession, ILC-484, is in on-farm tests in
Jordan. Cultivars have also been recommended for cultivation in Peru (Chancay) and Chile.

AVENUES OF COMMUNICATION

In addition to conventional reports in international and
national scientific journals and crop conference proceedings, there are various channels of communication either
specific to chickpeas or of which chickpeas form a major
component.

Workshops and conferences

International workshops have included one on 'Grain Legumes'
and another on 'Chickpea Improvement' held at ICRISAT in
1975 and 1979, respectively; and the workshop on 'Food Legumes' sponsored by ICARDA and IDRC at Aleppo in Syria in
1978. Proceedings of these workshops have been published
and form valuable, up-to-date accounts of the status of
chickpea improvement and production practices; we have
cited them widely in this chapter. Proceedings of the workshops on '*Ascochyta* Blight and Winter Sowing of Chickpea'
and on 'Faba Beans, Kabuli Chickpeas and Lentils in the
80s', held by ICARDA at Aleppo in 1981 and 1983, have just
been published or are in preparation, respectively. An
important national meeting is the Rabi (post-rainy season)

Pulses Workshop of the All India Coordinated Pulses Improvement Project held annually in India to consider previous
results and to formulate plans for the following season.
Also, ICRISAT annually invites chickpea breeders from other
institutes to its research centres at Hyderabad and (in collaboration with Haryana Agricultural University) at Hissar,
to select materials to grow at their own centres.

Publications

Three newsletters contain items of interest pertaining to
chickpeas. They are: *The Tropical Grain Legume Bulletin*,
published by the International Grain Legume Information
Centre of the International Institute of Tropical Agriculture (IITA) at Ibadan in Nigeria (see Chapter 18); the
International Chickpea Newsletter, distributed by the Pulses
Improvement Programme of ICRISAT; and the *Pulse Crop Newsletter* of the ICAR. One bibliography has been produced
[204], another is in preparation [199].

Training

Research and production-oriented training courses, held
annually at ICRISAT and ICARDA, are important for the dissemination of new ideas, techniques and seeds.

International trials and nurseries

Various international chickpea cultivar and agronomy trials
and disease and insect screening nurseries coordinated by
ICRISAT and ICARDA serve not only to disseminate seed materials but also to encourage contacts and visits between research centres and thus foster exchange of information.
These and other trials form the subjects of formal and informal publications from the two institutes and contain useful information regarding crop performance and responses to
environment.

PROSPECTS FOR LARGER AND MORE STABLE YIELDS

In common with other grain legumes, average and potential
seed yields of existing chickpea cultivars are smaller than
those of the cereals [225]. Various reasons for this have
been advanced, including unimproved cultivars, the relegation of chickpeas to marginal conditions [192] and the
larger seed protein (and sometimes oil) concentrations of
the legumes compared with the cereals [47]. It is also
frequently stated that research and breeding efforts have
been small but, while relative to the cereals this may be
so [191], it may be seen from the foregoing account that
efforts on chickpeas have been far from negligible in the
Indian subcontinent. Here, successive improved cultivars
have emerged from hybridisation programmes and have been
released for cultivation, and agronomic and crop protection
practices have received detailed attention.

Notwithstanding, improvements in chickpea seed yields
have been small and it is instructive to enquire why this
should be so. Two possible reasons for a disappointing rate
of genetic improvement are: first, the narrow genetic base
of existing cultivars; and secondly, the emphasis on breeding for yield. To illustrate the first point: apart from
cultivars released in Central Uttar Pradesh, most north
Indian desi cultivars trace back to either C 235 or S 26
and are virtually indistinguishable morphologically; and
kabuli cultivars trace back to Rabat (see Tables 8.2 and
8.3). These facts introduce the second possible reason:
that although the use of exotic germplasm in breeding has
been limited, their derivatives have not been selected, and
this probably relates to extreme specificity of adaptation
of chickpeas to the environments in which they have evolved.
Such specific adaptation is reflected in pronounced genotype × environment interactions for seed yield and associated characters which, coupled with their large non-additive
genetic components, have rendered conventional selection for
increased seed yield relatively ineffective. It is therefore

not surprising to find that progress in genetic improvement in productivity has been slow.

So what are the alternative breeding strategies? Considering the associations among characters, the observation that fruit and seed numbers per plant are significantly and positively correlated with seed yield has important physiological implications, but is of little utility for selection since these characters have as poor heritability and as large non-additive genetic components as yield *per se*, and they are more difficult to measure. Branching is also subject to considerable environmental modification and has poor heritability, but merits more detailed study. Seed size is extremely heritable and variation is predominantly additive. It is also usually correlated positively with seed yield, so yield improvement can be expected from selection for larger seeds. However, increases in seed size are usually accompanied by decreases in the numbers of fruits per plant and/or seeds per fruit; there is clearly a delicate source/sink balance.

Nevertheless, this balance may be manipulatable. Increased seed size may be achievable (within certain limits) without affecting fruit and seed numbers. Increased branch number is also important in this regard, since increases in fruiting nodes and fruits per plant can be accompanied by increased photosynthetic area, thereby not disturbing source/sink relations. Increased branching also implies changes in growth habit, requiring the development of new concepts of plant type.

The large non-additive component of genetic variation for seed yield presents problems. Although substantial heterosis has been reported, the breeding system and floral biology of chickpeas and the current lack of useful male sterility preclude the adoption of hybrid varieties or population improvement systems. Seed size (especially) and perhaps branching (provided satisfactory assessment methods can be developed) may be handled using conventional breeding techniques. However, the preponderance of non-additive

genetic variation for most yield components and the specific nature of adaptation suggest the need to consider
approaches which involve the incorporation of features
such as resistance to diseases and pests and tolerance of
physical stress factors into existing, heavy-yielding genetic backgrounds and the development of genotypes suited to
non-traditional situations.

The prospects for heavier and more stable yields, and
strategies to achieve them, vary among environments. In
peninsular India, where temperatures are relatively warm
during the cropping season, the heaviest yields without
irrigation are produced by genotypes which flower and commence fruiting in about 50 days and mature 80-90 days after
sowing [76], in response to rapidly increasing temperatures
and moisture stress. In these conditions seed yields of
about 1500 kg ha^{-1} are produced with a harvest index of
60% [184]. The principal modifiable constraint to production is soil moisture, and seed yields are substantially
increased by irrigation. Traditionally, in peninsular India,
rice has been grown when irrigation is available and latt-
erly, groundnuts; but pronounced increases in irrigated
area are projected and chickpeas may present an alternative
crop for these situations, though the economics of irrigated
chickpeas *vis-à-vis* other crops remains to be examined. The
incorporation of resistance to wilts, root-rots and to
Heliothis as well as tolerance of drought into adapted
genetic backgrounds will help to stabilïse production and
the latter may lead to slightly larger yields. Since
'source' is often limiting [184], increased biomass (one
property of the tall, erect genotypes) will be an advantage.
Cultivars which may be sown early or at the normal time
offer the possibilities of heavier seed yields (by extending
the growing period) and greater flexibility of sowing date
(to adjust to different cropping systems and seasonal variability in rainfall patterns). However, only minor contributions to improved and more stable yields are anticipated
from genetic improvement.

In northern areas of the Indian subcontinent, cooler temperatures during the growing season inhibit fruiting and extend crop duration to as long as 170 days. Long-duration cultivars produce the heaviest seed yields, which can exceed 4000 kg ha^{-1}. It is here that chickpeas have been displaced by modern productive wheat cultivars, capable of greater and more assured returns. The inherent instability of chickpeas has both pathological and physiological origins. *Ascochyta* blight and *Botrytis* grey mould cause serious losses when rainfall is excessive during the growing season. Established cultivars have only moderate resistance to *Ascochyta* blight and succumb to heavy disease pressure. There have been progressive small improvements in the resistance of successive cultivars and the resistance of the new cultivar H75-35 (derived from cultivar E 100Y from Greece) is apparently better. Sources of even greater resistance to *Ascochyta* blight [71] and moderate resistance to *Botrytis* grey mould [65] have been identified and incorporated in breeding programmes.

The agronomic requirements for heavy seed yields have not been satisfactorily defined. In the presence of only slightly more than adequate soil moisture, vegetative growth becomes excessive, lodging occurs, diseases (notably wet stem rot) are aggravated and seed yields decline [184]. The reasons are not well understood but germplasm is being screened for genotypes of compact growth habit, and those capable of fruiting in cool temperatures are also being examined. Delayed sowing (until December) would reduce vegetative growth and would also afford greater flexibility for rotation with crops such as rice. With existing cultivars, seed yields are reduced when sowing is delayed beyond mid-November, and genotypes which maintain their seed yields when sown late are being sought. Tall, erect plant types are better adapted to late sowing than conventional types [16].

Outside the Indian subcontinent, the predominantly spring-sown chickpeas experience progressively warming

378

temperatures and lengthening day, and produce seed yields of 1000-1500 kg ha^{-1} in 90-110 days, their duration being curtailed by increasing temperatures and moisture stress in June or July. In the Mediterranean region, winter-sown crops initially experience cooling temperatures and shortening days. Their growth is slow, flowering and fruiting are delayed, crop durations are extended to 160-180 days, and seed yields can exceed 5000 kg ha^{-1}. The growth durations and seed yields of these spring and winter-sown chickpeas are remarkably similar to those of crops of peninsular and northern India, suggesting at least some compatibility between the two situations.

When grown on the Indian subcontinent, genotypes adapted to the Mediterranean area flower very late (presumably due to different photo-thermal requirements for flowering [250]) and produce very poor seed yields (ICRISAT, unpublished). Conversely, chickpeas of Indian origin, when grown outside the Indian subcontinent, flower at least as early as locally adapted materials and are not conspicuously poorer yielding. Genotypes of Indian origin may therefore provide valuable sources of parental materials for breeding programmes elsewhere. In view of the similarities of the environments, north Indian materials may be especially useful for breeding for winter-sowing in the Mediterranean area, to which locally adapted materials, which have evolved in a spring-sowing situation, may be poorly adapted.

The introduction of Indian materials into such breeding programmes will inevitably involve crosses between desi and kabuli types. The evidence regarding the utility of desi/kabuli crossing *per se* is meagre but suggests that the variability generated by a cross is determined by the divergence in the origins [68] and characteristics of the parents, rather than whether they are desi or kabuli types (ICRISAT, unpublished). Furthermore, the recovery of segregants with kabuli type seeds from desi/kabuli crosses is very poor. Nevertheless, desi and kabuli types represent

divergent groups; both have characters which can be usefully introgressed into the other; established Indian
kabuli cultivars all derive from crosses between desi and
kabuli cultivars [68]; and wilt resistance has been successfully transferred from desi to kabuli types [96]. Crosses
between the two types are readily accomplished and the
recovery of kabuli type segregants can be improved by one
or two backcrosses to the kabuli type parent or intermating
among kabuli types in early generations. Thus, there seem
to be compelling reasons for using desi parents in crosses
for improvement of kabuli types.

Undoubtedly the most encouraging prospect for larger
yields is winter-sowing in areas where chickpeas are traditionally sown in spring. The most important constraints for
winter-sown chickpeas are *Ascochyta* blight and weed control.
Sources of good and stable resistance to *Ascochyta* blight
have been identified and are being incorporated into heavy-
yielding backgrounds [71]. For weed control, completely
satisfactory herbicides have not yet been isolated. However,
the sowing of the winter crop could probably be delayed
slightly to enable cultural weed control, without sacrific-
ing a significant proportion of the yield benefits resulting
from winter sowing *per se*.

Gains from other approaches are expected to be limited.
Phosphate and nitrogen requirements appear to be relatively
small and responses to their application are inconsistent.
Responses to inoculation with *Rhizobium* are also inconsistent but good quality inoculants will be necessary in situations where chickpeas have not been previously cultivated,
such as in the drier areas of the Mediterranean basin which
will become available to the crop with the advent of winter-
sowing.

ACKNOWLEDGEMENTS

We wish to acknowledge the contributions of many colleagues
at ICRISAT, notably those of Drs L.J.G. van der Maesen and

Jagdish Kumar, in reading earlier drafts of this chapter and for their valuable suggestions. We also thank Mr A.J. Rama Rao for his assistance in the typing and editing of the several drafts.

LITERATURE CITED

1. Abrol, I.P., Dargan, K.S. and Bhumbla, D.R. (1973), *Central Soil Salinity Research Institute Bulletin,* no. 2, New Delhi: Indian Council for Agricultural Research.
2. Abrol, I.P. and Fireman, M. (1977), *Central Soil Salinity Research Institute Bulletin,* no. 4. New Delhi: Indian Council for Agricultural Research.
3. Abrol, I.P. and Chatterjee, S.R. (1980), *Plant Biochemical Journal,* S.M. Siman Memorial Volume, 125-149.
4. Agarwal, S.C. and Sharma, C.P. (1979), *Recognising Micronutrient Disorders in Crop Plants on the Basis of Visible Symptoms and Plant Analysis.* Lucknow: Department of Botany, Lucknow University.
5. Ahlawat, I.P.S., Singh, A. and Saraf, C.S. (1981), *Experimental Agriculture* 17, 57-62.
6. Allen, D.J. (1983), *The Pathology of Tropical Food Legumes: Disease Resistance in Crop Improvement.* London: John Wiley.
7. Anonymous (1977), *Pulses in India (Production, Agriculture, Processing and Consumption).* Mysore: Central Food Technological Research Institute.
8. Anonymous (1982), *Unpublished Proceedings of the All India Coordinated Pulses Improvement Project Rabi Pulses Workshop, 3-5 October 1982.* New Delhi, India.
9. Argikar, G.P. (1970), in Kachroo, P. (*ed.*) *Pulse Crops of India.* New Delhi: ICAR, 54-135.
10. Aslam, M. (1984), in Saxena, M.C. and Singh, K.B. (*eds*) *Ascochyta Blight and Winter Sowing of Chickpeas.* The Hague: Martinus Nijhoff, 237-46.
11. Athwal, D.S. and Brar, H.S. (1964), *Journal of Research of Punjab Agricultural University* 1, 129-34.
12. Athwal, D.S. and Gill, G.S. (1964), *Journal of Research of Punjab Agricultural University* 1, 116-28.
13. Ayyar, V.R. and Balasubramanyan, R. (1936), *Indian Academy of Science Proceedings, Section B* 4, 1-26.
14. Aziz, M.A., Khan, M.A. and Shah, S. (1960), *Agriculture Pakistan* 11, 37-48.
15. Bahl, P.N. (1980), in *Proceedings of the International Workshop on Chickpea Improvement.* Hyderabad, India: ICRISAT, 75-80.
16. Bahl, P.N. (1983), *International Chickpea Newsletter* 8, 8-9.
17. Balasubrahmanyan, R. (1951), *Indian Journal of Agricultural Science* 21, 239-43.
18. Bapna, S.L., Binswanger, H. and Quizon, J.B. (1981), *Studies in Employment and Rural Development,* no. 73. Washington, D.C.: Development Economics Department, International Bank for Reconstruction and Development.
19. Behrman, J.R. and Murty, K.N. (1983), *Occasional Paper 52,* Hyderabad, India: Economics Program, ICRISAT.

381

20. Bhapkar, D.G. and Patil, J.A. (1962), *Science and Culture* 28, 441-2.
21. Bhat, N.R. and Argikar, G.P. (1951), *Heredity* 5, 143-6.
22. Bhatt, D.D. and Singh, D.P. (1980), *International Chickpea Newsletter* 3, 4-5.
23. Boorsma, P.A. (1980), *FAO Plant Protection Bulletin*, 28, 110-13.
24. Boulter, D., Evans, I.M. and Thompson, A. (1976), *Qualitas Plantarum* 26, 107-19.
25. Brar, H.S. and Athwal, D.S. (1970), *Indian Journal of Genetics and Plant Breeding* 30, 690-703.
26. Casas, A.T. and Diaz, R.M.Z. (1980), *International Chickpea Newsletter* 5, 10-11.
27. Central Food Technological Research Institute (1982), *Annual Report 1980-81*, Mysore, India.
28. Chahal, V.P., Joshi, P.K., Chahal, D.S. and Rewari, R.B. (1978), *Indian Journal of Microbiology* 18, 148-50.
29. Chandra, S. (1980), in *Proceedings of the International Workshop on Chickpea Improvement*. Hyderabad, India: ICRISAT, 97-105.
30. Chandra, S., Sohoo, M.S. and Singh, K.P. (1971), *Journal of Research of Punjab Agricultural University* 8, 165-8.
31. Chaubey, H.S., Beniwal, S.P.S., Tripathi, H.S. and Nene, Y.L. (1982), *International Chickpea Newsletter* 8, 20-21.
32. Chauhan, S.K. (1962), *Proceedings of Indian National Academy of Sciences India B* 32, 78-84.
33. Chauhan, S.K. (1962), *Proceedings of National Academy of Sciences India B* 32, 385-6.
34. Chauhan, S.K. (1963), *Proceedings of the National Academy of Sciences, India B* 33, 552-4.
35. Chopra, K. and Swamy, G. (1975), *Pulses: an Analysis of Demand and Supply in India 1951-1971*. Bangalore: Institute for Social and Economic Change.
36. Corbin, E.J., Brockwell, J. and Gault, R.R. (1977), *Australian Journal of Experimental Agriculture and Animal Husbandry* 17, 126-34.
37. Cubero, J. (1976), in *Proceedings of the International Workshop on Grain Legumes*. Hyderabad, India: ICRISAT, 117-22.
38. D'Cruz, R. and Tendulkar, A.V. (1970), *Research Journal of Mahatma Phule Agricultural University* 1, 121-7.
39. Dadarwal, K.R., Shashi Prabha and Tauro, P. (1976), in Sen, S.P., Abrol, Y.P. and Sinha, S.K. (eds) *Nitrogen Assimilation and Crop Productivity. Proceedings of a National Symposium*. New Delhi: Associated Publishing, 235-43.
40. Dahiya, B.S., Kapoor, A.C., Solanki, I.S. and Waldia, R.S. (1982), *Experimental Agriculture* 18, 289-92.
41. Dart, P.J., Islam, R. and Eaglesham, A. (1976), in *Proceedings of the International Workshop on Grain Legumes*. Hyderabad, India: ICRISAT, 63-83.
42. Davis, T.M. and Foster, K.W. (1982), *International Chickpea Newsletter* 7, 6-8.
43. Eberhart, S.A. and Russel, W.R. (1966), *Crop Science* 66, 36-40.
44. Echandi, E. (1970), *Phytopathology* 60, 1539.
45. Ekbote, R.B. (1937), *Current Science* 5, 548-9.
46. Esh, G.C., De, T.S. and Basu, U.P. (1959), *Science* 129, 148-9.
47. Evans, A.M. (1980), in Summerfield, R.J. and Bunting, A.H. (eds) *Advances in Legume Science*. London: HMSO, 337-48.

Chickpea (*Cicer arietinum* L.)

48. FAO (1950-1980), *Production Yearbooks*. Rome: FAO.
49. Gaur, Y.D. and Sen, A.N. (1979), *New Phytologist* 83, 745-54.
50. Govil, J.N. (1981), *Legume Research* 4, 23-6.
51. Gowda, C.L.L. (1981), *International Chickpea Newsletter* 5, 6.
52. Gowda, C.L.L. and Bahl, P.N. (1976), *Indian Journal of Genetics and Plant Breeding* 36, 265-9.
53. Gowda, C.L.L. and Kaul, A.K. (1982), *Pulses in Bangladesh*. Joydebpur, Dhaka: Bangladesh Agricultural Research Institute, and Rome: FAO.
54. Gowda, C.L.L. and Smithson, J.B. (1980), *International Chickpea Newsletter* 3, 10.
55. Gowen, S. (1982), *International Chickpea Newsletter* 7, 16-17.
56. Gupta, N. and Singh, R.S. (1982), *Indian Journal of Agricultural Research* 16, 113-17.
57. Gupta, R.F., Kazra, M.S. and Bhandari, S.C. (1982), in *Biological Nitrogen Fixation: Proceedings of the National Symposium held at Indian Agricultural Research Institute, New Delhi*, 544-5.
58. Gupta, S.P., Luthra, R.C., Gill, A.S. and Phul, P.S. (1972), *Journal of Research of Punjab Agricultural University* 9, 405-9.
59. Gupta, V.P. and Ramanujam, S. (1974), *Indian Journal of Genetics and Plant Breeding* 34a, 793-9.
60. Hadas, A. (1970), *Israel Journal of Agricultural Research* 20, 3-14.
61. Harjit, Singh, Kumar, J., Smithson, J.B. and Haware, M.P. (1984), *Journal of Heredity* (in press).
62. Haware, M.P. and Nene, Y.L. (1980), *Tropical Grain Legume Bulletin* 19, 38-40.
63. Haware, M.P. and Nene, Y.L. (1982), *Plant Disease Reporter* 66, 250-51.
64. Haware, M.P. and Nene, Y.L. (1982), *Plant Disease Reporter* 66, 809-10.
65. Haware, M.P. and Nene, Y.L. (1982), *International Chickpea Newsletter* 6, 17-18.
66. Haware, M.P., Nene, Y.L. and Rajeshwari, R. (1978), *Phytopathology* 68, 1364-7.
67. Hawtin, G.C. (1976), in *Proceedings of the International Workshop on Grain Legumes*. Hyderabad, India: ICRISAT, 109-16.
68. Hawtin, G.C. and Singh, K.B. (1980), in *Proceedings of the International Workshop on Chickpea Improvement*. Hyderabad, India: ICRISAT, 51-60.
69. Helbaek, H. (1970), in Mellaart, J. (*ed.*) *Excavations at Hacilar*. Edinburgh: University Press, 189-244.
70. ICARDA (1982), *Annual Report, 1980-81*. Aleppo, Syria.
71. ICARDA (1983), *Annual Report, 1982*. Aleppo, Syria.
72. ICRISAT (1979), *Annual Report, 1977-78*. Patancheru, India.
73. ICRISAT (1980), *Annual Report, 1978-79*. Patancheru, India.
74. ICRISAT (1981), *Annual Report, 1979-80*. Patancheru, India.
75. ICRISAT (1982), *Annual Report, 1980-81*. Patancheru, India.
76. ICRISAT (1983), *Annual Report, 1982*. Patancheru, India.
77. Indian Agricultural Research Institute (1971), *New Vistas in Pulse Production*. New Delhi: IARI.
78. Islam, R. (1979), in Hawtin, G.C. and Chancellor, G.J. (*eds*) *Proceedings of a Workshop held at the University of Aleppo, Syria*. Aleppo, Syria: ICARDA, 166-9.
79. Islam, R. (1981), *International Chickpea Newsletter* 5, 16.
80. Islam, R. and Afandi, F. (1980), *International Chickpea Newsletter*

$\underline{2}$, 18-19.
81. Jain, K.C., Pandya, B.P. and Pande, K. (1981), *Indian Journal of Genetics and Plant Breeding* $\underline{41}$, 220-25.
82. Jambunathan, R. and Singh, U. (1980), in *Proceedings of the International Workshop on Chickpea Improvement*. Hyderabad, India: ICRISAT, 61-6.
83. Jambunathan, R. and Singh, U. (1981), *Journal of Agriculture and Food Chemistry* $\underline{29}$, 1091-3.
84. Jambunathan, R., Singh, U. and Subramanian, V. (1984), in *Proceedings of the Workshop on Interfaces between Agriculture, Nutrition and Food Science*. Hyderabad, India: ICRISAT (in press).
85. Jeswani, L.M., Singh, S.P. and Raju, D.B. (1971), *Catalogue of World Genetic Stocks of Bengal Gram* (Cicer arietinum). New Delhi: Indian Council of Agricultural Research.
86. Johnson, H.W., Robinson, H.F. and Comstock, R.E. (1955), *Agronomy Journal* $\underline{47}$, 314-18.
87. Kaiser, W.J. (1981), *Economic Botany* $\underline{35}$, 300-20.
88. Kaiser, W.J. and Danesh, O. (1971), *Phytopathology* $\underline{61}$, 453-7.
89. Kay, D.E. (1979), *Crop and Product Digest No. 3 - Food Legumes*. London: Tropical Products Institute.
90. Keatinge, J.D.H. and Cooper, P.J.M. (1983), *Journal of Agricultural Science, Cambridge* $\underline{100}$, 667-80.
91. Khanvilkar, S.G. and Desai, B.B. (1981), *Journal of Maharashtra Agricultural University* $\underline{6}$, 226-8.
92. Krober, O.A., Jacob, M.K., Lal, R.K., and Kashkary, V.K. (1970), *Indian Journal of Agricultural Sciences* $\underline{40}$, 1025-30.
93. Kukula, S., Haddad, M.A. and Masri, H. (1984), in *Proceedings of the International Workshop on Faba Beans, Kabuli Chickpeas and Lentils in the 1980s*. Aleppo, Syria: ICARDA (in press).
94. Kumar, J., Gowda, C.L.L., Saxena, N.P., Sethi, S.C., Singh, U. and Sahrawat, K.L. (1983), *Current Science* $\underline{2}$, 82-3.
95. Kumar, J. and Haware, M.P. (1982), *Phytopathology* $\underline{72}$, 1035-6.
96. Kumar, J. and Haware, M.P. (1983), *International Chickpea Newsletter* $\underline{8}$, 7-8.
97. Kumar, J., Haware, M.P. and Nene, Y.L. (1980), *International Chickpea Newsletter* $\underline{3}$, 5.
98. Kumar, J., Smithson, J.B. and Singh, U. (1982), *International Chickpea Newsletter* $\underline{7}$, 20-21.
99. Labrousse, F. (1931), *Revue de Pathologie Végétale et d'Entomologie Agricole* $\underline{18}$, 226-31.
100. Ladizinsky, G. (1975), *Notes from the Royal Botanic Garden Edinburgh* $\underline{34}$, 201-2.
101. Ladizinsky, G. and Adler, A. (1975), *Israel Journal of Botany* $\underline{24}$, 183-9.
102. Ladizinsky, G. and Adler, A. (1976), *Euphytica* $\underline{25}$, 211-17.
103. Lal, B.M., Prakash, V. and Verma, S.C. (1963), *Experientiae* $\underline{9}$, 154-5.
104. Lal, B.M., Rohewal, S.S., Verma, S.C. and Prakash, V. (1963), *Annals of Biochemistry and Experimental Medicine* $\underline{23}$, 543-8.
105. Lal, S. and Singh, S.N. (1974), *Indian Journal of Genetics and Plant Breeding* $\underline{34a}$, 784-7.
106. Lal, S.S., Dias, C.A.R., Yadava, C.P. and Singh, D.N. (1980), *International Chickpea Newsletter* $\underline{3}$, 14-15.
107. Lateef, S.S. and Reed, W. (1979), *International Chickpea Newsletter* $\underline{1}$, 9.

Chickpea (*Cicer arietinum* L.)

108. Luthra, J.C., Sattar, A. and Bedi, K.S. (1935), *Agriculture and Livestock of India* 5, 489-98.
109. McKerral, A. (1923), *Agricultural Journal of India* 28, 608-13.
110. Maden, S., Singh, D., Mathur, S.B. and Neergaard, P. (1975), *Seed Science and Technology* 3, 667-81.
111. Malewar, G.U., Jadhav, D.K. and Ghonsikar, C.P. (1982), *International Chickpea Newsletter* 6, 13-14.
112. Malik, B.A. and Tufail, M. (1984), in Saxena, M.C. and Singh, K.B. (eds) Ascochyta *Blight and Winter Sowing of Chickpeas*. The Hague: Martinus Nijhoff, 229-35.
113. Manchanda, H.R. and Sharma, S.K. (1980), *International Chickpea Newsletter* 3, 7-8.
114. Mathur, R.S., Jain, J.S. and Atheya, S.C. (1960), *Current Science* 29, 403.
115. Mehra, R.B. and Ramanujam, S. (1979), *Indian Journal of Genetics and Plant Breeding* 39, 492-500.
116. Minchin, F.R., Summerfield, R.J., Hadley, P. and Roberts, E.H. (1980), *Experimental Agriculture* 16, 241-61.
117. Mishra, P.K., Pandey, R.L., Tomar, G.S. and Tiwari, A.S. (1974), *JNKVV Research Journal* 8, 290-91.
118. More, D.C. and D'Cruz, R. (1976), *Journal of Maharashtrah Agricultural University* 1, 11-14.
119. More, D.C. and D'Cruz, R. (1976), *Botanique* 7, 37-40.
120. Morena, M.T. and Cubero, J.I. (1978), *Euphytica* 27, 465-85.
121. Murty, B.R. (1976), in *Proceedings of the International Workshop on Grain Legumes*. Hyderabad, India: ICRISAT, 239-51.
122. Narayan, R.K.J. and Macefield, A.J. (1976), *Theoretical and Applied Genetics* 47, 179-87.
123. Nayeem, K.A., Kolhe, A.K. and D'Cruz, R. (1977), *Indian Journal of Agricultural Sciences* 47, 6068-611.
124. Nene, Y.L. (1980), in *Proceedings of the International Workshop on Chickpea Improvement*. Hyderabad, India: ICRISAT, 171-8.
125. Nene, Y.L. (1980), *Pulse Pathology, Progress Report 8*. Hyderabad, India: ICRISAT.
126. Nene, Y.L., Haware, M.P. and Reddy, M.V. (1979), *Information Bulletin No. 3*. Hyderabad, India: ICRISAT.
127. Nene, Y.L., Haware, M.P. and Reddy, M.V. (1981), *Information Bulletin No. 10*. Hyderabad, India: ICRISAT.
128. Nene, Y.L., Haware, M.P., Reddy, M.V. and Pundir, R.P.S. (1981), *Pulse Pathology Progress Report No. 15*. Hyderabad, India: ICRISAT.
129. Nene, Y.L., Kannaiyan, J., Haware, M.P. and Reddy, M.V. (1980), in *Proceedings of the Consultants' Group Discussion on the Resistance to Soil-Borne Diseases of Legumes*. Hyderabad, India: ICRISAT, 3-35.
130. Niknejad, M. and Khosh-Khui, M. (1972), *Indian Journal of Agricultural Sciences* 42, 273-4.
131. Niknejad, M. and Khosh-Khui, M. (1972), *Journal of Heredity* 63, 155-6.
132. Niknejad, M., Kheradnam, M. and Khosh-Khui, M. (1978), *Canadian Journal of Plant Science* 58, 235-40.
133. Nutman, P.S. (1976), in Nutman, P.S. (ed.) *Symbiotic Nitrogen Fixation in Plants*. Cambridge: Cambridge University Press, 211-37.
134. Okon, Y., Eshel, Y. and Henis, Y. (1972), *Soil Biology and Biochemistry* 4, 165-70.

135. Padwick, G.W. and Bhagwagar, P.R. (1943), *Indian Journal of Agricultural Science* 13, 289-90.
136. Pakistan Agricultural Research Council (1982), *Annual Report on Food Legumes Improvement in Pakistan in 1980-81*. Islamabad.
137. Pande, K., Jain, K.C. and Pandya, B.P. (1980), *Egyptian Journal of Genetics and Cytology* 9, 297-305.
138. Pandey, M.L., Beniwal, S.P.S., Pandya, B.P. and Arora, P.P. (1982), *International Chickpea Newsletter* 7, 13.
139. Pareek, R.P. (1979), *Indian Journal of Microbiology* 19, 123-9.
140. Pathak, M.M., Sindhu, J.S., Singh, K.B. and Srivastava, S.B.L. (1982), *International Chickpea Newsletter* 6, 9.
141. Patil, J.A. and Deshmukh, R.B. (1975), *Research Journal of Mahatma Phule Agricultural University* 6, 88-95.
142. Phadnis, B.A. (1976), *Indian Journal of Genetics and Plant Breeding* 36, 54-8.
143. Pieters, R. (1984), in Saxena, M.C. and Singh, K.B. (*eds*) *Ascochyta Blight and Winter Sowing of Chickpeas*. The Hague: Martinus Nijhoff, 89-93.
144. Polhill, R.M. and van der Maesen, L.J.G. (1985), Chapter 1 of this volume.
145. Popova, G.M. (1937), in Vavilov, N.I. (*ed.*) *Flora of the Cultivated Plants (of the USSR)*. Leningrad, 25-71.
146. Pundir, R.P.S. and van der Maesen, L.J.G. (1977), *Tropical Grain Legume Bulletin* 10, 26.
147. Pundir, R.P.S. and van der Maesen, L.J.G. (1983), *International Chickpea Newsletter* 8, 4-5.
148. Pundir, R.P.S. and van der Maesen, L.J.G. (1984), *Euphytica* (in press).
149. Rai, R. and Singh, S.N. (1979), *Journal of Agricultural Science, Cambridge* 93, 47-50.
150. Rajendran, S., Jha, D. and Ryan, J.G. (1982) *ICRISAT Economics Program, Progress Report 34*. Hyderabad, India: ICRISAT.
151. Ram, C., Chandra, S., Chaudhary, M.S. and Jatasra, D.S. (1978), *Indian Journal of Agricultural Research* 12, 187-90.
152. Ramanujam, S. (1976), in Simmonds, N.W. (*ed.*) *Evolution of Crop Plants*. London and New York: Longman, 157-9.
153. Ramanujam, S. and Gupta, V.P. (1974), *Indian Journal of Genetics and Plant Breeding* 34a, 757-63.
154. Ramanujam, S., Rohewal, S.S. and Singh, S.P. (1964), *Indian Journal of Genetics and Plant Breeding* 24, 239-43.
155. Randhawa, M.S. (1979), *History of the Indian Council of Agricultural Research, 1929-1979*. New Delhi: ICAR.
156. Rao, N.K., Pundir, R.P.S. and van der Maesen, L.J.G. (1980), *Indian Academy of Sciences Proceedings (Plant Sciences)* 89, 497-503.
157. Rao, P.P. and Subba Rao, K.V. (1981), *International Chickpea Newsletter* 5, 17.
158. Rao, P.U. (1980), *Indian Journal of Nutrition and Dietetics* 17, 408-9.
159. Rao, P.U. and Belavady, B. (1980), *Indian Journal of Nutrition and Dietetics* 17, 6-8.
160. Rastogi, K.B. and Singh, L. (1977), *Crop Improvement* 4, 191-7.
161. Reddy, M.V. (1980), *International Chickpea Newsletter* 3, 12.
162. Reddy, M.V. (1980), *International Chickpea Newsletter* 3, 13-14.
163. Reddy, M.V., Singh, K.B. and Nene, Y.L. (1982), *International*

Chickpea (*Cicer arietinum* L.)

Chickpea Newsletter 6, 18-19.

164. Reddy, V.G. and Chopde, P.R. (1977), *Journal of Maharashtra Agricultural University* 2, 224-6.

165. Reddy, V.G. and Chopde, P.R. (1977), *Journal of Maharashtra Agricultural University* 2, 228-31.

166. Reddy, V.G. and Nayeem, K.A. (1978), *Indian Journal of Heredity* 9, 27-35.

167. Reed, W., Lateef, S.S. and Sithanantham, S. (1980), in *Proceedings of the International Workshop on Chickpea Improvement*. Hyderabad, India: ICRISAT, 179-83.

168. Regional Pulses Improvement Project (1968), *Progress Report, 1968*. Beirut, Lebanon.

169. Regional Pulses Improvement Project (1969), *Progress Report, 1969*, Beirut, Lebanon.

170. Regional Pulses Improvement Project (1974), *Progress Report, 1974*. Appendix, Beirut, Lebanon.

171. Rembold, H. (1981), *International Chickpea Newsletter* 4, 18-19.

172. Rizvi, S.M.A. and Singh, H.M. (1980), *International Chickpea Newsletter* 3, 15.

173. Rupela, O.P. (1982), *International Chickpea Newsletter* 7, 9-10.

174. Rupela, O.P. and Dart, P.J. (1980), in *Proceedings of the International Workshop on Chickpea Improvement*. Hyderabad, India: ICRISAT, 161-7.

175. Rupela, O.P. and Dart, P.J. (1982), in Graham, P.H. and Harris, S.C. (eds) *Biological Nitrogen Fixation Technology for Tropical Agriculture*. Cali, Colombia: CIAT, 57-61.

176. Ryan, J.G. and Asokan, M. (1977), *Occasional Paper No. 18*. Hyderabad, India: Economics Program, ICRISAT.

177. Sah, R.P. (1982), PhD Thesis. Punjab Agricultural University, Ludhiana, India.

178. Sandhu, S.S., Keim, W.F., Hodges, H.F. and Nyquist, W.E. (1974), *Crop Science* 14, 649-52.

179. Sanoria, C.L. (1978), *Journal of Scientific Research, Banaras Hindu University* 28, 107-8.

180. Sarat, C.S. and Davis, R.J. (1969), in *Proceedings of a Symposium on Water Management*. Hissar, India: Punjab Agricultural University.

181. Saxena, M.C. (1979), in Hawtin, G.C. and Chancellor, G.J. (eds) *Food Legume Improvement and Development*. Ottawa, Canada: IDRC, 155-65.

182. Saxena, M.C. (1980), in *Proceedings of the International Workshop on Chickpea Improvement*. Hyderabad, India: ICRISAT, 89-96.

183. Saxena, M.C. and Yadav, D.S. (1976), in *Proceedings of the International Workshop on Grain Legumes*. Hyderabad, India: ICRISAT, 31-61.

184. Saxena, N.P. (1984), in Goldsworthy, P.R. and Fisher, N.M. (eds) *The Physiology of Tropical Field Crops*. London: John Wiley (in press).

185. Saxena, N.P., Kapoor, S.N. and Bisht, D.S. (1983), *International Chickpea Newsletter* 9, 12-14.

186. Saxena, N.P. and Sheldrake, A.R. (1980), *Field Crops Research* 3, 189-91.

187. Saxena, N.P. and Sheldrake, A.R. (1980), in *Proceedings of the International Workshop on Chickpea Improvement*. Hyderabad, India: ICRISAT, 106-20.

188. Saxena, N.P. and Sheldrake, A.R. (1980), *Field Crops Research* 3,

211-14.
189. Sen, A.N. (1966), *Indian Journal of Agricultural Science* 36, 1-7.
190. Shaikh, M.A.Q., Ahmed, Z.U., Majid, M.A., Bhuiya, A.D., Kaul,
A.K. and Miah, M.M. (1980), *Mutation Newsletter* 6, 1-3.
191. Sharma, B. and Mehra, R.B. (1981), *Pulse Crops Newsletter* 1(2), 8.
192. Sharma, D. and Jodha, N.S. (1983), in *Proceedings of a Symposium on Increasing Pulse Production in India: Constraints and Opportunities.* New Delhi, India: Hindustan Lever Research Organisation.
193. Sheldrake, A.R., Saxena, N.P. and Krishnamurthy, L. (1979), *Field Crops Research* 1, 243-53.
194. Shukla, U.C. and Yadav, O.P. (1982), *Plant and Soil* 65, 239-48.
195. Singh, G. and Bushan, L.S. (1979/1980), *Agricultural Water Management* 2, 299-305.
196. Singh, G., Kapoor, S. and Singh, K. (1982), *International Chickpea Newsletter* 7, 13-14.
197. Singh, K.B. and Dahiya, B.S. (1973), in *Proceedings of a Symposium on Wilt Problem and Breeding for Wilt Resistance in Bengal Gram.* New Delhi, India: IARI, 13-14.
198. Singh, K.B., Gridley, H.E. and Hawtin, G.C. (1984), in Saxena, M.C. and Singh, K.B. (eds) Ascochyta *Blight and Winter Sowing of Chickpeas.* The Hague: Martinus Nijhoff, 95-110.
199. Singh, K.B., Malhotra, R.S. and Muehlbauer, F.J. (1984), *An Annotated Bibliography of Chickpea Breeding and Genetics.* Patancheru, India: ICRISAT, (in press).
200. Singh, K.B., Malhotra, R.S. and Witcombe, J. (1983), *Kabuli Chickpea Germplasm Catalogue.* Aleppo, Syria: ICARDA.
201. Singh, K.B., Meyoeci, K., Izgin, N. and Tuwafe, S. (1981), *International Chickpea Newsletter* 4, 11-12.
202. Singh, K.B., Nene, Y.L. and Reddy, M.V. (1984), in Saxena, M.C. and Singh, K.B. (eds) Ascochyta *Blight and Winter Sowing of Chickpeas.* The Hague: Martinus Nijhoff, 67-87.
203. Singh, K.B. and Reddy, M.V. (1983), *Crop Science* 23, 9-10.
204. Singh, K.B. and van der Maesen, L.J.G. (1977), *Chickpea Bibliography 1930-1974.* Hyderabad, India: ICRISAT.
205. Singh, L. (1980), in *Proceedings of the International Workshop on Chickpea Improvement,* Hyderabad, India: ICRISAT, 191-6.
206. Singh, R.G. (1971), *Indian Journal of Agricultural Sciences* 41, 101-6.
207. Singh, S.P. and Mehra, R.B. (1980), *Tropical Grain Legume Bulletin* 19, 51-4.
208. Singh, U., Kumar, J. and Gowda, C.L.L. (1983), *Qualitas Plantarum* 32, 179-84.
209. Singh, U., Kumar, J., Jambunathan, R. and Smithson, J.B. (1980), *International Chickpea Newsletter* 3, 18.
210. Singh, U., Kumar, J., Sahrawat, K.L. and Smithson, J.B. (1982), *International Chickpea Newsletter* 7, 21-2.
211. Singh, U., Raju, S.M. and Jambunathan, R. (1981), *Journal of Food Science and Technology* 18, 86-8.
212. Sithanantham, S., Rao, V.R. and Reed, W. (1981), *International Chickpea Newsletter* 5, 14.
213. Sithanantham, S., Rao, V.R. and Reed, W. (1982), *International Chickpea Newsletter* 6, 21-2.
214. Sithanantham, S. and Reed, W. (1980), *International Chickpea Newsletter* 2, 15.
215. Sithanantham, S., Tahhan, O., Hariri, G. and Reed, W. (1984), in

Saxena, M.C. and Singh, K.B. (*eds*) Ascochyta *Blight and Winter Sowing of Chickpeas*. The Hague: Martinus Nijhoff, 179-87.

216. Sivakumar, M.V.K. and Virmani, S. (1979), *Experimental Agriculture* 15, 377-83.

217. Smartt, J. (1976), *Tropical Pulses*. London: Longman.

218. Smartt, J. and Hymowitz, T. (1985), Chapter 2 of this volume.

219. Sosulski, F.W. and Holt, N.W. (1980), *Canadian Journal of Plant Science* 60, 1327-31.

220. Subba Rao, N.S. (1976), in Nutman, P.S. (*ed.*) *Symbiotic Nitrogen Fixation in Plants*. Cambridge: Cambridge University Press, 255-68.

221. Subba Rao, N.S. (1982), in *Biological Nitrogen Fixation: Proceedings of the National Symposium Held at Indian Agricultural Research Institute, New Delhi*, 27-48.

222. Summerfield, R.J., Hadley, P., Roberts, E.H., Minchin, R.J. and Rawsthorne, S. (1984), *Experimental Agriculture* 20, 77-93.

223. Summerfield, R.J., Minchin, F.R., Roberts, E.H. and Hadley, P. (1980), in *Proceedings of the International Workshop on Chickpea Improvement*. Hyderabad, India: ICRISAT, 121-49.

224. Summerfield, R.J., Minchin, F.R., Roberts, E.H. and Hadley, P. (1981), *Tropical Agriculture* (Trinidad) 58, 97-113.

225. Summerfield, R.J. and Roberts, E.H. (1985), in Halevy, A.H. (*ed.*) *A Handbook of Flowering*. Florida: CRC Press (in press).

226. Sundara Rao, W.V.B., Venkatachari, A. and Madhava Reddy, V. (1975), *Indian Journal of Genetics and Plant Breeding* 35, 229-35.

227. Swaminathan, M.S., Naik, M.S., Kaul, A.K. and Austin, A. (1971), *Indian Journal of Agricultural Sciences* 41, 393-406.

228. Tahhan, O., Sithanantham, S., Hariri, G. and Reed, W. (1982), *International Chickpea Newsletter* 6, 21.

229. Tannous, R.I. and Singh, K.B. (1980), *International Chickpea Newsletter* 2, 20-21.

230. Thirumalachar, M.J. and Mishra, J.N. (1953), *Food and Agriculture Organisation Plant Protection Bulletin* 1, 145-6.

231. Thirumalachar, M.J. and Mishra, J.N. (1953), *Food and Agriculture Organisation Plant Protection Bulletin* 2, 11-12.

232. Thompson, J.A. (1982), in *Biological Nitrogen Fixation: Proceedings of the National Symposium held at Indian Agricultural Research Institute, New Delhi*, 49-64.

233. Tomar, G.S., Singh, L., Sharma, D. and Deodhar, A.D. (1973), *JNKVV Research Journal* 7, 35-9.

234. Toomsan, B., Rupela, O.P. and Dart, P.J. (1982), in *Biological Nitrogen Fixation: Proceedings of the National Symposium held at Indian Agricultural Research Institute, New Delhi*, 517-30.

235. Tyagi, P.S., Singh, B.D. and Jaiswal, H.K. (1982), *International Chickpea Newsletter* 6, 6-8.

236. Upadhyaya, H.D., Haware, M.P., Kumar, J. and Smithson, J.B. (1983), *Euphytica* 32, 447-52.

237. Upadhyaya, H.D., Smithson, J.B., Haware, M.P. and Kumar, J. (1983), *Euphytica* 32, 749-55.

238. van der Maesen, L.J.G. (1972), Cicer L., *a Monograph of the Genus, with Special Reference to the Chickpea* (Cicer arietinum L.), *its Ecology and Cultivation*. Wageningen: Mededelingen Landbouwhogeschool, 72-10.

239. van der Maesen, L.J.G. (1976), in *Proceedings of the International Workshop on Grain Legumes*. Hyderabad, India: ICRISAT, 229-37.

240. van der Maesen, L.J.G. and Pundir, R.P.S. (1984), *Research Bul-*

letin, Patancheru, India: ICRISAT (in press).

241. van der Maesen, L.J.G., Pundir, R.P.S., and Remanandan, P. (1980), in *Proceedings of the International Workshop on Chickpea Improvement*. Hyderabad, India: ICRISAT, 28-32.

242. Vavilov, N.I. (1951), *Chronica Botanica, New York* 13, 26-38; 75-8; 151.

243. Verma, M.M., Sandhu, T.S., Gill, A.S., Brar, H.S., Bhullar, B.S. and Sra, S.S. (1982), in *Unpublished Proceedings of All India Coordinated Pulse Improvement Project, Rabi Pulses Workshop, 3-5 Oct 1982*. New Delhi, India.

244. Verma, M.M., Sandhu, T.S., Sandhu, S.S. and Bhullar, B.S. (1981), *International Chickpea Newsletter* 4, 14-16.

245. Vir, S. and Grewal, J.S. (1974), *Indian Phytopathology* 27, 355-60.

246. Weltzein, H.C. and Kaack, H.J. (1984), in Saxena, M.C. and Singh, K.B. (*eds*) Ascochyta *Blight and Winter Sowing of Chickpeas*. The Hague: Martinus Nijhoff, 35-44.

247. Westphal, E. (1974), *Pulses in Ethiopia, Their Taxonomy and Agricultural Significance*. Wageningen: Centre for Agricultural Publishing and Documentation.

248. Williams, P.C. and Nakoul, H. (1984), in *Proceedings of the Workshop on Faba Beans, Kabuli, Chickpeas and Lentils in the 1980s*, Aleppo, Syria: ICARDA (in press).

249. Williams, P.C., Nakoul, H. and Singh, K.B. (1983), *Journal of Science of Food and Agriculture* 34, 492-6.

250. Roberts, E.H., Hadley, P. and Summerfield, R.J. (1985), *Annals of Botany* (in press).

9 Soyabean (*Glycine max* (L.) Merrill)

D.J. Hume, S. Shanmugasundaram and W.D. Beversdorf

INTRODUCTION

The soyabean (*Glycine max* (L.) Merrill) is the world's most important oilseed and grain legume crop. It is known colloquially by many trivial names including, in English, soya bean, soyabean, soya-bean or soybean. Soybean is the most common terminology in North America; soyabean is conventional in the UK and elsewhere (see Chapter 1) and has been used throughout this book.

History

Soyabeans originated in China [98,136] where they continue to be an important crop. They were first imported into France in 1740 and into England by 1790. Haberlandt [see 119] published detailed accounts of his work with soyabeans in Austria, beginning in 1875. The first studies on soyabeans in the USA [136] were reported by Mense in 1804 and by Nuttall in 1829. Major expansion outside of China began after numerous introductions into the USA around 1880. Soyabeans in the USA were grown initially as hay and forage crops [98,100], but then became increasingly important as an oilseed crop. Demand for oil during the Second World War stimulated the crushing of soyabeans for oil and meal [124].

Current status and future projections

Table 9.1 shows production trends around the world during the past decade. Between 1969-71 and 1982, world production more than doubled, mainly because of increases in output from the USA and Brazil. US production is concentrated in the central part of the country. Illinois and the surrounding Corn Belt states are the largest producers. Forty-five per cent of the 1983 US crop was exported.

In Brazil, rapid expansion of soyabeans occurred during the period 1973 to 1976, principally in the states of Rio Grande do Sul, Parana and Sao Paulo (i.e. at latitudes between 20°S and 30°S [124]).

Throughout the past decade, developing countries, including Brazil, have produced about 35% of the world's crop, but in Asia production has risen rapidly only in India [61]. Increased production in most countries has been due to an increased area planted. For example, US production increased by 101% during the period 1969-71 to 1982 and reflected an expansion of 68% in the area planted but only a 19% increase in average yield. In Asia, where production increased by 36% over this same period, average yields in 1969-71 (999 kg ha^{-1}) were equivalent to 67% of the world average at that time compared with 71% in 1982 (1244 and 1751 kg ha^{-1}, respectively). In comparison, average yields from all developing countries combined improved by 47% during the period 1969-71 to 1982 compared with just 18% for the world average. Clearly, there should be ample scope for improved cultivars and management practices to increase both yield and production in Asia in future years.

Although soyabeans are grown only once each year in temperate countries, in the tropics and subtropics they can be grown year-round if water is available [150]. Crop durations in the tropics are shorter [9,10,11,93,151] and productivity can be as large as 51 kg seed ha^{-1} day^{-1} (Table 9.2). Three consecutive crops can be grown successively on the same land [11] with a total annual yield of about

Table 9.1 Area (1000 ha), yield (kg ha^{-1}) and production (1000 t) of soyabeans from 1969-71 to 1982 [61]

	Area harvested			Yield			Production		
	1969-71	1980	1982	1969-71	1980	1982	1969-71	1980	1982
World	29247	50799	52209	1487	1595	1751	43487	81026	96103
Africa	193	339	377	413	941	1075	80	319	412
North America	17308	27900	29516	1831	1785	2176	31684	49799	63998
USA	17036	28700	28700	1830	1776	2181	31174	48772	62584
South America	1438	11468	10766	1218	1717	1647	1751	19687	17730
Brazil	1314	8774	8202	1178	1727	1562	1547	15156	12810
Asia	9334	9672	10210	999	1030	1244	9329	9957	12702
China	7873	7515	7712	1033	1052	1299	8131	7906	10017
Europe	108	510	526	1082	1095	1372	117	535	721
Oceania	5	57	45	1111	1450	1756	5	82	79
USSR	860	854	870	606	615	529	521	525	460

Table 9.2 Maturity durations and productivity potentials of soyabeans in selected countries [10,147]

Country/ Location	Cultivar	Latitude	Altitude (m)	Yield (t ha^{-1})	Maturity (days)	Productivity (kg seed ha^{-1} day^{-1})
Ecuador	Bossier	2°21'S	17	4.6	104	44
Sri Lanka	Davis	9°06'N	1	5.0	98	51
Bolivia	Jupiter	18°30'S	389	4.1	134	31
Taiwan	AGS 144	23°00'N	9	4.3	103	41

8 t ha^{-1}. The current world record yield for a crop grown in a temperate area is 7.32 t ha^{-1} in New Jersey, USA [8], compared with an average yield in the USA of 2.2 t ha^{-1}.

World soyabean demand is expected to double in the next twenty years [4], with US production accounting for 57% of the projected world total of 188 million tonnes. In this nation-wide survey, average yields in the USA were projected at 3.7 t ha^{-1} by the year 2002. Strong demands for meal and, particularly, soyabean oil are expected to continue, at least in the immediate future [153].

PRINCIPAL ECONOMIC YIELD AND USES OF CROP PRODUCTS

Soyabean crops are grown for their seeds in which proximate composition (in the USA) averages 40% protein, 21% oil, 34% carbohydrate and 4.9% ash [127]. Approximately 95% of the US crop is marketed for processing, either domestically or abroad. Processing usually involves dehulling (removing testas), flaking the cracked seeds, and extraction of oil in a hexane solvent which is recovered from the oil and reused.

Soyabean oil is used mainly for salad dressing, cooking, and in margarine. It is also used in whipped toppings, coffee whiteners, icings, ice cream, confections, shortenings, frozen desserts and soups. Industrial uses include the manufacture of soap, paints and varnishes, resins, and drying oil products.

After oil extraction, about 98% of the residue meal is used as a protein supplement in livestock feeds. The defatted, dehulled meal contains 49-50% protein. For fibre-rich feeds, the hulls are added back to the meal to produce a formulation containing 44% protein.

Other forms of soyabean protein products include flour, protein concentrates, protein isolates, and textured and modified proteins. The preparation of and uses for these products are discussed by Orthoefer [127]. The direct consumption of soyabean protein in foods is expected to increase in developed countries.

Soyabeans are one of the important food crops of the Chinese people. Agricultural experimental stations in Manchuria have developed cultivars for special purposes such as soyabean milk, bean curd, 'tofu' (Chinese soyabean cheese), bean sprouts, bean flour, immature green vegetable soyabeans, oil, soyabean meal and various fermented products [147].

In Japan, soyabeans are a principal ingredient in the production of 'shoyu' (soya sauce), 'miso' (fermented soyabean paste), 'tofu' (bean curd) and 'natto' (fermented whole soyabeans). The 'shoyu' and 'miso' preparations in Japan have evolved from that referred to as 'chiang' in China [69]. Nearly one third of all the soyabeans consumed in Japan are used directly as food [69].

'Tempeh' (fermented soyabean cake) is used primarily in Indonesia. It is fermented by *Rhizopus* and the resulting cake is deep-fried in vegetable oil.

Buddhist monks and other devout followers of Buddhism do not eat meat; soyabeans are one of the major protein sources in their diets. Among the Chinese, the proteins from soyabeans and mung beans (*Vigna radiata*) are combined and used to create various imitation (artificial) meats (e.g. chicken, fish, shrimp and sausage) for vegetarians.

In India, the Soya Production and Research Association in Pantnagar has produced vegetarian 'meat' from soyabean

meal using the trade name 'Nutri nugget'. Recipes have also
been developed to incorporate soyabeans into indigenous
food preparations [160].

In India, Pakistan, Bangladesh, Sri Lanka and other
developing African and Latin American countries where soyabeans have been introduced, the crop has found increasing
usage as a popular infant food. It is viewed as a crop that
will not only enhance the nutritive value of local diets
but also lessen national shortages of vegetable oil.

In a few countries, Taiwan, Hong Kong, Singapore,
Malaysia and Thailand, for example, soyabean milk as a
beverage is gaining as much popularity as any other soft
drink on the market.

Soyabean seeds are dry-roasted and consumed in a similar manner to groundnuts (*Arachis hypogaea*) in Nepal and
Indonesia.

Immature green soyabean fruits are boiled with salt and
consumed as a snack in Japan, Korea and Taiwan. Such 'vegetable soyabeans' are also popular in Thailand, Nepal and
other Asian countries. The shelled green seeds are also
cooked with various kinds of meat [113]. Black or brown-
coloured soyabeans are also cooked together with rice to
enrich the flavour, nutrition and appearance of that cereal
in Korea.

PRINCIPAL FARMING SYSTEMS

In North America, soyabeans traditionally have been grown
almost exclusively as a sole crop, and usually as the only
crop cultivated on a given field during the year. This traditional approach is, however, changing and soyabeans are
now increasingly double-cropped. This technique has been
used in the southern USA for several years and the practice
is gradually spreading into northern regions. In double-
cropping, soyabeans are planted immediately after a cereal
crop has been harvested and they mature in the same season
[167]. In 1982, a total of 3.2 million ha were double-

cropped with winter wheat, including more than one third
of the soyabeans planted in the southern states [137].

A small area of soyabeans is also grown by relay cropping, in which the beans are established in standing grain.
In the US Corn Belt states, soyabeans are most frequently
grown in rotation with maize, although production of continuous soyabeans is sometimes practised [132].

In the tropics and subtropics, soyabeans can be grown
all year round [145]. The diversity of Asian cropping systems into which soyabeans are integrated is evident from
recent reviews [151,152]. The cropping system in which soyabeans are included depends upon local environmental factors,
especially rainfall [67,139], and on socio-economic considerations [67]. Even though soyabeans are cultivated on
a limited scale during the wet season, most are grown during
the dry season in the tropics and subtropics and follow
rice, wheat, maize or sorghum [152]. Selected examples of
such cropping patterns in three Asian countries are shown
in Figure 9.1. Even within a country, for example in Taiwan,

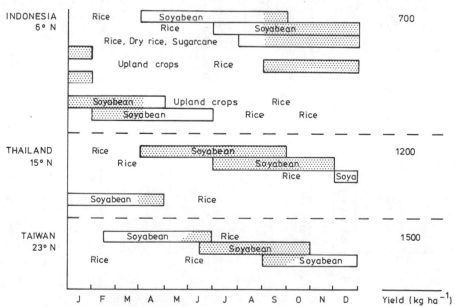

Figure 9.1 Diversity in soyabean growing seasons in three
selected Asian countries. Shaded and unshaded areas denote
wet and dry seasons, respectively [152]

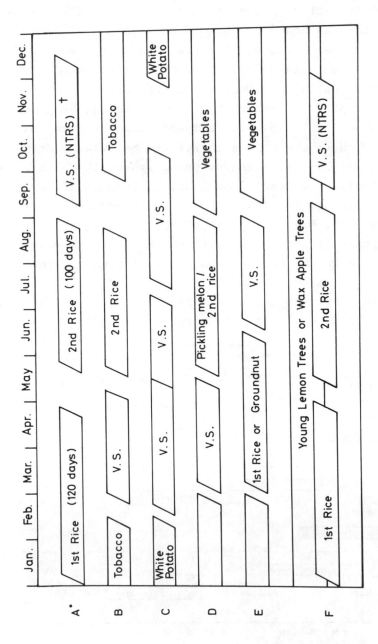

*Cropping patterns arranged according to the proportion of land devoted to each system, (A) being the most widespread.

†No-tillage, rice-stubble soyabean culture.

Figure 9.2 Major cropping patterns of vegetable soyabeans (V.S.) in Taiwan, 1980-81 [10]

the variety of cropping patterns in which vegetable soyabeans are included indicates the cropping system complexity
into which improved genotypes will have to fit (Figure 9.2).

In the subsistence farming systems of many warmer countries, major emphasis is given to the production of staple
foods. Either capital, or time, or both factors are critical for soyabean cultivation. Therefore, no-tillage, rice-
stubble culture (NTRS) is practised in China and Taiwan,
and is also considered to be of potential value in Thailand
and Indonesia [9,145,180].

Soyabeans are intercropped to a limited extent with
annuals such as maize and sorghum, or plantation crops such
as sugar cane, rubber, oil palm, banana, coconut and other
fruit trees [145]. However, most intercropping and mixed
cropping is still in the experimental stages and very little
is practised extensively by farmers. Nevertheless, soyabeans
are invariably intercropped with maize in Nepal, and in
Korea, Nepal and Japan the practice of growing soyabeans on
paddy rice bunds is also quite popular. In Korea, land and
suitable environments are limiting factors and so soyabeans
are cultivated in a number of different combinations; they
are the favourite crop for planting on newly reclaimed,
steep, hilly sites.

If they are to be successful in multiple cropping,
intercropping or mixed cropping systems, soyabean cultivars
with different characters to those developed for sole crops
may be needed. For example, photo-thermal insensitivity may
allow soyabeans to be introduced into the cropping cycle at
different times of year [67,164]. Tolerance of shading and
the ability to exploit limited environmental resources
(whatever these may be in different situations) in a manner
complementary to the needs of a companion crop may also be
advantageous. Of course, optimum plant populations and
planting patterns will need to be quantified too. Cultivars
developed for sole crop systems may be unsuitable for intercropping or mixed cropping [66] and it seems sensible that
cultivars should be selected specifically for the farming

system for which they are intended [66,152].

BOTANICAL AND AGRONOMIC FEATURES

The botany and physiology of soyabeans have been described
in detail in several reviews [27,36,90,157]. Germination is
epigeal and at 25°-30°C occurs within three or four days
[157]. The main stem node above the cotyledonary node has
opposite, simple, unifoliolate leaves. Above this, there may
be from about eight to twenty-four additional nodes with
alternate, trifoliolate leaves. The petioles are long, particularly on middle and lower leaves, so that most of the
crop canopy is within a narrow layer. Branches develop from
basal nodes, particularly at sparse plant population densities. Stems are erect, although lodging may still occur if
growth is luxuriant.

There are two common growth habits. Indeterminate types
begin to flower at lower nodes and while apices or terminal
buds continue to grow vegetatively. In these types, flowering normally begins at the fourth or fifth node above the
soil, and proceeds at successive nodes below and above that
point. Eventually, apices turn reproductive and flower.
Fruits form first close to the base of the plants. Indeterminate cultivars are grown in the northern USA and in the
soyabean-growing regions in Canada [157].

Determinate types produce flowers at all nodes over a
short period and each terminal bud forms a distinct raceme
with a dense cluster of fruits at the top of the plant.
Determinate cultivars are usually shorter and more branched
than indeterminate ones. Virtually all of the soyabeans
grown south of 36°N are determinate [163], as are most
tropical soyabeans.

Bernard [14] has described two major genes which control the stem termination characteristics. As well as the
common determinate (dt_1) and indeterminate (Dt_1, dt_2) types
(Plate 9.1), there are also semi-determinate types, such
as cultivar Will (Dt_1, Dt_2), which resemble the indetermin-

400

ates except that they are shorter and flower for a shorter period. Cooper [44] has developed 'semi-dwarf' cultivars from crosses between dt_1 and Dt_1 genotypes. Semi-dwarf determinates are used in the northern USA. Both semi-determinates and semi-dwarfs are used to reduce lodging in especially productive environments [40,44].

The stages of vegetative and reproductive development in soyabeans are most frequently described by adopting the system of Fehr *et al.* [64,65]. Rates of development are controlled primarily by photoperiod and temperature (see reviews by Shibles *et al.* [157] and Summerfield and Roberts [163]). Most soyabeans are short-day plants, although some early-maturing cultivars adapted to extreme latitudes [46, 134] and a few later-maturing [123] and tropical cultivars [149] are essentially insensitive to daylength during the period from planting to flowering. Very few genotypes, however, are completely insensitive to photoperiod throughout reproductive development [41,149]. Three major genes are known which influence times of flowering and maturity [13, 31]. Genes E_1 and E_2 have partial dominance for late maturity [157], with E_1 causing a greater effect in the field than E_2 [16]. Genotypes with E_3 also mature later than those with e_3, but the effect is less than that conditioned by E_1 and E_3. The three E genes are simply inherited but additional modifying genes make inheritance of time to maturity more complex [33,115].

Flowering in soyabeans has been reviewed in detail only recently [163]. The crop has typical papilionaceous flowers with a five-part corolla (a standard, two wing and two keel petals) which surrounds the pistil and ten stamens (nine fused and one free). The morphology of flower and fruit development is described by Carlson [36]. Flowering may begin within 25 to 50 days or even later after planting, depending on cultivars and environmental conditions. Soyabeans are virtually completely self-pollinated [63]. Abortion of flowers or subsequent fruits is common and ranges from 20% [78] to 80% [177]. Although rates of development

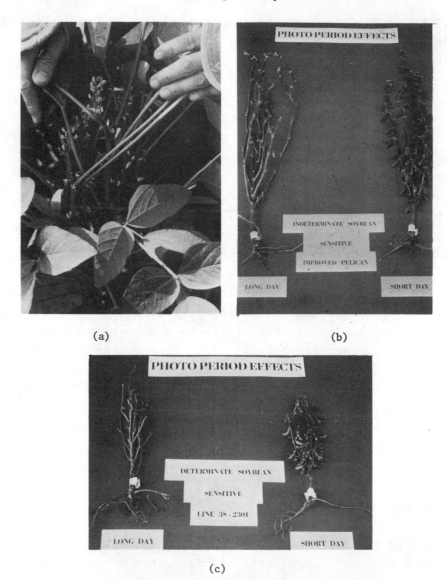

(a)

(b)

(c)

Plate 9.1 (a) Flowering habit of determinate soyabeans; and vegetative
framework and fruiting patterns of photo period-sensitive (b) indeterminate and (c) determinate genotypes (defoliated for clarity)
(Photographs: S. Shanmugasundaram (a) and H.C. Wien (b and c), New York
State College of Agriculture and Life Sciences, Cornell University)

are affected by environment and cultivar, fruits are usually
visible beyond the calyx within three days after blooming;
they reach their maximum length in 20-25 days, and are fully
expanded in 30 days [36]. Seeds reach their maximum size and
dry weight in 35-50 days. The duration of the linear phase
of seed-filling can vary from 21-52 days [19,53]. As seeds
reach maturity, they lose moisture and change from elongate
to spherical in shape.

Fruits (the pods) contain from one to five ovules but
most cultivars are three-ovule types [157]. Abortion of
ovules within fruits, especially at the basal position,
reduces the average number of seeds per fruit to about
2.0-2.5 [108,116].

Individual seeds reach maximum dry weight and then begin
to shrink, contract away from the pod wall [47] and turn
yellow [169]. However, the most useful indicator of whole-
plant physiological maturity in the field is when all
fruits have turned from green to yellow [70].

Soyabean yield components are traditionally thought of
as the numbers of plants per m^2, fruits per plant, seeds per
fruit and individual seed size. When a range of genotypes
or environments is investigated, number of fruits per plant
is the component most closely correlated with yield [89,
128]. Responses in this component may occur because of more
branch fruits, greater numbers of fruits per node, more
fruiting nodes, or combinations of all of these factors.
Number of seeds per fruit and individual seed size are
influenced both by environment and genotype, but individual
genotypes tend to have a characteristic seed weight which
ranges from 70-670 mg per seed at 13% moisture [9,11],
although most genotypes fall within the range 120-280 mg
per seed [157].

Soyabean yield components exhibit strong negative correlations. For example, in dense plant populations, the
numbers of fruits per plant are dramatically reduced. These
negative correlations allow plants to compensate for poor
stands, reduced fruit set, or changing environmental

403

conditions. In general, plant populations over the range of 15-50 plants per m^2 have little effect on seed yields of cultivars adapted to the Mid-West region of the USA, provided that the plants are uniformly distributed [112]. Dense plant populations increase yields of early-maturing, small-statured cultivars [51] by hastening canopy closure and ensuring more complete light interception. In contrast, the yields of later-maturing cultivars have often decreased at such dense populations [57,94] because of lodging [43].

Symbiotic nitrogen fixation potential

Symbiotic nitrogen fixation in soyabeans has been extensively studied (see reviews [54,79,179]), mainly because the species is a large-seeded annual with large nodules on its taproots and these attributes make the plants a convenient experimental tool. However, soyabeans are not the best nitrogen fixers among the grain legumes [125]. Weber [184, 185] estimated potential nitrogen fixation to be 160 kg N ha^{-1} in Iowa by comparing the N contents of nodulating and non-nodulating lines grown on soil amended with ground maize cobs to minimise available soil N. The fixed nitrogen represented 74% of the above-ground total N content. Field-grown soyabeans at several locations in Illinois obtained an average of only 25% of their total N by fixation, when this was measured by the natural ^{15}N abundance technique [107]. This estimate agrees with the approximate range of 60-80 kg N ha^{-1} per annum fixed under normal soil conditions in the USA [121]. In contrast, Hardy and Havelka [79] measured up to 450 kg N ha^{-1} fixed during a growing season when they enriched above-ground CO_2 to 1200 ppm.

Soyabeans characteristically fix nitrogen in symbiosis with *Rhizobium japonicum*. The amount of nitrogen fixed depends on a large number of factors. It is increased if soil N concentrations are small [46,107,184], if there are adequate populations of *R. japonicum* in the soil, especially of effective and efficient strains, and if environmental conditions are conducive to good plant growth and large yields. Good nodulation in new soyabean land requires 10^5-10^6 or more viable rhizobia per seed [26,28,29,74,110,182]. Strains of *R. japonicum* differ in their ability to effect nodulation and to fix nitrogen [178]. Interactions between soyabean host genotypes and strains of *R. japonicum* also occur, so that an individual strain may fix more nitrogen with some cultivars than with others [34]. Some strains of *R. japonicum* possess an uptake hydrogenase enzyme system which allows them to recapture the energy dissipated in hydrogen evolution during nitrogen fixation [88,143]. The presence of *R. japonicum* endowed with the hydrogen uptake (*Hup$^+$*) system appears to increase nitrogen fixation and seed yield [2,54] (and see Chapter 4).

Most soils in temperate regions and many of those in the tropics do not contain *R. japonicum* if they have not grown soyabeans before [28]. Inoculation with *R. japonicum* has traditionally been achieved using seed-applied, powdered peat inoculants. On new soyabean land, this inoculation procedure usually leads to nodule formation, but nodulation often improves the second and third time soyabeans are grown on the same land. The use of a granular inoculant, in which the *R. japonicum* are encased in peat granules and the inoculant is applied into the furrow with the seeds, has frequently improved nodulation and yield in first-time fields [17,120] (primarily because from as much as twenty to fifty times more inoculant can be added by this method than by that involving seed-dressing). Other types of inoculants have also been tested, including liquid formulations [167], concentrates [144] and those with charcoal, coir dust and other carriers [135,162]. However, peat-based inoculants are still the most widely used [28].

With any type of inoculant, it is important to ship and store the packages in cool and preferably dark conditions prior to use in order to maintain viability of the *R. japonicum* [68,187]. Even under good storage conditions, the viability of *R. japonicum* in peat carriers would be questionable after six months. This means that countries also need an effective distribution system for inoculants when soyabeans are introduced. A longer-term solution to the problem of ensuring effective inoculation in tropical countries may involve breeding soyabeans which nodulate with 'cowpea' rhizobia [138]. Rhizobia in this group are widely distributed in tropical soils and so planting soyabeans which can be nodulated by them effectively may circumvent inoculation or at least minimise the occasions on which inoculation is necessary.

Positive yield responses to inoculation in new soyabean land are well documented [17,20,144]. However, several studies have revealed little response to inoculation where soyabeans have been grown several times previously [35,55, 73,81,103,122,183]. On the other hand, positive yield responses have sometimes been reported following inoculation even in the presence of indigenous populations of *R. japonicum* [133]. The general lack of dramatic yield response to inoculation is attributed to the poor competitiveness of introduced strains and the poor recovery of the inoculant strains in nodules, with 5-10% recovery being common [35,73].

Recent advances in the ability to alter strains of *Rhizobium* using genetic engineering techniques suggest that 'superstrains' may be developed in the future. However, consistent field performance of new strains which is superior to the efficient, effective strains already available (such as USDA 110) has not yet been demonstrated. Even if superior strains are developed, they too must be able to compete with indigenous populations in the soil. Several techniques for enhancing the competitiveness of introduced rhizobia are under investigation [50,101,109] (and see Chapter 4).

Soyabean (*Glycine max* (L.) Merrill)

PRINCIPAL LIMITATIONS TO PRODUCTION AND YIELD

Developing countries

Limitations to national soyabean production and to yields
in farmers' fields will be treated here as separate topics,
although the two are related. For soyabean production to be
be successful in any country, four major facets need to be
developed simultaneously: production, marketing, processing
and utilisation [192]. The failure to establish soyabeans
successfully in some countries can be attributed to failure
in one or more of these areas.

In most developing countries and in South-east Asia,
soyabeans are not a major crop. Even in China and Indonesia, where the soyabean production areas are the most
extensive, small yields are attributed to the secondary
status of soyabeans in the rotation system [151,180], which
means that they are usually planted after the harvest of
winter crops or summer rice, and grown during the shortest
days and under adverse conditions.

The use of traditional local cultivars or introduced
cultivars with poor yield potentials also contributes to
small yields in developing countries. In the past decade,
the number of improved cultivars developed in many of these
countries has been negligible [151], and even those cultivars which have been developed are often not well adapted
to farmers' conditions. Seasonal variations with different
planting times in the cropping systems, environmental
fluctuations, and climatic variability between locations,
combine to limit the adaptability of environmentally sensitive genotypes [139]. Yields are further limited because
different types of cropping systems require specific cultivars which are seldom, if ever, available [67,87,147].

Poor seed quality is a common problem in tropical and
subtropical countries, particularly when soyabeans are
grown during the wet seasons [7,139]. Even when seeds of
good quality are harvested, viability of most cultivars is

407

lost rapidly in the tropics [49,56,167]. Crop stands are
often poor because of poor seed viability, lack of water,
hot soil temperatures and crusting [48,87].

Diseases such as soyabean rust (caused by *Phakopsora
pachyrhizi*) can reduce yields by as much as 90% [11].
Other yield-limiting diseases include bacterial pustule
(*Xanthomonas campestris* pv. *phaseoli*), downy mildew (*Peronospora manshurica*), anthracnose (*Colletotrichum dematium*
var. *truncata* and *Glomerella glycines*), purple seed stain
(*Cercospora kikuchii*), pod and stem blight (*Diaporthe
phaseolorum* var. *sojae*), Soybean Mosaic Virus, Yellow
Mosaic Virus, and various seedling diseases. (For illustra-
tions and details of these diseases see the recent compendium compiled by Sinclair [158]).

Insect pests are equally important risk factors for
tropical farmers. Some of the important insects are beanflies (also called stem miners: *Melanagromyza sojae* and
Ophiomyia centrosematis) which can cause 100% loss [165],
pod borers (e.g. *Maruca testulalis*), stink bugs (*Nezara
viridula* and *Piezodorus hybneri*) and a number of leaf
feeders. Detailed, illustrated descriptions are now available for tropical and subtropical soyabean insect pests
[155] and for soyabean insects in Brazil [129]. Beanflies
appear to be unique to the tropics and subtropics, whereas
most other insect pests of soyabeans are also found in
temperate regions.

Most soyabean growers in the tropics are subsistence
farmers who provide minimal inputs to their crops [154].
The 'yield gap' between fields grown with minimum and optimum inputs varies with season and environment, but the
yield reduction in the minimum input situation averages
more than 70% [11]. Nearly 56% of the variability in yield
among sites in the tropics could be attributed to agronomic
practices [189]. In Taiwan, yields were increased by 64%
following irrigation during the dry season and by 16.7%
through weed control. Genotypic interactions were observed
with all management factors tested, indicating the

potential for selecting cultivars adapted to specific situations [11].

The policies of national governments in many developing countries are such that major research emphasis is given to staple food crops. Soyabeans are a secondary crop, so research attention is minimal. The situation is aggravated further because many of these countries can import large quantities of soyabeans at prices lower than local production costs.

Developed countries

The principal limitations to yield in the USA are water, profitability and genotypes. Water often limits yields in the south [85] and Mid-West [168]. Profitability limits yield because small returns for farmers lead to fewer inputs in the form of fertiliser, protection against pests, possibly irrigation, and modern planting and harvesting equipment. Genotypes influence yield in two major ways. First, more recent cultivars outyield older ones, resulting in a gradual increase in yield, even under ideal conditions. Second, development of cultivars with resistance or tolerance to specific pests or diseases can result in rapid and large increases in yield when the pest or disease is prevalent. For example, the development of cultivars tolerant to root-knot (*Meloidogyne* spp.), soybean cyst (*Heterodera glycines*), and other nematodes has dramatically improved yields in infested fields in the southern USA [86,92]. For fine-textured soils in the USA and Canada, cultivars have been developed with tolerance to many races of *Phytophthora* root-rot [32,60].

Soyabean diseases have increased in number and severity as the area planted to the crop has expanded. More than one hundred diseases are known to affect soyabeans and about thirty-five of them are of economic importance [158]. In 1977, soyabean diseases caused losses estimated at 7 million tonnes [158], or 10% of the total world crop [4].

Soyabean damage by insects is more severe in warm climates; the northern USA states are not seriously affected [105]. With leaf-feeding insects, up to 33% defoliation can occur before chemical control is economic [172]. In a survey by US soyabean entomologists, reported in 1980, corn earworm (*Heliothis zea*) was ranked as the most important insect pest [105], followed by velvetbean caterpillars (*Anticarsia gemmatalis*), stink bugs (*Nezara viridula*), soya-bean loopers (*Pseudoplusia includens*), green cloverworms (*Plathypena scabra*) and Mexican bean beetles (*Epilachna varivestis*). In northern USA areas, grasshoppers (*Melanoplus* spp.) were the most serious pest. Various detailed descriptions are available for the insect pests of the American crop [106,172,173].

Weeds are estimated to decrease soyabean yield and quality by 10-15% in the USA [117] and to cause more losses than all other pests combined [175]. More than sixty species of weeds infest US soyabean fields, including both broadleaf and grass species [117]. In excess of 80% of the soyabean area is treated with herbicides [117], although combinations of herbicides and cultivation are also widely used, and may increase yields more than the use of herbicides alone [71].

Soyabean herbicides are applied at pre-planting, pre-emergence or post-emergence [167] and a combination of herbicides is usually used to increase the spectrum of weed control. The tolerance of soyabeans to many herbicide formulations is less than that desired, and there are only narrow ranges of herbicide rate between inadequate weed control and damaging or killing seedlings [167]. Some cultivars are particularly susceptible to metribuzin, a commonly applied herbicide, and must be avoided wherever it is used [6,12,52].

Another factor which limits yield is the row width in which soyabeans are planted. In the USA, most soyabean crops are grown in rows 75-100 cm apart but, in the northern USA, research indicates [1,91] that yield increases of

10-20% can be achieved by narrowing row widths to within 17-35 cm. In most studies, genotype x row spacing inter-actions for yield have been small or insignificant [40,72, 80,181,190], although some interactions have been reported [37,43]. The development of lodging-resistant, semi-dwarf cultivars [44] is promoting even greater interest in the use of narrow rows to increase yields [40,44,80]. At the northern fringes of production, soyabeans are usually planted in solid stands (rows 17 cm apart) to compensate for the small stature of the plants [141].

In much of Europe and Canada, as well as in parts of Russia, Australia and New Zealand, yields and production are limited by cool temperatures [161] and short growing seasons [156,171]. Although cultivars are being bred with greater ability to germinate [174], grow [95,161] and fruit [97] at cool temperatures, many areas are still too cold for successful soyabean production.

Factors which limit yield are active areas of current research. Indeed, there has been a remarkable increase in the research and breeding devoted to soyabeans in recent years [102,111]. Between 1961 and 1979, the annual number of scientist-years for soyabean production research in the USA increased from 38 to 230, while in addition the number of soyabean breeders increased from 39 to 266. Furthermore, the number of soyabean breeders employed by private com-panies in the USA increased from 36 in 1979 to 91 in 1982.

FERTILISER REQUIREMENTS

Soyabeans absorb nutrients from the soil more efficiently than most crops and so achieve maximum yields at smaller soil-test levels [132]. The nutrient composition per 1000 kg of seeds at 14% moisture content is approximately 65 kg N, 6 kg P, 18 kg K, 3 kg S, 3 kg Ca, 2 kg Mg, 0.3 kg Na, and 125 g Fe, 35 g Zn, 30 g B, 25 g Mn, 20 g Cu, 15 g Si and 8 g Al [76]. Patterns of N, P and K accumulation during the growing season differ [77,142]; N and P accumulate more

411

in the leaves than K does, and larger proportions of the total N and P are subsequently translocated to seeds. In experiments in Iowa, the average amounts of N, P and K accumulated in seven cultivars were 256, 23 and 93 kg ha^{-1}, respectively [77]. Intensive soyabean production seldom includes application of fertiliser N, but rather relies on nitrogen fixation to supplement uptake of residual N from the soil [132]. Phosphorus and K applications prior to growing soyabeans are recommended when available phosphorus values are below 30-40 kg ha^{-1} and exchangeable K is smaller than 200-240 kg ha^{-1}. If soil tests indicate residual P and K levels near to or greater than these values, then soyabeans are frequently grown without fertiliser.

Soyabeans are grown in rotation with maize more often than with any other crop [118]. Maize normally gives a greater response to direct fertiliser application, and soyabeans absorb nutrients from the soil more efficiently than maize. Therefore, the soil-test values are often maintained by fertiliser P and K applications to the maize and not to the soyabeans [20,118,132].

Soyabeans suffer from micronutrient deficiencies, usually caused by unavailability of the micronutrient, rather than small concentrations of the micronutrient in the soil. On acid soils with pH values below about 5.5, additions of Mo as a seed-dressing frequently improves yields because Mo is unavailable at these soil pH values [130]. Molybdenum is a necessary component of the nitrogenase enzyme involved in nitrogen fixation. Chlorosis caused by iron deficiency is a problem in alkaline soils and some breeding programmes select for tolerance to this deficiency [42]. Coarse-textured, alkaline soils also frequently cause Mn deficiency in soyabeans [3,126,140].

QUALITY OF SEED CONSTITUENTS

Seeds of most soyabean cultivars contain between 39.5% and 41.5% protein on a dry-weight basis [83]. The importance of soyabeans as a source of protein has stimulated breeding for increased protein concentration in seeds. Breeding lines with concentrations as large as 51% have been reported in Canada [58,59], and 52% protein has been reported from Minnesota [83], although these values were probably increased by the cool environments, presumably by extending the duration of protein accumulation.

The quality of the protein in soyabean flour depends primarily on the concentrations of various amino acids and on the digestibility of the protein [96]. The least concentrated amino acids are cysteine, methionine and tryptophan, with respective values of 0.8-0.9%, 1.1-1.4%, and 1.3-1.5% of the total protein [45]. However, methionine is the nutritionally limiting amino acid [39]. Soyabean meal is relatively rich in lysine (5.9-6.9% [45]) which makes it suitable for blending with cereals, which are poor in lysine, in foods and feeds (and see Chapter 3).

Soyabeans contain a number of antinutritional factors, including trypsin inhibitors, a haemagglutinin complex and a goitrogenic factor, but all are removed by moist heat or roasting and so soyabean meal is essentially free of them [45].

Soyabeans grown in the USA will normally contain from 20.5-21.5% oil, on a dry-matter basis [83]. The iodine number of the oil, which reflects its quality as a drying oil [83], ranges from 122 to 150, with most cultivars in the region of 130. The fatty acid composition of the oil is, approximately, 53% linoleic, 23% oleic, 11% palmitic and 8% linolenic [127]. When exposed to air or light the oil has a tendency to become rancid, a deterioration in flavour caused primarily by the instability of the linolenic acid. Several research programmes are attempting to breed for smaller linolenic acid concentration [30]. Breeding lines with linolenic acid concentrations as small as 3.4% of the total fatty acids have been obtained by mutation breeding [191].

413

GERMPLASM RESOURCES

Soyabean collections are maintained in a number of countries (Table 9.3). Large collections are maintained by the AVRDC in Taiwan (see Chapter 18), and in China, Japan, the USA, the USSR and Korea. In 1944, the total number of accessions in the USA collection [38] was 15000. However, during the process of screening, evaluation and selection for various latitudes, types judged unsuitable for seed production in the USA were unfortunately discarded [99, 147].

At a recent *ad hoc* working group meeting for the *Glycine* species, the International Board for Plant Genetic Resources (IBPGR; see Chapter 18) designated the People's Republic of China and the USA as the base collection centres for the long-term preservation of global *Glycine* collections, with Japan as a subsidiary storage centre. The base collection of the wild perennials of the sub-genus *Glycine* will be housed at CSIRO in Australia.

Although each one of the germplasm centres has evaluated its collection for different traits, a set of soyabean descriptors acceptable to all of them has only recently been finalised and will be published soon by the IBPGR.

Documentation and computerisation of the data collected on the germplasm has yet to be accomplished by most centres, with the probable exception of those in Japan.

On the basis of results from uniform test nurseries conducted over a period of years in numerous locations as far apart as Canada and Palmira in Colombia, the USA soyabean collection has been divided into various Maturity Groups (MG). Genotypes in MG00 are the earliest to mature; they are adapted to the longer daylengths and cooler temperatures at northernmost latitudes. Cultivars in MGX are late to mature and are adapted to the short days and warm temperatures of tropical regions [82,83]. This classification is very useful in the USA, Canada and in countries

Soyabean (*Glycine max* (L.) Merrill)

Table 9.3 *Glycine* species collections around the world*

Country	Location	Curator	No. of Accessions
Taiwan (AVRDC)	Tainan	S. Shanmugasundaram	10400[e]
Australia	Canberra	CSIRO	400[f]
China	Jilin	UNK[a]	3000[b]
	Hubei	UNK	3000[c]
	Shadong	UNK	2930[c]
	Beijing	UNK	2500
	Heilungjiang	UNK	960
France	Toulouse	R.M. Ecochard	500
Nigeria (IITA)	Ibadan	Q. Ng	1300
India	Pantnagar	P.S. Bhatnagar	4000
	Amravati	H.R. Bhatia	1800
Indonesia	Bogor	UNK	600
Japan	Tsukuba	K. Kumagai	3541[d]
	Morioka	J. Fukui	200
South Korea	Seoul	S.H. Kwon	2833
North Korea	Suweon	E.H. Hong	300
South Africa	Pretoria	J.W. Snyman	600
Sweden	Norrköping	UNK	1200
Thailand	Chieng Mai	Arwooth NaLampang	1686
USA	Urbana-Champaign and Stoneville, Mississippi	R.L. Bernard E.E. Hartwig	9648
USSR	Leningrad	N.F. Korsakov	3000

[a] Unknown

[b] About 1000 accessions are *G. soja*

[c] Also have additional wild soyabeans

[d] About 670 are possible duplicates and 1013 are introductions

[e] About 900 are duplicates

[f] Wild perennials

*Compiled from Report of the IBPGR *ad hoc* Working Group on the Genetic Resources of *Glycine* Species, held at INTSOY, Urbana, USA in 1982, and from references [15,99 and 151].

located within a similar range of latitudes. However, when
the germplasm has been evaluated at other tropical and subtropical locations, the maturity classifications and group-
ings break down, and a new classification system for the
tropics needs to be developed [146].

More than eighty specific gene loci of economic importance have been recorded for soyabeans [16,147]. A Soybean
Genetics Committee was formed in 1955. Beginning in 1983,
the committee will publish a revised quinquennial list of
all gene symbols, linkage groups, translocations, and tri-
somics in soyabeans in the *Soybean Genetics Newsletter* (see
below). The genetic Type Collections are being maintained
at the United States Regional Soybean Laboratory in
Illinois [16].

Sources of resistance to selected soyabean diseases
have been identified and reported [16,83,147,151,170].
Selected germplasm sources with resistance to insects and
diseases are listed in Tables 9.4 and 9.5, respectively.
In 1976, an International Working Group on Soybean Rust was
organised and each year the group publishes the *Soybean
Rust Newsletter* [148] to exchange information on this disease (see below).

Tremendous genetic diversity is available in the

Table 9.4 Sources of resistance among soyabeans to selected insect
pests [166,176]

Insect	Names of cultivars or PI numbers[+]
Beanflies (*Melanagromyza sojae* and *Ophiomyia centrosematis*)	AVRDC accession numbers G 3089; G 3091; G 3104; and G 3122*
Mexican bean beetles (*Epilachna varivestis* Mulsant)	PI 171451; PI 227687; PI 229358
Cotton boll worms (*Heliothis armigera*) .	As for Mexican bean beetles
Beet army worms (*Spodoptera exigua*)	As for Mexican bean beetles
Stink bugs (*Nezara viridula* and *Piezodorus hybneri*)	PI 227687

*All four accessions are *G. soja*.
[+]Pl denotes plant introduction (in the US National Collection).

Soyabean (*Glycine max* (L.) Merrill)

Table 9.5 Sources of resistance among soyabeans to selected diseases [16,21,83,147,151,159,170]

Disease	Names of Cultivars or PI numbers
Bacterial pustule (*Xanthomonas phaseoli* (Smith) Dawson var. *sojensis* (Hedges) Starr and Burk)	CNS; Lee; Hill; Bragg; FC 31592; PI 219656
Bacterial blight (*Pseudomonas glycinea* Coerper)	Norchief; Harosoy; PI 132207; PI 189968
Wild fire (*Pseudomonas tabaci* (Wolf and Foster) F.L. Stevens)	As for Bacterial pustule
Brown stem rot (*Cephalosporium gregatum* Allington and Chamberlain)	PI 84496-2
Frog-eye leaf spot (*Cercospora sojina* Hara)	Lincoln; Wabash; CNS; Dorman; Hood; Kent; Lee; Ogden; Roanoke
Downy mildew (*Peronospora manshurica* (Naum.) Syd. ex. Gaum)	Kanrich; Pine Dell Perfection; PI 166140; PI 171443; PI 174885; PI 18390; PI 200527; PI 201422
Phytophthora root-rot (*Phytophthora megasperma* Drechs. var. *sojae* Hild.)	Beeson; Calland; Semmes; CNS
Soyabean Mosaic Virus (SMV)	Buffalo; PI 341242; PI 324068; HLS 21
Cowpea Chlorotic Mottle Virus (CCMV)	CNS; Bragg; Dare; Dyer; Hardee; Hill; Lee; Pickett; Semmes
Yellow Mosaic Virus (YMV)	PI 171443
Soyabean rust (*Phakopsora pachyrhizi* Syd.)	Akiyoshi; Yoshida-2; Sanga; PI 70559; PI 70561; PI 79610; PI 68728; PI 230970; PI 230971; PI 200492; Ankur; PI 459025; PI 339871
Soyabean cyst nematode (*Heterodera glycines* Ichinohe)	Peking; Custer; Dyer; Pickett 71; Mack; Ilsoy; PI 84751; PI 88788; PI 89772; PI 87631-1; PI 90763; PI 209332
Root-knot nematode (*Meloidogyne incognita acrita*, Chitwood)	Laredo; Palmetto; FC 33243

417

soyabean germplasm. One hundred-seed-weight of the cultivars varies from 7 g to 67 g; their seed coat colour
includes black, brown, green, yellow, cream, chocolate and
various saddle patterns [16]. Leaf shapes, pubescence, colour and type, and flower, fruit and hilum colours also have
wide ranges of variation [16].

PRINCIPAL BREEDING STRATEGIES

Any discussion of soyabean breeding from a historical per-
spective must give credit to the Chinese, whose efforts
resulted in the domestication of *G. max*. Together with
other South-east Asian centres, the Chinese people provided
the world with the soyabean plant, essentially as we know
it today. The vast majority of soyabean cultivars grown in
North America or, for that matter, elsewhere up until 1940,
were introductions from Asia. Plant introductions provided
the initial variability for selection of improved homogeneous or pure lines. Introductions from China and several
other Asian countries, including Japan, have provided the
sources of genetic variability used by most intensive,
modern breeding programmes.

   As is the case with most self-pollinated seed crops,
soyabean breeding efforts until very recently have been
conducted in publicly supported breeding programmes. This
trend, however, has gradually changed and privately funded
breeding programmes now account for a very significant proportion of the total worldwide breeding effort on the crop.
This is particularly apparent in North America where many
large, privately funded soyabean breeding programmes have
evolved during the last twenty years.

   The primary objectives of the vast majority of soyabean
breeding programmes fall into the following categories:
improvement in yield potential; improved harvestability;
improved pest tolerance; improved adaptation to new geographic and climatic regions; improved breeding methodology;
and germplasm maintenance and evaluation. The success of

418

breeding programmes in achieving objectives in these areas
has made a major contribution to the expansion of world
soyabean production during the last twenty-five years.

Adaptation to new geographic areas

The massive increase in soyabean production in North America
(Table 9.1), coupled with the general acceptance of soyabean
oil and meal in world markets, has led to major increases
in breeding efforts in several soyabean-importing nations.
Soyabean production has moved rapidly into South America and
is spreading throughout Asia. Soyabean production is also
making inroads into Europe, and the Ukraine, Africa and
Australia. Adapting soyabeans to potential new production
areas has involved overcoming the limitations of sensitivities to photoperiod, heat, cold, drought and an array of
other stress factors.

Breeding methodology

Soyabeans are usually bred as pure lines or heterozygous
mixtures of almost pure lines. Therefore, the breeding
methods employed in soyabean improvement exploit additive
genetic variance and, to some extent, additive by additive
epistatic genetic variance. Breeding strategies applied in
most soyabean programmes involve the bulk method, the pedigree method (or modifications of it, including single-seed
descent and early generation testing) plus various recurrent
selection schemes. In addition, backcrossing schemes have
been used to improve cultivars, particularly for incorporation of genes for resistance to specific plant diseases,
insects, nematodes, and herbicides. In recent years, the
trend has been away from classical backcrossing (involving
between five and seven cycles) to selection within large F_2
populations following one or two cycles of backcrossing.

Single-seed descent, outlined by Brim [23], has been
particularly well suited to breeding programmes employing

winter (off-season) nurseries in tropical areas. Comparisons
of the modifications of the pedigree method have not demonstrated any superiority of one method over another [18].
Therefore, the breeding methods employed by particular
breeding programmes will be based primarily on tradition,
facilities and/or resources.

Selection in soyabeans is based primarily on selection
within populations synthesised from crosses between pairs
of homozygous lines [24], although the use of three-way and
four-way parental combinations is becoming more popular. The
process of advancing segregating populations to homozygosity
by selfing, after making single crosses between pure lines,
has been very restrictive in permitting recombination, particularly among linked genes. Hanson *et al*. [75] have suggested that recurrent selection, employing intermated populations, could improve recombination. Recurrent selection
has received relatively little attention until recently due
to the difficulty encountered in crossing soyabeans artificially. Application of genetic male sterility to recurrent
selection [25] has diminished the crossing limitations in
some programmes. Furthermore, single-seed descent and early
generation testing have commonly been used in recurrent
selection programmes [62]. The mechanics of recurrent
selection schemes depend on several factors, including
heritability of the traits under selection and sample sizes
or seed requirements for precise evaluation of those traits.

Soyabean hybrids cannot be purchased as seed stocks for
commercial production although a patent has been issued in
the USA for hybrid seed production [22]. Although heterosis
has been documented in some soyabean crosses [131,186] the
requirements for economic, commercial, single-cross hybrid
seed production (mating system control mechanisms and efficient pollen vectors) are not yet readily available.

Breeding objectives

Seed yield is the most important parameter in soyabean

420

breeding [62]. Yield is a quantitatively inherited parameter which is greatly influenced by the environment. Soyabean breeding efforts in the USA have resulted in considerable yield improvements compared with the introduced cultivars grown prior to 1940. Luedders [114] has documented these improvements, the early products of the North American breeding effort, and those more recent cultivars developed through public breeding institutes (and see [84]).

Improvements in harvestability have involved reductions in lodging and seed shattering. Better lodging resistance has involved selection for improved stem strength with traditional plant types and, more recently, morphological changes which have reduced plant height and hence lodging while maintaining or increasing yield. New, short-statured cultivars appear to be reasonably well adapted to particularly productive areas where lodging has often been a limiting factor to harvestable yield [44]. Improvements in resistance to shattering have also been significant since the instigation of organised breeding efforts in many areas of the world; shattering is no longer considered to be a major problem in most soyabean production areas.

Seed quality

Several breeding programmes, particularly publicly funded ones in North America, have placed emphasis on improving oil and protein concentrations and/or quality. Protein and oil concentrations are negatively related and simultaneous selection for increased quantities of both have usually been unsuccessful [24]. Genetic variability is available for linolenic acid concentration and some recent success in lowering traditionally expected values has been reported [30,191]. Improvements in protein quality, such as a major increase in methionine concentration, have limited potential because the necessary genetic variability has not yet been identified within available soyabean germplasm.

Improvements in other seed characteristics for specific

uses have been successful. Increases in seed size for the
manufacture of oriental food products, as well as the elimination of coloured hila, have resulted in premium markets
for certain cultivars in North America, particularly for
delivery to Japan and other eastern Asian countries. In
contrast, reductions in both seed size and oil concentration have led to the development of premium markets for
'natto-type' soyabeans in Japan.

Pest and disease tolerances

Breeding programmes have been reasonably successful in
improving pest and disease tolerances in the soyabean crop
in many countries. In North America, several major diseases,
including *Phytophthora* root-rot (*Phytophthora megasperma*
Drechs. var. *sojae* Hild.) and soyabean cyst nematodes
(*Heterodera glycines* Ichinohe), have been reasonably well
controlled using genes for specific resistance [83]. Several
other diseases and insect pests have also been controlled,
or at least partially controlled, through the development
of tolerant or resistant cultivars. Development of cultivars with horizontal or general resistance, which tends to
be more stable over time than resistance to a specific race,
has been less successful in soyabeans than in most other
self-pollinated crops.

Current trends in soyabean breeding

The past twenty years have seen a very dramatic shift in
the degree of mechanisation in soyabean breeding programmes.
This trend has resulted in huge increases in the volume of
breeding lines generated and evaluated within individual
programmes. The trend to larger volumes of material is
likely to continue, particularly in well-funded commercial
programmes competing for seed market shares in the North
and South American production areas.

   During the same period, there has also been a dramatic

422

increase in soyabean breeding efforts in North and South
America and a very substantial increase in several other
production and potential production areas. In North America,
the number of scientists employed in privately funded breeding programmes is escalating rapidly. It is probable that
as privately funded programmes evolve, publicly funded ones
will gradually shift their emphasis away from direct cultivar development to related research areas. These areas are
likely to include germplasm evaluation and improvement,
breeding methodology, and development of biotechnological
applications to soyabean breeding methodology and germplasm
improvement.

AVENUES OF COMMUNICATION AMONG RESEARCHERS

The major network assisting communication in soyabean
research has been the International Soybean Program (INTSOY),
a cooperative programme of the University of Illinois at
Urbana-Champaign and the University of Puerto Rico, Mayaguez
Campus (see Chapter 18). It has cooperated with international and national organisations to expand the use of
soyabeans for human food. INTSOY has coordinated cultivar
adaptation trials since 1973, following preliminary trials
instigated in 1970. The cultivar-testing programmes have
introduced improved soyabean germplasm into more than 110
countries [104,188]. INTSOY also organises extensive training programmes and has been a major publisher of information
on tropical and subtropical soyabeans. A current review
might lead to the development of INTSOY into an International Soybean Centre [104].

Soyabean development throughout much of Asia has been
assisted and coordinated by the Asian Vegetable Research
and Development Centre (AVRDC) in Taiwan (see Chapter 18).
The International Institute of Tropical Agriculture (IITA)
in Nigeria (see Chapter 18) has coordinated and promoted
research on the crop in Africa. A number of national,
cultivar-testing programmes exist, with the major ones in

Argentina, Brazil, China, Egypt, and Zimbabwe [104], as well as those in the USA, Canada, and Australia.

In Europe, the Food and Agriculture Organisation (FAO) established the European Cooperative Research Network on Soybean in 1976. The network involves twenty-six institutes from fourteen countries. FAO has also assisted in national programmes in other areas, as well as providing numerous publications and extension support in developing countries.

In the USA, the public soyabean breeders meet annually with breeders employed by commercial companies in a National Soybean Breeders Workshop. These workshops may also include researchers in agronomy, physiology, entomology and pathology. There have been World Soybean Research Conferences in 1976, 1979 and, most recently, in Ames, Iowa in 1984.

Communication among soyabean researchers is effected also through numerous scientific journals as well as an increasing number of books. A revision of the authoritative monograph entitled *Soybeans: Improvement, Production and Uses* [16] is underway. Preliminary research results on soyabean breeding and physiology are often reported in the *Soybean Genetics Newsletter*, and the *Soybean Rust Newsletter* publishes diverse reports on that disease. These publications are available from Dr R.G. Palmer, ARS-USDA, Department of Genetics, Iowa State University, Ames, Iowa 50011, USA and Mr S. Shanmugasundaram (see list of contributors), respectively.

PROSPECTS FOR LARGER AND MORE STABLE YIELDS

As indicated earlier, a group of experts (representing various disciplines) recently estimated that world soyabean production would double in the next twenty years. They also forecast [5] that average yields in the USA would increase from the current value of about 2.0 t ha^{-1} to 3.7 t ha^{-1} by the year 2002. This projected average yield is already achieved by a small number of US farmers. Raising average yields to this value will require more stable yields than

Soyabean (*Glycine max* (L.) Merrill)

are currently achieved. The rapidly increasing number of scientists working on soyabeans is expected to provide improved, larger-yielding germplasm, as well as better crop protection and more efficient production systems.

In developing countries, yields are also expected to increase and become more stable. Larger-yielding, better-adapted cultivars will become available and progress will be made in identifying the cultural practices necessary to allow these superior genotypes to yield reliably well in diverse regions and cropping systems.

LITERATURE CITED

1. Ablett, G.R., Schleihauf, J.C. and McLaren, A.D. (1984), *Canadian Journal of Plant Science* 64, 9-15.
2. Albrecht, S.L., Maier, R.J., Hanus, F.J., Russell, S.A., Emerich, D.W. and Evans, H.J. (1979), *Science* 203, 1255-7.
3. Alley, M.M., Rich, C.I., Hawkins, G.W. and Martens, D.C. (1978), *Agronomy Journal* 70, 35-8.
4. American Soybean Association (1981), *Soya Bluebook*. St Louis, Missouri: American Soybean Association.
5. American Soybean Association (1983), *2002: First the Questions, Now the Answers*. Supplement to *Soybean Update*. St Louis, Missouri: American Soybean Association.
6. Andersen, R.N. (1976), in Hill, L.E. (ed.) *World Soybean Research*. Danville, Illinois: Interstate Printers and Publishers, 444-52.
7. Andrews, C.H. (1982), in Sinclair, J.B. and Jackobs, J.A. (eds) (*Soybean Seed Quality and Stand Establishment*. INTSOY Series No. 22. Urbana-Champaign: University of Illinois, 19-25.
8. Anon. (1983), *Better Crops with Plant Food* 67, 4-5.
9. AVRDC (1976), *Soybean Report '75*. Shanhua, Taiwan: Asian Vegetable Research and Development Centre.
10. AVRDC (1981), *Progress Report for 1980*. Shanhua, Taiwan: Asian Vegetable Research and Development Centre.
11. AVRDC (1983), *Progress Report for 1982*. Shanhua, Taiwan: Asian Vegetable Research and Development Centre.
12. Barrentine, W.L., Hartwig, E.E., Edwards, C.J. and Kilen, T.C. (1982), *Weed Science* 30, 344-8.
13. Bernard, R.L. (1971), *Crop Science* 11, 242-4.
14. Bernard, R.L. (1972), *Crop Science* 12, 235-9.
15. Bernard, R.L., Nelson, R.L., Hartwig, E.E. and Edwards, C. (1983), *Soybean Genetics Newsletter* 10, 5.
16. Bernard, R.L. and Weiss, M.G. (1973), in Caldwell, B.E. *et al.* (eds) *Soybeans: Improvement, Production and Uses*. Agronomy Monograph 16. Madison, Wisconsin: American Society of Agronomy, 117-54.
17. Bezdicek, D.F., Evans, D.W., Abede, B. and Witters, R.E. (1978), *Agronomy Journal* 70, 865-8.
18. Boerma, H.R. and Cooper, R.L. (1975), *Crop Science* 15, 225-9.

19. Boote, K.J. (1981), *Agronomy Journal* 73, 854-9.
20. Boswell, F.C. and Anderson, O.E. (1976), *Agronomy Journal* 68, 315-18.
21. Bowers, G.R. (1980), PhD. thesis. Urbana-Champaign: University of Illinois.
22. Bradner, N.R. (1975), US Patent No. 3903 645.
23. Brim, C.A. (1966), *Crop Science* 6, 220.
24. Brim, C.A. (1973), in Caldwell, B.E. *et al.* (*eds*) *Soybeans: Improvement, Production and Uses.* Agronomy Monograph 16. Madison, Wisconsin: American Society of Agronomy, 155-86.
25. Brim, C.A. and Stuber, C.W. (1973), *Crop Science* 13, 528-30.
26. Brockwell, J. (1977), in Gibson, A.H. (*ed.*) and Hardy, R.W.F. (*gen. ed.*) *A Treatise on Dinitrogen Fixation. Sect. IV.* New York: Wiley Interscience, 227-310.
27. Brun, W.A. (1978), in Norman, A.G. (*ed.*) *Soybean Physiology, Agronomy, and Utilization.* New York: Academic Press, 45-76.
28. Burton, J.C. (1976), in Hill, L.D. (*ed.*) *World Soybean Research.* Danville, Illinois: Interstate Printers and Publishers, 170-79.
29. Burton, J.C. and Curley, R.L. (1965), *Agronomy Journal* 57, 379-81.
30. Burton, J.W., Wilson, R.F. and Brim, C.A. (1983), *Crop Science* 23, 744-7.
31. Buzzell, R.I. (1971), *Canadian Journal of Genetics and Cytology* 13, 703-7.
32. Buzzell, R.I. and Anderson, T.A. (1982), *Plant Disease Reporter* 66, 1146-8.
33. Byth, D.E. (1968), *Australian Journal of Agricultural Research* 19, 879-90.
34. Caldwell, B.E. and Vest, G. (1968), *Crop Science* 8, 680-82.
35. Caldwell, B.E. and Vest, G. (1970), *Crop Science* 10, 19-21.
36. Carlson, J.B. (1973), in Caldwell, B.E. *et al.* (*eds*) *Soybeans: Improvement, Production and Uses.* Agronomy Monograph 16. Madison, Wisconsin: American Society of Agronomy, 117-54.
37. Carter, T.E. Jr and Boerma, H.R. (1979), *Crop Science* 19, 607-10.
38. Cartter, J.L. (1944), *Soybean Digest* 4.
39. Cater, C.M., Cravens, W.W., Horan, F.E., Lewis, C.J., Mattil, K.F. and Williams, L.D. (1978), in Milner, M. *et al.* (*eds*) *Protein Resources and Technology: Status and Research Needs.* Westport, Connecticut: AVI Publishing, 278-301.
40. Chang, J.F., Green, D.E. and Shibles, R. (1982), *Crop Science* 22, 97-101.
41. Channer, G.W., Tanner, J.W. and Hume, D.J. (1982), *Agronomy Abstracts* 93.
42. Cianzio, S.R. and Fehr, W.R. (1982) *Crop Science* 22, 433-4.
43. Cooper, R.L. (1977), *Agronomy Journal* 69, 89-92.
44. Cooper, R.L. (1981), *Crop Science* 21, 127-31.
45. Cowan, J.C. (1973) in Caldwell, B.E. *et al.* (*eds*) *Soybeans: Improvement, Production, and Uses.* Agronomy Monograph 16. Madison, Wisconsin: American Society of Agronomy, 619-64.
46. Criswell, J.G., Tanner, J.W. and Hume, D.J. (1976), *Crop Science* 16, 400-4.
47. Crookston, R.K. and Hill, D.S. (1978), *Crop Science* 18, 867-70.
48. Dadson, R.B. (1982), in Sinclair, J.B. and Jackobs, J.A. (*eds*) *Soybean Seed Quality and Stand Establishment.* INTSOY Series No. 22. Urbana-Champaign : University of Illinois, 96-101.
49. Delouche, J.C. (1982), in Sinclair, J.B. and Jackobs, J.A. (*eds*)

Soyabean (*Glycine max* (L.) Merrill)

Soybean Seed Quality and Stand Establishment. INTSOY Series No. 22. Urbana-Champaign: University of Illinois, 57–66.

50. Devine, T.E. and Breithaupt, B.H. (1980), *Crop Science* 20, 269–71.
51. Dominguez, C. and Hume, D.J. (1978), *Agronomy Journal* 70, 801–5.
52. Eastin, E.F., Sij, J.W. and Craigmiles, J.P. (1980), *Agronomy Journal* 72, 167–8.
53. Egli, D.B. and Leggett, J.E. (1973), *Crop Science* 13, 220–22.
54. Eisbrenner, G. and Evans, H.J. (1983), *Annual Review of Plant Physiology* 34, 105–36.
55. Elkins, D.M., Hamilton, G., Chan, C.K.Y., Briskovich, M.A. and Vandeventer, J.W. (1976), *Agronomy Journal* 68, 513–17.
56. Ellis, R.H., Osei-Bonsu, K. and Roberts, E.H. (1982), *Annals of Botany* 50, 69–82.
57. Enyi, B.A.C. (1973), *Journal of Agricultural Science* 81, 131–8.
58. Erickson, L.R., Beversdorf, W.D. and Ball, S.T. (1982), *Crop Science* 22, 1099–101.
59. Erickson, L.R., Voldeng, H.D. and Beversdorf, W.D. (1981), *Canadian Journal of Plant Science* 61, 901–8.
60. Erwin, D.C., Bartnicki-Garcia, S. and Tsao, P.H. (1983), *Phytophthora, Its Biology, Taxonomy, Ecology and Pathology*. St Paul, Minnesota: American Phytopathology Society.
61. FAO (1983), *FAO Monthly Bulletin of Statistics* 6(1). Rome: FAO.
62. Fehr, W.R. (1978), in Norman, A.G. (*ed.*) *Soybean Physiology, Agronomy, and Utilization*. New York: Academic Press, 119–55.
63. Fehr, W.R. (1980), in Fehr, W.R. and Hadley, H.H. (*eds*) *Hybridization of Crop Plants*. Madison, Wisconsin: American Society of Agronomy, 589–99.
64. Fehr, W.R. and Caviness, C.W. (1977), *Stages of Soybean Development*. Special Report 80. Ames, Iowa: Iowa State University.
65. Fehr, W.R., Caviness, C.E., Burmood, D.T. and Pennington, J.S. (1971), *Crop Science* 11, 929–31.
66. Finlay, R.C. (1975), in Whigham, D.K. (*ed.*) *Soybean Production, Protection and Utilisation*. INTSOY Series No. 6. Urbana-Champaign: University of Illinois, 77–85.
67. Francis, C.A., Flor, C.A. and Temple, S.R. (1976), in *Multiple Cropping*. Madison, Wisconsin: American Society of Agronomy, 235–53.
68. Frederick, L.R. (1976), in Goodman, R.M. (*ed.*) *Expanding the Use of Soybeans*. INTSOY Series No. 10. Urbana-Champaign: University of Illinois, 44–5.
69. Fukushima, D. and Hashimoto, A. (1980), in Corbin, F.T. (*ed.*) *World Soybean Research Conference II: Proceedings*. Boulder, Colorado: Westview Press, 729–43.
70. Gbikpi, P.J. and Crookston, R.K. (1981), *Crop Science* 21, 469–72.
71. Gebhardt, M.R. (1981), *Weed Science* 29, 133–8.
72. Green, D.E., Burlamaqui, P.F. and Shibles, R.M. (1977), *Crop Science* 17, 335–9.
73. Ham, G.E., Cardwell, V.B. and Johnson, H.W. (1971), *Agronomy Journal* 63, 301–3.
74. Hamdi, Y.A., Abd-el-Samea, M.E. and Lofti, M. (1974), *Zentralblatt für Bakteriologie Abt. II* 129, 574–8.
75. Hanson, W.D., Probst, A.H. and Caldwell, B.E. (1967), *Crop Science* 7, 99–103.
76. Hanway, J.J. (1976), in Hill, L.D. (*ed.*) *World Soybean Research*. Danville, Illinois: Interstate Printers and Publishers, 5–15.

77. Hanway, J.J. and Weber, C.R. (1971), *Agronomy Journal* 63, 406–8.
78. Hardman, L.L. (1970), PhD. thesis, University of Minnesota [*Dissertation Abstracts* 31(5), 2401 B].
79. Hardy, R.W.F. and Havelka, U. (1975), in Nutman, P.S. (*ed.*) *Symbiotic Nitrogen Fixation in Plants*. Cambridge: Cambridge University Press, 421–39.
80. Hartung, R.C., Specht, R.C. and Williams, J.H. (1980), *Crop Science* 20, 604–9.
81. Hartwig, E.E. (1964), *Missouri Farm Research* 27(4) 1–3.
82. Hartwig, E.E. (1970), *Tropical Science* 12, 47–53.
83. Hartwig, E.E. (1973), in Caldwell, B.E. *et al.* (*eds*) *Soybeans: Improvement, Production and Uses*. Agronomy Monograph 16. Madison, Wisconsin: American Society of Agronomy, 187–210.
84. Hartwig, E.E. (1976), in Hill, L.D. (*ed.*) *World Soybean Research*. Danville, Illinois: Interstate Printers and Publishers, 217–21.
85. Hartwig, E.E. (1983), Personal communication.
86. Hartwig, E.E., Epps, J.M. and Buehring, N. (1982), *Plant Disease Reporter* 66, 18–20.
87. Harwood, R.R. and Price, E.C. (1976), in *Multiple Cropping*. Madison, Wisconsin: American Society of Agronomy, 11–40.
88. Haystead, A. and Sprent, J.I. (1981), in Johnson, C.B. (*ed.*) *Physiological Processes Limiting Plant Productivity*. London: Butterworth, 345–64.
89. Herbert, S.J. and Litchfield, G.V. (1982), *Crop Science* 22, 1074–79.
90. Hicks, D.R. (1978), in Norman, A.G. (*ed.*) *Soybean Physiology, Agronomy and Utilization*. New York: Academic Press, 17–44.
91. Hicks, D.R., Pendleton, J.W., Bernard, R.L. and Johnson, T.J. (1969), *Agronomy Journal* 61, 290–93.
92. Hiebsch, C.K., Kinloch, R.A. and Hinson, K. (1983), *Agronomy Abstracts* 107.
93. Hinson, K., Hartwig, E.E. and Minor, H.C. (1982), *Soybean Production in the Tropics*. Rome: FAO.
94. Hoggard, A.L., Shannon, J.G. and Johnson, D.R. (1978), *Agronomy Journal* 70, 1070–72.
95. Holmberg, S.A. (1973), *Agri Hortique Genetica* 31, 1–20.
96. Horan, F.E. (1976), in Hill, L.D. (*ed.*) *World Soybean Research*. Danville, Illinois: Interstate Printers and Publishers, 775–88.
97. Hume, D.J. and Jackson, A.K.H. (1981), *Crop Science* 21, 933–7.
98. Hymowitz, T. (1976), in Simmonds, N.W. (*ed.*) *Evolution of Crop Plants*. London and New York: Longman, 159–62.
99. Hymowitz, T. and Newell, C.A. (1980), in Summerfield, R.J. and Bunting, A.H. (*eds*) *Advances in Legume Science*. London: Royal Botanic Gardens, Kew, 251–64.
100. Hymowitz, T. and Newell, C.A. (1981), *Economic Botany* 35, 272–88.
101. Jones, R. and Giddens, J. (1983), *Agronomy Abstracts* 157.
102. Judd, R.E. (1979), *Soybean News* 31(1), 1 and 5.
103. Kapusta, G. and Rouwenhorst, D.L. (1973), *Agronomy Journal* 65, 916–19.
104. Kauffman, H.E. (1983), *Eurosoya* 1, 35–40.
105. Kogan, M. (1980), in Corbin, F.T. (*ed.*) *World Soybean Research Conferences II: Proceedings*. Boulder, Colorado: Westview Press, 303–25.
106. Kogan, M. and Kuhlman, D.E. (1982), *Soybean Insects: Identification and Management in Illinois*. Bulletin 773. Urbana-Champaign: University of Illinois.

Soyabean (*Glycine max* (L.) Merrill)

107. Kohl, D.H., Shearer, G. and Harper, J.E. (1980), *Plant Physiology*
66, 61-5.
108. Korte, L.L., Specht, J.E., Williams, J.H. and Sorensen, R.C.
(1983), *Crop Science* 23, 528-33.
109. Kvien, C., Ham, G.E. and Lambert, J.W. (1978), *Agronomy Abstracts*
142.
110. Lagacherie, B. and Obaton, M. (1973), *Comptes Rendus* 59, 67-79.
111. Leffel, R.C. (1983), *Eurosoya* 1, 17-20.
112. Leuschen, W.E. and Hicks, D.R. (1977), *Agronomy Journal* 69, 390-93.
113. Liu, Chiung-Pi and Shanmugasundaram, S. (1985), in *Proceedings of
the Symposium on Vegetables and Ornamentals in the Tropics*, Universiti Pertanian Malaysia (in press).
114. Luedders, V.D. (1977), *Crop Science* 17, 971-2.
115. McBlain, B.A., Alfich, R.A., Hesketh, D.J. and Bernard, R.L. (1985)
submitted to *Canadian Journal of Plant Science*.
116. McBlain, B.A. and Hume, D.J. (1981), *Canadian Journal of Plant
Science* 61, 499-505.
117. McWhorter, C.G. and Patterson, D.T. (1980), in Corbin, F.T. (*ed.*)
World Soybean Research Conference II: Proceedings. Boulder,
Colorado: Westview Press, 371-92.
118. Metcalfe, D.S. and Elkins, D.M. (1980), *Crop Production. Principles and Practices*. (4th edn), New York: MacMillan.
119. Morse, W.J. (1950), in Markley, K.S. (*ed.*) *Soybeans and Soybean
Products I*. New York: Interscience, 3-59.
120. Muldoon, J.F., Hume, D.J. and Beversdorf, W.D. (1980), *Canadian
Journal of Plant Science* 60, 399-409.
121. National Academy of Sciences (1979), *Microbial Processes: Promising Technologies for Developing Countries*. Washington, DC:
National Academy of Science.
122. Nelson, D.W., Swearingen, M.L. and Beckham, L.S. (1978), *Agronomy
Journal* 70, 517-18.
123. Nissly, C.R., Bernard, R.L. and Hittle, C.N. (1981), *Crop Science*
21, 833-6.
124. Norman, A.G. (1978), in Norman, A.G. (*ed.*) *Soybean Physiology,
Agronomy and Utilization*. New York: Academic Press, 1-15.
125. Nutman, P.S. (1976), in Nutman, P.S. (*ed.*) *Symbiotic Nitrogen
Fixation by Plants*. Cambridge: Cambridge University Press,
211-31.
126. Ohki, K., Wilson, D.O., Boswell, F.C., Parker, M.B. and Schuman,
L.M. (1977), *Agronomy Journal* 69, 597-600.
127. Orthoefer, F.T. (1978), in Norman, A.G. (*ed.*) *Soybean Physiology,
Agronomy and Utilization*. New York: Academic Press, 219-46.
128. Pandey, J.P. and Torrie, J.H. (1973), *Crop Science* 13, 505-7.
129. Panizzi, A.R., Correa, B.S., Gazzoni, D.C., de Oliveira, E.B.,
Newman, G.C. and Turnipseed, S.G. (1977), *Insetos da Soja no
Brazil*. EMBRAPA Boletim Tecnico No. 1.
130. Parker, M.B. and Harris, H.B. (1977), *Agronomy Journal* 69, 551-4.
131. Paschal, E.H. and Wilcox, J.R. (1975), *Crop Science* 15, 344-9.
132. Pepper, G.E. and many others (1982), *Illinois Grower's Guide to
Superior Soybean Production*. Circular 1200, Cooperative Extension
Service, Urbana-Champaign: University of Illinois.
133. Peterson, H.L., Switzer, R.E. and Peterson, I.R. (1977), *Agronomy
Abstracts* 150.
134. Polson, D.E. (1972), *Crop Science* 12, 773-6.
135. Pramanik, M. and Iswaran, V. (1973), *Zentralblatt für Bakterio-*

logie Abt. II **128**, 232-9.
136. Probst, A.H. and Judd, R.W. (1973), in Caldwell, B.E. *et al.*
(eds) *Soybeans: Improvement, Production, and Uses.* Agronomy
Monograph 16. Madison, Wisconsin: American Society of Agronomy,
1-15.
137. Prudential-Bache Securities (1982), *Consensus* **52**, 2, Kansas City,
Missouri: Consensus Inc.
138. Pulver, E.L., Brockman, F. and Wien, H.C. (1982), *Crop Science* **22**,
1065-70.
139. Rachie, K.O. and Plarre, W.K.F. (1975), in Whigham, D.K. *(ed.)*
Soybean Production, Protection and Utilization. INTSOY Series No.
6. Urbana-Champaign: University of Illinois, 29-47.
140. Randall, G.W., Schulte, E.E. and Corey, R.B. (1975), *Agronomy
Journal* **67**, 502-7.
141. Safo-Kantanka, O. and Lawson, N.C. (1980), *Canadian Journal of
Plant Science* **60**, 227-31.
142. Sale, P.W.G. and Campbell, L.C. (1980), *Field Crops Research* **3**,
157-63.
143. Schubert, K.R. and Evans, H.J. (1976), *Proceedings of the National
Academy of Sciences* (USA) **73**, 1207-11.
144. Scudder, W.T. (1975), *Soil and Crop Science Society of Florida
Proceedings* **34**, 79-82.
145. Shanmugasundaram, S. (1976), in Goodman, R.M. *(ed.) Expanding
the Use of Soybeans.* INTSOY Series No. 10. Urbana-Champaign:
University of Illinois, 44-5.
146. Shanmugasundaram, S. (1976), in Goodman, R.M. *(ed.) Expanding
the Use of Soybeans.* INTSOY Series No. 10. Urbana-Champaign:
University of Illinois, 191-5.
147. Shanmugasundaram, S. (1976), *Varietal Development and Germplasm
Utilization in Soybeans.* ASPCA/FFTC Technical Bulletin 30. Taipei,
Taiwan.
148. Shanmugasundaram, S. (1977), in Ford, R.E. and Sinclair, J.B.
(eds) *Rust of the Soybean - The Problem and Research Needs.*
INTSOY Series no. 12. Urbana-Champaign: University of Illinois,
71-2.
149. Shanmugasundaram, S. (1979), *Euphytica* **28**, 495-501.
150. Shanmugasundaram, S. (1981), *Bulletin of Institute of Tropical
Agriculture Kyushu University, Japan* **4**, 1-61.
151. Shanmugasundaram, S. (1982), in *Grain Legume Production in Asia.*
Tokyo: Asian Productivity Organisation, 137-66.
152. Shanmugasundaram, S., Kuo, G.C. and NaLampang, A. (1980), in
Summerfield, R.J. and Bunting, A.H. *(eds) Advances in Legume
Science.* London: Royal Botanic Gardens, Kew, 265-77.
153. Shanmugasundaram, S. and Selleck, G.W. (1985), in *Platinum Jubilee
Celebration of Tamil Nadu Agricultural University.* Coimbatore,
India: Tamil Nadu Agricultural University, (in press).
154. Shanmugasundaram, S., Yen, Chung-Ruey and Toung, T.S. (1982), in
Hsieh, Sung-Ching and Liu, Dah-Jiang *(eds) Proceedings of a Sym-
posium on Plant Breeding.* Taiwan: Agricultural Association of
China and Regional Society of SABRAO, 157-64.
155. Shepard, M., Lawn, R.J. and Schneider, M. (1983), *Insects on
Grain Legumes in Northern Australia.* St Lucia, Australia: University of Queensland Press.
156. Shibles, R.M. (1980), in Summerfield, R.J. and Bunting, A.H. *(eds)
Advances in Legume Science.* London: Royal Botanic Gardens, Kew,
279-86.

157. Shibles, R.M., Anderson, I.C. and Gibson, A.H. (1975), in Evans, L.T. (*ed.*) *Crop Physiology. Some Case Histories.* Cambridge: Cambridge University Press, 151–89.
158. Sinclair, J.B. (*ed.*) (1982), *Compendium of Soybean Diseases.* (2nd edn). St Paul, Minnesota: American Phytopathology Society.
159. Singh, B.B., Gupta, S.C. and Singh, B.D. (1974), *Indian Journal of Genetics and Plant Breeding* 34, 400–4.
160. Singh, Rajeswari (1970), *Soyahar Indian Recipes of Soybean.* India: U.P. Agricultural University.
161. Soldati, A., Keller, E.R., Brenner, H. and Schmid, J. (1983), *Eurosoya* 1, 27–34.
162. Strijdom, B.W. and Deschodt, C.C. (1976), in Nutman, P.S. (*ed.*) *Symbiotic Nitrogen Fixation in Plants.* Cambridge: Cambridge University Press, 151–68.
163. Summerfield, R.J. and Roberts, E.H. (1985), in Halevy, A.H. (*ed.*) *A Handbook of Flowering.* Boca Raton, Florida: CRC Press (in press).
164. Swaminathan, M.S. (1970), *Indian Farming* 29(7), 9–13.
165. Talekar, N.S. (1980), in *Proceedings of Legumes in the Tropics.* Malaysia: Universiti of Pertanian, Malaysia, 293–9.
166. Talekar, N.S. (1983), Personal communication.
167. Tanner, J.W. and Hume, D.J. (1978), in Norman, A.G. (*ed.*) *Soybean Physiology, Agronomy and Utilization.* New York: Academic Press, 157–217.
168. Taylor, H.M. (1980), in Corbin, F.T. (*ed.*) *World Soybean Research Conference II: Proceedings.* Boulder, Colorado: Westview Press, 161–78.
169. Tekrony, D.M., Egli, D.B., Balles, J., Pfeiffer, T. and Fellows, R.J. (1979), *Agronomy Journal* 71, 771–5.
170. Tisselli, O., Sinclair, J.B. and Hymowitz, T. (1980), *Sources of Resistance to Selected Fungal, Bacterial, Viral and Nematode Diseases of Soybeans.* INTSOY Series No. 18. Urbana–Champaign: University of Illinois.
171. Tollenaar, M. (1983), *Canadian Journal of Plant Science* 63, 1–10.
172. Turnipseed, S.G. (1973), in Caldwell, B.E. *et al.* (*eds*) *Soybeans: Improvement, Production and Uses.* Agronomy Monograph 16. Madison, Wisconsin: American Society of Agronomy, 545–72.
173. Turnipseed, S.G. and Kogan, M. (1976), *Annual Review of Entomology* 21, 247–87.
174. Unander, D.W., Lambert, J.W. and Orf, J.H. (1983), *Soybean Genetics Newsletter* 10, 59–62.
175. USDA (1976), *Report of the ARS, USDA Research Planning Conference on Weed Control in Soybeans.* Tifton, Georgia.
176. Van Duyn, J.W., Turnipseed, S.G. and Maxwell, J.D. (1971), *Crop Science* 11, 572–3.
177. Van Schaik, P.H. and Probst, A.H. (1958), *Agronomy Journal* 50, 98–102.
178. Vest, G., Weber, D.F. and Sloger, C. (1973), in Caldwell, B.E. *et al.* (*eds*) *Soybeans: Improvement, Production and Uses.* Agronomy Monograph 16. Madison, Wisconsin: American Society of Agronomy, 353–90.
179. Vincent, J.M. (1982), *Nitrogen Fixation in Legumes.* New York: Academic Press.
180. Wang, Chin Ling (1982), in Sinclair, J.B. and Jackobs, J.A. (*eds*) *Soybean Seed Quality and Stand Establishment.* INTSOY Series No.

22. Urbana-Champaign: University of Illinois, 136-9.
181. Weaver, D.B. and Wilcox, J.R. (1982), *Crop Science* 22, 625-9.
182. Weaver, R.W. and Frederick, L.R. (1972), *Agronomy Journal* 64, 597-9.
183. Weaver, R.W. and Frederick, L.R. (1974), *Agronomy Journal* 66, 233-6.
184. Weber, C.R. (1966), *Agronomy Journal* 58, 43-6.
185. Weber, C.R. (1966), *Agronomy Journal* 58, 46-9.
186. Weber, C.R., Empig, L.T. and Thorne, J.C. (1970), *Crop Science* 10, 159-160.
187. Weber, D.F. (1977) in Hollaender, A. (*ed.*) *Genetic Engineering for Nitrogen Fixation*. New York: Plenum Press, 433.
188. Whigham, D.K. (1975), *International Soybean Variety Experiment*. INTSOY Series No. 8. Urbana-Champaign: University of Illinois.
189. Whigham, D.K., Minor, H.C. and Carmer, S.G. (1978), *Agronomy Journal* 70, 587-92.
190. Wilcox, J.R. (1980), *Crop Science* 20, 277-80.
191. Wilcox, J.R., Cavins, J.F. and Nielsen, N.C. (1983), *Agronomy Abstracts* 85.
192. Williams, S.W., Hendrix, W.E. and von Oppen, M.K. (1974), *Potential Production of Soybeans in North Central India*. INTSOY Series No. 5. Urbana-Champaign: University of Illinois.

10 Common bean (*Phaseolus vulgaris* L.)

M.W. Adams, D.P. Coyne,
J.H.C. Davis, P.H. Graham and
C.A. Francis

HISTORICAL TRENDS AND CURRENT STATUS IN WORLD PRODUCTION

Common beans (*Phaseolus vulgaris* L.) are grown extensively in five major continental areas: eastern Africa, North and Central America, South America, eastern Asia, and western and south-eastern Europe. Production statistics for the crop published annually by the Food and Agriculture Organisation (FAO) [44] vary in reliability from reasonably accurate estimates to imprecise approximations [6].

A total of twenty-four African countries reportedly produce beans, seven of which (Angola, Burundi, Cameroon, Rwanda, Tanzania, Uganda, Zaire) devote large areas to the crop, ranging from nearly 120 000 ha annually in Angola to more than 350 000 ha in Uganda. Since 1961, about one half of the countries have increased the area devoted to beans. Annual variations in yield reflect favourable or adverse weather conditions, the critical factor usually being rainfall. Average yields over the period from 1960 to 1980 *appear* to have risen from 480 kg ha^{-1} to more than 600 kg ha^{-1}. In mixed cropping with maize under rainfed conditions and in monoculture, respectively, calculated daily productivity values [26] are 2.95 and 6.70 kg seed ha^{-1}.

In North and Central America, Mexico, with nearly two million hectares, dominates in its land area devoted to bean production [84]. Average yields for Mexico are less than 300 kg ha^{-1} with nearly 90% of the crop grown

433

under limited rainfall and often intercropped. With irrigation, yields larger than 1000 kg ha^{-1} are obtained. In the USA and Canada, beans are also grown under both rainfed and irrigated conditions. Production area and average yields are relatively stable, but variability is striking under dryland conditions. Daily productivity (yield efficiency) in one rainfed environment in the USA has been shown to range between 22.4 and 26.3 kg seed ha^{-1} (calculated from data in Burga-Mendoza [16]).

In the USA and Canada, where bean improvement and extension programmes have existed for many years, average yields have changed little in the past twenty years. One possible explanation is that improvement efforts have emphasised factors such as earlier maturity and seed quality which have only minor effects, if any, on yield *per se*. Research and extension efforts have, however, contributed greatly to stability of production.

While areas sown to beans have increased slightly in some Central American and Caribbean countries since 1961, there have been no corresponding increases in yield per unit area.

Brazil is by far the largest bean-producing nation in the world; its annual harvested area increased from 2.9 x 10^6 ha in the period 1961-65 to 4.3 x 10^6 ha in the period 1976-80. For South America as a whole, the bean area has increased in six countries, decreased in one, and remained static in three. In no single country, with the possible exception of Colombia, do the production statistics indicate a significant increase in yield per unit area. Indeed, in Brazil, the trend has been downward, from 650 kg ha^{-1} in 1961 to just over 500 kg ha^{-1} in 1979.

According to FAO statistics, China and India are very large producers of 'dry beans' but it is known that various species of *Vigna* account for a large proportion of this area. Within Asia, only Turkey, with 110 000 ha and an average yield of 1545 kg ha^{-1} in 1981 would be considered a significant common bean producing nation.

In Europe, a total of eighteen countries produce some dry beans; Romania and its neighbouring countries together with Portugal, Spain, and Italy are the major producers. Reported yields in 1981 ranged from less than 200 kg ha^{-1} for Portugal and Romania to more than 1500 kg ha^{-1} for each of France, Italy, Belgium, West Germany, the Netherlands, Hungary, Poland and Sweden.

PRINCIPAL CROPPING SYSTEMS

In North America, Europe, and in limited areas of other producing regions, beans are extremely commercialised; crops are grown on level lands with mechanisation, fertiliser and pesticide inputs, and sometimes with irrigation. In these areas, bush types (Plate 10.1) predominate since they are well suited to intensive cultivation including semi-mechanised harvest. A few areas, such as southern Brazil, have extensive mechanised culture of dry beans but use only modest inputs and so have correspondingly small yields. Most monoculture beans, however, are grown as a high-input crop with yield potentials ranging between 1000 and 3000 kg ha^{-1}.

In contrast to mechanised production systems, most dry bean producers in the developing countries cultivate unimproved varieties with different plant types and seed colours in complex multiple-cropping systems [9]. The centres of origin for *Phaseolus vulgaris* L. include the Central American and Andean zones [74,113] and beans have been grown in these regions in association with maize for centuries. The systems practised today evolved from those used by indigenous peoples, who had selected useful crop 'cultivars', developed relatively stable production systems, and chosen nutritious combinations of maize and beans for their diets [35,45]. Estimates of the proportion of beans produced in mixed-cropping systems are given in several publications: Mexico, 40% [83], Guatemala, 73%,

Plate 10.1 Determinate ('bush') *P. vulgaris* plants typical of the forms grown in intensive, commercialised production systems (Type I growth habit; see Table 10.6)

Colombia, 90% [64], and Brazil, 80%, [72,100]. It is estimated that between 75% and 80% of beans grown in this region are cultivated in some form of mixed-cropping [46,66]. The importance of these cropping practices cannot be ignored or underestimated.

Bean yields in these traditional cultural systems are small [11,20]. In the highlands, climbing types and landraces are grown with maize, potato, amaranth and other subsistence crops [35]. The beans often entwine the maize stems, or may be supported by stakes or tree branches placed next to the plants for that purpose. There is often direct competition for light, water, and nutrients, although temporal differences in the growth cycles of component species may make these complex associations more efficient than monocrops in the total exploitation of environmental resources. In lowland areas, determinate and indeterminate bush types predominate. These are grown in diverse planting patterns such as alternate rows, strip cropping, and sometimes random mixtures of beans with maize, cassava, sorghum, or other tropical subsistence crops.

Agronomic research on beans

The specific agronomic factors which most limit bean productivity vary between regions and the inputs available to local farmers [48]. In monoculture, most impressive gains in production have come from increasing plant density, from better weed control and where virus and bacterial blight problems have been most severe, from the use of disease-free seed. Timely dates of planting to make best use of available rainfall, and phosphorus fertilisation on some volcanic soils have also boosted yields. When grown in association with maize, bean density as well as appropriate relative planting dates of the two crops have been important.

Results from an experiment conducted on small farms in the Restrepo area in Colombia (4°N; 1500 m elevation; 1300 mm annual rainfall) illustrate several aspects of introducing new technology to traditional farmers. At this location, as shown in Table 10.1, the superiority of traditional black bean varieties as compared to the white and red types is clearly evident. Black beans are grown primarily for sale to other regions (if they are planted at all), while red beans are preferred for local consumption. In monoculture, improved technology increased bush bean yields by 33% and those of climbing beans by 57%,

437

Table 10.1 On-farm yields of bush and climbing beans at Restrepo, Colombia with two systems and two levels of technology: 'farmer technology' includes all traditional practices currently employed, while 'improved technology' includes greater plant density, application of granular insecticide at planting, and a modest input of chemical fertiliser

Bean genotype	Seed colour	Monoculture system		Associated with maize		Overall mean
		Farmer technology	Improved technology	Farmer technology	Improved technology	
Bush Beans (means of two trials)						
P459	Black	1807a*	2291abc	707abc	2073a	1720a
P302	Black	1589a	2439a	750ab	1663bc	1610ab
ICA Tui	Black	1711a	2331ab	652abc	1664bc	1590ab
P524	Cream	1698a	2032bcd	836a	1476bcd	1511bc
P756	White	1165bc	1744d	432c	1470bcd	1203d
P643	White	1475ab	1953d	605abc	1590bc	1406c
ICA Linea 17	Red	979cd	1291e	528bc	1200d	1000e
P758	Brown	1674a	1964cd	582abc	1798ab	1505bc
Calima	Red	780d	1086e	452c	576c	724f
Mean		*1431*	*1903*	*616*	*1501*	*1363*
Climbing Beans						
P525	Black	1176a	1570a	624a	1048a	1105a
P259	Brown	678bcd	1431a	351b	672ab	783bc
P364	White	1021bcd	1324ab	369b	832b	887b
P449	Brown	455cd	917bc	263b	800a	609c
P589	Cream	858abc	1243ab	424ab	673ab	800bc
Radical	Red	389d	682c	304b	227b	401d
Mean		*763*	*1195*	*389*	*709*	*764*

*Values within columns not followed by the same letter are significantly different (LSDs at $p = 0.05$).

and costs of the inputs were easily offset by these
greater yields. In association with maize, bush bean yields
were increased by 144% and climbing bean yields by 82%
as a result of introducing the technological package. A
similar economic advantage was achieved in the associated
cropping pattern. Improved technology in mixed culture
with maize raised yields above those achieved in monoculture with traditional technology [9]. All introduced
varieties were more productive than the 'Calima' bush
bean variety in common use in the zone, as were introduced
climbing varieties as compared with the traditional
variety 'Radical' at all levels of technology. The
important features in these experiments were the inexpensive cost and simplicity of the recommended technology,
the improved yields from introduced over common local
varieties, and the limited training of the extension
specialists who demonstrated and implemented the trials.

Similar encouraging results were obtained in a study
reported from El Salvador [7]. Much remains to be done,
however, to elaborate specific production packages for
each geographical zone, and to establish which combinations
of inputs are most cost-effective and reliable for farmers.

Economics of bean production

An economic survey conducted in Colombia in 1974 and 1975
identified the major constraints to productivity and production of beans at the farm level [21]. Field interviews
were conducted in four regions of the country, on a total
of 177 farms, and included three visits to each farm.
Associated maize was the factor which reduced bean yields
the most; two serious insect pests and Bean Common Mosaic
Virus were also major factors responsible for poor yields.
Small concentrations of potassium, four additional
diseases, and limited moisture for the crop were next in
importance. It was concluded that the development of more
disease-resistant and stress-tolerant varieties, and

improved agronomic practices to avoid some of these problems, could be very cost effective in increasing productivity of beans in the region.

An economic analysis of bean production in associated and monoculture cropping patterns, conducted at the Centro Internacional de Agricultura Tropical (CIAT) in 1975 and 1976, considered results from twenty experiments in which separate plantings of climbing beans and maize were compared with associated plantings of the two crops [47]. Production costs for the three systems were estimated from costs on the Station and from farms nearby. Monoculture beans were more profitable than associated cropping only if a suitable and inexpensive method was available to support the climbing beans. The advantage of the intercrop system was in the limited variation in crop yields from season to season, and thus less risk. The probability of net profit in this experiment was 0.73 for a bean-maize system, 0.53 for monoculture climbing beans, and 0.04 for monoculture maize. This illustrates the advantage of associated cropping for the subsistence farmer who is concerned with risk minimisation, and shows why these complex systems are maintained even when other technology and alternatives are available.

This discussion of agronomic and economic variables associated with dry bean culture highlights the complex decisions which must be made in growing the crop. Farmers must consider the potential productivity of the system and the risks which accompany any change in technology, as well as the nutritional needs of their families and the availability of markets. These factors are all critically important to the subsistence farmer with limited land, and somewhat distinct from the challenges facing commercial operators.

PRINCIPAL BOTANICAL CHARACTERISTICS

Phaseolus vulgaris L., is a member of the Leguminosae,

tribe Phaseoleae, subfamily Papilionoideae. Cultivated
forms are herbaceous annuals, determinate or indeterminate
in growth, and bearing papilionaceous flowers in axillary
and terminal racemes. Racemes may be one to many-flowered.
Flowers are zygomorphic, with a bi-petalled keel, two
lateral wing petals and a large, outwardly displayed
standard petal. Flower colours are genetically independent
of seed colour, though often associated, and may be white
or purple (also red in *P. coccineus*). The flower contains
ten stamens and a single multi-ovuled ovary, which is
normally self-fertilised, developing into a straight or
slightly curved fruit (the pod). Seeds may be round,
elliptical, somewhat flattened, or rounded-elongate in
shape, and have a rich assortment of coat colours and
patterns. They range in size from 50 mg seed^{-1} in ancestral
forms of *P. vulgaris* collected in Mexico, to more than
2000 mg seed^{-1} in some large-seeded Colombian varieties.
Close relatives, the Tepary (*P. acutifoliuc* A. Gray) and
Scarlet Runner (*P. coccineus* L.) beans, share many
morphological and anatomical characteristics with the
common bean. Tepary is better adapted to hot summers and
dry soils and Scarlet Runner to cool sites, particularly
in the uplands of Mexico and Central America, although
they are also grown commercially in England, northern
Europe and in South Africa. The Lima bean (*P. lunatus* L.)
appears less closely related to *P. vulgaris*, and differs
in leaf, fruit and seed characteristics and in adaptation
to hot temperatures (see Chapter 11). Leaves are trifoliolate and stipulate and may be pubescent.

Determinate plants of the common bean have a central
axis (the main stem) with five to nine nodes and from two
to several branches which arise from the more basal nodes.
Indeterminate plants (vegetative apices) have central
axes with twelve to fifteen nodes, or even more in
climbing, vine-types.

Germination is epigeal, (hypogeal in *P. coccineus*)
and requires 5-7 days at a soil temperature of 16°C. Time

to flowering varies with variety, temperature and photoperiod and is usually from 28-42 days. Flowering is usually
complete in 5-6 days in Type I genotypes or 15-30 days in
Types II, III, and IV (CIAT classification, see Table
10.6). As many as two-thirds of all flowers produced may
abscise, and under temperature or moisture stress young
fruits and/or developing seeds may also abort. Abcission
is greatest in flowers formed on the later nodes and
branches, and in the later flowers on racemes with multiple
flowers. Seed-filling periods may extend from as few as
23 days to nearly 50 days. Physiological maturity, the
stage where no further increase in dry-matter in seeds
takes place, may be reached in the earliest varieties in
only 60-65 days from planting; but some Type IV genotypes
in cooler upland sites may require 150 days.

P. vulgaris, *P. lunatus*, and *P. acutifolius* are
invariably self-fertilised, whereas *P. coccineus* is
normally cross-fertilised. Interspecific crossing, except
for *P. vulgaris* x *P. coccineus*, is rare in nature and,
if done by hand, the hybrid seeds survive only through
embryo-culture on synthetic media. Where growing naturally
side by side, *P. vulgaris* and *P. coccineus* have intercrossed and there is evidence that introgression of
P. coccineus into *P. vulgaris* has occurred.

Yield is measured as weight or volume of dry seeds
or, in 'snap beans', as green fruits harvested. Seed yields
may be expressed as the product of three primary components,
fruit number per plant or per unit land area, seed number
per fruit, and single-seed weight.

NITROGEN FIXATION

While common beans have often been regarded as weak in
their ability to fix nitrogen symbiotically [114],
surprisingly large rates of N_2 fixation can be obtained
under appropriate conditions [60,93,123]. Field studies
using the ^{15}N isotope dilution technique show maximum

rates of N_2 fixation equivalent to 64-121 kg N ha^{-1} per growth cycle [93,98,123], and give values for the proportion of plant N derived from fixation which are quite consistent across dissimilar cultural and environmental regimes, and which are as large as 68% [98]. In other studies, values for nodule mass, N_2 (C_2H_2) fixation and specific nodule activity have usually been comparable with those cited for soyabeans [56,57,58,60] (See Chapter 9).

Unfortunately, results such as these have rarely been obtained in farmers' fields, and fertilisation with applied N is still recommended in some countries. Among major problems which limit response to inoculation in the field are host genotype variation in ability to fix N_2, a lack of well tested strains of *Rhizobium* for inoculants, temperature effects on N_2 fixation, soil acidity problems and host plant nutrition. These topics were reviewed thoroughly by Graham [58]; only recent developments will be discussed here.

Variation among bean cultivars in ability to fix N_2 is well documented [39,58,60,98,122,123] and has been attributed to differences in growth habit, relative maturity, root mass and partitioning of assimilates. Fortunately, initial efforts to enhance rates of N_2 fixation in this crop have been very successful [90], with transgressive segregation for greater N_2 fixation activities obtained in a number of lines. Breeding programmes aimed at producing large-yielding and agronomically acceptable bean lines active in N_2 fixation are now underway in several centres.

Because common beans are a secondary crop in most developed countries, only limited production of bean inoculants has been necessary [33], and strain evaluation for this crop has not been intensive. Thus, the strain of *Rhizobium* CC 511 is still used in inoculant production in Australia despite increasing evidence of variation in its performance [61,92].

National bean programmes in Latin America, in collaboration with the bean programme at CIAT (see Chapter 18), have begun to emphasise strain selection and testing, with superior strains of *R. phaseoli* now available from CIAT or from the Microbiological Research Centre (MIRCEN), Porto Alegre, Brazil [39].

Temperature affects several processes associated with nodulation and N_2 fixation. These include survival of *Rhizobium* in the soil, root hair formation, root hair binding and infection thread formation, and nodule initiation, growth, leghaemoglobin content and nitrogenase activity. While Graham [57] reported little host-temperature interaction within the day/night temperature range 35°/25°C to 25°/15°C, Rennie and Kemp [92] have shown considerable interaction in the range 10°-16°C. Cultivar 'Aurora', a small white-seeded type, was superior to 'Kentwood' navy bean in nodulation, N_2 fixation and N yield at 10°C and 16°C; 'Kentwood' was superior at 14°C. Delayed nodulation by cool root temperatures would justify experimentation on the use of 'starter' N (and see Chapter 4).

Rhizobium phaseoli is normally considered to be sensitive to soil acidity, but recent studies suggest strain variation in tolerance to acid soil factors [62] with some, but not all, strains giving a strong inoculation response, even at pH 4.5. Tolerant strains have a strong competitive advantage in acid soils.

PRINCIPAL LIMITATIONS TO YIELD

The principal constraints to bean productivity are discussed below.

Weeds

Weeds must be controlled during the first thirty days of crop life, or until flowering, or yields are likely to be

substantially reduced. Seed-filling is also a critical
period, particularly for water. If plants are stressed
during seed-filling and weeds are excessive, yields can
be drastically reduced. Competition from other crop plants
and weeds is more likely to depress yields in some farming systems than in others in the developing countries.
Weed control in developing countries is usually by hand,
although in areas where labour is not readily available
during the early growth period, herbicides are sometimes
applied.

Abiotic stresses

The principal *non-biological stresses* which affect bean
productivity, both in developed and developing countries,
are drought, temperature extremes, salinity, poor soil
aeration, and toxicity or deficiency of certain elements
(notably P, Al and Zn). There appear to be genetic
differences among cultivars in response to each of these
conditions, with the possible exception of salinity,
although no genotype should be considered fully resistant
to any of these stresses.

Beans are not noted for their drought tolerance but
respond to water stress by leaf flagging, stomatal closure,
and shedding of leaves, flowers and young fruits. The
crop responds to irrigation with the greatest demand about
the time of flower initiation and early seed-filling. In
medium-textured soils in temperate climatic zones, the
crop requires about 2.5-3.0 cm of rainfall each week until
near to physiological maturity (estimated from yield and
rainfall records in Michigan [4]).

Among the more serious of the physiological disorders
are those due to chemicals in the atmosphere or to spray
drift. Some insecticides and herbicides can burn leaves
and stunt plants, while 2,4,D induces twisting and malformation of leaves and stems. Atmospheric pollutants are
also a problem in the industrial countries. Ozone (O_3)

445

causes leaf bronzing and often induces premature senescence. Peroxy-acetyl nitrate (PAN) produces shiny or silvery areas on leaves which become bronzed. Sulphur dioxide results in necrotic or bleached foliage. Genetic resistance to air pollutants has been found in beans, and benomyl applied as a foliar spray has reduced damage due to oxidants.

In addition to site-specific factors, temperature and photoperiod are major determinants of broad regional adaptation [54,55]. According to the recent reviews and synthesis by Wallace [118,120], there is for every genotype an optimum photoperiod-temperature regime in which that genotype will flower in the smallest possible number of days from emergence. This minimal period is termed by Wallace as the 'flowering tendency', and is closely related to, or identical with, the minimal (lowest) node at which flowering can occur under the most inductive conditions. Deviations in either temperature or photoperiod cause delays in flowering. Wallace distinguishes a 'below optimum temperature response' in which flowering is delayed at temperatures cooler than the optimum, and an 'above optimum temperature response' in which an increase in temperature above the optimum delays flowering proportionately. In addition, a photoperiod response delays flowering as the effective daylength becomes longer (for this short-day species). The photoperiod response may be positioned at, below or above the optimum temperature for flowering. There are, furthermore, interactions between temperature and photoperiod which delay flowering as either factor deviates from the optimum for minimal days to flower.

In all cases, Wallace expresses plant response in terms of rate of node development and lowest node on the central axis at which the first flower develops; there appear to be separate genic systems for these two responses.

Other researchers [126,127], whilst sharing many views in common with those summarised above [120], consider that rates of progress towards flowering (the reciprocals of times taken to flower) in common beans and other grain legumes in different photo-thermal regimes are amenable to quantitative description by a few, relatively simple linear relations in which interaction effects are seldom, if ever, significant.

Diseases and insects

Diseases.   Bean diseases are responsible for serious losses in yield and quality [105,125]. There is a greater array of disease-causing pathogens in tropical than in temperate production regions. The warm, often humid environments of the tropics favour pathogens, while continuous crop cycles in many regions provide a continuity of inoculum. Although yield losses to pathogens in the tropics are substantial, the local genetically variable landraces can have some tolerance which contributes to yield stability. In addition, the associated cropping systems used on small farms may reduce the rate of spread of pathogens and the amount of the disease caused by some of them. Cultivars in the developed temperate regions, grown in monoculture, are often homozygous and homogeneous, have a narrow genetic base and generally express vertical resistance, controlled by major genes. These cultivars often express immunity or at least extreme resistance to pathogens but if the resistance genes are overcome, they become very susceptible. The predominant diseases in the hot, humid tropics are web blight (*Rhizoctonia microsclerotia*), common blight (*Xanthomonas phaseoli*), and root-rot (*Fusarium* spp., *Macrophomina* spp. and *Rhizoctonia* spp.). In warm-hot, relatively dry climates, Bean Common Mosaic Virus (BCMV) [53], Golden Bean Mosaic Virus (GBMV), rust (*Uromyces phaseoli*), angular leaf spot (*Isariopsis griseola*), root-rots and common blight are especially troublesome. In cool climates, BCMV, rust, anthracnose (*Colletotrichum lindemuthianum*), angular leaf spot, root-rots, halo blight (*Pseudomonas phaseolicola*),

447

and white mould (*Sclerotinia sclerotiorum*) are of principal significance [105].

Bean diseases can be controlled or reduced by the following measures: use of clean seed, disease avoidance (appropriate plant architecture and planting date), disposal of infected debris, crop rotation, cultural practices, seed treatment, chemical sprays, and host genetic resistance. No successful chemical control has yet been developed for bacterial pathogens [103], but in the USA the severity of these pathogens and of anthracnose has been reduced by growing clean, bacteria-free seed of susceptible cultivars in arid areas using surface irrigation. Areas suited to disease-free bean seed production do not exist in many bean producing regions or, if they are present, the means of production and distribution of the disease-free seed have not yet been developed. In addition, subsistence farmers may not have sufficient funds to buy clean seed. Cultivars and genetic stocks with different levels of resistance to common blight and halo blight diseases are becoming available in many seed classes of dry beans [32].

Bean Common Mosaic Virus, transmitted by aphids and seeds, has been controlled successfully in the USA, Canada, Europe and, more recently, in CIAT materials by use of the dominant *I* gene and race-specific recessive genes for resistance [37], but much breeding is still needed to introduce resistance into beans grown elsewhere in the world [17] Bean Golden Mosaic Virus, transmitted by whiteflies, is serious in several countries in Latin America. The disease can be reduced by spraying to kill the vector. New cultivars and lines developed by CIAT have useful levels of genetic resistance. Resistance to 'summer death', a disease caused by mycoplasma in Australia, is found in cultivars of beans resistant to the Curly Top Virus in the USA [10].

Numerous races of the pathogens causing rust, anthracnose and angular leaf spot are known and new races continue to evolve and overcome known resistance genes. Despite this, genetic resistance is still the best method of control of these pathogens. Chemical sprays are also useful in the control of rust and angular leaf spot, especially where the early stages of plant growth are affected [105]. Root-rots caused by *Rhizoctonia, Fusarium, Macrophomina* and *Pythium* spp. are serious in some bean production areas [15]. No bean cultivars with outstanding resistance to these pathogens are known, but considerable progress has been made in incorporating useful levels of resistance into new breeding lines. Some measure of control can be achieved by cultural practices and by chemical treatments, but they are generally regarded as unsatisfactory [15].

Chemical control of white mould in bush snap beans has been achieved in the USA but is not generally satisfactory on indeterminate dry bean types [108]. The best means to reduce the incidence of white mould is to utilise dry bean types which possess plant architectural avoidance and physiological resistance mechanisms [50] combined with judicious irrigation after flowering [108].

Web blight causes large yield losses during the rainy season in the hot humid tropics. Systemic fungicides have proved to be a useful control measure [52].

Nematodes cause serious problems on beans; *Meloidogyne* spp. cause root galls, while *Pratylenchus* spp. produce brown or black lesions on roots [112]. Population levels can be reduced by crop rotation. Soil fumigation is a successful control measure but is expensive. Resistance to nematodes is available in a number of lines and is being incorporated into new cultivars [43].

Insects.   More than 200 different insects attack *P. vulgaris* [111]. Leafhoppers (*Empoasca* spp.), chrysomelids (*Ceratoma* spp. and *Diabrotica* spp.), bean pod weevils (*Apion godmani*), whiteflies (*Bemisia tabaci*), Mexican bean beetles (*Epilachna varivestis*), and bruchids (*Acanthosclelides obtectus*) are regarded as the most serious of the insects in South America. Mites (*Tetranychus* spp.) and slugs (*Vaginulus plebejus* and *Limax maximus*) are important in

some areas. An average yield loss of 47% was estimated
based on a literature review of sixteen insecticidal trials.
In North America, Mexican bean beetles, seed corn maggots
(*Hylemya cilicrura* and *Haliturata* spp.), leafhoppers and
mites are serious pests. Seed corn maggots and cutworms
(*Agrotis ipsilon* and *Spodoptera* spp.) attack seedlings;
chrysomelids (two species), lepidoptera (three species) and
Mexican bean beetles feed on leaves; leafhoppers, whiteflies, aphids, and thrips suck plant parts; weevils and
borers attack fruits, while two species of bruchids infest
stored seeds.

Yield loss due to leaf-eating insects is determined by
the amount of defoliation, the stage of plant development
when infestation occurs and the severity of infestation.
Considerable defoliation can occur in vegetative beans without significant reduction in subsequent seed yield and this
must be considered before applying insecticides. Chemicals
have been effective in the control of insect pests of beans.
Seed corn maggots are controlled by applying carbofuran,
diazinon and chlorpyrifas as granules in the furrow or as
a seed slurry [111]; Mexican bean beetles, leafhoppers, and
bean pod weevils are controlled by spraying plants with car-
baryl; and complete control of bruchids can be achieved by
coating stored seeds with cotton seed oil. Pyrethrins are
very effective in controlling weevils.

Useful resistance has been found in beans to leafhoppers, Mexican bean beetles and to bean pod weevils. Although
the task may be difficult, there are good prospects in
breeding commercial cultivars with satisfactory resistance
to these pests.

Associated cropping can be manipulated to reduce injury
from leafhoppers. This is of value to subsistence farmers
in South America who use this cultural system. When maize
was planted twenty days before beans the leafhopper population was significantly smaller than when both crops were
planted on the same day [110].

RESEARCH AND DEVELOPMENT

Researchers in twenty-nine countries of Africa and the
Americas (including Mexico and Canada) are currently working on dry beans. In the USA, the federal government, twelve
states, and several commercial companies maintain active
research programmes. At least six countries of western
Europe, and a smaller number of south-eastern European
nations also maintain bean research programmes. Activities
range from the testing of varieties and agro-chemicals, to
comprehensive genetical, physiological, pathological, and
variety development programmes. While considerable work is
underway on edible beans, both as a green vegetable and as
a pulse, much of the work is site (region)-specific, and
consists mostly of variety testing, and studies on fertility,
herbicides, insecticides and plant populations. The next
major category of research involves pathology since diseases
are almost universal and their control is one of the prime
objectives in many programmes; selection and breeding also
comprise a significant dimension in many programmes. Genetical, physiological, microbiological, cytogenetical and
taxonomic research, and those activities known as 'basic
research', are maintained in conjunction with variety
development programmes in a smaller number of locations.
CIAT (see Chapter 18) maintains a comprehensive programme
on dry bean research, and works cooperatively with most
Latin American and many African national programmes.

A special research initiative was launched in 1980 to
link American Universities with national programmes in
African and Latin American countries. Entitled 'The US/AID
Title XII Bean/Cowpea Collaborative Research Support Program', it deals with a broad spectrum of problems affecting
production and utilisation of beans and cowpeas on subsistence farms (see Chapter 18).

FERTILISER REQUIREMENTS

Nutrient requirements for different cultivars are usually very similar, except on problem soils where cultivars may differ specifically in their ability to grow and yield well. Table 10.2 shows the amounts of several major and minor nutrients contained in the above-ground components of a bean crop yielding 1000 kg seed ha^{-1}.

Table 10.2   Amounts of various nutrient elements contained in the above-ground (seeds and vegetative) parts of a bean crop yielding 1000 kg seed ha^{-1} [96]

Element		Amount (kg)	Element		Amount (kg)
Nitrogen	(N)	62	Copper	(Cu)	0.02
Phosphorus	(P)	48	Iron	(Fe)	0.40
Potassium	(K)	35	Manganese	(Mn)	0.07
Calcium	(Ca)	46	Zinc	(Zn)	0.04
Magnesium	(Mg)	13			

Fertile agricultural soils will generally contain adequate amounts of all of these elements except perhaps nitrogen. It is common practice, however, particularly in North America and in Europe, to apply fertiliser at planting time, usually N, P, and K. In mineral alkaline soils (pH > 7.2) additions of zinc and iron or manganese may be necessary.

Approximately 50% of the total dry-matter produced by a bean crop together with 75% of the N, 80% of the P, and 50% of the K, is partitioned into seeds.

Adequate nutrient concentrations, particularly of P and K, help crops to use water more efficiently. Adequate fertility, including the minor nutrients, also promotes normal phenology of bean plants [96].

YIELD AND QUALITY

There is a general expectation that seed yield and seed protein concentration (%) in beans are negatively correlated [65]. It was also reported [3] in a study of 198 plant introductions that percentage protein was negatively related to concentration of sulphur-containing amino acids. In that sample, seed protein concentration ranged from 19-31%. The 12% of the sample with least methionine (<0.9 g 100 g protein^{-1}) averaged 25.13% protein, whereas the 12% of the lines with most methionine (>1.05 g 100 g protein^{-1}) averaged only 21.56% protein. Data for cystine were similar. It was suggested that both yield and quality would be favoured by selecting lines averaging 20-22% protein. However, other studies [13,89] have shown that methionine-cystine concentrations as well as yield can be maintained by intensive selection at protein concentrations up to 28%.

Genotypic differences in several organic and mineral constitutents of seeds are evident in Table 10.3 which gives chemical analysis of seeds of twenty-eight standard American cultivars [117]. Apart from nitrogen, seeds were especially rich in potassium (averaging 1.47%, with a range of 1.12-1.94%), while iron was the most abundant micronutrient (at 92.9 ppm, ranging from 69-135 ppm). Oil is a very minor constituent in bean seeds, representing no more than 1% of the weight of seeds containing 3% moisture (see Chapter 3).

Table 10.3 Average and range in values of chemical constituents in seeds of 28 common dry bean varieties [117]

Constituent	Amount present	
	Mean	Range
Oil (%)	0.58	0.20 – 1.0
Total sugar (%)	6.2	4.40 – 9.20
Sucrose (%)**	46.4	33.6 – 57.0
Raffinose (%)**	10.4	2.4 – 16.0
Stachyose (%)**	43.0	33.6 – 64.0
Zn (ppm)	22.3	17 – 29
B (ppm)	15.1	12 – 18
Fe (ppm)	92.9	69 – 135
Mn (ppm)	11.7	6 – 20
Cu (ppm)	9.75	8 – 12
Mg (%)	0.16	0.13 – 0.18
Ca (%)	0.19	0.11 – 0.26
P (%)	0.40	0.28 – 0.50
K (%)	1.47	1.12 – 1.94
N (%)	3.70	2.84 – 4.28

**Values are percentages of total sugar.

Table 10.4 Essential amino acid content of bean seed protein
(g amino acid 16g N^{-1}) [41]

Amino acid	FAO* provisional standard	Commercial class of bean			
		Navy+	Red Kidney	Tropical Black	Small Red
Arginine	–	5.1	5.1	6.7	7.2
Histidine	–	2.0	2.6	2.9	3.2
Isoleucine	6.4	4.3	4.2	4.1	3.8
Leucine	11.2	7.5	8.1	7.7	7.7
Lysine	8.6	5.7	6.7	6.7	7.2
Methionine	–	1.3	0.9	2.0	1.3
Methionine plus cystine	5.6	2.0	1.9	3.0	2.2
Phenylalanine	9.8	5.0	5.3	5.1	5.4
Threonine	6.4	5.1	4.2	3.1	3.5
Tryptophan	1.6	1.8	1.5	1.5	1.7
Valine	8.0	4.7	5.1	4.9	5.0

*Food and Agriculture Organisation of the United Nations pattern of amino acid requirement (1973), FAO/WHO Technical Bulletin Series No. 522.
+Navy beans are white-seeded types with small (15-22 g 100 seeds^{-1}), round to oval seeds.

Common beans are a significant source of protein in human diets in many countries of Africa and Latin America. In conjunction with maize, for example, protein intake can be well balanced, the beans supplying the lysine and, to a lesser extent, the tryptophan deficient in maize, while maize provides the sulphur-amino acids. Table 10.4 provides data on mean amino acid values as determined for a standard variety in each of several commercial classes of dry beans.

Beans may be boiled, baked, canned in sauce, brine or syrup, refried, used in soups either alone or with other vegetables, meats and chili pepper, as a fresh salad, or curried. In many African countries immature leaves are harvested as a pot herb or 'spinach' or sun-dried and used as a base for soups during the dry season. Immature fruits are used as snap beans* or as a source of shelled beans.

Dry beans must be thoroughly cooked before being eaten

*Snap beans are those types grown for their edible fruits, harvested when the seeds are fully formed but immature.

since raw beans contain an array of antimetabolites that can adversely affect growth. Bean seeds contain significant amounts of thiamine, niacin, folic acid, as well as fibre [18]. In general, the deeply coloured beans, reds and blacks, are relatively rich in tannins as compared to the pale and lightly variegated types, and tannins are considered to reduce the dietary value of the protein (and see Chapter 3).

GERMPLASM RESOURCES

A world collection of *Phaseolus vulgaris* and related species has been assembled at CIAT (Table 10.5). A total of 28874 accessions of *P. vulgaris* had been accumulated up to 1982, inclusive of the large collections held by USDA and at Cambridge University, and by national programmes, particularly from Latin America, and some private organisations. Since 1978, CIAT and the International Board for Plant Genetic Resources (IBPGR; see Chapter 18) have funded collecting expeditions to Mexico, the Iberian Peninsula, Peru, Brazil and in several African countries. With such diverse sources, duplication in the collection is inevitable, but is not considered important so long as the material is readily available for use in breeding programmes.

Germplasm is stored at CIAT in a short-term working

Table 10.5   The number of accessions of *P. vulgaris* and related
species in the CIAT world germplasm collection [25,27]

Material	No. of Accessions
Cultivated *P. vulgaris*	28542
Ancestral *P. vulgaris*	332
P. coccineus subspecies *coccineus*	710
P. coccineus subspecies *polyanthus*	314
Ancestral *P. coccineus*	58
Cultivated *P. acutifolius*	89
Ancestral *P. acutifolius*	59
Non-cultivated *Phaseolus* sp	84

collection, designed to keep the seed 80% viable for five years, and in long-term storage where the seed can be pre- served for as long as twenty-five years. The short-term facilities provide a temperature of 5-8°C, with the seed kept in plastic jars at 12-15% moisture content and 60-65% relative humidity. Long-term material is sealed in laminated bags at 5-8% moisture content and stored at -2°C to -6°C.

Classifications of the variability within the collec- tion by seeking accessions which cluster when evaluated by twenty-seven morphological descriptors, as well as by chemo-taxonomy methods, are being investigated. A descrip- tion of the kinds of genetic variability available is pre- sented below.

Seed size

Evans [42] proposed a division of common beans into races based on seed size: medium and large-seeded types (≥300 mg) were considered to have originated and diversified in South America, small-seeded types (50 - 300 mg) in Middle America. Introgression of the large-seeded *P. coccineus* from the highlands of Mexico into smaller-seeded *P. vulgaris* from intermediate elevations was thought by Miranda [86] to have contributed to the genetic variability for seed size in common beans.

Small seeds are typical of germplasm from the lowland tropics, particularly Middle America. Wild *P. vulgaris* also tends to have very small seeds (50-150 mg) [86]. It has been assumed that the wild type was an indeterminate climber [87], but in the CIAT collection many of the wild accessions have determinate axes, whilst retaining a strong climbing tendency [110]. A dominant gene *F*, for indeter- minate growth, was proposed by Lamprecht [82], and the determinate character is thought to have been a recessive gene mutant that arose during domestication [69,107]. How- ever, extensive crossing between growth habits at CIAT does not always support these conclusions, which may have

456

been influenced by a linkage between photoperiod sensitivity and indeterminacy [29]. It seems the determinate raceme is dominant in the majority of crosses, but that a climbing tendency, associated with development of many long internodes before the terminal raceme is set, is also dominant.

Growth habit

For the purposes of classifying the collection, growth habits have been defined as in Table 10.6. Approximately equal numbers of each growth habit are found in the collection. Growth habit is plastic; temperature and photoperiod, and stresses such as drought and poor soil fertility, greatly affect vegetative stature. Growth habit can be related to the adaptation of a variety to different cropping systems, population densities [101], and to different stresses [80]. A growth habit instability (a change from indeterminate bush to an indeterminate climbing habit) has been related to a phytochrome response to differences in spectral quality [78,79].

Disease and pest resistance

A major reason for maintaining a world collection of germplasm is to preserve, and make available to breeders, different sources and mechanisms of resistance to diseases and insect pests. For extremely variable pathogens, such as rust, international germplasm nurseries have made it possible to identify accessions and improved lines with stable and possible non-race specific resistance. However, insufficient resistance to other diseases, such as common bacterial blight, has been found in the world germplasm collection. In order to increase levels of available resistance, interspecific hybrids from *P. acutifolius* [102] have been developed, and sources of field tolerance intercrossed. Web blight is another disease for which resistance sources are inadequate in existing germplasm. Finally,

457

Table 10.6 Growth habit classification and description of *Phaseolus* as defined at CIAT

Growth habit	Description
Type I	Determinate habit; reproductive terminals on main stem and no further node production on main stem after flowering.
Type II	Indeterminate habit (vegetative terminal on main stem); further node production on main stem after flowering; erect branches borne on lower nodes; erect plant with extremely variable guide development.
Type IIIa	Indeterminate habit; moderate node production on main stem after flowering; prostrate canopy with variable number of branches borne on lower nodes; main stem guide development extremely variable but generally showing poor climbing ability.
Type IIIb	Indeterminate habit; considerable node production on main stem after flowering; heavily branched with variable number of facultatively climbing branches borne on lower nodes; guide development variable; plants generally show moderate climbing tendency on supports with resulting cone-shaped canopy.
Type IVa	Indeterminate habit; heavy node production on main stem after flowering; branches not well developed compared to main stem development; moderate climbing ability on supports, with fruits load carried relatively uniformly along length of the plant.
Type IVb	Indeterminate habit; extreme node production after flowering; branches very poorly developed; strong climbing tendencies on supports, with fruit load borne on the upper nodes of main stem.

Ascochyta blight can be important in highland areas and satisfactory resistance has not been found in *P. vulgaris*.

Resistance is available, however, in *P. coccineus* sub-species *polyanthus*, and interspecific hybrid populations produced at the University of Gembloux in Belgium show potential for incorporating resistance into *P. vulgaris* [27].

Anthracnose resistance is found in some accessions from the highlands of Mexico and the Andean region [70]. In Colombia, anthracnose-resistant cultivars have given yields double those of commercial cultivars in on-farm trials [97]. Resistance to BCMV has been found in materials from the lowlands of Central America, but not in those from the Andean region. It has been difficult to incorporate *I*-gene resistance [36] into Andean region cultivars. Brazilian accessions are a common source of resistance to angular leaf spot [106]. These pathogens are also variable, and some accessions have been identified with field resistance to anthracnose which is not, however, expressed in glasshouse seedling tests.

Resistance to pests, particularly to *Empoasca kraemeri* and storage insects, has also been sought in the CIAT collection. Until the end of 1982, 11000 accessions had been evaluated for tolerance to *E. kraemeri*; with tolerance associated with reduced oviposition [27].

Tolerance to the bean fly, which is influenced by the degree of lignification and thickness of the basal portion of the stem, has been sought in Africa [63]. In collaborative work between the Asian Vegetable Research and Development Center (AVRDC) in Taiwan (see Chapter 18) and the University of Gembloux in Belgium, considerable tolerance has been detected in *P. coccineus* subspecies *coccineus* (CIAT accession G 35023), and progenies of crosses with *P. vulgaris* with much improved tolerance have now been selected [27].

The best resistance to bruchids has also been found outside the cultivated species, in this case in wild or regressive *P. vulgaris*, generally characterised by small seeds. The basis for resistance may be associated with the presence of trypsin inhibitors in seeds, as in cowpeas [71]. Two accessions from CIAT, G12952 and G12953, both originating from Mexico, are extremely resistant to both *Acanthoscel-ides obtectus* and *Zabrotes subfasciatus*.

Adaptation to photoperiod and temperature

Photoperiod-insensitive accessions originate mainly from extreme latitudes and tend to be found in growth habit Types I and II. Large-seeded climbing types, principally from the Andean region, are seldom insensitive [22,27]. Accessions which are specifically adapted to temperatures as cold as 13° C occur, but they are all sensitive to photo-period and most of them are climbing types from the high Andes [40]. Other accessions are specifically adapted to warm temperatures [67].

Drought tolerance

Most germplasm possessing drought tolerance originates from dry areas, particularly those of Mexico and coastal Peru. In material previously unselected for drought tolerance, there is a marked tendency for growth habit and vigour to be associated with drought tolerance (Table 10.7). The more vigorous plant types appear to be deeper rooting and so experience less stress in soils where deep root development is possible [27]. When selection for drought tolerance is practised, materials of growth habits other than Type IV can be obtained with good tolerance to drought.

Poor soil fertility

Variability in tolerance to acid-soil factors, particularly to Al and Mn toxicity and reduced P availability, has been reported, but lines selected for Al tolerance are not nec-essarily tolerant of excess Mn, and vice versa. Certain cultivars, such as 'Carioca' from Brazil, may be tolerant to soil acidity without necessarily being tolerant of Al. Other materials have shown tolerance to P-deficient, Al-rich

460

Common Bean (*Phaseolus vulgaris* L.)

Table 10.7 Yields under drought stress of breeding lines, previously unselected for drought tolerance, grouped according to growth habit [27]

Growth habit	Yield (kg ha^{-1}) under drought stress	No. of lines	LSD ($p < 5\%$)
Type I	417	20	80 (I/II)
Type II	482	41	84 (II/III)
Types IIIa and b	695	15	105 (III/IV)
Types IVa and b	1152	20	-
G 5059	1253	-	-
(tolerant control)		-	

soils (CIAT lines G 4000, BAT 26, BAT 28, EMP 28, and BAT 458) [23,24]. Because results can be affected by source of N (inorganic versus symbiotic fixation) and the presence of mycorrhizae, particular care must be exercised in screening germplasm for ability to grow well on poor soils.

Adaptation to intercropping

Germplasm has been screened for adaptation to row and relay intercropping with maize. For row intercropping, competitive ability varies with growth habit; Types I and II are the least competitive, and Types IIIb, IVa and IVb are progressively more competitive [81]. Despite severe reductions in the yield of maize, Type IVb accessions are found to give the most favourable combined production of beans and maize [34].

GENETICS AND BREEDING STRATEGIES

Genetics

Phaseolus vulgaris is normally a self-pollinated diploid ($2n=22$), but considerable outcrossing has been detected in some tropical locations, and attributed especially to carpenter

461

bees [49]. Floral characteristics and pollination methods have been described by Bliss [12]. Most new cultivars are homozygous and homogeneous while landrace populations may include a minor proportion of heterozygotes among the heterogeneous array of homozygous individuals. Some cultivars are derived from blends of closely related lines.

Considerable information exists on the inheritance of qualitative traits in *P. vulgaris* but it is scattered throughout the literature. Kooiman [77] and Yarnell (124] published substantial reviews on the inheritance of traits in the crop. There has, however, been no general review of the genetics of *P. vulgaris* since 1965 [124] although several excellent reviews on specific traits have been published. An up-dated gene list was published only recently [95]. There is much less information on the inheritance of quantitative traits available from the early literature, but a considerable amount has accumulated since 1970. Again, no general review has yet been published. Information on the linkage of genes in *P. vulgaris* is limited, especially when contrasted with the extensive information available, for example, in maize or tomatoes. Most breeders involved with beans have been interested largely in specific economic traits, with only a small number of them interested in genetics *per se*. Cytological and cytogenetic studies have been limited because of the difficulty of classifying individual chromosomes. The use and value of the Giesma staining technique has stimulated increased attention in this area [88]. Interspecific crosses have received considerable attention in some laboratories and their achievements are now being announced [109]. The cross *P. vulgaris* × *P. coccineus* has been made many times [124], while Honma [68] was the first to succeed in making the cross *P. vulgaris* × *P. acutifolius* through use of F_1 embryo culture. It has not yet been possible to regenerate whole plants from callus cultures of *P. vulgaris*. If the regeneration barrier is overcome, novel techniques for genetic modification of cells and tissue cultures are likely to receive increased application in both basic and applied research.

The genetics of seed-coat colour and disease resistance have been studied extensively. Lamprecht (reviewed by Yarnell [124] and Prakken [91]) developed the most extensive information on the former trait. Seed-coat pigmentation is determined by major genes in the presence of the dominant gene *P*; white seed-coat is expressed in the presence of recessive *p* no matter what alleles for colour are present at other loci. Whiteness can also be expressed in the presence of recessive alleles which prevent pigment formation. Modifying genes, which regulate the expression of genes for colour, i.e. genes for seed-colour patterns, have also been identified.

Variation in time to flowering results from differences in sensitivity of genotypes to changes in photo-thermal regimes [28,40,120,126,127]. This response affects days to maturity, adaptation, and yield in diverse environments [120]. The variation encountered in any particular cross appears to be determined by relatively few genes [120], but the inheritance pattern can seemingly change with the environment [55,85]. A number of major genes have been identified which control plant habit and leaf abnormalities in various crosses [124]. These can easily be removed from populations through progeny testing and selection.

Major genes are important in providing resistance to nematodes and several fungal, bacterial, and virus diseases of beans [105,124]. Numerous sources of major dominant genes which confer resistance to rust have been reported but none has provided generalised durable resistance [31]. The *Are* gene provided resistance to the prevailing races of *Colletotrichum lindemuthianum* (Sacc. and Mag.) Scrib. for several years but it does not protect against some new races of the pathogen in Europe and South America [19]. Major genes have now been identified which provide resistance to these new races. Major gene resistance has been identified for the bacterial pathogens causing halo blight and bacterial wilt diseases [32]. Different linked genes have been found to

control the leaf water soaked and leaf systemic chlorosis
phases of the disease (the latter being due to mobility of
bacterial toxin), while the genes which determine fruit
reaction were inherited independently of the former ones
[32].

Two types of genetic resistance to various strains of
Bean Common Mosaic Virus have been identified. The host/
virus genetic interrelations have been clarified in a
thorough systematic study by Drijfhout [36]. Cultivars or
lines which are resistant to systemic mosaic and express
no local or systemic necrosis, possess recessive *i i* alleles
and one or more recessive genes for resistance. Cultivars
which possess dominant *I I* do not express systemic mosaic
but are susceptible to some virus strains which cause systemic necrosis.

Reactions to the pathogens causing white mould, root-
rots incited by numerous pathogens, common blight, and
brown spot incited by *Pseudomonas syringae* pv. *syringae*,
are quantitatively inherited [32,50]. Moderately small to
moderately large estimates of heritability have been reported depending upon the inoculation method used and degree of environmental control imposed. Linkage of genes
which control resistance to common blight and delayed
flowering was reported in the temperate zone but not in
the tropics [32]. No association was found between the
reactions of fruits and leaves. Genotypes with resistance
to common blight in the long-day temperate regions were
found to be susceptible when grown under short days in the
tropics but their resistance was expressed when daylength
was extended artificially using lamps in the field [121].

Tolerances to mineral deficiencies and toxicities,
although seldom studied, are generally quantitatively inherited but, in some cases, one or a few major genes have
been postulated to explain responses to small P and K concentrations in soil [30,51].

Seed yield and its components, protein concentration
in seeds, and nitrogen fixation were found to be

quantitatively inherited, with small to moderately small
heritability estimates, except for single-seed weight,
which is large [14,119]. Adams [1] reported negative correlations between major yield components and postulated
that these correlations were a consequence of development
rather than due to genetic causes. Negative associations
were found between yield and seed protein concentration;
however, lines which combine large yield and protein concentration have now been developed [14].

Heterosis for yield and other complex traits has been
reported [124]. In one case Duarte and Adams [38] showed
that heterosis was due to component interaction at the
developmental level where variation in the components was
regulated by additively acting and/or partially or completely dominant genes. Additive gene action appears to be
the predominant type of gene action determining variation
in quantitative traits in *P. vulgaris*. It is recommended
that breeding systems should be adopted that best use this
genetic effect - such as structured or non-structured recurrent selection schemes.

The use of induced mutations for genetic and breeding
studies, emphasised in the past, is not now widely adopted
as a major means of genetic improvement. Two useful traits
were, however, derived from induced mutations; a white-
seeded mutant, 'Nep-2', derived from the black-seeded 'San
Fernando' bean in Costa Rica, and upright plant architectural types in Michigan [2,5].

Breeding strategies

As an extremely autogamous species, beans are bred by
methods appropriate to such crops, of which wheat, barley,
oats, rice and soyabeans are typical examples. Such methods
are adequately described in plant breeding texts, and do
not need repetition here. Introduction, mass selection,
pedigree selection from simple two-parent crosses, and
standard backcrossing are the methods currently used to

some extent in many breeding programmes. Recurrent phenotypic selection in two-parent and multiple-parent hybrid
populations is probably the most favoured method at the
present time.

The classical backcross method has been thought to be
appropriate primarily to transfer a single desirable gene,
an extremely penetrant dominant or recessive gene, from an
otherwise undesirable 'donor' parent to a standard cultivar
or breeding line as recipient recurrent parent, defective
only in the trait to be transferred. Voysest [115] and
Knott and Talukdar [75] have shown that the backcross
method can also be effective in transferring a polygenic
trait if that trait is strongly heritable, their examples
being seed size in beans and wheat, respectively. Bliss and
his colleagues [89] have successfully employed the back-
cross-inbred method to improve nitrogen fixation in navy
and tropical black beans. This method was originally devised to isolate individual polygenes and to estimate their
effects and was not intended as a breeding method *per se*.

Diallel crosses have been used frequently in bean
improvement programmes [42]. It should be recognised, however, that the diallel is most often employed for combining
parents for biometrical genetic studies but it may lead to
populations for selection purposes. In this latter context,
a modified version, the diallel selective mating design,
has been suggested by Jensen [73] as a method of breeding
autogamous crops that overcomes several disadvantages of
the more traditional methods.

All breeding strategies for self-fertilising species
involve six fundamental tasks:

> (a) selection of the target environment and farming
>   system, and development of a sound understanding
>   of their characteristics [80,116];
> (b) identification of goals reasonable in the chosen
>   environment and farming system;
> (c) parent selection;

(d) combining parents in a mating design that will
yield superior populations;

(e) employing population management strategies that
either preserve and add to favourable parental
genic combinations, or that promote the formation
of new favourable genic recombinations;
and

(f) designing and using screening techniques that identify reliably the genic combinations desired.

With respect to objectives, many breeding programmes
must deal with multiple objectives, whose genetic bases may
range from simple to complex in terms of gene numbers, gene
action, and genic linkage. Even in programmes with a single
major objective, the trait to be improved may be regulated
by both major and polygenic factors, together with significant environmental influences. At this stage of bean
improvement, multiple disease resistance is a major objective. Additional goals include insect resistance, resistance
to various abiotic stresses, improved plant architecture,
biological nitrogen fixation and nutritional and processing
quality, various agronomic characteristics and, nearly
always, yield *per se*.

Certain programmes select one or two major goals, for
example, multiple disease resistance, or improved nitrogen
fixation, and concentrate upon producing, not new cultivars
per se, but elite breeding populations or lines to be released to other breeders. Given the task of breeding for
multiple objectives or for single traits regulated by complex quantitative factors, a breeder will most advisedly
adopt a strategy involving numerous parents and many
crosses, employed perhaps initially in a parental set or
sets of bi-parental matings (set A with set B), in a cyclic
or recurrent design, with the use of single seed (pod)
descent and early generation selection where possible.
After a one-generation advance and testing, the selections
re-enter the crossing design to initiate the next cycle.

This strategy uses either a narrow or broad genetic base, as desired; it permits the use of the breeder's skill in selecting upon phenotypic merit, and, in the progeny testing year, on genotypic merit; the opportunities for genetic recombination are improved by the early selection and recurrent intercrossing. Populations can be expanded in the recombining generations and contracted as the successive groups of parents are chosen. The system also permits the incorporation of fresh germplasm periodically into the crossing phases. It is important in a system such as this that genotype x environment interactions are minimised by ensuring, as far as possible, a succession of stable environments through fertility, cultural, and moisture controls. If wide adaptation is a breeding goal, then region-wide testing will have to be conducted, either in the phase of early testing of lines as potential parents, or when advanced selections are propagated and tested.

Although heterosis exists in crosses among genetically contrasting parental lines, its exploitation in beans is not feasible.

Creation and exploitation of bulk hybrids in beans is not practised to any extent, although some of the characters of behaviour of bulks in Lima beans (see Chapter 11) were studied by Allard and co-workers [8] several years ago. Plant and seed types that are important to commercial growers and consumers tend not to be the types that have superior fitness in a hybrid bulk population. Populations could be stratified to minimise that constraint, at the risk, however, of narrowing the gene base in other respects. This has not been done in any current programme.

PROSPECTS FOR LARGER AND MORE STABLE YIELDS

There are now developing global bean research and development networks involving international centres, regional centres or institutes, and multi-university programmes, in collaboration with national ministries of agriculture [6,

94]. In addition, in Europe and North America, various commercial seed companies and agro-chemical organisations also engage in varietal development and research in management technologies that result in improved yields or protection from losses due to insects or diseases. The positive achievements of these programmes and activities are beginning to make their mark in individual countries, states, and regions, as improved cultivars are accepted by growers, and as new or more effective environmental and pest management techniques are employed.

Insensitivity or reduced sensitivity to photoperiod can broaden potential adaptability of common beans and improved vigour in stress environments should help to increase or stabilise yields. Changes in plant architecture have contributed to improved yields [5,76] and may also lead to escape from diseases [104]. It is likely that greater progress, at least in the short-term, will be achieved by lessening the difference between farmers' yields (a global average of 514 kg ha^{-1} during the period 1974-76 [99]) and potential yields under good management (commonly regarded to be 4000-5000 kg ha^{-1} [56]), rather than seeking to improve this potential.

Improvements in yield and yield stability, although widely expected to materialise from the many programmes now underway, will nevertheless be measured, not by any large or sudden global increase in production per unit area, but by more gradual and subtle processes, first in one location, then elsewhere, as marginally better cultivars are produced and management techniques are developed. Furthermore, it must be recognised that economic yield, as viewed on a regional or countrywide scale, depends to a large extent upon such factors as rainfall, purchasing capacity of growers for agro-chemicals, production incentives, and upon what may be termed as deteriorating environments for growing beans - examples being increasing air pollution, soil compaction, soil erosion and salt accumulation, and expansion of cultivation to less hospitable soils and into less

469

suitable climatic zones.

In the final analysis, then, regional or national or even global yields will result from the interplay of both promoting and detracting forces, the balance being different from region to region, and possibly even from farm to farm. What is generally clear, however, is that *yield potential* and *stability potential* will continue to improve, albeit by small and irregular increments, and not necessarily simultaneously.

AVENUES OF COMMUNICATION BETWEEN RESEARCHERS

It is imperative to the advancement of the subject that grain legume researchers communicate with each other. This is accomplished in numerous ways, ranging from more formal, refereed publications to informal telephone conversations. To some extent, language differences impede the full and free exchange of information, and political restraints may also impede the movement of personnel and/or germplasm among some of the world's bean scientists. Instances of deliberate withholding of information or materials are rare if, indeed, they occur at all. But there have been instances of confiscation by quarantine agencies of bean seeds and shipments of bean soils between countries.

In the scientific community of bean researchers, the following list illustrates those avenues which provide ample opportunity for scholarly and germplasm exchanges, and sharing of information.

(a) Abstracts, such as *Abstracts on Field Beans*, CIAT Series; *Plant Breeding Abstracts*, Cambridge.
(b) Bean Improvement Cooperative (BIC); biennial meeting, and annual publication of research notes.
(c) Books and conference publications. Examples of the latter are *The Manipulation of Genetic Systems in Plant Breeding*, published in 1981 by The Royal Society, London, and *Genetic Engineering in the Plant Sciences*,

also published in 1981 by Praeger, New York.

(d) Centro Internacional de Agricultura Tropical (CIAT); an international network of exchange of trainees, germ-plasm, and reports - including annual reports and abstracts (see Chapter 18).

(e) Congresses, such as *The Xth Eucarpia Congress*, at Wageningen in the Netherlands in 1983; and the *International Genetics Congress* in Moscow in 1978.

(f) Informal exchanges of annual reports, pre-publication manuscripts, seed requests, and personal contacts.

(g) Newsletters, such as *Phaseolus - Beans Newsletter for Eastern Africa*, Nairobi, Kenya and *Pulse Beat*, the newsletter from the Bean/Cowpea CRSP Management Office, East Lansing, Michigan, USA.

(h) Professional society annual meetings, such as those of the PCCMCA (Programa Cooperative Centroamericano para el Frejoramiento de Cultivos Alimenticios), the American Society of Agronomy and the American Society for Horticultural Science.

(i) Special grain legume workers conference and symposia (see the literature cited below).

(j) Standard professional journals.

(k) Title XII Bean/Cowpea Collaborative Research Support Program (CRSP). An international network involving exchanges of students, investigations, reviewers, materials and reports (see Chapter 18).

(l) Workshops, such as *Potential for Field Beans in Eastern Africa*, Proceedings of a Regional Workshop held in Lilongwe, Malawi, 1980; *Bean Rust Workshop*, held at the University of Puerto Rico, Mayaguez, 1983, and CIAT-sponsored workshops on various subjects held periodically at Cali, Colombia.

Clearly, there are many avenues for potentially fruitful communications between bean researchers, although those working in the developing regions may not always have sufficient funds from national resources to ensure that they keep abreast of latest developments.

LITERATURE CITED

1. Adams, M.W. (1967), *Crop Science* 7, 505-10.
2. Adams, M.W. (1973), in *Potentials of Field Beans and Other Food Legumes in Latin America*. CIAT Seminar Series 2E. Colombia: CIAT, 266-78.
3. Adams, M.W. (1973), in Milner, M. (*ed.*) *Nutritional Improvement of Food Legumes by Breeding*. New York: PAG, 143-50.
4. Adams, M.W. (1978), Unpublished data.
5. Adams, M.W. (1981), *Iowa State Journal of Research* 56, 225-54.
6. Adams, M.W. (1984), 'Beans-Cowpeas: Production Constraints and National Programs'. *Bean/Cowpea Collaborative Research Support Program*, Michigan State University, East Lansing, Michigan.
7. Aguirre, J.A. and Miranda, M.H. (1973), in *Potentials of Field Beans and Other Food Legumes in Latin America*. CIAT Seminar Series 2E. Colombia: CIAT, 161-87.
8. Allard, R.W., Jain, S.K. and Workman, P.L. (1969), *Advances in Genetics* 14, 55-131.
9. Andrews, D.J. and Kassam, A.H. (1976), in *Multiple Cropping*, ASA Special Publication No. 27. Madison, Wisconsin: American Society of Agronomy, 1-10.
10. Ballantyne, Barbara, (1970), *Plant Disease Reporter* 54, 903-5.
11. Bastidas, G. (1977), in *I Curso Intensivo de Adiestramiento en Produccion de Frijol*. Colombia: CIAT.
12. Bliss, F.A. (1980), in Fehr, W.R. and Hadley, H.H. (*eds*) *Hybridization of Crop Plants*. Madison, Wisconsin: American Society of Agronomy-Crop Science Society of America, 273-84.
13. Bliss, F.A. and Hall, T.C. (1977), *Cereal Foods World* 22, 106-13.
14. Bliss, F.A. and Brown, John W.S. (1983), in Janick, J. (*ed.*) *Plant Breeding Reviews*. Westport, Connecticut: AVI Publishing, 59-102.
15. Bolkan, H.A. (1980), in Schwartz, H.H. and Galvez, G.E. (*eds*) *Bean Production Problems*. Colombia: CIAT, 65-100.
16. Burga-Mendoza, C. (1978), PhD. thesis, Michigan State University, East Lansing, Michigan.
17. Cafati, C. and Alvarez, M. (1975), *Agricultura Téchnica* (Chile) 35, 152-7.
18. Carpenter, K.J. and Wolfe, J.A. (1973), in Milner, M. (*ed.*) *Nutritional Improvement of Food Legumes by Breeding*. New York: PAG, 313-18.
19. Chaves, G. (1980), in Schwartz, H.F. and Galvez, G.E. (*eds*) *Bean Production Problems*. Colombia: CIAT 37-54.
20. CIAT (1973), *Potentials of Field Beans and other Food Legumes in Latin America*. Series Seminars No. 2E. Colombia: CIAT.
21. CIAT (1976), *Annual Report 1976*. Cali, Colombia.
22. CIAT (1977), *Annual Report 1977*. Cali, Colombia.
23. CIAT (1979), *Bean Program 1979 Annual Report*. Cali, Colombia.
24. CIAT (1980), *Bean Program 1980 Annual Report*. Cali, Colombia.
25. CIAT (1981), *Bean Program 1981 Annual Report*. Cali, Colombia.
26. CIAT (1981), *Potential for Field Beans in Eastern Africa*. Proceedings of Regional Workshop, Lilongwe, Malawi, CIAT Series No. 03EB-1.
27. CIAT (1982), *Bean Program 1982 Annual Report*. Cali, Colombia.
28. Coyne, D.P. (1966), *Proceedings of the American Society for*

472

Common Bean (*Phaseolus vulgaris* L.)

Horticultural Science 89, 350–66.
29. Coyne, D.P. (1967), *Journal of Heredity* 58, 313–14.
30. Coyne, D.P. (1982), *Journal of Plant Nutrition* 5, 575–85.
31. Coyne, D.P. and Schuster, M.L. (1975), *Euphytica* 24, 795–803.
32. Coyne, D.P. and Schuster, M.L. (1979), in Summerfield, R.J. and Bunting, A.H. (*eds*) *Advances in Legume Science*. London: Royal Botanic Gardens, Kew, 225–32.
33. Date, R.A. (1969), *Journal of the Australian Institute of Science* 35, 27–37.
34. Davis, J.H.C. and Garcia, S. (1983), *Field Crops Research* 6, 59–75.
35. Donoso Puga, A. (1980), *Bulletin*, Ministerio de Agricultura y Ganaderia, Ecuador.
36. Drijfhout, E. (1978), *Agricultural Research Report (versl. Land-bouwk. Onderz.)* 872.
37. Drijfhout, E. (1982), *Bean research at IVT for CIAT, Report 1982* (mimeo). Colombia: CIAT.
38. Duarte, R. and Adams, M.W. (1963), *Crop Science* 3, 185–6.
39. Duque, F.F., Salles, L.T.G., Pereira, J.C. and Dobereiner, J. (1982), in Graham, P.H. and Harris, S.C. (*eds*) *Biological Nitrogen Fixation Technology for Tropical Agriculture*. Colombia: CIAT, 63–6.
40. Enriquez, G.A. and Wallace, D.H. (1980), *Journal of the American Society Horticultural Science* 105, 583–91.
41. Evans, R.G. and Bandemer, S. (1967), Personal Communication.
42. Evans, A.M. (1974), in *Potentials of Field Beans and other Food Legumes in Latin America*. CIAT, Seminar Series 2E. Colombia: CIAT, 279–86.
43. Fassuliotis, G., Deakin, J.R. and Hoffman, J.C. (1970), *Journal of the American Society for Horticultural Science* 95, 640–45.
44. Food and Agriculture Organisation (1975), *Production Yearbook*. Rome: FAO.
45. Francis, C.A. (1978), *HortScience* 13, 12–17.
46. Francis, C.A., Flor, C.A. and Prager, M. (1977), *Fitotecnia Latinoamericana*.
47. Francis, C.A. and Sanders, J.H. (1978), *Field Crops Research* 1, 319–35.
48. Freytag, G. (1973), in *Potentials of Field Beans and Other Food Legumes in Latin America*. CIAT Seminar Series 2E. Colombia: CIAT, 199–217.
49. Freytag, G. (1975), 'Breeding beans for the tropics'. *Programa Cooperativo Centroamericano para el Mejoramiento de Cultivos Alimenticos* (XXI Annual Meeting), 97–104.
50. Fuller, P. (1983), PhD. thesis, University of Nebraska, Lincoln.
51. Gabelman, W.H. and Gerloff, G.C. (1978), *HortScience* 13, 682–4.
52. Galvez, G.E., Guzman, P. and Castono, M. (1980), in Schwartz, H.F. and Galvez, G.E. (*eds*) *Bean Production Problems*. Colombia: CIAT, 101–10.
53. Games, R. (1977), *Fitopatologia* 12, 24–7.
54. Gniffke, P.A. (1981), MSc. thesis, Cornell University, Ithaca, New York.
55. Gniffke, P. (1982), 'Daylength-temperature adaptation study in bean flowering, preliminary report'. Colombia: CIAT.
56. Graham, P.H. (1978), *Field Crops Research* 1, 295–317.
57. Graham, P.H. (1979), *Journal of Agricultural Science, Cambridge* 93, 365–70.
58. Graham, P.H. (1981), *Field Crops Research* 4, 93–112.

59. Graham, P.H. (1982), *Phaseolus vulgaris: proposed studies*. Instituto de Investigacion Nutricional, Lima, Peru and CIAT, Colombia.
60. Graham, P.H. and Rosas, J.C. (1977), *Journal of Agricultural Science, Cambridge* 88, 503-8.
61. Graham, P.H., Apolitano, C., Ferrera-Cerrato, R., Halliday, J., Lepiz, R., Menendez, O., Rios, R., Saito, S.M.T. and Viteri, S.E. (1982), in Graham, P.H. and Harris, S.C. (*eds*) *Biological Nitrogen Fixation Technology for Tropical Agriculture*. Colombia: CIAT, 223-9.
62. Graham, P.H., Viteri, S.E., Mackie, F., Vargas, A.T. and Palacios, A. (1982), *Field Crops Research* 5, 121-8.
63. Greathead, D.J. (1969), *Bulletin of Entomology Research* 59, 541-61.
64. Gutierrez, U., Infante, M. and Pinchinat, A. (1975), Centro Internacional de Agricultura Tropical, Cali, Colombia, Boletin Informe ES-19.
65. Hamblin, J. (1973), *Bean Improvement Cooperative* 16, 27-9.
66. Hernandez-Bravo, G. (1973), in *Potentials of Field Beans and Other Food Legumes in Latin America*. CIAT Seminar Series 2E. Colombia: CIAT, 144-50.
67. Hershey, C., Miles, J. and Davis, J.H.C. (1982), Discussion Paper, CIAT Annual Review, December 1982.
68. Honma, S. (1955), *Proceedings of the American Society for Horticultural Science* 65, 405-8.
69. Hutchinson, J.B. (1970), *Proceedings of the Nutrition Society* 29, 49-55.
70. ICA (1982), *ICA Plegable de Divulgacion* No. 166, Ministerio de Agricultura, Colombia.
71. IITA (1982), *Research Highlights*. Ibadan, Nigeria: International Institute of Tropical Agriculture, 43-5.
72. Instituto Interamericano de Ciencias Agricolas (1969), 'Reunion tecnica sobre programacion de la investigacion y extension de frijol y tres leguminosas de grano para America Central'. IICA, Turrialba, Costa Rica, Publ. ZN 11-65, Vol. 2.
73. Jensen, N.F. (1970), *Crop Science* 10, 629-35.
74. Kaplan, L. (1981), *Economic Botany* 35, 240-54.
75. Knott, D.R. and Talukdar, B. (1971), *Crop Science* 11, 280-83.
76. Konzak, C.F., Kleinhofs, A. and Ullrich, S.E. (1985), in Janick, J. (*ed.*) *Plant Breeding Reviews*. Westport, Connecticut: Avi Publishing. (In press)
77. Kooiman, H.N. (1931), *Bibliographia Genetica* 8, 259-409.
78. Kretchmer, P.J., Ozbun, J.L., Kaplan, S.L., Laing, D.R. and Wallace, D.H. (1977), *Crop Science* 17, 797-9.
79. Kretchmer, P.J., Laing, D.R. and Wallace, D.H. (1979), *Crop Science* 19, 605-7.
80. Laing, D.R. (1978), Workshop on International Bean Yield and Adaptation Nurseries. CIAT, January 1978.
81. Laing, D.R., Jones, P.G. and Davis, J.H.C. (1983), in Goldsworthy, P.R. and Fischer, N.M. (*eds*) *The Physiology of Tropical Field Crops*. Chichester: Wiley, 305-52.
82. Lamprecht, H. (1935), *Hereditas* 20, 71-93.
83. Lepiz, R. (1972), *Agricola Tecnico Mexico* 3, 90-101.
84. Lepiz, R. (1974), Instituto Nacional de investigaciones Agropecuarios, Secretaria de Agricultura y Ganaderia, Mexico. Bolleto Tecnico No. 58.

Common Bean (*Phaseolus vulgaris* L.)

85. Leyna, H.K., Korban, S.S. and Coyne, D.P. (1982), *The Journal of Heredity* 73, 306-8.
86. Miranda, C.S. (1967), Colegio de Postgraduados, Chapingo, Mexico. Serie de Investigaciones No. 9.
87. Miranda, C.S. (1979), in Engleman, M. (*ed.*) *Contribuciones al Conocimiento del Frijol* (Phaseolus) *en Mexico*. Colegio de Post-graduados, Chapingo, Mexico.
88. Mok, D.W.S. and Mok, M.C. (1976), *The Journal of Heredity* 67, 187-8.
89. McFerson, J., Bliss, F.A. and Rosas, J.C. (1982), in Graham, P.H. and Harris, S.C. (*eds*) *Biological Nitrogen Fixation Technology for Tropical Agriculture*. Colombia: CIAT, 39-44.
90. Munevar, F. and Wollum, A.G. (1982), in Graham, P.H. and Harris, S.C. (*eds*) *Biological Nitrogen Fixation Technology for Tropical Agriculture*, Colombia: CIAT, 173-82.
91. Prakken, R. (1970), *Mededelingen Landbouwhogeschool Wageningen* 70-23, 1-38.
92. Rennie, R.J. and Kemp, G.A. (1982), *Canadian Journal of Botany* 60, 1423-7.
93. Rennie, R.J. and Kemp, G.A. (1983), *Agronomy Journal* 75, 645-9.
94. Roberts, L.M. (1970), 'The food legumes: recommendations for expansion and acceleration of research'. Rockefeller Foundation, New York. (Mimeo Report).
95. Roberts, Mary-Howell E. (1982), *Annual Report of the Bean Improve-ment Cooperative* 25, 109-27.
96. Robertson, L.S. and Frazier, R.D. (1978), *Dry Bean Production - Principles and Practices*. Extension Bulletin E-1251, Cooperative Extension Service, Michigan State University, East Lansing, Michigan.
97. Roman, A., Davis, J.H.C., Garcia, S., Graham, P.H., and Temple, S.R. (1979), Informe de Trabajos, Convenio ICA-CIAT, Programa de Frijol. CIAT, Colombia.
98. Ruschel, A.P., Vose, P.B., Matsui, E., Victoria, R.L. and Saito, S.M.T. (1982), *Plant and Soil* 65, 397-407.
99. Sanders, J.H. and Alvarez, C. (1978), CIAT Technical Series.
100. Santa Cecilia, F.C. and Vieira, C. (1978), *Turrialba* 28, 19-23.
101. Scarisbrick, D.H. and Wilkes, J.M. (1973), *Nature* 242, 619-20.
102. Schuster, M.L. (1955), *Phytopathology* 45, 519-20.
103. Schuster, M.L. and Coyne, D.P. (1981), in Janick, J. (*ed.*) *Horticultural Reviews*. Westport, Connecticut: Avi Publishing, 28-58.
104. Schwartz, H.F. and Steadman, J.R. (1978), *Phytopathology* 68, 383-8.
105. Schwartz, H.F. and Galvez, G.E. (*eds*) (1980), *Bean Production Problems*. Colombia: CIAT.
106. Schwartz, H.F., Pastor Corrales, M.A. and Singh, S.P. (1982), *Euphytica* 31, 1-14.
107. Smartt, J. (1969), in Ucko, P.J. and Dimbleby, G.W. (*eds*) *The Domestication and Exploitation of Plants and Animals*. London: Duckworth, 451-62.
108. Steadman, J.R. (1983), *Plant Disease* 67, 346-50.
109. Thomas, C.V. and Waines, J.G. (1985), *Crop Science* (in press).
110. Vanderborght, T. (1982), *CIAT-Gembloux Project, Report, July 1982*, Colombia: CIAT.
111. Van Schoonhoven, A. and Cardona, C. (1980), in Schwartz, H.F.

and Galvez, G.E. (*eds*) *Bean Production Problems*. Colombia: CIAT, 363-412.

112. Varon de Agudelo, F. (1980), in Schwartz, H.F. and Galvez, G.E. (*eds*) *Bean Production Problems*. Colombia: CIAT, 915-26.

113. Vavilov, N.I. (1951), *The Origin, Variation, Immunity and Breeding of Cultivated Plants*. New York: Ronald Press.

114. Vincent, J.M. (1974), in Quispel, A. (*ed.*) *The Biology of Nitrogen Fixation*. Amsterdam: North Holland Publishing, 266-341.

115. Voysest-Voysest, O. (1970), PhD. thesis, Michigan State University, East Lansing, Michigan.

116. Voysest, O.V. (1982), CIAT Seminar Series SE-7-82, Cali, Colombia.

117. Walker, W.M. and Hymowitz, T. (1972), *Communications in Soil Science and Plant Analysis* 3, 505-11.

118. Wallace, D.H. (1980), in Summerfield, R.J. and Bunting, A.H. (*eds*) *Advances in Legume Science*. London: HMSO, 349-58.

119. Wallace, D.H. (1980), in *Proceedings of the 14th International Congress of Genetics* 1(2), Moscow: MIR Publishers, 306-17.

120. Wallace, D.H. (1985), in Janick, J. (*ed.*) *Plant Breeding Reviews*. Westport, Connecticut: Avi Publishing (in press).

121. Webster, D.M., Temple, S.R. and Galvez, G. (1983), *Plant Disease* 67, 394-6.

122. Westermann, D.T. and Kolar, J.J. (1978), *Crop Science* 18, 986-90.

123. Westermann, D.T., Kleinkopf, G.E., Porter, L.K. and Leggett, G.E. (1981), *Agronomy Journal* 73, 660-64.

124. Yarnell, S.H. (1965), *The Botanical Review* 31, 247-330.

125. Zaumeyer, W.J. and Thomas, H.R. (1957), USDA Technical Bulletin 868.

126. Hadley, P., Summerfield, R.J. and Roberts, E.H. (1983) in Jones, D.G. and Davies, D.R. (*eds*) *Temperate Legumes: Physiology, Genetics and Nodulation*. London: Pitman, 19-42.

127. Summerfield, R.J. and Roberts, E.H. (1985) in Halvey, A.H. (*ed.*) *A Handbook of Flowering*. Florida: CRC Press. (In press.)

11 Lima bean (*Phaseolus lunatus* L.)

J.M. Lyman, J.P. Baudoin and R. Hidalgo

TRENDS IN PRODUCTION

The Lima bean *(Phaseolus lunatus* L.) is an underdeveloped crop with potential for good yields in the tropics. Current production, although undoubtedly underestimated, represents less than 1% of total dry pulse output worldwide. Although the species originated in the humid tropics [54,169], it is cultivated intensively only in temperate regions in the USA and in arid regions in Malagasay and Peru. Production statistics from other regions are fragmentary and often aggregated with those of other pulses. Harvests from traditional mixed cropping systems and garden cultivation are casual and seldom recorded officially.

Intensive farming systems

The USA is the world's largest producer of Lima beans, primarily fresh beans for processing and domestic consumption. The eastern coastal states dominated the area harvested and total production until the late 1940s, when California became the major producer. In 1980, the four major production regions were in California, Delaware, Maryland and Wisconsin. Area and production have both declined steadily in recent years, from 42400 ha harvested producing 104 600 t green beans in 1968 to 20600 ha harvested and 55600 t in 1980. This trend was related to limited consumer demand and

unfavourable economic factors. Yields of green beans remained stable at about 2.7 t ha^{-1} during this period [112]
(dry seed weights can be estimated roughly as 40-50% of
fresh weights). Production of large-seeded cultivars exceeded that of small-seeded ones until 1975, when production
of both types became more or less equal [55]. Despite decreases in total production, Lima bean exports (as dry
seeds) from the USA increased from 5800 t per annum during
the period 1965-69, to 9800 t per annum from 1970 to 1974 [55

Malagasy is the second largest commercial producer of
Lima beans; almost exclusively large, white-seeded types are
produced for export. Lima beans are grown in the south-west
part of the island and yields range from 500-600 kg ha^{-1} in
mixed cropping systems to 1000-2000 kg ha^{-1} in sole crops.
Growing seasons extend from 200-270 days [18]. Production
of dry seed increased from 17200 t per annum in 1965-69 to
22250 t per annum in 1970-74 [55]. However, area harvested
and total production declined from 19100 ha and 19375 t in
1975 to only 6790 ha and 8050 t in 1978 [32]. Exports averaged 15624 t per annum during 1965-69 and 19910 t per annum
during 1970-74. The UK and Japan are the largest importers
of dry Lima beans [55].

In Peru, Lima beans are grown under extremely arid conditions, mostly in the Ica and Pisco valleys, in the coastal
region south of Lima. Large-seeded vining types are harveste
after 270-300 days. Small-seeded native types and introduced
varieties are also grown. Dry seed production (1973-79) aver
aged 5295 t from 5652 ha (a mean yield of 936 kg ha^{-1}) compared with 5713 t, 1567 ha and 3638 kg ha^{-1}, respectively,
for the green (fresh vegetable) seed crop during the same
period [64 and *Anuario Estadistico del Ministerio de Agricultura de Peru*, 1973-79].

Traditional farming systems

Lima beans are cultivated as a pulse crop in traditional
farming systems, particularly in arid regions. In the Yucatan Peninsula of Mexico, for example, approximately 20000 ha

of small-seeded landraces are cultivated in the traditional
Mayan system of intercropped maize and Lima beans [61].
Dry seed yields average 200-400 kg ha^{-1} in sub-humid to humid conditions and growing seasons are 180-270 days. In
Ecuador, small-seeded Lima beans are grown for sale in local
markets, either as a vegetable or pulse [66]. Local vining
types are tended in household gardens throughout the Philippines. Seeds are used principally as a fresh vegetable
and are dried only for planting. In India, vining types are
hand-harvested frequently [39] and yield 5-8 t green fruits
ha^{-1} (200-300 kg dry seed).

The area devoted to Lima beans in the sub-humid and
humid tropics of Africa is estimated to be between 120 000
and 200 000 ha, with total production varying from 50000 to
100 000 t [87]. Local types are predominantly small-seeded
vining forms which yield 200-1500 kg dry seed ha^{-1}. In
Nigeria, Lima beans are cropped in compounds (areas about
500 m^2) in pure or mixed stands and produce yields of
450-600 kg dry seed ha^{-1} in 120-180 days.

Current Lima bean production in Africa is sufficient
only to satisfy local consumption. Local types are adapted
for mixed cropping systems in gardens. Improved production
would require more intensive management practices and
shorter-season varieties resistant to disease infestations
and insect depredations.

CROP USES

Lima beans are cultivated primarily for immature or dry
seeds and, in some regions, for immature sprouts, leaves and
fruits. In the USA, green Lima beans are canned or frozen
as a vegetable. In the tropics, the mature dry beans are a
valuable pulse that enriches diets composed largely of
cereals or starchy vegetables. The beans are added to soups
or stews in South America, and boiled, fried or mixed with
other vegetables in Africa [55,122]. Leaves, fruits and
seeds are eaten as vegetables in Asia and India.

Proper preparation of primitive types requires

changing and discarding the cooking water because raw seeds of some wild types may taste bitter if they contain potentially toxic concentrations of a cyanogenic glucoside. However, this is not a problem in improved cultivars (see below).

Dry seeds produce a protein-rich flour which is added to bread or noodles in the Philippines and used for bean paste in Japan [15,55]. The flour contains 345 calories, 10.5% moisture, 21.5 g protein, 1.4 g oil, 63.0 g total carbohydrates, 20 g fibre and 3.6 g ash per 100 g flour [39].

Seeds and leaves are valued for their astringent qualities in traditional Asian medicine. A diet of Lima bean seeds is prescribed for fevers; a poultice of the seeds for stomach-ache; and a decoction of green fruits, leaves and stems for Bright's disease, diabetes, dropsy and eclampsia [39].

Lima bean vines are sometimes used as cattle fodder after the fruits have been harvested but green vines need to be ensiled to avoid toxicity. Lima bean silage contains 27.3% dry matter, 3.3% protein, 2.1% digestible protein, 14.2% total digestible nutrients, and has a nutritive ratio of 5:8 [55]. In Malagasy, yields of 15 t green matter ha^{-1} have been obtained for use as hay or fodder for cattle during the dry season, [18]. Similar practices are reported from Malaysia and Indonesia.

In Malaysia, Lima beans are also grown as a short-duration cover or green manure crop. Dwarf cultivars perform best as cover crops since vining forms tend to smother the main crop. Nevertheless, vining forms are grown successfully as cover crops in citrus plantations in Mexico. They are preferred to other legumes because of their drought tolerance and long duration of growth [61].

FARMING SYSTEMS

Farming systems which include Lima beans vary from the

intensive monocrop systems of the temperate zones to the
mixed-cropping systems of humid tropical regions. Determinate, compact, 'bush' varieties are better suited to intensive cropping systems, while vining types predominate in
the traditional farming systems of the tropics. Optimum
management practices and varieties have been developed for
temperate conditions, but not for humid tropical regions.

USA

Lima beans are grown as monocrops for commercial processing and, to a limited extent, as a garden vegetable. In
northern regions with short growing seasons, small-seeded,
early-maturing 'bush' varieties are prevalent. Green fruits
and seeds are harvested for the fresh vegetable market or
for the processing industry. In southern and western regions,
both 'bush' and climbing forms are cultivated for their
green or mature seeds. In California, successive Lima bean
crops are grown either without rotation, or are rotated
with alfalfa, tomatoes, sugar beets, melons or winter
vegetables [1]. Lima beans have been cropped continuously
for many years on fertile soils in California without a
decline in yields.

Areas between 40 ha and 400 ha are seeded mechanically.
Sowing rates are 90-170 kg ha^{-1} for large-seeded cultivars
and 45-80 kg ha^{-1} for small-seeded ones. Seeds are planted
2.5-10 cm deep, 5-30 cm apart within-rows and 60-100 cm
between-rows. Vining types grown for their green fruits are
planted in hills 90-120 cm apart, with three or four seeds
per hill [55]. Indeterminate forms are staked in humid areas
or left prostrate under hot, dry conditions in California.

The choice of varieties and appropriate crop management
practices depends upon local conditions. In California, the
large-seeded, indeterminate varieties grown in the southern
coastal valleys are usually irrigated once or twice after
planting. Small-seeded, heat-tolerant 'Sieva' types are
grown on considerable areas in the warmer interior valleys,
where large transpiration losses necessitate at least one

irrigation before sowing and as many as four irrigations thereafter. Efficient weed, pest and disease control measures are practised during the growing season [1].

Green and mature fruits are combine-harvested. Mature fruits are left to dry on the vines in windrows before threshing. Commercial producers of green seeds for processing are changing to mobile fruit-stripping and shelling machines which reduce time and labour requirements. The new machines permit a reduction in row width from 100 cm to 50 cm, or even 33 cm. Increased irrigation will be required in these systems and the incidence of diseases may become more severe [102].

Malagasy (Madagascar)

Lima beans are grown under semi-arid conditions (Plate 11.1); maximum temperatures vary from 27°-35°C and relative humidity is usually very low (although cool nights and proximity to the Indian Ocean can lead to more humid conditions during some periods of the year). Following a brief rainy season, seeds are sown as a monocrop or are interplanted with cassava, groundnuts, maize or cotton. They are planted on alluvial soils in mounds or ridges as the flood waters from a nearby river recede, or are irrigated via canals from the river. Plants are not staked. Weeding is necessary during the initial weeks of crop life due to the wide spacing used and the slow growth rate of the crop. At harvest, whole plants are cut and left to dry in the field. After removing the fruits, the vines are fed to domestic livestock [18].

Peru

Lima bean crops are cultivated in coastal valleys where annual rainfall averages only 2.2 mm. Temperatures average 13°-19°C due to the cooling effects of cloudy skies and ocean currents, and dense fogs provide additional moisture

Lima Bean (*Phaseolus lunatus* L.)

Plate 11.1 'Intensive' production of Lima beans in Malagasy
(Photograph by J.P. Baudoin)

for growth. Large-seeded, 'Big Lima' types are planted far
apart (0.6-2 m between-rows, 0.3-2 m within-rows) and are
not staked. Plants are irrigated with approximately 3500
m^3 water ha^{-1} during the initial period of vegetative
growth. Lima beans are most often grown as monocrops but
occasionally they are intercropped with maize or cotton
[64].

Farming systems in the humid and subhumid tropics

The area devoted to Lima beans in the tropics is small
compared with that planted to starchy staple and export
crops. Indeterminate 'Potato' and 'Sieva' types are planted
in traditional farming systems ranging from shifting to
permanent cultivation. In drier savanna areas, Lima beans
are frequently cultivated in small pure stands, or intercropped with cereals. In the rainforest regions, cereals,
roots and tubers, grain legumes, vegetables and condiments

483

are grown in mixed or relay intercropping systems, often beneath the canopy of tree crops. This scheme reduces risk of crop failure and provides food throughout the year. Nevertheless, crop management is rudimentary. When the rains start, three or four seeds per hill are planted close to maize, shrubs, hedges or fences. Plants may even climb on the same stakes as used for the yam crop. Lima beans may also be planted in homestead gardens where the soil is enriched by domestic refuse and animal manure. Weeds are removed at the beginning of the growing season. Harvest begins 120-150 days after planting, and continues for several weeks.

In the Yucatan Peninsula of Mexico, Lima beans have always been associated with maize in the ancient system of shifting cultivation known colloquially as 'milpa'. On the Jos Plateau of Nigeria, plants are grown in pure stands or intercropped with maize or sorghum and supported by robust trellises [18].

PRINCIPAL FEATURES OF THE SPECIES

Taxonomy

The monospecific status of the Lima bean has been recognise(
for more than fifty years [83], despite the integration of wild and cultivated forms. Baudet [17] proposed the follow-ing divisions within the species, based on evolutionary trends described by Mackie [69]:

P. *lunatus* var. *silvester*: the wild forms; and
P. *lunatus* var. *lunatus*: with three 'cultigroups', i.e.
    Sieva - medium-sized, flat seeds,
    Potato - small, globular seeds, and
    Big Lima - large, flat seeds.

The Lima bean is more closely related to the wild species P. *metcalfei*, P. *ritensis*, P. *polystachyus*,

Lima Bean (*Phaseolus lunatus* L.)

P. pachyrrhizoides and *P. tuerckheimii*, than it is to the
cultivated *P. vulgaris* - *P. coccineus* complex [72].

Cultivars

Many cultivars have been developed in the USA. Commercial
'bush' cultivars are the small-seeded Bridgeton, Kingston,
Jackson Wonder, Dixie Butter Pea and related selections,
and the large-seeded Fordhook 242 and 1042. Commercial vining cultivars are the large-seeded Ventura and small-seeded
Wilbur and Westan [85,102]. King of the Garden, Carolina
and Florida Speckled Butter are vining cultivars suitable
for production of green fruits and seeds in the tropics
[39,55]. Tropical landraces are predominantly small-seeded,
vining types.

Botany

Phaseolus lunatus, an herbaceous species, includes annual,
determinate 'bush' types (to 0.6 m tall) and indeterminate
climbers (2-4 m tall) with robust tap roots and which are
occasionally perennial. Germination is epigeal. Leaves are
trifoliolate; leaflets are ovate and acuminate; stipellae
are glandular. Inflorescences comprise long, axillary racemes with many nodes and flowers. Calyx-bracts are glandular and about one half as long as the calyx. Standard petals are usually pubescent and coloured pale-green or violet, whereas the wing petals are white or violet. The keel
terminates in a spiral and is usually white or occasionally
pigmented. There are ten stamens in a diadelphous arrangement. The style is coiled; its apical region is pubescent
on the inner face. The stigma is introrse, slightly sub-
terminal and adaxially situated. Pollen grains are tricol-
porated with distinct pseudocolpus and exine devoid of
structure and reticule. Fruits are oblong and often curved,
terminating in a sharp beak orientated towards the dorsal
suture; they contain between two and four seeds and

eventually dehisce. Seeds vary in size from about 30-300 g
100 seeds^{-1} and are kidney-shaped, rhomboid or round. Testa
patterns may be solid, speckled or mottled and colours vary
from white, green and yellow to brown, red, purple or black.
The cotyledons are white or green [70]. Transverse lines
often radiate from the hilum (Plate 11.2).

Wild forms are always indeterminate and perennial; they
are readily distinguished from cultivated forms by their
smaller leaves, fruits and seeds. Seeds are often grey with
black speckles; they generally contain large concentrations
of the cyanogenic glucoside linamarin. Flowers are always
pigmented. Fruits shatter at maturity. Natural crossing
between wild and cultivated forms and between wild forms
and regressive mutations have given rise to weedy forms
which persist in many tropical environments.

Ecology

Lima beans occur from the lowland humid tropics to the
subtropical steppes and in warm temperate regions of higher
latitudes. They are found from sea level to 2000 m altitude
and thrive in regions with monthly average temperatures of
16°-27°C. Cool temperatures (below 13°C) slow growth and hot
temperatures (days and nights warmer than 32° and 21°C,
respectively) are detrimental to fruit set and seed yield.
The crop requires a frost-free growing season and dry conditions during fruit maturation but tolerates wetter weather better than common beans (*P. vulgaris*) during active
growth. Growth duration varies from 75-90 days for early
'bush' cultivars to 90-110 days for small-seeded indeterminate cultivars, and 105-130 days for large-seeded indeterminate ones.

The crop tolerates annual rainfall in excess of 1500
mm to as little as 500-600 mm, provided soil moisture is
adequate and air is humid during flowering and fruit set.
Once established, it is reasonably tolerant of drought.
Well-drained soils of pH 6.0-6.8 are optimal but some

Plate 11.2 Seeds of Potato, Big Lima, Sieva and 'unclassified' Lima bean types (clockwise from top left) showing variations in size, shape, colour and pattern (GRU, CIAT, Colombia; A. Cuellar)

cultivars are able to tolerate acid soils (see below).

Germination and seedling establishment

Lima beans require soil temperatures of 18°-21°C for optimum
rate of germination. They may be injured if exposed to
temperatures below 15°C during imbibition [84], although
some cultivars germinate satisfactorily at 15°C. No relation has been established between seed size and cold temperature tolerance, although large-seeded types are said
to be more sensitive to cold [37].

Poor germination may be caused by impermeable seed coats
or damaged seeds [98]. White seeds are more easily damaged
mechanically due to their thinner testas which contain relatively little lignin [53]. Post-harvest seed handling techniques can affect seed viability [119].

Flower initiation

Lima bean genotypes have different critical daylengths for
flowering [46,47]. Day-neutral genotypes from both temperate
and tropical regions flower in more or less the same time
in daylengths of 9-18 h. Qualitative short-day types respond
sharply to photoperiod and their critical daylength for
flower initiation lies between 11 h and 12.5 h. Quantitative
short-day plants respond gradually and flower at progressively higher nodes as days become longer [46]. Photoperiod-
temperature interactions, important in many legumes, have
not been investigated extensively in Lima beans, but photo-
thermal effects on flowering are almost certain to be
common [127].

Flower development and pollination

Buds open early in the day. Once open, flowers do not close
and the corolla is shed after two or three days. Pollen and
stigma mature synchronously and in close proximity within

the unopened bud, thus favouring self-pollination. The keel
of small-seeded varieties is more tightly wrapped around
the style than in large-seeded ones, bringing pollen into
closer contact with the stigma. Pollen of small-seeded
types germinates on the same day of anthesis while that of
large-seeded types germinates one or two days later [13].
These phenomena partly account for wider adaptation of
small-seeded forms and the restricted distribution of 'Big
Lima' cultivars to humid coastal regions.

Lima beans are said to be primarily self-pollinated.
However, pressure on the wings of fully-open flowers (as
when insects alight) forces the stigma and style to protrude
through the keel. The exposed stigma remains receptive to
pollen for several hours, in contrast to the common bean,
where the stigma dries soon after anthesis [115]. Thus,
frequency of cross-pollination ranges from almost absent to
almost complete depending upon variety, plant spacing,
direction of prevailing wind and local insect populations
[3,31]. Bees visit the flowers for both pollen and nectar,
secreted by discoid nectaries at the base of each ovary.
The mechanical extrusion of the stigma can be exploited as
a rapid crossing method without emasculation [118].

Anthesis and fruit set

The most favourable conditions for fruit set and development
are humid air, cool nights and adequate soil moisture [43].
Humid conditions promote good pollen germination; adequate
soil water is necessary to maintain sufficient moisture on
the stigmatic surface. Lima bean pollen grains germinate
over a wide range of temperatures (16°-31°C), but pollen
tube growth and fertilisation have more precise requirements. At least one ovule in each pistil must be fertilised
and begin development in order to prevent abscission of the
flower [59]. Numerous (76-83%) flowers and/or immature
fruits abort under field conditions [88]. Early blooming
racemes are more productive than later ones, and basal

nodes of the racemes are potentially more fruitful than
terminal ones. Fruit setting proceeds until a 'capacity
set' is attained; remaining reproductive structures then
abscise. Leaf area duration and irradiance have important
effects after this initial stage [34,59].

Components of seed yield

Studies of physiological and morphological determinants of
seed yield have been conducted with only a few Lima bean
genotypes and without regard to the farming systems in
which the crop might be used. However, from experiments
under temperate [34] and tropical conditions [18,66], it
is clear that the number of racemes (reproductive sites)
and fruits per plant, plant size or vigour (i.e. node number and leaf area), and crop longevity are positively correlated with seed yield (Plate 11.3). Specific traits which
contribute to better yielding ability are small, thin
leaves with good phototropic orientations, open plant canopies and large root:shoot ratios. These attributes, combined with greater chlorophyll and carbohydrate concentrations, greater osmotic pressure of the leaf sap, and more
open stomata account for the better capacity set of small-
seeded over large-seeded types [13,14]. Light penetration
into the canopy is considerably improved with lanceolate
leaves, but yields are small despite a greater leaf area
index [124]. Variations in numbers of seeds per fruit are
small. Under subhumid conditions in Nigeria, seed and fruit
size were negatively correlated with total dry seed production, especially in 'Sieva' and 'Big Lima' cultivars [18].

PLANT NUTRITION

Fertiliser rates

Slightly acid, well-drained, loams are most suitable for
Lima bean cultivation, but commercial production on soils

Lima Bean (*Phaseolus lunatus* L.)

Plate 11.3 Many racemes per plant (a) and fruits per raceme (b)
contribute to improved yields of Lima beans
(Photographs J.M. Lyman.)

of pH 5.0-5.6 is common in the eastern USA [36]. Indeed,
Lima beans grow on soils ranging in pH from 4.2 to 7.0 in
Colombia, Brazil and Nigeria [18,40,68].

Recommended compound fertiliser rates for Lima beans
range from 25-280 kg ha^{-1} of N, P_2O_5 and K_2O (hereafter
denoted N-P-K), the ratio of components depending on soil
fertility [39,55]. In California, fertiliser application
did not increase the economic returns from Lima beans grown
on fertile soils [1], but good yields have been achieved
with rates of 10-90 kg N-P-K ha^{-1} in Pennsylvania [95],
Wisconsin [24] and in Washington [33].

Larger fertiliser rates may be required on acid, sandy
soils where efficiency of uptake of applied nutrients may
be 50% or less [74]. Rates of 135-224 kg P ha^{-1} were
recommended in Virginia [81] and on Long Island, New York
[36] although cultivars Fordhook and Fordhook 242 responded

491

to large phosphorus applications only in potassium-rich soils on Long Island [36].

Fertiliser practices for Lima beans are said to be similar to those for common beans in tropical regions, but largely because of a lack of specific recommendations for the crop. Rates of 17-90-120 kg N-P-K ha^{-1} were used in Africa and applications of up to 15000 kg manure ha^{-1} were also suggested [55]. However, optimum fertiliser rates remain to be established.

Fertiliser application and cultivation practices

Fertiliser is applied at planting, often in bands 5.0-7.5 cm below and adjacent to the seeds [39,55]. Lima bean seeds are sensitive to soluble salts and direct contact with fertilisers greatly reduces the proportion which germinate [81]. Supplemental nitrogen and phosphorus may be side-dressed at the early bud stage and again at fruit-filling [81]. Soil moisture conditions influence the effectiveness of side-dressing; adequate rainfall or carefully timed irrigation is required to achieve optimum results [96].

Cultivation practices affected nutrient distribution on sandy loams in Tennessee. Nutrients were distributed to greater depth by conventional tillage than by reduced till or no-till practices, but plant uptake of N, P, K and Ca was not significantly different. However, yields were best with conventional tillage [79].

Mineral nutrition

Lima bean growth varies with different sources of nitrogen fertiliser. Total plant dry weight was greater when nitrates were the primary source of nitrogen in solution culture [74]. Peak absorption of nitrate occurred during flower initiation and pod filling. Ammonium concentrations greater than 50% were toxic and reduced growth [74]. Physiological maturity of field-grown plants was delayed by side-dressing of nitrogen during blossoming and fruit set [59]. Nitrate nitrogen fertiliser delayed maturity of cultivar Fordhook 242 more than ammonium-based sources [95].

Leaf tissue analyses of plant nutrient status did not accurately predict dry-matter yields of plants grown in solution culture [74]. This was confirmed in field tests, when respective foliar concentrations of N, P, and K were the same in fertilised and unfertilised plants, regardless of treatment and yield differences [95].

The response of Lima beans to calcium is important where the crop is grown on acid soils. Calcium applications did not increase the availability of N, P, K or Mg in three Virginia soils once the optimum soil pH had been achieved [49]. Calcium added to acid soils in Georgia raised the pH from 4.7 to 6.1 and significantly altered the nutrient status of cultivar Nemagreen, but had no effect on plant growth or dry weight, indicating considerable tolerance of soil acidity in that cultivar [48]. Differences in varietal response to liming were observed in Maryland and correlated with region of varietal development: those developed in the south and east were most tolerant of acid soils rich in exchangeable aluminium, whereas varieties developed in the Mid-West or western USA were the least tolerant [44].

Liming affected foliar mineral concentrations of culti-var Nemagreen more than those in the roots. Increasing lime from 0 to 2242 kg ha^{-1} increased foliar concentrations of calcium, barium and copper but decreased those of phosphorus, potassium, magnesium, zinc, manganese, molybdenum, iron and aluminium [48]. An application of 5-10 kg Mn ha^{-1} was recommended when the soil pH exceeded 6.5 [39], and 5-16 kg boric acid ha^{-1} applied one week before flowering significantly increased the yields of large-seeded 'bush' types grown on sandy loams of pH 4.8-5.0 [120].

NITROGEN FIXATION

Artificial inoculation of Lima bean seed with strains of

Rhizobium is not recommended because of typically poor
responses [1]. Strains reliably beneficial to Lima beans
on particular soil types are not available commercially.
Testing and evaluation of effective rhizobia for Lima beans
have been limited, but are now underway at the USDA Nitro-
gen-Fixation Laboratory [56].

Lima beans and at least ten other agronomically important legumes belong to the cowpea cross-inoculation group
[30]. Some of these host species, including Lima beans,
demonstrate specificity for those strains effecting beneficial associations [12].

All eighty-seven strains of *Rhizobium* isolated in Indonesia from diverse hosts in the cowpea group produced nodules following inoculation of Lima beans in sand culture,
but only thirty-three of them improved the growth of the
host plant [91]. Similarly, only two out of fifteen strains
from the cowpea group improved Lima bean growth in the
south-eastern USA [29]. Lima beans were placed in the same
cross-inoculation subgroup as *Canavalia ensiformis* and *C.
gladiata, Macrocarpus pruriens* and *Vigna unguiculata*,
although *V. unguiculata* had been excluded from this subgroup in an earlier classification [30,91].

Specificity and recognition phenomena between a given
strain of *Rhizobium* and its legume host occur at the surface of root hair tips [22]. Infection of Lima bean root
hairs by rhizobia can be observed within two hours after
inoculation [92] (and see Chapter 4).

Root nodule bacteria may affect plant growth by synthesising hormones or altering plant hormone metabolism,
but the role of hormones in the growth and development of
nodulated Lima beans is unclear, as it is in other legumes
[41,108].

On the few occasions when Lima beans have responded to
inoculation, the effects have been comparable to chemical
fertiliser treatments in glasshouse trials with nine infertile Indonesian soils. However, competition from indigenous
rhizobia reduced the effectiveness of some strains [90].

Seed yield and total seed protein concentration can be enhanced by effective rhizobial associations. Seed number, weight, and nitrogen concentration were all increased in eight Lima bean introductions obtained from Africa and South America in glasshouse inoculation tests with selected strains on infertile soils. Total seed protein concentration was 300-500% greater than in the uninoculated controls and improvements in seed production and protein concentration were greater in long-season introductions than in early-maturing ones [35].

In summary, when compatible strains of *Rhizobium* are selected for effectiveness with Lima beans on particular soil types, the increase in plant growth can be comparable to that which results from inorganic nitrogen fertilisers, and significant increases in seed yield and total seed protein concentration can be achieved. The problem is to better ensure a reliable and economic response to inoculation, especially on small farms in the warmer regions.

SEED QUALITY

Lima beans are an important component of the diets in many tropical regions. The nutritional value of seeds depends upon their chemical composition and amino acid content, the presence of antimetabolic factors, cooking characteristics, and on the composition of the diet (Chapter 3). Unfortunately, very few Lima beans have been evaluated for these traits.

Chemical composition and amino acid spectrum

The chemical composition of green and dry seeds (Table 11.1) is similar to that of other pulses. Only soyabeans and winged beans are richer in oil and protein. Total sugars consist of the monosaccharides glucose, fructose and galactose, the disaccharide sucrose and the trisaccharides raffinose and stachyose [55]. The glyceride fractions include

495

Table 11.1 Chemical and amino acid composition of dry mature seeds of *P. lunatus*

Chemical composition [42]			Amino acid composition [39]		
Calories	335 kcal 100 g^{-1}		Crude protein	20.0%	
Water	12.7		Asparagine	11.9	
Protein	21.4		Glutamic acid	14.9	
Oil	1.4	%	Alanine	4.7	
Carbohydrate	61.1		Arginine	6.3	
Fibre	4.3		Cystine	1.1	
Ash	3.4		Glycine	4.4	
			Histidine	3.2	
Ca	116.0		Isoleucine	5.3	
Fe	4.9		Leucine	8.9	g 16g N^{-1}
Thiamine	0.33	mg 100 g^{-1}	Lysine	7.5	
Riboflavin	0.16		Methionine	1.2	
Niacine	2.1		Phenylalanine	6.4	
			Proline	4.6	
			Serine	8.1	
			Threonine	5.1	
			Tyrosine	3.4	
			Valine	5.8	

principally palmitic, linoleic and linolenic acids, with smaller amounts of stearic and oleic acids. Lipid composition is similar in both immature and mature beans and is characterised by a relatively large proportion of linolenic acid [58].

The seeds of *P. lunatus* average 20% crude protein, with variations reflecting genotypic differences and ecological conditions. The principal proteins are α- and β-globulin, plus a small quantity of albumin. Like most other grain legumes, Lima bean seeds are deficient in sulphur-containing amino acids and, to a lesser extent, in tryptophan, but they contain relatively large amounts of lysine (Table 11.1). Disparities in amino acid profiles are due not only to environmental and genotypic influences but also to the analytical methods used. However, the following generalisations seem to be reliable [80]:

(a) the amino acid balance remains fairly uniform
     throughout all wild and cultivated forms;

496

(b) ecological conditions markedly influence the total nitrogen content of the seeds, and consequently the relative proportions of essential amino acids, particularly of lysine and methionine;

(c) total nitrogen content is greater in wild and weedy forms than in cultivars, but the latter are relatively richer in sulphur-containing amino acids;

(d) a slight increase in total nitrogen content is often associated with a relative decrease in lysine and methionine values;

(e) methionine concentration averages about 1% within the amino acid spectrum, which is similar to values recorded for common beans but inferior to those of *Vigna radiata* (1.15%), *Vigna unguiculata* (1.25%) and *Vigna mungo* (1.5%).

Like other pulses, raw Lima beans are poorly digestible but this can be improved by cooking [27]. The beneficial effect of sulphur-containing amino acid supplementation on the protein quality of Lima bean foods has been well established in feeding tests with rats [25,71]. Lima beans are usually consumed with cereal grains or with cassava and plantains in many tropical regions. Cereal and Lima bean proteins can complement each other effectively (Chapter 3) and the legume is especially important in starchy food-bean production systems.

Antimetabolic factors

Toxic or antinutritional compounds in Lima beans which are known to have harmful effects on humans or animals include protease inhibitors, phytohaemagglutinins or lectins, and cyanogens.

Protease inhibitors.   Lima bean seeds contain heat-labile proteins which inhibit proteolytic enzymes, principally trypsin and chemotrypsin [45,57,100,101,116].

Compared to other pulses, Lima beans contain relatively large amounts of trypsin inhibitors, approaching those values recorded in soyabeans and winged beans [89]. Rats fed with raw seeds suffer hypertrophy of the pancreas, causing loss of sulphur-containing amino acids from the body [63]. This explains the beneficial effect of methionine added to raw Lima bean seeds in rat-feeding trials [57].

Phytohaemagglutinins or lectins.   Lima bean seeds contain phytohaemagglutinins or lectins, a group of heat-labile substances which can agglutinate red blood cells by virtue of their specificity for glycoprotein receptor sites on the cell surface. The lectins, mainly concentrated in the cotyledons, decrease in concentration during germination and seedling development [73]. Lima bean lectins show specificity for the agglutination of erythrocytes types A and AB in haemagglutination tests against human blood groups [26]. Inter-varietal differences in agglutinating activities may be genetically controlled [93,106].

Cyanogens.   Some wild forms of Lima bean are potentially toxic because they contain glucosides (phaseolunatin or linamarin) from which HCN may be released by hydrolysis through the action of the enzyme linamarase or β - glucosidase [9,78]. Hydrolysis occurs quite rapidly when the macerated beans are cooked in water and most of the liberated HCN is lost by volatilisation. It has been postulated that cooking destroys the enzyme linamarase (thermo-labile at 80°C) but not the glucoside linamarin (thermo-labile at 141°C). If linamarase is inactivated before all of the cyanogen has been hydrolysed, the residual cyanogen may be decomposed by stomach enzymes or bacteria with a consequent release of HCN [62,123].

Lethal doses of ingested cyanide for an adult human have been estimated at 30-200 mg. A maximum content of 20 mg HCN per 100 g Lima bean seeds (i.e. 200 ppm) is permitted by USA legislation [78]. The distribution of cyanogenic glucosides in Lima bean seeds is extremely variable, ranging from 10 ppm in some lines to more than 4000 ppm in others [38,66, 113]. Wild forms have large HCN contents, markedly greater than those found in any of the cultivated forms so far ana- lysed [113]. Unlike sorghum, cassava and flax, considerably more HCN is released from Lima bean seeds than from other plant components, especially the roots [113].

Cooking characteristics

Cooking time, volume of the whole beans following rehydra- tion, taste, texture and appearance of local dishes made from the bean flour are among the most important factors which determine the choice of pulses by consumers. Although these problems have been investigated by the processing industry in the USA, very few studies of this type have been carried out in the tropics.

In Nigeria, cooking times of twelve Lima bean cultivars varied from 60 to 86 minutes, with an average of 76 minutes [65]. In comparison, cowpeas required an average cooking time of only 52 minutes. The relatively hard testas of Lima beans might account for the longer time required to soften them.

PRINCIPAL YIELD CONSTRAINTS

Research and development

Lima beans, like most other grain legumes, have not received the benefits of intensive research programmes such as those devoted to the cereals and other major crops. Research efforts have been limited primarily to the USA, where environmental conditions are quite different from those of the tropical regions from where the Lima bean originated. Rigorous evaluations of the Lima bean's range of adaptation and genetic diversity have not been under- taken, and there is little information concerning its potential, particularly in the tropics. These factors

constitute the most serious constraint to the improvement
and wider use of the crop; attempts to grow temperate 'bush'
varieties in the lowland humid tropics have usually failed.

Crop management

Neither good agronomic practices nor improved cultivars
suitable for more intensive production systems have been
developed for the tropics, as they have for temperate
regions. Timing of planting is critical to conform to local
climatic patterns, and is influenced by photoperiod and
temperature sensitivity [127] and duration of growing season. Poor knowledge about Lima bean tolerance of acid soils
and ability to nodulate reliably hampers recommendations
for fertilisation and inoculation. Weed control is essential
during the early stages of growth. Disease and pest control
practices may be critical, especially during the flowering
and seed-filling stages. Trellising reduces disease and
improves yields of vining varieties, but increases time
and labour requirements. Sequential ripening patterns make
mechanical harvesting difficult and prolong the harvest
season. Premature fruit dehiscence may necessitate frequent
harvesting. Poor management of any of these factors,
exacerbated by incomplete knowledge of varietal traits,
contributes to poor yields.

Environmental adaptation

Lima beans may be more exacting than many other legumes in
respect of the temperature and humidity conditions affecting seed germination, photoperiodic and temperature effects
on flowering, and environmental effects on fruit-setting.
Serious flower shedding and fruit abortion occur under arid
conditions. In contrast, certain cultivars have good tolerance of cold, heat, drought, acidity or salinity, and some
are insensitive to photoperiod. Recent trials also indicate
a range in efficiency of symbiotic nitrogen fixation.

Lima Bean (*Phaseolus lunatus* L.)

Disease and pest constraints

In traditional farming systems, Lima beans appear to be
less affected by diseases and pests than some other pulses.
Nevertheless, these depredations increase in dense cultivation. More than fifty species of fungi, eleven bacteria,
eleven viruses and sixteen nematodes have been reported to
attack Lima beans in the USA [126]. These, or related,
diseases and pests are also common in the tropics.

Among the fungal diseases, web blight, caused by *Thanatephorus cucumeris* (known as *Rhizoctania solani* in the
imperfect stage), is perhaps most devastating to Lima beans
in the humid tropics of America and Africa. 'Bush' types
developed in temperate regions are particularly susceptible
[18]. Root-rots caused by *Fusarium, Pythium, Rhizoctonia,
Thielaviopsis* and *Macrophomina* spp. are important in cool,
humid areas. Downy mildew, caused by *Phytophthora phaseoli*
and specific to Lima beans, was first discovered in eastern
USA but is now known to be distributed worldwide. Other
fungi of potential importance, and seed-transmitted, are
*Diaporthe phaseolorum, Colletotrichum truncatum, C. linde-
muthianum, Cercospora canescens, Ascochyta phaseolorum* and
Nematospora phaseoli. Of the bacterial diseases, common
blight (*Xanthomonas* sp.) may be severe in the tropics if
warm, moist conditions prevail at flowering.

Virus diseases are probably the most limiting diseases
of Lima beans in tropical regions. They can be categorised
by their insect vectors: Cucumber Mosaic, Common Mosaic,
Lima Bean Green Mottle and Veinal Necrosis are all transmitted by aphids. Bean Rugose Mosaic and Bean Curly Dwarf
Mosaic are transmitted by beetles. Double Bean Yellow Mosaic and Lima Bean Golden Mosaic are transmitted by whiteflies (*Bemisia* spp.). Golden Mosaic Virus (Plate 11.4) is
the major constraint to intensified production in the
humid, tropical forest regions of Africa and Latin America
[18]. Root-knot nematodes (principally *Meloidogyne* spp.)
are serious pests and cause considerable yield reductions

in commercial-production areas of the USA and in the tropics.

In general, insect pests that attack common beans also attack Lima beans. Leafhoppers (*Empoasca kraemeri*), Mexican bean beetles (*Epilachna varivestis*), bean pod weevils (*Apion godmani*), and bruchids (*Acanthoscelides obtectus* and *Zabrotes subfasciatus*) are important in Latin America. Bud thrips (*Megalurothrips sjostedi*) and bruchids (*Callosobruchus maculatus*) are severe in Africa. Pod borers (*Maruca testulalis, Etiella zinckenella* and *Cydia* spp.) are important worldwide.

Plate 11.4 A Lima bean plant infected with Golden Mosaic Virus
(IITA, Nigeria; Photograph by J.P. Baudoin)

Consumer preference

Commercial processors of Lima beans in the USA have

developed exacting specifications for varieties and processing procedures to meet consumer demands. Varieties with
green cotyledons are preferred for fresh processing, while
white cotyledons are preferred for dry seeds. Speckled seeds
are popular for consumption as dry seeds in southern states.

Regional preferences may also restrain the spread of
the crop throughout the tropics. Popular pulses such as cowpeas, common beans, pigeonpeas or mung beans may have longer
traditions of use. Prolonged cooking times and a bitter
taste due to the cyanogenic glucosides found in some primitive forms may also limit the popularity of the crop.

GERMPLASM RESOURCES

Centres of origin and domestication

The Lima bean has a neotropical origin with two primary
centres of domestication in Central America and Peru.
Mackie [69] suggested that the species originated in Guatemala and was dispersed along three Indian trade routes in
pre-Columbian times: the 'Sieva' types along the Hopi, or
northern Indian route, extending from Mexico to the southern part of the United States; the Potato types along the
Caribbean-West Indian route toward the Amazon basin; and
the 'Big Limas' along the Inca, or southern Indian route,
from Central America to Peru.

A common origin for the 'Sieva' types of Central America
and the 'Big Limas' of Peru has never been substantiated.
Recent archaeological evidence suggests that the 'Big Limas'
were domesticated east of the Andean highlands in the warm
humid tropics of Peru, while the 'Sievas' were domesticated
independently in the Pacific coastal foothills of Mexico
[54]. Other seed types do not conform to these groups and
complicate the understanding of the species' domestication.

Lima beans were carried later from Brazil to Africa by
the slave trade; from Peru to Madagascar (Malagasy), and
from Mexico to the Philippines and Asia (Java, Burma and

Mauritius) by the Spaniards. West Africa, Malagasy, and South-east Asia are considered secondary centres of diversification for cultivated and weedy forms.

The gene-pool of *P. lunatus* is comprised of three major sections [18,94]: a primary gene-pool, represented by cultivars and landraces of *P. lunatus* var. *lunatus*; a secondary gene-pool, constituted by wild forms of *P. lunatus* var. *silvester*; and a tertiary gene-pool, which includes those wild species that can exchange genes with cultivated forms despite genetic barriers.

World germplasm collections

The International Board for Plant Genetic Resources (IBPGR; see Chapter 18) has given *Phaseolus* beans priority among the food legumes for collection and preservation [50]. In 1976, the IBPGR designated CIAT (see Chapter 18) as the world repository for *Phaseolus* bean germplasm. To date, CIAT's collection of *Phaseolus lunatus* totals 2311 accessions, including both cultivated (97%) and wild forms (3%). Lima bean germplasm represents 7% of the total *Phaseolus* collection. Donations were received from Brazil, Peru, Mexico, Colombia, Nigeria, the USA and elsewhere. Collections were made on recent expeditions to Ghana, Ivory Coast, Liberia, Sierra Leone, Togo and Mexico.

North American and African countries provided the majority of CIAT's germplasm accessions (34% and 40%, respectively). However, when the sources of these donations were traced, 14% of the germplasm was originally obtained from Central America and 31% from South America. Wild forms came almost exclusively from Central America. 'Sieva' and 'Potato' types came predominantly from lowland areas of Central and South America. 'Big Lima' types came mostly from South American Andean areas and also from Africa (Table 11.2). The geographical distribution of CIAT's collection of Lima beans supports Mackie's theory on the origin and dispersal of seed types [69].

Lima Bean (*Phaseolus lunatus* L.)

Table 11.2 Geographical distribution of Lima bean seed types by
provenance (*Phaseolus lunatus* collection at CIAT,
Colombia, 1982)

Zone	Wild forms (%)	Cultivated forms (%)		
		Sieva	Potato	Big Lima
North America	0	5.7	2.8	3.1
Central America	75.5	17.1	13.1	1.8
Caribbean	0	1.0	0.9	0
South America:				
Andes	7.5	4.7	11.2	46.5
Lowlands	1.9	21.9	46.7	8.0
Europe	1.9	0.5	0	0.3
Africa	0	4.5	11.2	13.2
Middle East	0	0.1	0	0
Asia-Oceania	0	0.7	0	0.3
Unknown	13.2	43.8	14.0	26.8
Total no. of accessions	53	1468	107	325

Variability in seed-coat colour and pattern is as great
as that found in common beans: white, cream-beige, yellow,
brown, pink, red, purple and black seeds are available.
However, white, cream-beige, and red are the predominant
colours. Testa patterns include mottles, spots, stripes and
speckles; this variation may have been increased by outcrossing.

Variability for growth habit is limited. Indeterminate
prostrate or climbing types predominate. Determinate, 'bush'
forms show little variability for erectness, branching
patterns or duration of growth.

CIAT's collection of Lima beans shows ample variability
in physiological-agronomical characteristics of interest
to breeders. Evaluation is proceeding for numbers of racemes, fruits, and seeds per plant, as well as for other
traits included in the list of standard descriptors prepared by the IBPGR [21].

Smaller Lima bean collections exist in Latin America,
North America, Europe, Africa and in Asia [15]. Institutions maintaining Lima bean collections in Latin America
are: INIA* in Mexico, EMBRAPA* in Brazil, CATIE* in Costa

Rica, and ICA* in Columbia. In North America the USDA
maintains an important collection at Pullman, Washington.
In Europe, the N.I. Vavilov All Union Institute of Plant
Industry in the Soviet Union has a collection of landraces.
In Belgium, the Faculty of Agricultural Sciences at Gembloux holds the basic world collection of the wild *Phaseolus*, containing species of the tertiary gene pool. In Japan,
there are *Phaseolus* collections at the National Institute
of Agricultural Sciences and at Kyoto University.

Evaluation of these collections has been extremely
limited, but some screening has been carried out for traits
such as tolerance to cold [37,107], drought [14,59,99] and
heat [5], and for photoperiod response [47] and ability to
grow well on acid soils [44]. A wide range in cyanide content was reported in seeds of fifty-six wild accessions and
cultivars [113]. Variation for many of these traits appears
to be greater in the indeterminate forms than in determinate
ones.

Tolerance to downy mildew was found in the 'bush' varieties Thaxter and Early Thorogreen [117], and to web-blight
in some indeterminate varieties [114]. Other indeterminate
varieties are tolerant of Golden Mosaic and Green Mottle
Viruses [20,51], and tolerance to stem-rot [97] and to root-
knot nematodes [4,51] has also been reported. Some large-
seeded cultivars are tolerant of leafhoppers [67] (Plate
11.5(a)).

BREEDING STRATEGIES

Objectives

Lima bean improvement programmes seek to develop cultivars
which combine improved productivity, stable yields, disease
and pest resistance, and desirable agronomic traits. In the

---

*Instituto Nacional de Investigaciones Agrícolas; Empresa Brasiliera
de Pesquisa Agropecuaria; Centro Agronómico Tropical de Investigación
y Enseñanza; and Instituto Colombiano Agropecuario, respectively.

Lima Bean (*Phaseolus lunatus* L.)

Plate 11.5 (a) Tolerance and (b) susceptibility to leafhoppers
(*Empoasca kraemeri*) in a large-seeded indeterminate and a
small-seeded, determinate Lima bean cultivar, respectively
(Photographs J.M. Lyman)

USA, mechanical cultivation and industrial processing have
led to the development of varieties of short duration,
determinate 'bush' habit, synchronous fruit maturation and
uniformity in plant height, seed shape and size. In tropical
regions, crop improvement should focus on incorporating
tolerance to pests and diseases, and adverse climatic and
edaphic conditions into a range of plant types suited to
various farming systems. New cultivars should be developed
continually as farming systems intensify to meet increasing food requirements.

Genetic systems ($2n = 22$)

Knowledge of genetic systems and diversity in the Lima
bean is fragmentary.

However, twenty-three monogenic plant and seed characters and four independent linkage groups have been identified [10]. Inheritance of seed size involves complex genotype x environment interactions [82]. Two dominant gametophytic factors were identified which prevent self-fertilisation by inhibiting pollen tube growth on incompatible stylar tissue [7,23]. A genetic male sterile factor was identified in cultivar Henderson [2]. Yarnell [125] provides a detailed review of cytogenetics in the Lima bean.

The inbreeding mating system of the Lima bean restricts genetic recombination, but several mechanisms ensure that genetic variability is retained during the evolution of selfed progenies. These include the self-incompatibility factors described above, the occurrence of outcrossing at rates which vary between seasons and populations [11], the adjustment of recombination rates by selection pressures [8], and changes in selective values for heterozygotes [9,11].

Mono- or oligogenic inheritance has been postulated for resistance to downy mildew (caused by *Phytophthora phaseoli*) [105,121], to stem-rot (caused by *Rhizoctonia solani*) [97], to Cucumber Mosaic Virus [104] and to root-knot nematodes [4,75]. Polygenic inheritance controls resistance to bacterial blight (caused by *Xanthomonas phaseoli*) [66] to Lima Bean Golden Mosaic [20] and to the leafhopper, *Empoasca kraemeri* [67].

Intraspecific hybridisation

Improvement of Lima beans in the USA has been achieved through the selection of pure-line cultivars and by conventional breeding methods. Pedigree and backcross methods were convenient for incorporating simply-inherited characters into the best horticultural types, such as resistance to root-knot nematodes [4,75] and to downy mildew [103,117]. Bulk breeding methods were effective in

improving yield stability [6,9], but not in developing the
most productive cultivars [110,111]. Resistance to Golden
Mosaic Virus was successfully transferred from 'Big Lima'
to 'Sieva' types in the Lima bean improvement programme
at IITA in Nigeria (see Chapter 18). Early flowering and
other desirable traits were also incorporated [18].

Recurrent selection techniques have greater potential
for improving quantitatively inherited traits such as seed
yield or horizontal resistance to diseases and pests than
pedigree or backcross breeding methods. The major constraint in adapting recurrent selection methods to self-
pollinated crops has been the difficulty in effecting
massive gene flow during recombination. This problem can
be partially overcome by exploiting genetic male sterility,
and by enhancing natural outcrossing in the field. The
recessive male sterile factor identified by Allard [2] was
closely linked with compact, dark green foliage in the
seedling stage. A dual population scheme was proposed [86]
to exploit natural outcrossing. Alternate rows of two
populations are planted, each possessing a distinctive
recessive trait for which the other is dominant. Hybrid
progeny are identified by the presence or absence of
appropriate marker traits. Selection for quantitative
traits is carried out in subsequent generations.

Interspecific hybridisation

Interspecific crosses offer the opportunity to broaden
genetic variability by introducing useful characteristics
poorly expressed or non-existent in the primary gene-pool
of *P. lunatus*, such as thicker stems, rigid racemes,
hairiness and resistance to disease. Three wild taxa
(*P. polystachyus*, *P. metcalfei* and a newly recognised,
as yet un-named species have been crossed successfully
with the Lima bean [19,60]. Interspecific hybrids were
obtained without embryoculture. The first generations
tested both in a glasshouse in Belgium and in the field in

509

Nigeria displayed satisfactory fertility restoration and a gradual introgression of desirable genes from the wild species [18]. Genotypes combining good productivity and disease resistance were selected in advanced progenies. Such individuals should be integrated in conventional populations through a recurrent breeding scheme to break up unfavourable linkage blocks.

Future prospects

Steady progress has been made in Lima bean improvement for temperate regions, but much remains to be done in the tropics in order to develop improved cultivars for intensified farming systems.

Climbing types perform satisfactorily in tropical regions but need to be staked and so are unsuitable for intensive mechanised cropping systems. Cereals, root or tuber crops could provide support for the grain legume in intercropping schemes. A large number of genotypes should be evaluated to assess their compatibility and aggressiveness with companion crops.

'Bush' types are more appropriate materials for intensive production systems. However, poor plant architecture and susceptibility to diseases contribute to very small yields in the humid tropics. An intermediate type (e.g. the indeterminate 'bush' type) is very productive in the common bean (see Chapter 10) but was not found in limited collections of Lima beans evaluated in Colombia and Nigeria [18,66]. Developing such a form would require exploitation of the total species gene-pool (both primary and secondary) and realisation of wide crossing programmes to combine genes for adaptation and plant architecture.

At present, many more efforts are devoted to the improvement of common beans (poorly adapted in many regions) and cowpeas for the tropics, than to Lima beans. Notwithstanding, promising results at CIAT in Colombia and at IITA in Nigeria indicate that the Lima bean offers broad

adaptability and a potential for good yields, yet the
potential for improvement of this crop remains to be more
fully exploited.

YIELD POTENTIAL

Average yields for different production intensities can be
estimated from production data and compared with yields of
improved varieties on experimental farms to estimate yield
potential. Stabilising better average yields will depend
on the dissemination of improved varieties and management
practices, as well as favourable climatic conditions and
economic policies.

Yields of Lima beans grown commercially in the USA for
green seeds average 2000-3000 kg ha^{-1}. Assuming 100 days
to harvest for small-seeded Limas, daily productivity can
be calculated as 20-30 kg ha^{-1} day^{-1}. Dry seed yields
average 2500-4000 kg ha^{-1}. Assuming 120 days to harvest
for large-seeded Limas, daily productivity is 21-33 kg ha^{-1}
day^{-1}. Experimental yields as large as 6000-7000 kg ha^{-1}
(50-70 kg ha^{-1} day^{-1}) have been reported for both green
and dry beans [33,76].

Yields may increase by as much as 50% at the greater
plant densities made feasible by modern mobile fruit-
stripping and shelling machines. More compact varieties
bearing fruits above the foliage will be required [102],
and are currently under development by private seed companies [76]. The USDA programme is concentrating on increasing resistance to anthracnose and downy mildew,
adding heat tolerance and developing lines with green
cotyledons [102]. Other programmes at State Experiment
Stations are focussed on resistance to nematodes and *Lygus*
bugs, earliness, herbicide and fungicide testing, and
variety trials [102,109].

In Peru, where yields of dry beans averaged 850 kg ha^{-1}
in 1974, North American varieties with growing seasons of
120-160 days have been introduced. The national agricultural

programme (INIPA) is developing improved types with shorter growing seasons and which yield 1500-2900 kg ha^{-1} at close spacings [64].

Average yields of small-scale, less intensive systems range from 200-1500 kg dry seed ha^{-1}. Since growing seasons vary from 100-270 days, productivity is difficult to estimate. Good productivity values under these conditions may reach 15 kg ha^{-1} day^{-1}. These regions may have the greatest potential for developing varieties with dramatic-ally improved yields, since so little work has been done with tropical germplasm. Traits such as drought tolerance will be especially valuable.

In Mexico, where yields of Lima beans grown in mixed cropping systems average 200-400 kg ha^{-1}, the national agricultural programme (INIA) is developing improved varieties from local and introduced parental lines. Yields of experimental lines range from 1700-1900 kg ha^{-1} for 'bush' types of 100 days duration to 2000-3000 kg ha^{-1} for vining types of 160-180 days duration [61]. Trials at the humid Guayaquil Station of the Ecuadorian agricultural programme (INIAP) have given seed yields in the range of 3800-4700 kg ha^{-1} for improved vining types [28]. Maximum yields as large as 4800 kg ha^{-1} were achieved in Colombia from trials of selected vining lines [68]; productivity ranged from 15-44 kg ha^{-1} day^{-1}. In preliminary trials under humid conditions at Manaus, Brazil, yields of up to 4000 kg ha^{-1} were estimated; early-maturing, short-day 'bush' types show promise there [40].

Yields from household gardens in the Philippines are small. Selections from introduced varieties were made at the University of the Philippines at Los Banos. Four cultivars were recommended for commercial production [16], ranging from short-season, determinate types with yields of 8.0-12.7 t green fruits ha^{-1} (1.5-2.0 t green seeds) to long-season, indeterminate types with green fruit yields of 12.8-15.7 t ha^{-1}.

In Mauritius, the Ministry of Agriculture conducted

Lima Bean (*Phaseolus lunatus* L.)

trials on eight US varieties under 'optimum' conditions. Dry seed yields ranged from 1500-2000 kg ha^{-1} for 'bush' types and 4400-9200 kg ha^{-1} for trellised, vining types from multiple harvests over a two-month period [77].

An intensive Lima bean improvement programme has been carried out at IITA in Nigeria. Experimental yields exceeding 5000 kg dry seed ha^{-1} were obtained from vining types under 'optimum' conditions [51]. However, Lima bean resistance to disease and insect pressures (especially thrips, soil-borne fungi and Golden Mosaic Virus) decreased in dense stands. Indeterminate varieties grown on trellises performed better and had the potential to produce very large seed yields, but the staking was expensive. The Lima bean improvement programme at IITA was discontinued in 1979 when efforts were diverted to cowpeas [52]. Nevertheless, it is clear that more intensive production could be realised from mixed cropping systems. Such systems can protect the soil and give stable yields, but few data are available on the performance of Lima beans when intercropped.

AVENUES OF COMMUNICATION

Communication between researchers studying Lima beans is limited primarily to personal contacts between interested individuals and to occasional reports in publications with restricted circulation. The lack of communication is particularly marked between temperate and tropical researchers, and reports published by tropical research organisations can be difficult to obtain in temperate regions.

There is no international newsletter devoted exclusively to the Lima bean. However, reports on Lima bean research appear occasionally in the *Tropical Grain Legume Bulletin*, published by IITA. IITA also published summaries of Lima bean research in its annual reports from 1972-79. CIAT publishes very little information about the Lima bean germplasm collection; the Institute has no Lima bean improvement programme.

Only a few institutions in temperate countries publish
reports on tropical legume research. These include the
Centre for Agricultural Publishing and Documentation at
Wageningen, the Netherlands; the *Bulletin des Recherches
Agronomiques* from the University of Gembloux, Belgium; and
the *Crop and Product Digest* from the Tropical Development
and Research Institute in London, England. Research efforts
in the USA are confined solely to temperate varieties and
have declined in recent years. The USDA and certain state
university agricultural programmes publish occasional
research bulletins on Lima beans in agronomic and horticultural journals. Plant breeders and agronomists in the
seed and processing industries maintain an informal network of contacts among themselves and with state extension
agents.

The scarcity of publications and other organised forms
of communication appears to reflect the limited production
of Lima beans worldwide, and the limited priority given to
improvement programmes. However, interest in the grain
legumes as a group is expanding and may generate increased
attention to the Lima bean; its food value, wide adaptation
and potential for good seed production in tropical environments merit this expanded effort.

LITERATURE CITED

1. Allard, R.W. (1953), *University of California Agriculture
   Experiment Station Circular 423*.
2. Allard, R.W. (1953), *Proceedings of the American Society of
   Horticultural Science* 61, 467-71.
3. Allard, R.W. (1954), *Proceedings of the American Society of
   Horticultural Science* 64, 410-16.
4. Allard, R.W. (1954), *Phytopathology* 44, 1-4.
5. Allard, R.W. (1954), *California Agriculture* 8(3), 5.
6. Allard, R.W. (1961), *Crop Science* 1, 127-33.
7. Allard, R.W. (1963), *Der Zuchter* 33(5), 212-16.
8. Allard, R.W. (1963), *Genetics* 48, 1389-95.
9. Allard, R.W. and Bradshaw, A.D. (1964), *Crop Science* 4, 503-8.
10. Allard, R.W. and Clement, W.M. (1959), *Journal of Heredity*
    50, 63-7.
11. Allard, R.W. and Workman, P.L. (1963), *Evolution* 17(4), 470-80.
12. Allen, O.N. and Allen, E.K. (1939), *Soil Science* 47, 63-76.

Lima Bean (*Phaseolus lunatus* L.)

13. Andrews, F.S. (1936), *Proceedings of the American Society of Horticultural Science* 34, 498–501.
14. Andrews, F.S. (1939), *Proceedings of the American Society of Horticultural Science* 37, 752–8.
15. Ayad, G. and Anishetty, M. (1980), *Directory of Germplasm Collections. I. Food Legumes*. Rome: International Board for Plant Genetic Resources.
16. Aycardo, H. (1982), Personal communication. College of Agriculture, University of the Philippines, Los Banos, Philippines.
17. Baudet, J.C. (1977), *Bulletin Société Royale Botanique de Belgique* 10, 65–76.
18. Baudoin, J.P. (1981), Doctoral thesis, Faculté des Sciences Agronomiques de l'Etat, Gembloux, Belgium.
19. Baudoin, J.P. (1981), *Bulletin des Recherches Agronomiques, Gembloux* 16, 273–86.
20. Baudoin, J.P. and Allen, D.J. (1978), in Maraite, H. and Meyer, J.A. (*eds*) *Diseases of Tropical Food Crops*. Proceedings of the International Symposium, Louvain-La-Neuve, Belgium, 237–49.
21. Baudoin, J.P. and Maréchal, R. (1982), *Descriptors for Lima Bean*. Rome: International Board for Plant Genetic Resources.
22. Bauer, W.D. (1981), *Annual Review of Plant Physiology* 32, 407–49.
23. Bemis, W.P. (1959), *Genetics* 44, 555–62.
24. Binning, L.K., Wyman, J.A., Schulte, E.E. and Stevenson, W.R. (1982), *Commercial Lima Bean Production*. University of Wisconsin Coop. Ext. Prog. No. A2338.
25. Boulter, D., Evans, A.M. and Thompson, A. (1976), *Qualitas Plantarum Plant Foods for Human Nutrition* 26, 107–19.
26. Boyd, W.C. and Reguera, R.M. (1949), *Qualitas Plantarum Plant Foods for Human Nutrition* 62, 333–9.
27. Bressani, R. and Elias, L.G. (1980), in Summerfield, R.J. and Bunting A.H. (*eds*) *Advances in Legume Science*. London: HMSO, 135–55.
28. Buestan, H. (1978), 'Uniform Lima bean trial'. INIAP, Guayaquil, Ecuador. Unpublished.
29. Burton, J.C. (1952), *Soil Science Society of America Proceedings* 16, 356–8.
30. Burton, J.C. (1967), in Peppler, H.J. (*ed*.) *Microbial Technology*. New York: Reinhold, 1–33.
31. Cetas, R.C. and Wester, R.E. (1956), *Proceedings of the American Society of Horticultural Science* 68, 392–3.
32. Chatel, M. (1981), *Agronomie Tropicale* 36(3), 294–8.
33. Clore, W.J. and Stanberry, C.O. (1951), 'Growing Lima beans in irrigated central Washington'. *University of Washington Agriculture Experiment Station Bulletin 530*.
34. Cordner, H.B. (1933), *Proceedings of the American Society of Horticultural Science* 30, 571–6.
35. Cornet, D., Meulemans, M. and Otoul, E. (1980), *Bulletin des Recherches Agronomiques, Gembloux* 15(1), 17–29.
36. Dallyn, S.L. and Sawyer, R.L. (1959), *Proceedings of the American Society of Horticultural Science* 73, 355–60.
37. Dickson, M.H. (1973), *HortScience* 8(5), 410.
38. Dilleman, G. (1958), in Ruhland, W. (*ed*.) *Handbuch der Pflanzenphysiologie 8*. Berlin: Springer-Verlag, 1050–75.
39. Duke, J.A. (1981), in *Handbook of Legumes of World Economic Importance*. New York: Plenum, 191–5.

40. Erickson, H.T. (1983), Personal communication, Purdue University, Indiana.
41. Evensen, K.B. and Blevins, D.G. (1981), *Plant Physiology* 68, 195-8.
42. FAO (1968), *Food Composition Table for use in Africa*. FAO Nutrition Division and US Department of Health, Education and Welfare Public Health Service.
43. Fisher, V.J. and Weaver, C.K. (1974), *Journal of the American Society of Horticultural Science* 99, 448-50.
44. Foy, C.D., Armiger, W.H., Fleming, A.L. and Zaumeyer, W.J. (1967), *Agronomy Journal* 59, 561-3.
45. Friedman, M., Zahnley, J.C. and Wagner, J.R. (1980), *Analytical Biochemistry* 106(1), 27-37.
46. Harding, J., Tucker, C.L. and Barnes, K. (1981), *Journal of the American Society of Horticultural Science* 106, 69-72.
47. Hartmann, R.W. (1969), *Journal of the American Society of Horticultural Science* 94, 437-40.
48. Hegwood, D.A. (1972), *Journal of the American Society of Horticultural Science* 97, 232-5.
49. Hester, J.B. (1934), *Proceedings of the American Society of Horticultural Science* 32, 600-3.
50. IBPGR (1981), *Revised Priorities among Crops and Regions*. Rome: International Board for Plant Genetic Resources.
51. IITA (1979), *Annual Report for 1978*. Ibadan, Nigeria: International Institute of Tropical Agriculture, 44-6.
52. IITA (1980), *Annual Report for 1979*. Ibadan, Nigeria: International Institute of Tropical Agriculture, 103.
53. Kannenberg, L.W. and Allard, R.W. (1964), *Crop Science* 4, 621-2.
54. Kaplan, L. (1965), *Economic Botany* 19(4), 358-68.
55. Kay, D.E. (1979), in Food Legumes. *Crop and Product Digest*, no.3. London: Tropical Products Institute, 224-45.
56. Keyser, H. (1983), Personal communication, US Department of Agriculture, Beltsville, Maryland.
57. Klose, A.A., Hill, B., Greaves, J.D. and Fevoid, H.L. (1949), *Archives of Biochemistry* 22, 215-23.
58. Korytnyk, W. and Metzler, E.A. (1963), *Journal of Science of Food and Agriculture* 14, 841-4.
59. Lambeth, V.N. (1950), Some factors influencing pod set and yield of the Lima bean. *University of Missouri Agriculture Experiment Station Research Bulletin*. no.466.
60. Le Marchand, G., Maréchal, R. and Baudet, J.C. (1976), *Bulletin des Recherches Agronomiques, Gembloux* 11(1-2), 183-200.
61. Lepiz, R. (1982), Personal communication. Instituto Nacional de Investigaciones Agropecuarias, Mexico.
62. Liener, I.E. (1962), *American Journal of Clinical Nutrition* 11, 281-98.
63. Liener, I.E. (1976), in Norton, G. (*ed.*) *Plant Proteins*. London: Butterworth, 117-40.
64. Lozano, C. (1982), Personal communication. Instituto Nacional de Investigaciones y Promocion Agropecuaria, Peru.
65. Luse, R.A. (1975), in *Proceedings of the IITA Collaborators' Meeting on Grain Legume Improvement*, Ibadan, Nigeria, 101-4.
66. Lyman, J.M. (1980), PhD. thesis, Cornell University, Ithaca, New York.
67. Lyman, J.M. and Cardona, C. (1982), *Journal of Economic Entomology* 75(2), 281-6.

Lima Bean (*Phaseolus lunatus* L.)

68. Lyman, J.M. (1983), *Journal of the American Society of Horticultural Science* 108(3), 369-73.
69. Mackie, W.W. (1943), *Hilgardia* 15(1), 1-24.
70. Magruder, R. and Wester, R.E. (1940), *Proceedings of the American Society of Horticultural Science* 38, 581-4.
71. Maneepun, S., Lum, B.S. and Rucker, R.B. (1974), *Journal of Food Science* 39, 171-4.
72. Maréchal, R., Mascherpa, J.M. and Stainier, F. (1978), *Boissiera* 28, 146-7.
73. Martin, F.W., Waszczenko-Zacharczenko, E., Boyd, W.C. and Schertz, K.F. (1964), *Annals of Botany N.S.* 28(110), 319-24.
74. McElhannon, W.S. and Mills, H.A. (1978), *Agronomy Journal* 70, 1027-32.
75. McGuire, D.C., Allard, R.W. and Harding, J.A. (1961), *Proceedings of the American Society of Horticultural Science* 78, 302-7.
76. Mendoza, A. (1983), Personal communication. Ben Fish Seed Co., California, USA.
77. Ministry of Agriculture (Mauritius) (1980), *Annual Report of the Ministry of Agriculture and Natural Resources and the Environment (Mauritius) for 1978*, Port Louis, Mauritius, 44-6, 155 and 170.
78. Montgomery, R.D. (1969), in Liener, I.E. (*ed.*) *Toxic Constituents of Plant Foodstuffs*. New York: Academic Press, 143-57.
79. Mullins, C.A., Tompkins, F.D. and Parks, W.L. (1980), *Journal of the American Society of Horticultural Science* 105, 591-3.
80. Otoul, E. (1976), *Bulletin des Recherches Agronomiques, Gembloux* 11, 207-20.
81. Parker, M.M. (1939), *Proceedings of the American Society of Horticultural Science* 37, 737-42.
82. Parsons, P.A. and Allard, R.W. (1960), *Heredity* 14, 115-23.
83. Piper, C.V. (1926), *Contributions from the US National Herbarium* 22, 663-701.
84. Pollock, B.M. and Toole, V.K. (1966), *Plant Physiology* 41, 221-9.
85. Purseglove, J.W. (1968), *Tropical Crops: Dicotyledons*. London: Longman, 296-301.
86. Rachie, K.O. and Gardner, C.O. (1975), *Proceedings of the International Workshop on Grain Legumes*, Andhra Pradesh, India: International Crops Research Institute for the Semi-Arid Tropics, 285-300.
87. Rachie, K.O. and Silvester, P. (1977), in Leakey, C.L.A. and Wills, J.B. (*eds*) *Food Crops of the Lowland Tropics*. Oxford: Oxford University Press, 41-74.
88. Rappaport, L. and Carolus, R.L. (1956), *Proceedings of the American Society of Horticultural Science* 67, 421-8.
89. Rosario, R.R. del, Lozano, Y., Pamorasamit, S. and Noel, M.G. (1980), *Philippines Agriculturalist* 63(4), 339-44.
90. Saono, S. and Karsono, H. (1979), *Annales Bogoriensis* 7(1), 21-33.
91. Saono, S., Karsono, H. and Suseno, D. (1976), *Annales Bogoriensis* 6(2), 83-95.
92. Shantharam, S. and Wong, P.P. (1982), *Applied and Environmental Microbiology* 43(3), 677-85.
93. Schertz, K.F., Jurgelsky, W. and Boyd, W.C. (1960), *Proceedings of the National Academy of Science, USA* 46, 529-32.
94. Smartt, J. (1980), *Economic Botany* 34(3), 219-35.
95. Smith, C.B. (1980), *Journal of the American Society of Horticultural Science* 105, 472-5.

96. Smittle, D.S. (1979), *Journal of the American Society of Horticultural Science* 104, 176-8.
97. Steinswat, W., Pollard, L.H. and Anderson, J.L. (1967), *Phytopathology* 57, 102.
98. Steinswat, W., Pollard, L.H. and Campbell, W.E. (1971), *Journal of the American Society of Horticultural Science* 96, 312-15.
99. Tan, B.S. (1978), *Bulletin des Recherches Agronomiques, Gembloux* 13(1), 73-81.
100. Tan, C.G.L. and Stevens, F.C. (1971), *European Journal of Biochemistry* 18, 503-14.
101. Tan, C.G.L. and Stevens, F.C. (1971), *European Journal of Biochemistry* 18, 515-23.
102. Thomas, C.A. (1983). Personal communication, US Department of Agriculture, Beltsville, Maryland, USA.
103. Thomas, C.A. and Fisher, V.J. (1979), *Crop Science* 19, 419.
104. Thomas, H.R., Zaumeyer, W.J. and Jorgensen, H. (1951), *Phytopathology* 41, 231-4.
105. Thomas, H.R., Jorgensen, H. and Wester, R.E. (1952), *Phytopathology* 42, 43-5.
106. Toms, G.C. (1981), in Polhill, R.M. and Raven, P.H. (*eds*) *Advances in Legume Systematics*. London: HMSO, 561-77.
107. Toole, V.K., Wester, R.E. and Toole, E.H. (1951), *Proceedings of the American Society of Horticultural Science* 58, 153-9.
108. Triplett, E.W., Heitholt, J.J., Evensen, K.B. and Blevins, D.G. (1981), *Plant Physiology* 67(1), 1-4.
109. Tucker, C. (1983), Personal communication, University of California, Davis, USA.
110. Tucker, C.L. and Harding, H. (1974), *Euphytica* 23(1), 135-9.
111. Tucker, C.L. and Webster, B.D. (1970), *Crop Science* 10, 314-15.
112. USDA (1982), *Agricultural Statistics, 1981*. Washington DC: Government Printing Office, 156-7.
113. Vanderborght, T. (1979), *Annales de Gembloux* 85, 29-41.
114. Warren, H.L., Helfrich, R.M. and Blount, V.L. (1972), *Plant Disease Reporter* 56, 268-70.
115. Webster, B.D., Lynch, S.P. and Tucker, C.L. (1979), *Journal of the American Society of Horticultural Science* 58, 153-9.
116. Weder, J.K.P. (1981), in Polhill, R.M. and Raven, P.H. (*eds*) *Advances in Legume Systematics*. London: HMSO, 533-60.
117. Wester, R.E. (1967), *Phytopathology* 57, 648-9.
118. Wester, R.E. and Jorgensen, H. (1950), *Proceedings of the American Society of Horticultural Science* 55, 384-90.
119. Wester, R.E. and Magruder, R. (1938), *Proceedings of the American Society of Horticultural Science* 36, 614-22.
120. Wester, R.E. and Magruder, R. (1941), *Proceedings of the American Society of Horticultural Science* 38, 472-4.
121. Wester, R.E., Fisher, V.J. and Blount, V.L. (1972), *Plant Disease Reporter* 56, 65-6.
122. Westphal, E. (1974), *Pulses in Ethiopia: Their Taxonomy and Agricultural Significance*. Wageningen, the Netherlands: Centre for Agricultural Publishing and Documentation.
123. Winkler, W.O. (1958), *Journal of the Association of Official Agricultural Chemists* 41, 282-7.
124. Williams, W.A., Tucker, C.L. and Guerrero, F.P. (1978), *Crop Science* 18(1), 62-4.
125. Yarnell, S.H. (1965), *Botanical Review* 31(3), 247-330.

Lima Bean (*Phaseolus lunatus* L.)

126. Zaumeyer, W.J. and Thomas, H.R. (1957), 'Bean diseases and methods for their control'. *US Department of Agriculture Technical Bulletin* 868, 65–74.
127. Summerfield, R.J. and Roberts, E.H. (1985), in Halevy, A.H. (*ed.*) *A Handbook of Flowering*. Florida: CRC Press, (in press).

12 Cowpea (*Vigna unguiculata* (L.) Walp.)

W.M. Steele, D.J. Allen and
R.J. Summerfield

INTRODUCTION

Cowpeas, *Vigna unguiculata* (L.) Walp. (syn. *V. sinesis* (L.) Savi ex Hassk), although an ancient African domesticate, remain predominantly a nutritionally important but minor component of subsistence agriculture in the semi-arid and subhumid tropics of Africa and, to a lesser extent, of India and Asia. The crop is grown for its mature seeds, and/or for its immature fruits and leaves (which are used as vegetables); the haulms are also fed to livestock. In recent times it has become an important pulse crop in Brazil but elsewhere in the advanced agriculture of the tropics and subtropics it is used chiefly as a forage or cover crop.

Much has been written about cowpeas during the past eighty years (e.g. see the reviews [7,21,66,208,232,233,234, 256]) but the research reported so far has had little impact upon the crop in subsistence agriculture; in most of the African centres of production today, cowpea landraces are cultivated in traditional farming systems much as they have been for centuries. However, this situation is expected to change in the foreseeable future as the results of intensive research at the International Institute of Tropical Agriculture (IITA) at Ibadan in Nigeria (see Chapter 18) and at the numerous research organisations with which it has collaborated over the past decade begin to have impact. A world collection of germplasm exceeding 12000 accessions has now

been established, evaluated and used at IITA in a breeding
programme (see [220] and below) which has already had
a significant effect on cowpea production in parts of
Nigeria, Brazil, the Yemen and, most recently, in the rice
fallows of South-east Asia [102].

ORIGIN, DISPERSAL AND CLASSIFICATION

Although much of the evidence is circumstantial, there is
now general agreement about the origin and dispersal of cowpeas, and a confused classical taxonomy of the genus *Vigna*
has also been clarified [247]. Linnaeus received seeds of
cowpea from Barbados [39] and named the plants he grew from
them *Dolichos unguiculatus*. In 1924, Savi renamed the genus
Vigna, after Dominic Vigna, a seventeenth-century professor
of botany at Pisa, and then, in 1842, Walpers established
the binomial *Vigna unguiculata* which has precedence over
V. sinensis (L.) Hnook (1844).

Vigna belongs to the tribe Phaseoleae of the subfamily
Papilionoideae, and is a pantropical genus comprising about
160 species, most of them from Africa, where 66 are endemic,
22 from India, and a few from Australia and America [65].
With the proposed transfer of Bambarra groundnut, *Vigna subterranea* (syn. *Voandzeia subterranea*), to the genus [134,248],
there are seven cultivated species, five of which are Asian.
Verdcourt [247] recognises five inter-fertile subspecies of
V. unguiculata, distinguished by the lengths of their calyx
lobes and by the lengths and dehiscence of their fruits. Two
are wild subspecies with short, scabrous, dehiscent fruits
and small seeds whose relatively impermeable testas confer
a degree of dormancy rare in the cultivated forms [132]. The
wild types are confined to Africa, and are the progenitors
of the cultivated subspecies [65,197,226]. Subspecies
dekindtiana comprises the common wild and weedy cowpeas of
the African savanna zones and both perennial and annual
forms occur in West Africa; wild subspecies *mensensis* differs little from subspecies *dekindtiana*, but extends into

521

the wetter forest zones.

Verdcourt's nomenclature for cultivated and wild cowpeas [247] is now in common usage, and is adopted here. However, it is important to note also the work of Marechal *et al.* [134] who, following a taxonometric study of the *Phaseolus-Vigna* complex, proposed a single cultivated subspecies and three wild subspecies for *V. unguiculata* (and see Chapter 1) Marechal's cultivated subspecies *unguiculata* comprises cultivar groups *unguiculata* (cowpea), *biflora* (syn. subsp. *cylindrica*, catjang), and *sesquipedalis* (yard-long bean). This arrangement differs little from Verdcourt's treatment of the cultivated forms [247], except that it emphasises the close affinity between them. On the other hand, Marechal's classification of the wild forms of *V. unguiculata* is of great significance to cowpea breeders because it suggests that the gene-pool available for hybridisation is much larger than was previously anticipated. The three wild subspecies are: subsp. *dekindtiana* (Harms) Verdc., with the botanical varieties *dekindtiana, mensensis, pubescens* and *protracta*; subsp. *tenuis* (E. Mey) M.M.& S.; and subsp. *stenophylla* (Harvey) M.M.& S.

Vigna nervosa (Markotter) may also belong to the secondary gene-pool (wild forms) of the cowpea. Marechal's nomenclature for *V. unguiculata* has been summarised by The International Board for Plant Genetic Resources [90] (IBPGR; see Chapter 18) with the synonyms for wild subspecies and their varieties which are important in plant exploration for wild germplasm. IBPGR gives urgent priority to exploration for the wild forms in South Africa and Zimbabwe, and in the East African and Zambezian phytogeographical zones.

Subspecies *unguiculata*, the common cowpea, is the most widespread, economically important and genetically diverse of the cultivated subspecies. It was domesticated in Africa, probably around the third millenium BC, and within the ancient sorghum (*Sorghum bicolor*) and pearl millet (*Pennisetum americanum*) farming systems of the savanna zone [224, 226]. West Africa is the centre of genetic diversity of subsp. *unguiculata*, but whether domestication occurred only in that region, or throughout the sorghum/pearl millet belt of Africa north of the equator, remains in doubt [227]. Subspecies *unguiculata* first reached the Indian subcontinent with trade along the Sabaean Lane, probably with sorghum and pearl millet, and more recently than 1500 BC [48]. Subsequently, more cowpeas were taken to India with human migrations from Africa. The two other cultivated subspecies were selected from *unguiculata* after it reached India, and were later dispersed to South-east Asia.

Subspecies *cylindrica*, the catjang (Jerusalem pea or marble pea), was selected in India as a forage and fodder crop, and remains important for that use today. It is characterised by short fruits and small seeds, and by a frequently determinate and erect growth habit. These latter traits have recently been exploited in cowpea breeding.

Subspecies *sesquipedalis*, the yard-long or asparagus bean (sitao, bodi bean or snake bean), was selected from subsp. *unguiculata* for its very long (up to 1 m), flaccid, fleshy fruits which are eaten as a vegetable. The modern centres of production are in Asia and Oceania, but the yard-long bean has spread widely throughout the tropics as a minor vegetable crop.

The first European reference to cowpeas was made by Theophrastus in 300 BC [39]. The Romans called cowpeas 'phaseolus', and by the end of the Roman era they were established as a minor legume in southern Europe. They reached the New World with Spanish and Portugese settlements in the seventeenth century and, later, many cultivars must have been carried there with the slave trade from West Africa. In more recent times many cultivars have been bred in the USA [150].

PRODUCTION AND USES

Unofficial statistics for 1981 record the world area of cowpeas harvested as 6 million hectares, with an average

seed yield of just 240 kg ha^{-1} and total production of 1.4
million tonnes. However, more recent official statistics
[64], and the unreported production of cowpeas for home consumption, suggest at least that amount is produced in Africa
alone, and that world seed production is significantly
greater than 2 million tonnes annually.

Summerfield *et al.*[237] comment upon the unreliability
of official estimates of cowpea seed production, and it
remains impossible to obtain precise data because an unknown,
but very large, proportion of the world crop is grown in the
subsistence agriculture of developing countries, and goes
unrecorded. Cowpeas do not enter world trade in significant
amounts, although for some years they have been available
from retail outlets in Europe. Nor are they recorded in the
published statistics of world food production except in so
far as they are included in the FAO data on pulses or, in
the case of the large Brazilian crop, under the heading of
'dry beans' ('frijoles secos') which are mostly *Phaseolus
vulgaris*. Not even these statistics are available from mainland China, and those which report pulse production from the
Indian subcontinent include data about so many grain legumes
that no inference can be made about the proportion that
might be cowpeas.

The major producers and consumers of cowpeas are the
subsistence farmers of the semi-arid and subhumid lowlands
of Africa whose crops provide leaves and fruits for vegetables, fodder and dry seeds. In much of eastern Africa,
where *Phaseolus vulgaris* is the predominant food legume,
cowpeas are grown primarily as a leafy vegetable. The importance and nutritional value of cowpea leaves and fruits may
be greatly underestimated, for the crop is also grown mainly
for these products in South-east Asia. Rachie [188] has
emphasised this point, and he estimates that the protein
content of those cowpea plant parts which are eaten is equivalent to 5 million tonnes of dry seeds annually, a figure
equivalent to almost 30% of food legume production in the
lowland tropics.

Cowpea (*Vigna unguiculata* (L.) Walp.)

Seed yields vary enormously [237], from crop failure to around 3800 kg ha^{-1}. In Africa, they seldom exceed 400 kg ha^{-1} and are commonly much less depending upon the primary product for which the crop is grown, and upon husbandry, the incidence of extremely destructive diseases and pests, and upon the amount and duration of rainfall [7]. It is not surprising that official production statistics for the crop are unreliable, and that the past and present status of cowpeas has been under-rated.

The world centre of cowpea production is West Africa, where pulse production in 1982 was estimated to be about 1.65 million tonnes [64]. It is unlikely that the only other important food legume in the region, Bambarra groundnut, accounts for more than one quarter of this amount. Nigeria is the world's leading cowpea seed producer with at least 0.75 million tonnes annually; significant amounts are also produced in Niger and Upper Volta (Berkina Faso). Steele [225] estimated that about 80% of the Nigerian crop is grown in the ancient cereal farming system of the savanna zones, where unimproved, spreading, photosensitive and locally adapted landraces are interplanted with cereals. Throughout the West African centre of production, most cowpeas come from this traditional farming system where they are nutritionally important for their seed protein (commonly 23% by weight) and for their lysine (5.5-7.8% of the protein) in these cereal-based diets. Their protein is important also in the root and tuber diets of the West African coastal forest zone to which there is a substantial trade in cowpeas from the inland savannas.

Brazil is second to Nigeria in world cowpea production. In a survey of pulse production in that country, François and Sizaret [70] found that 23% (0.5 million tonnes) of the pulses eaten are cowpeas. They are produced in the seasonally dry north-east states (mostly in Pernambuco, Ceara and Piaui) where they are grown in association with maize, and are dominant to it except on the most fertile soils [7]. Watt [252] suggests that the Brazilian crop may be as large

as 0.6 million tonnes annually, but that the rest of Latin
America combined produces only about 50000 tonnes, or even
less. About 80000 ha of cowpeas are grown annually in the
USA, most of them in Texas, Georgia and California [67]
where the crop is entirely mechanised and is grown for its
immature seeds ('peas') which are commercially canned or
frozen for food.

In the most common recipes for the preparation of cowpeas in the West African centre of production [50], the
seeds are soaked to soften them and to facilitate removal
of their testas. Cultivars with thin testas, especially
those which are wrinkled or 'rough' [130], imbibe water more
rapidly than those with thicker, smooth testas [207]; those
seeds with large pectin and phytin concentrations soften
and cook most quickly [170]. Soaking and the removal of
testas renders cowpea food products more digestible, but
also leaches from the seeds some of their thiamin, niacin
and riboflavin [49]. Cooking seeds inactivates heat-labile
antinutritional factors and decreases available lysine concentration [11], but their nutritional value is little
affected [239] because the sulphur-amino acids, not lysine,
are those which limit the nutritional value of cowpea
protein [24].

After removal of the testas, in West African recipes the
softened seeds are ground into a raw paste, spices and often
onions are then added, and balls of the mixture are either
deep-fried in vegetable oil [142], or are wrapped in leaves
(usually those of *Sarcotherynium brachystachys*) and steamed.
The latter product, called 'moin-moin' in south-west Nigeria,
is also suitable for canning [2]. Whole seeds are often
boiled and eaten in soups and stews. Many other recipes
have been published [92].

In the developing countries of Africa, subsistence farmers do not grow cowpeas specifically as a cover crop or for
fodder, but most (80%) of those questioned in a survey in
northern Nigeria fed the haulms of cowpeas grown primarily
for seeds to their livestock [225]. Seed yield is often

sacrificed for fodder yield when plants are uprooted and
dried for this purpose while they still bear green leaves
and immature fruits. In contrast, the principal use of cowpeas (subsp. *cylindrica*) in India is as a forage and fodder
crop [227], and in the developed countries of the tropics
and subtropics cowpeas are grown as cover crops, green
manures and for fodder. Green fodder contains 15-20% crude
protein and about 50% digestible carbohydrate at the early
fruiting stage [249]. Air-dried cowpea hay contains about
11% crude protein [25] and in feeding trials with cattle it
proved to be more digestible than *Stizolobium* hay [145].
Many forage cultivars were grown in the USA at the turn of
the century; among them were such well known ones as
'Brabham', 'Victor' and 'Iron' [51,71].

In West Africa, rare cowpeas were once grown by fishing
communities along the banks of the Benue and Niger rivers
for the fibre (called 'yawa') in their penduncles (which can
be as long as 1 m). The fibre was used to make fishing lines,
and has also been considered as a source of pulp to make
good quality paper [1,15]. Chevalier [39] described such
cultivars as *V. sinensis* var. *textileis*. Rare examples of
these forms in the world collection of germplasm at IITA
have been ascribed to subsp. *unguiculata*. However, they may
merit subspecific status equal to the other cultivated
groups, even though they have a very restricted distribution
and are no longer used as a source of fibre.

BOTANY

Cultivated cowpeas are annual herbs whose remarkable morphological diversity, and the physiological basis for it, have
been studied intensely over the past decade in both controlled environments and field crops [256]. This work has greatly
increased our understanding of cowpeas in agriculture [31,
227,255], and has provided plant breeders with useful in-
sights into those genotype and environment interactions
which are the principal determinants of vegetative growth,

527

reproductive ontogeny and, ultimately, seed yield. It is largely in this important context that the botany of cowpeas is discussed here.

Germination is epigeal. During the rapid, vigorous, early growth of seedlings before emergence, the cotyledons lose as much as 91% of their dry weight, and do not persist [158]. Characteristically, the strong, deep tap root supports a much-branched, well-nodulated lateral root system in the surface soil. The large, globular nodules have flattened surface lenticulations and this may account for the relative intolerance of cowpeas to ephemeral anoxia associated with water logged soils [146]. The first pair of simple, opposite leaves is succeeded by alternate, pinnately trifoliolate leaves on stout, grooved and sometimes pigmented petioles 5-15 cm long, and subtended by large, cordate, spurred stipules which are characteristically inserted far above their bases. The leaflets are ovate to lanceolate, sometimes hastate, 5-18 cm long and 3-16 cm wide, entire or lobed and subtended by inconspicuous stipels. Rare mutants have sessile, unifoliolate leaves without stipels [198].

In the axil of each leaf there are three buds. Only the central bud normally expands to produce either a potentially indeterminate, monopodial branch, or a racemose inflorescence. Consequently, the number of branches is the complement of the number of inflorescences. Rare determinate genotypes have terminal inflorescences on their main stems and branches, and such types may have more than one inflorescence, or an inflorescence and a reduced branch, at a single node [233].

The plant architecture and growth habits of cowpeas are diverse because genotypes and environments interact to influence both the numbers and the lengths of their branches. At one extreme are the bushy, erect types with few short branches; at the other are prostrate, spreading or sometimes twining and climbing forms with five or more orders of branching, and first and second order branches up to 5 m long (Plate 12.1). Since the number of branches is the

Cowpea (*Vigna unguiculata* (L.) Walp.)

Plate 12.1.(a)   Range of traditional, variously indeterminate
cowpea forms available in the world germplasm collection

(b)   An example of the more productive, 'new generation' of
shorter-duration, more compact and 'determinate' types now
available as a result of the efforts of cowpea breeders
(Photographs : R.J. Summerfield and IITA, Nigeria, respectively)

complement of the number of inflorescences, the sooner and more profusely and synchronously a cowpea plant flowers and bears fruit, the fewer are its branches, and the shorter the duration of their growth before reproductive demands deprive them of carbohydrates and nitrogen [167]. While soil fertility [208], soil moisture [235] and crop husbandry [168] can influence the growth and length of branches, the chief determinants of plant architecture and growth habits appear to be those interactions between genotypes on the one hand, and daylength and air temperature on the other, which influence the rates, durations and relative synchrony of successive stages of reproductive ontogeny [87,88,233,255].

Most cowpeas are quantitative short-day plants [255,259] and those which have been studied in detail have distinct photo-thermal requirements for inflorescence initiation and expansion [131,225]. In these genotypes, the duration of vegetative growth, and so the number and lengths of branches, depend upon sowing date in relation to seasonal changes in daylength and temperature and their effects on the phenology of flowering. Other genotypes are less sensitive to photo-period over the range which prevails in tropical and sub-tropical latitudes. In both photosensitive and photoinsensitive genotypes, flowering is hastened by warmer temperatures [87,230]. Indeed, simple, linear relations have now been established between the reciprocals of the times taken to flower (rates of progress towards flowering) and both photo-period and mean diurnal temperature. When the equations which describe these two responses are solved, the time to flower in any given photo-thermal regime is predicted accurately by whichever solution calls for the greater delay in flowering. Thus, in different circumstances flowering is controlled exclusively by *either* photoperiod *or* mean temperature. The value of the critical photoperiod (that which, if exceeded, causes a delay in flowering) is now known to be temperature-dependent and an equation derived from the linear relations mentioned above predicts this value. Considered together as a quantitative model, these relations

suggest simple field methods for screening genotypes to
determine photo-thermal response surfaces. These relations
and their implications are discussed in detail elsewhere
[79].

Steele and Mehra [227] distinguish three groups of cowpeas in a simple classification based upon genetic differences in response to photoperiod, and upon growth habits.
Their scheme may be summarised as follows:

Type IA: Reproductively photoinsensitive, determinate genotypes with erect growth and few branches (Plate
12.1 (b)). They flower at most nodes as early as
five weeks after sowing, and die as their fruits
mature. Such genotypes are important in breeding
programmes because they tend to be phenotypically
stable in diverse environments, and because their
fruits mature synchronously within as few as sixty
days after sowing (and see below).

Type IB: Reproductively photoinsensitive, indeterminate
genotypes whose growth and plant habit are much
influenced by photo-thermal environment. They
flower at many nodes within five or six weeks
after sowing, while branching from others. The
apices of main stems and branches remain vegetative, and while fruits develop the plants grow to
form a more or less dense ground cover depending
upon the influence of soil fertility and moisture,
air temperature and daylength. Typical of such
types are the forage cowpeas bred in the USA.

Type II: Reproductively photosensitive, indeterminate genotypes typified by the ancient landraces grown
with cereals in the traditional agriculture of the
West African savanna zones. They have a range of
short-day requirements for inflorescence initiation [161], and some of them may require even
shorter days before initiated inflorescences expand and flower. This photoperiodic mechanism

confers adaptation to local environments whereby
West African cultivars tend to flower close to the
average date when the rains end in their own locality [225,255]. In days longer than about 12 h
45 min these cowpeas are entirely vegetative and
develop many long diageotropic or twining branches.
As days shorten, inflorescences are initiated in
an unsystematic sequence with branches, but may
lie dormant (or even die) without increasing the
size of the sink for assimilates until the second
photoperiodic trigger stimulates flowering and
synchronous fruit development. These genotypes are
phenotypically unstable in diverse environments.
If they are grown continuously in days shorter
than 11h 30min they appear to be photoinsensitive,
flowering within five or six weeks after sowing,
being erect and producing only a few, short
branches.

The floral characteristics and reproductive biology of
cowpeas have recently been reviewed [233]. Each axillary
inflorescence is a compound raceme of several simple racemes
[166] carried on a grooved peduncle 5-60 cm long, occasionally up to 1 m long in those forms described as var. *textilis*
by Chevalier [39]. Each simple raceme has between six and
twelve flower buds, but only the lower, first-formed pair
develops while the rest degenerate to form extra-floral
nectaries between the paired flowers. Each flower is subtended by a deciduous bract and has a calyx of five wrinkled
acute lobes whose length (which varies from 2-16 mm) has
been used by Verdcourt [247] to distinguish wild subsp.
mensensis (calyx lobes longer than the tube) from the other
four subspecies.
The cleistogamous flowers are typically papilionaceous
and large, with a standard petal 2-3 cm wide and which is
either white or with anthocyanin pigmentation in shades of
pale mauve or pink to dark purple, with or without a 'V'

shaped distribution at the top centre of the petal. The
flowers open in the early morning and then wither, fade and
are shed the same day. Outcrossing is uncommon [187,260]
although two outcrossing mechanisms, either simple recessive
genetic male sterility or mechanical sterility conferred by
constricted petals, have now been described [191]. It is
uncommon for an inflorescence to carry more than two mature
fruits. Various estimates suggest that less than 20% of the
flowers which open produce fruits [166,171,229], a loss of
reproductive and yield potential discussed in detail elsewhere [233,256] and variously ascribed to competition for
assimilates (perhaps mediated by hormones), hot temperatures,
wide diurnal temperature fluctuations, drought stress and
adverse photoperiods. More research is required to determine
the influence of these physiological phenomena on the seed
yield of field crops, but there is no doubt about the depression of yield which results from the depredations of
post-flowering insect pests which cause fruits to be shed
or which feed on them voraciously [28,218,241,245,254].

Mature fruits vary widely in size, shape, colour and
texture. The wild subspecies have short (about 4 cm), scabrous, dehiscent, straight, black fruits which are held erect
at maturity. Among cultivars, fruit length varies between
approximate extremes of 6 cm (subsp. *cylindrica*) to 1 m
(subsp. *sesquipedalis*), with most cultivars of subsp.
unguiculata having fruits 12-20 cm long which are straight,
curved or coiled (the 'ramshorn' varieties of the USA). They
may lack pigmentation (tan or straw coloured) or have varying intensities of anthocyanin pigmentation from pink through
red, to purple or almost black. Any colouring is most often
uniformly distributed, but sometimes occurs in a mottled
pattern, or is confined to, or is only absent from, the
sutures and beak. In texture, fruits of the cultivars are
brittle or soft, and they are more or less easy to thresh
depending upon whether the valves separate at the sutures
or fracture between the seeds. They contain between six and
twenty-one kidney shaped, oval or rarely almost spherical

seeds whose longest dimension is within the range 5-12 mm, and with individual weights between 50 mg and 340 mg, in contrast to those of the wild subspecies which are 2-6 mm long and weigh 20-40 mg. The testas of wild cowpeas and most cultivars are smooth, but among cultivars from the West African savanna zones finely wrinkled, rough testas are common and are the ones preferred by consumers (see earlier comments). Rare cultivars have testas which are split, exposing the cotyledons. Diverse seed colours and colour patterns have been classified by Porter *et al.* [184]. Testas without pigment are white and such seeds are popular with consumers. When present, pigment may be confined to a narrow 'eye' around the hilum in an otherwise white seed; or it may spread over the seed from the hilum to varying extents, and in various combinations of speckled, mottled or solid distributions. Common seed colours are buff, brown, red and black. Green seeds are rare.

Cowpeas are diploid (2 n = 22) with little genetic and no chromosomal divergence of the cultivated subspecies from their wild progenitors [65,226]. Gene exchange between the five subspecies of *V. unguiculata* (*sensu* Verdcourt) is readily achieved by artificial hybridisation, and occurs to a limited extent naturally, but there are no reports of successful wild crosses with other species [197].

FARMING SYSTEMS AND THEIR CONSTRAINTS

Traditional systems and their stability

Traditional cultivars of crops, and traditional farming systems, are often adapted with great precision both to their social uses and to the time-tables imposed by the environment. For example, in the semi-arid regions of northern Nigeria, the effective dates for the onset of the rains is uncertain but those for the end of the rainy season are predictable. Here, farmers have two overriding purposes when the rains begin: to secure the earliest possible crop from part of their land in order to break the 'hungry gap', and to sow and weed as large an area as possible of the staple cereal crops on which they and their families depend for survival, before newly mineralised nitrogen is leached. The first crops to be sown, on the first light rains, are day-neutral types of bulrush millet (*Pennisetum typhoides*); these mature quickly and, being photoinsensitive, are independent of the calendar. The main crop, sorghum, is sown when the rains are established. Whenever they are sown, the sorghum cultivars flower at dates precisely related to the average date of the end of the rains in their home locality. If they were to flower too soon, insects and fungi would be likely to devastate the grains; if, on the other hand, they were to flower too late, there would be insufficient water left in the soil profile for a large proportion of the grains to fill. The adaptation mechanism is a short-day photoperiodic response for flowering.

After weeding their crops of sorghum, farmers usually sow into them a long-season, spreading, photoperiod-sensitive cowpea. The spreading intersown legume helps to control late weeds; it may also protect the soil from erosion during intense rain and supply edible leaves and protein-rich seeds for the diet of the farm family. The cereals may also derive some nitrogen from the associated cowpea late in the season [53], especially with these traditional, spreading cultivars from which more nitrogen (up to 50 kg ha^{-1}) may be turned back into the soil than from less vigorous erect forms.

Photoperiodic sensitivity initiates flowering in the cowpeas at a time when the sorghum is heading and its leaf area is beginning to decline. So, the phenological time-table of the sorghum and of the associated cowpea are closely, even elegantly, interrelated [31,227].

However, it seems probable that co-evolved crops in traditional farming systems have achieved equilibrium not only with each other and with their environment, but also with their parasites. Supporting evidence comes from genetic analyses of obligate parasites and their hosts, suggesting that such systems are the relics of an ancient and balanced polymorphism which was a necessary condition for their co-evolution, from the concentration of resistance genes present in centres of origin, and from the relatively low levels of disease usually found in traditional farming systems [7].

Too little is known as yet of the effects of mixed cropping on populations of insect pests and pathogens, but the few studies that have been made do suggest that the damage they cause is often, but not invariably, less in complex crop mixtures than it is in many sole crops. For example, in cowpeas associated with maize, the severity and rate of spread of *Ascochyta* blight have been found to be greater in the sole-cropped cowpea than in the intercrop, apparently because the maize foliage impedes spore dispersal. In contrast, powdery mildew (*Erysiphe polygoni*) may initially develop more rapidly in sole crop cowpea than when intercropped with maize, but when maize populations are sufficiently dense to shade the cowpea, powdery mildew incidence may be much greater in the intercrop. Then again, beetle-transmitted virus diseases of cowpeas may occur less often when the legume is grown in association with maize, either through smaller vector populations in the mixture or through their reduced mobility [7,149]. Similarly, studies on the effects on insect pests of intercropping of cowpeas with cereals have suggested that the cereal may act as a barrier to the entry and dispersal of some insects although, again, there are exceptional cases where larger numbers of pests occur in the mixture [13,116,140,180]. Matteson [140] has reported a 42% reduction in numbers of flower bud thrips (*Megalurothrips sjostedti*) on cowpea in association with maize relative to a cowpea monocrop in southern Nigeria where this pest is responsible for very substantial crop losses. Infestation by other species, such as the legume pod borer (*Maruca testulalis*), was unaffected by cropping system.

Cowpea (*Vigna unguiculata* (L.) Walp.)

Stresses as limitations to yield

In the savannas of Africa, where most cowpeas are grown, cowpeas are most commonly found as a subordinate companion crop in complex mixtures to which few inputs are added. It has been emphasised that such traditional systems tend to be relatively stable, but they also tend to be poor yielding. Although soil moisture may be limiting, it seems that the traditional practice of late-sowing *per se* forfeits some cowpea yield and it is evident that pests and diseases are the principal restraints to seed yield in Africa. In north-eastern Brazil, the principal restraints to cowpea seed yield appear to be poor soil fertility and drought, especially the irregularity of rainfall; pests and diseases seem not to be the principal determinants of crop losses, as they are in Africa.

Pests and pathogens

Cowpeas are susceptible to very wide ranges of pests and pathogens which attack the crop at all stages of growth; without some protection seed yields in Africa are often negligible [215]. The economic importance of cowpea diseases varies considerably with ecological zone. Thus, web blight, *Cercospora* leaf spot, anthracnose, rust, Cowpea (Yellow) Mosaic Virus and bacterial pustule are considered the major diseases of the West African forest belt, whereas scab (*Sphaceloma* sp.), *Septoria* leaf spot, brown blotch (*Colletotrichum capsici* and *C. truncatum*), bacterial blight (*Xanthomonas campestris* pv. *vignicola*) and Cowpea Aphid-Borne Mosaic Virus (CABMV) are the principal diseases of the African savannas, where the parasitic angiosperm *Striga gesnerioides* is also locally important. *Ascochyta* blight can be destructive under wet conditions at elevations above 1000 m. In the dryland areas of north-eastern Brazil, the most important diseases are Cowpea Severe Mosaic Virus, CABMV and its relatives, scab, leaf smut (*Entyloma vignae*),

Cercospora leaf spot and powdery mildew; while in California and the southern states of the USA, *Fusarium* wilt and root-knot nematodes have long been considered important. For greater detail about the major fungal and virus diseases of cowpea the reader is referred to a recent review by Allen [7] where, in his Tables 20 and 21, an attempt is made to categorise them by their symptoms, giving details also of their geographical distribution and economic importance.

Insect pests not only damage cowpeas at all stages of growth of the crop but also seeds in storage. In Africa, the most important pests are: (a) beetles, leafhoppers and aphids, which feed on foliage before flowering; (b) flower bud thrips (*Megalurothrips sjostedti*), pod borers (particularly *Maruca testulalis*) and pod sucking bugs, which are the principal pests post-flower; and (c) bruchids (especially *Callosobruchis maculatus*), which cause major losses of seed during storage. The extent to which the relative economic importance of these pests varies with ecological zone is uncertain. There is some indication that flower bud thrips may be rather less damaging in the semi-arid savannas of Africa than they are in the more humid zones, but this species and the damage it causes may often have been over-looked. It is rather more clear that pod-sucking bugs, notably *Clavigralla tomentosicollis*, reach greater population densities under savanna conditions than in the forest zone. Conversely, the cowpea seed moth (*Cydia ptychora*) is apparently confined to the more humid zones of West Africa. In north-east Brazil and other cowpea-growing areas of warm temperate and tropical America, the cowpea curculio (*Chalcodermus aeneus*) is the most important insect pest. Leafhoppers are also said to be important in Brazil but this assertion appears to warrant substantiation. The lesser cornstalk borer (*Elasmopalpus lignosellus*) is a locally important cause of damage to seedlings, and various chrysomelid beetles (*Ceratoma* spp. and *Diabrotica* spp.), and a weevil (*Pantomorus* sp.), cause foliar damage. Various groups of insects are not only important as pests but are also vectors

Table 12.1 Insect pests of cowpeas

Pest	Insect species	Distribution	Importance/crop loss	References
I Pre-flowering pests				
Leafhoppers	*Empoasca dolichi* Paoli, *E. christiani* Dworak, *E. kerri* Pruthi and *E. kraemeri* Ross & Moore	Widespread in Africa (*E. dolichi* and *E. christiani*), India (*E. kerri*) and Latin America (*E. kraemeri*)	Considered a major pest in north-east Brazil. In Nigeria, 27% crop loss recorded from infestation (15 adults per plant) at two weeks after sowing.	[178,195,203,215]
Aphids	*Aphis craccivora* Koch and *A. fabae* Scop.	Cosmopolitan	*A. craccivora* is the more important and is locally and seasonally damaging (yield loss up to 40%). Especially important as a vector of Cowpea Aphid-Borne Mosaic Virus.	[214,215]
Whiteflies	*Bemisia tabaci* Genn.	Pantropical	Damage probably negligible. Important as vector of virus disease.	[14,250]
Foliage thrips	*Sericothrips occipitalis* Hood	Africa	Minor foliar feeder at seedling stage. Crop loss 0–15%.	[164,215,218]
Bean stem flies	*Ophiomyia phaseoli* (Tryon), *O. centrosematis* (de Meij), and *O. spencerella* Greathead	*O. phaseoli* and *O. centrosematis* are widespread throughout the Old World tropics; *O. spencerella* is known only from East Africa.	Usually minor on cowpea though locally important as cause of stem and branch break.	[78,126,203]

Table 12.1 contd

Pest	Insect species	Distribution	Importance/crop loss	References
Leaf miners	*Liriomyza trifolii* (Burgess)	North and East Africa	Minor	[82,209]
Lesser cornstalk borers	*Elasmopalpus lignosellus* (Zeller)	America	Seedling damage to cowpea in north-east Brazil.	[8]
Hairy caterpillars	*Amsacta moorei* (Butler)	Senegal, India	Locally serious seedling defoliator.	[157,203]
Foliage beetles	*Ootheca mutabilis* Sahlb. and *O. bennigseni* Weise	Africa; *O. bennigseni* known only from East Africa	Cause considerable damage to seedlings but, because of plants' ability to withstand leaf damage, effects on yield usually minor. *O. mutabilis* is an important vector in West Africa of Cowpea (Yellow) Mosaic, Cowpea Mottle and Southern Bean Mosaic Viruses.	[9,26,27,37,59, 119,165]
Striped foliage beetles	*Medythia quaterna* (Fairmaire)	Africa	Minor. Vector of Cowpea (Yellow) Mosaic and Cowpea Mottle Viruses.	[9,218,253]
Bean leaf beetles	*Ceratoma ruficornis* (Oliver) and *C. trifurcata* (Forster)	Tropical America	Minor. Chief vectors of Cowpea Severe Mosaic Virus. Also transmit Cowpea Chlorotic Mottle and Southern Bean Mosaic Viruses, and bacterial blight.	[19,108,115,251]

540

Common name	Scientific name	Distribution	Remarks	References
Mexican bean beetle and allies	*Epilachna varivestis,* Muls., *E. vigintiocto-punctata* (Fab.)	Tropical America (*E. varivestis*) and Asia (*E. vigintiocto-punctata*)	Minor. *E. varivestis* transmits Cowpea Severe Mosaic Virus.	[109,114]
Striped bean weevils	*Alcidodes leucogrammus* (Erichson)	Africa	Locally and seasonally important as a stem girdler. Larval feeding inside stem causes stunting and yield loss.	[28,192]
II Post-flowering pests				
Legume pod borers	*Maruca testulalis* (Geyer)	Pantropical	Probably the principal cowpea pest worldwide. Damage to flower buds, flowers and green fruits.	[28,104,105,106, 113,172,266]
Lima bean pod borers	*Etiella zinckenella* (Trietschke)	America, Asia and Australia	Minor feeder on developing seeds.	[212,215]
Cowpea seed moths	*Cydia ptychora* (Meyrick)	Africa	Locally important in southern Nigeria as feeder on seeds within mature fruits.	[62,63,174,181, 182,240,241]
African bollworms	*Heliothis armigera* (Hb.)	Widespread in Old World tropics	Sporadic, as a pod borer.	
Long-tailed blues	*Lampides boeticus* (L.)	Widespread in Old World.	Minor. Locally important as a flower feeder on the Accra plains of Ghana.	[5,203]
Flower bud thrips	*Megalurothrips sjostedti* (Tryb.), *Frankliniella schultzei* (Tryb.) and *F. tritici*	Tropical Africa (*M. sjostedti* and *F. schultzei*) and southern USA (*F. tritici*)	*M. sjostedti* is a major pest, responsible for substantial crop loss (up to 100%) in Nigeria. Other spp. are of minor importance.	[159,173,214,244, 254]

Table 12.1 contd

Pest	Insect species	Distribution	Importance/crop loss	References
Pod flies	*Melanagromyza chalcosoma* Spencer	Uganda	Locally important as a pod borer leading to damage of seed with subsequent loss of germinability.	[120]
Pod sucking bugs	*Clavigralla tomentosicollis* Stal., *C. elongata* Signoret, *C. shadebi*, *Anoplocnemis curvipes* F., *Mirperus jaculus* Thunb., *Riptortus dentipes* F. and *Creontiodes purgatus* (Stal.)	Widespread in Africa, except *C. elongata* which is confined to East Africa. *C. purgatus* is reported from cowpeas in Brazil.	A group of major pests which cause shedding, drying and shrivelling of young fruits, and discoloration and shrivelling of seeds.	[56,101,102,128, 137,139]
Green stink bugs	*Nezara viridula* (L.)	Cosmopolitan	Minor in Africa but important in southern USA where infestation (three adults per plant) at first flower causes complete loss of seed. Infestation at late flowering leads to 74% loss.	[160,204]
Blister beetles and pollen beetles	*Mylabris* spp. and *Coryna* spp.	Tropical Africa and India	Both are flower feeders but the damage they cause is unquantified. Locally important, especially in East Africa.	[214]
Pod weevils	*Piezotrachelus varius* Wagner	Africa	Minor feeder on seeds within immature fruits.	[214,215]

542

Cowpea curculio	*Chalcodermus aeneus* Bohem.	Warm temperate and tropical America	The principal cowpea pest in southern USA and north-east Brazil. [35,36,42,246]
III *Storage pests*			
Cowpea seed beetles	*Callosobruchus maculatus* F., *C. chinensis* (L.), *C. rhodesianus* (Pic.) and *Bruchidius atrolineatus* Pic.	*C. maculatus* and *C. chinensis* pantropical; *C. rhodesianus* in West and East Africa.	Major pests of stored seeds. In Nigeria, it has been estimated that each emergence hole of *C. maculatus* in a seed represents 5% loss. *C. maculatus* is the most important species; the infestation begins just before harvest and develops rapidly in store. Conversely, infestation of *B. atrolineatus* starts in the field when fruits are ripening, but later dies out in early weeks of storage. [12,29,32,33,186, 216,221]

543

of virus, notably whiteflies (*Bemisia tabaci*), aphids and beetles (*Ootheca mutabilis* in Africa; *Ceratoma ruficornis* and other species in tropical America). The principal pests of cowpeas worldwide have been described in several reviews, notably those of Singh *et al.* [218], Singh and Allen [214] and Singh and Allen [215], and are summarised in Table 12.1.

Edaphic factors and mineral nutrition

Cowpeas are grown on a wide variety of soils, from sands to heavy, expandable clays; they are said to grow best on well-drained, 'sandy substrates'. Whether grown in rotation with maize or cotton in the southern USA, or variously inter-planted with millet, sorghum, cassava or maize in the tradi-tional farming systems of the tropics, the cowpea crop is often subjected to tillage practices developed primarily for the companion crop: 'cowpeas succeed on a seed bed as pre-pared for maize' [136]; 'ordinary maize cultivator equipment is satisfactory' [52]; 'tillage is normally done for the crops with which the cowpea is planted' [223]. Whilst zero tillage in conjunction with an appropriate herbicide can improve nodulation, vegetative growth, water-use efficiency and seed yield of cowpeas in the humid forest zone of Nigeria [124] the effect is, apparently, by no means univer-sal even at nearby sites [60]. It also needs to be evaluated in semi-arid environments.

Cowpeas share with most other green plants the ability to improve their nutrient uptake by mycorrhizal associations between their roots and soil fungi [94]. Along with other legumes, the potential for assimilation of nitrogen through symbiosis with *Rhizobium* creates special demands, notably for molybdenum and cobalt but also for boron, copper, phos-phate and zinc [153]. However, neither mycorrhiza nor root nodules are essential for cowpea growth and since both can be partially or completely replaced by appropriate fertili-sers it seems absolutely essential to evaluate and to quan-tify the changes in microbial dependence which are likely

to result from applications of inorganic fertilisers to
crops. Unfortunately, this is seldom done and so it is
hardly surprising that estimates of the nutrient requirements of cowpea crops differ widely (Table 12.2) and that
a plethora of responses to various inorganic fertilisers
have been reported. Consequently, there seems little justi-
fication for the statement [136] that 'fertiliser requirements (for cowpeas) are similar to those for soyabeans' (and
see Table 12.2).

Table 12.2 Estimates of the nutrients removed (kg ha^{-1}) in cowpea
seeds for crops producing an economic yield of 1000 kg ha^{-1} (data for
soyabeans included for comparison)

Nutrient (kg 1000 kg seeds^{-1})*						Comments	References
N	P	K	Ca	Mg	S		
40.0	7.4	39.8	11.4	9.0	4.0	Unqualified statement	[190]
56.7	4.4	16.6	NI	NI	NI	Unqualified statement	[118]
42.9	5.0	13.0	NI	NI	NI	Calculated from nutri- tional data	[103]
71.0	6.1	20.3	3.0	3.0	1.7	Soyabeans (USA crops)	[206]
66.8	6.3	17.8	NI	NI	NI	Soyabeans; calculated as above	[103]

*All data converted to elemental equivalents to facilitate comparisons;
NI, no information given.

Other recommendations are often questionable. For example, it seems strange to recommend applications of 165 kg N,
67 kg P and 73 kg Mg ha^{-1} to achieve a seed yield of 807 kg
ha^{-1} from non-nodulated cowpeas on the acid soils of Sierra
Leone [76] when a judicious application of lime may allow
nodule-dependent crops to yield well [93,94,202,208]. However, the role of lime in tropical legume culture has proved
a most controversial subject [47] and debates continue. Then
again, appropriate additions of sulphur can circumvent iron
chlorosis on alkaline, calcareous soils [20] but there are
complex interactions between iron and phosphate in similar
substrates elsewhere [117]. Although large amounts of applied
P may become fixed when added to calcareous soils, it cannot

be wise to advocate foliar applications for cowpeas when, it seems, control plots in the experiments on which the advice was based were not sprayed with equivalent volumes of water [147]. Similarly, it is difficult to assess the agronomic significance of genotypic differences in P absorption by cowpea roots when neither nodulation nor mycorrhizal status can be assessed from the data presented [3]. Many other examples could be cited. It is clear that the recent discussions by Fox and Kang [68], Munns [152] and Munns and Mosse [153] on the type and magnitude of various problems associated with the current and potential fertility of tropical soils, and their collective recommendations for experiments needed to evaluate these problems and the agronomic practices which are required to overcome them, must be heeded in the future, particularly as the productivity of cowpea crops is improved.

Despite the dearth of reliable information on the mineral nutrient requirements of cowpea crops we can be confident that cultivars of subsp. *unguiculata* are potentially extremely productive grain legumes: that is they can symbiotically fix quantities of nitrogen sufficient for very large seed yields (Table 12.3).

Table 12.3 Annual symbiotic fixation (kg N ha^{-1} annum^{-1}) of cowpeas and, for comparison, soyabeans in (a) properly conducted field experiments and (b) 4-year average on-farm values in different countries (after Nutman [162]).

Attribute		Cowpeas	Soyabeans
(a)	Range	73–354	1–168
	Average	198	103
(b)	USA farms	25	88
	Egyptian farms	192	NI

NI, no information

Perhaps it is not surprising that the selection history of cowpeas and the edaphic conditions with which most crops have been forced to contend have proved indirect pressures for 'symbiotic potential'. Clearly, little of this potential

can be expressed against the fertile backcloth of crop rotations in the USA (Table 12.3) and, as we have argued before
[231], cowpea breeders will need to select for nodulation
and nitrogen fixation [75] which means they will need to
restrict applications of inorganic fertilisers to their
breeding plots. Otherwise, the symbiotic potential of cowpeas could decline in proportion to the breeding effort put
into the crop. It is clear that if the amount of nitrogen
removed when the crop is harvested is greater than the
amount fixed during growth then a legume crop will deplete
soil reserves which, of course, negates their assumed beneficial role in crop rotations [4].

Water relations

Most cowpea crops are rainfed, a small proportion are irrigated (e.g. in the southern USA and in Iraq) and others utilise mainly water residual in the soil after a crop of rice
has been harvested [190].

Dry-matter production (DM) by crops grown in semi-arid
environments can be thought of in relatively simple terms:
the total amount of water transpired from the canopy during
a given period of time (T), the efficiency with which the
crop uses this water (the ratio of assimilation to transpiration, WUE) and the distribution (F) of each increment of
dry-matter between various plant organs (leaves, roots,
seeds and so on) viz [68]:

$$DM = f (T, WUE, F)$$

Unfortunately, there have been few attempts to generate
these data for cowpea crops and much research needs to be
done in order both to substantiate similar results and,
especially, to resolve the markedly different conclusions
which have been reported.

Cowpeas with an appropriate crop duration have, it seems,
a wide range of adaptation in the tropics; from short-duration

cultivars in the semi-arid regions (< 600 mm annual rainfall) through the subhumid to the humid forest belt (1000-
1500 mm). In humid southern Nigeria they are said to require
between 51 mm and 127 mm rainfall each month from sowing
through to flowering, but 'much less' thereafter so that
ripe seeds of good quality can be harvested [95,155,169].
Rachie [188,189] has emphasised that without fertiliser applications or irrigation, but with all major pests controlled,
a cowpea crop in the semi-arid or subhumid tropics can be
expected to produce seed yields between 1600 kg ha^{-1} (good)
and 3000 kg ha^{-1} (maximum) within 85 days from sowing; in
doing so, the crop is said to need 'adequate water' for only
55 days (65%) of its life. The problem is not only, 'How
much water is adequate?' but also, 'On which days should it
be available?'

Weekly mean maximum evapotranspiration (ETm) from a
well-watered crop of cultivar TVu 6207 grown in drainage
lysimeters at Ibadan was closely related to the prevailing
weather - small values of ETm (2-4 mm d^{-1} depending on crop
development) were associated with small values of each of
net radiation, mean saturation vapour pressure deficit, and
wind speeds, but the seasonal trend was closely related to
leaf area index [125]. Total maximum evapotranspiration
during a crop duration of 77 days was 275 mm and 311 mm on
two representative soils of the region and the two crops
produced 4.3 g and 4.9 g seed m^{-2} cm water transpired^{-1}
(WUE), respectively. Some of these data can be compared witl
others for irrigated cowpea crops (assumed to have optimal
water supply) in Senegal (Table 12.4). Data for the three
short-duration crops agree reasonably well whereas the
longer-duration crop in the hotter, drier, northernmost site
transpired more than twice as much water despite an increase
in crop duration of only 64%. Dancette and Hall [44] have
stressed that maximum water requirements of annual crops in
one region are strongly dependent upon crop duration; indeed,
the two variables may be linearly correlated. These limited
data for cowpeas are inevitably confounded with differences

in phenology and plant habit; they neither prove nor dis-
prove a linear relationship between the duration of a crop
and its maximum water requirement.

Table 12.4 Comparative attributes of the water requirements for cowpea crops at different locations (calculated or extracted from Dancette
and Hall [44] and Lawson [125])

Cultivar	Crop duration (days)	Location	CWR (mm)	Ep (mm d^{-1})	ETm	ETm/Ep
TVu 6207	77	*Nigeria* Ibadan (7°N)	275	5.0	3.6	0.72
TVu 6207	77	Ibadan	311		4.1	0.82
B 21	75	*Senegal* Bambey (14°N)	336	5.9	4.5	0.76
Unknown	125	Guede (16.5°N)	715	8.8	5.7	0.65

CWR: total crop water requirement
Ep: Class A pan evaporation (mm d^{-1})   } average values
ETm: Mean weekly maximum evapotranspiration (mm d^{-1}) } for cropping
                                                             season

A series of experiments during the dry season at Ibadan,
when there is little chance of rain for several weeks but
temperatures are similar to those which prevail during the
main cropping season, involved irrigating crops in the
field at weekly intervals except for two weeks either just
before or just after the onset of flowering [258]. Stomatal
conductance (measured between 13.00 and 15.00 h on the basal
leaflets of a recently expanded trifoliolate at the top of
the canopy) declined from a value of $0.6 \, cm \, s^{-1}$, typical for
the non-stressed controls, to a minimum of $0.2 \, cm \, s^{-1}$ after
seven days of water stress: the stomata were closed and probably remained so for most of each subsequent day without
water. Although leaf water potential (pressure bomb) was
consistently smaller on the stressed compared with the control plants, it remained more or less stable throughout the
stress period (between -1.0 and -0.8 MPa) whereas values for
soyabean decreased from -1.1 to -2.2 MPa during this time.
Wien and his colleagues considered that the resistance to
water movement through the plant might decline as the

severity of water stress increased although, as they discussed, in well-watered plants grown in containers leaf water potential is linearly and negatively related to transpiration rate - which implies a constant resistance to water movement through the plant [80]. Clearly, further research is needed, especially on stressed plants grown in the field.

Withholding water during vegetative growth decreased the rate of initiation of main stem leaves but Wien *et al*. [258] were not able to show a significant relation between rate of leaf appearance and leaf water potential (range -0.7 to 1.2 MPa) and concluded that leaf water potential would not serve as a good indicator of plant water stress. On the other hand, Cutler [43] described the seasonal changes in midday leaf water potentials for rainfed crops at Niomo, Mali and noted changes in leaf temperature and leaflet orientation when values were between -0.81 and -0.82 MPa for all five cultivars; he concluded that a midday value of approximately -0.82 MPa 'could be considered a critical leaf water potential and that cowpeas are very sensitive to relatively mild (-1.0 MPa) midday deficits'. Furthermore, after comparing the temporal changes in leaf water potential with the rate of elongation of terminal leaflets, Cutler [43] also concluded that differences in the rate of leaflet elongation provide a very sensitive indication of the onset of detrimental water deficits in vegetative plans (and see Elston and Bunting [57]).

One of the cultivars studied in Mali, TVu 662 (an indeterminate type classified as intermediate between semi-erect and semi-prostrate in habit and originating from southern Nigeria [184]) was by far the best 'drought avoider'. Plants of this cultivar were able to maintain a more favourable water status in a drought environment; pre-dawn, midday and diurnal changes in leaf water potential were consistently more positive (the latter by an average of 0.2 MPa) than in the other four cultivars. On the other hand, TVu 37 (perhaps better known colloquially as Pale Green; an erect, indeterminate vigorous cultivar with a tendency to twine and

originating from the Republic of South Africa [184]) was
described as the most sensitive cultivar to drought of the
five tested. We had screened these two cultivars previously
in controlled daylength, day and night temperature glasshouses [236] for their ability to produce seeds in hot days
- a major feature which influences adaptability to the
tropical aerial environment [237]. On average, economic
yields of the twenty-two cultivars tested were 42.3% smaller
in hot (33°C) than in warm (27°C) days and the responses of
TVu 662 and TVu 37 were at opposite extremes (Table 12.5).
The former was the only cultivar to produce dramatically
larger yields in hot than in warm days, the latter was the
most sensitive and yield declined by 65% in the warmer daytime regime. Although neither tissue temperatures nor plant
water status were measured, these data may have exciting
implications. The fact that such large genotypic differences
in response to adverse environments were detected among even
such a small collection of germplasm must offer encouragement to the physiologist and plant breeder alike. Secondly,
and perhaps more important, there seems to be a potential
for using precisely-controlled glasshouses in which to
screen nodule-dependent, well-watered, pot-grown plants not
only for adaptability to temperature *per se* but also, indirectly, for genotypic differences in ability to withstand
the consequences of drought.

Table 12.5 Relative changes (%) in seed yield (g plant^{-1}) of selected
cowpea cultivars grown in factorial combinations of hot and warm days
(33° and 27°C) with warm and cool nights (24° and 19°C) in a glasshouse
[236]

Cultivar	*Mean seed yield per plant in:*		Relative effect of
	Hot days (33°/19°-24°C)	Warm days (27°/19°-24°C)	hot temperature %
Overall average of 22 cultivars	17.9	31.0	−42.3
TVu 662	42.3	27.9	+51.6
TVu 37	19.2	55.2	−65.2

Probably the most extensive data on the consequences of water stress in cowpea have been generated for cultivar TVu 4552, an erect, acutely branched, non-twining, non-vigorous, more or less determinate cultivar with a typical crop duration of 60-80 days [258]. Withholding weekly irrigations to induce a ten-day period of progressively more severe water stress immediately before the onset of flowering increased leaf temperatures so that they were at least as warm as ambient air during the middle of the day, reduced net photosynthesis by at least 27% at full irradiance, and restricted leaf area development so that the maximum leaf area index achieved was only about 3 compared with 5 in the non-stressed plots. The stressed plants had shorter main stems which, together with peduncles, contributed a smaller proportion to total plant dry weight at maturity, when fruits represented 55% and 45% of the total in the stressed and control plants, respectively (seed yields of 203 and 209 $g\ m^{-2}$). Drought stress significantly reduced nodule dry weight per plant and symbiotic activity during the stress period but, after drought was alleviated, nitrogen fixation efficiency (specific activity) became significantly greater than the controls. Overall, the stressed plants accumulated about 25% less total reduced nitrogen during a crop duration of 63 days (about 12 and 16 $g\ m^{-2}$, respectively) but remobilised a larger proportion from vegetative components into fruits. However, although a reduction of 15.3% in the number of fruits per m^2 was offset by a 14.8% increase in mean seed weight, so that the stressed plants produced seed yields equally as good as the controls, their fruits contained almost 20% less nitrogen; presumably the seeds of the stressed plants contained a smaller concentration of protein.

Wien and his colleagues [258] have discussed the problems involved in screening for drought resistance and have urged that it is important to use field conditions whenever possible and that several traits will probably need to be measured. Plants grown in pots may develop atypically large leaf water potentials; they probably develop water deficits

at rates which are abnormally rapid and which preclude the
acclimatisation responses which seem to be typical of field-
grown plants [81]. Indeed, the only 'safe' generalisations
which might be made to date are that cowpeas seem better
able than soyabeans to withstand drought in tropical temperature conditions and although water stress is undoubtedly
an important selective force there is little evidence for
any species that it has a direct regulatory action on flower
initiation [154]. Whether or not severe drought stress has
a greater adverse effect on phenological potential [235] or
on the subsequent realisation of this potential [81] will
probably depend on previous and subsequent environmental
conditions (e.g. the effects will be less severe when post-
drought temperatures and vapour pressure deficits favour
recovery), on whether or not the rate of a particular process (e.g. leaf initiation or leaf expansion) is changed
permanently or only temporarily [148], and whether the cul-
tivar(s) concerned are effectively nodulated or largely
dependent on inorganic nitrogen [222] (see Chapter 4).

PEST AND DISEASE MANAGEMENT POTENTIAL

Host plant resistance

Among the control strategies available, host plant resistance is now widely recognised as the pivot of an integrated
management of both pests and diseases, against which not
only chemical control measures but also cultural and biological methods may be appropriate. Resistant cultivars cost
the farmer nothing, nor does their adoption necessarily dis-
rupt his farming system.

The potential for controlling diseases by host plant
resistance was first realised more than eighty years ago
when Orton [176] reported differences in reaction to *Fusarium*
wilt among cowpea cultivars. Combined resistance to wilt
and root-knot nematodes was found in cultivar Iron, which
was used by Mackie [133] in crosses with the susceptible

cultivar California Blackeye to produce disease-resistant blackeye types for California, and later by Hawthorn [83] in breeding resistant 'crowder' types in Louisiana. However, despite these early efforts, relatively little subsequent work was devoted to cowpeas until the late 1960s and early 1970s, when an increased awareness of the need to improve food legumes for the tropics led to the establishment of various national and international research and breeding programmes.

As a consequence of intensive work during the last twenty years, perhaps most notably at IITA in Nigeria (see Chapter 18), resistance is now available to most of the major diseases of cowpea [7,215]. Indeed, many genotypes possess combined disease resistance [10,73,264] and several culti- vars are known to combine resistance to as many as nine or ten diseases [7]. The emphasis placed by cowpea improvement programmes on disease resistance has largely obviated the need for chemical protection. Resistance to *Striga gesne- rioides* has now also been identified [101].

Pest resistance has also been successfully identified in cowpeas. Resistance of various types has been found agai- nst leafhoppers *(Empoasca dolichi)* [195], aphids *(Aphis craccivora)* [38,213], thrips *(Megalurothrips sjostedti)* [254], the legume pod borer *(Maruca testulalis)* [106], the cowpea seed moth *(Cydia ptychora)* [62,181], the cowpea cur- culio *(Chalcodermus aeneus)* [36,41], various pod bugs [101, 129,159] and bruchids *(Callosobruchis maculatus)* [72,101, 163].

There is obviously considerable variation in the degree, or level, of resistance expressed in different host-parasite combinations as well as in the way in which that particular resistance is expressed. In some cases, for example, there is a continuous gradation between resistance and susceptibi- lity, typical of a quantitative interaction, whereas in others there is a clear-cut distinction between the two classes, indicative of a qualitative interaction between the host and its parasite. Of the two types of interaction,

qualitative resistance more frequently confers complete protection although lower levels are often sufficient to reduce
damage to below the economic threshold. Qualitative interactions appear to be more common with diseases than with
pests, and pest resistance more often requires some supplementation by other components of an integrated management
system.

The ways in which resistance is expressed may be divided
into three broad categories:

(a)   decreased infection frequency or degree of infestation,
      conferred by avoidance ('escape') mechanisms prior to
      invasion, including resistance to penetration or
      probing;

(b)   inhibition of parasite development, expressed as a de-
      creased rate of reproduction or a slower development
      of symptoms and governed by mechanisms operating after
      invasion;

(c)   decreased damage, or tolerance.

Several, possibly independent, mechanisms may be involved in each of these main components of resistance to a
particular pest or disease. Whereas more attention has been
devoted to mechanisms of 'escape' and non-preference among
pests relative to their equivalents against pathogens (resistance to spore deposition and to penetration), the biochemical bases for 'antibiosis' against insects remain
remarkably poorly understood despite the relatively long
history of research into plant chemical defences against
pathogens [110].

Examples of non-preference among cowpea pests include
the apparent ovipositional non-preference of *Bemisia tabaci*
for cultivars resistant to Cowpea Golden Mosaic Virus [250],
the ovipositional non-preference shown by *Cydia ptychora*
for fruits of certain wild types [62], and the ovipositional
non-preference of *Callosobruchus maculatus* for cultivars
with particularly rough seed coats [163]. Thick pod walls

of certain cultivars afford a mechanical resistance to oviposition by *Chalcodermus aeneus* [41] and possibly also by pod bugs and bruchids [129]. Mechanisms of escape from pests include early maturity, the hosts' capacity for profuse and early flowering, the height at which fruits are borne above the canopy, and the angle at which they are subtended from their peduncles [213,215]. Among cowpea pathogens, there is evidence of an inhibition of spore germination in *Cercospora* leaf spot [205] and of failed cuticular penetration by *Septoria vignicola* [199]. In other host-parasite combinations, resistance to spore deposition may be attributed to leaf waxiness, and insect feeding is often impeded by pubescence, but neither has been found among cowpea cultivars.

Histological studies of the mechanisms of resistance to rust and to anthracnose suggest that, in each case, there are 'switching points' in the infection process beyond which further fungal development may, or may not, continue, depending on the host cultivar concerned. Ultrastructural examination of the interaction of various cowpea cultivars and rust *(Uromyces appendiculatus)* has revealed that there are at least four histologically distinct resistance mechanisms, including three 'immune' responses and a necrotic fleck reaction [84,85]. A similar range of mechanisms is involved in restricting the development of *Colletotrichum lindemuthianum* in host tissue. These mechanisms include penetration leading to encapsulation of hyphae, failure of the primary vesicle to develop, and cell death which leads to the cessation of fungal growth associated with the accumulation of phytoalexins [219]. Vignafuran and methyl phaseollidin isoflavan appear to be the most important antifungal compounds associated with resistance to anthracnose [185]. Similarly, kievitone seems to be the basis for the expression of race-specific resistance to stem rot *(Phytophthora vignae)* [179], and the differential synthesis of chlorogenic acid may account for verticillium wilt resistance [141]. Immunity to Cowpea (Yellow) Mosaic Virus has been investigated at the cellular level by Beier *et al.* [22,23] who found that proto-

plasts from immune plants either prevented or delayed virus
replication, or died after inoculation, depending on the
host cultivar.

Mechanisms of antibiosis against insect pests are much
less well known. Studies with non-host legumes suggest that
saponins [16], L-DOPA [201] and canavanine and phytohaemagglutinins [111,112] provide bases for resistance and host-
specificity among seed-eating beetles *(Callosobruchus* spp.).
More recent work [72] has shown that an abnormally large
concentration of tripsin inhibitor is the basis of resistance to *Callosobruchis maculatus* in the cowpea cultivar
TVu 2027. Antibiosis could be the basis also of resistance
of cultivar TVu 123 to *Empoasca dolichi*, being expressed as
decreased survival of nymphs and increased susceptibility
to insecticide. Although associated with strong phenolase
activity, the biochemistry of leafhopper resistance remains
largely unknown [58,195]. Poor nutritional quality, as
opposed to toxicity, may underlie larval mortality and reduced fecundity of *Cydia ptychora* on resistant cultivars [62,
182]. Resistance to damage caused by *Megalurothrips sjostedti* is related in some cases to a resistance to floral abscission induced by ethylene, which is known to be produced by
infested cowpea peduncles [254]. But, it remains unknown
whether differential synthesis of, or sensitivity to ethylene is one basis of thrips resistance in cowpeas, in addition to an antibiosis [101].

Tolerance is also recognised in cowpeas against infestation with insects, including leafhoppers and aphids, but
the mechanisms involved are not known. However, it is clear
that cultivars differ in their ability to compensate for
damage, depending on their leafiness and vigour as well as
upon the growth stage at which that damage is inflicted
[257].

Chemical protection

Early work on the control of insect pests [28,46,113,121,

557

193,242] and, to some extent, also on the control of diseases [151,177,262,263] of cowpeas, focussed on the use of
chemicals. Insecticides such as DDT, lindane, dieldrin,
malathion and monocrotophos applied on as many as six occasions increased seed yield by as much as ten-fold compared
to unprotected plots. Since the most critical stage for pest
attack was the period of early flower formation and fruit
development, the application of insecticide was recommended
to begin just before flowering. However, there are several
important limitations to the wider adoption of chemical
control measures in cowpeas, including the expense and difficulty of obtaining not only the chemicals themselves but
also the equipment to apply them, the technology and labour
required for repeated and timely applications, and the scarcity of water in many regions. Despite these limitations,
which apply particularly to subsistence farmers with limited
resources in Africa, there is considerable potential for a
minimum use of insecticides as part of an integrated management system, particularly if selective compounds and formulations with a low toxicity to the predators and parasites
of pests are used. Ultra-low volume application of synthetic
pyrethroids may have such a potential [34,61,194]; also, in
certain cases, seed dressings may have a similarly useful
future in the control of seedling diseases [262,263] and
seed-eating beetles [216]. Irrespective of the chemical and
its mode of application, chemical control measures should
seek to supplement other components of an integrated management of pests and diseases; unfortunately, too little is
known at present about the specific effects of insecticides
on beneficial insects.

Biological control

There are many examples of antagonism, parasitism and predation of the pests and, to a lesser extent, the pathogens of
cowpeas, and yet almost nothing is known of their contribution to the natural suppression of insect and pathogen popu-

lations on cowpeas in traditional agriculture in the tropics.
A potential is sometimes suggested by the accidental demonstration that the use of a pesticide exacerbates a pest or
disease by eliminating its natural enemies [265]. Attempts
to harness this sort of potential, and introductions of
exotic parasites and predators to control cowpea pests, have
not been reported.

Against this background, there is a growing body of information on the parasites and predators of insect pests.
For example, brachonid wasps of the genera *Braunsia* and
Bracon are larval parasites of *Maruca testulalis* and *Cydia
ptychora*, respectively [175,218], whereas the anthocorid
Orius is a predator of *Megalurothrips sjostedti* [74]. Pod
sucking bugs are parasitised by tachinid flies [137,138] and
their eggs are parasitised by encyrtid and scelinid wasps
[55,137,139]. Certain fungi are parasites of cowpea pests,
for example, *Entomophthora fresenii* on *Aphis craccivora*
in Nigeria (D.J. Allen, unpublished). Studies in Brazil
[45] have revealed that fungal pathogens play a potentially
important role in the natural regulation of cowpea pests:
both *Beauveria bassiana* and *Metarhizum anisopliae* were associated with *Chalcodermus aeneus* and with chrysomelid beetles;
Paecilomyces fumosoroseus caused a serious disease of *Lagria
villosa*, and an *Entomophaga* sp. was obtained from the leafhopper, *Empoasca kraemeri*. Almost nothing is known about the
potential for the biological control of cowpea pathogens.
And yet, there is some evidence that certain interactions
between pathogens, or between pathogens and non-pathogens,
can lead to some protection from diseases [17,123,127].
Gall-forming weevils (*Smicronyx* sp.) are parasites of the
parasitic weed *Striga gesnerioides* on cowpeas in the savannas of West Africa [261]. More work is warranted on the
potential of biological control, particularly of insect
pests, in an integrated management system.

Cultural control practices and the potential of crop mixtures

The management of pests and diseases of cowpeas by cultural means has received very little attention, despite evidence from other legume crops that sanitation measures, the use of clean seed and the modification of agronomic practices could each contribute to their suppression. Potentially useful sanitation measures include the burning of crop residues and the destruction of alternate hosts, whereas rotation, altered tillage regime, adjustment to the time of sowing or plant spacing, the use of trap crops, the timing of harvest, and post-harvest treatment are agronomic practices that could be modified so as to lessen the effects of a pest or pathogen. Numerous pathogens and a few pests are seed-borne in cowpea, and the production and use of clean seed through certification schemes is potentially very important although difficult to implement in developing countries.

Despite the preponderance of intercropping practices in subsistence farming in the tropics, there is a dearth of information on the effects of intercropping on pests and diseases [7,101,180]. The extent to which a particular association affords protection obviously depends on many factors, including the temporal and spacial arrangements of the cropping system, the relative susceptibility of its components, and the biology of the insects and pathogens themselves. Although it is not safe to assume that damage is necessarily less in mixtures than in monocrops, it seems probable that there is considerable potential for the manipulation of cropping systems so as to maximise their productive effects.

Integrated systems of pest and disease management

In 1969, Taylor [243] reported an attempt to introduce an integrated approach to pest control of cowpeas in Nigeria where the combination of a bacterial preparation (Bacillus thuringiensis), the use of insecticides and sanitation were

560

considered promising. Subsequent work has been reviewed by Singh *et al.* [218]. That the adoption of resistant cultivars might reduce dependence on insecticides was demonstrated by Nangju *et al.* [156], and later work by Matteson [140] explored the feasibility of combining a 'minimum' of insecticide with resistant cultivars in cropping systems which favoured the maintenance of natural enemies while providing some protection from pest damage. Although resistance to disease in cowpeas is, in most cases, sufficient to obviate reliance on chemical protection, it seems certain that there are important lessons to be learnt from existing systems of traditional agriculture themselves, for it is here that stability between co-evolved hosts and parasites has often been attained [7].

GERMPLASM RESOURCES

The primary centre of genetic diversity of subsp. *unguiculata* is in West Africa, and within that region especially the savanna zones of Nigeria north of about latitude 9°N. The Indian subcontinent is a secondary centre. Variation in the wild progenitors in Africa has yet to be adequately studied, but they appear to be equally diverse in both West and eastern Africa. India is the primary centre of diversity of subsp. *cylindrica*, and South-east Asia of subsp. *sesquipedalis*.

The present status of cowpea germplasm resources has been greatly influenced by the activities of the International Board for Plant Genetic Resources (IBPGR; see Chapter 18), and is dominated by a world collection of more than 12000 accessions, including landraces, gathered from eighty-five countries and preserved at IITA at Ibadan in Nigeria (see Chapter 18). Genetic resource centres in fifteen countries are listed by the IBPGR [18] as holding significant cowpea collections. Among the largest of them are:

(a) Southern Region Plant Introduction Station,

Experiment, Georgia, USA.

(b) N.I. Vavilov All-Union Institute of Plant Industry, Leningrad, USSR.

(c) National Bureau of Plant Genetic Resources, New Delhi, India.

(d) National Plant Genetic Laboratory, University of the Philippines, Los Banos, Philippines.

(e) Chitedze Agricultural Research Station, Lilongwe, Malawi.

The earliest published accounts of cowpea germplasm collections (see [30,40,150,183]) described the agricultural and horticultural varieties grown in the USA. More recently, studies of diversity in collections have been conducted in India [6,122,143,211], and in Nigeria [225], but the great majority of published work relating to genetic diversity in cowpeas has been concerned with inheritance studies. This work has been reviewed by Fery [66].

The systematic collection of cowpea germplasm in the West African centre of diversity began in the 1950s. The cultivars and wild subspecies gathered then were added to the Nigerian Grain Legumes Gene Bank [54] and eventually, with other collections donated from India and the USA, were the foundation in 1972 of IITA's world collection. Exploration for cowpea germplasm in West Africa was renewed by IITA in 1972 [196] and was greatly expanded between 1976 and 1980 when the Institute organised thirty-eight plant collecting missions to eighteen African countries and added 4672 accessions to the world collection [228]. In a revision of its priorities for exploration and germplasm collection among crops and regions, the IBPGR [89] accorded cowpea (with other crops) first priority in the West African and Southeast Asian regions, and second priority elsewhere.

During IITA's exploration activities, priority has been given to gathering seeds from mature crops but, inevitably, many samples come from farm stores and village markets. Theoretical sampling strategies [135] designed to maximise

562

the probability of capturing alleles present in populations
at predetermined frequencies are hard to implement when
gathering germplasm from African subsistence agriculture.
The strategy adopted has been to sample the widest possible
range of environments, using both a random and selective
approach to individual crops within which, at least in West
Africa, there is often great diversity. With each sample
collected a comprehensive description of the source is recorded and kept as part of the documentation of the germplasm.

An evaluation of fifty characters of the first 6837 accessions in IITA's world collection has been published as a
germplasm catalogue [184]. Although the collection has since
doubled in number (to more than 12700), there has been no
significant shift in the means and ranges of quantitative
characters, nor of the class frequencies of qualitative
characters reported in that catalogue. On the other hand,
the number of sources has been increased. The most important
sources, and the numbers of accessions from them, are:

Afghanistan	61	Malawi	171	Zambia	822
Benin	372	Niger	587	Brazil	161
Cameroon	430	Nigeria	2846	Upper Volta (Berkina Faso)	190
Egypt	340	Philippines	50	United States	869
Ghana	278	South Africa	81	34 Other Countries	341
India	1638	Tanzania	411	No information	2684
Ivory Coast	127	Togo	94		
Mali	194	Turkey	47		

Important sources of germplasm not yet adequately represented in the IITA world collection have been listed in
order of priority for exploration by the IBPGR [89] as:

Cultivated forms:
    (a)   Africa
          - Fernado Po., Mozambique, Zimbabwe

- Botswana, Lesotho, South Africa, Swaziland
- Central African Republic, Zaire
- Angola, Congo, Gabon

(b) South-east Asia and China.

Wild forms:

- South Africa (Natal and Transvaal), Zimbabwe
- East African and Zambezian phytogeographical zones.

While the morphological diversity in the IITA collection is most remarkable, its principal value has been as a source of resistances to several major diseases [7] and to some pests [217] which have now been incorporated into otherwise elite breeding lines (see below). These achievements illustrate the importance and practical value for crop improvement of a very large and genetically diverse germplasm collection, and fully justify its costly acquisition and preservation. An outstanding example was the discovery of resistance (conferred by an unusually large concentration of tripsin inhibitor) to the cowpea storage weevil *(Callosobruchus maculatus)* in only one accession from the entire collection [217]. The major deficiency in the IITA world collection is that wild cowpeas are insufficiently represented; nor has their genetic diversity been adequately studied, and it has yet to be used in cowpea improvement.

The descriptors and descriptor states used for the routine evaluation of cowpea germplasm collections have recently been reviewed [91] and now provide an internationally accepted standard for the exchange of information. The descriptors are in three categories:

(a) *passport data* which consist of accession identifiers and the information recorded by collectors;

(b) *characterisation data* which describe those characters which are strongly heritable and expressed in all environments; and

(c) *preliminary evaluation data* about a limited number

of traits thought desirable by a consensus of
users.

Using sixty-two descriptors in these categories, and a
total of 305 descriptor states, the documentation of the
IITA world collection is computerised for selective retrieval, and is freely available, with seeds, on request.

The IITA germplasm is preserved as a 'working' collection from which seeds are drawn in response to about 10000
requests each year, and in a 'base' collection for long-term
preservation. The working collection contains up to one kg
of seed of each accession. For the base collection, freshly-
harvested seeds with viability greater than 90% are dried
to 5-6% moisture content in dry air without heat and sealed
in aluminium cans for storage at -20°C. Three cans, each of
0.33 litre capacity, are stored for each accession. The
number of seeds stored therefore varies greatly depending
upon seed size.

There are numerous problems associated with the management of such a large and much-used collection. Principal
among them are the need to maintain the genetic integrity
of each accession through initial seed increase and eventually rejuvenation, and the related need to provide reliable
documentation which enables plant breeders to select material with confidence that the traits of interest to them
are accurately described and will be faithfully reproduced.
The widespread and effective use of the germplasm depends
upon the documentation, and the seeds it describes, fulfil-
ling these criteria.

When new accessions are received they often arrive in
small quantities (100 seeds or less) and may be mixtures of
pure lines. As a practical expedient to achieve reliable
documentation and to promote effective use, such mixtures
are separated on the basis of seed characters and the derived lines are subsequently maintained as distinct accessions. On the rare occasions when accessions consist of
hybrid seed, they are maintained as such.

BREEDING STRATEGIES AND PROGRESS

Breeding objectives and strategy

Recent progress in the genetic improvement of *Vigna unguiculata* has been thoroughly reviewed by Smithson *et al*. [220] and by Goldsworthy and Redden [77], and the role that disease-resistance breeding has played in that improvement has been discussed by Allen [7]. The genetics of disease resistance has been reviewed by Fery [66] and Meiners [144], and recent work on the inheritance of pest resistance has been described from IITA [101].

The objectives of earlier breeding work emphasised the improvement of yield potential by the incorporation of photoperiod insensitivity, an erect growth habit with fruits borne well above the leaf canopy, and pest and disease resistance, while maintaining seed characteristics acceptable to consumers. However, with the recognition of insect pests and pathogens as the principal factors which limit productivity throughout Africa, and the importance of both photoperiod sensitivity and indeterminate plant types in traditional agriculture, a more recent strategy at IITA has attempted to incorporate pest and disease resistance into a range of plant types and photoperiodic reactions suited to different cropping systems and environments [220]. The rate at which these objectives may be expected to be achieved obviously depends on a complex of factors, including the size and diversity of the available germplasm, the breeding procedure, the potential for advancing several generations each year, and the efficacy of selection.

Although bulk and pedigree selection, and backcross procedures have led to significant improvements (especially with regard to those characters which are controlled by relatively few genes), genetic progress tends to be restricted by continued selfing and by the small numbers of parents which can be conveniently handled in crossing, so that the probability of novel recombinations is reduced. These limitations may be lessened by the use of population improvement and recurrent selection methods as have been applied to cowpeas, either with or without the use of genetic male sterility, and some evidence of genetic advance by such an approach has been obtained [220].

The principal limitations to conventional breeding strategies for resistance appear to be twofold: that the durability of resistance in some cases is inadequate; and that levels of resistance are sometimes insufficient. Although it is premature to assess whether population improvement and related strategies are likely to overcome these limitations, it is evident that there are cases in which resistance is known, or suspected, to be a quantitative character to which these breeding methods are better suited. Effective selection for resistance depends to a large extent on the reliability and reproducibility of the screening technique, and on the method of selection and its intensity. The former depends heavily upon the ability to regulate pest populations and manage disease epidemics [7].

The limitations of concentrating selection in early generations at a single location were recognised at IITA in 1976 (and see Fig. 12.1, later). In an attempt to solve the problems of genotype by environment interactions, a set of testing locations was established by IITA in 1977 to take advantage of the diversity of environments within Nigeria, selecting sites representative of the major ecological zones of West Africa from the very humid lowland tropics to the dry sahel. Information obtained from trials across locations has since provided the bases on which to stratify breeding materials by the environment(s) for which they are ultimately intended. Recent work on cowpeas at IITA has therefore shifted its emphasis from the exclusive development of lines which perform well across environments to a strategy of also harnessing environment specificity through which there is an ecological deployment of specific combinations of characters [7,77,220].

567

Breeding progress and impact on agriculture

Traditional cultivars of cowpea in Africa are usually prostrate, leafy and photosensitive. Though they are often
adapted with considerable precision to the environments in
which they have evolved [255], they have inherently small
yield potentials and are widely susceptible to pests and
diseases. Selection among locally adapted types has led to
the identification of somewhat superior cultivars, such as
SVS 3 in Tanzania, and crosses between local cultivars and
introduced germplasm have produced the improved cultivars
Westbred, Ife Brown, N'Diambour and Baol in West Africa
[220].

Work at IITA in the early 1970s led to the identification of four lines which combined good yield performance
with resistance to a range of diseases and insect pests and
a fifth, comprising an F_2 population segregating for male
sterility [191], which have been described as 'VITA' lines.
The identification of suitable breeding materials was accompanied by the intensification of recombination procedures
and, by 1975, 3000 crosses had been achieved using conventional methods. Comparisons of lines from this programme
across environments have indicated a progressive improvement
in yield of the breeding material relative to the VITA lines
[220]. New VITA lines were described, possessing good combined resistance to disease and, in favourable environments,
capable of seed yields larger than 2.5 t ha^{-1}, which represents almost a ten-fold increase over average yields in
Africa. Breeding progress is also reflected in the composition of the Cowpea International Trials from 1974-79. In
1974, only three of the sixteen entries were the products
of IITA's breeding efforts whereas, by 1979, twenty-three
of the twenty-eight entries were so derived [77].

The early emphasis on selection of erect, photoperiod-
insensitive cultivars tended to eliminate much of the northern Nigerian material which is characterised by large,
rough, white seeds. In consequence, improved genotypes were

often unacceptable to consumers. To overcome this, a series
of backcrosses were begun in 1976 to increase the seed size
of the otherwise superior VITAs 4 and 5; preferred seed-
types have now been successfully combined with heavy yield
potential and with pest and disease resistance.

Progress made with the incorporation of combined resistance to diseases, largely obviating the need for any reliance on chemical control [7], has revealed more clearly the
extent of losses attributable to insects. Since 1979, the
cowpea programme at IITA has intensified its efforts in
searching for sources of resistance to the major insect
pests. Work on the biology and behaviour of *Maruca testulalis* has subsequently provided a first step in the development of control measures based on host plant resistance
[105,106], and a *Maruca* colony has been established so as
to provide insects needed to infest field nurseries from
which to identify resistance among breeding lines [107]. The
partial resistance to *Maruca* possessed by TVu 946 has been
used in breeding, and TVx 3890-010F is among the most promising of its derivatives [106]. Much attention has also been
devoted to breeding for bruchid resistance based on TVu 2027,
into which resistance to Cowpea (Yellow) Mosaic Virus has
been crossed from TVu 408-P_2 [99]. Advanced lines have now
been developed that combine bruchid and disease resistance,
heavy yield potential and good seed quality. With the identification of further sources of resistance, including a
resistance to pod-wall penetration in cultivar Worthmore,
there are opportunities for both raising the level and increasing the durability of bruchid resistance [101]. Lines
with partial resistance to flower bud thrips, such as TVu
1509, have also been used as parents in breeding. Line TVx
3236, derived from a cross between TVu 1509 and Ife Brown,
has maintained the level of thrip resistance found in the
resistant parent [238], and attempts are now being made to
incorporate aphid resistance into TVx 3236 [101. Thus, combined resistance to the major insect pests is beginning to
become a reality.

During the last few years, the conduct of international trials, the establishment of national programmes and of scientific training are together beginning to have some impact. Genetically superior material from IITA has been evaluated widely, and selections have been used either directly as new cultivars or as parents in further adaptive breeding. For instance, the national programme in Tanzania has released two new cultivars derived from the locally accepted SVS 3 into which resistance to two viruses and a heavier yield potential has been incorporated [210]. Subsequently, the cultivars Hope and Pride have also been released to Tanzanian farmers [101]. Releases have also been made in Guyana, where ER 7, VITA 8, TVx 2907 and VITA 6 have been re-named Minica 1 to 4, respectively. The national programme in Brazil is making a considerable impact. VITAs 3 and 6 have been released as EMAPA 822 and 821, respectively, and more recently the cultivars CNC 0434 and Manaus have overcome a net deficit in grain legume seed production in at least one region of that country. In Upper Volta (Berkina Faso), VITA 7 has been released as KN 1, and in the north of Nigeria TVx 3236 is being multiplied on a large scale in Kano state where farmers planned to grow more than 4000 ha in 1983. Farmers' yields from TVx 3236 ranged from 1000-2000 kg ha^{-1} with two or three applications of insecticide, so emphasising the contribution that is to be made from host plant resistance [101].

AVENUES OF COMMUNICATION

From a modest beginning in the early 1970s, the Documentation and Information Services maintained at IITA have transformed the dissemination and exchange of information on the cowpea crop worldwide. The world literature on the crop has been collated and published in three volumes of abstracts [96,98,100]. Most of these publications are in the collection of the International Grain Legume Information Centre (IGLIC), which is based at IITA and sponsored jointly by

the Institute and the International Development Research
Centre of Canada. IGLIC will provide photocopies of publications on request.

The *Tropical Grain Legume Bulletin*, published quarterly
by IGLIC, provides a forum for researchers and extension
workers to exchange information and ideas about tropical
grain legume research and development. The *Bulletin* attracts
many contributions on cowpeas.

Annual Reports and *Research Highlights* are published and
distributed widely by IITA each year; the Institute has also
organised numerous international conferences and workshops
on diverse subjects, including the breeding and improvement
of cowpeas and the farming systems in which the legume is
cultivated (see Literature Cited).

A World Cowpea Research Conference, sponsored by IITA
and the Bean/Cowpea Collaborative Research Support Program
(see Chapter 18) at Michigan State University, was organised in November 1984. The objectives of the conference
were threefold:

(a) to collate information on cowpea research related
to cropping systems, crop physiology, breeding,
agronomy, entomology, pathology, microbiology and
quality improvement;
(b) to establish collaborative links between various
national and international organisations for cooperative research and exchange of materials; and
(c) to attract donor attention for funding of cowpea
research.

The conference itself, and the published proceedings
from it, represent the latest links in a communication network which continues to expand to the benefit of all who
are interested in cowpeas, whatever their discipline and
sphere of concern.

PROSPECTS AND FUTURE CHALLENGES

There are opportunities for further substantial advances in cowpea improvement through the fuller exploitation of the germplasm diversity of *V. unguiculata* currently available. The substantial progress referred to in this chapter has been based on a much narrower germplasm [77]. Specific characters have not been exploited and so it seems possible, for example, that attempts to transfer the pubescence available in *V. unguiculata* subsp. *dekindtiana* var. *protracta* (*sensu* Marechal, *et al.* [134]), with which the cultigen is interfertile, could confer some advantage with respect to insect resistance.

There are also opportunities for both stabilising and improving the level of partial resistance by intercrossing lines known to possess distinct components of resistance to a particular pest or pathogen, a strategy tantamount to a 'pyramiding of mechanisms' [7]. Such an opportunity has been appreciated at IITA where pod-wall resistance and seed resistance are being combined to combat bruchids [99]. One of the greatest challenges now facing plant pathologists is to provide guidance to breeders on the means of selecting durable resistance.

We have emphasised that genetically superior cultivars of cowpeas are now available, and that they are beginning to make an impact on productivity in some places. The magnitude and extent to which such impact is likely to be sustained in a predominantly subsistence crop depends, we suggest, on the ability to match cultivar development to the environment and cropping system for which they are intended. Increased attention to the definition of agro-ecological zones, based on analyses of climatic and edaphic data as well as on cropping systems and their principal constraints, seems a vital prerequisite for the rational deployment of cultivars. The large interactions that occur between cultivars and environments [220] led to an appreciation at IITA for stratifying the cowpea breeding programme, as we have discussed earlier.

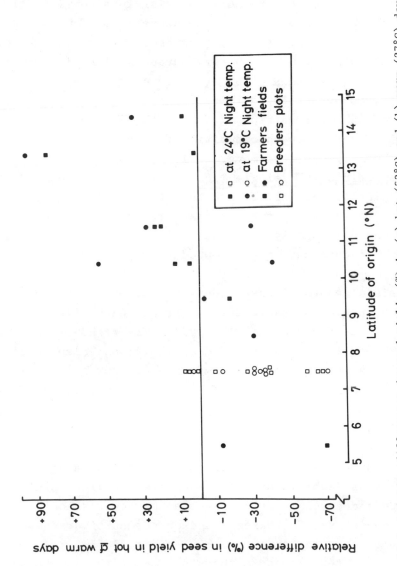

Figure 12.1 Relative differences in seed yields (%) in (a) hot (33°C) and (b) warm (27°C) days for cowpeas grown by subsistence farmers in West Africa and for those 'improved' genotypes bred or selected at Ibadan (7°30'N) during the early 1970s (relative differences calculated as [(a)−(b)]/(b)×100)

Such a stratification allows for the deployment of characters of relevance to specific agro-ecological zones. For
instance, accessions from hot, arid environments are more
tolerant of heat, and are able to maintain a consistently
larger plant water status throughout periods of hot temperature and soil drying, than accessions from the forest zone
[97]. Cultivars bred for the dry savannas of Africa would
therefore require this combination of characters or, alternatively, to be sufficiently early-maturing so as to escape
severe drought stress. The advent of the 'sixty-day' cultivars [101] opens new opportunities for these dry environments. Furthermore, the move, in 1976, away from concentrating selections at a single location (i.e. at Ibadan; 7°30'
N) was significant too; before that time, cowpea breeders
seeking to improve pest and disease resistance had concurrently and unknowingly selected against an ability to yield
well in hot days during the reproductive period (see Fig.
12.1)

Resistances to diseases such as *Septoria* leaf spot,
scab, brown blotch, bacterial blight and CABMV [10], to
pests such as the pod bug *Clavigralla tomentosicollis*, and
to witchweed (*Striga gesnerioides*) [101], are also required
for these arid regions. There may be room for greater emphasis on screening for adaptation to intercropping, as well as
to the production of 'dual purpose' types for eastern Africa,
where leaf plucking is practised. In the forest zone of West
Africa, vegetable types harvested for their green fruits
have potential [200]. Here, resistance to Cowpea Golden
Mosaic Virus, web blight and *Maruca* are required. Aluminium
toxicity and calcium deficiency are characteristic of soils
in these high rainfall areas, and cowpea cultivars have been
found with some tolerance of soil acidity [86].

If the materials and methods now emerging from international programmes are used adequately by national research
teams to adapt them to local requirements, and if international breeding efforts become increasingly decentralised,
then we may expect greater and more enduring productivity
of cowpeas in the future.

LITERATURE CITED

1. Adeleke, A. and Holgate, R.A. (1964), Technical Memorandum No. 21, Federal Institute for Industrial Research, Nigeria.
2. Adeniji, A.O. and Potter, N.N. (1980), *Journal of Food Science* 45, 1359-62.
3. Adepetu, J.A. and Akapa, L.K. (1977), *Agronomy Journal* 69, 940-43.
4. Agboola, A.A. and Fayemi, A.A.A. (1972), *Agronomy Journal* 64, 409-12.
5. Agyen-Sampong, M. (1978), in Singh, S.R., van Emden, H.F. and Taylor, T.A (eds) *Pests of Grain Legumes: Ecology and Control*. London : Academic Press, 85-92.
6. Ahmad, S.T., Magoon, M.L. and Mehra, K.L. (1972), *SABRAO Newsletter* 4, 103-06.
7. Allen, D.J. (1983), *The Pathology of Tropical Food Legumes : Disease Resistance in Crop Improvement*. Chichester : Wiley.
8. Allen, D.J., Redden, R. and Jackai, L.E.N. (1980), Unpublished report, International Institute of Tropical Agriculture. Ibadan : IITA.
9. Allen, D.J., Anno-Nyako, F.O., Ochieng, R.S. and Ratinam, N. (1981), *Tropical Agriculture, Trinidad* 58, 171-5.
10. Allen, D.J., Emecheve, A.M. and Ndimande, B. (1981), *Tropical Agriculture, Trinidad* 58, 267-74.
11. Almas, K. and Bender, A.E. (1980), *Journal of the Science of Food and Agriculture* 31, 448-52.
12. Alzouma, I. (1981), *Series Entomologica* 19, 205-13.
13. Alzouma, I. and Huignard, J. (1981), *Oecologica Applica* 2, 391-400.
14. Anno-Nyako, F.O., Vetten, H.J., Allen, D.J. and Thottappilly, G. (1983), *Annals of Applied Biology* 102, 319-23.
15. Anon. (1953), *Colonial Plant and Animal Products* 3, 331-41.
16. Applebaum, S.W., Marco, S. and Birk, Y. (1969), *Journal of Agriculture and Food Chemistry* 17, 618-22.
17. Asafo-adeji, B. (1982), MSc. thesis, University of Guelph, Ontario, Canada.
18. Ayad, G. and Murthi Anishetty, N. (1980), *Directory of Germplasm Collections I. Food Legumes*. Rome : IBPGR Secretariat.
19. Bancroft, J.B. (1971), *Cowpea Chlorotic Mottle Virus*, CMI/ABB Descriptions of Plant Viruses no. 49.
20. Bansal, K.N. and Singh, H.G. (1975), *Soil Science* 120, 20-24.
21. Baradi, T.A.El (1975), *Abstracts on Tropical Agriculture* 1, 9-19.
22. Beier, H., Siler, D.J., Russell, M.L. and Bruening, G. (1977), *Phytopathology* 67, 917-21.
23. Beier, H., Bruening, G., Russell, M.L. and Tucker, C.L. (1979), *Virology* 95, 165-75.
24. Bender, A.E., Mohammadiha, H. and Almas, K. (1979), *Qualitas Plantarum Plant Foods for Human Nutrition* 29, 219-26.
25. Bhaid, B.D. and Talapatra, S.K. (1965), *Indian Journal of Dairy Science* 18, 153-5.
26. Bock, K,R., Guthrie, E.J. and Meredith, G. (1978), *Annals of Applied Biology* 89, 423-8.
27. Booker, R.H. (1965), *Samaru Miscellaneous Paper No. 10*; Samaru, Zaria, Nigeria.
28. Booker, R.H. (1965), *Bulletin of Entomological Research* 55, 663-72.

29. Booker, R.H. (1967), *Journal of Stored Products Research* 3, 1-15.
30. Brittingham, W.H. (1946), *Proceedings of the American Society of Horticultural Science* 48, 478-80.
31. Bunting, A.H. (1975), *Weather* 30, 312-25.
32. Caswell, G.H. (1959), *Bulletin of Entomological Research* 50, 617-80.
33. Caswell, G.H. (1961), *Tropical Science* 3, 154-8.
34. Caswell, G.H. and Raheja, A.K. (1977), *Samaru Agriculture Newsletter* 19, 46-9.
35. Chalfant, R.B. (1973), *Journal of Economic Entomology* 66, 727-9.
36. Chalfant, R.B., Suber, E.F. and Canerday, T.D. (1972), *Journal of Economic Entomology* 65, 1679-82.
37. Chant, S.R. (1959), *Annals of Applied Biology* 47, 565-72.
38. Chari, M.S., Patel, G.J., Patel, P.N. and Raj, S. (1976), *Gujarat Agricultural University Research Journal* 1, 130-32.
39. Chevalier, A. (1944), *Revue de Botanique Appliquée et d'Agriculture Tropicale* 24, 128.
40. Connell, J.H. (1894), *Texas Agriculture Experiment Station Bulletin* 34, 582-4.
41. Cuthbert, F.P. and Davis, B.W. (1972), *Journal of Economic Entomology* 65, 778-81.
42. Cuthbert, F.P. and Fery, R.L. (1976), *Tropical Grain Legume Bulletin* 5, 43-4.
43. Cutler, J.M. (1979), Unpublished report, International Institute of Tropical Agriculture, Ibadan : IITA.
44. Dancette, C. and Hall, A.E. (1979), *Ecological Studies* 34, 98-118.
45. Daoust, R.A., Roberts, D.W. and Soper, R.S. (1983), *Annual Review of the Bean Improvement Cooperative* 26, 86-7.
46. Dina, S.O. (1977), *Experimental Agriculture* 13, 155-9.
47. Dobereiner, J. (1974), *Proceedings of the Indian National Academy of Science B.* 40, 768-94.
48. Doggett, H. (1970), *Sorghum.* London : Longman.
49. Doughty, J. and Oracca-Tetteh, R. (1966), in Stantion, W.R. (*ed.*) *Grain Legumes in Africa.* Rome; FAO, 9-32.
50. Dovlo, F.E., Williams, C.E. and Zoaka, L. (1976), *Cowpeas: Home Production and Use in West Africa.* IDRC Publ. 055e. Ottawa: IDRC.
51. Duggar, J.F. (1900), *Alabama College Station Bulletin* 118, 1-40.
52. Duke, J.A. (1980), *Handbook of Legumes of World Economic Importance.* New York: Plenum Press.
53. Eaglesham, A.R.J. (1981), *Soil Biology and Biochemistry* 13, 169-71.
54. Ebong, U.U. (1971), *Samaru Agriculture Newsletter* 13, 21-4.
55. Egwuatu, R.I. and Taylor, T.A. (1977), *Bulletin of Entomological Research* 67, 31-3.
56. Egwuatu, R.I. and Taylor, T.A. (1977), *Bulletin of Entomological Research* 67, 249-57.
57. Elston, J. and Bunting, A.H. (1980), in Summerfield, R.J. and Bunting, A.H. (*eds*). *Advances in Legume Science.* London: HMSO, 37-42.
58. Emden, H.F.van (1980), in Summerfield, R.J. and Bunting, A.H. (*eds*) *Advances in Legume Science.* London: HMSO, 187-97.
59. Enyi, B.A.C. (1974), *Experimental Agriculture* 10, 87-95.
60. Ezedinma, F.O.C. (1964), *Nigeria Agricultural Journal* 1, 21-5.
61. Ezueh, M.I. (1976), *Tropical Grain Legume Bulletin* 4, 15-18.
62. Ezueh, M.I. (1981), *Annals of Applied Biology* 99, 313-21.
63. Ezueh, M.I. and Taylor, T.A. (1981), *Annals of Applied Biology* 99, 307-312.

Cowpea (*Vigna unguiculata* (L.) Walp.)

64. FAO (1983), *Production Yearbook for 1982*. Rome: FAO.
65. Faris, D.G. (1965), *Canadian Journal of Genetics and Cytology*, 7, 433-52.
66. Fery, R.L. (1980), in Janick, J.L. (*ed.*) *Horticultural Reviews*. *Vol. 2*. Westport, Connecticut: Avi Publishing, 311-93.
67. Fery, R.L. (1981), *HortScience* 16, 474.
68. Fischer, R.A. and Turner, N.C. (1978), *Annual Review of Plant Physiology* 29, 277-317.
69. Fox, R.L. and Kang, B.T. (1977), in Vincent, J.M., Whitney, A.S. and Bose, J. (*eds*) *Exploiting the Legume - Rhizobium Symbiosis in Tropical Agriculture*. University of Hawaii, College of Tropical Agriculture, Miscellaneous Publication no. 145, 183-210.
70. François, P. and Sizaret, F. (1981), *Food and Nutrition* 7, 11-18.
71. Garrison, W.D. (1905), *South Carolina Agriculture Experiment Station Bulletin* 123, 1-15.
72. Gatehouse, A.M.R., Gatehouse, J.A., Dobie, P., Kilminster, A.M. and Boulter, D. (1979), *Journal of the Science of Food and Agriculture* 30, 948-58.
73. Gay, J.D. (1976), *HortScience* 11, 621-2.
74. Ghauri, M.S.K. (1980), *Bulletin of Entomological Research* 70, 287-91.
75. Gibson, A.H. (1980), in Summerfield, R.J. and Bunting, A.H. (*eds*) *Advances in Legume Science*. London: HMSO, 69-75.
76. Godfrey-Sam-Aggrey, W. (1975), *Zeitschrift für Acker-und Pflanzenbau* 141, 169-77.
77. Goldsworthy, P.R. and Redden, R. (1982), *Tropical Grain Legume Bulletin* 24, 9-15.
78. Greathead, D.J. (1968), *Bulletin of Entomological Research* 59, 541-61.
79. Hadley, P., Roberts, E.H., Summerfield, R.J. and Minchin, F.R. (1983), *Annals of Botany* 51, 531-43.
80. Hailey, J.L., Hiler, E.A., Jordan, W.R. and van Bavel, C.H.M. (1973), *Crop Science* 13, 264-7.
81. Hall, A.E., Foster, K.W. and Waines, J.G. (1979), *Ecological Studies* 34, 148-79.
82. Hammad, S.M. (1978), in Singh, S.R., van Emden, H.F. and Taylor, T.A. (*eds*) *Pests of Grain Legumes: Ecology and Control*. London: Academic Press, 135-7.
83. Hawthorne, P.L. (1943), *Proceedings of the American Society of Horticultural Science* 42, 562-4.
84. Heath, M.C. (1971), *Phytopathology* 61, 383-8.
85. Heath, M.C. (1974), *Physiological Plant Pathology* 4, 403-14.
86. Horst, W.J., Wagner, A. and Marchner, H. (1983), *Zeitschrift für Pflanzenphysiologie* 109, 95-103.
87. Huxley, P.A. and Summerfield, R.J. (1974), *Plant Science Letters* 3, 11-17.
88. Huxley, P.A. and Summerfield, R.J. (1976), *Annals of Applied Biology* 83, 259-71.
89. IBPGR (1981), *Revised Priorities Among Crops and Regions*. Rome: IBPGR Secretariat.
90. IBPGR (1982), *Report of the IBPGR Ad Hoc Working Group on* Vigna *Species*. Rome: IBPGR Secretariat.
91. IBPGR (1983), *Cowpea Descriptors*. Rome: IBPGR Secretariat.
92. IDRC (1982), *Cowpeas - Soups to Snacks*. IDRC Nutrition Report No. 11. Ottawa: IDRC.

93. IITA (1977), *Grain Legume Improvement Program In-House Report for 1976*. Ibadan: IITA.
94. IITA (1977), *Annual Report for 1976*. Ibadan: IITA.
95. IITA (1979), *Annual Report for 1978*. Ibadan: IITA.
96. IITA (1979), *Cowpeas (Vigna unguiculata L. Walp.): Abstracts of World Literature, 1900-1949*. Ibadan: IITA.
97. IITA (1980), *Annual Report for 1979*. Ibadan: IITA.
98. IITA (1980), *Cowpeas (Vigna unguiculata L. Walp.): Abstracts of World Literature, 1950-1973*. Ibadan: IITA.
99. IITA (1981), *Annual Report for 1980*. Ibadan: IITA.
100. IITA (1981), *Cowpeas (Vigna unguiculata L. Walp.): Abstracts of World Literature, 1974-1980*. Ibadan: IITA.
101. IITA (1983), *Annual Report for 1982*. Ibadan: IITA.
102. IITA (1983), *Research Highlights for 1982*. Ibadan: IITA.
103. Isom, W.H. and Worker, G.F. (1979), *Ecological Studies* 34, 200-23.
104. Jackai, L.E.N. (1981), *Insect Science and its Application*, 2, 205-7.
105. Jackai, L.E.N. (1981), *Annals of the Entomological Society of America* 74, 402-8.
106. Jackai, L.E.N. (1982), *Bulletin of Entomological Research* 72, 145-56.
107. Jackai, L.E.N. and Raulston, J.R. (1982), *IITA Research Briefs* 3, 1-6.
108. Jager, C.P.De (1979), *Cowpea Severe Mosaic Virus*. CMI/AAB Descriptions of Plant Viruses no. 209.
109. Jansen, W.P. and Staples, R. (1970), *Journal of Economic Entomology* 63, 1719-20.
110. Janzen, D.H. (1981), in Polhill, R.M. and Raven, P.H. (eds), *Advances in Legumes Systematics*. London: HMSO, 951-77.
111. Janzen, D.H., Juster, H.B. and Liener, I.E. (1976), *Science* 192, 795-6.
112. Janzen, D.H., Juster, H.B. and Bell, E.A. (1977), *Phytochemistry* 16, 223-7.
113. Jerath, M.L. (1968), *Journal of Economic Entomology* 61, 413-16.
114. Kabir, A.K.M.F. (1978), in Singh, S.R., van Emden, H.F. and Taylor, T.A. (eds). *Pests of Grain Legumes: Ecology and Control*. London: Academic Press, 33-6.
115. Kaiser, W.J. and Vakili, N.G. (1978), *Phytopathology* 68, 1057-63.
116. Karel, A.K., Lakhani, D.A. and Ndunguru, B.J. (1982), in Keswani, C.L. and Ndunguru, B.J. (eds), *Intercropping*. Ottawa: IDRC 102-9.
117. Kashirad, A., Bassiri, A. and Kheradnam, M. (1978), *Agronomy Journal* 70, 67-70.
118. Kassam, A.H. (1976), *Crops of the West African Semi-Arid Tropics*. Hyderabad: ICRISAT.
119. Kayumbo, H.Y. (1975), *Proceedings IITA Collaborators Meeting on Grain Legume Improvement*, 89-94.
120. Koehler, C.S. and Mehta, P.N. (1970), *East Africa Agriculture and Forestry Journal* 36, 83-7.
121. Koehler, C.S. and Mehta, P.N. (1972), *Journal of Economic Entomology* 65, 1421-7.
122. Kohli, K.S., Singh, C.B., Singh, A., Mehra, K.L. and Magoon, M.L. (1971), *Genetica Agraria* 25, 231-42.
123. Kuhn, C.W. and Dawson, W.O. (1973), *Phytopathology* 63, 1380-5.
124. Lal, R., Maurya, P.R. and Osei-yeboah, S. (1978), *Experimental Agriculture* 14, 113-20.

Cowpea (*Vigna unguiculata* (L.) Walp.)

125. Lawson, T.L. (1979), in Lal, R. and Greenland, D.J., (*eds*), *Soil Physical Properties and Crop Production in the Tropics*. New York: Wiley, 227-34.
126. Litsinger, J.A., Quirino, C.B., Lumban, M.D. and Bandong, J.P. (1978), in Singh, S.R., van Emden, H.F. and Taylor T.A. (*eds*), *Pest of Grain Legumes: Ecology and Control*. London: Academic Press, 309-20.
127. Long, D.W. (1963), *Phytopathology* 53, 881 (abstract).
128. Lukefahr, M.J. (1981), *Tropical Grain Legume Bulletin* 23, 29.
129. Lukefahr, M.J. (1981), *Annual Report for 1980, International Institute of Tropical Agriculture*. Ibadan: IITA.
130. Lush, W.M. and Evans, L.T. (1980), *Field Crops Research* 3, 267-86.
131. Lush, W.M. and Evans, L.T. (1980), *Annals of Botany* 46, 719-25.
132. Lush, W.M., Evans, L.T. and Wien, H.C. (1980), *Field Crops Research* 3, 173-87.
133. Mackie, W.W. (1934), *Phytopathology* 24, 1135 (abstract).
134. Marechal, R.J., Mascherpa, J.M. and Stainier, F. (1978), *Bossiera* 28, 10-225.
135. Marshall, D.R. and Brown, H.D. (1975), in Frankel, O.H. and Hawkes, J.G. (*eds*), *Genetic Resources for Today and Tomorrow*. Cambridge: Cambridge University Press, 53-80.
136. Martin, J.H., Leonard, W.H. and Stamp, D.L. (1976), *Principles of Field Crop Production*. London: Collier Macmillan.
137. Materu, M.E.A. (1971), *East African Agriculture and Forestry Journal* 36, 361-83.
138. Matteson, P.C. (1980), *Entomologist's Monthly Magazine* 66.
139. Matteson, P.C. (1981), *Bulletin of Entomological Research* 71, 547-54.
140. Matteson, P.C. (1982), *Tropical Pest Management* 28, 372-80.
141. McNeill, K.E. (1976), PhD. thesis, Mississippi State University (Dissertation Abstracts Int. B. Sci. Eng. 36, 1005).
142. McWatters, K.H. and Brantley, B.B. (1982), *Food Technology* 36, 66-8.
143. Mehra, K.L., Singh, C.B. and Kohli, K.S. (1969), *India Journal of Heredity* 1, 53-8.
144. Meiners, J.P. (1981), *Annual Review of Phytopathology* 19, 189-209.
145. Miller, T.B., Rains, A.B. and Thorpe, R.J. (1964), *Journal of the British Grassland Society* 19, 77-80.
146. Minchin, F.R. and Summerfield, R.J. (1976), *Plant and Soil* 25, 113-27.
147. Mohta, N.K. and De, R. (1971), *Indian Farming* 20, 27-8.
148. Monteith, J.L. (1977), in Alvim, P. (*ed.*) *Ecophysiology of Tropical Crops*. London: Academic Press, 1-27.
149. Moreno, R. (1975), *Turrialba* 25, 361-4.
150. Morse, W.J. (1920), *USDA Farmers' Bulletin* no. 1148, 26.
151. Mukiibi, J. (1969), 'Control of leaf diseases of cowpeas caused by fungal pathogens'. Contribution to the 12th Meeting of Specialists, Committee for Agricultural Botany, Kampala, September 1969.
152. Munns, D.N. (1977), in Vincent, J.M., Whitney, A.S. and Bose, J. (*eds*) *Exploiting the Legume – Rhizobium Symbiosis in Tropical Agriculture*. University of Hawaii, College of Tropical Agriculture, Miscellaneous Publication no. 145, 211-36.
153. Munns, D.N. and Mosse, B. (1980), in Summerfield, R.J. and Bunting, A.H., (*eds*) *Advances in Legume Science*. London: HMSO, 115-25.
154. Murfet, I.C. (1977), *Annual Review of Plant Physiology* 28, 253-78.

155. Nangju, D. (1976), *Journal of Agricultural Science, Cambridge* 87, 225-35.
156. Nangju, D., Flinn, J.C. and Singh, S.R. (1979), *Field Crops Research* 2, 373-85.
157. Ndoye, M. (1978), in Singh, S.R., van Emden, H.F. and Taylor, T.A. (*eds*), *Pests of Grain Legumes: Ecology and Control*. London: Academic Press, 113-15.
158. Ndunguru, B.J. and Summerfield, R.J. (1975), *East African Agriculture and Forestry Journal* 41, 65-71.
159. Nilakhe, S.S. and Chalfant, R.B. (1982), *Journal of Economic Entomology* 75, 223-7.
160. Nilakhe, S.S., Chalfant, R.B. and Singh, S.V. (1981), *Journal of Economic Entomology* 74, 21-30.
161. Njoku, E. (1958), *Journal of the West African Science Association* 4, 99-111.
162. Nutman, P.S. (1976), *Journal of the Royal Agricultural Society of England* 137, 86-94.
163. Nwanze, K.F., Horber, E. and Pitts, C.W. (1975), *Environmental Entomology* 4, 409-12.
164. Nyiira, Z.M. (1971), *PANS* 17, 194-7.
165. Ochieng, R.S. (1978), in Singh, S.R., van Emden, H.F. and Taylor, T.A. (*eds*), *Pests of Grain Legumes: Ecology and Control*. London: Academic Press, 187-91.
166. Ojehomon, O.O. (1968), *Journal of the West African Science Association* 13, 227-34.
167. Ojehomon, O.O. (1970), *Journal of Agricultural Science, Cambridge* 74, 375-81.
168. Ojehomon, O.O. and Bamiduro, T.A. (1971), *Tropical Agriculture, Trinidad* 48, 11-19.
169. Ojomo, O.A. (1967), Mimeo Report, Ministry of Agriculture and Natural Resources, Ibadan, Nigeria, No. 30.
170. Ojomo, O.A. and Chheda, H.R. (1972), *Journal of the West African Science Association* 17, 3-10.
171. Okelana, M.A.O. and Adedipe, N.O. (1982), *Annals of Botany* 49, 485-91.
172. Okeyo-Owuor, J.B. and Ochieng, R.S. (1981), *Insect Science and its Application* 1, 263-68.
173. Okwakpam. B.A. (1967), *Nigerian Entomological Magazine* 1, 45-6.
174. Olaifa, J.A. and Akingbohungbe, A.E. (1981), *Insect Science and its Application* 1, 151-60.
175. Olaifa, J.I. and Akingbohungbe, A.E. (1982), *Bulletin of Entomological Research* 72, 567-72.
176. Orton, W.A. (1902), *USDA Agricultural Bureau and Plant Industry Bulletin* 17, 9-20.
177. Oyekan, P.O. (1979), *Plant Disease Reporter* 63, 574-7.
178. Parh, I.A (1983), *Bulletin of Entomological Research* 73, 25-32.
179. Partridge, J.E. and Keen, N.T. (1976), *Phytopathology* 66, 426-9.
180. Perrin, R.M. (1977), *Agroecosystems* 3, 93-118.
181. Perrin, R.M. (1978), *Bulletin of Entomological Research* 68, 47-56.
182. Perrin, R.M. (1978), *Bulletin of Entomological Research* 68, 57-63.
183. Piper, C.V. (1912), *USDA Agricultural Bureau and Plant Industry Bulletin* 29, 1-160.
184. Porter, W.M., Rachie, K.O., Rawal, K.M., Wien, H.C., Williams, R.J. and Luse, R.A. (1974), *Cowpea Germplasm Catalog No. 1*. Ibadan: IITA.

Cowpea (*Vigna unguiculata* (L.) Walp.)

185. Preston, N.W., (1975), *Phytochemistry* 14, 1131-2.
186. Prevett, P.F. (1961), *Bulletin of Entomological Research* 52, 635-45.
187. Purseglove, J.W. (1968), *Tropical Crops. Dicotyledons*. London: Longman.
188. Rachie, K.O. (1977), in Ayanaba, A. and Dart, P.J. (*eds*) *Biological Nitrogen Fixation in Farming Systems of the Tropics*. New York: Wiley, 45-60.
189. Rachie, K.O. (1978), in Jung, G.A. (*ed*) *Crop Tolerance to Sub-Optimal Land Conditions*. American Society of Agronomy, Special Publication no. 32, 71-96.
190. Rachie, K.O. and Roberts, L.M. (1974), *Advances in Agronomy* 26, 1-132.
191. Rachie, K.O., Rawal, K.M., Franckowiak, J.D. and Akinpelu, M.A. (1975), *Euphytica* 24, 159-63.
192. Raheja, A.K. (1975), *Tropical Grain Legume Bulletin* 1, 8.
193. Raheja, A.K. (1976), *PANS* 22, 229-33.
194. Raheja, A.K. (1976), *PANS* 22, 327-32.
195. Raman, K.V., Singh, S.R. and van Emden, H.F. (1980), *Journal of Economic Entomology* 73, 484-8.
196. Rawal, K.M. (1973), in *Proceedings of the First IITA Grain Legume Improvement Program Workshop*, October 1973. Ibadan: IITA, 15-16.
197. Rawal, K.M. (1975), *Euphytica* 24, 699-707.
198. Rawal, K.M., Porter, W.M., Franckowiak, J.D., Fawole, I. and Rachie, K.O. (1976), *Journal of Heredity* 67, 193-4.
199. Rawal, R.D. and Sohi, H.S. (1981), *Current Science* 50, 188-9.
200. Redden, R.J. (1981), *Tropical Grain Legume Bulletin* 23, 13-17.
201. Rehr, S.S., Janzen, D.H. and Feeny, P.P. (1973), *Science* 181, 81-2.
202. Rhodes, E.R., Evenhuis, B. and Taylor, W.E. (1979), *Tropical Agriculture, Trinidad* 56, 241-3.
203. Saxena, H.P. (1978), in Singh, S.R., van Emden, H.F. and Taylor, T.A. (*eds*) *Pests of Grain Legumes: Ecology and Control*. London: Academic Press, 15-23.
204. Schalk, J.M. and Fery, R.L. (1982), *Journal of Economic Entomology* 75, 72-5.
205. Schneider, R.W. and Sinclair, J.B. (1975), *Phytopathology* 65, 63-5.
206. Scott, W.O. and Aldrich, S.R. (1970), *Modern Soybean Production*. Illinois: S & A Publications.
207. Sefeh-Dedeh, S. and Stanley, D.W. (1979), *Cereal Chemistry* 56, 379-86.
208. Sellschop, J.P.F. (1962), *Field Crop Abstracts* 15, 259-66.
209. Singh, B.B. and Merrett, P.J. (1980), *Tropical Grain Legume Bulletin* 21, 15-17.
210. Singh, B.B. and Mligo, J.K. (1981), *Tropical Grain Legume Bulletin* 22, 14-16.
211. Singh, C.B., Mehra, K.L. and Kohli, K.S. (1971), *SABRAO Newsletter* 3, 11-16.
212. Singh, H. and Dhooria, M.S. (1971), *India Journal of Entomology* 33, 12-130.
213. Singh, S.R. (1978), in Singh, S.R., van Emden, H.F. and Taylor, T.A. (*eds*) *Pests of Grain Legumes: Ecology and Control*. London: Academic Press, 267-79.
214. Singh, S.R. and Allen, D.J. (1979), *Cowpea Pests and Diseases*. Manual Series no. 2. Ibadan: IITA.
215. Singh, S.R. and Allen, D.J. (1980), in Summerfield, R.J. and

Bunting, A.H. (*eds*) *Advances in Legume Science*. London: HMSO, 419–43.

216. Singh, S.R., Luse, R.A., Leuschner, K. and Nangju, D. (1978), *Journal of Stored Products Research* 14, 77–80.

217. Singh, S.R., Singh, B.B., Jackai, L.E.N. and Ntare, B.R. (1983), *Cowpea Research at IITA*. IITA Information Series no. 14. Ibadan: IITA.

218. Singh, S.R., van Emden, H.F. and Taylor, T.A. (*eds*) (1978), *Pests of Grain Legumes: Ecology and Control*. London: Academic Press.

219. Skipp, R.A. (1975), 'Progress in understanding the nature of resistance to anthracnose disease in cowpea'. *Proceedings of the IITA Collaborators' Meeting on Grain Legume Improvement*. Ibadan: IITA.

220. Smithson, J.B., Redden, R. and Rawal, K. (1980), in Summerfield, R.J. and Bunting, A.H. (*eds*), *Advances in Legume Science*. London: HMSO, 445–57.

221. Southgate, B.J. (1958), *Bulletin of Entomological Research* 49, 591–9.

222. Sprent, J.I. (1976), in Kozlowski, T.T. (*ed.*) *Water Deficits and Plant Growth. Volume 4*. London: Academic Press, 291–315.

223. Stanton, W.R., Doughty, J., Orraca-Tetteh, R. and Steele, W.M. (1966), *Grain Legumes in Africa*. Rome: FAO.

224. Steele, W.M. (*ed.*) (1963), *Proceedings of the 1st Nigerian Grain Legumes Conference*, Samaru, Zaria. Zaria: Institute of Agricultural Research, Ahmadu Bello University, 86–90.

225. Steele, W.M. (1972), PhD. thesis, University of Reading.

226. Steele, W.M. (1976), in Simmonds, N.W. (*ed.*) *Evolution of Crop Plants*. London: Longman, 183–5.

227. Steele, W.M. and Mehra, K.L. (1980), in Summerfield, R.J. and Bunting, A.H. (*eds*), *Advances in Legume Science*. London: HMSO, 393–404.

228. Steele, W.M., Perez, A.T. and Ng, N.Q. (1980), *Genetic Resources Unit Exploration 1980*. Ibadan: IITA.

229. Stewart, K.A., Summerfield, R.J., Minchin, F.R. and Ndunguru, B.J. (1980), *Tropical Agriculture, Trinidad* 55, 117–25.

230. Summerfield, R.J. (1976), in McFarlane, N.R. (*ed.*) *Crop Protection Agents – Their Biological Evaluation*. London: Longman, 251–7.

231. Summerfield, R.J. (1980), in Hurd, R.G., Biscoe, P.V. and Dennis, C. (*eds*) *Opportunities for Increasing Crop Yields*. London: Pitman, 51–69.

232. Summerfield, R.J. and Bunting, A.H. (*eds*) (1980), *Advances in Legume Science*. London; HMSO.

233. Summerfield, R.J. and Roberts, E.H. (1985), in Halevy, A.H. (*ed.*) *A Handbook of Flowering*. Florida: CRC Press. (In press.)

234. Summerfield, R.J., Huxley, P.A. and Steele, W.M. (1974), *Field Crop Abstracts* 27, 301–12.

235. Summerfield, R.J., Huxley, P.A., Dart, P.J. and Hughes, A.P. (1976), *Plant and Soil* 44, 527–46.

236. Summerfield, R.J., Minchin, F.R. and Roberts, E.H. (1978), *Reading University – IITA Internal Communication No. 17*, 18pp.

237. Summerfield, R.J., Minchin, F.R., Roberts, E.H. and Hadley, P. (1983), in Yoshida, S. (*ed.*) *Potential Productivity of Field Crops under Different Environments*. Los Banos: IRRI, 249–80.

238. Ta'ama, M. (1983), *Tropical Grain Legume Bulletin* 27, 26–8.

239. Tannus, R.I. and Ullah, M. (1969), *Tropical Agriculture, Trinidad* 46, 123–9.

240. Taylor, T.A. (1965), *Bulletin of Entomological Research* 55, 761–73.
241. Taylor, T.A. (1965), *Proceedings of the Agricultural Society, Nigeria* 4, 50–3.
242. Taylor, T.A. (1968), *Nigerian Agricultural Journal* 5, 29–37.
243. Taylor, T.A. (1969), *Journal of Economic Entomology* 62, 900–2.
244. Taylor, T.A. (1969), *Bulletin of the Entomological Society of Nigeria* 2, 142–5.
245. Taylor, T.A. (1974), *Revue Zoologique Africaine* 88, 689–702.
246. Todd, J.W. and Canerday, T.D. (1968), *Journal of Economic Entomology* 61, 1327–30.
247. Verdcourt, B. (1970), *Kew Bulletin* 24, 507–69.
248. Verdcourt, B. (1978), *Taxon* 27, 219–22.
249. Verma, J.S. and Mishra, S.N. (1982), *Indian Farmers Digest* 15, 32–4.
250. Vetten, H.J. and Allen, D.J. (1983), *Annals of Applied Biology* 102, 219–27.
251. Walters, H.J. and Henry, D.G. (1970), *Phytopathology* 60, 177–8.
252. Watt, E.E. (1983), *The IITA/EMBRAPA/IICA Cowpea Program in Brazil.* Internal report of the International Institute of Tropical Agriculture. Ibadan: IITA.
253. Whitney, W.K. and Gilmer, R.M. (1974), *Annals of Applied Biology* 77, 17–21.
254. Wien, H.C. and Roesingh, C. (1980), *Nature, London* 283, 192–4.
255. Wien, H.C. and Summerfield, R.J. (1980), in Summerfield, R.J. and Bunting, A.H. (eds), *Advances in Legume Science.* London: HMSO, 405–17.
256. Wien, H.C. and Summerfield, R.J. (1984), in Goldsworthy, P.R. and Fisher, N.M. (eds), *The Physiology of Tropical Field Crops.* London Wiley, 353–83.
257. Wien, H.C. and Tayo, T.O. (1978), in Singh, S.R., van Emden, H.F. and Taylor, T.A. (eds) *Pests of Grain Legumes: Ecology and Control* London: Academic Press. 241–52.
258. Wien, H.C., Littleton, E.J. and Ayanaba, A. (1979), in Mussell, H. and Staples, R. (eds) *Stress Physiology in Crop Plants.* New York: Wiley, 294–301.
259. Wienk, J.F. (1963), *Mededelingen Landbouwhogeschool, Wageningen* 63, 1–82.
260. Williams, C.B. and Chambliss, O.L. (1980), *HortScience* 15, 179.
261. Williams, C.N. and Caswell, G.H. (1959), *Nature, Lond.* 184, 1668.
262. Williams, R.J. (1975), *Plant Disease Reporter* 59, 245–8.
263. Williams, R.J. (1975), in Bird, J. and Maramorosch, K. (eds), *Tropical Diseases of Legumes.* New York: Academic Press 139–46.
264. Williams, R.J. (1977), *Tropical Agriculture, Trinidad* 54, 53–60.
265. Williams, R.J. and Ayanaba, A. (1975), *Phytopathology* 65, 217–18.
266. Woolley, J.N. and Evans, A.M. (1979), *Journal of Agricultural Science, Cambridge* 92, 417–25.

13 Mung bean (*Vigna radiata* (L.) Wilczek/ *Vigna mungo* (L.) Hepper)

R.J. Lawn and C.S. Ahn

INTRODUCTION

The Asiatic *Vigna* or 'mung bean' group consists of several related species (Table 13.1) which have long been culti-vated in regions throughout central, southern and eastern Asia. All are annual, short-duration plants and are grown primarily for their edible dry seeds although the immature fruits and seeds are occasionally used as green vegetables and the plants as fodder or green manure. Green gram (*Vigna radiata* (L.) Wilczek) and black gram (V. *mungo* (L.) Hepper) are the most widely utilised species of the group.

Despite their long history in Asian agriculture, green gram and black gram received limited agricultural research interest prior to the last decade [8,100]. For many years, seedling physiology was the most documented aspect of both crops [9], mainly because their readily sprouting seeds provide convenient laboratory material for the studies of physiologists and biochemists.

In 1972, green gram was established as one of the main crops of the Asian Vegetable Research and Development Centre (AVRDC) in Taiwan [13] (See Chapter 18). Several countries, particularly India, the Philippines, Thailand, the USA and Australia, have augmented or initiated active research programmes on one or both of the species and the number of recent reviews and bibliographies [8,35,62,78,89, 102,118] is indicative of current research interest. Yet,

Table 13.1   Cultivated Asiatic *Vigna* species

Species	Synonyms	Common names
Vigna aconitifolia (Jacq.) Maréchal	*Phaseolus aconitifolius* Jacq.	moth bean, mat bean
Vigna angularis (Willd.) Ohwi and Ohashi	*Phaseolus angularis* (Willd.) W.F. Wright	adzuki bean
Vigna mungo (L.) Hepper	*Phaseolus mungo* L.	black gram, urd, mash, mungo bean, woolly pyrol, black mapte
Vigna radiata (L.) Wilczek	*Phaseolus radiatus* L. *Phaseolus aureus* Roxb.	green gram, golden gram, mung bean, Oregon pea, moong
Vigna trilobata (L.) Verdc.	*Phaseolus trilobatus* (L.) Schreb.	pillipesara bean, jungli bean
Vigna umbellata (Thunb.) Ohwi and Ohashi	*Phaseolus calcaratus* Roxb.	rice bean, red bean, Pegin bean

green gram and black gram are among the least researched of the major grain legume crops and their agricultural potential remains substantially under-exploited.

TAXONOMY, ORIGIN AND DISTRIBUTION

The taxonomic history of the Asiatic *Vigna* species has been confused at both the generic and specific levels [17,21,26, 130]. The group was formerly placed in the genus *Phaseolus* (see Table 13.1) but subsequent nomenclatural proposals [54,130,133], now generally accepted [35,50,62], have resulted in its transfer to the genus *Vigna* Savi, subgenus *Ceratotropis* (Piper) Verdc. The debate continues as to whether green gram and black gram, which share many morphological similarities, warrant their present status as separate species [31,62,130,131] (and see Chapter 1).

Attempts at interspecific hybridisations and cytogenetic studies with hybrids from successful crosses indicate that relations between the cultivated Asiatic *Vigna* species are complex, and hypotheses on the progress of speciation

585

within the group remain conjectural and controversial [62].
A substantial degree of chromosomal homology persists among
the cultivated species, each of which can be crossed
successfully with at least one other; although, in most
instances, the hybrids are more or less completely sterile
and the success of the cross depends on the use of one
particular species as the pollen receptor [2,36,37,40].
All species have a somatic chromosome number of
$2n = 22$ [38].

There are many reports of successful crosses between
black gram and green gram [2,3,13,36,40]. Hybrids have
differed in degree of fertility, from poorly to moderately
fertile. Chromosomal homology between green gram and black
gram is very substantial; one study indicates that the
species are separated only by one translocation, one
deletion and one duplication [40].

It has long been believed [26,130,137], largely on
morphological similarities, that green gram, and most
probably black gram as well, are domesticates which were
derived by selections from within a base population similar
to the wild species *Vigna sublobata* (syn. *V. radiata* var.
sublobata (Roxb.) Verdc.), which is distributed throughout
the Indian subcontinent, Indo-China and Australasia.
Hybridisation and cytogenetic studies are interpreted as
supporting *V. sublobata* as the presumed wild progenitor of
both black gram and green gram [62] but chemotaxic evidence
has been advanced to refute that claim for black gram [131].
Two distinct races of *V. sublobata* bearing some morphological similarity to either green gram or black gram
have been described [4,108]. A recent suggestion is that
the two races are the respective wild progenitors of the
two cultivated species, but are themselves distinct taxa
[31]. Viable but sterile hybrids have been obtained with
crosses between *Vigna radiata* and *V. umbellata, V. angularis*
[3] and *V. trilobata* [37] when green gram was used as the
female parent.

Green gram and black gram are believed to be of Indian

586

[16,41,137] or Indo-Burmese [62,110,129] origin. Vavilov
[129] also suggested the central Asian region as a primary
centre of diversity for green gram and a secondary centre
for black gram.

Green gram is said to have been widely cultivated in
the Indian subcontinent and adjacent regions for several
thousand years, and to have spread at an early time into
other Asian countries and to northern Africa [26,41,137].
Its present wide distribution throughout the tropics and
subtropics of Africa, the West Indies, North America and
Australasia is comparatively recent. Currently, green gram
is the most important grain legume in Thailand and the
Philippines; it ranks second in Sri Lanka and third in each
of India, Burma, Bangladesh and Indonesia [89]. It is a
minor crop in Australia, China, Iran, Kenya, Korea, Malaysia,
the Middle East, Peru, Taiwan and in the USA.

Black gram is considered to be of more recent deriv-
ation than green gram. It is an important pulse crop in
India, Burma, Bangladesh, Pakistan and Thailand and of
minor but increasing importance in Sri Lanka. It is grown
on a very limited scale in other parts of South-east Asia,
Australia and Fiji.

PRODUCTION AND USES

There are no comprehensive and accurate production statistics compiled on an international basis for either green
gram or black gram. Nationally, compilation of accurate
statistics is made difficult by the nature of production
and trade in these crops. Most are grown on small holdings,
often as mixed or intercrops, and are utilised and/or
traded locally. Furthermore, it is difficult to obtain
current statistics for some countries (e.g. Burma) and for
others (e.g. China and the USSR) statistics are not available. Within these constraints however, sufficient information can be summarised from various sources to reveal a
broad outline of production of the two crops (Table 13.2).

At least 5.8 million ha are sown annually to the two crops combined, with a larger area devoted to green gram than black gram. This total, based on the *FAO Production Yearbook for 1981* statistics, would, if accurate, represent around 9% of the total area sown to all pulses, and a quarter of the area sown to 'dry beans'. Total production of the two crops is of the order of 2.2 million t, which represents about 5% of total pulse production and 16% of total 'dry bean' production. The fact that the two crops represent a smaller proportion of the total production than of the area sown to all pulses and 'dry beans' indicates that average yields of green gram and black gram are generally smaller than those of the other categories.

About 70% of total world production of green gram and black gram occurs in India, which devotes around 12% of its total pulse area to these crops [86]. As with most of the producing countries, virtually all of the Indian crop is consumed domestically. World trade in green gram and black gram is dominated by Thailand, which exports about 40% of its total production [29,94], mainly to Japan, Taiwan, the Philippines, Malaysia, Singapore, Europe and North America. Production in Thailand has increased six-fold over the past twenty years so that it is now the second major producer of both crops.

Throughout Asia, green gram and black gram are used primarily for food, and in many countries they provide a major source of protein in cereal-based diets. The dried seeds may be eaten whole or split, cooked or fermented, or parched, milled and ground into flour. Whole or split, they are used to make *dhal*, soups, and curries, and are added to various spiced or fried dishes. Popular fermented foods in India, such as *idli* and *dosa*, are made from mixtures of rice with green gram or black gram ground together. The flour is used to make noodles, breads, and biscuits. An important green gram by-product is starch noodle, which is transparent, easy to cook, and stores well. Air-classification has recently been proven feasible

Mung Bean (*Vigna radiata/Vigna mungo*)

Table 13.2   Green gram and black gram production in selected countries in recent years

Country	Year	Area (1000 ha)	Production (1000 t)	Yield (kg ha^{-1})	Source
Green Gram					
Australia[+]	1977-81[++]	8.9	4.1	463	‡
Bangladesh	1976/77	15.6	9.5	611	106
Burma	1975	35.0	9.8	280	10
India	1978	2525.0	854.0	338	11*
Indonesia	1978	193.0	111.0	575	11
Iran	1977	27.5	15.1	550	5
Korea	1979	6.2	5.5	889	**
Pakistan	1978/79	66.0	30.0	455	11
Philippines	1978	45.1	26.2	580	11
Sri Lanka	1979	12.2	9.7	794	11
Taiwan	1979	4.7	3.6	764	11
Thailand	1979	275.0	157.0	571	***
Total Green Gram		3214.2	1235.5		
Overall average yield				384	
Black gram					
Bangladesh	1976/77	52.3	38.4	733	106
Burma	1975	74.5	38.0	510	10
India	1978	2260.0	698.0	309	11
Pakistan	1978/79	49.0	25.0	504	11
Sri Lanka	1979	8.7	6.0	694	11
Thailand	1979	149.3	93.7	628	***
Total Black Gram		2593.8	899.1		
Overall average yield				347	

+ Green gram and black gram combined.
++Annual averages over the five-year period.
* Asian Productivity Organisation 1982; various authors.
**E.H. Hong (1980), personal communication.
***A. NaLampang (1983), personal communication.
‡Based on combined State Departments of Agriculture estimates.

for extracting starch from green gram [13] while at the same time recovering the protein fraction which was previously lost in the traditional 'wet-milling' process. The recovered protein is now available for other uses (e.g. for fortifying cereal flours).

Both green gram and black gram seeds can be germinated to produce bean sprouts, the nutritional value of which is comparable with asparagus or mushrooms [123]. With sprouting, there is said to be an increase (on a dry weight basis) in thiamine, riboflavin, niacin, and ascorbic acid concentrations [66]. In Japan, black gram is favoured for the production of bean sprouts and the recent growth of the Thai and Australian industries has been directed towards production for this market. Throughout most of Asia, Europe and North America, green gram is preferred for sprouting.

In village agriculture, fruits are usually hand-picked as they ripen, and plant residues are used either as animal fodder or chopped into the soil as a green manure. Immature green fruits are sometimes used as vegetables. Cracked or weathered seeds can be used as feed. Crops have been used as forage [81] or as a green manure [52].

The composition of green gram and black gram seeds is similar, with approximately 25.0-28.0% protein, 1.0-1.5% oil, 3.5-4.5% fibre, 4.5-5.5% ash and 62.0-65.0% carbohydrate on a dry weight basis [110,128]. However, depending on genotype and environment, seed protein concentration ranges from 19-29% [128,136]. Amino acid analyses [18,124] indicate that, as with most grain legume crops, the concentrations of the sulphur-containing amino acids methionine and cystine are small. Methionine concentration is larger in black gram than in green gram [128]. Lysine values are comparatively large, which is why the protein of both crops is an excellent complement to rice in terms of balanced human nutrition [25,128]. Green gram and black gram contain trypsin and other growth inhibitors but these are denatured by heating and sprouting [43,80].

DESCRIPTION

Morphology

Green gram (Plate 13.1) is a branching, pubescent, erect or semi-erect, herbaceous annual between 25 cm and 125 cm tall. Leaves are alternate and trifoliolate (rarely pentafoliolate) with large, usually ovate leaflets (the laterals

somewhat asymmetric), although deltoid, ovate-lanceolate
and lobed [56] leaflet forms exist. Petioles are long with
large ovate stipules peltately attached. Branching usually
occurs from the lower and intermediate nodes, rarely from
either the cotyledonary or unifoliolate nodes. Flowers are
large (standard petals are 10-18 mm across), yellow or
greenish-yellow, and borne on axillary inflorescences, 2-10
cm long on both the main stem and branches, with five to
fifteen flowers clustered at the top of each peduncle which
is slightly thickened at the point of flower insertion. The
keel is coiled with a basal horn-like protuberance. Inflorescences remain meristematic and if flowers are progressively removed, more than thirty new flowers may develop
successively. Flowers are primarily self-fertilised. Cleistogamy is not unusual [61], particularly during periods of
drought and/or cool temperatures. Fruits are long, slender,
and straw, tan, brown or black at maturity; they may be
constricted between seeds. Shatter-resistant types exist.

Fruits may be borne above or within the canopy; they
are characteristically spreading and pendant and may be
pubescent (hairs as long as 2.0 mm [134]) or glabrous, with
nine to seventeen globose or eliptical seeds in each one.
The seed testas are usually green or yellow, occasionally
brown or blackish, and may be smooth and shiny, or dull,
with fine wavy ridges. Differences in seed-coat texture are
thought to be associated with the adherence or otherwise of
pod wall remnants [132]. The hilum is white, and may be flat
or convex. Mean seed weight ranges from 20-80 mg. Germination is epigeal, and in deep, friable soils, pronounced tap-
root development may occur.

Black gram (Plate 13.1) is a pubescent herbaceous annual
with an erect, decumbent or trailing habit of somewhat similar morphology to green gram. Branches arise from the basal
nodes. Inflorescences may occur at all nodes but do so
rarely from the cotyledonary nodes. Flowers are bright yellow and four to six are clustered on short axillary peduncles which elongate after fruit set has occurred. Fruit-

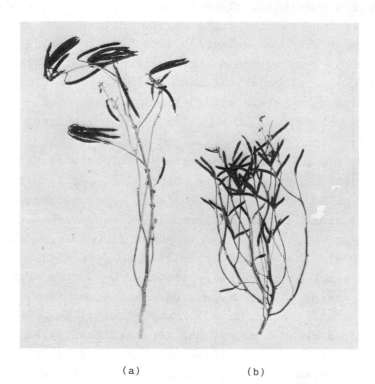

(a)   (b)

Plate 13.1 Mature shoots of (a) green gram (*Vigna radiata*)
cultivar Berken and (b) black gram (*V. mungo*) cultivar Regur

ing habit is characteristically erect or suberect with
fruits borne within the canopy. The fruits are usually
strongly pubescent (hairs as long as 4.5 mm [134]), although
glabrous forms exist. They are cylindrical and contain six
to ten oblong seeds with somewhat flattened ends. The seed
testas are usually grey, black or brown with a smooth but
dull surface, although shiny, dark olive-green types exist.
The hilum is white and usually concave. Individual seed
weight usually ranges between 40 mg and 60 mg. Germination
is epigeal.

On the basis of existing germplasm collections, the
genetic variations in vegetative and reproductive morphological traits are substantially less in black gram than
in green gram [13,56,60,134].

Mung Bean (*Vigna radiata/Vigna mungo*)

Crop growth and development

Green gram and black gram are short-duration crops, usually
flowering within 30-60 days after sowing, depending on
photo-thermal regime [71,99] and maturing within 60-120
days depending on photo-thermal regime [71] and water
supply [73]. Prior to canopy closure, crop growth rates
can be directly related to leaf area indices [68] and the
amount of photosynthetically active radiation intercepted
[90]. Early leaf area development may be slow, so that
ultimate seed yield potential is limited by inadequate
vegetative dry-matter accumulation prior to flowering [69].
In favourable conditions however, early vegetative growth
at a range of densities may exceed that needed to ensure
canopy closure prior to seed filling [76,90,91]. Mean crop
growth rates prior to flowering are usually in the range
50-95 kg ha^{-1} d^{-1} depending on density and genotype [68,90],
with peak rates greater than 200 kg ha^{-1} d^{-1} under favourable conditions.

Green gram is usually determinate [70] so that formation
of new leaves effectively ceases at flowering. However, as
much as one third of the total dry-matter accumulated during
seed-filling may still be partitioned into non-reproductive
structures [91]. Under favourable conditions, successive
flushes of fruits will form on each inflorescence [115].
In contrast, black gram is usually indeterminate [70] and,
in favourable conditions, growth of new leaves may continue
for several weeks after the onset of flowering, with relatively more of the current dry-matter accumulation (in the
region of 45%) being partitioned to non-reproductive structures after flowering than in green gram [91]. In both
crops, fruits ripen as they reach physiological maturity,
usually three to four weeks after flower opening, so that
the period of crop ripening may extend over many weeks in
favourable conditions [72] (Plate 13.2).

Abscission of flowers is prevalent in both crops [65]
(up to 90% in green gram [68]) even under apparently

593

Plate 13.2 Under favourable conditions, green gram and black
gram may produce successive flushes of flowers and fruits.
Asynchronous fruiting poses various problems, including
difficulties with harvest

favourable conditions. Within cultivars, the numbers of
fertile nodes and fruits per plant are usually more environmentally sensitive than the number of seeds per fruit or
seed size [68,70,75]. In green gram, flowering is restricted
to the uppermost four or five nodes on the stems and
branches but many more nodes are potentially fertile in
black gram [70]. Thus, an increase in plant size in green
gram either through reduced plant density [76] or later
maturity [72] leads to the accumulation of relatively more
sterile nodes than occurs with black gram. This difference,
together with differences in expression of determinance,
contributes to a yield potential per plant which is much
less responsive to environment in green gram than black
gram [70].

ADAPTATION

Regional and multi-environment evaluation tests indicate
large environmental effects on the performance of green

gram and black gram with substantial genotype x environment
(G x E) interaction [19,59,60,70,92,98]. Phenology and crop
yields are particularly sensitive.

Climatic factors affecting adaptation

The climatic factors identified as primarily involved in
conditioning adaptation of green gram and black gram are
daylength, temperature, rainfall and humidity. More quantitative information on the effects of these environmental
factors on growth and development is available for green
gram but the need remains for substantially more information
on both crops.

Daylength.   Rate of reproductive development in both species is sensitive to daylength [1,71,83,105]; most genotypes show quantitative short-day responses. Genotypes which
are insensitive over a wide range of daylengths are available [53,83] but absolute day neutrality in either species
has yet to be confirmed.

The primary effect of a quantitative short-day response
to photoperiod is for flowering to be progressively delayed
as daylength increases beyond the critical photoperiod.
Since summer daylengths increase with latitude, flowering
tends to be progressively delayed as a given genotype is
sown on the same date further from the equator [1,98,99].
At extreme latitudes, genotypes particularly responsive to
daylength may be exposed to cool temperatures or even frost
before maturity. Differences in times to flowering among
genotypes are minimised in short days (e.g. in the tropics
or in winter sowings in the subtropics [99]). Differential
daylength sensitivity can also result in large genotype x
sowing date interaction effects on phenology, particularly
in subtropical locations [71]. In the tropics, where seasonal variations in daylength are smaller, the contributions
of differential photoperiodic responses to genotype x sowing
date interactions are reduced [19].

Daylength can also affect development of both species subsequent to the onset of flowering [71]. In each case, long days prolong flowering and delay maturity. Indeed, long days stimulate successive flushes of flowers in green gram so that, in extreme cases, individual racemes may bear flower buds, green fruits and ripe fruits simultaneously.

Some progress has been made in analysing the genetic basis of photoperiodic responses in green gram. A major gene (*Ps*) with dominance or partial dominance for lateness under long daylengths has been identified [121]. In the absence of expression of this gene in short days, complex epistatic effects of several genes appear to control time to flowering. Surprisingly, 'insensitivity' is reported to be dominant to 'sensitivity' to photoperiod [125]. However, because phenotypic expression was not tested in controlled photoperiods, doubt remains as to whether the observed 'insensitivity' was apparent or real.

Temperature.   Both green gram and black gram can be described as 'warm season' crops. Growth is adversely affected by cool temperatures and plants are killed by frost. In the International Mungbean Nursery (IMN) evaluations conducted by AVRDC, green grams survived over a wide range of temperatures but were generally unproductive where mean temperatures were cooler than 20°-22°C [98]. Optimum mean temperatures were in the range 28°-30°C. For this reason, cropping is generally restricted to the summer and autumn months in warm temperate and subtropical regions, and to altitudes below 2000 m in the tropics.

Germination, emergence and seedling growth are all temperature-sensitive. Rates of germination in laboratory tests are slower at temperatures cooler than 25°C and are particularly slow below 14°-15°C [6,103]. The threshold (base) temperature for emergence in the field is around 10.5°C [6], while seedling growth rates are substantially reduced below 15°C [103]. Field studies have revealed differences between genotypes in rates of vegetative growth

596

and seed dry-matter accumulation at given mean temperatures [72]. Among genotypes, sensitivity of vegetative growth to temperature was positively correlated with their latitude of origin: genotypes from the tropics were more sensitive than those from the subtropics, which implies that differential temperature responses have played a part in determining adaptation across latitudes.

Rates of development in green gram and black gram are also strongly modulated by temperature, with controlled environment [1] and field experience [71] suggesting that in most genotypes the effects are additive to those of daylength. In general, warmer temperatures hasten development, so that flowering is most rapid and crop duration is shortest in warm-temperature, short-day conditions. However, seemingly quite complex daylength x temperature interactions can occur [1]. Field data indicate differential genotypic responses in both species to daily maximum and minimum temperatures. Rates of progress towards flowering of genotypes originating in the tropics were more sensitive to variations in minimum temperatures, while those from the subtropics were more sensitive to changes in maximum temperatures, when grown in a subtropical environment [71]. Overall, the effects of temperature in that environment were greater in the tropical genotypes.

Rainfall and humidity.   Most of the world production of green gram and black gram is concentrated in the 600-1000 mm (semi-arid to subhumid) annual rainfall zone [89,102]. The upper rainfall limit exists in part because other more productive crops are economically more attractive, and partly because of specific problems with green gram and black gram in humid environments. Both crops are susceptible to weather damage [57], although black gram much less so than green gram [79]. Weather damage results from the occurrence of rainy and humid conditions at fruit maturity. Inhibition of water by seeds in the fruits can cause them to sprout (Plate 13.3), or seeds become discoloured and

Plate 13.3 Mature green gram fruits in which the seeds have sprouted during rainy weather; this is one aspect of 'weather damage' limiting production of the crop in humid regions

shrivelled, or they lose viability and suffer spoilage by fungi.

Both crops are extremely sensitive to waterlogging prior to emergence which can rot seeds and reduce crop stands. There is some tolerance to waterlogging after emergence, particularly within black gram germplasm [79]. In both crops, prolonged waterlogging induces senescence of nodules leading to nitrogen deficiency which ultimately reduces growth [55]. Wet humid conditions also stimulate vining and lodging [78] and favour the incidence and spread of foliar diseases [135].

On the other hand, both species have a reputation of being 'drought resistant' [62,89,102]; they show considerable developmental plasticity and dehydration avoidance when grown in drought conditions [73]. When stressed, the period to flowering remains largely unaffected, but the period from flowering to maturity can be curtailed dramatically and so the crops 'escape' drought. Dehydration avoidance is achieved by stomatal regulation of water loss: as leaf water potentials fall, stomata close and leaf conduct-

ances to water are reduced [73]. Green gram and black gram
stomata are less sensitive to leaf water potential than
those of cowpea (*Vigna unguiculata*) but much more sensitive
than those of soyabeans (*Glycine max*). Stomatal control of
water loss restricts soil water use [74] but also limits
growth and yield potential [75]. Water use efficiency has
proved remarkably constant, (at around 5 kg ha^{-1} mm^{-1}),
over a range of values of water availability [7,33,75].

In comparative studies with a large range of accessions
in eastern Australia, green gram and black gram performed
better in drought conditions than cowpeas [92], but were
less productive than cowpeas in favourable conditions.
Nonetheless, both green gram and black gram are quite
responsive to water [33,70].

It has been stated [62,118] that black gram requires
wetter conditions than green gram, but it is not certain
whether this view is based on observations of crop responses
or on the pattern of traditional crop production. The latter
is likely to be influenced by the superior weather resistance, waterlogging tolerance, and ability of black gram to
respond to better growing conditions due to its indeterminate growth habit [70]. In some areas, black gram is also
the more 'highly esteemed' pulse [22]. Detailed studies [70,
75,92] have indicated similar yields from both crops under
severe drought conditions.

Edaphic factors affecting adaptation

Soil physical properties.   Green gram and black gram are
grown primarily as rainfed crops on a very wide range of
soils, but general experience indicates that they perform
best on deep, well-drained loams and sandy loams [89,102,
110]. In general, black gram is better suited to heavy clay
soils, perhaps in part because of better tolerance to ephemeral waterlogging. Neither crop is particularly well-suited
to shallow or infertile sandy soils although green gram is
often grown on them. Both crops are said to tolerate soils

that are moderately alkaline and saline [102] but experiments in Australia indicate they are less well suited to such conditions than either soyabeans or cowpeas (B.A. Keating, pers. comm.). Soil type and depth influence both water-holding capacity, which is important for rainfed cropping in semi-arid areas [75], and rooting depth [7,74,89].

Emergence of both species when sown deeper than 5-6 cm is variable, particularly on heavy textured soils and on poorly structured soils which tend to cap after rainfall. Apart from the physical obstruction to emergence provided by the soil crust, injury and death of tissues adjacent to the crust can be caused by extremely hot (> 55°C) temperatures occurring in the vicinity of the emerging seedling.

Fertility. Chemical analyses of green gram and black gram seeds (Table 13.3) indicate that with each tonne (dry weight) of seed harvested, the following quantities of nutrients are removed from the field: 40-42 kg of nitrogen, 3-5 kg of phosphorus, 12-14 kg of potassium, 1.5-2.0 kg of each of sulphur and magnesium, and 1.0-1.5 kg of calcium. Additional requirements are of course necessary to sustain vegetative dry-matter production and in many cropping systems these nutrients are recycled as the crop stubble is incorporated into the soil. But, where crop residues are fed to animals, significant further quantities of mineral nutrients may be removed from the field at harvest (Table 13.3).

Many of the tropical soils on which green gram and' black gram are grown are phosphorus-deficient, so that

Table 13.3 Mineral composition (%) of seeds and forage residues of green gram cv. Berken and black gram cv. Regur grown on black earths in Queensland, Australia

Cultivar	Component	Mineral element					
		N	P	K	Ca	Mg	S
Berken	Seeds	3.95	0.26	1.16	0.09	0.16	0.19
	Residues	0.60	0.20	2.20	1.11	0.66	0.07
Regur	Seeds	4.21	0.48	1.40	0.14	0.23	0.20
	Residues	1.59	0.29	2.96	1.86	0.40	0.35

responses to added phosphate are common [42,64,117]. In
general, given current yield potentials, the most economic
P application rates have been of the order of 20-40 kg
P_2O_5 ha^{-1} although on lateritic soils of intense P fixation
capacity, yield responses up to 100 kg P_2O_5 ha^{-1} have been
reported [88].

As with other legumes, both crops can assimilate mineral
and symbiotically fixed nitrogen. Nitrogen fertilisation
tends to suppress nodulation, and does not increase seed
yields except in the absence of effective nodulation [23].
There are relatively few reports of responses by either
species to other nutrients. Positive responses to zinc,
iron and magnesium in pot experiments with calcareous soils
have been reported [13], and in pot/sand culture the concentrations of sulphur-containing amino acids in the seeds
have been improved as the sulphur concentration of the
nutrient medium was increased up to 90 ppm [12]. Zinc deficiency symptoms have been observed on some heavy clay soils
in eastern and northern Australia. Descriptions of foliar
symptoms of nutrient disorders (both deficiencies and exces-
ses) are available for green gram [116].

Part of the reason for the comparatively few reports of
significant nutrient responses in green gram and black gram
lies in their small yield potentials, and hence smaller
nutrient requirements, relative to those of more developed
grain legume crops such as soyabeans and groundnuts (*Arachis
hypogaea*). Furthermore, in many of the environments where
green gram and black gram are grown, water rather than
nutrients is likely to be the major factor limiting growth.
However, as the yield potential of both crops is improved,
and as agronomic and production techniques are refined,
responses to fertilisation will become more common.

SYMBIOTIC NITROGEN FIXATION

Seasonal profiles of nodule growth and activity

Under favourable conditions, crown nodulation on primary

roots of green gram and black gram is apparent within one week of emergence [39,96] and measurable acetylene reduction activity can be detected within ten to twenty-five days after sowing [39,114,122]. Nodule numbers and fresh weight increase rapidly during vegetative growth, reach a maximum at flowering [39] or during early fruit filling [32] and decline thereafter as nodules senesce. The onset of nodule senescence has been observed as early as four weeks from sowing [96] but why this should have happened remains obscure. Nodules of both species are globose, smooth, and usually in the range of 1-4 mm in diameter on actively grow-ing plants.

Acetylene reduction assays [32,39,122] and bleeding sap analyses [114] indicate that symbiotic fixation in green gram increases to a maximum during flowering and early fruit filling, but declines rapidly thereafter. The onset of nod-ule senescence and decline in fixation activity can be advanced or delayed by source-sink manipulations [32], indi-cating a close relation between nodule growth and function, and supply of photo-assimilates. As in cowpeas, transport of xylem nitrogen in nodulated green gram occurs principally in the form of ureides [95].

Host x strain of *Rhizobium* interactions and inoculation

Green gram and black gram are generally promiscuous and nodulate readily in the presence of a wide range of strains from the cowpea cross-inoculation group [24,77,93] (see Chapter 4). Nonetheless, there are substantial host geno-type x strain of *Rhizobium* interactions in terms of both infectiveness of strains and effectiveness of the symbiosis. Partitioning genotypic effects of the host into root and shoot components has indicated that infectiveness is cont-rolled primarily by the root but that effectiveness of the symbiosis is influenced by both root and shoot genotype [77].

Because native strains of the cowpea cross-inoculation group of rhizobia are widespread in many areas of the

tropics and subtropics, positive responses by either species to inoculation with effective strains can be difficult
to obtain in the field (see [28]). This lack of response to
inoculation can be accentuated by difficulties in establishment or poor persistence of inoculant strains in competition
with indigenous ones [27]. Many of the reports of positive
responses to inoculation have been from India [89], which
is perhaps surprising given the long history and wide distribution of both crops in the region, and the likelihood
of the widespread occurrence of effective native strains.

Because trials which fail to show positive responses to
inoculation are less likely to be reported formally, it is
difficult to assess realistically how widespread the need
is for inoculation of either green gram or black gram crops
on the basis of current information, and it is impossible
to evaluate the reasons for lack of response to inoculation.
What is needed therefore is not only a critical evaluation
on a regional scale of the need for inoculation, but also
research to identify why inoculation with apparently superior strains has failed to induce positive responses. Where
competition from an abundance of native strains precludes
the ready introduction of superior inoculant strains,
attempts to manipulate the host genetically may prove more
useful [77].

Environmental factors affecting the symbiosis

Soil acidity (pH ≤ 5.0) is detrimental to nodulation in
both green gram [93] and black gram [27] although with green
gram, differential sensitivity to soil pH exists among
strains, and there are host genotype x strain interactions [93]. Effects of soil acidity appear to be related
directly to the initiation of nodules, rather than to the
survival of rhizobia in the rhizosphere, and can be ameliorated by liming [27,93]. Lime pelleting of seeds has enhanced crown nodulation in acid soils [27]. As with most
legumes, nodulation is stimulated and nitrogen fixation is

603

enhanced by small amounts of 'starter' nitrate in the soil
solution, but both are progressively depressed as nitrate
concentration increases [39].

It is likely (see Chapter 4) that nodulation and nitrogen fixation in green gram and black gram will prove sensitive to a range of environmental influences, but few data
are yet available for either crop. Factors such as temperature, fertility and water (too much or too little) can be
expected to influence nitrogen fixation either directly or
indirectly through effects on crop growth. The effects of
shading and other factors on assimilate supply and nodule
activity of green gram and black gram when grown as intercrops, remain to be critically investigated.

Nitrogen fixation and transfer

There are few reliable estimates of total fixation of nitrogen by either green gram or black gram. One estimate for
green gram derived from a two-year rotation study in Thailand [48] was 58-107 kg N ha^{-1} crop^{-1}. This study was conducted on a nitrogen-deficient soil but the estimate can
be expected to be somewhat larger than that actually fixed
in farmers' fields where stands are often poor and crop
growth is limited by soil fertility, weed competition and
pests. Another estimate of fixation by a green gram crop
was derived from long-term studies of cereal-legume rotations in Oklahoma [52] which indicated that a green gram
crop turned in as green manure increased the yield of a
subsequent oat crop by an amount equivalent to the addition
of 70 kg N ha^{-1}. Other suggestions [89] estimate fixation
in the range of 50-100 kg N ha^{-1} crop^{-1}.

The potential value of both green gram and black gram
as nitrogen-contributing components in crop rotations or
intercrops is widely recognised [13,35,89] but little
definitive information exists on the likely *net* input of
nitrogen (if any) from crops harvested for seeds. Clearly,
for a net nitrogen contribution to be made by the legume,

the amount of nitrogen removed in plant material must be less than the crop fixes symbiotically. Where seeds are harvested and crop residues are removed it is unrealistic to expect a net nitrogen input into the soil from either black gram or green gram. Indeed, in a seven-year continuous cropping study in eastern Australia, there was an average net loss of 74 kg N ha^{-1} per annum from the top 15 cm of soil in black gram plots where plant residues were removed after each crop [30]. Only part of these losses were accounted for by runoff and movement of nitrogen down the soil profile.

CULTURAL SYSTEMS

Green gram and black gram are flexible crop plants grown in diverse cropping systems and seasons, subjected to a wide range of cultural management techniques and more or less well supported by technological inputs [35,102,119]. Perhaps because of its greater genetic diversity, cultural systems which involve green gram are rather more diverse than those for black gram.

Cropping systems

Both species are grown as mixed-, inter- and relay-crops [89,110]. With mixed-cropping, seeds of the pulse and one (or more) other crops, often a cereal (e.g. millet or sorghum), are mixed and broadcast. Mixed-cropping is usually associated with subsistence agriculture in marginal cropping areas, involving few agronomic and technological inputs. Stands of the component crops of the mixture are variable, often patchy, and yields tend to be small.

Intercropping with green gram and black gram usually involves the systematic interplanting of one or more rows of the pulse alternating with rows of an intercrop. Their short crop durations suit green gram and black gram to intercropping with other longer duration crops such as sugar

605

cane, maize, sorghum, sunflower or cotton. The pulses can
be harvested before competition from the longer season
intercrop is thought to become excessive. As with many
pulses, green gram and black gram are favoured as components for intercropping with non-legumes because of their
potential to fix part of their own nitrogen requirements
and so lessen nitrogen competition with the non-legume
component.

Relay-cropping involves the sequential growing of crops,
often in rapid succession, to maximise the total income
from cropping which can be attained from the land within
local time and climatic constraints. As short-season crops,
green gram and black gram are well-suited to relay-cropping,
being able to produce a crop on the limited resources available following a main crop or within the short time availabl
between two other crops [20,126].

Cropping seasons

Most of the production of green gram and black gram in the
monsoonal tropics occurs in three seasons: the monsoon or
rainy season, the post-monsoon or cool-dry season and the
pre-monsoon or hot season [35,62,110]. Monsoon crops are
sown after the first rains and grow through the rainy season, often as mixed-crops or intercrops. This system of
cropping is common in northern India, much of South-east
Asia, Africa, and in northern Australia where long duration
crops are preferred [19] so that maturation occurs after
the monsoon and the risk of weather damage to seeds is
minimised.

Post-monsoon or dry season crops are sown around the
end of the monsoon, often as a relay-crop following rice,
maize or sorghum, thus utilising residual water and fertility. Shorter duration varieties are usually preferred to
ensure maturity before the onset of serious water stress.
Post-monsoon cropping of green gram and black gram is common in southern and eastern India [110], Bangladesh [20],

Thailand [94], the Philippines [29] and in Indonesia.

Pre-monsoon or 'summer' crops of green gram and black gram are grown during the period of warming temperatures prior to the monsoon. Harvest has to be completed before the onset of heavy rains to avoid weather damage to the seeds. Traditionally, pre-monsoon cropping has been limited to those few areas (e.g. Thailand [94]) where early rains preceding the monsoon provide adequate water for crop growth. However, irrigation facilities are increasing the opportunities for producing short-duration crops of green gram or black gram during the hot dry period before the monsoon as relay-crops following wheat or perhaps vegetables, and preceding rice [20,62,89].

In the subtropics and warm temperate areas, cropping of green gram and black gram is restricted to the summer-autumn period, because temperatures are too cool at other times of the year. In the southern USA, green gram is sown in mid-summer as a short-season 'catch' crop following wheat [126]. In subtropical Australia, green gram and black gram are grown as rainfed crops during the summer and autumn, often following a winter cereal [78]. Short-season crops are used to minimise crop water requirements and to increase the probability that rainfall and stored soil water reserves will be adequate [75].

PESTS AND DISEASES

Green gram and black gram are susceptible to a wide range of insect pests [82,107,113] many of which are pan-tropical in distribution, but comprehensive information on the various species, their biology and relative economic importance is limited. A one-year survey in the Philippines identified twenty-six pest species on green gram [82] while a recent survey in northern Australia [107] reported more than 130 species of grain legume pests, most of which attack green gram and/or black gram. Use of chemicals is more common with mechanised production in the USA and in Australia [89] than it is in Asia [29,82]. Current recommendations usually stress the need to control insects once flowering commences [97,101].

Economically, the most serious pests are those which feed on flowers, fruits and seeds. Foremost among these are the pod-boring caterpillars such as *Heliothis* spp. and *Etiella* spp., the bean pod-borer *Maruca testulalis* and the pod-sucking bugs such as *Nezara viridula* (Plate 13.4). Apart from direct feeding losses, pod-feeding insects can substantially reduce the sprouting quality of remaining seeds, either by exposure to weather damage, as with the pod-borers [107], or through loss of viability as with seeds attacked by sucking bugs [78].

Another serious pest of both species is the bean fly (*Ophiomia phaseoli*), the larvae of which tunnel through stems and petioles, causing wilting and stunting and predisposing plants to secondary attack by pathogens. Problems

Plate 13.4 Damage from insect pests can be severe in both green gram and black gram. Most serious are the pod-sucking bugs, such as these green vegetable bug (*Nezara viridula*) nymphs, and the pod-boring caterpillars. The sucking bugs cause damage by piercing seeds through the pod walls. Pod-boring caterpillars chew holes through the pod walls (as in the fruit on the right) allowing the entry of water and pathogens.

with bean fly damage are usually most serious during the
seedling stage when stands can be destroyed completely [13].
Other pests of generally localised economic importance are
aphids, leafhoppers and whiteflies [13]; the last can also
be important as a virus vector. Nematodes can also be of
economic significance in localised areas [135]. Numerous
insect species, most notably caterpillars, beetles and
locusts, feed on the leaves of green gram and black gram
crops [82,107] but there seems to be little evidence that
yields are depressed except after severe defoliation. The
consequences of artificial defoliations [46] support this
view.

A number of widespread and serious diseases caused by
fungal, bacterial and viral pathogens limit production of
both crops [13,49,89,135]. Indeed, diseases may be one of
the major constraints to yield throughout much of Asia [60,
98]. Most serious of the fungal diseases are *Cercospora*
leaf spot (*Cercospora canescens*) and powdery mildew (*Erysiphe polygoni*). *Cercospora* leaf spot is primarily a disease of the humid tropics and therefore tends to be most
prevalent in crops grown during the monsoon. The disease
results in brown, necrotic spots on the leaves which abscise prematurely. With severe outbreaks, lesions occur also
on stems and fruits.

Powdery mildew is very widespread in green gram (Plate
13.5) and black gram areas and is favoured by a combination
of dry weather and cool temperatures. It is therefore most
commonly a disease of post-monsoon crops, and then often
only as the crop approaches maturity. However, where crops
become infected before they flower, yield reductions can be
severe. Other fungal diseases of localised importance include damping-off (*Rhizoctonia* spp. and *Pythium* spp.), wilts
(*Fusarium* spp.), several rusts (*Uromyces* spp.), scab (*Elin-
soe iwatae*) and anthracnose (*Colletotrichum lindemuthianum*)
[49,135].

Two serious bacterial leaf diseases, bacterial leaf spot
or blight (*Xanthomonas phaseoli*) and halo blight (*Pseudo-*

Plate 13.5 An extensive sowing of irrigated green gram in the Ord
River area of Western Australia during early fruiting. The greyish/
white colour of the lower leaves (most evident on the rows to the
left) is due to powdery mildew, a serious disease of green gram
and black gram in some regions.

monas phaseolicola) are favoured by warm, wet weather.
Bacterial leaf spot produces small, dry necrotic areas
which may gradually coalesce to form large necrotic patches
which ultimately disintegrate. Halo blight produces water-
soaked lesions on the leaves surrounded by chlorotic
margins.

Most serious of the numerous virus diseases to affect
green gram and black gram is Yellow Mosaic Virus, which is
widespread on the Indian subcontinent and in adjacent regions [49,104]. Symptoms of the disease are small yellow
spots on the young leaves which coalesce into larger yellow
patches or completely yellow leaves in advanced stages.
Infected plants are stunted and fruit set is sparse or
inhibited completely. Other virus diseases of modest importance are Leaf Crinkle Virus, which mainly affects black
gram, Leaf Curl, Mosaic Mottle Virus and Mungbean Mosaic
Virus [135].

Fortunately, sources of resistance have been identified

for all of the current major diseases of green gram and black gram except Yellow Mosaic Virus, for which sources of only moderate resistance are available [13,104]. Substantial progress is being made by the AVRDC (see Chapter 18) in incorporating resistances into genotypes with desirable agronomic traits and local adaptation [13].

CROP IMPROVEMENT

Potential for improvement

The major constraints to performance of green gram and black gram derive from their traditional role as secondary crops whereby they have been grown in areas marginal for fertility and water, often following a staple crop, with minimal cultivation or weed control and without inputs of fertiliser, pesticides or irrigation. These constraints involve, on the one hand, the immediate limitations posed by environmental stresses on the expression of the yield potential of existing varieties and, on the other, genetic limitations deriving from adaptation, over thousands of years of cultivation, to stability of performance in marginal environments.

It is clear that the yield potential of current cultivars substantially exceeds that realised in farmers' fields. Yields larger than 3 t ha^{-1} have been reported in many trials [19,70,99]. In the first four years, mean yields in more than one quarter of the IMN trials [98] were 1000-1771 kg ha^{-1}, which is three or four times larger than the world average yield for green gram (Table 13.2). Furthermore, only one fifth of the trials reported mean yields as small as world average yields. The significance of the IMN data is that, although yields were only assessed from small plots, site means included the yields of non-adapted as well as adapted genotypes.

It is reasonable to attribute the superior performance of genotypes in cultivar trials such as the IMN to the impact of agronomic management and, on this basis, Poehlman [98] concluded that there was a need for more production-orientated research as well as more effective utilisation of current agronomic knowledge in green gram production. As with most crops, the extent to which agronomic inputs are currently being used in production depends on many social and economic factors (see [29,82,87]) not the least of which are farmers' perceptions of the magnitude of likely economic benefits relative to the potential risks of losses incurred through subsequent crop failure and from unstable market prices. Optimal inputs and appropriate 'input packages' in green gram and black gram production therefore vary substantially between regions [29].

In this context, both green gram and black gram generally compare unfavourably with many other crops: although they respond positively to agronomic inputs such as sowing practice [89], water [33,70,75], weed control [45] and herbicides [87], pesticides [82] and agronomic management [29,72,75,76], both crops are less responsive than many alternatives. Ironically, it is probably those homeostatic physiological attributes which contribute to stability of performance in marginal environments (e.g. long duration, good biological productivity, poor harvest index, lack of synchronous flowering and production of many more flower buds than could ever be retained as fruits) which limit the potential for responsiveness in better environments.

It is probable that in both species, alleles for large yields exist, but are infrequently distributed throughout landrace populations. Although characteristically poor yielding and less responsive to management inputs than is desirable, these materials are known to possess a number of desirable traits such as tolerance to environmental stresses, and resistance or tolerance to important diseases and insects [62].

Mung Bean (*Vigna radiata/Vigna mungo*)

Germplasm resources

More than 20000 accessions of *Vigna radiata* are maintained
by various institutes in Australia [58], India [15,112,119],
Indonesia [112], Iran [11], Korea [11], the Philippines
[112], Taiwan [11] and in the USA [15,112] but these involve
substantial duplication. The largest collections are maintained by the AVRDC in Taiwan (5108 accessions), the United
States Department of Agriculture (3494 accessions) and the
Punjab Agricultural University (3000 accessions). The AVRDC
collection represents germplasm from fifty-one countries
although about one half of it is of Indian origin. Less
information is available on the extent of the germplasm
resources of *Vigna mungo* but it is clear that they are
substantially fewer than for *V. radiata*. Current collections
are held in Australia [58], India [119], Taiwan [11] and
in the USA [15,112].

Various correlation, path coefficient and heritability
studies [14,34,51,84,136] indicate an optimistic outlook in
green gram with regard to genetic variability for the principal yield components, fruit number per plant, seed number
per fruit and mean seed weight. Less information is available on black gram, but yield components, fruit length,
plant height and branch number per plant may be of value,
either alone or in combination, in indirect selection
schemes for larger seed yield.

Although sufficient genetic variation is believed to be
already available to sustain adequate genetic advance,
mutation breeding may be useful to generate variability
for specific characters such as synchronous senescence or
improved protein concentration [10]. Increased variability
in the treated populations has been reported for yield and
yield components [109]. Once identified, desirable mutant
characteristics still remain to be transferred into
adapted backgrounds by traditional recombination breeding
procedures.

Current knowledge on the genetics of green gram and

black gram has recently been summarised [47].

Strategies and objectives

Good yield potential with stability of performance is the ultimate breeding objective with both crops but specific objectives vary with crop and region. Early and uniform maturity, resistance to diseases, insects, weather damage and shattering are usually specific breeding objectives for green gram while for black gram, common specific objectives include erect and determinate growth habit, and photo-thermal insensitivity.

Green gram and black gram are self-pollinating, with little natural outcrossing (0-0.75% in green gram) [13,44]. Like other self-pollinated species, the principal breeding approach for both crops is selection among available germ-plasm for direct release or use as parents, and hybridisa-tion followed by pedigree selection. Introduction of exotic germplasm is usually necessary to augment the limited vari-ability of local collections, and to widen the genetic base. Hybridisation is used to recombine the desired charac-ters. Multiple crossing is often used to combine the characteristics of several parents, or to obtain a wider spectrum of gene recombinations for characters inherited quantitatively. Backcross procedures are useful for improv-ing an otherwise useful genotype which may lack one or two strongly heritable characteristics. The characteristically large G x E interaction for yield with the two crops [60, 63,99] has to be considered in evaluating lines for large yield potential.

Seed yields of F_1 green gram hybrids range from 37% [120] to 74% [111] larger than the better-yielding parent. With the closed pollination system, it is unlikely that heterosis will be exploited commercially unless effective outcrossing mechanisms can be established. Autotetraploids have been induced in both crops [47,67], but have usually exhibited gigantism in vegetative and reproductive parts

and do not seem to have commercial value because of late
maturity and poor fertility.

Green gram and black gram, although closely related,
differ in relation to several important characteristics.
Black gram traits which might be useful in green gram
improvement include better resistance to several major diseases and insects (e.g. the sucking bugs [107]), better
tolerance to environmental stresses such as waterlogging
and seed weather damage [79], less fruit shattering, and
larger methionine concentration [128]. The determinate
growth habit, erect plant type, short duration, less photo-
thermal sensitivity [83] and more seeds per fruit of green
gram may be useful traits in black gram.

Although hybrid breakdown has been reported as a major
reproductive barrier between the two species, the strength
of the isolating mechanisms varies with the parental genotypes, especially the green gram female line used in the
crosses [2,3], indicating the possibility of locating green
gram genotypes which may be effectively used as genetic
bridges between the two species. Thus, interspecific hybridisation is desirable as a long-term programme to bring the
two species into a common gene-pool. *Vigna sublobata* is
another possible gene source, especially for disease resistance [4].

International improvement programmes.   An International
Mungbean Nursery (IMN), was initiated in 1972 by the University of Missouri. Through the IMN, the extent of G x E
interaction was assessed and genotypes with large yield
potential and generally wide adaptation were identified.
In 1976, the IMN was transferred to the AVRDC, which has
an intensive green gram improvement programme supported by
an integrated multi-disciplinary research group. The AVRDC
programme involves breeding green gram material for adaptation to different regions, preliminary testing and
evaluation of germplasm, and dissemination of accessions,
segregating populations and advanced breeding material to

national programmes (see Chapter 18).

Breeding objectives at AVRDC include the development of widely adapted, stable and high-yielding genotypes with early and synchronous maturity, resistance to important diseases and insects, and improved seed quality. Breeding populations at AVRDC are handled as dynamic gene-pools for integrated population improvement, into which new sources of germplasm are added whenever feasible; frequencies of favourable alleles are progressively increased through cyclic selection, genetic recombination is enhanced by hybridisation among selected genotypes, and promising lines are extracted at any cycle [13]. Special consideration is given to the genetic diversity of parental lines. Segregating populations are grown in three successive seasons for rapid generation advance and identification of widely adaptable genotypes through 'disruptive seasonal selection' [127] under distinctly different environmental and biological pressures at various sites in Asia.

The AVRDC programme has included screening for: (a) resistance to major diseases (e.g. *Cercospora* leaf spot, powdery mildew, root-rot and virus) and pests (e.g. beanflies, pod borers, aphids and nematodes) [13]; (b) photoperiod insensitivity [83]; (c) improved protein concentration and quality [128]; and (d) desirable agronomic traits [13]. Many AVRDC breeding lines have been named and released by national programmes. For example, in Costa Rica, VC 1089 A was released as ASVEG '78 in 1978 and VC 1628 A was released as Tainan (Selection) No. 3 in Taiwan in 1981. Releases in 1982 included a selection from VC 1000-45-B as Station 46 in Fiji, and VC in 1973A as Shanhua in Korea. No international programme exists for improvement of black gram.

National improvement programmes.   Varietal improvement programmes have been conducted in several countries with most emphasis on green gram. In India, improvement of both green gram and black gram initially involved the introduction of, and selection from within, limited local collections

by the Indian Agricultural Research Institute, New Delhi,
research stations under the Directorate of Pulses Development, and various agricultural universities. Active breeding began with the establishment of the All-India Coordinated Pulse Improvement Research Project in 1969.

In the Philippines, varietal improvement of green gram
is mainly conducted by the Bureau of Plant Industry and the
University of the Philippines, Los Banos, with field experiment stations and other agricultural colleges conducting
regional tests. Smaller improvement programmes exist in
Indonesia, Thailand, Sri Lanka, Pakistan and Bangladesh.
Although some active breeding is involved, much of the
emphasis in these countries is on germplasm introduction
and selection for local adaptation. In the USA, three varieties have been released by the Oklahoma Experiment Station
[85]. In Australia, limited introduction and selection by
the State Departments of Agriculture and the CSIRO led to
the release of one black and two green gram varieties [78].
Active breeding supported by multi-disciplinary research
was initiated in 1980 by the CSIRO Division of Tropical
Crops and Pastures.

Avenues of communication.   There are no international
newsletters or bulletins uniquely available for either formal or informal communication of research developments in
mung bean. As the bibliography below shows, mung bean researchers avail themselves of a wide range of publications
for the formal communication of their research data. Most
of them rely on direct personal communication for informal
exchanges of information. The IMN trials [99] have provided
an important avenue for the exchange of information and
germplasm, particularly in the developing countries of
South-east Asia. In turn, this has enabled a more comprehensive overview of the adaptation of green gram cultivars
to be formulated [60,98].

As the only international research centre with green
gram as a principal crop, the AVRDC plays an important role

in communicating research developments. Green gram is an important component of the Centre's Outreach Programme in Thailand, and of the various bilateral arrangements formulated between the AVRDC and a number of countries, including Korea, the Philippines and Taiwan. AVRDC achievements in research on green gram are communicated regularly in *Annual Reports* [13], and in the Centre's newsletter, *Centerpoint*.

One of the more significant events in the communication of mung bean research data was the First International Mungbean Symposium held in 1977 at Los Banos in the Philippines. That meeting, jointly sponsored by the University of the Philippines, Los Banos, the South-east Asian Regional Centre for Graduate Study and Research in Agriculture, the Asia Foundation, the Philippines Government, and the AVRDC, provided an opportunity for 135 participants from sixteen countries to establish contacts and to exchange information [87].

ACKNOWLEDGEMENTS

The authors thank Dr B.C. Imrie for reviewing this chapter and Ms Anne Ting for typing the manuscript.

LITERATURE CITED

1. Aggarwal, V.D. and Poehlman, J.M. (1977), *Euphytica* 26, 207-18.
2. Ahn, C.S. and Hartmann, R.W. (1978), *Journal of the American Society of Horticultural Science* 103, 3-6.
3. Ahn, C.S. and Hartmann, R.W. (1978), in Cowell, R. (*ed.*) *Proceedings of the First International Mungbean Symposium*. Taiwan: AVRDC, 240-46.
4. Ahuja, M.R. and Singh, B. (1977), *Indian Journal of Genetics and Plant Breeding* 37, 133-6.
5. Amirshahi, M.C. (1978), in Cowell, R. (*ed.*) *Proceedings of the First International Mungbean Symposium*. Taiwan: AVRDC, 233-5.
6. Angus, J.F., Cunningham, R.B., Moncur, M.W. and Mackenzie, D.H. (1981), *Field Crops Research* 3, 365-78.
7. Angus, J.F., Libbon, S.P., Hsiao, T.C. and Hasegawa, S. (1979), *Proceedings of the Philippines Crop Science Society*. Los Banos, Philippines: April 1979.
8. Anon.(1970), Mung Bean (*Phaseolus aureus*), *Annotated Bibliography No. 1210*. Farnham Royal: Commonwealth Agricultural Bureau; and see Supplements No. 1210A (1974), G210B (1975) and G210C (1979).
9. Anon.(1980), Germination of *Vigna radiata* and *V. mungo, Annotated*

Bibliography No. G510. Farnham Royal: Commonwealth Agricultural Bureau.

10. Anon.(1977), *Induced Mutations for the Improvement of Grain Legumes in South East Asia (1975)*. IAEA-203. Vienna: International Atomic Energy Agency.
11. APO (1982), *Grain Legumes Production in Asia*. Tokyo: Asian Productivity Organisation.
12. Arora, S.K. and Luthra, Y.P. (1971), *Plant and Soil* 34, 91-96.
13. AVRDC (1974-83), *Annual Reports for 1972/73-1982*. Taiwan: The Asian Vegetable Research and Development Centre.
14. AVRDC (1977), *Mungbean Progress Report for 1975*. Taiwan: The Asian Vegetable Research and Development Centre.
15. Ayad, G. and Anishetty, M.N. (1980), *IBPGR Directory of Germplasm Collections I. Food Legumes*, AGP/IBPGR 180/45. Rome: IBPGR.
16. Bailey, I.H. (1949), in *Manual of Cultivated Plants*. New York: Macmillan.
17. Baker, J.G. (1876), in Hooker, J.D. (*ed.*) *The Flora of British India*. Ashford, Kent: L. Reeve. Reprinted 1961, Ashford, Kent: Headley.
18. Bandemer, S.L. and Evans, R.J. (1963), *Journal of Agricultural and Food Chemistry* 11, 134-7.
19. Beech, D.F. and Wood, I.M. (1978), in Cowell, R. (*ed.*) *Proceedings of the First International Mungbean Symposium*. Taiwan: AVRDC, 107-11.
20. Begum, S., Shaikh, M.A.Q., Kaul, A.K., Ahmed, Z.U., Oram, R.N. and Malifant, K. (1983), *Field Crops Research* 6, 279-92.
21. Bose, R.D. (1932), *Indian Journal of Agricultural Science* 2, 607-24.
22. Bose, R.D. (1932), *Indian Journal of Agricultural Science* 2, 625-37.
23. Brockwell, J. (1971), *Plant and Soil* (Special Volume), 265-72.
24. Brockwell, J. and Gault, R.R. (1973), *Plant Introduction Review* 9, 30-40.
25. Bunce, G.E., Modie, J.A., Miranda, C., Gonsales, J. and Salon, D.T. (1970), *Nutrition Report International* 1, 325-36.
26. Burkill, I.H. (1935), *A Dictionary of the Economic Products of the Malay Peninsula. II*. London: Crown Agents for the Colonies.
27. Bushby, H.V.A. (1981), *Soil Biology and Biochemistry* 13, 241-5.
28. Bushby, H.V.A., Date, R.A. and Butler, K.L. (1983), *Australian Journal of Experimental Agriculture and Animal Husbandry* 23, 43-53.
29. Calkins, P.H. (1978), in Cowell, R. (*ed.*) *Proceedings of the First International Mungbean Symposium*. Taiwan: AVRDC, 54-63.
30. Catchpoole, V.R. (1983), *Australian Society of Soil Science: Soil News* 55, 7.
31. Chandel, K.P.S. (1982), *IBPGR Regional Committee for Southeast Asia Newsletter* 6, 7-8.
32. Chen, C.L. and Sung, F.J.M. (1982), *Field Crops Research* 5, 225-31.
33. Chiang, M.Y. and Hubbell, J.N. (1978), in Cowell, R. (*ed.*) *Proceedings of the First International Mungbean Symposium*. Taiwan: AVRDC, 93-6.
34. Chowdhury, J.B., Chowdhury, R.B. and Kakar, S.N. (1971), *Journal of Research of the Punjab Agricultural University* 8, 169-72.
35. Cowell, R. (*ed.*) (1978), *Proceedings of the First International Mungbean Symposium*. Taiwan: AVRDC.
36. Dana, S. (1966), *Genetica* 37, 259-74.

37. Dana, S. (1966), *Cytologia* 31, 176–87.
38. Darlington, C.D. and Janaki Ammal, E.K. (1945), *Chromosome Atlas of Cultivated Plants*. Aberdeen: Allen and Unwin.
39. Das, G. (1982), *Canadian Journal of Botany* 60, 1907–12.
40. De, D.N. and Krishnan, R. (1966), *Genetica* 37, 588–600.
41. De Candolle, A. (1886), *Origin of Cultivated Plants*. New York: Hafner Publishing. Reprinted 1959.
42. Deshpande, A.M. and Bathkal, G.B. (1965), *Indian Journal of Agronomy* 10, 271–8.
43. Devadas, R.P., Leela, R. and Chandrasekaran, K.N. (1964), *Journal of Nutrition and Dietetics* 1, 84–6.
44. Empig, L.T., Lantican, R.M. and Escuro, P.B. (1970), *Crop Science* 10, 240–41.
45. Enyi, B.A.C. (1973), *Journal of Agricultural Science, Cambridge* 81, 449–53.
46. Enyi, B.A.C. (1975), *Annals of Applied Biology* 79, 55–66.
47. Fery, R.L. (1980), *Horticultural Review* 2, 311–94.
48. Firth, P., Thitipoca, H., Suthipradit, S., Wetselaar, R. and Beech, D.F. (1973), *Soil Biology and Biochemistry* 5, 41–6.
49. Grewal, J.S. (1978), in Cowell, R. (*ed.*) *Proceedings of the First International Mungbean Symposium*. Taiwan: AVRDC, 165–8.
50. Gunn, C.R. (1973), *Crop Science* 13, 496.
51. Gupta, M.P. and Singh, R.B. (1969), *Indian Journal of Agricultural Science* 39, 482–93.
52. Harper, H.J. and Gray, F. (1957), *Agronomy Journal* 49, 293–6.
53. Hartman, R.W. (1969), *Journal of the American Society of Horticultural Science* 94, 437–40.
54. Hepper, F.N. (1956), *Kew Bulletin* 10, 113–34.
55. Herrera, W.A.T. and Zandstra, H.G. (1979), *Philippines Journal of Crop Science* 4, 146–52.
56. IBPGR (1980), *Descriptors for Mungbean*. Rome: International Board for Plant Genetic Resources Secretariat.
57. Imrie, B.C. (1983), in *Proceedings of the Australian Plant Breeding Conference*. Adelaide, 348–50.
58. Imrie, B.C., Beech, D.F., Blogg, D., and Thomas, B. (1981), 'Mungbean Catalogue'. *Genetic Resources Communication No. 2*. Brisbane: CSIRO Division of Tropical Crops and Pastures.
59. Imrie, B.C. and Butler, K.L. (1982), *Australian Journal of Agricultural Research* 33, 523–30.
60. Imrie, B.C., Drake, D.W., De Lacy, I.H. and Byth, D.E. (1981), *Euphytica* 30, 301–11.
61. Jain, H.K. (1977), in *Food Legume Crops - Improvement and Production*. FAO Plant Production and Protection Paper No. 91. Rome: FAO, 132–6.
62. Jain, H.K. and Mehra, K.L. (1980), in Summerfield, R.J. and Bunting, A.H. (*eds*) *Advances in Legume Science*. London: HMSO, 456–68.
63. Joshi, S.N. (1969), *Indian Journal of Agricultural Science* 39, 1010–12.
64. Kaul, J.N. and Sekhon, H.S. (1976), *Indian Journal of Agronomy* 21, 83.
65. Kaul, J.N., Singh, K.B. and Sekhon, H.S. (1976), *Journal of Agricultural Science, Cambridge* 86, 219.
66. Kylen, A.M. and McCready, R.M. (1975), *Journal of Food Science* 40, 1008–9.

Mung Bean (*Vigna radiata/Vigna mungo*)

67. Krishnan, R. and De, D.N. (1968), *Indian Journal of Genetics and Plant Breeding* 28, 12-22.
68. Kuo, C.G., Tsay, J.S., Liou, T.D. and Hsu, F.H. (1979), *Philippine Journal of Crop Science* 4, 102-6.
69. Kuo, C.G., Wang, L.J., Cheng, A.C. and Chou, M.H. (1978), in Cowell, R. (ed.) *Proceedings of the First International Mungbean Symposium*. Taiwan: AVDRC, 205-9.
70. Lawn, R.J. (1978), in Cowell, R. (ed.) *Proceedings of the First International Mungbean Symposium*. Taiwan: AVDRC, 24-7.
71. Lawn, R.J. (1979), *Australian Journal of Agricultural Research* 30, 855-70.
72. Lawn, R.J. (1979), *Australian Journal of Agricultural Research* 30, 871-82.
73. Lawn, R.J. (1982), *Australian Journal of Agricultural Research* 33, 481-96.
74. Lawn, R.J. (1982), *Australian Journal of Agricultural Research* 33, 497-509.
75. Lawn, R.J. (1982), *Australian Journal of Agricultural Research* 33, 511-21.
76. Lawn, R.J. (1983), *Australian Journal of Agricultural Research* 34, 505-15.
77. Lawn, R.J. and Bushby, H.V.A. (1982), *The New Phytologist* 92, 425-34.
78. Lawn, R.J. and Russell, J.S. (1978), *Journal of the Australian Institute of Agricultural Science* 44, 28-41.
79. Lawn, R.J., Russell, J.S., Williams, R.J. and Coates, D.B. (1978), *Journal of the Australian Institute of Agricultural Science* 44, 112-14.
80. Liener, I.E. and Kakade, M.L. (1969), in Liener, I.E. (ed.) *Toxic Constituents of Plant Foodstuffs*. New York: Academic Press.
81. Ligon, L.L. (1945), 'Mungbeans: a legume for seed and forage production'. *Oklahoma Agricultural Experiment Station Bulletin* B-284.
82. Litsinger, J.A., Price, E.C., Herrera, R.T., Bandong, J.P., Lumaban, M.D., Quirino, C.B. and Castillo, M.D. (1978), in Cowell, R. (ed.) *Proceedings of the First International Mungbean Symposium*. Taiwan: AVRDC, 183-91.
83. MacKenzie, D.R., Ho, L., Liu, T.D., Wu, H.B.F. and Oyer, E.B. (1975), *HortScience* 10, 486-7.
84. Malhotra, V.V., Singh, S. and Singh, K.B. (1974), *Indian Journal of Agricultural Science* 44, 136-41.
85. Matlock, R.S. and Oswalt, R.M. (1963), 'Mungbean varieties for Oklahoma'. *Oklahoma Agricultural Experiment Station Bulletin* B12.
86. Mehta, T.R. (1970), in Kachroo, P. (ed.) *Pulse Crops of India*. New Delhi: ICAR.
87. Moody, K. (1978), in Cowell, R. (ed.) *Proceedings of the First International Mungbean Symposium*. Taiwan: AVRDC, 132-6.
88. Moolani, H.K. and Jana, M.K. (1965), *Indian Journal of Agronomy* 10, 43-4.
89. Morton, J.F., Smith, R.E. and Poehlman, J.M. (1982), *The Mungbean*. Mayaguez, Puerto Rico: University of Puerto Rico Department of Agronomy and Soils Special Publication.
90. Muchow, R.C. and Charles-Edwards, D.A. (1982), *Australian Journal of Agricultural Research* 33, 41-51.
91. Muchow, R.C. and Charles-Edwards, D.A. (1982), *Australian Journal*

of Agricultural Research 33, 53-61.

92. Mungomery, V.E., Byth, D.E., and Williams, R.J. (1972), *Australian Journal of Experimental Agriculture and Animal Husbandry* 12, 523-7.

93. Munns, D.N., Keyser, H.H., Fogle, V.W., Hohenberg, J.S., Righetti, T.L., Lauter, D.L., Zaroug, M.G., Clarkin, K.L. and Whitacre, K.W. (1979), *Agronomy Journal* 71, 256-60.

94. Nalampang, A. (1978) in Cowell, R. (*ed.*) *Proceedings of the First International Mungbean Symposium.* Taiwan: AVRDC, 12-14.

95. Pate, J.S., Atkins, C.A., White, S.T., Rainbird, R.M. and Woo, K.C. (1980), *Plant Physiology* 65, 961-5.

96. Pawar, N.B. and Ghulghule, J.N. (1980), *Tropical Grain Legume Bulletin* 17/18, 3-5.

97. PCARR (1977), 'The Philippines recommendations for mungo in 1977'. *Philippine Council for Agriculture and Resources Research Report.*

98. Poehlman, J.M. (1978), in Cowell, R. (*ed.*) *Proceedings of the First International Mungbean Symposium.* Taiwan: AVRDC, 97-100.

99. Poehlman, J.M., Sechler, D.T., Swindell, R.E. and Sittiyos, P. (1976), 'Performance of the fourth international mungbean nursery'. *Special Report 191,* Agriculture Experiment Station, University of Missouri-Columbia.

100. Poehlman, J.M. and Yu-Jean, F.F.M. (1972), 'Bibliography of mungbean research'. *Missouri Agricultural Experimental Station Bulletin* 991.

101. Putland, S., Strickland, G. and Conde, B. (1982), 'Mung bean recommendations for the N.T.' *Northern Territory Department of Primary Production Agnote,* No. 82/54.

102. Rachie, K.O. and Roberts, L.M. (1974), *Advances in Agronomy* 26, 1-132.

103. Raison, J.K. and Chapman, E.A. (1976), *Australian Journal of Plant Physiology* 3, 291-9.

104. Sandhu, T.S. (1978), in Cowell, R. (*ed.*) *Proceedings of the First International Mungbean Symposium.* Taiwan: AVRDC, 176-9.

105. Sen, N.K. and Chedda, H.R. (1960), *Indian Journal of Agricultural Science* 30, 250-55.

106. Shaikh, M.A.Q., Majid, M.H., Ahmed, Z.U. and Shamsuzzaman, K.M. (1982), in *Induced Mutations for Improvement of Grain Legumes Production II.* IAEA-TECDOC-260. Vienna: International Atomic Energy Agency.

107. Shephard, M., Lawn, R.J. and Schneider, M. (1983), *Insects on Grain Legumes in Northern Australia - A Survey of Potential Pests and Their Enemies.* St Lucia: University of Queensland Press.

108. Singh, B.V. and Ahuja, M.R. (1977), *Indian Journal of Genetics and Plant Breeding* 37, 130-32.

109. Singh, D.P. (1981), *Theoretical and Applied Genetics* 59, 1-10.

110. Singh, H.B., Joshi, B.S. and Thomas, T.A. (1970), in Kachroo, P. (*ed.*) *Pulse Crops of India.* New Delhi: ICAR.

111. Singh, K.B. and Jain, R.P. (1970), *Indian Journal of Genetics and Plant Breeding* 30, 251-60.

112. Singh, R.B. (1980), *IBPGR Regional Committee for Southeast Asia Newsletter* 4, 5-6.

113. Singh, S.R., van Emden, H.F. and Ajibola Taylor, T. (*eds*) (1978), *Pests of Grain Legumes: Ecology and Control.* London: Academic Press.

114. Sinha, S.K., Khanna-Chopra, R., Chatterjee, S.R. and Abrol, Y.P. (1978), *Physiologia Plantarum* 42, 45-8.

115. Sinha, S.K. and Savithri, K.S. (1978), in Singh, S.R., van Emden, H.F. and Ajibola Taylor, J. (*eds*) *Pests of Grain Legumes: Ecology and Control*. London: Academic Press, 233–40.

116. Smith, F.W., Imrie, B.C. and Pieters, W.H.J. (1983), 'Foliar symptoms of nutrient disorders in mungbean (*Vigna radiata*)'. *CSIRO Division of Tropical Crops and Pastures Technical Paper*, No. 24.

117. Sreenivas, L., Upadhyay, U.C. and Warokar, R.T. (1968), *Indian Journal of Agronomy* 13, 137–41.

118. Summerfield, R.J. and Roberts, E.H. (1985), in Halevy, A.H. (*ed.*) *A Handbook of Flowering*. Florida: CRC Press, (in press).

119. Swaminathan, M.S. and Jain, H.K. (1975), in Milner, M. (*ed.*) *Nutritional Improvement of Food Legumes by Breeding*. New York: Wiley.

120. Swindell, R.E. and Poehlman, J.M. (1976), *Tropical Agriculture* 53, 25–30.

121. Swindell, R.E. and Poehlman, J.M. (1978), *Euphytica* 27, 325–33.

122. Talekar, N.S. and Kuo, Y.C. (1979), *Tropical Grain Legume Bulletin* 15, 9–14.

123. Tan, S.Y. (1973), *Food Industries (Taiwan)* 5, 20–26.

124. Tawde, S. and Cama, H.R. (1962). *Journal of Scientific and Industrial Research* 21C, 212–19.

125. Tiwari, A.S. and Ramanujam, S. (1976), *Indian Journal of Genetics* 36, 418–19.

126. Tomlinson, J. and Plaxico, J.S. (1962), 'An economic analysis of mungbeans as a crop for sandy soils of central Oklahoma'. *Oklahoma Agricultural Experiment Station Bulletin* B-595.

127. Tsai, K.H., Lu, L.C. and Oka, H.I. (1967), *Botanical Bulletin of Academia Sinica* 8, 209–20.

128. Tsou, C.S., Hsu, M.S., Tan, S.T. and Park, H.G. (1979), *Acta Horticulturae* 93, 279–87.

129. Vavilov, N.I. (1926), *The Origin, Variation, Immunity and Breeding of Cultivated Plants*. Translation by Chester, K.S., *Chronica Botanica* 13, 1951.

130. Verdcourt, B. (1970), *Kew Bulletin* 24, 507–69.

131. Watt, E.E. and Maréchal, R. (1977), *Tropical Grain Legume Bulletin* 7, 31–3.

132. Watt, E.E., Poehlman, J.M. and Cumbie, B.G. (1977), *Crop Science* 17, 121–5.

133. Wilczek, R. (1954), in *Flora Congo Belge et Ruanda-Urundi Vol. VI*. Brussels.

134. Williams, R.W., Lawn, R.J., Imrie, B.C. and Byth, D.E. (1983), in *Proceedings of the Australian Plant Breeding Conference*. Adelaide, 298–9.

135. Yang, C.Y. (1978), in Cowell, R. (*ed.*) *Proceedings of the First International Mungbean Symposium*. Taiwan: AVRDC, 141–6.

136. Yohe, J.M. and Poehlman, J.M. (1972), *Crop Science* 12, 461–4.

137. Zukovskij, P.M. (1950), *Cultivated Plants and Their Wild Relatives*. Translation by Hudson, P.S. (1962), Farnham Royal: Commonwealth Agricultural Bureau.

14 Winged bean (*Psophocarpus tetragonolobus* (L.) DC.)

G.E. Eagleton, T.N. Khan and
W. Erskine

INTRODUCTION

The winged bean (*Psophocarpus tetragonolobus* (L.) DC.) is a
semi-domesticated legume distributed throughout the humid
tropics of Asia and Melanesia. In Burma and Papua New Guinea
there are landraces which produce edible tubers on the scale
of a field crop. More commonly, winged beans are grown for
their 'winged' fruits and, then, one or two vines are sufficient to meet individual household needs.

In recent years, researchers have become interested in
the winged bean as a potential grain legume crop. This
attention was prompted after recognition by the US National
Academy of Sciences of the winged bean's capacity to fix
carbon and nitrogen in difficult humid tropical environments
and to transform these elements into seed protein and oil
composed of a nutritionally valuable configuration of amino
and fatty acids, respectively [98].

This review traces the natural history of the winged
bean as a cultivated plant, examines its 'potential' to
become a major grain legume and considers the transforma-
tions which we believe will be necessary to realise this
potential. It does not address the question as to whether
another grain legume is required, or whether this is the
best emphasis for the development of the winged bean.

THE PLANT

Taxonomy and distribution

It was around 1690, on the small Dutch outpost of Amboina, that Georg Eberhard Rumpf recorded the leguminous species *Lobus quadrangularis* which is now known as *Psophocarpus tetragonolobus* (L.) DC. [129]. His description of the species and the drawing of it (Figure 14.1, possibly attributable to his son, with its quaint depiction of nodules or perhaps tubers) are models of pre-Linnean observation. He noted its viny, climbing habit, tuberous roots ('like turnips'), papilionaceous flowers and various parallels with other species of what is now known as the tribe Phaseoleae.

Although Lackey [84] placed *Psophocarpus* Neck ex DC. within the subtribe Phaseolineae, it is not typical of those genera, having a chromosome number of $n = 9$ instead of 11 [51,74,103,145]. The two species of *Psophocarpus* examined by Pickersgill [103], *P. scandens* and *P. tetragonolobus*, also have $n = 9$.

There are nine species in the genus of which all but the winged bean are African [151]. *Psophocarpus scandens*, *P. palustris* and *P. grandiflorus* are the species most closely related to the winged bean. On present evidence, *P. scandens* is reproductively incompatible with the winged bean [33,37] and there are differences between their karyotypes [103]. There is no cytogenetic information for *P. grandiflorus* (and see Chapter 1).

Winged beans have not been recorded in the wild [37,95, 151] and there are suggestions of either an African [9] or an Asian origin [9,32,61,103].

When Burkill [9] recorded the species, the distribution of winged beans stretched from Mauritius in the Indian Ocean (but not Malagasy and the African mainland) across southern India and South-east Asia into New Guinea. Northern Burma, at 25°N, is the highest latitude where winged beans have been traditionally grown, although there is a taxonomic record from Yunan [151]. There is no evidence of a natural overlap between winged beans and the African species of *Psophocarpus*, although both winged beans and *P. scandens* were widely dispersed during colonial times.

Ecology

Winged beans are found in well-drained lowland environments with a mean temperature range of 15°-28°C and an annual rain fall in excess of 700 mm [29,117,118]. In Papua New Guinea

Figure 14.1 *Lobus quadrangularis (= Psophocarpus tetragonolobus)* according to Rumphius (1747) [129]

and Burma they grow at altitudes up to 2000 m [9,74] but do not survive frosts [80]. Controlled-environment experiments suggest that day temperatures in the region of 27°C and nights warmer than 18°C are optimal for growth and reproductive development [33,55,56,147,155], although tuber initiation is favoured by somewhat cooler temperatures [128,140].

Winged beans require large amounts of water [3,8,69,97, 127] but, at the same time, they do not tolerate waterlogging [125,127]. Short days are needed to initiate both flowers and tubers [33,34,127,147]. Wong [154] determined the critical daylength for flowering to be between 11.25 and 12.5 hours for one Malaysian accession. Daylength and temperature interact in their influence on flowering [127,147], but the relative importance of these two factors and their interactions are poorly understood [160].

Although winged beans grow and yield less well in soils more acidic than pH 5.5, the plants and nodules can survive at a pH of 4.5 [63,156, S. Zulkifli, pers.comm.]. There is usually a positive response to lime application [158,159] although interactions with aluminium concentration are an important factor in this response on acid soils [156, S. Zulkifli, pers.comm.].

General morphology and development

Individual winged bean seeds vary in weight from 50 mg to 600 mg and in colours from white through to black, depending on genotype [17,74,131]. If artificially dried to less than 11% water content at harvest, winged beans develop hard-seededness which, however, can be broken by mechanical scarification or by soaking seeds in concentrated sulphuric acid [87,101,142]. Germination is hypogeal; seedlings usually emerge within eight days of sowing but their initial growth is slow. Thus, even after fourteen days, plants are still less than 10 cm tall and it is not until about three weeks that leaf area begins to increase more or less linearly; this phase then continues until at least eleven weeks after sowing [56,139]. By this time, most genotypes at equatorial latitudes have begun to flower although some require as long as 150 days, irrespective of latitude or altitude [130].

Winged beans are indeterminate, climbing, annual or perennial herbs. The flowers are borne in axillary racemes comprising from three to twelve flower buds but seldom do more than one or two fruits develop on any one raceme [89, 105]. Individual flowers take 21-25 days to open from the time the buds are macroscopically visible. Anthers begin to dehisce in the evening, flowers open the following morning, and stigmas are most receptive at around the time of flower opening [41,89,133]. Pollinating insects, particularly bees of *Xylocarpa* spp., are frequent visitors and cross pollination may reach 7.6% [37,38,68,89].

The individual winged fruits show a rapid sigmoidal growth pattern whereas seed growth is diauxic [24]. Fruits reach their maximum length after about 20 days from anthesis and maximum weight at about 25 days. Seeds and fruits do not reach full maturity until 50-60 days after anthesis. The pod walls are very fibrous and so for use as vegetables the fruits must be picked within 15-20 days after anthesis [15,23,24]. There is genetic diversity for flower colour and in fruit colour, shape and size [74]. There is variation too in the branching habit and flowering pattern of cultivars which leads to different patterns of dry-matter accumulation [139].

The fibrous root systems with their large nodules (up to 1.5 cm in diameter) grow in proportion to leaf area until about 100 days after planting [2,158]. Then, root growth either levels off as the reproductive sinks sequester photosynthate, or accelerates in those cultivars, environments and management circumstances which favour continued vegetative growth and initiation of tubers [2]. In tuberous cultivars, increases in root dry weight continue undiminished beyond the sixth month after planting, although nodule dry weight does not increase further after about three months (M.A.

628

Zainah and N.F. Chai, unpublished). Tubers are ready to harvest by the seventh to eighth month, when the shoots senesce.

If tubers are not harvested, such cultivars are able to regenerate from the old root stocks when conditions are favourable. Vegetative propagation is more reliable from stem cuttings or from tissue culture than from separated tubers or tuber sections [33,87,150].

Plate 14.1 gives a general impression of fruiting plants and of genotypic differences in tuber production.

Plate 14.1   (a) Staked, fruiting plants of cultivar UPS 32 and
(b) tuberous roots harvested from four winged bean genotypes

Genetic variability

With the exception of Burma, Laos and Kampuchea, all countries which traditionally grow winged beans are represented in national germplasm collections [16,17,50,56,66,74,120, 131]. Efforts are now underway to describe these collections systematically and to catalogue them [134].

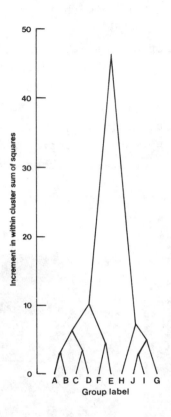

Number of accessions in the group

Group	Papua New Guinea	Indo-nesia	Malaysia	Brunei	Thailand	Burma	India	Sri Lanka	Miscel-laneous
A	11	0	0	0	0	0	0	0	1
B	7	0	0	0	0	0	0	0	1
C	24	0	0	0	0	0	0	0	0
D	6	0	0	0	0	0	0	0	0
E	12	0	0	0	0	0	0	0	6
F	2	2	1	0	2	0	0	0	1
G	0	2	12	0	0	0	0	1	0
H	0	6(Bogor)	6	1	10	0	1	0	0
I	0	3	2	0	16	0	2	0	0
J	0	0	0	0	16	1	0	0	0

Figure 14.2 Pattern analysis of 147 accessions of winged bean scored for forty-seven characters

Winged Bean (*Psophocarpus tetragonolobus* (L.) DC)

The winged bean germplasm contains striking diversity in morphological and agronomic characters [74,131]. Erskine and Khan [44] have demonstrated that a gene which affects stem colour is linked on the same chromosome with one affecting calyx colour. There is linkage too between major genes which affect pod wall colour and one affecting the colour of the fruit wings. Using the technique of diallel analysis, Erskine [40] demonstrated significant additive genetic components of variation for the characters of seed weight, number of seeds per fruit, fruit length, individual leaf area, leaf area index, time to first flower, shelling percentage and seed yield, amongst eight genotypes from Papua New Guinea and Indonesia.

The genetic variability is not distributed uniformly across the geographic range of the species. Figure 14.2 presents a pattern analysis carried out on 147 accessions from several countries scored for forty-seven characters. This classifies the accessions into ten groups using a strategy based on minimising incremental sum of squares applied to standardised Euclidean distances within a 47-dimensional space. This figure demonstrates the validity of a taxonomic distinction between the Papua New Guinea ecotype on the one hand and the South-east Asian on the other. Subdivision within these two ecotypes is of a much smaller order of magnitude. Nevertheless, the concept can be sustained of a predominantly Malaysian grouping, as distinct from two Thai groups and from a more general group centred on the late-maturing Indonesian genotypes in the Bogor collection [130]. Group F is a bridging category gathering together the accession UPS 121 from lowland Papua New Guinea with the early-maturing South-east Asian accessions such as UGM 1, UGM 19 and UGM 100 of Thompson and Haryono [144]. Of principal importance in differentiating highland New Guinea accessions from lowland South-east Asian ones are the characters' number of days to first flower, number of branches arising from the first (lowermost) ten nodes and calyx colour.

Evidence suggests that there have been at least two

centres for the domestication of winged bean - one in the
New Guinea Highlands and another in the central plains of
Burma [9,85,109]. Winged beans from Papua New Guinea are
non-branching, early maturing and morphologically diverse
annuals. Northern Thai and Burmese accessions are branching,
late-maturing, facultative perennials with a strong tendency
to develop edible root tubers [31]. In New Guinea there is
considerable differentiation between landraces collected
just a few kilometres apart and heterogeneity is a common
feature of these accessions [44,74].

There is an unfortunate lack of representative germplasm
from Burma. However, in related material from north-east
Thailand the signs of selection towards tuberous cultivars
are unmistakable.

Winged beans in Indonesia and Malaysia show less evidence of selection towards a domesticated plant type (and
see Chapter 2).

TRADITIONAL CULTIVATION AND USE

It is possible that winged beans have been cultivated in
South-east Asia for some considerable time. However, the
evidence is evasive and, at least in the case of Papua New
Guinea, suggestive of a quite late origin - as recent as
three or four centuries ago [85,109]. Whatever the time-
scale, three broad patterns of cultivation and use can be
discerned.

Field crop tuber production - the Burmese example

Cultivation of winged beans for edible tubers in the Irra-
waddy basin of Burma was described in detail by Burkill [9,
119]. In his time, at the turn of the century, they were
grown unstaked in plots as large as one or two hectares, as
a commercial crop with hired labour to carry out weeding and
harvesting. Sandy loam soils were preferred and flood irrigation was carried out at least three times every two months,

depending on rainfall. Crops were planted on ridges using seeds brought in from villages in the Shan Hills, since the crops grown in the plains were unsuitable for good seed production. Tubers were harvested before the crops reached the mature-seed stage.

No other species produces tubers with a greater concentration of crude protein than those of winged bean, although Claydon's [20] estimate of 10.5% protein and 30.5% carbohydrate now seems excessive in the light of the more recent findings of Evans *et al.* [45] and Poulter [107] that Kjeldahl nitrogen determinations overestimate true protein concentration by a factor of 20%.

Nevertheless, there is genotypic variation in protein concentration, fibre concentration and probably in overall nutritional value [60, M.A. Zainah and N.F. Chai, unpublished]. Claydon's [21] observations that tubers are still a popular food commodity in Burmese markets attest to their nutritional usefulness.

In experimental plots, fresh-weight tuber yields of 1500 g m^{-2} (15000 kg ha^{-1}) have been obtained from one staked northern Thai winged bean accession [31]. Without staking, yields are reduced [69,153] and in Burma, tuber yields from commercial crops [9] varied from 2500-6200 kg ha^{-1}. Perhaps after noting similar yields, Burkill [9] commented:

'The margin left to *bona fide* cultivators after trouble and expenses would by itself not be encouraging but the cultivators ... generally grow a bumper crop of sugar cane in the year after.... It is said that the cane crop if preceded by Goa beans (i.e. winged beans), yields half as much again as usual.'

Multipurpose horticultural crop - the New Guinea example

Winged bean is the most important leguminous crop grown in the settled horticultural valleys of the New Guinea Highlands, at altitudes of 1400-2200 m [108]. In the valley pockets, 10-20% of the cultivated area may be used to grow winged beans during June to December, a relatively dry period of the year [78]. Its role is both nutritional and ecological since the staple crop, sweet potato, is poor in protein and demanding of soil fertility.

Winged beans may be grown as a monocrop (usually 0.3-1.0 hectares) in rotation with sweet potato, or in separate blocks within mixed gardens. Well-drained, raised beds are provided, seeds are planted at around 150 000 ha^{-1} and the plants are always staked [78].

All parts of the plant are eaten: shoots, leaves, flowers, young fruits, green seeds and root tubers. When grown for tubers, the flowers and young fruits are removed and eaten either raw or after steam-roasting in a ground-oven style of cooking (*mumu*). Trials have shown that tuber yields can be increased by as much as four-fold by pruning reproductive sinks [2,54,69]. Tuber yield estimates in Highland gardens [78,108] have ranged from 1800-11000 kg ha^{-1}. Experimental yields in the Highlands [140] have been as large as 17000 kg ha^{-1}.

Experiments with New Guinea cultivars in Port Moresby produced a mean green fruit yield of 10290 kg ha^{-1} over nine weeks of regular harvest, with a 'best yield' of 17820 kg ha^{-1} from an experimental hybrid [42,72]. On a fresh-weight basis, the fruits contain 1-3% protein [21] and compare favourably with other legumes. The fruits are best picked when they are about one half to three-quarters fully grown [15, 23]. In older fruits, the green ripe seeds, which contain 5-10% protein, are removed and eaten as 'peas' usually after whole fruits have been steam-roasted in the *mumu*. In Papua New Guinea, fully mature seeds are used for sowing and are not consumed.

Claydon's [22] nutritional survey of two areas in the Highlands estimated that in the tuber season winged beans provide up to 25% on a dry weight basis of the total nutrient intake. Poulter's [107] recent studies, however, would suggest that the nutritional value of this tuber consumption

may be quite small because of an unbalanced amino acid composition. Winged bean fruits and unripe seeds, while comprising only 4% of the total nutrient intake [22], may be valuable because of their better balance of nutrients.

In addition to their direct nutritional value to the
cultivators, winged beans are also valued as a cash crop.
Fruits and tubers are a popular food item at the local and
even central markets [141]. The winged bean's 'soil restora-
tive properties' are also well recognised by the Highlanders
[78], if not yet well understood by researchers.

Minor garden vegetable - the South-east Asian example

Throughout much of lowland South and South-east Asia winged
beans play a minor role as an occasional vegetable in household gardens [17,53,54,58,106,129,130,136,153]. One or two
vigorous perennial or semi-perennial vines are planted at
the base of fences, stumps or trees over which they sprawl,
sometimes to a height of 3-7 metres. The preference is for
a wet-season planting and the vigorous growth of such cultivars enables them to persist with minimal management.

In Vietnam, fresh fruit yields of 10000 kg ha^{-1} have
been obtained over a forty-day harvest period [106]. In the
cool winter season, Vietnamese cultivars become dormant but
regenerate from their tuberous root bases with the advent
of warm, wet conditions. By contrast, in Malaysia, vigorous
perennial cultivars can supply fruits over a prolonged
period. Wong [153] has quoted a fresh vegetable yield of 35
t ha^{-1} from the best Malaysian cultivar.

Only rarely are any plant components other than the
immature (vegetable) fruits consumed in lowland South-east
Asia. In Java, the mature seeds are sometimes eaten after
prolonged soaking and cooking, or in fermented bean cake
preparations known as 'tempeh'. Elsewhere in South-east
Asia, winged bean seeds find little use.

One of the striking features of traditional winged bean
cultivation is that the potential nutritional value of the

plant's mature seeds seems not to have been exploited. This is probably due to the difficulty of removing the antinutritional factors contained in the seeds without decreasing their nutritional value in other ways.

GRAIN LEGUME POTENTIAL OF WINGED BEAN

Masefield [95] first drew attention to four features of winged beans which suggested a potential for an expanded agricultural role. These features: (a) the good nutritional value of winged bean seeds; (b) the ready nodulation and nitrogen fixing capacity of its nodules; (c) its strong tropical adaptation; and (d) a demonstrated capacity to produce large seed yields, are examined below.

Nutritional qualities of the seed

The nutritional value of winged bean seeds has been recognised since the turn of the century [75]. Mature seeds contain 20-46% protein and 17-22% oil [60]. The amino acid profile compares favourably with that of soyabeans (*Glycine max*) having comparable concentrations of sulphur-containing amino acids and a larger concentration of lysine (Table 14.1 and see Chapter 9) [11,35,48]. In fatty acid profile, winged bean seeds resemble that of groundnuts (*Arachis hypogaea*) rather than soyabeans, with larger oleic acid and behenic acid concentrations and smaller linoleic acid concentrations than in soyabeans [14], and a ratio of saturated to unsaturated fatty acids of 1:3 in contrast to a ratio of 1:6 for soyabeans [6].

Nodulation and nitrogen fixation

Early reports suggested that winged bean is exceptional in its capacity to nodulate, fix nitrogen and to survive on nitrogen-poor soils [9,95,98]. In Malaysia, Masefield [93] obtained fresh nodule yields of 700 kg ha^{-1}. Then again,

Winged Bean (*Psophocarpus tetragonolobus* (L.) DC)

Table 14.1 Composition of winged bean seeds

(a) Proximate analysis

Component	Concentration (%)	
	Range	Mode
Protein	21–46	37
Oil	7–22	19
Crude fibre	9–11	10
Nitrogen-free extract	24–28	25
Moisture	5–10	9

(b) Amino acids, fatty acids, and antinutritional factors

Amino acid	Concentration (%)	Fatty acid	Concentration (%)
Ile	4.0–4.9	Myristic	0–0.1
Leu	6.8–9.0	Palmitic	6.3–10.2
Lys	6.8–8.0	Stearic	2.6– 7.7
Met	0.8–1.2	Oleic	34.0–40.2
Cys	0.6–1.6	Linoleic	23.0–31.6
Phe	3.8–5.8	Arachidic	1.3– 2.3
Thr	3.3–4.3	Linolenic	2.0– 5.6
Trp	0.9	Behenic	12.5–15.9
Val	4.6–5.5	Antinutritional factors	
Tyr	2.6–4.7		
Arg	6.4–7.5	Lipoxygenase activity : 31–44 units mg protein^{-1}	
His	2.7–4.2	Trypsin inhibitor : 22.2–42.5 mg g seed meal^{-1}	
Ala	2.4–4.3	Chymotrypsin inhibitor: 30.1–47.6 mg g meal^{-1}	
Asp	6.9–11.5	*Haemagglutinin activity*	
Glu	10.1–15.3	Human 0+ erythrocytes : 615–2460 units mg meal^{-1}	
Gly	3.6– 4.3	Rabbit erythrocytes : 77–308 units mg meal^{-1}	
Pro	4.5–6.9		
Ser	3.8–4.9		

Harding *et al.* [52] reported greater numbers and weights of
nodules from winged beans than from four other tropical
legume crops with which it was sown, uninoculated, on an
Oxisol with no previous history of legume cultivation. On
the other hand, there have been reports of nodulation difficulties in the crop [118, M. Aslam, pers. comm.].

These different experiences could perhaps be explained
by the non-specific rhizobial requirements of winged bean
[62]. It forms symbioses with the broad-spectrum cowpea
group of rhizobia which inhabit most tropical soils [36].
But, the effectiveness of the symbiosis varies with strain
of *Rhizobium* and host genotype [62,64]. In newly planted
areas the formation of nodules does not necessarily trans-
late into effective nitrogen fixation. (see Chapter 4.)

Effectively nodulated winged beans have shown considerable potential to fix nitrogen, whether this be measured in
terms of acetylene reduction or in terms of plant nitrogen
concentration [65,94,157]. In pot trials, nitrogen fixation
equivalent to 140 kg N ha^{-1} in a nine-week period has been
reported [65]. All authors point to the forage and green
manure qualities of well-nodulated winged bean genotypes.

Humid tropical adaptation

Nangju and Baudoin [97] found that winged beans gave the
second poorest seed yield (bush Lima beans being the least
productive) of six 'tropical' grain legumes (the other
species being cowpeas, jack beans, pigeonpeas and soyabeans)
when grown at a subhumid site in Nigeria (120 cm rainfall
per annum). But, in a wet region in that country (240 cm
rainfall per annum), they produced the largest yield of all
six species. The vigorous growth, daylength and temperature
sensitivity and late-maturity characteristics of winged
beans are typical of humid tropical species which must contend with strong weed competition.

Winged bean seeds retain viability better under tropical
conditions than do those of soyabeans but not as well as

mungbeans [57,91,142]. The hard-seededness in winged bean
seeds dried to small moisture contents is perhaps beneficial
to seed storage. The seeds are reported to be relatively
resistant to *Bruchid* storage pests (T.V. Price, pers.comm.)

Winged beans show resistance to several insects and
diseases to which many other legumes succumb when grown in
the tropics, e.g. *Phytophthora* root-rots [112]. They are
reported to be relatively resistant to bean fly borers of
the *Agromizidae*; and, perhaps, to pod-sucking bugs such as
Nezara and *Riptortus* spp. [100,112,125] against which it
is possible that the large fibrous pod walls offer some
protection.

Seed yield

Seed yields from experiments and observation plots [54,69,
77,97,99,125] regularly exceed 2000 kg ha^{-1}. In some recently concluded International Trials (Table 14.2) [76] one
cultivar achieved a mean seed yield across sixteen sites in
excess of 1.0 t ha^{-1}, and at one site two cultivars exceeded
3 t ha^{-1} (T.N. Khan, unpublished). The largest seed yield
recorded in the literature, 4.5 t ha^{-1} obtained from an
experimental crop in Malaysia [153], has recently been repeated, also with Malaysian material, in Java [144]. Tham
[142] reported a yield of 2.8 t ha^{-1} over three consecutive
seasons in Malaysia. Clearly, there is potential for large
yields, but it is important to realise that such yields have
been obtained from trellised plants, regularly hand-harvested
over long periods. In few reports do authors detail the
duration of growing season required to obtain such yields.
An exception is the report of Nangju and Baudoin [97] in
Nigeria, who obtained a seed yield of 2400 kg ha^{-1}, or 11 kg
ha^{-1} day^{-1}. In many trials, it is probable that between six
and ten months were required to achieve the yield figures
quoted.

Table 14.2 Performance of thirteen cultivars included in the First
International Winged Bean Variety Trial (T.N. Khan, unpublished), means
across sixteen locations

Cultivar	Days from sowing to:		Mean seed yield (kg ha^{-1})	Threshing percentage	Best seed yield (kg ha^{-1})
	first flowering	last ripe fruit			
LBNC 1	87	190	881	44.3	3493
LBNC 3	86	193	854	41.7	2699
Nakhon Sawan	77	179	1082	39.3	3655
UPS 31	63	158	794	40.1	1690
UPS 32	64	153	661	41.3	1305
UPS 45	64	157	602	41.3	1563
UPS 47	62	157	720	44.8	1451
UPS 53	64	153	784	45.0	1776
UPS 62	64	153	713	47.0	1578
UPS 99	60	155	681	44.7	1429
UPS 102	62	159	824	44.1	1829
UPS 121	76	165	899	39.8	2279
UPS 122	68	162	924	41.0	2017

PRINCIPAL LIMITATIONS TO GRAIN LEGUME POTENTIAL

Growth habit

Wider use of winged beans as a grain legume crop is discouraged by their exclusively climbing habit. Substantial trellising and hand-harvesting represent significant costs to
the cultivator.

There are two schools of thought on these issues. One
is that the large labour requirement of winged beans is an
asset in the context of some developing economies, and that
the way to 'pay' for this is to aim for good productivity
per unit labour. K.C. Wong (pers.comm.) believes that in
Malaysia it ought to be possible to pay for the capital cost
of trellising by aiming for a perennial plant type capable
of ratooning up to four times from the same root stock. Pate

and Minchin [102] have reported that the nodulated tuberous roots of winged bean continue to fix nitrogen even after the tops have been removed, and work in Vietnam suggests that vigorous regeneration arises from such root stocks [106].

The second, more conventional, approach is to mount a breeding programme to produce plants with bush habits. The characteristics associated with such a habit are: abrupt termination of stem elongation; shortening and thickening of internodes; self-supporting stem and branches; and flowering at basal nodes. In soyabeans, these attributes are conferred by the interaction of a few major genes with environment and with minor gene modifiers [4,5] (and see Chapters 2 and 9).

A significant difference in growth habit, which is strongly heritable (narrow-sense heritability of 65%), has been demonstrated between winged bean ecotypes from Papua New Guinea and those from South-east Asia (G.E. Eagleton, unpublished). This difference is expressed mainly by a shift from apical stem dominance to axillary stem dominance, but also in terms of a shift from early to late flowering.

In none of the several hundred accessions which have been screened so far has trace of either a bush habit or of determinacy been detected. Nevertheless, the Law of Homologous Series [148] leads to the expectation that simply-inherited determinacy ought to be obtainable in this species. The search for determinate types is of singular importance and if they are not forthcoming soon, then it may be necessary to resort to mutation breeding [73].

Maturity and fruiting characteristics

Rapid, synchronous reproductive development in a monocropped grain legume ensures economical use of land and harvest labour. Most winged bean cultivars require at least forty days to flower, and many considerably longer. The first mature fruit is rarely ready for harvest in less than 110 days. Accessions from Papua New Guinea are generally earli-

est to flower (Table 14.2); however, most cultivars are sensitive to photoperiod and, even in tropical latitudes, a
delay in flowering of several days occurs in response to
small seasonal variations in daylength [40]. At subtropical
latitudes the delay is greater (Table 14.3).

Table 14.3 Performance of winged beans in eight environments included
in the First International Winged Bean Variety Trial ([76] and unpublished data), means of thirteen cultivars

Location	Latitude	Date of sowing*	No. of days to flowering	Mean seed yield (kg ha^{-1})	Best seed yield (kg ha^{-1})
Singapore	1°N	8 March	70	678	1588
Sri Lanka	7°N	1 May	58	1110	1717
Philippines	8°N	–	44	528	990
India (Chethali)	11°N	15 Sept	68	699	1440
India (Bangalore)	12°N	–	68	634	1002
Taiwan	23°N	21 Jan	101	1914	3655
Bangladesh	24°N	–	90	1082	2173
Nepal	27°N	19 June	71	417	829

*Dashes denote 'not available'

Late development is not necessarily a problem in well-
watered equatorial environments since the extra growth can
be translated into larger seed yields. It is the late-
maturing South-east Asian accessions which give the largest
seed yields [76,144,152]. However, associated with late
maturity is usually a very drawn-out harvest period. To
obtain the full yield from climbing winged beans, as many
as ten weeks of hand-harvesting may be required.

Amongst winged bean germplasm, there is, in fact, considerable genotypic variation in the respective times to
flowering and from flowering to first mature fruit [39,40,
144] as well as in the sensitivity of phenology to daylength
and temperature [127,147]. In diallel cross analyses involving several accessions, narrow-sense heritability for time

to first flower has been estimated at 68% in Port Moresby [40] and 17% in the subtropical location of Perth, where daylength sensitivity dominates phenological development (G.E. Eagleton, unpublished). In both environments, genetic advances can be made through selection in segregating generations of appropriate crosses.

In the phase from flowering to first mature fruit, heritability is poor [40]. There is little published information as to the degree of genetic variability in the duration of the fruit maturation phase *per se*.

Disease and insect pests

Although winged beans show resistance to some tropical legume diseases and insects, there are several specific diseases and pests which pose problems to expanded use of the crop.

The false rust (*Synchytrium psophocarpi* (Rac.) Gaumann) is a widely distributed and potentially destructive disease [26,27,143]. In Java, where yields can be severely reduced, resistant genotypes have, however, been detected [143]. At least two genes are involved [90].

Leaf spots (caused by *Pseudocercospora psophocarpi* (Yen) Deighton [113]) may be damaging in humid climates. Powdery mildew (*Erysiphe cichoraceum*) [110] and collar-rots caused by various ubiquitous pathogens, including *Macrophomina phaseolina*, *Rhizoctonia solani* and *Fusarium* species [114], have been reported, but their importance has not been assessed.

Virus diseases have been widely reported [25,47,135, 143]. In the Ivory Coast, Fauquet *et al.* [47] encountered three diseases, two of which have been confirmed to be, and the third one suspected of being, caused by viruses. There is evidence of genetic resistance to a virus disease (probably Cowpea Mosaic) in Indonesian germplasm [143].

Root-knot nematodes (*Meloidogyne incognita*, *M. javanica* and *M. arenaria*) have been reported as serious pests [46,

74,78,111,123,137]. Price [112] estimated reductions in
tuber yield by as much as 50% in Papua New Guinea due to
root-knot nematodes. There appears to be little genetic
resistance against nematode infection [30,88,111] and so the
prospects of sustaining winged bean yields over a series of
successive ratoons do not look promising.

A large number of insect pests have been reported to
attack winged beans [75]. Two classes, however, seem to be
of particular importance: bud- and flower-boring larvae of
the *Lepidoptera* and *Coleoptera*, and a range of leaf-eaters,
particularly from the *Lepidoptera*. *Maruca testulalis* (Geyer)
is the most significant flower-borer in Papua New Guinea;
Lamb and Price [86] described how 31% of flowers examined
in a planting in Port Moresby were infested, but that there
was little fruit damage. In India, Srinivasan *et al.* [138]
reported that 20-30% of fruits were infected by the pod-
borer *Heliothis armigera*. Nangju and Baudoin [97] found that
bi-weekly sprayings of monocrotophos at 0.5 kg ha^{-1} marginally improved yields in Nigeria, but there has been no systematic work to determine economic thresholds for spraying.

Fruit structure and seed losses

Since winged beans have not traditionally been grown for
their seeds, it is not surprising that the large awkward
fruits are sources of inefficiencies. Fruit length varies
from 10-60 cm. Fruit structure also varies: most are rectan-
gular in cross section whereas others are flatter (i.e. with
reduced 'wings'). This variation is controlled by a single
major gene [43].

Despite variations in fruit size and structure, shelling
percentage is consistently below 50% (Table 14.2) and, in
most cultivars, fruit shattering at maturity is a significant source of seed loss [125].

There is evidence from diallel analyses that fruit
length is a strongly heritable character [40, G.E. Eagleton,
unpublished]. N. Chomchalow (pers.comm.) has detected

genotypes with resistance to shattering. Shelling percentage, despite a narrow phenotypic range, has an estimated heritability of 77% [40]. It is evident that there is scope for genetic advance towards a more favourable fruit type in the species.

Seed processing difficulties

Despite similarities with soyabeans in protein and oil concentrations and in amino acid profile, Gillespie and Blagrove [48] have characterised three storage protein fractions in winged bean seeds which have different properties from those of soyabeans. The solubility characteristics of these proteins and their extraction and commercial recovery will present more difficult problems than for soyabeans.

Another problem in the use of winged bean seeds is that the seed coats, which represent 12% of their total weight [1], are difficult to remove [11]. Unless preceded by soaking in sodium bicarbonate, rate of water uptake and cooking times are much slower than in other edible legumes such as *Phaseolus vulgaris*, *Cicer arietinum* and *Vigna unguiculata* [92,126].

Winged beans, unlike soyabeans, contain a specific chymotrypsin inhibitor in addition to the usual trypsin inhibitor [82,83]. In contrast to soyabeans, which contain only rabbit erythrocyte agglutinating lectins, winged beans also contain human erythrocyte agglutinins [49,67,115,124,132]. It is essential that these antinutritional factors be de-activated before seeds are eaten by humans or animals, but to do this requires at least thirty minutes of autoclaving which renders the proteins insoluble and decreases lysine availability [35]. For these reasons, in traditional planting areas, mature winged bean seeds have never been a popular food item in spite of their nutritional value.

There is useful genetic variation in seed composition: in protein and oil concentrations [59,121], in haemagglutinin and proteolytic enzyme activity (A.A. Kortt, in press), and in seed coat colour [17]. However, there is negligible variation in protein composition, solubility or processing quality [7].

It is likely that if use of winged bean seeds is to expand, it will be for vegetable oil, and as protein meal in animal feeds and processed foods [18,28,79]. In Ghana, Czechoslovakia and Vietnam, trials with children suffering from kwashiorkor fed a diet in which winged bean seeds provided most of the protein, have demonstrated that they can safely be used as vegetable protein substitute for milk [12,13]. Autoclaving is essential to render the protein meal safe [10,11,152]. Processed winged bean meal is unlikely to have any inherent disadvantage when compared with other legumes [18,19] and its oil may even be preferred to the comparable groundnut oil because of its larger tocopherol concentration [11,28].

Narrow adaptation

The poor average seed yields recorded in the International Winged Bean Trials (Table 14.2) point to an important limitation - the lack of broad adaptation both of the species and of its cultivars. These figures were averaged over several sites which ranged in latitude from 1°N to 27°N (Table 14.3). In only two sites did the best genotype reach 2000 kg ha^{-1}. The cultivar Nakhon Sawan, from Thailand, with the best average yield across sites, was also the most unstable in yield and in time to maturity. Photoperiodic sensitivity seemed to be one of the important factors in this instability, as was also the case in a similar study of ten accessions in Western Australia [34].

Even where photoperiod sensitivity is of little relevance, such as in Khan and Erskine's [77] trial in Papua New Guinea, instability is still characteristic of winged bean seed yields. Thirty cultivars from Papua New Guinea were involved in this trial which was planted at two Highland and three Lowland locations. Environmental variation was of overwhelming importance, accounting for 69.9% of total variation, while genetic differences represented a minute 0.3% of the total and were not significant when tested against the genotype x environment component. Because of the large differences in environments, the authors concluded that different strategies would be required in varietal selection for the Lowlands and Highlands. Similar conclusions were reached by Kesavan [71] for fruit yield and by Stephenson *et al.* [140] for tuber yield, who also suggested that temperature sensitivity was the primary determinant of the different responses between Highland and Lowland sites.

Controlled-environment studies have emphasised this narrow adaptability of winged beans to environmental variation. In relation to moisture deficit, temperature extremes and daylength, winged beans have shown the kinds of responses expected from a species adapted to the well-buffered climate of equatorial latitudes [8,33,127,155]. The repeated failure of the crop in waterlogged soils has, however, been somewhat surprising [125,128]. There is, unfortunately, little information as to the soil requirements of winged bean, apart from one study of nutritional requirements on acid tropical soils [116].

Undoubtedly there is genetic variation in response to most factors limiting winged bean's response to environment. Cultivars from Papua New Guinea showed greater stability in seed yield than those from South-east Asia in the International Trial because they were early to mature [76]. However, under the hot, wet, alkaline soil conditions of north-western Australia, cultivars from Papua New Guinea died, while those from South-east Asia thrived [125] to produce hand-harvested seed yields of 2-3 t ha^{-1}. Genotypic variations in response to daylength, temperature and water deficit [32,34,70,127] suggest that through judicious selection and plant breeding, the development of cultivars with greater flexibility or with adaptation to specific environmental niches ought to be possible.

647

PROSPECTS FOR WINGED BEAN AS A GRAIN LEGUME CROP

The need

In the lowland humid tropics, several grain legumes are cultivated but none of them has the potential of winged beans.
Vigna species grow well in humid tropical environments but
their seeds do not contain as much protein or oil as those
of winged bean nor do they have as favourable a composition
of amino and fatty acids. The popular *Phaseolus* species are
better adapted to the highland tropics than to the lowlands
but, again, do not have as valuable seed composition as in
winged beans. Soyabeans, while of excellent nutritional
value, have their centre of origin in the subtropics and
efforts to disseminate the crop throughout the warmer
regions have been frustrating (but see Chapter 9).

'Ideotypes'

An attempt to identify those characteristics which would be
required in winged beans designed to meet particular economic circumstances, suggests three 'ideotypes':

(a) For village economies: Climbing perennial habit; vigorous
   vegetative development prior to fruit set; resistance
   to *Synchytrium psophocarpi*, viruses and nematodes; sequential, regular seed harvest; good seed-storage characteristics; non-shattering fruits; protein-rich seeds.
(b) For intermediate economies: Climbing ratooning habit;
   short harvest period; non-shattering fruits; cream-
   coloured processable seeds.
(c) For developed economies: Bush-type, annual growth habit;
   rapid development to full seed yield; synchronous maturity; small non-shattering fruits; cream-coloured processable seeds.

Priorities

Village economies.   For village-based economies, suitable
winged bean cultivars could be selected from amongst existing germplasm. The important research priority is to develop
appropriate agronomic and nutritional packages. The prospect
for intercropping late-maturing cultivars with crops such
as maize and sorghum needs close examination, and must be
accompanied by a vigorous extension effort.

K. Cerny and colleagues in the early seventies demonstrated that winged bean seeds could contribute significantly to the correction of nutritional deficiency in children
under medical supervision in Ghana [12,14,81]. Concurrently,
vigorous agronomic research and extension efforts were centred on the University of Ghana [69,104,105]. By 1978, S.K.
Karikari reported that 166 farmers in the Kade area were
growing winged beans in a country where previously it was
not known.

In the Ivory Coast, the approach was more specific in
its objectives and largely extension-oriented. Over a four-
year period, agronomic and nutritional procedures were introduced directly into a single village which was to be compared with a neighbouring 'control' village [122,123]. At
the outset, one third to one half of the children in the
target village were assessed as suffering from marginal
malnutrition.

The introduction of winged bean cultivation into the
village was slow. All farmers in the first year, 1977, received seeds and instruction, but only a fraction planted
them. The initial labour requirement involved in trellising
was perceived as an unnecessary burden. By 1979 however,
50% of villagers grew the crop through to harvest. In 1980,
twenty-six of the 110 families in the village cooked winged
bean products an average of more than three times each week.
Recipes in which a proportion of the normal ingredients such
as cassava and cornflour were replaced by winged bean flour
had better nutritional value than traditional recipes, with-

649

out impairing the flavour and popularity of the original dish. The winged bean flour was prepared by 'cooking, parching and grinding'. Since autoclaving was not used, it may still have contained some antinutritional factors but the authors did not comment on this possibility.

Intermediate economies.   There are no traditional examples of winged beans grown as field crops for seed. A significant research effort will be required if winged beans are to be developed into a commercial pulse crop.

In Malaysia, Wong [152] and Tham [142] have regularly recorded seed yields larger than 2 t ha^{-1} from trellised, late-maturing South-east Asian cultivars. In Vietnam, the feasibility of ratooning winged bean has been demonstrated [106] and K.C. Wong (pers.comm.) in Malaysia is examining whether this approach to seed production can recover the initial expenditure on trellising.

In Thailand, a large-scale planting of trellised winged bean has produced quantities of seed for processing research (N. Chomchalow, pers.comm.). With this, Verangoon et al. [149] have shown that winged bean milk, though comparable in appearance and nutritional value with commercial soyabean milk, has a strong 'beany' flavour. The lypoxygenase activity of winged bean seeds, which is thought to be responsible for the liberation of the volatile beany flavour, is, however, no greater than that in soyabeans [146]. The flavour of winged bean milk can be improved by adding chocolate essence, as is often done with commercially bottled soya milk. Cream-coloured seeds are preferable for milk production.

Ramanvongse and Munsakul [121] have shown that commercial quantities of good quality oil can be obtained using an efficient expeller and simple bleaching, or by solvent extraction and conventional refining. Oleic and linoleic acids represent 39% and 27%, respectively, of the fatty acids and there are no extremely toxic fatty acids present.

On the other hand, the protein meal is relatively rich

in antinutritional factors and further research is required
to define accurately the appropriate heat or other processing conditions which are required to bring the nutritional
value up to its potential.

Developed economies.   In the absence of genotypes with a
suitable growth habit there has been little interest in
winged beans for large-scale mechanised farming.

In northern Australia, Robertson *et al.* [125] chemically
desiccated and mechanically harvested winged beans grown
without trellising. Fruit shattering rather than plant habit
per se reduced yields by more than 50%. Accessions from
South-east Asia, with their characteristic branching habit
(see earlier), yielded better when unsupported than did
those from Papua New Guinea.

It is clear that substantial breeding and agronomic
research will be required before large-scale, mechanised
cultivation of winged bean crops becomes possible [96].

Prospects

Despite limitations as a grain legume, winged beans offer a
range of possibilities for utilisation in the humid and
semi-humid tropics [76]. With the encouragement of the US
National Academy of Sciences and the International Council
for Development of Underutilised Plants, researchers have
expanded the basic knowledge of winged bean biology, agronomy and nutritional value to the point where a coordinated,
mission-oriented programme of research is now possible. In
acknowledgement of this, an International Dambala (Winged
Bean) Institute was established in Sri Lanka in 1982. The
Institute intends to conduct applied research which seeks
to overcome the agronomic limitations of winged beans for
vegetable and pulse production.

Research units for winged bean crop improvement have
also been organised in Thailand and Malaysia, and an international repository for winged bean germplasm has been

651

established in the Philippines. This centre (the International Documentation Centre for the Winged Bean, Agricultural Information Bank for Asia, College, Laguna 3720, Philippines) publishes a newsletter, *The Winged Bean Flyer*, which is widely distributed to subscribers throughout the world.

The proceedings of the First International Symposium on Winged Beans held at Manila in 1978 have been published (see below), but it is unfortunate that those of the Second Symposium, held in Sri Lanka in 1981, have been delayed by financial limitations. Without adequate funds, the international efforts begun with such promise in the late seventies will flounder. The time for hyperbole about the valuable characteristics of the winged bean is now well past. So begins the task of bringing these potentials to fruition.

LITERATURE CITED

1. Agcaoili, F. (1929), *Philippine Journal of Science* 40, 513-14.
2. Bala, A.A. and Stephenson, R.A. (1978), in *The Winged Bean*. Papers presented at the First International Symposium on Developing the Potentials of the Winged Bean, January, 1978. Los Banos: Philippine Council for Agriculture and Resources Research, 63-70, (*Los Banos Symposium*).*
3. Balasubramanian, S., Thiagarajah, M.R., Udugampola, N. and de Silva, M. (in press), in Herath, H.M.W. (ed.), *Proceedings of the Second International Seminar on Winged Bean*, January, 1981, held at Colombo, Sri Lanka. Orinda, California USA: International Council for Development of Underutilised Plants (*Colombo Seminar*).*
4. Bernard, R.L. (1971), *Crop Science* 11, 242-4.
5. Bernard, R.L. (1972), *Crop Science* 12, 235-9.
6. Berry, S.K. (1977), *Malaysian Applied Biology* 6, 33-8.
7. Blagrove, R.J. and Gillespie, J.M. (1979), *Australian Journal of Plant Physiology* 5, 371-5.
8. Broughton, W.J. and Lenz, F. (in press), *Colombo Seminar*.
9. Burkill, I.H. (1906), *The Agricultural Ledger (Calcutta)* 4, 51-64.
10. Caygill, J.C. and Jones, N.R. (in press), *Colombo Seminar*.
11. Cerny, K. (1978), *Los Banos Symposium*, 281-9.
12. Cerny, K. and Addy, H.A. (1973), *British Journal of Nutrition* 29, 105-7.**
13. Cerny, K., Hoa, D.Q., Dinn, N.L. and Zelona, H. (in press), *Colombo*

*Many other references are contained in the proceedings of these two conferences; for brevity they will be referred to hereafter as *Los Banos Symposium* or *Colombo Seminar*, as appropriate.
**Although this paper deals with *P. palustris*, K. Cerny later pointed out at the Los Banos Symposium that the species was in fact *P. tetragonolobus*.

Seminar.
14. Cerny, K., Kordylas, M., Pospisil, F., Svabensky, O. and Zajic, B. (1971), *British Journal of Nutrition* <u>26</u>, 293-9.
15. Chai, N.F., Jalani, B.S. and Singah, R. (in press), *Colombo Seminar.*
16. Chandel, K.P.S., Arora, R.K., Joshi, B.S. and Mehra, K.L. (1978), *Los Banos Symposium,* 393-5.
17. Chamchalow, N., Supputtitada, S. and Peyachoknagul, S. (1978), *Los Banos Symposium,* 46-62.
18. Chubb, L.G. (1978), *Los Banos Symposium,* 340-49.
19. Chubb, L.G. (in press), *Colombo Seminar.*
20. Claydon, A. (1975), *Science in New Guinea* <u>3</u>, 103-14.
21. Claydon, A. (1978), *Los Banos Symposium,* 263-80.
22. Claydon, A. (1979), *Science in New Guinea* <u>6</u>, 144-53.
23. Claydon, A. (in press), *Colombo Seminar.*
24. Data, E. and Pratt, K.K. (1980), *Tropical Agriculture, Trinidad* <u>57</u>, 309-16.
25. Dale, W.T. (1949), *Annals of Applied Biology* <u>36</u>, 327-33.
26. Drinkall, M.J. (1978), *PANS* <u>24</u>, 160-66.
27. Drinkall, M.J. and Price, T.V. (1979), *Transactions of the British Mycological Society* <u>72</u>, 91-8.
28. Duff, D.H. (1978), *Los Banos Symposium,* 350-57.
29. Duke, J.Λ. (1981), *Handbook of Legumes of World Economic Importance.* New York: Plenum Press.
30. Duncan, L.W., Caveness, F.E. and Parez, A.T. (1979), *Tropical Grain Legume Bulletin* <u>15</u>, 30-34.
31. Eagleton, G.E., Halim, A.H. and Chai, N.F. (1980), in *Proceedings of Legumes in the Tropics.* Serdang, Malaysia: Universiti Pertanian, 133-44.
32. Eagleton, G.E., Khan, T.N. and Chai, N.F. (in press), *Colombo Seminar.*
33. Eagleton, G.E., Thurling, N. and Khan, T.N. (1978), *Los Banos Symposium,* 110-20.
34. Eagleton, G.E., Thurling, N. and Khan, T.N. (in press), *Colombo Seminar.*
35. Ekpenyong, T.E. and Borchers, R.L. (1978), *Los Banos Symposium,* 300-12.
36. Elmes, R.P.T. (1976), *Papua New Guinea Agriculture Journal* <u>27</u>, 53-7.
37. Erskine, W. (1978), *Los Banos Symposium,* 29-35.
38. Erskine, W. (1980), *SABRAO Journal* <u>12</u>, 11-14.
39. Erskine, W. (1981), *Journal of Agricultural Science, Cambridge* <u>96</u>, 503-8.
40. Erskine, W. (in press), *Colombo Seminar.*
41. Erskine, W. and Bala, A.A. (1976), *Tropical Grain Legume Bulletin* <u>6</u>, 32-5.
42. Erskine, W. and Kesavan, V. (1982), *Journal of Horticultural Science* <u>57</u>, 209-13.
43. Erskine, W. and Khan, T.N. (1977), *Euphytica* <u>26</u>, 829-31.
44. Erskine, W. and Khan, T.N. (1980/81), *Field Crops Research* <u>3</u>, 359-64.
45. Evans, I.M., Boulter, D., Eaglesham, A.R.J. and Dart, P.J. (1975), *Qualitas Plantarum Plant Foods for Human Nutrition* <u>27</u>, 275-85.
46. Fajardo, T.G. and Palo, M.A. (1933), *Philippine Journal of Science* <u>51</u>, 457-84.
47. Fauquet, C., Lamy, D. and Thouvenel, J.C. (1979), *Plant Protection Bulletin* <u>27</u>, 81-7.

48. Gillespie, J.M. and Blagrove, R.J. (1978), *Australian Journal of Plant Physiology* 5, 357-69.
49. Gillespie, J.M., Blagrove, R.J. and Kort, A.A. (in press), *Colombo Seminar*.
50. Haq, N. (1982), *Zeitschrift für Pflanzenzüchtung* 88, 1-12.
51. Haq, N. and Smartt, J. (1978), *Los Banos Symposium*, 96-102.
52. Harding, J., Lugo-Lopez, M.A. and Perez-Escolar, R. (1978), *Tropical Agriculture, Trinidad* 55, 315-24.
53. Haryono, S.K., Soedarsono, J. and Thompson, A.E. (1978), *Los Banos Symposium*, 40-45.
54. Herath, H.M.W. and Fernandez, G.C.J. (1978), *Los Banos Symposium*, 161-72.
55. Herath, H.M.W. and Ormrod, D.P. (1979), *Annals of Botany* 43, 729-36.
56. Herath, H.M.W., Dharmawansa, E.M.P. and Ormrod, D.P. (1978), *Los Banos Symposium*, 84-6.
57. Hew, C.S. and Lee, Y.H. (in press), *Colombo Seminar*.
58. Heyne, K. (1927), *Nuttige Planten von Ned. Indie* 2, 849-50.
59. Hildebrand, D.F., Chaven, C., Hymowitz, T. and Bryan, H.H. (in press), *Colombo Seminar*.
60. Hildebrand, D.F., Chaven, C., Hymowitz, T., Bryan, H.H. and Duncan, A.A. (in press), *Colombo Seminar*.
61. Hymowitz, T. and Boyd, J. (1977), *Economic Botany* 31, 180-88.
62. Ikram, A. and Broughton, W.J. (1980), *Soil Biology and Biochemistry* 12, 77-82.
63. Ikram, A. and Broughton, W.J. (1980), *Soil Biology and Biochemistry* 12, 203-9.
64. Iruthayathas, E.E. and Vlassak, K. (1982), *Scientia Horticulturae* 16, 313-22.
65. Iruthayathas, E.E., Vlassak, K. and Reynders, L. (1982), *Zeitschrift für Pflanzenernaehrung und Bodenkunde* 145, 398-410.
66. Jalani, B.S. (1978), *Los Banos Symposium*, 135-9.
67. Kantha, S.S. and Hettiarachchy, N.S. (in press), *Colombo Seminar*.
68. Karikari, S.K. (1972), *Ghana Journal of Agricultural Science* 5, 235-9.
69. Karikari, S.K. (1978), *Los Banos Symposium*, 150-60.
70. Karikari, S.K. (in press), *Colombo Seminar*.
71. Kesavan, V. (in press), *Colombo Seminar*.
72. Kesavan, V. and Erskine, W. (1978), *Los Banos Symposium*, 211-4.
73. Kesavan, V. and Khan, T.N. (1978), *Los Banos Symposium*, 105-9.
74. Khan, T.N. (1976), *Euphytica* 25, 693-706.
75. Khan, T.N. (1982), *Winged Bean Production in the Tropics*, Plant Production and Protection Paper 38. Rome: FAO.
76. Khan, T.N. and Edwards, C.S. (in press), *Colombo Seminar*.
77. Khan, T.N. and Erskine, W. (1978), *Australian Journal of Agricultural Research* 29, 281-9.
78. Khan, T.N., Bohn, J.C. and Stephenson, R.L. (1977), *World Crops* 29, 208-16.
79. Khor, H.T., Tan, N.H. and Wong, K.C. (1980), in *Proceeding of the 6th Annual Conference of the Malaysian Biochemistry Society*. Kuala Lumpur, Malaysia.
80. Kim, K. (1978), Los Banos Symposium, *435-6*.
81. Kordylas, M., Dsei, Y.D. and Asibey-Berko, E. (1978), *Los Banos Symposium*, 363-70.
82. Kortt, A.A. (1979), *Biochemica et Biophysica Acta* 577, 371-82.

83. Kortt, A.A. (1980), *Biochemica et Biophysica Acta* 624, 237-48.
84. Lackey, J.A. (1977), *Botanical Journal of the Linnean Society* 74, 163-79.
85. Lam, H.J. (1945), *Observations of a Naturalist in Netherlands New Guinea*, Sargentia V Fragmenta Papuana, Harvard, Massachusetts, USA.
86. Lamb, K.P. and Price, T.V. (1978), *Los Banos Symposium*, 281-5.
87. Lawhead, C.W. (1978), MSc. thesis, University of California, Davis, USA.
88. Linge, E. (1976), BSc. thesis, University of Papua New Guinea, Port Moresby.
89. Lubis, S.H.A. (1978), *Los Banos Symposium*, 121-3.
90. Lubis, S.H.A. and Sastrapradja, S. (in press), *Colombo Seminar*.
91. Lubis, S.H.A., Hanson, J. and Mumford, P.M. (in press), *Colombo Seminar*.
92. Martin, F.W. and Ruberte, R.M. (1978), *Los Banos Symposium*, 371-2.
93. Masefield, G.B. (1957), *Empire Journal of Experimental Agriculture* 25, 139-50.
94. Masefield, G.B. (1961), *Tropical Agriculture, Trinidad* 38, 225-9.
95. Masefield, G.B. (1973), *Field Crop Abstracts* 26, 157-60.
96. Mayo,10. (1980), *The Theory of Plant Breeding*. Oxford: Clarendon Press.
97. Nangju, D. and Baudoin, J.P. (1979), *Journal of Horticultural Science* 54, 129-36.
98. NAS (1975), *The Winged Bean, a High Protein Crop for the Tropics*. Washington DC: National Academy of Sciences.
99. Odor, D.R.B. (1978), *Los Banos Symposium*, 147-59.
100. Otanes, Y. and Quesales, F. (1918), *Philippine Agriculturist* 7, 1-27.
101. Padmasiri, I.S. and Pinto, M.E.R. (in press), *Colombo Seminar*.
102. Pate, J.S. and Minchin, F.R. (1980), in Summerfield, R.J. and Bunting, A.H. (*eds*), *Advances in Legume Science*, London: HMSO, 105-14.
103. Pickersgill, B. (1980), *Botanical Journal of the Linnean Society* 80, 279-91.
104. Pospisil, F., Karikari, S.K. and Boamah-Mensah, E. (1971), *World Crops* 23, 260-4.
105. Pospisil, F., Hlava, B. and Buresova, M. (1978), *Los Banos Symposium*, 124-35.
106. Pospisil, F., Buresova, M., Hlava, B., Hrachova, B., Michl, L., The-Tue, T., Diu, T.Z., Manh, N.X., Quynh, N.T. and Hoan, N.V. (in press), *Colombo Seminar*.
107. Poulter, N.H. (1982), *Journal of the Science of Food and Agriculture* 33, 107-14.
108. Powell, J. (1976), *New Guinea Vegetation*. Canberra, Australia: CISRO, 106-84.
109. Powell, J., Kulunga, A., Moge, R., Pono, C., Zimike, F. and Golson, J. (1975), *Agricultural Traditions of the Mt Hagen Area*. Department of Geography, Occasional Paper No. 12, University of Papua New Guinea, Port Moresby.
110. Price, T.V. (1977), *Plant Disease Reporter* 61, 384-5.
111. Price, T.V. (1978), *Los Banos Symposium*, 236-47.
112. Price, T.V. (in press), *Colombo Seminar*.
113. Price, T.V. and Munro, P.E. (1978), *Transactions of the British Mycological Society* 70, 47-55.
114. Price, T.V. and Munro, P.E. (1978), *PANS* 24, 53-6.

115. Pueppke, S. (1979), *Biochemica et Biophysica Acta* 581, 63-70.
116. Purcino, H.M.A., Purcino, A.A.C. and Lynd, J.Q. (in press), *Colombo Seminar*.
117. Purseglove, J.W. (1968), *Tropical Crops. Dicotyledons Vol. I.* London: Longman.
118. Rachie, K.O. and Roberts, L.M. (1974), *Advances in Agronomy* 26, 1-132.
119. Rajan, S.S. (1978), *Notes on Winged Bean in Burma.* Unpublished report. Rome: FAO.
120. Rajendran, R., Satyanarayana, A., Selvaraj, Y. and Bhargava, B.S. (1978), *Los Banos Symposium*, 71-82.
121. Ramanvongse, S. and Munsakul, S. (in press), *Colombo Seminar*.
122. Ravelli, G.P., N'zi, G.K., Diaby, L., Ndri, K.B., Mayer, G.G. and Sylla, B.S. (1978), *Los Banos Symposium*, 313-21.
123. Ravelli, G.P., N'zi, G.K. and Sylla, B.S. (in press), *Colombo Seminar*.
124. Renkonen, K.O. (1948), *Annales Medicinae Experimentalis et Biologiae Fenniae* 26, 66-72.
125. Robertson, G.A., Warren, J. and Khan, T.N. (1978), *Los Banos Symposium*, 173-82.
126. Rockland, L.B. (1979), *Journal of Food Science* 44, 1004-7.
127. Ruegg, J. (1981), *Journal of Horticultural Science* 56, 331-8.
128. Ruegg, J. (in press), *Colombo Seminar*.
129. Rumphius (1747), *Herbarium Amboinensis* 5, 374, t.133.
130. Sastrapradja, S. (1978), *Los Banos Symposium*, 36-9.
131. Sastrapradja, S. and Lubis, S.H.A. (1976), in *South-East Asian Plant Genetic Resources.* Proceedings of a symposium held at Bogor, Indonesia, 20-22 March 1975, Lembaga Biologi Nasional, Bogor, 147-51.
132. Schertz, K.F., Boyd, W.L., Jugelsky, W.J.R. and Cabanillas, E. (1960), *Economic Botany* 14, 323-40.
133. Senanayake, Y.D.A. and Thiruketheeswaran, A. (1978), *SABRAO Journal* 10, 116-19.
134. Singh, R.B. (in press), *Colombo Seminar*.
135. Sinnadurai, S. (1977), *Tropical Grain Legume Bulletin* 10, 14-15.
136. Soriano, J.M. and Batugal, P.A. (1978), *Los Banos Symposium*, 432-4.
137. Sornay, de P. (1913), *Green Manures and Manuring in the Tropics.* London: John Bale.
138. Srinivasan, K., Rajendran, R. and Satyanarayana, A. (1978), *Los Banos Symposium*, 255-60.
139. Stephenson, R.A. (1978), *Los Banos Symposium*, 191-6.
140. Stephenson, R.A., Drake, D.W. and Kesavan, V. (in press), *Colombo Seminar*.
141. Strathern, A. (1978), *Los Banos Symposium*, 12-18.
142. Tham, W.F. (1980), in *Proceedings of Legumes in the Tropics.* Serdang, Malaysia: Universiti Pertanian, 21-8.
143. Thompson, A.E. and Haryono, S.K. (1979), *HortScience* 14, 532-3.
144. Thompson, A.E. and Haryono, S.K. (1980), *HortScience* 15, 233-8.
145. Tixier, P. (1965), *Revue de Cytologie et de Biologie Végétales* 28, 133-5.
146. Truong, V.D., Raymundo, L.C. and Mendoza, E.M.T. (in press), *Colombo Seminar*.
147. Uemoto, S., Fujieda, K., Nonaka, M. and Nakamoto, Y. (in press), *Colombo Seminar*.
148. Vavilov, N.I. (1949/50), *Chronica Botanica Vol. 13*, Massachusetts,

Winged Bean (*Psophocarpus tetragonolobus* (L.) DC)

USA: Chronica Botanica Co.
149. Verangoon, P., Paklamjeak, M., Srisawat, S. and Pathomyothin, W. (in press), *Colombo Seminar.*
150. Venketeswaran, S. (in press), *Colombo Seminar.*
151. Verdcourt, B. and Halliday, P. (1978), *Kew Bulletin* **33**, 191-227.
152. Wong, Kai Choo (1976), in *Malaysian Food Self-Sufficiency, Proceedings of a Conference.* UMAGA, University of Malaya, Kuala Lumpur, 103-5.
153. Wong, Kai Choo (1978), *Los Banos Symposium,* 220-26.
154. Wong, Kai Choo (in press), *Colombo Seminar.*
155. Wong, Kai Choo and Schwabe, W.W. (1980), in *Proceeding of Legumes in the Tropics,* Serdang, Malaysia: Universiti Pertanian, 73-86.
156. Woomer, P., Guevarra, A. and Stockinger, K. (1978), *Los Banos Symposium,* 197-204.
157. Yap, T.N., Van Soest, P.J. and McDowell, R.E. (1979), *Malaysian Applied Biology* **8**, 119-23.
158. Zulkifli, S. and Aini, Z. (1978), *Los Banos Symposium,* 215-9.
159. Zulkifli, S. and Othman, Y. (1977), in *Proceeding of the Symposium on Soil Microbiology and Plant Nutrition,* Kuala Lumpur: University of Malaya Press.
160. Summerfield, R.J. and Roberts, E.H. (1985), in Halevy, A.H. (*ed.*) *A Handbook of Flowering.* Florida, USA: CRC Press. (In press.)

15 Pigeonpea (*Cajanus cajan* (L.) Millsp.)

P.C. Whiteman, D.E. Byth and
E.S. Wallis

INTRODUCTION

Although little of the crop enters world trade, pigeonpea
(*Cajanus cajan* (L.) Millsp.) is the fifth most important
pulse crop in the world. It is mainly a subsistence crop in
the tropics and subtropics of India, Africa, South-east Asia
and the Caribbean, but also an important cash crop in the
West Indies. The crop is most commonly grown for its dry,
split seeds (dhal), which have a protein concentration of
approximately 20-25%; but seeds are also eaten as a green
vegetable. In these ways it is an important component of
human nutrition, particularly in vegetarian diets. Dry seeds
and the by-products of dhal manufacture, as well as leaf and
pod-wall residues after harvest, can provide suitable feed
for ruminants, which may also graze the standing crop [100].
Pigeonpea stems are an important source of fuel in rural
India [47]; other minor uses are reviewed by Morton [45].
Production is concentrated in India, where the crop is
second only to chickpeas (*Cicer arietinum*) in area and output (and see Chapter 8), but substantial areas also occur
in Africa (Table 15.1).

Pigeonpeas have been grown as a traditional crop in the
semi-arid tropics over many thousands of years. Many production systems are practised, some of which are discussed
by Byth *et al*. [11]. Although mechanised harvesting is technically possible [95], the seeds are almost invariably hand-

Pigeonpea (*Cajanus cajan* (L.) Millsp.)

Table 15.1 World pigeonpea production 1970–1980 [55]

	1970	1974	1979	1980*
WORLD				
Area (1000 ha)	2982	2999	3000	2951
Yield (kg ha^{-1})	684	541	703	684
Production (1000 t)	2039	1622	2111	2017
AFRICA				
Area (1000 ha)	214	241	252	255
Yield (kg ha^{-1})	593	565	589	599
Production (1000 t)	127	136	149	153
NORTH AND CENTRAL AMERICA				
Area (1000 ha)	24	28	2	9
Yield (kg ha^{-1})	1603	1411	2500	2222
Production (1000 t)	38	40	5	20
ASIA – (Including India)				
Area (1000 ha)	2723	2723	2718	2656
Yield (kg ha^{-1})	703	530	713	687
Production (1000 t)	1913	1442	1938	1824
INDIA				
Area (1000 ha)	2655	2646	2663	2600
Yield (kg ha^{-1})	709	532	719	692
Production (1000 t)	1883	1408	1914	1800

*FAO estimates (includes China)

harvested and there are few commercial mechanised production systems.

Many factors have contributed to the continued and widespread use of pigeonpeas in the semi-arid tropics. The ability of the crop to survive and reproduce in environments characterised by severe moisture stress and poor soil fertility is especially important. Other attractive characteristics of the crop include resistance to lodging, shattering and pre-harvest weathering, a perennial habit, suitability for intercropping with tall cereals, human preferences for the seeds, and the use of its by-products for fuel and fodder.

Despite its international importance, and although

Borlaug [10] considered that increased research on this crop
was justified, sustained scientific studies of its production, improvement and utilisation have been limited until
recently. Considerable research has occurred in India since
the 1920s [57] but it had limited impact on production prior
to the development of a coordinated national programme [65].
Concerted international research on pigeonpea improvement
began in 1973 at the International Institute for Tropical
Agriculture (IITA) at Ibadan in Nigeria (see Chapter 18) and
is now centered at the International Crops Research Institute for the Semi-Arid Tropics (ICRISAT) at Hyderabad in
India (see Chapter 18). Furthermore, active pigeonpea research is ongoing in India, Kena, various Caribbean countries
and in Australia. Morton *et al*. [46] have listed current programmes and documented contacts in pigeonpea research.

BOTANY

Taxonomy

The genus *Cajanus* is classified in the subtribe Cajaninae
within the tribe Phaseoleae [93]. The subtribe Cajaninea is
separated in the Phaseoleae on the basis of a glabrous
style, nodes of the racemes not swollen and leaves copiously
gland-dotted [64]. Within the Cajaninae, there are thirteen
closely-related genera, separated initially on the basis of
fruits containing one or two versus two or more ovules [72].
Within the latter group, the genus *Atylosia* is separated on
the basis of the seed having a strophiole while seeds of
Cajanus do not. However, even this distinction is blurred
as some *Cajanus* genotypes do have small strophioles [93].
Other genera within the Cajaninae include *Dunbaria*, *Rhynchosia* and *Flemingia* (and see Chapter 1).

The genus *Cajanus* is usually described as monospecific,
but it is closely related to *Atylosia*. Some taxonomists
suggest *Atylosia* and *Cajanus* should be combined into a single genus [17,93]. However, Reynolds and Pedley [73] com-

660

pleted a botanical revision of Australian *Atylosia*, and considered that while *Atylosia* and *Cajanus* may not be distinct, the groups are probably best maintained as separate genera since *C. cajan* is known only in cultivation and exhibits a wide variation in characters such as height, morphology, photoperiodism, and in resistance to insects and diseases (but see Chapter 1).

Origin

Earlier literature suggested that the crop originated in Africa [62]. However, the genus *Cajanus* does not show the diversity expected if this suggestion were valid [17]. Dc [17] suggests that the wild progenitor of the cultivated *Cajanus* was derived from an erect *Atylosia* species such as *A. cajanifolia*, and that the most likely centre of origin is the upper western Ghats area of India.

Morphology

Pigeonpeas are usually described as perennial shrubs [58, 62,87]. True annual forms are not known, but it is widely grown as an annual crop with a single harvest or perhaps allowed to ratoon after harvest.

Traditional landraces are shrubby and up to 4 m tall (Plate 15.1). More recently, photoperiod-insensitive cultivars have been grown as standard field crops 70-150 cm tall, with or without ratooning [77,96] (Plate 15.2).

Two botanical varieties were recognised in an older classification [17,58], viz. variety *flavus* D.C. or 'Tur' types and variety *bicolor* or 'Arhar' types. However, the classification has doubtful validity since there is a full range of phenotypes across the species.

Growth habit ranges from erect with acutely angled branches (30° or less) to more spreading types with branch angles as large as 60°. Some cultivars tend to produce long primary branches with leaves along their entire length and

Plate 15.1   A traditional, photo-sensitive pigeonpea
cultivar grown in Viti Levu, Fiji

Plate 15.2   Experimental plots of an early-flowering
photo-insensitive cultivar after 60 days from
planting at Ubon, Thailand

with fruits concentrated in the terminal one-third or one-
half of each branch. Some genotypes branch very little and
produce flowering racemes directly on the main axis. The
extent and position of branching, and the position of inflorescences, are also greatly influenced by plant density.
Forms with mainly terminal flowering have been termed determinate types. While this may be botanically correct, the
term has limited meaning in a perennial plant in which regrowth and development of new inflorescences effectively
ensures reproductive indeterminacy.

Trifoliolate leaves are arranged spirally in a 2/5 phyllotaxy; petioles are 2-8 cm long and grooved above; stipules
are about 4 mm long and small, ovate and hairy; leaflets are
entire, lanceolate to narrow elliptical, acute at both ends,
pubescent on both surfaces, with minute resin glands beneath.

Inflorescences are 4-12 cm long and borne either terminally or at axillary nodes. Flowering can be diffuse over
the whole plant throughout a long period, or synchronous,
depending upon the genotype and on the photoperiod and temperature regime. Flowers are about 2.5 cm long with four
calyx lobes (the upper two are united) and a large and broad
standard which is auricled at the base. Petal colour varies
from yellow to red or purple, with some tinged, striped or
mottled with red or purple. Stamens are arranged 9 + 1,
while the style is beardless. The stigma is terminal, and
the ovary and base of style are hairy. Fruits are flattened
pods with diagonal depressions between each of the two to
nine locules, and are up to 10 cm long, beaked and often
hairy. Pod wall colour can be green, brown, dark maroon to
dark purple or blotched, with a greasy or waxy surface
when immature. Fruits contain between two and nine seeds
and do not shatter. Seeds are orbicular, oval to flattened;
testa colour varies from white, brown, red, to purple, and
sometimes speckled; the hilum is small and white. Seed size
varies from 6-28 g per 100 seeds [58].

Phenology

Phenology can be regarded as the sequence of developmental
events involving times to germination, emergence to flowering, flowering to fruit fill, fruit fill to harvest and
ratoon crop development.

Germination

Germination of pigeonpea seeds is not normally limited by
hardseededness although some genotypes have well-developed
hard seeds. Germination is hypogeal [2]. P.L.M. De Jabrun
(unpublished) found that seeds harvested fifteen days after
anthesis and dried in the fruits had only 10-20% germination, whereas those collected after twenty-two days had 90%
and at thirty days, 90-100% germination. Older seeds also
had greater rates of germination and hypocotyl elongation.
    Pigeonpeas germinate over a wide range of temperatures
from 19°C to 43°C, but do so most rapidly at 20°-30°C and
only poorly at temperatures below 19°C [18]. Similar responses have also been observed in the field (D. van Cooten,
pers.comm.).

Emergence to flowering

The duration of the vegetative phase is extremely variable,
depending upon genotype, photoperiod, temperature and, perhaps, moisture status, all of which interact to determine
time of flowering. Depending on the genotype, days to
flowering can range from about 60 to more than 200 days for
sowings made prior to the longest day at 17° N (Figure 15.1)
[25]. Most photo-sensitive cultivars flower more rapidly
when sown after the longest day (Table 15.2).
    These data are consistent with a quantitative short-day
response and this is supported by limited controlled environment studies (H.G. McPherson, pers.comm.) [90,92]. Where
the daylength regime for a region is shorter than the

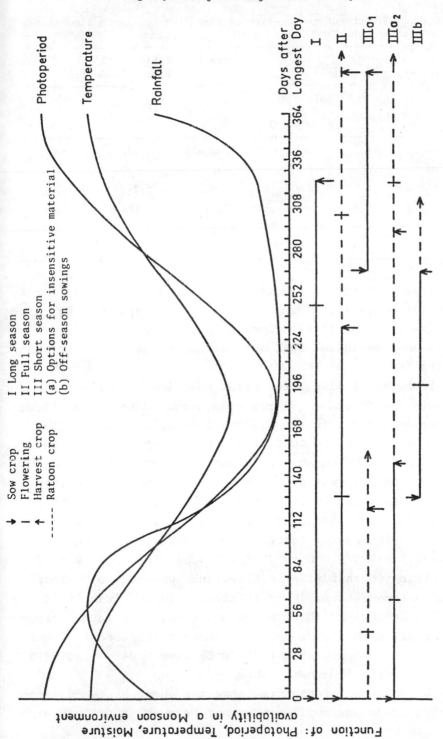

Figure 15.1 Generalised classification of production systems in pigeonpeas

665

Table 15.2 Effects of sowing date on time from sowing to flowering and on plant height

(a) Australia - 27°S Variety - Royes [95]			(b) Puerto Rico - 18° N Variety - Florido [75]		
Month of planting	Days to flower	Height at first flower (m)	Month of planting	Days to flower	Height at first flower (m)
October	140	2.3	January	322	4.6
December	118	1.7	March	265	4.5
March	94	0.7	May	212	3.8
			July	164	2.4
			September	143	1.3
			November	142	1.0

critical daylength for a genotype, it may be insensitive to photoperiod [5,76]. Turnbull *et al*. [92] reported that time to flowering was little changed in daylengths of 16 and 24 hours and considered that several genotypes approached day neutrality.

Temperature also affects the phenology of pigeonpeas. Akinola and Whiteman [5] reported slower development under cooler conditions in the field. Field studies at ICRISAT (see Chapter 18) and controlled environment work in New Zealand (H.G. McPherson, pers.comm.) suggest that both cold and hot temperatures delay flowering. Turnbull *et al*. [92] found that floral initiation of photoperiod-insensitive genotypes was delayed, and floral bud development was accelerated by increasing temperature of alternating diurnal regimes from 24°/16°C to 32°/24°C. The effects were self-cancelling so that time to flowering appeared to be insensitive to temperature. There is clear evidence from field studies at ICRISAT (D. van Cooten, pers. comm.) that flowering of all maturity groups of pigeonpea [25] occurs sooner in moderate temperatures (22°-30°C) even under relatively long days (12h 30 min to 16h).

The duration of the pre-flowering phase has a fundamental impact on agronomic management because it influences

666

biomass accumulation. In the absence of other limitations, the optimum sowing density/arrangement should enable full light interception by the canopy immediately prior to commencement of reproductive growth [43]. Since phenology is a function of genotype and environment (photoperiod and temperature), marked genotype x sowing date or latitude x density interactions occur. Optimal agronomic systems have been developed for exploitation of short-season pigeonpea culture [88,96] and similar systems have been reported by others [15,29,82].

Growth rates of pigeonpeas from emergence to canopy closure are relatively small. In early and mid-season cultivars, Sheldrake and Narayanan [82] recorded growth rates of $1.4 - 1.6\,g\,m^{-2}\,d^{-1}$ during the first 63 days from sowing and maximum growth rates of $10.0-17.1\,g\,m^{-2}d^{-1}$ during flowering and early fruit filling. Pandey [53] recorded similar values for the period until 60 days from sowing. Small early crop growth rates are due to small leaf area indices (LAI) and maximum growth rates are not achieved [31,53,77] until LAI reaches 4.0 to 5.0.

Flowering to fruit fill

Sheldrake and Narayanan [82] reported that in the cultivar Pusa Ageti, which flowered in 70 days, some 73% of the total stem weight was added during the reproductive phase. Pandey [53] also reported that 44-77% of total dry-matter was accumulated after flowering in four early maturing cultivars. The fact that a relatively large proportion of current photosynthesis is being translocated into vegetative growth during fruit filling appears to be related to the fact that pigeonpea is a perennial crop.

Fruit fill to maturity

The duration of fruit filling and ripening depends on the synchrony of flowering, and on climatic conditions. Akinola and Whiteman [5] found the fruit filling period varied from 25 days during the summer to 61 days during the cool winter period at Redland Bay, Queensland (27°S). Fruit ripening also varied from 14-39 days, again dependent upon irradiance, temperature and humidity. Sheldrake and Narayanan [82] found that for the early cultivar T-21, time from flowering to harvest at Hyderabad, India was 46 days while in the inter-mediate cultivar ICP-1 it was 86 days. In short-statured, photoperiod-insensitive cultivars, time from first flowering to harvest averages 50 days in favourable (25°-30°C) temper-atures [96].

In the interval from flowering to harvest, Sheldrake and Narayanan [82] noted an almost linear rate of abscission of leaves such that 2.16 t ha^{-1} of fallen material was col-lected (out of a total of 6 t ha^{-1} of above-ground dry-mat-ter) of which 86% was leaves, 12% flowers and 2% shrivelled fruits.

Ratooning

Because of their perennial habit, pigeonpea plants remain vegetative at fruit maturity. This is of little consequence with hand-harvesting but mechanical harvesting requires either prior defoliation of the canopy or seed drying, or both. Conversely, the perennial habit enables ratoon growth after harvest of the plant crop, and two or more subsequent harvests are possible [96]. Furthermore, the reproductive indeterminacy conditioned by perenniality represents an important homeostatic mechanism in that it allows great plasticity of response to biotic or environmental challenge. Unlike annual crops, pigeonpea plants can regrow or reflower after alleviation of stress, which can result in a success-ful, albeit deferred crop.

Sharma et al. [80] compared the ratooning ability of early-, medium- and late-maturing cultivars. Early cultivars produced ratoon yields equal to the plant crop, but the plant-crop yields were relatively small. Medium cultivars produced seed yields in the ratoon crop equal to about 50% of the plant crop, while late cultivars (238 days to first harvest) grew vegetatively after plant crop harvest but did not produce seeds in the ratoon. In the medium and late cultivars, there were marked differences among genotypes in ability to ratoon. In India, the incidence of *Fusarium* wilt disease is an important factor determining survival of the ratoon [82].

In Australia, ratoon yields equal to or greater than those of the plant crop have been achieved with the cultivar Royes and with short-season photo-insensitive genotypes [96]. Genotypes differ in ratoon characteristics, including some which commence their ratoon growth before harvest of the plant crop, and others which survive winter frosting and can begin ratoon growth in spring.

YIELD AND YIELD ACCUMULATION

Vegetative yield

When pigeonpeas are grown as a green manure crop or for grazing, both the yield and quality of foliage are important. The greatest dry-matter yields reported, 51 t ha^{-1} in northern Australia [54] and 57 t ha^{-1} in Colombia [30], were based on two cuts per year. Akinola and Whiteman [5] suggested cutting at 8-12 week intervals to give adequate yields and good feeding quality. Liveweight gains as large as 1 kg head^{-1} d^{-1} have been recorded for beef cattle grazing fruiting pigeonpea stands [100]. The utilisation and feeding value of by-products such as pod walls, harvest trash, seed and dhal mill by-products have also been reviewed by Whiteman and Norton [100].

Seed yield

Seed yield per hectare in grain legume crops has traditionally been divided into the following broad components:

number of fruits plant^{-1} and number of seeds fruit^{-1}; mean dry weight seed^{-1}; and number of plants ha^{-1}.

In general, the number of seeds fruit^{-1} and mean seed weight are characteristic of a genotype and influenced relatively little by environmental conditions. Rowden et al. [77] reported that sowing density had no effect on either component but Akinola and Whiteman [4] found that average seed weight was slightly greater at those densities which had few fruits per plant. Times of sowing and harvest had relatively small effects on mean seed weight, but seeds maturing in the hot wet period were slightly smaller than those produced in cooler, drier conditions of greater insolation [3]. Number of seeds fruit^{-1} varied depending on sowing date and harvest time. Artificial shading treatments reduced the number of fruits plant^{-1} but had little effect on the number of seeds fruit^{-1} or on mean seed weight [53, 82, A. Thirathon, pers. comm.]. Removal of either 33% or 66% of leaves at flowering had little effect on either component of yield, but removal of leaves at the flowering nodes led to a small decrease in the number of seeds fruit^{-1} and a large decrease in mean seed weight [53].

In most situations, economic yield is determined largely by fruit number plant^{-1} which is related to plant size and duration of the crop [3,35,69]. In particular [5], the number of fruit-bearing branches and the length of stem over which inflorescences are produced are clearly related to fruit number plant^{-1} and are affected by crop density, time of sowing (in photo-sensitive genotypes) and climatic factors. As in other grain legumes, a large proportion of pigeonpea flowers abort. Pandey [53] recorded 80-85% flower drop regardless of the number of flowers set, which he induced by different amounts of defoliation. Sheldrake [81] suggests that pigeonpeas develop only sufficient fruits that individual plants are capable of filling completely, but the realisation of this capability depends on maintaining a critical threshold of assimilate supply, the use of which is dependent upon maintaining a mass flow from the sources to

the sinks. However, the effects of removal of a proportion
of the fruits during filling also suggests that the presence
of fruits *per se* tends to prevent others from forming (N.
Venkataratnam and A.R. Sheldrake, pers.comm.).

Crop yield depends upon harvesting the maximum number
of fruits per unit area. As population density increases,
so fruit number plant^{-1} declines, essentially asymptotically
[5,19,23,35,77]. Since the number of seeds fruit^{-1} and mean
seed weight are relatively stable, seed yield per plant
declines approximately in parallel with changes in fruit
number. With increasing density, the reduction in fruit number per plant is more than compensated by increasing plant
number, leading to a quadratic relation in which maximum
yield is achieved at an optimum or plateau of plant densities. With further increases in density, yields will decline
[4]. The density ranges over which these functions operate
depend upon plant size, and particularly the duration from
sowing to flowering which is influenced by sensitivity to
photoperiod and temperature and time of sowing. These are
modified by other factors, including moisture stress [35],
and interact with development of leaf area index (LAI). The
time taken to reach LAI values at which 95% of sunlight is
intercepted, which is in the range 4.0-5.0 [31,77], and the
duration over which large intercept values are maintained,
have important effects on fruit set and seed development.

CROPPING SYSTEMS

Pigeonpeas are grown in a wide range of cropping systems;
Willey *et al.* [101] describe several, including the following broad categories: sole crops; intercrops with cereals
(often with sorghum, maize or pearl millet), or with other
legumes (such as groundnuts, cowpeas, mung beans or soyabeans), or with long-season annuals such as castor, cotton
or cassava. These cropping systems were reduced to three
general classes based on the phenology of pigeonpea, viz.
long-season crops, full-season crops and short-season crops
[11].

Long-season crops

This classification is considered to include those crops
sown at or around the longest day of the year but in which
flowering generally occurs after the shortest day of the
year. Commonly, the crops are sown at sparse densities, are
almost invariably intercropped, and produce massive vegetative growth throughout the wet season. The intercrop (often
sorghum or millet) is sown at the same time and matures at
or before the time of flowering of pigeonpea. The intercrop
is harvested and the pigeonpea flowers and produces its
seeds relying on residual soil moisture, and is harvested ·
between nine and eleven months after sowing.

Willey *et al.* [101] have described intercropping systems
involving pigeonpea in which the yield of the companion crop
is not significantly reduced compared to that of a sole
crop, and in which pigeonpea yields up to 70% of a sole
pigeonpea crop can be obtained. Such cropping systems, which
are being researched in some detail, appear to have potential in particular environmental conditions and especially
in subsistence agriculture.

Full-season crops

Production systems that utilise the entire duration of the
available growing season are included here. In some situations, intercrops, ratoon crops, or rotations with a short-
season spring-sown crop may be possible.

An example of this system is that developed by Wallis
et al. [95] for cultivar Royes. Sowings at or after the
longest day have been coupled with appropriate agronomic
management practices to produce canopies which flower more
rapidly and are shorter in stature than long-season crops
sown at the same time. Substantial genotype x sowing date x
density interactions are common [28,75,88,96].

Pigeonpea (*Cajanus cajan* (L.) Millsp.)

Short-season crops

Two distinct subclasses of this production system have been
defined.

Early-maturing crops.   This subclass includes all genotypes that are early flowering and insensitive to the photoperiod regime at the latitude of testing. Wallis *et al.* [97]
reported a method for screening under artificially extended
photoperiods in the field to identify genotypes which will
flower in approximately 60 days where temperature does not
limit growth or influence flowering.

These production systems require large plant populations
$(3\text{-}5 \times 10^5$ plants ha^{-1}) for maximum yields and do not include an intercrop. With appropriate agronomic practices
[99], yields in replicated small plots have often exceeded
6 t ha^{-1} and some have been greater than 8 t ha^{-1}. Farmers
in northern India have rapidly adopted early-maturing cropping systems to provide a wet-season crop which can be harvested before sowing winter wheat [22,33]. Ratoon cropping
is feasible if environmental conditions are favourable [96].

Off-season crops.   This subclass involves sowings made two
or more months after the longest day. Such crops are only
possible where winter temperatures are favourable and frost
does not occur. Due to rapid floral induction modulated by
short days, both photoperiod-sensitive and insensitive genotypes may be used in this system. However, only insensitive
material can provide subsequent short-season ratoon crops.
Flowering in photo-sensitive genotypes would be delayed by
the lengthening days during ratoon growth.

The application of this system in Bihar, India is discussed by Roy Sharma *et al.* [78]. Yields of off-season
pigeonpeas in that environment have been promising, with up
to 3400 kg ha^{-1} in 209 days from the late-sown cultivar
Bahar. Earlier-maturing cultivars have also shown potential.

These production systems are represented in Fig. 15.1

with respect to days after the longest day. Generalised distributions of photoperiod, temperature and moisture availability likely to occur in a monsoon environment are also shown. This classification is general and will not explicitly identify all production systems. However, it is useful in understanding the diversity of adaptation available in this species.

NITROGEN FIXATION

Nodulation

Pigeonpeas are generally regarded to be well-adapted to soils of moderate to poor fertility [1]. However, the nitrogen nutrition of pigeonpea crops, and particularly the role of nitrogen fixation, are poorly understood. Field surveys suggest that yield is limited by poor nodulation in farmers' fields in north India [38] where individual plants produced fewer than ten nodules in most soils. However, nodule weight and number per plant were not closely correlated with the effectiveness of the strain of *Rhizobium* in increasing seed yield [74].

Nodulation in pigeonpeas can proceed rapidly in most soils. Thompson *et al.* [91] recorded about twenty-five nodules per plant fifteen days after sowing in an alfisol at Hyderabad. Rate of nodulation is affected by soil nitrogen concentration. Rao *et al.* [63] reported that nodulation and nitrogenase activity were depressed by soil nitrogen concentrations greater than 25 ppm N (as NO_3). Quilt and Dalal [60] found negligible nodulation in plants up to ten weeks old in soils with 50 ppm N whereas normal nodule formation occurred at soil N concentrations of around 20 ppm. Applications of nitrogen fertiliser at planting reduced nodule weight per plant by 74% at twenty days, but by sixty days no differences in nodulation were apparent [63]. While early N application increased plant size, seed yield was increased in only one year out of three.

674

Pigeonpea (*Cajanus cajan* (L.) Millsp.)

Responses to inoculation

Marked increases in seed yield have been reported following inoculation with an efficient and dominant strain of *Rhizobium*. In trials conducted within the All-India Coordinated Pulse Improvement Project on Improvement of Pulses [74], there was a general increase in yield in eight of the eighteen trials conducted in different climatic zones. Increases in seed yield by as much as 79% (730 kg ha^{-1}) over uninoculated controls were recorded. Other authors have recorded increases ranging from 9-45% [15,27,38,39,60].

Given the diversity of pigeonpea cultivars, production systems and cultural environments (including soils), host genotype x strain of *Rhizobium* x location interaction effects on plant productivity would not be unexpected and have been reported [74]. The cause of these interactions was not determined since nodule collection and typing was not carried out, but they may reflect in part the competitive advantage of the inoculum strains in different soils. If this is so, then introduction of inoculants of specific strains for particular cultivars and regions may be justified. Thompson *et al*. [91] considered that the inoculum must provide at least 1000 and preferably close to 10000 *Rhizobia* seed^{-1} if adequate opportunity for infection by an applied strain is to exist.

Fixation of nitrogen

In general, nitrogen accumulation in the young crop is slow, and maximum nitrogen uptake does not occur until immediately before flowering. Khanna-Chopra and Sinha [37] reported that the greatest rate of nitrogen accumulation in cultivar Prahbat occurred between eight and ten weeks, and maximum crop growth rate between ten and twelve weeks. Similarly, Sheldrake and Narayanan [82] detected maximum nitrogen uptake in the cultivar ICP-1 of 1.7 kg N ha^{-1} d^{-1} between days 60 and 90, while maximum crop growth rate occurred between 90

675

and 120 days. In a study of the transport of ureides (allantoin) in xylem sap, used as an indirect measure of nitrogen
fixation, Rao *et al.* [63] found maximum concentration occurred some two weeks before flowering in the cultivar UWI-17.
Net uptake of nitrogen may continue throughout the reproductive phase, although various proportions of nitrogen for
seed yield originate from retranslocation from elsewhere in
the plant. Khanna-Chopra and Sinha [37] found that 68% of
total N uptake was achieved before flowering. Under field
conditions, nitrogen fixation may decline during the reproductive phase due to nodule degeneration, lack of soil moisture or nodule damage by the larvae of a platystonated fly
(*Rivellia augulate*) [84].

The nitrogen yield in above-ground organs varies greatly
with crop duration and dry-matter yield in particular seasons. Akinola and Whiteman [5] reported a harvest of 460 kg
N ha^{-1} in total biomass for a long-season cultivar cut at
16-week intervals over a 48-week period. Sheldrake and Narayanan [82] recorded nitrogen yields at harvest (including
fallen material) of cultivar ICP-1 of 113 and 89 kg N ha^{-1}
in separate years on a vertisol, and 79 kg N ha^{-1} on an
alfisol. At ICRISAT, Rao *et al.* [63] estimated nitrogen
fixation by subtracting the amount of nitrogen accumulated
by a 175-day sorghum crop from the total nitrogen yield of
the pigeonpea (Table 15.3). For cultivars of similar crop
duration, nitrogen fixation varied from 13-55 kg N ha^{-1}.

Nitrogen return to the cropping system

Approximately 35 kg N ha^{-1} were removed with each tonne of
seeds harvested [24]. Assuming return of all non-seed materials to the soil, returns of 26-100 and 56-70 kg N ha^{-1}
have been estimated [63,82]. The harvest index for nitrogen
ranged from 21% to 52% and was greatest in the early cultivars. These values are small relative to other grain legumes
[56], viz. cowpeas 61%, soyabeans 75%, groundnuts 80%, faba
beans 76% and chickpeas 73%.

Table 15.3 Total nitrogen uptake and fixation by some pigeonpea cultivars on an alfisol at ICRISAT [63] (rainy season 1977)

Cultivar	Plant growth habit*	Maturity (days from sowing)	N yield (kg ha^{-1})				Total N uptake (kg ha^{-1})	Balance against sorghum (N fixed)
			Fruits	Shoots	Roots + nodules	Fallen plant parts		
Prabhat	D	115	43.4	11.6	2.7	11.4	69.1	+ 4.4
Pant A-3	D	115	44.3	14.9	3.9	8.5	71.6	+ 6.9
T-21	I	130	55.7	31.1	4.6	16.5	107.9	+ 43.2
UPAS-120	I	125	46.1	24.9	5.3	15.5	91.8	+ 27.1
BDN-1	I	130	52.3	37.2	4.1	24.6	118.2	+ 53.5
No. 148	I	150	53.2	40.9	8.1	17.6	119.8	+ 55.1
JA-275	I	170	33.7	19.0	7.5	17.7	77.9	+ 13.2
ICP-7035	I	170	33.4	34.7	11.9	21.0	101.0	+ 36.3
ICP-7065	I	175	49.0	23.0	7.6	28.1	107.7	+ 43.0
T-7	I	215	33.3	64.2	15.3	21.3	134.1	+ 69.4
NP (WR)-15	I	240	34.0	54.1	11.5	14.7	114.3	+ 49.6
Sorghum		175	9.5	51.6	3.6	0	64.7	

*I = Indeterminate; D = Determinate

677

In most traditional production systems, whole plants are cut and threshed after removal from the field, the stems being used as firewood and the threshings as fodder. Sheldrake and Narayanan [82] estimated that root systems contribute about 10 kg N ha^{-1} and fallen material 30-35 kg N ha^{-1}, making an average return of 40 kg N ha^{-1}. Similar estimates were reported by Rao *et al.* [63]. These values are similar to comparisons with nitrogen-fertilised maize crops following pigeonpea [63], where the residual effect of pigeonpea was also equivalent to the application of about 40 kg N ha^{-1}.

NUTRIENT REQUIREMENTS

While pigeonpea crops are commonly grown in, and regarded to be well adapted to, soils of poor fertility, plant and seed yields may be limited under these conditions. Relatively few critical studies of the mineral nutrition of pigeonpea have been reported. As with most grain legumes, responses to fertiliser applications have been recorded but may be small because of other limitations. Determination of the effect of improved mineral nutrition on seed yield is complicated by the long duration and large biomass of most traditional cultivars.

Geervani [24] estimated that a harvest of 1000 kg of seeds involved the removal of 35 kg N, 2.8 kg P, 14 kg K, 2 kg Ca, and 0.1 kg Fe. The total nutrient removal was greater where the shoots and pod walls were also harvested [1], with 1000 kg of shoots containing approximately 6 kg N, 0.8 kg P and 4-5 kg K.

Critical concentrations for nutrient elements in the leaves of pigeonpea have not yet been defined. However, Nichols [51] compared nutrient concentrations in the leaves of deficient and well-fertilised pigeonpeas grown in sand culture (Table 15.4), and the critical concentrations probably fall within the ranges shown. Nichols [50] has also presented deficiency symptoms for N, P, K, Ca, Mg and Fe. Deficiency symptoms for S, Mn, B and Zn, and those of Mn toxicity, have been produced by D.G. Edwards and C.J. Asher (unpublished; and see Table 15.6).

Table 15.4 Concentrations of six elements in leaves of pigeonpea cultivar 02/58 grown in sand culture irrigated with either a complete nutrient solution or a solution deficient in the element under study [51]

| Element | *Concentration (%) in dry-matter at:* | | | |
| | *9 weeks* | | *16 weeks* | |
	Deficient	Adequate	Deficient	Adequate
Nitrogen	1.3	3.2	2.0	3.3
Phosphorus	0.05	0.38	0.06	0.22
Potassium	0.34	2.60	1.37	1.95
Calcium	0.69	2.00	0.50	1.19
Magnesium	0.06	0.23	0.15	0.42
Iron	0.027	0.028	0.017	0.016

Responses to nitrogen (N)

Applications of large doses of N have had small or negligible effects on pigeonpea growth, and have also depressed the rate of nodulation early in the vegetative stage [40,63, E.S. Wallis, unpublished]. However, applications of small amounts of N (20-25 kg N ha^{-1}) have increased seed yields by as much as 60-670 kg ha^{-1} in farmers' fields over a wide range of soils in India. Nitrogen responses were greater where adequate phosphate had also been applied [40]. Applications of N increased the number of fruits plant^{-1}, with only a small effect on mean seed weight (and see Chapter 4).

Responses to phosphorus (P)

In general, pigeonpea is not thought to be responsive to large concentrations of soil phosphate. While Kulkarni and Panwar [40] reported increases in seed yield ranging from 180-600 kg ha^{-1} in trials and in farmers' fields in India, all these responses were obtained in the range of 8-15 kg P

679

ha^{-1}. The greatest responses were obtained when 20-25 kg N ha^{-1} was also applied, and were related to increases in branching and numbers of fruits per plant.

Responses to potassium (K)

Kulkarni and Panwar [40] reported either no response or increases in seed yield of up to 200 kg ha^{-1} after applications of 16 kg K ha^{-1} in trials in farmers' fields in India. On more acid tropical soils, K release is usually sufficient to sustain adequate pigeonpea growth and production [44].

Responses to calcium (Ca)

Calcium has two roles in plant nutrition, one as a nutrient element, the other as lime which affects soil pH. The critical concentration of Ca in pigeonpea leaves is between 0.7% and 1.0% [51] (Table 15.4). In acid tropical soil, Ca concentrations can be small and additions are necessary to improve Ca status. Edwards limed an acid tropical ultisol with an initial pH of 4.3 and obtained marked responses in plant growth and nodulation (Table 15.5). In the unlimed soil, Ca deficiency was evident and Ca concentration in leaves was small (0.17%). However, when given 1.0-1.6 t lime ha^{-1}, with pH still at 4.6, pigeonpeas grew and nodulated well. Following greater applications of lime up to 5 t ha^{-1}, growth and nodulation declined due to zinc deficiency.

Responses to sulphur (S)

No clear evidence is available on the responses by pigeonpea to S, but responses are likely to occur on S-deficient soils. Many reports on the effects of phosphate fertilisation (e.g. [40]) do not stipulate the form of phosphate applied and so where single superphosphate has been used, any effects may reflect response to either P or S, or both.

Table 15.5 Effect of lime on soil pH, relative above ground dry-matter yields, and relative dry-matter yields of nodules of pigeonpea cultivar 3D-8103G grown for 45 days in ultisol from south-eastern Nigeria (D.G. Edwards and B.T. Kang, unpublished)

Lime application (t ha^{-1})	Soil pH (H_2O)	Relative yield of shoots (%)	Relative yield of nodules (%)
0	4.30	31.0	0.0
0.25	4.48	46.1	0.9
0.5	4.60	70.1	52.0
1.0	4.88	85.5	95.5
1.6	5.21	100.0	100.0
2.5	5.68	96.2	89.1
3.75	6.30	78.3	20.4
5.0	7.03	44.8	0.3

Responses to micronutrients

There are few reports of field studies of the responses of pigeonpeas to micronutrients. The availability of Zn, Fe, Mn, Cu and B is restricted in alkaline soils but that of Mo is improved. In India, where a large proportion of pigeonpeas are grown on alfisols and vertisols, 47% of 22400 soil samples analysed were categorised as deficient in zinc [34]. Deficiency of Zn reduced seed yield by as much as 67% and there appeared to be differences between cultivars in their tolerance to small concentrations of this element.

Some published values of micronutrient concentrations in pigeonpea leaves [21] are given in Table 15.6.

WATER USE

Pigeonpea is classified as a species well adapted to the semi-arid tropics, largely because of its ability to tolerate dry conditions. The plants have deep roots which can extend to more than 150 cm, and root growth can continue during the reproductive period [82].

Expanding leaves of pigeonpea transpire at small rates

Table 15.6 Micronutrient concentrations in pigeonpea leaves [21]

Micronutrient	Range (ppm)	Concentrations below which deficiency symptoms were observed (ppm)*	Type of culture
Zn	14–45	14	Soil, pot culture
	10–38	15	Soil, pot culture
	25–52	30	Sand culture
	26–30	SNO	Field experiment
Cu	18–21	SNO	Field experiment
Mn	4–6	SNO	Field experiment
	5–31	8	Sand culture
Fe	108–160	SNO	Field experiment
	61–193	61	Solution culture

*SNO = Symptoms not observed

in comparison with fully expanded leaves. Over a given range of vapour pressure deficit, pigeonpea leaves had rates of water loss similar to those of sorghum, and smaller than those in sunflower, soyabean and groundnut [66]. In terms of CO_2 uptake relative to water loss from leaves, pigeonpea efficiency was similar to that in groundnut and greater than in sunflower or cotton. Its photosynthetic efficiency in dim light is also greater than in these crops and this may help to explain the usefulness of pigeonpeas as an intercrop.

Under water stress, leaf photosynthesis rates began to decline at $1.0-1.1 \times 10^6$ Pa leaf water potential, which was somewhat greater than that for sunflower, sorghum and soya-bean [66] which did not decline until 1.5×10^6 Pa. At 2.0×10^6 Pa, pigeonpea leaves stopped photosynthesising whereas the other species were fixing carbon at 40-60 ng CO_2 cm^{-2} s^{-1}. However, M.M. Ludlow (pers.comm.) suggests that, in field-grown pigeonpea, photosynthesis continues at water potentials of at least 3.0×10^6 Pa. In part, its ability to control water loss is related to a capability for osmotic adjustment caused by the accumulation in the cells of ions, sugars and organic acids. This may allow continued photo-synthesis in stress situations, thus enabling root extension to exploit additional water at depth in the soil profile.

Singh and Russell [85] reported a detailed study of water use in which maize and pigeonpeas were grown as a mixed crop. Water use was assessed for the maize/pigeonpea crop during the monsoon, and continued for the pigeonpea following harvest of the maize. Pigeonpea roots extracted water throughout the whole profile to a depth of 187 cm. In the period from 84 to 200 days after sowing, pigeonpea transpired 200 mm of water. The water-use efficiencies of the maize and pigeonpeas are shown in Table 15.7. In terms of seed yield per unit of water transpired (WUE1), maize was more efficient than pigeonpea. The second value (WUE2) is the ratio of seed produced per unit of available water to which the crop had access in the growing season. Here, the values for pigeonpea were greater. The third value (WUE3) is the fraction of seasonally available water that the crop transpired. Pigeonpea was more efficient in the post-rainy season than maize was in the monsoon.

Although pigeonpeas can tolerate considerable moisture deficit, Singh [86] showed that the early-maturing cultivar Prabhat responded well to irrigation. One irrigation increased vegetative dry-matter by 51% and seed yield by 94-108%. Increased seed yield was due to increases in the numbers of fruits plant^{-1} and seeds fruit^{-1}, and in average seed weight.

LIMITATIONS TO YIELD

In general, seed yields obtained from pigeonpeas in traditional farming systems are small (Table 15.1). The major production system (intercropping of late-maturing pigeon-peas) necessarily restricts the yield potential in farmers' fields. The major factors limiting yields in these situations are moisture stress during fruit and seed development, cool temperatures, depredations by pests and diseases, and unfavourable edaphic factors. The outstanding ability of

Table 15.7 Seed yields and water-use efficiencies of maize and pigeon-pea crops on a deep vertisol at ICRISAT [85]

Crop	Yield (kg ha^{-1})	T_m (cm)	AW (cm)	*Water-use efficiency (WUE)*		
				kg ha^{-1} cm^{-1} WUE 1	kg ha^{-1} cm^{-1} WUE 2	cm cm^{-1} WUE 3
1977–78						
Rainy season Maize/pigeonpea	2490	21.2	57.3	117.5	43.5	0.37
Post-rainy season Pigeonpea	1580	19.6	29.3	80.6	53.9	0.67
1978–79						
Rainy season Maize/pigeonpea	2530	24.6	114.0	102.8	22.2	0.22
Post-rainy season Pigeonpea	1070	14.1	27.1	75.9	39.5	0.52

T_m = Transpiration

AW   = Seasonal available water = $M + P$, where M = available profile water to 187 cm at planting, and P = rainfall during growth.

WUE 1 = Yield per unit of water transpired

WUE 2 = Yield per unit of water available

WUE 3 = Fraction of seasonal available water transpired

pigeonpeas to survive in marginal environments, their drought tolerance and ability to recover after severe environmental or biotic stress, are the major reasons for the crop's success in subsistence agricultural systems in the semi-arid tropics.

Yields in improved production systems in more favourable environments with inputs of fertiliser, pesticides and management skills demonstrate that pigeonpeas have extremely large yield potentials [99]. The adoption of these improved production systems in subsistence agricultural systems, however, is problematical.

Two major limitations to yield that can be manipulated genetically are disease and pest susceptibility. The major diseases of pigeonpea include wilt (*Fusarium udum*), Sterility Mosaic Virus, *Phytophthora drechsleri* F.sp. *cajani*,

witches broom (an unclassified, mycoplasma-like organism),
rust (*Uredo cajani*) and leaf spots (*Cercospora cajani*) [49].
Nene [48] listed diseases recorded on pigeonpeas throughout
the world.

Wilt disease is widely spread throughout central and
northern India, and in Kenya and Malawi. Resistant genotypes
have been identified [49] and a coordinated testing programme is underway. Multi-location trials in India have indicated that races of the fungus exist and that further study
is urgently required to identify resistant germplasm.

Sterility Mosaic disease is thought to be caused by a
virus and is transmitted by an Eriophyid mite, *Aceria cajani*
[49]. This disease is particularly important since infected
plants are partially or completely sterile. Resistance has
been identified by field screening techniques [49] but has
not been complete at all locations, suggesting the existence
of different strains of the mite or causal agent.

Phytophthora blight is common where long wet periods
occur during the growing period. Researchers at ICRISAT have
screened a wide range of germplasm and identified sources
of resistance.

Resistance to all or combinations of these three major
diseases is an obvious requirement. Screening at ICRISAT
[49] has identified promising material for further evaluation and incorporation in plant improvement programmes.
Other diseases are of less importance and have been reviewed
elsewhere [9,49].

Insect damage to seed yield can be devastating. The
major insect pests include the moth *Heliothis armigera* (a
larval pod borer) and *Melanagromyza obtusa* (the pod fly)
[71]. Severe damage can also be caused by many other insect
pests and these are discussed by Davies and Lateef [16].

Widespread use of pesticides in subsistence agricultural
systems is unlikely in the foreseeable future. In more intensive production systems, such as the short-season cropping patterns of north India, and in some proposed mechanised
systems [96], pesticide applications are economically justi-

fied and appropriate if large yields are expected. Research is required to develop effective integrated pest management strategies for all pigeonpea production systems. The potential and applicability of such systems have been reviewed [8,71].

Host plant resistance to insect attack is an exciting possibility in the search for resistance to pest attacks. Despite intensive research at ICRISAT [42] genotypes immune to the major pests have yet to be identified. However, significant progress has been made and, in 1983, some progenies of plants selected for resistance to *Heliothis* attack looked promising (W. Reed, pers.comm.). Additional testing and research are required in this area.

A further limitation to productivity in many cropping systems is weed growth. Pigeonpea seedlings grow quite slowly and plants take a considerable period to cover the ground completely. Shetty [83] suggests that weed control during the first five to seven weeks of crop growth is important in attaining acceptable yields. Mechanical control is not feasible in some intensive production systems which utilise narrow rows and large plant populations, so that effective chemical control must be identified. Shetty [83] reviewed aspects of chemical weed control. In traditional production systems, weed control can be achieved by thorough appropriate seed bed preparation and subsequent cultural control.

QUALITY

Numerous aspects of quality in pigeonpea are considered to be important nutritionally or economically. These include visual appearance, cooking time, milling characteristics and yield, protein quality and quantity, digestibility and mineral content. The relations between quality, preference and price are complex and poorly understood. Von Oppen [94] considered that market price in India was a function of seed colour (white being preferred), swelling capacity of the seed and losses during milling to produce dhal. However,

Pigeonpea (*Cajanus cajan* (L.) Millsp.)

Pushpamma and Rao [59] reported that consumers in Andhra Pradesh preferred the local landrace over new cultivars, and also large red seeds with thin testas.

Despite considerable research into aspects of chemical, nutritional, milling, cooking and storage quality, a clear understanding of the variability of these criteria, their interrelations and their fundamental importance in pigeonpea improvement has not yet materialised. Cultivars differ in milling efficiency, and this may be a complex function of seed size and shape, testa composition, gum content and chemistry, moisture content and hardness of the cotyledons [41]. Large differences in cooking time exist among cultivars, but the cause is unknown. Whole-seed protein concentrations range from 15.5-26.8% while some wild relatives and some breeding lines from intergeneric crosses contain more than 30% protein [32]. Protein concentration varies greatly between environments but the nature and cause of genotype x environment interactions has received little study. Aspects of analysis for, and variability in, amino acid composition (especially variations in methionine, cystine and tryptophan concentrations) and antinutritional factors (trypsin and chymotrypsin inhibitors, and tannins) have been reviewed [32]. T. Visitpanich (unpublished) reported that heat treatment increased the nutritional value of pigeonpea seeds in pig rations, whereas untreated seeds were considered to be an acceptable protein source for poultry at inclusion rates up to 30% [20,89]. Relatively little is known about the quality requirements for vegetable use, or for processing as canned or frozen products.

Considerable research is required to define objective criteria of quality that justify improvement, and to document the relative importance of genetic and environmental influences on these criteria. In the interim, improvement and stabilisation of seed yield should remain the primary objectives in plant improvement and advanced lines should be confirmed to be of acceptable quality prior to release.

GERMPLASM RESOURCES

Formal collection of pigeonpea germplasm commenced in the 1920s at the Imperial Agricultural Research Institute at Pusa, India. That collection (approximately 5000 accessions) now forms the basis of the world collection of some 8900 accessions maintained at ICRISAT [93]. Approximately 90% of the accessions have been collected in India which is the principal centre of diversity of pigeonpea. Only limited collections have been made so far from other countries in Asia, Africa, Central and South America and Australia.

Continued, widespread culture of landraces of pigeonpea within traditional agricultural systems internationally sug- gests that considerable opportunity exists for further col- lections of germplasm, and such activities are continuing [93]. The first geographical priority has been given to South and South-east Asia, East Africa and Central America [6]. By contrast, the wild relatives of pigeonpea are rela- tively poorly represented in the current collection (129 accessions representing only 5 genera and 35 species of the 13 genera and 450 species in Cajaninae), and are perhaps subject to greater threat of genetic erosion worldwide [72]. In view of their potential importance to future pigeonpea improvement, active collection of wild relatives is justi- fied and necessary.

Systematic descriptions of the germplasm on the basis of standard descriptors [6] and the results of screening for specific characteristics (e.g. resistance to insects, dis- eases, waterlogging, photoperiod response, annuality and plant habit) are now stored on computer for rapid retrieval and use [93]. The collection is maintained by ICRISAT using controlled self-pollination of plants within each accession. Although germplasm is actively exchanged, more effective procedures to exploit the germplasm in pigeonpea improvement are required.

Some potentially useful characters have been identified in the progeny of pigeonpea x *Atylosia* spp. crosses, including cytoplasmic male sterility, improved protein concentration and partial resistance to *Heliothis* spp. [68]. Transfer of other characteristics is also possible [72]. However, there is limited knowledge of intergeneric compatibilities and of the cytogenetics of intergeneric hybrids.

Exploitation of wild relatives in pigeonpea improvement requires better understanding of intergeneric hybridisation and introgression, and of the physiological mechanisms present in those relatives which may broaden the adaptation of the crop.

STRATEGIES OF PLANT IMPROVEMENT

Some regional perceptions of relevant improvement strategies and methodology have been reviewed recently for India [65], Kenya [52], and for the Caribbean [7]. Green *et al.* [26] have discussed pigeonpea breeding strategies at ICRISAT, and some broader aspects of pigeonpea adaptation and improvement internationally have been reviewed [11].

Relatively little formal plant improvement has been attempted in pigeonpea compared with the major cereal crops. This implies that relatively large improvements in productivity and adaptation should be possible in the short term.

Because of its great diversity of habit and use in quite contrasting production systems, greater differences exist in growth and development among genotypes adapted to the various production systems than exist in many other crops. Thus, it is probable that quite different physiological limitations will exist in materials adapted to the various systems, and that expression of genetic variability and heritability of characters and interrelations among them will also differ between systems. Byth *et al.* [11] suggested that improvement within the different systems will pose quite different problems, and that separate improvement programmes for particular systems were necessary.

A logical strategy for pigeonpea improvement therefore involves determination, by modelling and experimentation, of the relevance and productivity of alternative production systems within a region, and subsequent definition of those improvements necessary within the relevant systems.

Current scientific knowledge of factors which influence pigeonpea growth and development, and their inheritance, is limited so that a clear definition of objectives in quantitative improvement is difficult. The remainder of this section should be read in this context.

Limited tests of different breeding methods have been made. Reasonable discrimination among crosses was demonstrated on the basis of F_1 performance [25], but Green *et al.* [26] preferred the use of F_2 and F_5 bulk performance in multi-location trials in order to account for population x environment interactions. Line x tester matings have been used effectively to determine the breeding value of potential parents, and triple crossing of the hybrid of two divergent parents to a well-adapted parent has proved successful.

In general, Green *et al.* [26] concluded that mass selection for seed yield was relatively ineffective due to poor heritability, and that progeny testing of lines for yield was complicated by line x environment interactions over years and locations. However, pedigree selection for simply inherited and strongly heritable characters such as disease resistance, phenology, seed and fruit size, and plant height was highly effective. Single seed descent has been used effectively for recombination and population advancement.

Outcrossing, mainly by bees [57], can exceed 50% in some circumstances [11]. This imposes considerable costs, difficulties and inefficiencies in breeding and genetic tests, and in the maintenance of cultivars. Two floral modifications which enforce or predispose genotypes towards self-pollination have been reported [12,67] which would greatly simplify pure seed production. However, natural outcrossing does enable exploitation of recombination through open pollination. A number of population improvement schemes have been implemented [26,52,61], or proposed [11,36], with and without incorporation of genetic male sterility.

The literature regarding gene action in pigeonpeas is conflicting [79] and this complicates definition of appropriate breeding procedures and objectives. Most breeding is directed towards the development of homozygous cultivars. However, some varietal hybrids based on genetic male sterility [70,98] have exhibited considerable heterosis for seed yield [25,26]. The cause of such heterosis is not yet known but the use of hybrid cultivars can be considered in some circumstances.

Considerable advances have been made in the last decade in identifying the major diseases of pigeonpeas and sources of genetic resistance to them [49]. Pot and/or field screening procedures have been developed, and sources of resistance identified, for the wilt, mosaic, blight and leaf spot diseases [49]. Pedigree and backcross schemes have been used to incorporate single and multiple resistances into agronomically desirable backgrounds [26].

By contrast, relatively little progress has been made in identifying and incorporating genetic resistance to insect pests. Considerable differences exist among selections in susceptibility to the pest complex (mainly *Heliothis armigera* and *Melanagromyza obtusa*) in India [42]. These differences may reflect resistance, tolerance and/or compensation, and their inheritance and relevance to different production systems are unknown.

The long generation interval of many pigeonpea genotypes restricts breeding and genetic studies. Implementation of accelerated generation turnover would greatly facilitate genetic improvement. However, knowledge of the control and inheritance of floral induction and development is limited. Recent studies suggest that the inheritance is relatively complex [13] and that genotypes of all maturity groups [25] flower relatively quickly under short daylengths (11 h 30 min to 12 h 30 min) and moderate temperatures (22°-30°C) [92, D. van Cooten and Y.L. Nene, pers. comm.].

Breeding and selection of pigeonpea are normally conduc-

ted in pure stands [26]. While this is justifiable for sole
crop production systems, the implications of selection in
pure stands for performance in intercrop systems is of concern. Limited data indicate different genotype yields in
pure stands and intercrops, and that selection in pure
stands rejected a large proportion of genotypes which were
elite in intercrops [26]. Green *et al.* [26] recommended that
selection should be practised in the target environment.

FUTURE PROSPECTS FOR IMPROVEMENT

In common with other tropical and subtropical grain legumes,
pigeonpeas are both under-researched and relatively primitive in their development for domesticated agriculture.
Despite its long history of cultivation, it remains a funda-
mentally wild species. While substantial genetic and environmentally controlled gains in productivity may be possible,
sustained advance in productivity and adaptation, we believe,
will only result from development of an adequate scientific
understanding of its growth and development.

Byth [14] considered that the production system is the
central issue in pigeonpea improvement, and that because of
the large differences between systems, much scientific knowledge should be considered system-specific until proven
otherwise. Thus, advances may arise from comparative evaluation of systems and improvements within particular systems,
together with deliberate introgression between systems as
necessary.

Pigeonpeas exhibit both remarkable tolerance of harsh
environments (e.g. moisture stress and poor soil fertility)
and extremely large seed yield potential in favourable
environments. Both forms of adaptation need to be exploited
and, if possible, combined.

The development of short-season, photoperiod-insensitive
(or nearly so) genotypes considerably broadens the adaptations of systems and improvements within particular systems,
rotations and allows mechanical harvesting. Entirely new

692

production systems are possible, and very large seed yields
in both the plant and ratoon crops can be achieved in favourable environments with effective crop management [99].
However, relatively little is known of the genetic and
physiological limits to productivity in such canopies, or
how they vary in different environments.

At the other extreme, yield accumulation in traditional
production systems involving long-season (nine to eleven
months) cultivars grown in sparse stands either in pure culture or in intercrops with little or no inputs, is complex,
and the prospects for yield improvement are perhaps limited;
yield stabilisation by incorporation of insect and disease
resistance may be more realistic. Research into systems of
intercropping [101] are likely to contribute to increases
in productivity.

COMMUNICATION BETWEEN RESEARCHERS

The creation of ICRISAT in 1972 provided a focal point for
contacts between research workers interested in pigeonpea
improvement. ICRISAT's charter is to act as a world centre
dedicated to improve the yield and nutritional quality of
pigeonpea (and of other crops; see Chapter 18), to develop
appropriate farming systems for these crops in the seasonally dry semi-arid tropics, and to assist national and regional
programmes wherever possible.

The most significant contributions to worldwide cooperation in pigeonpea research have been two International Workshops held at ICRISAT in 1975 and 1980. In addition, scientists from throughout India (and from other countries, where
possible) meet annually in informal, practically-orientated
sessions to select material and to discuss progress. Training programmes for post-graduate and technical personnel
from several countries are also provided by ICRISAT. A
Pigeonpea Newsletter is published twice yearly by ICRISAT,
and provides a forum for exchange of ideas and rapid communication of recent advances in pigeonpea improvement.

Various national programmes have fostered communication between pigeonpea research workers. The All-India Coordinated Pulse Improvement Project provides an annual forum for Indian scientists to develop collaborative testing and evaluation of material from throughout India. The University of West Indies sponsored an International Meeting in 1979 at which problems especially related to the Caribbean region were discussed. The University of Puerto Rico, Mayaguez Campus has also undertaken research on pigeonpea and has published a recent report on the crop [46].

As a result of these national, and especially the international initiatives, direct channels of communication have now been established between individual scientists and institutions for the exchange of information, concepts and materials on the production, improvement and utilisation of this important crop.

LITERATURE CITED

1. Ahlawat, I.P.S. (1981), in *Proceedings of the International Workshop on Pigeonpeas. Volume 1.* ICRISAT, 227-37.
2. Akinola, J.O. and Whiteman, P.C. (1972), *Australian Journal of Agricultural Research* 23, 995-1005.
3. Akinola, J.O. and Whiteman, P.C. (1974), *Australian Journal of Agricultural Research* 26, 43-56.
4. Akinola, J.O. and Whiteman, P.C. (1974), *Australian Journal of Agricultural Research* 26, 57-66.
5. Akinola, J.O. and Whiteman, P.C. (1975), *Australian Journal of Agricultural Research* 26, 67-79.
6. Anon.(1981), *Descriptors for Pigeonpea.* AGP: IBPGR/80/74. International Board for Plant Genetic Resources and International Crops Research Institute for the Semi-Arid Tropics.
7. Ariyanayagam, R.P. (1981), in *Proceedings of the International Workshop on Pigeonpeas. Volume 1.* ICRISAT, 415-26.
8. Blood, P. (1981), in *Proceedings of the International Workshop on Pigeonpeas. Volume 1.* ICRISAT, 106-16.
9. Brathwaite, C.W.D. (1981), in *Proceedings of the International Workshop on Pigeonpeas. Volume 1.* ICRISAT, 129-36.
10. Borlaug, N.E. (1972), in *Nutritional Improvement of Food Legumes by Breeding.* Proceedings, Conference of Protein Advisory Group. Rome: FAO, 7-11.
11. Byth, D.E., Wallis, E.S. and Saxena, K.B. (1981), in *Proceedings of the International Workshop on Pigeonpeas. Volume 1.* ICRISAT, 450-65.
12. Byth, D.E., Saxena, K.B. and Wallis, E.S. (1982), *Euphytica* 31, 405-8.

Pigeonpea (*Cajanus cajan* (L.) Millsp.)

13. Byth, D.E., Saxena, K.B., Wallis, E.S. and Lambrides, C. (1983), in *Proceedings of the Australian Plant Breeding Conference*. Adelaide, 14-18 February, 136-8.
14. Byth, D.E. (1981), in *Proceedings of the International Workshop on Pigeonpeas*. *Volume 1*. ICRISAT, 487-95.
15. Dahiya, J.S., Khurana, A.L. and Dudeja, S.S. (1981), in *Proceedings of the International Workshop on Pigeonpeas*. *Volume II*. ICRISAT, 373-9.
16. Davies, J.C. and Lateef, S.S. (1975), in *Proceedings of the International Workshop on Grain Legumes*. ICRISAT, 319-31.
17. De, D.N. (1974), in Hutchinson, J. (*ed.*) *Evolutionary Studies in World Crops - Diversity and Change in the Indian Subcontinent*. Cambridge: Cambridge University Press, 79-87.
18. de Jabrun, P.L.M., Byth, D.E. and Wallis, E.S. (1981), in *Proceedings of the International Workshop on Pigeonpeas*. *Volume II*. ICRISAT, 181-7.
19. Dhingra, K.K., Satnam Singh and Tripathi, H.P. (1981), in *Proceedings of the International Workshop on Pigeonpeas*. *Volume II*. ICRISAT, 229-34.
20. Draper, C.I. (1944), 'Algarroba beans, pigeonpeas and processed garbage in the laying mash'. *University of Hawaii Agricultural Experiment Station Progress Notes* 44. Honolulu, Hawaii, USA.
21. Edwards, D.G. (1981), in *Proceedings of the International Workshop on Pigeonpeas*. *Volume 1*. ICRISAT, 205-11.
22. Faroda, A.S. and Johri, J.N. (1981), in *Proceedings of the International Workshop on Pigeonpeas*. *Volume 1*. ICRISAT, 45-50.
23. Faroda, A.S. and Singh, R.C. (1981), in *Proceedings of the International Workshop on Pigeonpeas*. *Volume II*. ICRISAT, 223-8.
24. Geervani, P. (1981), in *Proceedings of the International Workshop on Pigeonpeas*. *Volume II*. ICRISAT, 427-34.
25. Green, J.M., Sharma, D., Saxena, K.B., Reddy, L.J. and Gupta, S.C. (1979), 'Pigeonpea breeding at ICRISAT'. Presented at the *Regional Workshop on Tropical Grain Legumes*. University of the West Indies, 18-22 June 1979, St Augustine, Trinidad.
26. Green, J.M., Sharma, D., Reddy, L.J., Saxena, K.B., Gupta, S.C., Jain, K.C., Reddy, B.V.S. and Rao, M.R. (1981), in *Proceedings of the International Workshop on Pigeonpeas*. *Volume 1*. ICRISAT, 437-49.
27. Gupta, B.R. (1981), in *Proceedings of the International Workshop on Pigeonpeas*. *Volume II*. ICRISAT, 387-9.
28. Hammerton, J.L. (1971), *Tropical Agriculture, Trinidad* 48, 341-50.
29. Hammerton, J.L. (1976), *Journal of Agricultural Science, Cambridge* 87, 649-60.
30. Herrera, P.G., Lotero, C.J. and Crowder, L.V. (1966), *Agricultura Tropicale* 22, 473-83.
31. Hughes, G., Keatinge, J.D.H. and Scott, S.P. (1981), *Tropical Agriculture, Trinidad* 58, 191-9.
32. Jambunathan, R. and Singh, U. (1981), in *Proceedings of the International Workshop on Pigeonpeas*. *Volume 1*. ICRISAT, 351-6.
33. Kanwar, J.S. (1981), *International Pigeonpea Newsletter* 1, 6-7.
34. Katyal, J.C. (1981), in *Proceedings of the International Workshop on Pigeonpeas*. *Volume 1*. ICRISAT, 221-5.
35. Keatinge, J.D.H. and Hughes, G. (1981), *Tropical Agriculture, Trinidad* 58, 45-51.
36. Khan, T.N. (1973), *Euphytica* 22, 273-7.
37. Khanna - Chopra, Renu and Sinha, S.K. (1981), in *Proceedings of the*

International Workshop on Pigeonpeas. Volume II. ICRISAT, 397-401.

38. Khurana, A.L. and Deduja, S.S. (1981), in *Proceedings of the International Workshop on Pigeonpeas. Volume II.* ICRISAT, 381-6.

39. Khurana, A.S. and Phutela, R.P. (1981), in *Proceedings of the International Workshop on Pigeonpeas. Volume II.* ICRISAT, 391-5.

40. Kulkarni, K.R. and Panwar, K.S. (1981), in *Proceedings of the International Workshop on Pigeonpeas. Volume I.* ICRISAT, 212-20.

41. Kurien, P.P. (1981), in *Proceedings of the International Workshop on Pigeonpeas. Volume I.* ICRISAT, 321-8.

42. Lateef, S.S. and Reed, W. (1981), in *Proceedings of the International Workshop on Pigeonpeas. Volume II.* ICRISAT, 315-22.

43. Lawn, R.J. (1981), in *Proceedings of the International Workshop on Pigeonpeas. Volume I.* ICRISAT, 151-64.

44. Lugo-Lopez, M.A. and Abrams, R. (1981), *Journal of Agriculture of the University of Puerto Rico* 65, 21-8.

45. Morton, J.F. (1976), *Hortscience* 11, 11-19.

46. Morton, J.F., Smith, R.E., Lugo-Lopez, M.A. and Abrams, R. (1982), *Pigeonpeas* (Cajanus cajan *(L.) Millsp.*). *A Valuable Crop for the Tropics.* Special Publication of the College of Agricultural Sciences, Department of Agronomy and Soils, University of Puerto Rico, Mayaguez, Puerto Rico.

47. National Academy of Sciences. (1980), *Firewood Crops: Shrub and Tree Species for Energy Production.* Washington, DC: NAS, 118-19.

48. Nene, Y.L. (1980), 'A World List of Pigeonpea (*Cajanus cajan* (L.) Millsp.) and Chickpea (*Cicer arietinum* L.) Pathogens'. *ICRISAT Pulse Pathology Progress Report 8.* Patancheru, Andhra Pradesh, India.

49. Nene, Y.L., Kannaiyan, J. and Reddy, M.V. (1981), in *Proceedings of the International Workshop on Pigeonpeas. Volume I.* ICRISAT, 121-8.

50. Nichols, R. (1964), *Plant and Soil* 21, 377-87.

51. Nichols, R. (1965), *Plant and Soil* 22, 112-16.

52. Onim, J.F.M. (1981), in *Proceedings of the International Workshop on Pigeonpeas. Volume I.* ICRISAT, 427-36.

53. Pandey, R.K. (1981), in *Proceedings of the International Workshop on Pigeonpeas. Volume II.* ICRISAT, 203-8.

54. Parbery, D.B. (1967), 'Pasture and fodder crop plant introductions at Kimberley Research Station, Western Australia, 1963-64. Part 1 - Perennial legumes'. *Technical Memorandum, CSIRO, Australian Division, Land Resources. No. 67/6.*

55. Parpia, H.A.B. (1981), in *Proceedings of the International Workshop on Pigeonpeas. Volume 1.* ICRISAT, 484-6.

56. Pate, J.S. and Minchin, F.R. (1980), in Summerfield, R.J. and Bunting, A.H. (*eds*) *Advances in Legume Science.* London: HMSO, 105-25.

57. Pathak, G.N. (1970), in Kachroo, P. (*ed.*) *Pulse Crops of India,* New Delhi, India: ICAR, 4-53.

58. Purseglove, J.W. (1974), *Tropical Crops: Dicotyledons, Volume 1.* London: Longman, 236-41.

59. Pushpamma, P. and Chittemma Rao, K. (1981), in *Proceedings of the International Workshop on Pigeonpeas. Volume II.* ICRISAT, 435-42.

60. Quilt, P. and Dalal, R.C. (1979), *Agronomy Journal* 71, 450-52.

61. Rachie, K.O. and Gardener, C.O. (1975), in *Proceedings of the International Workshop on Grain Legumes.* ICRISAT, 285-300.

62. Rachie, K.O. and Roberts, L.M. (1974), *Advances in Agronomy* 26, 1-132.

Pigeonpea (*Cajanus cajan* (L.) Millsp.)

63. Rao, Kumar J.V.D.K., Dart, P.J., Matsumoto, T. and Day, J.M. (1981), in *Proceedings of the International Workshop on Pigeonpeas. Volume I*. ICRISAT, 190-99.
64. Raju, D.C.S. (1981), in *Proceedings of the International Workshop on Pigeonpeas. Volume II*. ICRISAT, 15-22.
65. Ramanujam, S. and Singh, S.P. (1981), in *Proceedings of the International Workshop on Pigeonpeas. Volume I*. ICRISAT, 403-14.
66. Rawson, H.M. and Constable, G.A. (1981), in *Proceedings of the International Workshop on Pigeonpeas. Volume I*. ICRISAT, 175-89.
67. Reddy, L.J. (1979), 'Pigeonpea breeding'. *ICRISAT Pulse Breeding Progress Report 3*. Patancheru, Andhra Pradesh, India.
68. Reddy, L.J., Green, J.M. and Sharma, D. (1981), in *Proceedings of the International Workshop on Pigeonpeas. Volume II*. ICRISAT, 39-50.
69. Reddy, R.P., Singh, D. and Rao, N.G.P. (1975), *Indian Journal of Genetics* 35, 119-22.
70. Reddy, B.V.S., Reddy, L.J. and Murthi, A.N. (1977), *Tropical Grain Legume Bulletin* 7, 11.
71. Reed, W., Lateef, S.S. and Sithanantham, S. (1981), in *Proceedings of the International Workshop on Pigeonpeas. Volume I*. ICRISAT, 99-105.
72. Remanandan, P. (1981), in *Proceedings of the International Workshop on Pigeonpeas. Volume II*. ICRISAT, 29-38.
73. Reynolds, S.T. and Pedley, L. (1981), *Austrobaileya* 1, 420-28.
74. Rewari, R.B., Vinod Kumar and Subba Rao, N.S. (1981), in *Proceedings of the International Workshop on Pigeonpeas. Volume I*. ICRISAT, 238-48.
75. Riollano, A., Perez, A. and Ramos, C. (1962), *The Journal of Agriculture of the University of Puerto Rico* 46, 127-34.
76. Riollano, A. (1964), *The Journal of Agriculture of the University of Puerto Rico* 48, 232-5.
77. Rowden, R., Gardiner, D., Whiteman, P.C. and Wallis, E.S. (1981), *Field Crops Research* 4, 201-13.
78. Roy Sharma, R.P., Thakur, H.C. and Sharma, H.M. (1981), in *Proceedings of the International Workshop on Pigeonpeas. Volume I*. ICRISAT, 26-36.
79. Saxena, K.B., Byth, D.E., Wallis, E.S. and De Lacy, I.H. (1981), in *Proceedings of the International Workshop on Pigeonpeas. Volume II*. ICRISAT, 81-92.
80. Sharma, D., Saxena, K.B. and Green, J.M. (1978), *Field Crops Research* 1, 165-72.
81. Sheldrake, A.R. (1979), *Indian Journal of Plant Physiology* 22, 137-43.
82. Sheldrake, A.R. and Narayanan, A. (1979), *Journal of Agricultural Science, Cambridge* 92, 513-26.
83. Shetty, S.V.R. (1981), in *Proceedings of the International Workshop on Pigeonpeas. Volume I*. ICRISAT, 137-46.
84. Sithanantham, S., Kumar Rao, J.V.D.K., Reed, W. and Dart, P.J. (1981), in *Proceedings of the International Workshop on Pigeonpeas. Volume II*. ICRISAT, 409-16.
85. Singh, Sardar and Russell, M.B. (1981), in *Proceedings of the International Workshop on Pigeonpeas. Volume I*. ICRISAT, 271-82.
86. Sinha, S.K. (1981), in *Proceedings of the International Workshop on Pigeonpeas. Volume I*. ICRISAT, 283-8.
87. Smartt, J. (1976), *Tropical Pulses*. Tropical Agriculture Series.

London: Longman.

88. Spence, J.A. and Williams, S.J.A. (1972), *Crop Science* 12, 121-2.

89. Springhall, J., Akinola, J.O. and Whiteman, P.C. (1974), *Proceedings of 1974 Australian Poultry Science Convention.* 117-19.

90. Summerfield, R.J. and Wein, H.C. (1980), in Summerfield, R.J. and Bunting, A.H. (*eds*) *Advances in Legume Science.* London: HMSO, 17-36.

91. Thompson, J.A., Kumar Rao, J.V.D.K. and Dart, P.J. (1981), in *Proceedings of the International Workshop on Pigeonpeas. Volume I.* ICRISAT, 249-53.

92. Turnbull, L.V., Whiteman, P.C. and Byth, D.E. (1981), in *Proceedings of the International Workshop on Pigeonpeas. Volume II.* ICRISAT, 217-22.

93. van der Maesen, L.J.G., Remanandan, P. and Murthi, Anishetty N. (1981), in *Proceedings of the International Workshop on Pigeonpeas. Volume I.* ICRISAT, 385-92.

94. Von Oppen, M. (1981), in *Proceedings of the International Workshop on Pigeonpeas. Volume I.* ICRISAT, 332-43.

95. Wallis, E.S., Whiteman, P.C. and Byth, D.E. (1979), *Queensland Agricultural Journal* 105, 487-92.

96. Wallis, E.S., Byth, D.E. and Whiteman, P.C. (1981), in *Proceedings of the International Workshop on Pigeonpeas. Volume I.* ICRISAT, 51-60.

97. Wallis, E.S., Byth, D.E. and Saxena, K.B. (1981), in *Proceedings of the International Workshop on Pigeonpeas. Volume II.* ICRISAT, 143-50.

98. Wallis, E.S., Saxena, K.B. and Byth, D.E. (1981), in *Proceedings of the International Workshop on Pigeonpeas. Volume II.* ICRISAT, 105-8.

99. Wallis, E.S., Byth, D.E., Whiteman, P.C. and Saxena, K.B. (1983), in *Proceedings of the Australian Plant Breeding Conference.* Adelaide. 14-18 February, 142-5.

100. Whiteman, P.C. and Norton, B.W. (1981), in *Proceedings of the International Workshop on Pigeonpeas. Volume I.* ICRISAT, 365-77.

101. Willey, R.W., Rao, M.R. and Natarajan, M. (1981), in *Proceedings of the International Workshop on Pigeonpeas. Volume I.* ICRISAT, 11-25.

16 Lupin (*Lupinus* spp.)

J.S. Pate, W. Williams and P. Farrington

INTRODUCTION - LUPIN SPECIES AND THEIR HISTORY OF CULTIVA-
TION AND SELECTION AS CROP PLANTS

Lupinus, with some 200 New World and 12 Old World species
[108], is the only genus of the predominantly tropical tribe
Genisteae (Fabaceae) from which species have been exploited
as temperate annual grain legume crops. Three major crop
species, *Lupinus albus* L. (white lupin), *L. luteus* L.
(yellow lupin) and *L. angustifolius* L. (narrow-leafed lupin),
and one with potential for the future, *L. cosentinii* Guss.
(sandplain lupin), have originated from Europe. A single New
World species, *L. mutabilis* Sweet (tarwi), came from the
Andean Highlands of South America [50]. Distinguishing features of the five species are shown in Table 16.1.

Use of lupins as food or forage can be traced to ancient
times. The progenitor types of 'wild' species were probably
all endowed with characters undesirable for immediate acceptance as seed crop plants, including seeds and foliage rich
in alkaloids, asynchronous ripening of impermeable seeds,
and shattering of fruits. However, the large seeds of wild
lupins must have rendered them easily harvested by hand,
while man learned that the bitter principles of the seeds
could be removed by prolonged soaking. Also in their favour,
lupins proved markedly resilient and persistent in primitive
agriculture, largely because of their unpalatability to
grazing animals, their great seed longevity and their

Table 16.1 Distinguishing features of current and potential crop lupin species [42,47,83,84,109,155]

Species	Chromosome number (2n)	Plant height (cm)	*Flowers*		Fruit size (length x width) (mm)	Shape	*Seeds*	
			Colour	Arrangement on inflorescence			Testa surface	Mean seed weight (mg)
L. albus (white lupin)	50	30–120	White tinged with violet or blue	Spiral	70–150 x 12–20	Square, compressed	Smooth	360
L. angustifolius (narrow-leafed lupin)	40	20–150	Normally blue, but white in cultivated species	Spiral	35–50 x 7–10	Globular	Smooth	170
L. luteus (yellow lupin)	52	20–80	Yellow	Verticillate	40–60 x 10–14	Orbicular-quadrangular, compressed	Smooth	130
L. cosentinii (sandplain lupin)	32	20–120	Blue	Subverticillate-verticillate	40–55 x 13–16	Orbicular-quandrangular, compressed	Rough	230
L. mutabilis (tarwi)	48	50–150	Purple, blue, pink, white	Partially verticillate	70–100 x 18–24	Ovoid, compressed	Smooth	200

ability to thrive in disturbed or roughly cultivated areas of poor fertility. Indeed, the name lupin (*Lupinus*) derives from *Lupus* (the wolf), attesting to a supposed ability of the species to 'rob' the soil of nutrients.

Lupins have been cultivated in the Mediterranean region for probably more than 3000 years [45]. *Lupinus albus*, the oldest European domesticate, has been used since Greek and Roman times as a green manure, as forage for stock, or as a pulse for human consumption when alternative foods were scarce. Cultivation of *L. mutabilis* in subsistence agriculture in South America dates back to the Incas, who had long known how to rid the seeds of bitter alkaloids by boiling and leaching them in water [18]. The earliest cultivations of *L. luteus* for forage or green manure, were probably in the Iberian Peninsula (North Africa) and Madeira. Later, the species was carried to northern Europe as an ornamental. Cultivation of *L. angustifolius* is reported for southern Europe in the eighteenth century, when it was grown as a forage crop or for seeds to use as a substitute for coffee.

During the nineteenth century, *L. albus* and *L. luteus* became widely cultivated in northern Germany for fodder and as green manure. Thereafter, following the collapse of the Saxony wool industry in the early twentieth century, use of lupins declined markedly. However, during the First World War (1914-18) methods were devised in Germany to remove the alkaloids on a commercial scale and 'debittered' seeds were then incorporated into stock feed.

Lupinus albus was taken overseas in the nineteenth century by southern European migrants to many New and Old World countries. *Lupinus angustifolius* was already being grown in Suffolk, England by 1859 [88]; both this species and *L. luteus* were firmly established by the 1930s as fodder crops in the Cape Province of South Africa [117], the Canterbury Plains of New Zealand [73,133] and in Australia [43,50], and by the 1940s in the south-eastern coast region of the USA [17]. *Lupinus luteus* was probably introduced into the USSR and Portugal at the beginning of the twentieth century [131], while *L. cosentinii* was introduced, probably accidentally, into Western Australia in the mid-nineteenth century - quickly to become naturalised on the coastal sand plain surrounding Perth [37]. *Lupinus mutabilis* was imported into England from Colombia in about 1825, principally for use as an ornamental [83] in view of its showy flowers.

Development of lupins as commercial seed crops dates to the first selections of low-alkaloid or 'sweet' mutants in the various species. The chronology of selection for these and other desirable characters is given in Table 16.2.

Alkaloids in crop lupins comprise at least ten different compounds, of which lupanine is predominant in most species [151]. *Lupinus mutabilis* is exceptional for its large proportion (up to 20% of total alkaloid) as sparteine [151]. All low-alkaloid varieties are homozygous for recessive or partly-recessive alleles which almost completely block alkaloid synthesis in green tissues, so that only small amounts accumulate in seeds. In contrast, all wild forms and unselected farm stocks contain dominant alleles for large alkaloid concentrations. The main mutant alleles which govern reduced alkaloid synthesis in *L. albus*, *L. angustifolius* and *L. luteus* are listed in Table 16.3. Research and breeding of low-alkaloid line continues in *L. mutabilis* [150] (see Table 16.2).

In the 1920s and 1930s characters such as permeable testas (seed coats) and non-dehiscent fruits were incorporated into the new 'sweet' lines of *L. luteus* (e.g. as seen in cultivars Weiko II and III), while parallel breeding programmes on *L. angustifolius* developed 'sweet', soft-seeded lines such as Borre [61]. Further selections of *L. angustifolius* in Western Australia (see Table 16.2) eventually led to the discoveries of early flowering mutants with reduced fruit shattering and white flowers and seeds. Crosses with existing lines of permeable-seeded sweet cultivars or varieties led to the release of forms suitable for seed production (e.g. 'Uniwhite' in 1967, 'Uniharvest' in 1970 and 'Unicrop' in 1973). Concurrently, selections in

702

Table 16.2 Chronology of the development of varieties of lupins suitable as seed crops [45,52,64,149,150]

Species	Character	Gene	(date reported)	Variety/cultivar	(date)
Lupinus albus	low-alkaloid	*nutricius*	(1932)	Nahrquell	(1949)
	low-alkaloid	*pauper*	(1932)	Kraftquell	(1949)
	early flowering	*brevis*	(?)	Ultra	(1950)
	(All genotypes contained mutant alleles of ancient origin for non-dehiscent fruits and permeable testas prior to 1949)				
Lupinus luteus	low-alkaloid	*dulcis*	(1928)	Muncheberger St. 8	(1933)
	low-alkaloid	*dulcis*	(1928)		
	permeable testas	*w*	(1928)	Weiko II	(1943)
	non-dehiscent fruits	*invulnerabilis*	(1935)		
	white seeds	*coloratus niveus*	(1932)		
	rapid early growth	*crescens celer*	(1938)	Weiko III	(1952)
Lupinus angustifolius	low-alkaloid	*iucundus*	(1928)	Muncheberger	(1933)
				Blaue Susslupine	
	low-alkaloid	*iucundus*	(1928)	Muncheberger	(1944)
	permeable testas	*mollis*	(?)	Blaue Susslupine II	
	permeable testas	*mollis*	(?)		
	non-dehiscent fruits	{*tardus*	(1960)	Uniharvest	(1970)
		{*lentus*			
	white flowers and seeds	*leucospermus*	(?)		
	large seeds	*retardans*	(?)		
	early flowering	*Ku*	(1961)	Unicrop	(1970)
	anthracnose resistance	*An*	(1951)	Illyarrie	(1979)
	grey leaf spot resistance	g^1_1	(1953)		

Table 16.2 contd

Species	Character	Gene (date reported)		Variety/cultivar (date)	
Lupinus cosentinii	low-alkaloid	*sw*	(1963)	Eregulla	(1977)
	reduced fruit-dehiscence	*coniunctus*	(1961)		
		macer			
	white flowers and seeds	*wfs*	(1963)		
	early flowering	*xe*	(1955)		
Lupinus mutabilis	non-dehiscent fruits	not known	(prehistory)	Diverse farm stocks	
	permeable testas	?*	?	?	
	low-alkaloid	*mutal*	(1983)	(unnamed)	

*? denotes 'not recorded'

Table 16.3 The main mutant alleles conditioning reduced alkaloid concentrations in crop lupins [64,140]

	Lupinus albus		*Lupinus angustifolius*			*Lupinus luteus*		
Mutant allele	Alkaloid concentration (%)	Principal variety/ cultivar	Mutant allele	Alkaloid concentration (%)	Principal variety/ cultivar	Mutant allele	Alkaloid concentration (%)	Principal variety/ cultivar
Pauper (Synonyms *Primus* and *Tertius*)	0.03–0.05	Kievskij mutant, Kraftquell, Blanca, Ultra	*Iucundus*	0.02–0.05	Borre, Blanco, Frost, Uniwhite, Uniharvest	*Dulcis*	0.04	Weiko II, Weiko III, Alteria
Nutricius	0.04–0.05	Nahrquell	*Depressus*	–*	–	*Amoenus*, *Liber*	0.01	–
Exiguus	0.03–0.05	Neuland	*Esculentus*	–	–			
Reductus	–	–	*Tantalus*	–	–			
Mitis	–	–						
Bitter type	1.5–2.2	Lupini bean and all unselected farm stocks**	Bitter types	1.5–1.8	'Fest' and all wild and unselected farm stocks	Bitter types	0.9–1.2	'Schwako' and all wild and unselected farm stocks

*Dashes denote 'no information available'.
**Lupini bean is a large-seeded form of unknown origin and cultivated in Italy.

the USA produced a cultivar ('Rancher') resistant to anthracnose (*Glomerata cingulata*) and grey leaf spot (*Stemphylium
vesicarum*) [33], which was used in Western Australia to
develop the disease-resistant cultivars 'Marri' (1976),
'Illyarie' (1979), 'Yandee' (1980) and 'Chittick' (1982).

CURRENT STATUS AND FUTURE PROJECTIONS AS CROP SPECIES

Lupins are as yet insignificant in the world league of grain
legume crops. Figures reported by the Food and Agriculture
Organisation (FAO) in 1982 (FAO, pers. comm.) show a total
of 518 000 tonnes of harvested seed - a mere 0.3% of the
estimated total world grain (seed) legume production. Production is currently centred on eastern Europe, notably in
the USSR and Poland, the only major lupin growing area outside Europe being in Australia (Fig. 16.1(a)). Seed production declined markedly in the USSR and other European countries in the 1970s (Fig. 16.1(b)), largely because of widespread losses due to *Fusarium* wilt disease and the replacement of lupins by cereal crops [150]. However, there is a
current resurgence of interest in Europe, with the prospect
of seed lupin crops being grown within countries of the
European Economic Community (EEC) to reduce the need to
import soyabean meal from abroad. Production in South
Africa declined from 60000 tonnes in the 1960s to a mere
3000 tonnes today - a situation attributed to the better
profitability of wheat over wool, a succession of dry
seasons, and severe outbreaks of powdery mildew (*Erysiphe*
spp.) [143].

The development of 'sweet', non-shattering varieties of
L. angustifolius suited for dryland farming led to lupin
seed production in Australia expanding rapidly, from 260
tonnes in 1966 to 135 000 tonnes in 1981-82 (see Fig. 16.1
(b)). Western Australia is the major producer (82000 tonnes),
followed by Victoria (20000 tonnes), South Australia (19000
tonnes) and New South Wales (14000 tonnes).

Production statistics (FAO, pers. comm.) do not specify

706

Lupin (*Lupinus* spp.)

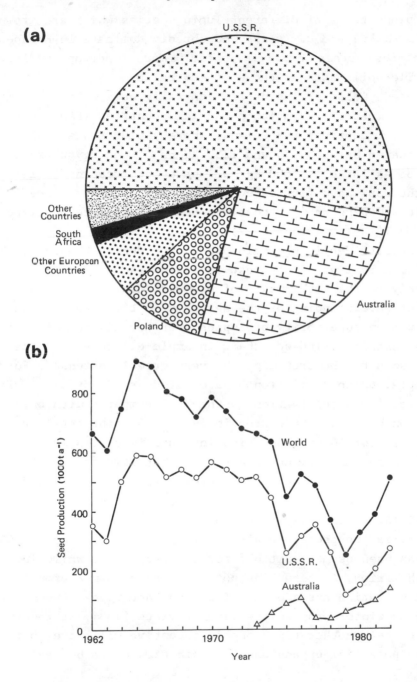

Figure 16.1 World production of lupin crops (a) Contributions by
different countries to production in 1981–82 (b) Annual seed
production 1962–82

707

the proportions of different lupin species which are grown, but *L. albus* and *L. luteus* undoubtedly dominate in European countries whereas *L. angustifolius* is more important on less fertile soils such as those of Australia.

Seed yields of lupins are generally well below 1000 kg ha^{-1} in most of the major producing countries (FAO, pers. comm.). Exceptions are Poland, where yields since 1962 have fluctuated between 950 and 1250 kg ha^{-1}, and Egypt, where yields as large as 1600 kg ha^{-1} have been recorded. Yields in the USSR have fluctuated very widely (a range since 1962 of 265-1040 kg ha^{-1}); those for South Africa are consistently small (below 300 kg ha^{-1}).

Despite declining world production during the 1960s and 1970s, a revival of interest in lupins is under way, based largely on the performance of new varieties and cultivars, better understanding of the agronomy of the species, and the proven superiority of lupins over cereals on soils of poor fertility. Indeed, there is ample evidence that the area sown to the crop might be very greatly extended. For example, up to 30 million ha are rated as potentially suitable for lupin cultivation in the USSR compared with only 2.5 million ha sown to the crop in 1973 [130]; the total area suitable for lupin production in Spain is estimated at 0.8 million ha, many times larger than that used at present [118], and a doubling of the current area is considered possible in South America [127], including an expansion of tarwi (*L. mutabilis*) by at least 1000 ha [135]. In Western Australia prospects are also good, since an area of 250 000 ha has been deemed suitable for the crop [63], while in South Africa, *L. albus* has potential as a summer crop in high elevation areas of the Transvaal and Orange Free State, or as a winter crop in the Cape Province [143]. It remains to be seen whether the economic incentive will develop for these potential expansions of lupin growing to be realised.

Lupin (*Lupinus* spp.)

PRINCIPAL FARMING SYSTEMS

Lupins fit into farming systems most frequently as summer
crops in cool temperate climates with summer rainfall, but
they also perform particularly well as winter crops in milder
Mediterranean climates of reliable winter rainfall (i.e. in
conditions close to those obtaining where the European
species of lupins evolved). In northern Europe, lupins are
normally grown as full-season forage crops, planted early
(April/May) to ensure maximum vegetative growth [45,60].
They are also used as 'catch crops' for green manure, sown
immediately after harvest of a winter cereal and then ploughed in during early winter. They are sometimes grown in
eastern Europe in mixtures with rye, oats, or barley, when
the crops are grazed green or made into silage.

    Lupinus luteus has been used extensively as a green manure,
silage or hay crop in north-central Europe and the USSR. It
is also grazed in parts of the Iberian Peninsula.

    Lupinus angustifolius is grown in Europe as forage for sheep
This species, and *L. albus*, are best cut for silage during
mid-fruiting, when plants are still leafy and of near maximum nutrient content. Crops may then yield as much as 11 t
dry-matter ha^{-1}, the resulting silage being of good digestibility with a large crude protein concentration [129].

    Seed production of lupins in northern Europe is extreme-
ly unreliable because of the long time required for fruits
to mature and the possibility of damp summer weather conducive to severe fungal infestations.

    In New Zealand, spring-sown lupins are used for forage
and green manuring in rotation with cereals or, increasingly,
as seed crops for export or local consumption [152].

    In Australia, lupins are almost exclusively winter-grown
seed crops and, with the current trend away from pastured
livestock, they are being increasingly included in all-
arable systems as 'break crops' in rotations with wheat and
other cereals. Wheat crops following lupins frequently produce larger yields than those following wheat [9,10], but it

is too early to judge the long-term viability of continuous arable cropping of this type, especially in relation to nitrogen carryover from the lupins to succeeding crops [145]. Lupins can noticeably improve soil structure, principally due, perhaps, to their deep-rooting habit.

In the Cape Province of South Africa, bitter lupins (mainly *L. angustifolius*) rotated with cereals are traditionally used for summer grazing or for harvesting of seeds followed by grazing of stubble [143]. The seeds are either fed to local livestock or exported to Europe for sowing as green manure or seed crops. In the south-eastern USA, *L. angustifolius* is grown as a winter forage crop for beef cattle, often in rotation with summer-grown maize [33].

Lupins are being evaluated in Brazil as alternative winter crops to cereals, especially where prevalence of root diseases necessitates a break of three years or longer between successive cereal crops. Elsewhere in South America subsistence farming of *L. mutabilis* is still practised in the high Andean regions of Ecuador, Chile and Peru [83], where it is grown in small plots close to homesteads and harvested as a pulse [8,126].

MAJOR MORPHOLOGICAL FEATURES OF VEGETATIVE AND REPRODUCTIVE PLANTS

Crop lupins are erect, self-supporting, herbaceous to woody annual plants, which are 20-150 cm tall when mature (see Table 16.1). Plants develop a strong main stem bearing a terminal racemose inflorescence and a series of first order lateral branches (Figure 16.2). Some of these laterals remain vegetative and of strictly limited growth, while others produce terminal inflorescences and second order lateral branches. Provided the growing season is long enough, third and even fourth order lateral branches may then develop from the second order laterals.

Species, and cultivars and varieties of a given species, differ in the number of nodes developed on a main shoot

710

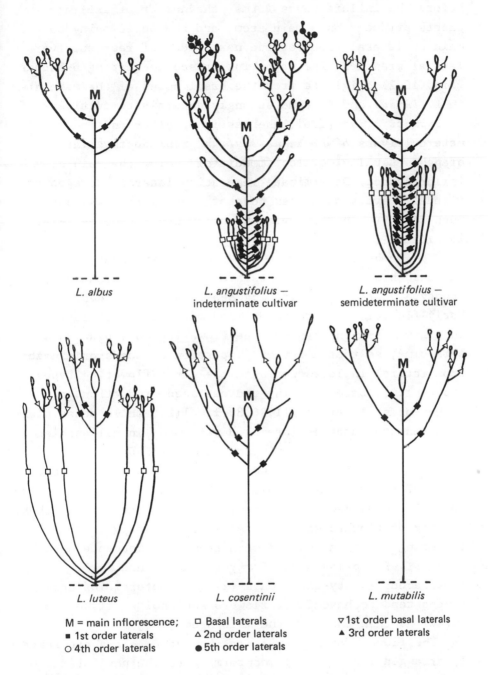

Figure 16.2 Diagrammatic representations of individual lupin crop plants showing differences in reproductive branching structure

before the inflorescence forms, the number of axillary
shoots produced on a main stem, and the positioning on a
main or lateral shoot of the next order of reproductive
lateral branches. The major inflorescence-bearing branches
(Fig. 16.2) originate from the basal nodes of the main stem
in *L. luteus* and in some strongly-determinate cultivars of
L. angustifolius [138], whereas in *L. albus* and indeterminate cultivars of *L. angustifolius*, reproductive lateral
branches usually develop immediately below the main stem
inflorescence. Determinate vegetative laterals develop at
virtually every main stem node of *L. angustifolius* but are
much less frequent or even absent on other species (Fig.
16.2).

The number, final size and reproductive status of lateral branches of lupin plants are strongly influenced by
genotype, duration of growing season and plant density.
Early-flowering genotypes develop more orders of laterals
in a given season than do late-flowering ones planted at the
same time, so that a larger proportion of seeds originates
from lateral inflorescences in the early-flowering types
[26,29,110]. Late planting reduces seed production on
higher-order laterals [11,26,39,110,112], while dense plantings tend to suppress lateral development in all species
except *L. luteus* (P. Farrington, unpublished data).

Flowering of lupins usually commences soon after emergence of the terminal raceme on the main stem and continues
for as long as new inflorescences form on lateral branches.
Opening of flowers within a raceme is strictly acropetal,
flowering duration for the main stem inflorescence of *L.
angustifolius* being about forty-two days, and for *L. cosentinii* about twenty-eight days [27,31]. Lateral inflorescences tend to have fewer flowers and shorter flowering
durations than those on the main stem [134].

The flowers of *L. angustifolius* and *L. albus* are spirally arranged (Table 16.1), conforming to the phyllotaxis of
the main stem or lateral on which the inflorescence is borne
[28,89,107]. *Lupinus cosentinii* and *L. luteus* exhibit a verti-

712

cillate floral arrangement with five or more flowers in each
successive whorl [47]. *Lupinus mutabilis* is partially verticil-
late - the lower groups of flowers on an inflorescence
clearly so, the upper ones mostly alternate. Flower number
on individual plants varies from about 50 in certain varieties of *L. albus* to almost 200 in some genotypes of *L.
angustifolius* [27].

The lupin fruit is a single carpel (legume), having two
main longitudinal blocks of vascular tissue strands along
its dorsal suture, a single one along the ventral suture,
and a network of veins in its lateral walls. Seeds are
attached alternately to the opposing halves of the vasculature of the dorsum [107]. Fruit length varies from 35-150
mm, depending on species (Table 16.1). Fruits are thick and
fleshy, due mainly to the multilayered parenchyma of the
mesocarp of the pod wall [102]. The outer layers of the
mesocarp and the inner epidermis of the pod wall are photosynthetic, allowing most, or all, of the CO_2 respired by
seeds during the daytime to be reassimilated [101,106]. The
fleshiness of fruit walls is considered to be responsible
for the delayed ripening and for increasing susceptibility
to fungal attack in cool climates [147].

Reduction in seed shedding at maturity has been achieved
by selecting mutants with modified fruit wall anatomy [44].
In *L. angustifolius* and *L. luteus*, the paired strips of
sclerenchyma associated with the vasculature of the dorsal
suture of the fruits are normally separate, but in mutants
with reduced fruit shattering, these strips are fused. In
other mutants of *L. angustifolius*, *L. albus* and *L. mutabilis*,
the normally parallel alignment of endocarp sclerenchyma in
the pod walls is irregular so that twisting and explosive
shattering of fruits during drying is prevented.

The average weight, shape and surface texture of seeds
vary widely between species (Table 16.1). The proportion
of testa is large in comparison with other grain legumes:
e.g. values of 28.0% of seed dry weight in *L. cosentinii*;
26.1% in *L. luteus*; 23.6% in *L. angustifolius*; and 18.6%

in *L. albus* [155].

Lupin leaves are compound, comprising between five and eleven leaflets (depending on species) which radiate from the apex of a long petiole. Leaflet shape varies from narrow and linear in *L. angustifolius* to oblong-obovate in *L. albus* and *L. mutabilis* [42,47]. Flexible pulvini at the base of the leaflets engage in sun-tracking movements, thus enhancing photosynthetic performance when solar elevations are small. Some species (e.g. *L. angustifolius*) are particularly prone to shed their leaflets when moisture-stressed, diseased or severely shaded [46,58].

Lupins typically develop a robust tap root, from which arises numerous, relatively thick, lateral roots. When grown on deep sands, lupin roots may penetrate to below 2 m, thus enabling plants to acquire water and nutrients from depths below that possible for most other crops [20,25,90]. Nodules are usually confined to the upper part of the tap root, provided that the symbiosis is an effective one. Each nodule is spherical and possesses a peripheral, potentially perennial meristem which may extend such that the nodule eventually forms a collar-like structure completely encircling the root [93]. Most nodules remain active and continue growth throughout the life cycle of the host species. Truly perennial nodules of massive size (up to 50 g) have been recorded for perennial species of lupin (e.g.*L. arboreus* L.) [93].

REPRODUCTIVE BIOLOGY

Crop lupins are quantitative long-day plants, and many genotypes are responsive to vernalisation [45,50,134]. Floral initiation in *L. albus* and *L. angustifolius* is controlled more by vernalisation requirement than by photoperiod [124], while *L. luteus* responds strongly to both factors, with long photoperiods substituting markedly for vernalisation. In *L. cosentinii*, a species with only a modest vernalisation requirement, flowering is accelerated by lengthening photoperiods and increasing temperatures during the post-vernalisation stage [52,120,122].

The gene *Ku* (the '*Julius*' characteristic) eliminates the requirement for vernalisation in *L. angustifolius* [52] and, when incorporated into early flowering lines, it advances flowering by up to five weeks [49]. Two early flowering forms of *L. cosentinii* are known; one, a natural mutant, shows no response to vernalisation, the other, an artificially-induced mutant, is less sensitive to short days [42, 52].

Most lupin species are renowned for premature abscission of their flowers and fruits. For example, in *L. albus* and *L. luteus*, up to 70% of flowers may fail to form fruits [66, 138], while up to 80% can fail to set in certain cultivars of *L. angustifolius* [27,58,66,110], especially under temperature and moisture stress [6]. In contrast, *L. mutabilis* is much less sensitive to floral abscission. Determinate genotypes of a species are usually much less prone to floral abscission than are indeterminate ones.

The physiological mechanisms involved in floral abscission have been studied in both *L. luteus* and *L. angustifolius*. In *L. luteus*, abscission of upper flowers on an inflorescence was suggested to result from synthesis of an inhibitor in the basal fruits [138]. The inhibitor was identified as abscisic acid [116], although a direct role of this substance in floral abscission has since been questioned [114,115].

Studies on the *L. angustifolius* cultivar Unicrop [28,31, 99] have indicated that flower-shedding on a given inflorescence coincides with the successful setting of the first-formed fruits and with outgrowth of laterals immediately below that inflorescence. Flowers fail to import assimilates shortly prior to their abscission. The overriding inhibitory influence of lateral shoots on fruit set is demonstrated by the fact that a virtually complete complement of twenty to twenty-five fruits will form on an inflorescence if the laterals below it are removed at an early stage of development, whereas only the lowermost three to five fruits normally set on inflorescences of plants in which lower laterals have been left intact [31]. Similar consequences resulting from artificially manipulating laterals have been reported for *L. albus* and *L. mutabilis* [115].

Following fertilisation, lupin fruits import assimilates sparingly and grow at negligible rates for a few days, after which their 'sink' strength increases and growth accelerates markedly [31]. As with other grain legumes, the fruit wall develops precociously to attain maximum dry weight just as the seeds enter their exponential phase of embryo growth [70,106]. The testa of a seed matures well in advance of its embryo. Liquid endosperm is transient and its reserves are completely resorbed (at least in *L. albus* and *L. angustifolius*) before the embryo has attained one-quarter of its final dry weight [5,70]. Dry-matter losses from the fruit wall and testa occur mainly during the last third of the exponential phase of embryo growth. These organs contribute, at most, only one-quarter of the total requirements of the embryos for dry-matter, and so seed-filling is heavily dependent on current photosynthesis, or mobilisation of previously-stored assimilates from vegetative organs [97, 100].

Field crops of *L. angustifolius* in Australia frequently suffer severe moisture stress during mid- to late-fruit ripening, and, if leaf shedding ensues, yields may be drastically reduced and crop maturity delayed. Fruits on all orders of inflorescences mature almost synchronously in the *L. angustifolius* cultivar Unicrop, although a main stem inflorescence may have set its fruits several weeks before those on lateral inflorescences [27,31,58,67,110].

Assimilate translocation and budgets for carbon have been studied in fruits and whole plants of both *L. albus* [100,104,154] and *L. angustifolius* [97,99]. Translocation patterns for photosynthate indicate an initial tendency for upper leaves to feed fruits on the same orthostichy of a shoot but, as fruiting progresses and secondary thickening of vascular tissues permits lateral exchange of assimilates, fruits may develop the capacity to acquire assimilates from a wide range of photosynthetic surfaces. Vegetative organs of the shoot and the nodulated root also have access to this 'common' pool of assimilates, so competition between these structures and fruits is evident well into fruit ripening. The nodulated root of *L. albus* still commands about one-half of the plant's net photosynthate during the last few weeks of crop life, and of the remaining half of the net photosynthate, less than 40% is directly committed to seeds. With demands of such magnitude from roots, it is not surprising that seed yields are small. But, the benefits from maintaining a large root system are all too evident when one considers how successful lupins are in withstanding water and nutrient stresses in marginal dryland agriculture.

AGRONOMIC CONSIDERATIONS INCLUDING REQUIREMENTS FOR WATER AND MINERALS

Cultivated lupins are cool temperate species generally requiring a five-month growing season free from water stress and with mean monthly maximum temperatures within the range 15°-25°C [45].

Climatic requirements differ between the species, probably reflecting the basic adaptations of the progenitor types [45]. *Lupinus angustifolius*, which is generally the most tolerant of frost [45], germinates and grows more rapidly at cooler temperatures than the other species [121]. *Lupinus cosentinii* tends to be better adapted to milder growing conditions, is very frost sensitive, and slow to germinate at cold temperatures [121]. *Lupinus luteus* and *L. albus* are intermediate between *L. angustifolius* and *L. cosentinii*, both in temperature response and frost tolerance [121].

Lupins are poor competitors with weeds during early growth, particularly when seedling growth is slow at cool temperatures [58]. Early weed infestations can result in poor stand establishment, or even crop failure [2], so

717

there is incentive to sow early, even into dry soil in anti-
cipation of the onset of seasonal rains. The most serious
weeds of lupins in mainland Australia are annual ryegrass
(*Lolium rigidum*) and capeweed (*Arctotheca calendula*); in
Tasmania [38] and New Zealand [68] other broad-leafed weeds
assume greater importance. Control of weeds is best achieved
by early sowing; reduction of the weed seed residual from
previous crops can be accomplished by heavy grazing and
burning of stubble. Useful pre-emergence herbicides include
simazine, divron and trifluralin. A good post-emergence
herbicide is 'Hoegrass'. Future research should aim at developing broad-spectrum herbicides capable of total control
from emergence until the crop forms a closed leaf canopy
[40].

As with any seed crop, the components most likely to
influence economic yield are plant density, the numbers of
fruits on individual plants, the numbers of seeds contained
in each fruit, and the average weight of seeds. In general,
economic yields of seed crops improve with increasing plant
density until a plateau is reached, but studies on lupins
have recorded near maximum yields over quite wide ranges of
sowing densities. Thus, seed yields of *L. angustifolius* in
New Zealand [57,152] are maximal as density approaches 100
plants m^{-2}, whereas at other locations maximum yields occur
at half this value [153]. Similarly, *L. albus* seed yields
may peak at between 50 and 93 plants m^{-2} in certain situations, but not until twice this density has been reached
elsewhere [65,85,86].

Differences in growth habit between species appear to
have marked effects on crop responses to density. For example, seed yields of strongly indeterminate varieties and
cultivars of *L. angustifolius* are generally maximal at much
smaller densities than in the semi-determinate *L. albus*
[65].

At any given density, seed yield is stongly correlated
with the number of fruits able to mature within a particular
season [26]. Late sowings of indeterminate cultivars, for

718

example, give small seed yields because fruits on the
higher order lateral shoots fail to mature [11,26,39,65,
110].

The number of ripe seeds formed within individual fruits
is relatively homeostatic, as abortion of seeds is not a
common event in lupins even under severe moisture or heat
stress [26,110]. However, mean seed weight can vary by as
much as 20% in *L. albus* and *L. angustifolius*, and is thus
an important element modulating seed yield between sites
and seasons [26,110]. Much of the variability in mean seed
weight in *L. angustifolius* results from variation in size
of seeds of the later-formed fruits on high order laterals
[110].

Lupins can be grown on a wide range of soils but are
markedly intolerant of waterlogged or extremely alkaline or
saline substrates [13,45,123,132,139,141]. Coarse-textured,
moderately acid to neutral soils (pH 5.0 to 7.0) seem most
suitable for most species [45], although *L. luteus* appears
to be best adapted to very acid conditions.

The species have been generally rated in order of decreasing tolerance of poor fertility, in the sequence: *L.
cosentinii = L. luteus > L. angustifolius > L. albus* [45].
The tolerance of lupins to particularly small concentrations
of phosphate may relate to the ability of certain species to
proliferate specialised proteoid-like root clusters under
mineral-deficient conditions [136]. *Lupinus luteus* is said to be
especially susceptible to iron and manganese deficiency,
L. angustifolius to cobalt deficiency and *L. cosentinii* to
a deficiency of molybdenum [45,56]. *Lupinus angustifolius* is also
apparently more vulnerable to potassium, cobalt and copper
deficiencies than either *L. luteus* or *L. cosentinii* [53,56],
while the split-seed syndrome caused by manganese deficiency
is most prevalent in 'sweet' cultivars of *L. angustifolius*
[111,142]. This disorder is characterised by small concentrations of manganese in phloem sap supplying seeds, despite
large concentrations of the element in leaves supplying
assimilates to the fruits [72]. The relatively substantial

requirements of *L. albus* for nutrients may explain why cultivation is largely restricted to fertile soils of heavy texture [9,10,48].

Pot studies of *L. albus* and *L. angustifolius* have shown broadly similar patterns of uptake and distribution of mineral elements to those recorded for other crop plants. However, some lupin species may accumulate certain elements to larger concentrations than do other members of the genus when grown under the same mineral regime. For example, greater accumulations of Ca, K, Mg, Fe and Cu and lesser accumulations of Na and Mn occur in *L. angustifolius* than in *L. albus* (see Table 16.4). Nevertheless, these two species show similar patterns of specialisation by vegetative organs in storage of specific nutrients, and similar ratings of elements in relation to the efficiency of mobilisation from vegetative parts to seeds during fruiting [70,71].

Field studies on the accumulation of mineral elements provide evidence of significant differences between species grown at the one site and, within a species, when grown on different soil types [51,54,70,92,142].

The amounts of mineral elements absorbed by a crop will obviously depend on its productivity of dry-matter and on the availability of specific nutrients in the soil in which it is rooted. Element concentrations in dry-matter of *L. albus* and *L. angustifolius* given in Table 16.4 provide a basis for estimating the likely uptake of nutrients by lupin crops, provided that information is available on yield of crop dry-matter at harvest.

The water requirements of field-grown lupins have been poorly investigated. However, recent studies in Western Australia suggest that a lupin crop transpires considerably more of the intercepted rainfall than does a pasture of subterranean clover. Thus, a cereal/lupin rotation is likely to reduce percolation of rain to groundwater more than a cereal/clover pasture rotation, and so lessen salinity problems caused by capillary translocation of subterranean salt

Table 16.4 Mineral concentrations in dry-matter of *L. albus* and *L. angustifolius* grown in mineral-sufficient conditions in sand culture [71]

Element	Whole plant at beginning of fruiting*		Non-reproductive parts at plant maturity+		Fruits at plant maturity+	
	L. albus	*L. angustifolius*	*L. albus*	*L. angustifolius*	*L. albus*	*L. angustifolius*
	Major elements (mg g dry wt^{-1})					
Nitrogen	20.4	20.9	8.3	11.9	32.3	27.3
Potassium	8.8	13.7	4.4	10.4	11.8	14.7
Phosphorus	4.3	4.7	1.7	2.9	7.0	6.1
Calcium	4.6	15.2	5.1	27.5	4.5	6.4
Sodium	5.4	2.4	6.9	6.1	4.5	1.1
Magnesium	1.9	3.4	1.2	3.8	2.6	3.1
	Minor elements (μg g dry wt^{-1})					
Iron	105.8	171.5	143.6	274.6	56.4	90.9
Manganese	59.6	43.3	58.5	73.1	53.1	24.1
Zinc	34.2	38.2	25.6	43.1	42.6	39.1
Copper	9.1	12.3	9.2	16.9	9.5	11.4

*At end of exponential phase of dry-matter increase in whole plants (after 120 days in *L. albus* and 90 days in *L. angustifolius*).

+At final harvest (after 158 days in *L. albus* and 124 days in *L. angustifolius*).

to the soil surface [87].

Judging from the behaviour of potted plants of L. albus, water-use efficiency (transpiration ratio) for whole plants varies during growth within the range 210-480 g H_2O g dry-matter^{-1} produced [104]. Assuming a mean value of 350 g H_2O g dry-matter^{-1} and a crop production of 10 t dry-matter ha^{-1}, the volume of water utilised by the crop over a complete cycle of growth would be of the order of 3500 kl ha^{-1}.

Many aspects still need to be investigated of the water relations of fruiting plants, especially of the factors inducing leaf shedding following water stress. Fruits themselves use water remarkably efficiently [94,106], transpiring only 20-25 g H_2O g dry-matter accumulated^{-1}. Almost leafless plants translocating previously accumulated assimilates to their fruits during late crop growth are therefore likely to be extremely economical of water.

NODULATION, NITROGEN FIXATION AND NITROGEN METABOLISM

Artificial inoculation of lupin seeds with Rhizobium is not widely practised, except where crops are sown into areas in which lupins have not previously been grown or where appropriate strains are unlikely to be present. Indigenous strains of Rhizobium of the appropriate cross-inoculating specificity normally promote effective symbiosis, as gauged by extent of nodulation, haemoglobin pigmentation of nodules, and rates of acetylene-reducing activities by nodules [30,62]. The strains of Rhizobium isolated from lupin nodules are usually slow growing and tolerant of acidic conditions in culture on artificial media.

Nodulation of the seedling tap root occurs promptly under normal conditions and there is no evidence of a 'nitrogen hunger' stage in seedling growth - as evident, for instance, in some species of Vigna and Phaseolus (see Chapter 4).

Studies monitoring nitrogen fixation during the life of field-grown crops of L. angustifolius have suggested that fixation (as assessed by C_2H_2 reduction) starts some five weeks after sowing and continues almost until reproductive maturity [30,137]. Maximum specific fixation rates are achieved during mid-vegetative growth, the greatest total fixation rates per plant occurring at or close to flowering. Unlike the situation in many other legumes, C_2H_2-reducing activity shows little or no diurnal rhythm, and nodulated roots continue to reduce acetylene for many days after shoots have been removed [62,137]. These findings suggest that the root and nodules carry ample reserves of available carbohydrate.

Marked adverse effects of waterlogging [30] and cobalt deficiency [15] have been recorded for the symbiosis of lupins.

In view of the extremely poor nitrogen status of many of the infertile deep sands on which lupins are often grown it is likely that most crops depend heavily on root nodules for their nitrogen. However, definitive studies on the proportion of total plant nitrogen coming from symbiotic fixation have yet to be undertaken under field conditions, but, judging from comparisons of the nitrogen accumulation of nodule-dependent and NO_3-dependent *L. albus* plants in pot culture, symbiosis can provide nitrogen at almost equal rates, and at almost as little cost in terms of photosynthate, as when plants receive optimal amounts of nitrate [4 104].

An important aspect of the nitrogen economy of both *L. albus* and *L. angustifolius* is that only one-quarter, or less, of the total nitrogen fixed during growth comes from nodular activity before flowering [30,71,100]. Symbiotic nitrogen fixation therefore makes heavy demands upon the plant for photosynthate during flowering and fruiting [95, 100]. Nodules require, on average, 17 g carbohydrate g N_2 fixed^{-1} and at times they consume as much as 15-20% of the net photosynthate currently available from the plant [95, 98].

Balance sheets for nitrogen in field- or pot-grown

lupins have revealed great efficiency (70-80%) in withdrawal of nitrogen from vegetative parts of the shoot to fruits during seed filling [30,71], despite evidence in some studies of net losses of nitrogen from fruiting plants [55]. Harvest indices for nitrogen vary from 0.61 to 0.84 depending on variety or cultivar and species [30,71,100]. These values compare favourably with those for other grain legumes [105].

Lupinus albus and *L. angustifolius* offer classic examples of species with a root-dominated nitrate reductase system. Indeed, substantial spillover of nitrate to shoots in *L. albus* occurs only under concentrations of nitrate in excess of 200 ppm NO_3-N, a situation unlikely to obtain for long, if at all, during growth of a crop. Under such nitrate regimes, stems and petioles, not the photosynthesising leaflets, are responsible for reducing the nitrate received from the root [4]. Two amides, asparagine and glutamine, are the principal organic nitrogenous solutes transported in xylem and phloem of plants dependent on nitrate or symbiotically-derived nitrogen [4,95]. Synthesis of other amino acids occurs predominantly in young leaves, shoot apices and fruits following enzymatic breakdown of amides in these organs [96]. Complex interchanges of solutes between organs via xylem and phloem are involved in the provision of carbon and nitrogen to root and shoot apices [103] as well as to developing seeds [95,106].

DISEASES, PESTS AND LUPINOSIS

The following disease agencies are especially serious for lupin crops

Fungal diseases

Wilts. Wilting diseases can be due to *Fusarium oxysporum* subspecies *lupini* or *Verticillium albo-atrum*. Varietal resistance to certain physiological races of *F. oxysporum*

subspecies *lupini* are known, but best overall control is achieved by appropriate crop rotations.

Brown leaf spot (*Pleiochaeta setosa*).   This fungus is particularly aggressive on winter-grown crops. Prevalent under cool, moist conditions, it is transmitted mainly from residues of a previous crop and, to a lesser extent, as infected seeds. Brown leaf spot infects all above-ground parts of the plant, forming dark brown to purple necrotic spots which increase in size and number with intensity of infection. Infected leaflets are shed prematurely. *Lupinus albus* is the most susceptible species with *L. angustifolius*, *L. luteus* and *L. mutabilis* relatively less affected. *Lupinus cosentinii* is fairly resistant [45]. Control measures include removal of crop residues by grazing, wider rotations and early sowing [157].

Mildew (*Erysiphe* spp.).   This common fungal disease forms wefts of mycelium on leaves, but seldom causes severe necrosis. Debilitation by *Erysiphe* can cause severe losses of *L. angustifolius*, especially in South Africa [143].

Grey leaf spot (*Stemphylium vesicarum*).   Symptoms of this disease are grey or grey-brown necrotic spots, each surrounded by a dark brown halo [157]. Stems, fruits and leaves are attacked. The fungus is particularly aggressive on *L. angustifolius* in Western Australia, where it can be sufficiently severe to kill large numbers of plants in a crop. Rotation, crop hygiene and restriction of cultivation to well-drained soils are the best control measures currently available, but resistant forms of *L. angustifolius* (such as Ritchie, Illyarie and Tifblue 78) offer promise [157].

Seedling diseases.   Fungal pathogens causing 'damping-off' of lupin seedlings include species of *Pythium* and *Phytophthora*, *Botrytis cinerea* and *Rhizoctonia solani*. Their effects are most noticeable at relatively cool temperatures. Seed dressing with fungicides can give partial control.

Other fungal diseases.   Fungal pathogens locally important
in lupins include anthracnose (*Glomerata cingulata*), to
which a source of resistance is now available, stem canker
(*Sclerotium rolfsii*) and stem rot (*Sclerotinia sclerotiorum*
and *S. minor*). The latter is very important on *L. angusti-*
folius in Western Australia [157].

Bean Yellow Mosaic Virus (BYMV)

Many seed stocks of crop species are infected with this
virus, but seed transmission amongst lupins has been demonstrated only for *L. luteus*. Infected vegetative plants initially show wilting of apical leaves and stem tips; the
whole plant then turns black and dies [46]. Late infections
cause dwarfing of axillary branches and blackening of roots.
Virus spread is confined largely to 'sweet' genotypes since
the sap of alkaloid-rich types is toxic to sucking insects
which may be vectors of the virus.

Insect pests

Budworm moths (*Heliothis* sp.) are a most destructive pest
of winter-grown lupin crops in Australia, South Africa,
Portugal and the south-eastern USA [45]. Attacks normally
occur in spring when young caterpillars burrow into young
fruits and then feed on developing seeds. Damage is confined
to soft immature fruits, though heavy infestations may still
devastate a maturing crop. Losses can be minimised by early
sowing, use of early-maturing varieties, and by aerial
spraying or misting with insecticides.

Other significant arthropod pests include the red-legged
earth mites (*Halotydeus destructor*) and the lucerne fleas
(*Smynthurus viridis*), both of which are especially damaging
to young plants of alkaloid-free cultivars.

Aphid attacks on 'sweet' varieties or cultivars are

another common cause of crop losses. The aphid *Macrosiphon albifrom* Essig., first recorded in the UK in 1981, has quickly become widespread on both 'sweet' and 'bitter' forms of *L. albus* and *L. mutabilis*.

Lupinosis

Experience with this important disorder derives largely from Australia [36,49], where it mainly afflicts sheep grazing mature crops of lupin. Symptoms of acute attack are enlarged 'fatty' livers, swollen gall bladders and kidney damage. Most severely affected animals eventually die. In milder cases, principal symptoms are loss of appetite and body weight. If detected sufficiently early, affected animals will recover if they are withdrawn from lupin pasturage. Cattle and horses also suffer lupinosis.

Although certain epidemiological features of lupinosis remain unclear, primary causality is due to ingestion of toxins produced by the fungus *Phomopsis leptostromiformis*. This pathogen infects green tissues of both 'sweet' and 'bitter' genotypes, but can also persist and proliferate saprophytically on the stubble, particularly in warm humid conditions. The fungus usually causes little damage to its lupin host.

Differences exist between lupin species in susceptibility to *Phomopsis* infection. *Lupinus luteus* and *L. cosentinii* are usually, but not always, readily infected, and therefore more likely to be responsible for lupinosis than *L. angustifolius* or *L. albus* [3,156]. The extent of susceptibility of *L. mutabilis* to the fungus is unclear. Incidences of lupinosis involving *L. albus* have been infrequent, but sufficiently unpredictable for farmers to need to take care when committing stock to grazing the species.

Control measures centre mainly on daily monitoring of animal stock grazing lupin stubble, and shifting them to alternative pasture immediately symptoms appear. Other measures include the use of rotational grazing and lupin/cereal mixtures to reduce lupin intake, and reduction of stubble residues by raking and burning. Breeding for *Phomopsis*-resistant lupin cultivars is in progress in Western Australia using *L. albus* [1].

YIELD, COMPOSITION, QUALITY AND NUTRITIONAL VALUE OF SEEDS

Lupin seeds are richer in protein than most other grain legumes, with the notable exception of soyabeans (see Chapter 3). So, if seed yields are expressed as biomass of protein ha^{-1}, lupin crops, with an average production (1965-74) of 286 kg seed protein ha^{-1}, emerge as second to soyabeans (499 kg seed protein ha^{-1}) amongst the eleven categories of grain legumes whose production is recorded by the FAO [94].

Reported values for seed composition vary widely but the overall ranking of lupin species in respect of oil and protein (see Table 16.5) is fairly constant [69,75, 146]. The protein in whole seeds varies from a mean value of about 31% in *L. angustifolius* to 44% in *L. mutabilis*. The lipid concentration (mainly fatty acids) exceeds 20% in some samples of *L. mutabilis*, comparing favourably with that in soyabeans. *Lupinus albus* is the only other crop lupin that possesses substantial amounts of seed oil (10-14%).

Table 16.5 The composition of whole lupin seeds (range, % dry-matter) [69,75]

Species	Crude protein	Ether extract (lipid)	Nitrogen-free extract	Crude fibre
L. albus	34–45	10–15	35–46	3–10
L. angustifolius	28–38	5–7	37–46	13–17
L. luteus	36–48	4–7	29–39	15–18
L. cosentinii	28–40	3–4	34–45	19–22
L. mutabilis	32–46	13–23	25–26	7–11

Seeds of *L. luteus*, *L. cosentinii* and *L. angustifolius* contain relatively large amounts of fibre (Table 16.5). The seeds are characterised by very small amounts of starch, varying from zero in *L. luteus* and some cultivars of *L. albus*, to 7% in *L. angustifolius* [12]. The non-pectic poly-saccharides of seeds are mainly wall-bound in cotyledon cells and composed of galactose, arabinose and galacturonic acid residues with little cellulose [12]. The wall reserves are utilised as an energy source when the seeds germinate [12].

Lupin seed protein resembles that of other grain legumes in having sub-optimal concentrations of the sulphur-containing amino acids, methionine and cystine (Table 16.6). Percentage composition of these two amino acids is also lowered substantially when lupins are grown on sulphur-deficient soils [7].

Methionine concentration is small in all species, attaining values greater than 1 g 16g N^{-1} only in *L. mutabilis*. Values recorded for cystine (4.8 g 16g N^{-1} in *L. luteus* and 3.8 g 16g N^{-1} in *L. cosentinii*) are outstandingly large for grain legumes and exceed the optimal combined value (3.54 g 16g N^{-1}) prescribed for methionine plus cystine by the FAO (1974) [32]. Given that cystine and methionine are fully interconvertible alternative sources of sulphur-amino acid to most animals, *L. luteus* and *L. cosentinii* would appear to provide valuable potential material for genetic programmes aimed at improving the dietary value of seed protein.

The major protein in lupin seed is termed conglutin and is of the type generally classified as globulin [6]. This protein comprises about 80% of the total insoluble nitrogen in seeds, the remaining protein being largely of the albumin type. Electrophoresis shows conglutin to comprise three separate components, α, β and γ, which occur in *L. angustifolius* [6,7,41] in the respective molar ratios 40:50:10. Two additional components, conglutin ∂ and fraction 9b, have been identified in *L. albus* [7,21,22].

Conglutin α and ∂ are deemed equivalent to the legumin fraction, and conglutin β to the vicilin fraction of seeds of *Vicia faba* and *Pisum sativum*. Ultracentrifugation reveals

729

Table 16.6 Range of amino acid composition (g 16g N^{-1}) of lupin protein [32,69,75]

Species	Iso-leucine	Leucine	Lysine	Methionine	Cystine	Phenyl-alanine	Threonine	Tryptophan	Valine
Lupinus albus	4-6	7-9	5-6	0.3-0.5	1-2	4-5	4-5	1-2	4-5
Lupinus angustifolius	4-5	6-7	4-6	0.4-1.0	0.5-0.9	3-4	3-4	1-2	3-5
Lupinus luteus	3-5	7-10	3-6	0.3-1.0	0.7-4.8	3-5	2-4	0.7-1.2	3-5
Lupinus cosentinii	3.0	6-7	4-5	0.6-0.8	3-4	3-4	3	-	3
Lupinus mutabilis	4.0	7.0	5	0.4-1.4	1.4	3-4	3-4	1	4
FAO 'Ideal Protein'	4.0	7.04	5.44		3.52	6.08*	4.0	0.96	4.96

*Value includes tyrosine

a 7S fraction (equivalent to conglutin β), an 11S fraction
(comprising conglutin α and ϑ), and a small molecular weight
(2S) fraction equivalent to the 9b protein. The latter is
gauged to be distinct from either legumin or vicilin [125].

Concentrations of the sulphur-amino acids are large in
the conglutin γ of lupin seeds, relatively small in conglutin α (legumin), and absent from conglutin β (vicilin) [6,
7]. The possibility therefore exists for improving the biological value of protein by altering the genetic regulation
of the synthesis of these conglutin fractions in favour of
those rich in sulphur [6,7] (and see Chapter 3).

Fatty acid composition of the two species with appreciable amounts of oil in their seeds (*L. albus* and *L. mutabilis*) is recorded in Table 16.7. Both species have oleic
acid (C18:1) as the predominant fatty acid, but differ significantly in relative concentration of the extremely unsaturated fatty acid, linolenic acid (C18:3). Since the
latter causes instability of extracted oil, the smaller concentrations found in *L. mutabilis* seeds are of particular
importance. In view of pathological effects attributed to
diets rich in erucic acid (C22:1), absence of this compound
in *L. mutabilis* would be an important consideration when
evaluating the species as a potential oil crop.

Table 16.7 Fatty acid composition of lupin oil (%) [69,75]

Species	*Fatty acid*						
	16:0	18:0	18:1	18:2	18:3	22:1	Other
L. albus	8	2	53–61	17–23	3–9	2–7	8
L. mutabilis	9–13	2–7	44–56	28–30	2	0	1

Of the many antinutritional factors recorded for cultivated grain legumes [81], the presence of large concentrations of alkaloids is the principal problem likely to be
associated with use of 'bitter' lupin seeds [149]. There is
also the more general problem in both 'bitter' and 'sweet'
genotypes of indigestibility of seeds due to their large

fibre concentrations and the presence within them of unsuitable forms of carbohydrate. Small concentrations of
starch, for instance, give small potential energy values to
certain monogastric animals, while large concentrations of
stachyose and verbascose (especially in seeds of *L. albus*)
are undesirable in view of the implication of these oligosaccharides as flatus-inducing materials [82].

Small amounts of cyanophoric compounds, haemagglutinins
and antitryptic activity have been detected in lupin seeds
[128] but, generally, at too small concentrations to be considered of antinutritional significance. The same probably
applies to the small amounts of sapogenins in seeds (less
than 1.7% in seed dry-matter [74] and also to flavonoids
[23] (less than 0.37 mg 100 g seed^{-1}).

UTILISATION OF SEEDS AS FOOD

Because the concentration and quality of protein in crop
lupin seeds are essentially similar to those of soyabeans,
the greatest economic prospect for lupin seeds is as a
protein-rich substitute for soyabeans, particularly for the
intensive feeding of pigs, poultry and ruminants. An export
trade based on this usage is being established in Australia
[14].

Energy values of white and narrow-leafed lupins in pig
rations exceed 17 Mj kg^{-1} which compare very favourably with
soyabean meal, but detrimental effects have been reported
when a diet contains more than 50% lupin seed [49]. Below
this amount, and given supplementation of limiting amino
acids, the responses have been satisfactory. Similarly, with
adult poultry, despite a small energy value due to non-
utilisable forms of carbohydrate, 'sweet' cultivars of both
white and narrow-leafed lupins have given adequate perfor-
mances in rations containing up to 20% lupin seed [49],
though supplementation with methionine and tryptophan enhanced production in some trials. By contrast, young chicks,
during the first week after hatching, are very sensitive to

lupin-containing diets, especially those based on 'bitter' species [79]. Heating to improve feed efficiency in young chicks has given variable results, mostly negative. However, after the first week, rations containing as much as 35% lupin meal effect a satisfactory weight gain and conversion efficiency.

Good results have been recorded with ruminants (including milk cows) fed on concentrates containing up to 30% of lupin. However, lupin protein is poorly protected against microbial degradation in the rumen and so its biological efficiency is small. On the other hand, digestibility of the crude fibre is good [46,49]. The safe level to which 'sweet' lupin may be incorporated into a feed ration often ultimately depends on the concentration of alkaloid in the currently used stocks of lupin seeds. Since alkaloid concentrations of sweet cultivars may vary with season and site, accurate assays of seeds need to be undertaken frequently and feed ration formulae adjusted accordingly.

Use of 'sweet' lupins as food for humans is an economically attractive proposition, the potential of which has yet to be fully explored. The traditional use of seeds for food in subsistence agriculture in southern Europe, Africa and South America suggests that even alkaloid-containing species could be eaten after proper debitterisation on a local or commercial scale. Interest has developed recently in Europe, Peru and Chile in the potential of alkaloid-free lupins as direct sources of protein isolates [76,113] while kernel flour or hydrolysed lupin protein have also been successfully substituted for soyaflour as a meat-extender in ground meat products such as sausages, canned meats and hamburgers [34,76]. Nevertheless, medical health authorities in some countries have expressed concern about the possible long-term toxic effects from continued ingestion of the small amounts of alkaloids (usually less than 0.1% of seed dry weight) present in seed flour of 'sweet' genotypes [14]. Lupin seeds have also proved equally effective to those of soyabeans for making traditional Asian foods such as 'tempeh' [78].

733

CYTOGENETICS AND PRINCIPAL BREEDING OBJECTIVES TOWARDS
ACHIEVING LARGER AND MORE STABLE SEED YIELDS

Although frequently visited by pollinating insects, crop
lupins are predominantly self-fertilised [19], with no evidence of incompatibility mechanisms or inbreeding depression. Inter-crossing at the intraspecific level occurs in
the field, but at varying proportions between species and
locations. *Lupinus luteus* is reported to have the largest degree
of cross-pollination, *L. cosentinii* the smallest [47,50].
Records of outcrossing in *L. albus* range from 3.7-8.2%,
while *L. mutabilis* produces 4-11% naturally outcrossed progeny in its native Peru [8,24]. Some introductions of this
species have shown 17-35% incidence of male sterility [91].
A small measure of cross-pollination by insects and other
arthropods has been recorded for field crops of *L. angustifolius* [35,80,119].

The large chromosome numbers of crop lupins (see Table
16.1) indicate polyploidy, yet genetic segregation and
chromosome pairing suggest the species to be functional
diploids. The reported survival of a haploid of *L. luteus*
and the univalent disposition of its twenty-six chromosomes
strongly suggest an amphidiploid origin, although the putative diploid parents have yet to be named.

Genetic isolation is virtually complete between the crop
lupin species, as it is also between these cultivated species and wild undomesticated members of the genus, so plant
breeders are likely to have access to only limited ranges
of variability for use in selection studies and mutation
programmes. Nevertheless, based on similarities in alkaloid
patterns [151], some affinities appear to exist between
certain of the Mediterranean species. Likewise, pollen tube
growth and carpellary stimulation in crosses between *L.
mutabilis* and European species suggest that gene exchange
might prove possible even between Old and New World species.
Partial compatibility is also suggested from the reports of
sterile hybrids between *L. albus* and *L. angustifolius*,

between female *L. luteus* and male *L. rothmaleri* (Klinkowskii)
[77], and from the production of several F_1 hybrids in
crosses between female *L. luteus* and male *L. hispanicus*
(Boiss and Reuter). In particular, possibilities might exist
for using *L. rothmaleri* to extend the gene pool of *L. luteus*
towards improved cold tolerance.

There are many ways in which lupins might be bred for
improved seed yields and the objectives to be pursued will
obviously vary according to the species and the geographical
region in which it is to be grown. Listed below are some
attributes which might be suggested to merit special attention in future selection programmes.

Improved resistance to diseases

Foremost among the most damaging diseases currently affecting lupins are the various damping-off diseases of seedlings, other fungal diseases such as *Pleiochaeta setosa*,
powdery mildew, grey leaf spot (*Stemphylium vesicarium*) and
Bean Yellow Mosaic Virus. In all cases, positive selection
programmes for resistance are warranted in one or more vulnerable lupin species. Experience with other crops has shown
that where resistance is controlled by a few major genes,
protection is ephemeral due to the development of new physiological races of the pathogen with altered patterns of
pathogenicity. Durable resistances should therefore be based
on polygenic control.

Reduction in or abolition of vernalisation requirements

A requirement for vernalisation in a cultivar is obviously
an undesirable characteristic if it leads to variable rates
of development to flowering in spring-sown crops at various
sites and in different seasons [148]. Elimination of the
requirement would ensure more predictable phasic development
and earlier crop maturity in such circumstances.

Decreased sensitivity to cold and frost

Winter hardiness is an obviously important character for
those geographical areas where crops are winter-sown. Its
expression should involve resistance to seedling diseases
as well as tolerance of frost and desiccating cold winds.
Genes for cold resistance are already available in the
American cultivars Tifblue 78 (*L. angustifolius*) and Tif-
white 78 (*L. albus*) [144].

Reduced fruit and seed loss prior to harvest

Seed yields of lupins are often small and unreliable because
of excessive flower abscission from the mid and upper positions of all inflorescences. Flower drop is particularly
yield-limiting in certain cultivars of *L. albus* and *L. angustifolius*, but genotypes of *L. mutabilis* and *L. cosentinii*
normally set fruits along the length of their racemes. Identification of genotypes with increased capacity for fruit
retention on the primary inflorescences would constitute a
major improvement in *L. angustifolius* and *L. albus*, especially for use in areas where the season is short and terminated abruptly by adverse temperatures or water stress.
Selections might be made on the basis of later, less vigorous outgrowth of lateral shoots following flowering of the
main stem inflorescence.

Reduced fruit maturation time and optimisation of seed size

Expansion of lupin cultivation for seed production into more
northerly latitudes will demand cultivars which flower early
and ripen their fruits under relatively cool and damp conditions. This is particularly important in *L. albus* and *L.
angustifolius* which both require a protracted period of
development between seed set and fruit maturation [16]. The
breeding of genotypes with thin fruit walls capable of desiccating during early fruiting would be extremely desirable

in view of the proneness of slow-ripening fleshy fruits to fungal attack. Selections of this nature have already been made in *Phaseolus vulgaris* when developing this species as a dry seed crop (and see Chapter 10).

Seed size in *L. albus* is extremely variable, but most currently-available cultivars are relatively large-seeded (Table 16.1). Since large seeds add materially to costs of sowing, selection for smaller unit size (e.g. not greater than 150 g 1000 seeds^{-1}) would be advantageous. Seed size is not a problem in other lupin crops.

Improved seed quality

Attempts to improve either oil or protein concentrations in the embryos of grain legumes have been disappointing due to the strong negative correlations between these characters [12,59]. However, considerable scope does exist in *L. albus*, *L. angustifolius*, *L. cosentinii* and *L. luteus* for reducing the proportion of testa in the seed, in which case fibre content would be reduced and oil and protein concentrations increased proportionately.

Increased oil concentration is a particularly important objective in *L. albus* and *L. mutabilis*, were these species to be developed for production of vegetable oil and protein concentrates. Development of alkaloid-free forms of *L. mutabilis* is an urgent objective. The possibility of improving protein quality by selecting for seed protein fractions rich in limiting sulphur-containing amino acids has been considered earlier (and see Chapter 3).

Improved physiological efficiency in terms of conversion of plant photosynthate to seed yield

All too rarely is the possibility explored in crop improvement programmes of deliberately selecting for physiological features such as larger rates of photosynthate production, better economy of photosynthate use in vegetative growth,

and partitioning profiles for assimilates which favour
fruits rather than vegetative organs during later stages of
development. Table 16.8 lists some of the physiological
criteria which might be examined and provides relevant data
for two contrasting genotypes of *L. albus*, one an unimproved
indeterminate landrace used principally as a green manure
crop, the other selected from this parental stock on the
basis of large seed yield. Of the seven items listed, the
first three relate to efficiency of utilisation of net
photosynthate by roots and shoots. These features do not
differ significantly between genotypes and, indeed, might
not be amenable to improvement in the species as a whole.
The remaining four criteria (items 4-7) concern the conversion of photosynthate and fixed nitrogen into dry-matter and
protein in seeds. Each criterion differs greatly between
landrace and cultivar, due mainly, one would suspect, to the
selection in the improved cultivar of a more determinate
habit and, consequently, a lesser investment of carbon and
nitrogen in non-harvestable products such as woody stems
and roots. Most noticeable is the much better conversion
efficiency of photosynthate to seed protein in the improved
cultivar, this being reflected also in the much larger harvest index for nitrogen than in the parental landrace. It
remains to be seen how physiologically-based features of this
type vary between genotypes and how they might most easily be
assessed in the mass progeny of breeding programmes.

AVENUES OF COMMUNICATION BETWEEN RESEARCHERS

The International Lupin Association has organised an international lupin conference biennially since 1980. The third
conference was held at La Rochelle, France in June 1984. The
Association also produces a newsletter for distribution to
members; it summarises new literature on crop lupins and
updates information on germplasm resources. The Association's
address is: International Lupin Association, (Professor Luiz
Lopez Bellido, General Secretary), Avenida Menendez Pidal s/n,

Table 16.8 Functional economy of carbon and nitrogen of two genotypes of white lupin (*Lupinus albus*) [94,100, 104] (J.S. Pate, unpublished data)

Criterion	Parent landrace used as green manure crop	Cultivar selected from parental landrace for large seed yield	Comments
1. Translocate utilised by roots over whole growth cycle ($mg\ CH_2O\ mg\ N\ fixed^{-1}$)	55.0	48.1	Similar budgets for C in nodulated roots
2. Net photosynthate produced per unit protein in vegetative parts by end of flowering ($g\ CH_2O\ g\ protein^{-1}$)	24.7	27.1	Similar economies of photosynthate usage for protein synthesis during pre-fruiting stages of growth
3. Proportion of plant's net photosynthate lost in respiration by roots (day and night) and shoots (night); (data for the 10-day period before flowering)	45	46	
4. Harvest index for carbon	0.27	0.46	Vastly dissimilar economies of photosynthate usage in relation to final yield of seed due mainly to selection of determinate habit
5. Net photosynthate produced over growth cycle per unit seed dry-matter ($g\ CH_2O\ g\ seed\ dry\text{-}matter^{-1}$)	9.9	5.3	
6. Harvest index for nitrogen	0.61	0.84	
7. Net photosynthate produced over growth cycle per unit seed protein ($g\ CH_2O\ g\ seed\ protein^{-1}$)	31.0	16.8	

PO Box 3048, Cordoba, Spain.

There are twelve major international centres which carry substantial numbers of accessions of crop lupins. Populations of wild species are also grown or stored at some of these locations. The addresses of the centres and details of their collections are available from: IBPGR Executive Secretariat, Crop Genetics Resources Centre, Plant Production and Protection Division, FAO, Via delle Terme di Caracalla, 0100 Rome, Italy (and see Chapter 18).

LITERATURE CITED

1. Allen, J.G., Wood, P.M., Croker, K.P. and Hamblin, J. (1980), *Farmnote 80/80*. Western Australia Department of Agriculture.
2. Allen, J.M. (1978), *Australian Journal of Experimental Agriculture and Animal Husbandry* 17, 112-17.
3. Arnold, G.W., Hill, J.L., Maller, R.A., Wallace, S.R., Carbon, B.A., Nairn, M., Wood, P.M. and Weeldenburg, J. (1976), *Australian Journal of Agricultural Research* 27, 423-5.
4. Atkins, C.A., Pate, J.S. and Layzell, D.B. (1979), *Plant Physiology* 64, 1078-82.
5. Atkins, C.A., Pate, J.S. and Sharkey, P.J. (1975), *Plant Physiology* 56, 807-12.
6. Biddiscombe, E.F. (1975), *Journal of the Australian Institute of Agricultural Science* 41, 70-72.
7. Blagrove, R.J. and Gillespie, J.M. (1982), in Gross, R. and Bunting, E.S. (*eds*) *Agricultural and Nutritional Aspects of Lupins*. Eschborn, West Germany: Deutsche Gesellschaft für Technische Zusammenarbeit, 447-63.
8. Blanco Galdos, O. (1982), in Gross, R. and Bunting, E.S. (*eds*) *Agricultural and Nutritional Aspects of Lupins*. Eschborn, West Germany: Deutsche Gesellschaft für Technische Zusammenarbeit, 33-49.
9. Boundy, K.A. (1978), *Journal of Agriculture, Victoria* 76, 8-9.
10. Boundy, K.A. (1978), *Journal of Agriculture, Victoria* 76, 292-4.
11. Boundy, K.A., Reeves, T.G. and Brooke, H.D. (1982), *Australian Journal of Experimental Agriculture and Animal Husbandry* 22, 76-82.
12. Brillouet, J.M. (1984), in *Proceedings of the 3rd International Lupin Conference*. La Rochelle, France (in press).
13. Broue, P., Marshall, D.R. and Munday, J. (1976), *Australian Journal of Experimental Agriculture and Animal Husbandry* 16, 549-54.
14. Brown, N.F. and Curtis, F. (1982), *Journal of Agriculture, Western Australia* 23, 100-5.
15. Chatel, D.L., Robson, A.D., Gartrell, J.W. and Dilworth, M.J. (1978), *Australian Journal of Agricultural Research* 29, 1191-202.
16. Davies, S. and Williams, W. (1982) in Thompson, R. and Casey, R. (*eds*) *Perspectives for Peas and Lupins as Protein Crops*. World Crops: Production, Utilization and Description, Vol. 8. The Hague, Boston and London: Martinus Nijhoff, 87-93.
17. Decker, P., Webb, T.E. and Edwardson, J.R. (1953), *Soil and Crop*

Lupin (*Lupinus* spp.)

Science Society of Florida Proceedings <u>13</u>, 75-78.

18. de Mayolo, S.A. (1982), in Gross, R. and Bunting, E.S. (*eds*) *Agricultural and Nutritional Aspects of Lupins*. Eschborn, West Germany: Deutsche Gesellschaft für Technische Zusammenarbeit, 1-11.
19. Duke, J.A. (1981), *Handbook of Legumes of World Economic Importance*. New York: Plenum Press, 345.
20. Dukhanin, A.A. (1974), cited in *Field Crop Abstracts* <u>29</u>, No. 2999.
21. Duranti, M. and Cerletti, P. (1982), in Thompson, R. and Casey, R. (*eds*) *Perspectives for Peas and Lupins as Protein Crops*. World Crops: Production, Utilization and Description, Vol. 8. The Hague, Boston and London: Martinus Nijhoff, 227-40.
22. Duranti, M., Restani, P., Poniatowska, M. and Cerletti, P. (1981), *Phytochemistry* <u>20</u>, 2071-5.
23. El-Difrawi, E.A. and Hudson, B.J.F. (1979), *Journal of the Science of Food and Agriculture* <u>30</u>, 1168-70.
24. Faluyi, M.A. and Williams, W. (1981), *Zeitschrift für Pflanzenzuchtung* <u>87</u>, 233-9.
25. Farrington, P. (1972), MSc. (Agric.) thesis, University of Western Australia.
26. Farrington, P. (1974), *Australian Journal of Experimental Agriculture and Animal Husbandry* <u>14</u>, 539-46.
27. Farrington, P. (1976), *Australian Journal of Experimental Agriculture and Animal Husbandry* <u>16</u>, 387-93.
28. Farrington, P. (1979), PhD. thesis, University of Western Australia.
29. Farrington, P. and Gladstones, J.S. (1974), *Australian Journal of Experimental Agriculture and Animal Husbandry* <u>14</u>, 742-8.
30. Farrington, P., Greenwood, E.A.N., Titmanis, Z., Trinick, M.J. and Smith, D.W. (1977), *Australian Journal of Agricultural Research* <u>28</u>, 237-48.
31. Farrington, P. and Pate, J.S. (1981), *Australian Journal of Plant Physiology* <u>8</u>, 293-305.
32. Food and Agriculture Organisation (1974), *Report of the Joint FAO/ WHO Ad Hoc Expert Committee*. Nutrition Meetings Report Series No. 52. Rome: FAO.
33. Forbes, I. and Wells, H.D. (1966), *Proceedings of the 10th International Grassland Congress, Helsinki*. 708-11.
34. Fountain, R.D. (1974), 'Grain Pool of Western Australia'. Unpublished report, Perth, Western Australia.
35. Free, J.B. (1970), *Insect Pollination of Crops*. London: Academic Press.
36. Gardiner, M.R. (1975), *Journal of Agriculture, Western Australia* <u>16</u>, 26-30.
37. Gardner, C.A. and Elliott, H.G. (1929), *Journal of Agriculture, Western Australia* <u>6</u>, 414-20.
38. Garside, A.L. (1977), *Journal of Agriculture, Tasmania* <u>48</u>, 7-11.
39. Garside, A.L. (1979), *Australian Journal of Experimental Agriculture and Animal Husbandry* <u>19</u>, 64-71.
40. Gilbey, D.J. (1982), *Farmnote 35/81*. Western Australia Department of Agriculture.
41. Gillespie, J.M. and Blagrove, R.J. (1975), *Australian Journal of Plant Physiology* <u>2</u>, 29-39.
42. Gladstones, J.S. (1958), *Journal of the Royal Society of Western Australia* <u>41</u>, 29-33.
43. Gladstones, J.S. (1959), *Empire Journal of Experimental Agriculture* <u>27</u>, 333-42.

44. Gladstones, J.S. (1967), *Australian Journal of Experimental Agriculture and Animal Husbandry* 7, 360-66.
45. Gladstones, J.S. (1970), *Field Crop Abstracts* 23, 123-48.
46. Gladstones, J.S. (1972), *Bulletin 3834*. Western Australia Department of Agriculture.
47. Gladstones, J.S. (1974), *Technical Bulletin 26*. Western Australia Department of Agriculture.
48. Gladstones, J.S. (1976), *Journal of Agriculture, Western Australia* 17, 70-74.
49. Gladstones, J.S. (1984), in *Proceedings of the 3rd International Lupin Conference*. La Rochelle, France (in press).
50. Gladstones, J.S. (1980), in Summerfield, R.J. and Bunting, A.H. (eds) *Advances in Legume Science*. Kew, England: Royal Botanic Gardens, 603-11.
51. Gladstones, J.S. and Drover, D.P. (1962), *Australian Journal of Experimental Agriculture and Animal Husbandry* 2, 46-53.
52. Gladstones, J.S. and Hill, G.D. (1969), *Australian Journal of Experimental Agriculture and Animal Husbandry* 9, 213-20.
53. Gladstones, J.S., Leach, B.J., Drover, D.P. and Asher, C.J. (1964), *Australian Journal of Experimental Agriculture and Animal Husbandry* 4, 158-64.
54. Gladstones, J.S. and Loneragan, J.F. (1970), *Proceedings of the 11th International Grassland Congress, Surfers Paradise*. 350-4.
55. Gladstones, J.S. and Loneragan, J.F. (1975), *Australian Journal of Agricultural Research* 26, 103-12.
56. Gladstones, J.S., Loneragan, J.F. and Goodchild, N.A. (1977), *Australian Journal of Agricultural Research* 28, 619-28.
57. Goulden, D.S. (1976), *New Zealand Journal of Experimental Agriculture* 4, 181-4.
58. Greenwood, E.A.N., Farrington, P. and Beresford, J.D. (1975), *Australian Journal of Agricultural Research* 26, 497-510.
59. Gross, R. and Von Baer, E. (1982), *Plant Research and Development* 15, 54-65.
60. Hackbarth, J. and Husfeld, B. (1939), *The Sweet Lupin*. Berlin: Paul Parey.
61. Hagberg, A. and Josefsson, A. (1947), cited in *Plant Breeding Abstract* 18, No. 1661.
62. Halliday, J. (1975), PhD. thesis, University of Western Australia.
63. Halse, N.J. (1973), in *Proceedings of the 45th ANZAAS Congress, Section 13, Perth*. 58-60.
64. Harrison, J.E.M. and Williams, W. (1982), *Euphytica* 31, 357-67.
65. Herbert, S.J. (1977), *Proceedings of the Agronomy Society of New Zealand* 7, 69-73.
66. Herbert, S.J. (1979), *Annals of Botany* 43, 55-63.
67. Herbert, S.J. and Hill, G.D. (1978), *New Zealand Journal of Agricultural Research* 21, 475-81.
68. Herbert, S.J., Lucas, R.J. and Pownall, D.B. (1978), *New Zealand Journal of Experimental Agriculture* 6, 299-303.
69. Hill, G.D. (1977), *Nutrition Abstract and Review, Series B*, 47, 511-29.
70. Hocking, P.J. and Pate, J.S. (1977), *Annals of Botany* 41, 1259-78.
71. Hocking, P.J. and Pate, J.S. (1978), *Australian Journal of Agricultural Research* 29, 267-80.
72. Hocking, P.J., Pate, J.S., Wee, S.C. and McComb, A.J. (1977), *Annals of Botany* 41, 677-88.

Lupin (*Lupinus* spp.)

73. Hudson, A.W. (1934), *New Zealand Journal of Agriculture* 40, 346-51.
74. Hudson, B.J.F. and El-Difrawi, E.A. (1979), *Journal of Plant Foods* 3, 181-6.
75. Hudson, B.J.F., Fleetwood, J.G. and Zand-Moghaddam, A. (1976), *Plant Foods and Man* 2, 81-90.
76. Junge, I. (1973), *Publication 1, Anales Escuela da Ingenieria*. University Concepcion, Chile.
77. Kazmierski, T. (1982), in Gross, R. and Bunting, E.S. (*eds*) *Agricultural and Nutritional Aspects of Lupins*. Eschborn, West Germany: Deutsche Gesellschaft für Technische Zusammenarbeit, 51-68.
78. Kidby, D.K., McComb, J.R., Snowdon, R.L., Garcia-Webb, P. and Gladstones, J.S. (1977), *Proceedings of the UNEP/UNESCO/ICRO Symposium - Workshop on Indigenous Fermented Foods*. 5th International Conference on Global Aspects of Applied Microbiology, Bangkok.
79. Lacassagne, M. (1983), in Thompson, R. and Casey, R. (*eds*) *Perspectives for Peas and Lupins as Protein Crops*. World Crops: Production, Utilization and Description, Vol. 8. The Hague, Boston and London: Martinus Nijhoff, 365-71.
80. Langridge, D.F. and Goodman, R.D. (1977), *Australian Journal of Experimental Agriculture and Animal Husbandry* 17, 319-22.
81. Liener, I.E. (1980), in Summerfield, R.J. and Bunting, A.H. (*eds*) *Advances in Legume Science*. Kew, England: Royal Botanic Gardens, 157-77.
82. Macrae, R. and Zand-Moghaddam, A. (1978), *Journal of the Science of Food and Agriculture* 29, 1083-6.
83. Masefield, G.B. (1975), *Experimental Agriculture* 11, 113-18.
84. Mora, G.S. (1980), *Agro Sur* 8, 43-56.
85. Muntean, L. (1971), cited in *Field Crop Abstracts* 26, No. 6645.
86. Muntean, L. (1972), cited in *Field Crop Abstracts* 28, No. 6584.
87. Nulsen, R.A. (1982), *Journal of the Department of Agriculture, Western Australia* 23, 91.
88. Oldershaw, A.W. (1920), *Agriculture* 26, 982-91.
89. O'Neill, T.B. (1961), *Botanical Gazette* 123, 1-9.
90. Ozanne, P.G., Asher, C.J. and Kirton, D.J. (1965), *Australian Journal of Agricultural Research* 16, 785-800.
91. Pakendorf, K.W. (1974), *Zeitschrift für Pflanzenzuchtung* 72, 152-9.
92. Pakendorf, K.W., Coetzer, F.J. and Van Schalkwyk, D.J. (1970), *Zeitschrift für Pflanzenzuchtung* 63, 125-36.
93. Pate, J.S. (1977), in Hardy, R.W.F. and Silver, W.S. (*eds*) *A Treatise of Dinitrogen Fixation*. Section III - Biology. New York: Wiley, 473-517.
94. Pate, J.S. (1979), *Journal of the Royal Society of Western Australia* 62, 83-94.
95. Pate, J.S. (1984), in *Proceedings of the 3rd International Lupin Conference*. La Rochelle, France (in press).
96. Pate, J.S., Atkins, C.A., Herridge, D.F. and Layzell, D.B. (1981), *Plant Physiology* 67, 37-42.
97. Pate, J.S., Atkins, C.A. and Perry, M.W. (1980), *Australian Journal of Plant Physiology* 7, 283-97.
98. Pate, J.S., Atkins, C.A. and Rainbird, R.M. (1981), in Gibson, A.H. and Newton, W.E. (*eds*) *Current Perspectives in Nitrogen Fixation*. Australia: Griffin Press, 105-16.
99. Pate, J.S. and Farrington, P. (1981), *Australian Journal of Plant Physiology* 8, 307-18.
100. Pate, J.S. and Herridge, D.F. (1978), *Journal of Experimental Botany*

29, 401-12.
101. Pate, J.S. and Hocking, P.J. (1978), *Annals of Botany* 42, 911-21.
102. Pate, J.S. and Kuo, J. (1981), in Polhill, R.M. and Raven, P.H. (eds) *Advances in Legume Systematics*. Kew, England: Royal Botanic Gardens, 903-12.
103. Pate, J.S. and Layzell, D.B. (1981), in Bewley, J.D. (ed.) *Nitrogen and Carbon Metabolism*. The Hague: Martinus Nijhoff, 94-134.
104. Pate, J.S., Layzell, D.B. and Atkins, C.A. (1980), *Bericht der Deutschen Botanischen Gesellschaft* 93, 243-55.
105. Pate, J.S. and Minchin, F.R. (1980), in Summerfield, R.J. and Bunting, A.H. (eds) *Advances in Legume Science*. Kew, England: Royal Botanic Gardens, 105-14.
106. Pate, J.S., Sharkey, P.J. and Atkins, C.A. (1977), *Plant Physiology* 59, 506-10.
107. Pate, J.S., Sharkey, P.J. and Lewis, O.A.M. (1974), *Planta* 120, 229-43.
108. Pazy, B. (1982), PhD. thesis, Hebrew University of Jerusalem.
109. Pazy, B., Heyn, C.C., Herrnstadt, I. and Plitmann, V. (1977), *Israel Journal of Botany* 26, 115-27.
110. Perry, M.W. (1975), *Australian Journal of Agricultural Research* 26, 809-18.
111. Perry, M.W. and Gartrell, J.W. (1976), *Journal of Agriculture, Western Australia* 17, 20-25.
112. Perry, M.W. and Poole, M.L. (1975), *Australian Journal of Agricultural Research* 26, 81-91.
113. Pompei, C. (1983), in Thompson, R. and Casey, R. (eds) *Perspectives for Peas and Lupins as Protein Crops*. World Crops: Production, Utilization and Description, Vol. 8. The Hague, Boston and London: Martinus Nijhoff, 344-57.
114. Porter, N.G. (1977), *Physiologia Plantarum* 40, 50-54.
115. Porter, N.G. (1982), *Australian Journal of Agricultural Research* 33, 957-65.
116. Porter, N.G. and Van Steveninck, R.F.M. (1966), *Life Science* 5, 2301-8.
117. Preller, J.H. (1949), *Farming in South Africa* 24, 25-9.
118. Puerta, J. (1982), in Gross, R. and Bunting, E.S. (eds) *Agricultural and Nutritional Aspects of Lupins*. Eschborn, West Germany: Deutsche Gesellschaft für Technische Zusammenarbeit, 856-63.
119. Quinlivan, B.J. (1974), *Journal of Agriculture, Western Australia* 15, 54.
120. Rahman, M.S. and Gladstones, J.S. (1972), *Australian Journal of Experimental Agriculture and Animal Husbandry* 12, 638-45.
121. Rahman, M.S. and Gladstones, J.S. (1973), *Journal of the Australian Institute of Agricultural Science* 39, 259-62.
122. Rahman, M.S. and Gladstones, J.S. (1974), *Australian Journal of Experimental Agriculture and Animal Husbandry* 14, 205-13.
123. Reeves, T.G. (1974), *Journal of Agriculture, Victoria* 72, 285-9.
124. Reeves, T.G., Boundy, K.A. and Brooke, H.D. (1977), *Australian Journal of Experimental Agriculture and Animal Husbandry* 17, 637-44.
125. Restani, P., Duranti, M., Cerletti, P. and Simonetti, P. (1981), *Phytochemistry* 20, 2077-83.
126. Reyes, J., Gross, U., Ponce, L. and Serpa, M. (1982), in Gross, R. and Bunting, E.S. (eds) *Agricultural and Nutritional Aspects of Lupins*. Eschborn, West Germany: Deutsche Gesellschaft für Tech-

Lupin (*Lupinus* spp.)

nische Zusammenarbeit, 707-20.
127. Romero, J.J. (1982), in Gross, R. and Bunting, E.S. (*eds*) *Agricultural and Nutritional Aspects of Lupins*. Eschborn, West Germany: Gesellschaft für Technische Zusammenarbeit, 805-18.
128. Schoeneberger, H., Ildefonso, C., Cremer, H.-D., Gross, R. and Elmadfa, I. (1982), in Gross, R. and Bunting, E.S. (*eds*) *Agricultural and Nutritional Aspects of Lupins*. Eschborn, West Germany: Deutsche Gesellschaft für Technische Zusammenarbeit, 553-83.
129. Sheldrick, R.D., Tayler, R.S., Maingu, Z. and Pongkao, S. (1980), *Grass and Forage Science* 35, 323-7.
130. Shutov, G.K. (1982), in Gross, R. and Bunting, E.S. (*eds*) *Agricultural and Nutritional Aspects of Lupins*. Eschborn, West Germany: Deutsche Gesellschaft für Technische Zusammenarbeit, 248-61.
131. Silva, G.M. and De Oliviera, A.J. (1959), *Agronomia Lusitana* 21, 43-74.
132. Sim, J.T.R. (1958), *Scientific Bulletin 373*. South Africa Department of Agriculture.
133. Stoker, R. (1974), *New Zealand Journal of Agriculture* 129, 40-43.
134. Summerfield, R.J. and Roberts, E.H. (1985), in Halevy, A.H. (*ed.*) *A Handbook of Flowering*. Florida: CRC Press, (in press)
135. Tapia, M. (1982), in Gross, R. and Bunting, E.S. (*eds*) *Agricultural and Nutritional Aspects of Lupins*. Eschborn, West Germany: Deutsche Gesellschaft für Technische Zusammenarbeit, 819-34.
136. Trinick, M.J. (1977), *New Phytologist* 78, 297-304.
137. Trinick, M.J., Dilworth, M.J. and Grounds, M. (1976), *New Phytologist* 77, 359-70.
138. Van Steveninck, R.F.M. (1957), *Journal of Experimental Botany* 8, 373-81.
139. Van Vuvren, P.J. (1964), *Farming in South Africa* 39, 44-6.
140. Waller, G.R. and Nowacki, E.W. (1978), *Alkaloid Biology and Metabolism in Plants*. New York: Plenum Press.
141. Walton, G.H. (1980), *Farmnote 41/80*. Western Australia Department of Agriculture.
142. Walton, G.H. and Francis, C.M. (1975), *Australian Journal of Agricultural Research* 26, 641-6.
143. Wasserman, V.D. (1981), *Lupine Newsletter* 2, 18-20.
144. Wells, H.D. and Forbes, I. (1982), in Gross, R. and Bunting, E.S. (*eds*) *Agricultural and Nutritional Aspects of Lupins*. Eschborn, West Germany: Deutsche Gesellschaft für Technische Zusammenarbeit, 77-89.
145. White, D.H., Elliott, B.R., Sharkey, M.J. and Reeves, T.G. (1978), *Journal of the Australian Institute of Agricultural Science* 44, 21-7.
146. Williams, W. (1979), *Euphytica* 28, 481-8.
147. Williams, W. (1982), *Lupine Newsletter* 4, 16.
148. Williams, W. and Brocklehurst, T. (1983), in Thompson, R. and Casey, R. (*eds*) *Perspectives for Peas and Lupins as Protein Crops*. World Crops: Production, Utilization and Description, Vol. 8. The Hague, Boston and London: Martinus Nijhoff, 59-71.
149. Williams, W. and Harrison, J.E.M. (1983), *Phytochemistry* 22, 85-90.
150. Williams, W., Harrison, J.E.M. and Jayasekera, S. (1984), *Euphytica* 33 (in press).
151. Withers, N.J. (1973), *Proceedings of the Agronomy Society of New Zealand* 3, 63-8.
152. Withers, N.J. (1975), *Proceedings of the Agronomy Society of New*

Zealand <u>5</u>, 13–16.
153. Withers, N.J., Baker, C.J. and Lynch, T.J. (1974), *Proceedings of the Agronomy Society of New Zealand* <u>4</u>, 4–8.
154. Withers, N.J. and Forde, B.J. (1979), *New Zealand Journal of Agricultural Research* <u>22</u>, 463–74.
155. Withers, N.J., King, S. and Hove, E.L. (1975), *New Zealand Journal of Experimental Agriculture* <u>3</u>, 331–2.
156. Wood, P.M. and Allen, J.G. (1980), *Australian Journal of Experimental Agriculture and Animal Husbandry* <u>20</u>, 316–18.
157. Wood, P.M. and McLean, G.D. (1982), *Journal of Agriculture, Western Australia* <u>23</u>, 86–8.

17 Groundnut (*Arachis hypogaea* L.)

A.H. Bunting, R.W. Gibbons and J.C. Wynne

INTRODUCTION

The groundnut or peanut (*Arachis hypogaea* L.) is grown as an annual crop on about 19 million hectares in the warmer regions of the world, principally for its edible, oil- and protein-rich seeds (kernels), borne in fruits (pods) which develop below the soil surface. The dry pericarp of the mature fruit is known as the shell or husk. In this chapter, the terms seed, fruit and shell are used.

HISTORICAL TRENDS

Both the cultivated groundnut and the genus *Arachis* are exclusively South American. The genus is thought to have originated in central Brazil, and the cultivated forms were probably domesticated in northern Argentina and eastern Bolivia (and see Chapter 2). The earliest archaeological evidence of cultivated groundnuts is from graves in Peru, dated at 3200-3500 BP. When the Spanish and Portuguese first entered the American continent, the crop was being cultivated in Mexico, the Caribbean basin, north-eastern and eastern Brazil and throughout the coastal regions of Peru [78].

Few people believe that the crop had reached other continents before the time of Columbus [112]. It was probably taken from Brazil to Africa, India and the Far East by the

Portuguese, and from the West Coast to the Western Pacific and to Indonesia and China by the Spaniards, early in the sixteenth century. Groundnuts had probably girdled the globe well before the middle of that century [112].

Small-seeded 'runner' types (see below) may have reached the south-eastern coast of North America from Africa in slave ships, but it is at least equally possible that the crop entered the USA from Mexico or the Caribbean. The 'Spanish' types (see below) were brought to Europe in 1794, and introduced into the USA from Spain in 1871. 'Valencia' cultivars (see below) were introduced into the USA about thirty years later, also from Spain. The sources of the large-seeded 'Virginia' forms (see below), which are commonly grown in the states of Virginia and the Carolinas in the USA, are not yet clear.

Oil crushing mills were established in Europe early in the nineteenth century. By the early 1840s, many thousands of tons were imported annually from the Gambia into the UK for oil and into the USA for roasting. Groundnuts became a commercial crop in the USA after the Civil War, and cultivation increased in the states of Alabama, Georgia and the Carolinas as depredations by the boll weevil (*Anthonomus grandis*) compelled farmers to abandon cotton.

After soyabeans (*Glycine max*), the groundnut is the second most important source of vegetable oil in the world. In 1982, the recorded worldwide output of soyabeans (about 93 million t) contained about 19.5 million t of oil. The 13 million t of groundnut seeds, corresponding to the reported production of about 19 million t in shell (at a shelling percentage of 70%) contained about 6.5 million t of oil. The total cotton output of the world in 1982 corresponded to about 5.8 million t of oil [53], and in that year the other main sources (sunflower, brassicas, oil palm and copra from coconut) each contributed 3.2-5.1 million t [53].

From 1969/71 to 1980/82 the reported annual world production of groundnuts in shell remained fairly constant at about 18-19 million t from a similar number of hectares

(Table 17.1). In individual countries, however, there have been some significant changes. During 1969-71, Nigeria produced over 1.6 million t per annum and was an important exporter; by 1980-82 the annual output was less than 600 000 t and Nigeria had become an importer [53]. Many technical reasons (including the arrival of groundnut rust [117]) have been suggested for this change, but in the background lies a general decline in agricultural output, resulting from insufficiently effective national policies for agriculture and the rural areas in a rapidly changing economy and society.

In Senegal, where groundnuts are the main cash crop, the area sown has remained fairly constant at about one million ha. The government plans [57] to stabilise production at an annual maximum of 1.2 million t. The recorded increases in both area and yield in Sudan form part of the diversification of cropping in the irrigated cotton areas. In Brazil, both area and output of groundnuts have declined since 1969-71 (from 0.67 to 0.27 million ha and from 0.88 to 0.39 million t, respectively), while the area and output of soyabeans have increased from 1.3 to 8.5 million ha and from 1.5 to 14.3 million t, respectively [53].

In the USA, the area of the crop has been relatively stable since 1938 but, by 1977, the new runner-type Virginia cultivars (preferred in the edible trade to the smaller-yielding Spanish types) had increased yields so much that a system of quotas with a minimum support price, to be set every year, was introduced to limit supplies in the domestic market. Non-quota groundnuts, supported at a much smaller price, have enabled the USA, which had previously been a supplementary supplier in world markets, to become the largest exporter (seeds, cake and meal, and oil equivalent to about 360 000 t in shell out of a total production of 1.56 million t in 1982) [115].

Table 17.1 Production of groundnuts in shell in various regions and various years [53]

Region	Area harvested (ha x 10³)			Yield (kg ha⁻¹)			Output (t x 10³)		
	1969–71	1974–76	1980–82	1969–71	1974–76	1980–82	1969–71	1974–76	1980–82
World	19747	18974	18859	926	929	996	18293	17631	18787
Developed market economies	1068	932	854	1721	2338	2205	1838	2180	1883
North America	591	606	561	2182	2810	2621	1289	1701	1470
Western Europe	9	7	6	2015	2449	2853	17	17	17
Oceania	35	26	30	873	1249	1538	30	32	47
Other*	434	294	256	1157	1459	1364	502	429	350
South Africa	370	247	220	984	1360	1218	364	333	268
Developing market economies	16415	16035	15416	836	828	841	13720	13284	12972
Africa	6206	5740	5063	778	745	700	4827	4277	3543
Nigeria	1846	937	600	900	420	972	1660	393	583
Senegal	1006	1267	1019	789	969	683	794	1228	696
Latin America	1127	923	725	1225	1115	1262	1381	1029	915
Argentina	255	337	213	1099	993	1192	280	334	254
Brazil	670	363	265	1307	1288	1453	876	468	385
Near East	546	897	1031	883	1048	894	483	941	922
Sudan	490	838	967	756	984	830	370	825	803
Far East	8531	8470	8589	824	830	883	7027	7034	7583
India	7287	7109	7295	797	803	821	5807	5710	5986
Centrally planned economies	2264	2007	2590	1208	1080	1518	2735	2167	3931
East Europe and USSR	2	2	6	980	1549	1267	2	4	7
Asia	2262	2004	2584	1208	1079	1519	2733	2163	3924
China	2165	1909	2471	1216	1083	1547	2634	2068	3822
Developed (all)	1070	934	860	1720	2338	2198	1840	2184	1890
Developing (all)	18677	18039	18000	881	856	939	16953	15447	16896

*South Africa, Israel and Japan

Groundnut (*Arachis hypogaea* L.)

Yields per unit of area and time (productivity)

Of the main producing countries the USA has the largest average yield per hectare (Table 17.1), exceeding 3 t ha^{-1} in 1982 [53]. Yields in developing countries are usually considerably smaller. The current area of the crop in China (about 2.5 million ha) is similar to that of Nigeria fifteen years ago, and is probably at least partly irrigated. Yields, at about 1.5 t ha^{-1}, are about half those of the USA average. In all remaining areas in Table 17.1, reported or estimated average yields are smaller than 1 t ha^{-1}, though yields obtained by the best farmers or those on experiment stations commonly exceed 3 t ha^{-1}. In India, the world's largest producer, the average yield is around 900 kg ha^{-1}.

The largest recorded commercial yield, 9.6 t ha^{-1}, was harvested in Zimbabwe from a crop of a long-season Virginia cultivar established with irrigation before the onset of the rains [84]. The crop occupied 182 days, which represents 36.9 kg seed ha^{-1} day^{-1}. A crop of 3000 kg ha^{-1} grown in Florida in 167 days produces 12.6 kg seed ha^{-1} day^{-1}. An average crop in the developing world, grown under limited rainfall and with little or no protection, and yielding 800 kg ha^{-1} in 120 days, produces about 4.6 kg seed ha^{-1} day^{-1}.

Future prospects

The future prospects for the crop around the world depend on competition with other sources of vegetable oil, particularly soyabeans, on the growth of market demands and decreases in the unit cost of production in the developing countries, on the control of aflatoxin (see below) in the product, and on the extent to which prices of groundnuts and competing products are supported, particularly in the developed countries. They are consequently difficult to predict.

751

PRINCIPAL ECONOMIC PRODUCTS AND USES

Groundnut crops are grown for their seeds, the oil and meal derived from them, and the vegetative residues (haulms) [117]. About two-thirds of the world crop is crushed for oil, which is used mainly for cooking. It may also be used for margarines and vegetable ghee, for shortening in pastries and bread, in soaps, pharmaceutical and cosmetic products, lubricants and emulsions for insecticides, and as a fuel for diesel engines [47]. The residue, after the oil has been extracted (press cake), contains 40-50% of protein and provides valuable food and feed. Groundnut flour, produced from the cake, is used in many foods and food products.

The entire seeds are consumed fresh, dry, boiled, or roasted. In the USA, about half the groundnut harvest is used to make peanut butter; salted groundnuts and confectionery products are also important.

Groundnut haulm is nutritionally comparable to grass hay and is used as feed in many countries [60]. The shells may be used for fuel, as a soil conditioner, as a filler in feeds and as a source of furfural [4], processed as a substitute for cork or for hardboard, or composted with the aid of lignin-decomposing bacteria.

PRINCIPAL FARMING SYSTEMS

Developed countries

In most developed countries, groundnuts are a rainfed and at least partly mechanised row-crop, though supplementary irrigation is increasingly used [60]. In the USA it is recommended that groundnuts are not planted in the same field more than once in three years, in order to use fertiliser more efficiently, to lessen losses from pests and diseases, and to help control weeds [83]. Seedling diseases are more prevalent in groundnuts following cotton, tobacco or soyabeans than after grain crops such as sorghum, maize or millets. The grain crops also respond more profitably to fertilisers, and the succeeding groundnuts can benefit from the fertiliser residues.

In Australia, where production is heavily capitalised and mechanised, larger yields have been achieved where groundnuts are grown on land recently ploughed from pasture [113]. Although the area of groundnuts in Israel is very small, it is completely irrigated and yields are large (5.6 t ha^{-1} in 1982). The crop follows wheat, potatoes, carrots or radishes [55].

Developing countries

Sole crops.   In Senegal, small-scale farmers grow groundnuts as sole crops in a two-year rotation with pearl millet [57]. In the Sudan, 75% of the area of the crop is rainfed in shifting systems on sandy soils west of the White Nile. No fixed rotations have been recorded. In the irrigated areas of the lower Blue and White Niles, on alkaline vertisols, sole groundnuts are grown in two-, three- or four course rotations [92], including cotton, wheat, sorghum, rice and fallow. Average yields are double those harvested in the rainlands. In Zimbabwe, there are two distinct production systems. On mechanised commercial farms on the highveld, Virginia cultivars, sown with irrigation before the onset of the rains, may take nearly seven months to mature. In the hotter, lower-lying areas, and in subsistence systems at both lower and higher elevations, early-maturing cultivars are rainfed [84].

The East African Groundnuts Scheme of the late forties and fifties in what was then Tanganyika was intended to produce groundnuts and other crops, in fully mechanised solecropping systems, on a very large scale. It failed spectacularly, because large-scale investment and physical development were started without adequate knowledge of the proposed production environments, and before appropriate methods of land development and management, and of crop production, had been established. In certain areas, however, and under a very different system of administrative management, a

753

general mechanised farming development including groundnuts could probably have succeeded.

Mixed and sequential cropping.   In both Asia and Africa, groundnuts are very often mixed with cereals such as pearl millet, sorghum or maize. It has been estimated that 95% of the groundnuts in Nigeria and 56% of those in Uganda are grown in mixtures [131]. In studies at ICRISAT, Hyderabad, India (see Chapter 18), three rows of groundnuts and one of pearl millet (which matured first), gave an output advantage of 26% (land equivalent ratio (LER) of 1.26), presumably because the mixture intercepted more solar radiation (28%), over the season, than the sole crops [149]. Output advantages of 6-16% have been reported for maize-groundnut mixtures in Cameroon, and of 9-54% in Tanzania [51,122]. The advantages are probably obtained because the groundnuts are able, in the early part of the season, to use resources (such as time, radiation and water) which would otherwise be wasted or less effectively used [149]. In addition, when groundnuts are intercropped with taller cereals, diseases may be less severe than in sole crops [91].

Groundnuts are also commonly intercropped with long-season annual or biennial crops. In south India, considerable output advantages have been measured experimentally with from six to eight rows of groundnuts to one row of pigeon-peas [11,94,183], and output advantages of 53-65% have recently been obtained with 1:1 and 5:1 combinations [149]. In both India and Africa, mixed crops of cotton, castor or cassava with groundnuts appear to be more profitable than sole crops of these species [10,52,141,148,182].

Groundnuts are important in rice-based sequential cropping systems in South-east Asia. They can either be planted with irrigation before a rice crop or grown with residual water after the rice has been harvested. In Thailand, rice is more vigorous after groundnuts, which also provide extra income [123]. In the rainfed rice area of Malaysia, groundnuts are rotated with rice and there is considerable potential for expansion of this system [76]. However, groundnuts require more water than some alternative crops (e.g. soya-beans, mungbeans, sorghum or maize [201]). In the Guangdong Province of China, where three crops a year are often grown, early rice is followed by a second rice crop and then a crop of groundnuts [201].

Groundnuts are often grown beneath perennial tree crops such as coconut, oil palm or rubber [149]. In smallholder rubber schemes in South-east Asia, groundnuts may provide the only cash income until the trees are old enough to tap. It is thought that they may also provide sufficient fixed nitrogen to the rubber to save expenditure on nitrogen fertilisers.

PRINCIPAL BOTANICAL, MORPHOLOGICAL, PHENOLOGICAL AND AGRONOMIC FEATURES OF THE CROP

Botany and morphology

Arachis ($2n = 2x = 20$; $4n = 40$), a papilionate legume of the tribe Aeschynomeneae, subtribe Stylosanthineae, consists of perennial or annual herbs. There are twenty-two described and possibly forty or more undescribed species, all located in South America east of the Andes, south of the Amazon and north of la Plata [74]. Only one species, *A. hypogaea*, is cultivated as a field crop, though others are used for pasture or ground cover.

Vegetative and reproductive morphology.   Two types of branches, 'vegetative' and reproductive (inflorescences), are formed on plants of the genus *Arachis* [152]. On the main axis, primary ($n + 1$) 'vegetative' branches are borne in the axils of the cotyledons and of the first few normal leaves (which are stipulate and compound, with three or four leaflets). In some forms, inflorescence branches may appear at higher nodes on the main axis. Each node on a 'vegetative' branch produces either a 'vegetative' branch of ($n + 2$) or

higher order, or an inflorescence, up to the point in the
life of the plant at which all further branches are suppressed. In the cultivated forms, the branches which arise
in the axils of the cotyledons are strongly developed and
produce a substantial fraction of the crop (Plate 17.1).

Plate 17.1 Branching patterns in *Arachis hypogaea*: (top) cotyledonary
branch of subspecies *hypogaea* showing alternate branching (pairs of
'vegetative' branches alternate at successive nodes with pairs of
inflorescences); (bottom) cotyledonary branch of subspecies
fastigiata showing sequential branching (inflorescences in
sequence at successive nodes, no 'vegetative' branches).
In both cases, the subtending leaves have been removed
to reveal these patterns more clearly (ICRISAT
Information Services and Dr R.W. Gibbons)

The inflorescence branches are inconspicuous among the
stipules and petioles of their subtending leaves, except
that a few inflorescences may grow on to produce a few normal leaves [143]. Each inflorescence produces several

flowers, in sequence in time, in the axils of small bracts (cataphylls). Flowers are sessile and consist of a long (up to 5 cm) tubular hypanthium (fused lower parts of calyx, corolla and staminal tube), at the top of which are borne the expanded lobes of the sepals and petals (coloured white, through yellow, to orange-red), a tube of 9 stamens, and the anthers, each on a short filament. The superior, unilocular ovary of one carpel is enclosed inside and at the bottom of the hypanthium tube, within which the free style ascends, to bear a single capitate stigma, located among the anthers when the flower opens.

On each inflorescence, only one or two flowers are visible at a time. Each flower opens early in the morning, but the anthers usually burst even earlier. Hence self-fertilisation before the flower opens is common, at least in the cultivated forms [73]. However, the papilionate corolla can attract wild bees, and perhaps other pollinators. Up to 2.56% of flowers have been found to be outcrossed in Florida [129]. Though several flowers in an inflorescence may be fertilised, the development of the first-fertilised ovary or ovaries prevents or delays that of later ones. Thus, if all fruits and all new flowers are removed from a plant, a number of new fruits may develop from fertilised but suppressed ovaries. Normally, only a small proportion of the flowers which open produce fruits [167].

After fertilisation, an intercalary meristem at the base of the ovary generates the 'peg', a sort of stalk, morphologically unique in plants. In flowers borne above the ground, the peg grows upward at first, carrying with it the ovary (which develops a sharp, hardened tip), but it soon curves to grow more or less vertically downwards towards the soil surface. The tip penetrates the surface and then usually turns horizontally, away from the main axis of the plant. In flowers which lie below the soil surface (because they have been initiated there or because they have been covered by soil wash or cultivation), the peg grows horizontally from the outset.

In many forms, both cultivated and wild, one or more additional intercalary meristems become active in the ovary wall between the ovules, as the ovary starts to develop underground. The mature fruit may consequently be constricted (lomentiform) or even remarkably extended to form a long, slender 'isthmus', from a few centimetres to more than one metre long, between the seed-containing segments. A fleshy endocarp surrounds the ovules in the expanding fruit, but it decreases as the ovules enlarge, and has virtually disappeared by the time the seeds are mature [156]. Tannins accumulate and darken the inner surface of the shell. Fruits may contain from two to as many as seven seeds (in some cultivated forms), enclosed in papery testas with a wide range of colours.

The seeds contain two large cotyledons and a straight 'embryo' [74] which consists of the upper stem axis plus leaf and bud primordia (the epicotyl), and the radicle, which consists of the hypocotyl and the primary root [73]. The initials of the first inflorescences on the cotyledonary branches are already present in the seeds of at least some early-maturing (sequentially branched) forms [106] (and personal observations).

In many of the longer-season cultivated forms, the seeds are dormant for a longer or shorter time after they are morphologically mature. Non-dormant seeds germinate within 24 h at 25°C and the radicle extends rapidly [197]. In the soil, the outer layers of the primary root are sloughed or abraded off, so that normal root hairs can be seen only on plants grown so that the roots are not abraded, for example *in vitro* [73]. At the points where lateral roots may emerge (in four rows opposite the angles of the stele), structures resembling root hairs are formed. The roots may be tap-rooted or tuberous, and the tuberous-rooted types are further divided into those in which the hypocotyl and the primary root become tuberous and those in which the lateral roots do so [166]. (The nodules in which nitrogen is fixed are considered later.)

Classification within the genus.   Earlier classifications of *Arachis* were based on incomplete collections and inadequate herbarium material. New explorations and collections, and studies of living plants (including an extremely large number of experimental cross-pollinations) have led to the division of the genus into seven sections [72,74,166]. Unfortunately, a formal taxonomic treatment of the genus has not yet been published, and several commonly cited species names are *nomina nuda* (and see Chapter 1).

Arachis hypogaea is assigned to the section *Arachis*. It is a tetraploid, like the closely related *A. monticola*, but all the remaining species in the section are diploid. *Arachis hypogaea* is an amphidiploid [88], and it is thought that species similar in structure to *A. cardenasii* and *A. batizocoi* are involved in its ancestry [73,165].

Classification within *A. hypogaea*.   In the past, many entries in collections have been crude mixtures of genotypes. To develop and use a classification it is essential to separate the components of such mixtures at harvest and sometimes to screen them through progeny rows [23].

Classification within the species is based on branching pattern and plant habit. Characters of the fruit wall and seeds are then used to divide the species into agronomic groups or variety clusters [23,24,63,75].

Branching pattern.   Virtually all cultivated groundnuts fall into one of two types which differ in branching pattern. In the alternately branched forms, pairs of 'vegetative' branches alternate, at successive nodes of the ($n + 1$) and higher order branches, with pairs of inflorescence (reproductive) branches; no reproductive branches are formed (or at any rate expanded) on the main (n) axis. In the sequentially branched forms, the first node, and sometimes a few much higher nodes, on the ($n + 1$) branches may bear ($n + 2$) 'vegetative' branches, but most nodes bear reproductive branches, which consequently appear in sequence at successive nodes. In these forms, reproductive branches are also expanded on the main axis.

This difference in branching pattern divides *A. hypogaea* into two subspecies, *hypogaea* (the alternately branched Virginia and the less-common *hirsuta* forms), and *fastigiata* (the sequentially branched Spanish and Valencia forms) [74, 100,166]. There are several associated differences [23,144]. The alternately branched forms are darker green in colour, the plants are longer-lived, their seeds are usually dormant for a significant time after they are morphologically mature, and in very many the testas are russet-brown in colour. The sequentially branched forms are paler green in colour, they are shorter-lived, seed dormancy is absent or brief, and there is a wide range of testa colours at maturity in different genotypes, from pale cream through pink, red and brown to very dark blue.

The two subspecies have frequently been crossed. The crosses segregate in later selfed generations. Although a few apparently stable cultivars have been created by breeding, or by natural outcrossing, intermediate forms are very rare in collections. This could suggest a deeper-seated difference between the subspecies, even though they are interfertile. Perhaps the two subspecies differ in at least one diploid ancestor.

Plant habit.   A series of differences in branch posture and plant habit is superimposed on the difference in branching pattern. In all forms, the main (n) axis is usually erect. The 'vegetative' branches, of ($n + 1$) and higher orders, grow in one of three main ways. They may ascend at relatively small but constant angles to the main axis, so that the plants are fastigiate in appearance (upright bunch). This is the characteristic habit of the Spanish cultivars, and it is not known in the alternately branched forms. In the second habit type, the branches may start to grow outwards and then curve upwards, giving a basket-like or spreading bunch appearance. This is the habit of the

Valencia types of the sequentially branched subspecies, and of the Virginia bunch types in the alternately branched subspecies. The latter bear so many 'vegetative' branches of (n + 2) and higher orders that they are very much denser in appearance than the Valencia bunch types, which may have no more than four 'vegetative' (n + 1) branches. In the third habit type, the branches grow horizontally on the surface of the soil, so that the plants are of the runner type. Sequentially branched runners are very rare: virtually all runners are alternately branched.

These patterns of variation are summarised in Table 17.2. Evidently the older view that there are only two habit types in the cultivated groundnuts, bunch and runner, is over-simple, even though it persists in some recent texts.

Table 17.2 Variation and taxonomy in *Arachis hypogaea* (not including var. *hirsuta*)

| Character | Subspecies | |
	fastigiata	*hypogaea*
Branching pattern	sequential	alternate
Colour of foliage	paler green	darker green
Dormancy of seeds	none to brief	weeks to months
Duration of crops	shorter	longer
Common testa colours	creamy white, pink, red, brown, deep blue	mostly russet brown, but many other colours, and variegated
Fruit size (range)	small to medium	small to very large
Region of origin	central and southern Brazil	Bolivia, Amazonia

| | Variety | | |
	vulgaris	*fastigiata*	*hypogaea*
Common English name	Spanish	Valencia	Virginia bunch	Virginia runner
Branch posture	upright	curved	curved	prostrate
Habit	upright bunch	open bunch	spreading bunch	runner

Agronomy

Climate and soil. Groundnuts are grown in warm tropical, subtropical and winter rainfall environments, and in temperate humid climates with sufficiently long warm summers. In these conditions, groundnuts can succeed wherever the wet season is sufficiently long or irrigation can be provided, between 40°N and 40°S. Mean day temperatures in such areas lie between 24°C and 33°C [155].

The time from sowing to harvest is independent of photoperiod but dependent on temperature, variety and population density. At a mean day temperature of 30°C in the Sudan rainlands (400-500 metres above sea level), or at similar altitudes in West Africa, and in the densely sown crops necessary for acceptable yields, short-season varieties of the sequentially branched group are ready for harvest in 90-95 days, whereas alternately branched forms require 120-130 days. In cooler conditions, at 1200 metres in Central Africa, these longer-season varieties may need 150-160 days [60].

The time of initiation, and perhaps the node number, of the first-formed inflorescence [91,193], and the rate at which flowers appear, are affected by temperature but do not appear to be affected by photoperiod, at least in the strict sense More flowers are produced at temperatures and daylengths that allow most vegetative growth [17,41,167]. It may be that in some studies effects of photoperiod have been confounded with those of radiation receipt. The harvest indices of some cultivars increase in longer days, whereas in others they are not affected [15

Seeds will germinate at temperatures of 5°-40°C, but more rapidly at 20°-35°C. It may be possible to select forms that tolerate cold soils [2]. The growth of fruits is most rapid at soil temperatures of 30°-34°C but competition from other fruits may decrease its potential rate and/or duration [45, 186].

The amount of water required for the groundnut (or any other) crop is not fixed: it depends on the duration of the crop (which varies between cultivars and is affected by plant density), the rate of expansion of the canopy, and the evaporative demand of the environment. A rainfed crop occupying 95 days on the deep vertisols of the Sudan rainlands

Groundnut (*Arachis hypogaea* L.)

(with a mean evaporation from the crop of about 5 mm day^{-1})
would require less than 475 mm if the rain were perfectly
distributed; in practice at least 600 mm of rain are required. On a light soil with smaller storage capacity in the
same climate, still more would be needed.

The groundnut is well suited to regions with a marked
dry season, where it has to be timed to mature soon after the
rains are expected to end. In wet conditions, harvesting and
the subsequent collection, threshing and drying of the crop
are difficult. In addition, the fruits and seeds may be
invaded by *Aspergillus flavus* (see below), and seeds which
are not dormant may germinate. At the start of the season,
wet conditions make sowing more difficult. In temperate
climates, cool temperatures in spring and autumn may damage
the crop. So, the selection of the cultivar and the time-
table of the crop have to be carefully planned in relation
to climate and expected weather [60]. However, since the
crop is of indeterminate habit, and the first fruits are
formed at early nodes, some yield may be harvested even if
the season should end early (unlike a cereal crop, which,
being determinate, forms yield only during the later part
of its life and so may yield little or nothing if the season
ends too soon). Hence, like other grain legumes of similar
habit, groundnuts may be grown as a catch crop to give a
small yield in marginal regions.

Groundnuts are said to grow best on soils limed to pH
5.8-6.2, provided nutrient elements are available in balanced supply, but the crop grows well under both irrigated and
rainfed conditions in the Sudan on vertisols as alkaline as
pH 8.5, so long as the sodium content is small.

On light soils, the crop is easier to harvest and the
fruits are largely free of adhering earth. It can grow
satisfactorily on heavy soils, but unless they are self-
mulching, harvesting may be difficult [60], particularly
of the alternately branched spreading bunch and runner
varieties. (Many of the sequentially branched Spanish and
Valencia varieties form their fruits in a compact pattern

763

close to the tap root and are much easier to harvest). With all varieties on heavy soils the produce may contain many small clods of clay, which are not easy to remove, so light soils are preferred by the grower, rather than by the crop.

Nutrition and fertiliser needs.   Groundnuts do not seem to have any peculiar nutritional needs, except that calcium (which can be absorbed by the pegs) is particularly important for the growing fruits and seeds, especially in large-fruited Virginia cultivars. The uptake of the most important elements for each 1000 kg of fruit produced is shown in Table 17.3.

Table 17.3 Approximate quantities (kg) of some elements in groundnut crops for each 1000 kg of fruit produced [142]

| Element | Quantities of elements (kg) in: | |
	foliage and stems	fruits
N	10–12	30–35
P_2O_5	1.5–2	6
K_2O	10–12	6–10
CaO	8–12	1–2
MgO	8–10	2

Phosphate, potassium, sulphur, magnesium or minor elements may have to be added on some soils; molybdenum, cobalt, boron, copper and zinc are needed to support the symbiotic fixation of nitrogen [120]; but responses to potassium are uncertain [151]. Groundnut roots appear to absorb potassium efficiently; and moreover in many tropical soils the crop seldom responds to potassium. In the USA, if groundnuts follow another crop which has been well-fertilised, a further supply of fertiliser will seldom increase yield.

The large nitrogen needs of the crop are usually satisfied by symbiotic fixation [138,157]. Nevertheless, there are many reports (as in other annual legume crops) of important responses to 'starter' nitrogen fertiliser, particularly in short-season cultivars grown in warmer locations. These effects have not been unequivocally explained. They may be associated with more vigorous growth during the first three weeks (nearly one-quarter of the life of a thirteen-week crop), before symbiotic fixation of nitrogen is well-established, but during which the number and size of the reproductive initials which will produce most of the fruits are being determined - see Chapter 4.

Production methods.   In the USA, land preparation for groundnuts has two purposes. One is to form a deep, friable, even, level and slightly raised seed bed. The second is to bury completely both weed seeds and crop residues (to limit infection by *Sclerotium rolfsii* [83]). Seeds are routinely dressed with protective chemicals. The sowing of fruits has often been investigated and usually rejected. The seeds have to be stored and prepared with great care to minimise damage to the delicate testas. Sowing is by precision planter, usually with a picker wheel rather than a plate, to minimise mechanical damage. Pre- and post-emergence herbicide applications, and routine pest control and cultivations, are mechanised. Harvesting, curing and drying are very sophisticated, partly because they are conducted at a time of year when temperatures are decreasing and humidity is increasing. The procedures include lifting (digging), shaking, wind-rowing and threshing with a pick-up combine before drying. The harvested fruits are rapidly dried, in bins or wagons, to less than 10% water content, to control insects and moulds which would otherwise soon affect quality [200]. Once-over combines which both lift and strip the fruits have been constructed in both the UK and the USA but have not succeeded commercially.

In developing countries, the seed bed, which is sometimes ridged, is usually made by hand-hoes or by animal-drawn ploughs. It tends to be cloddy and uneven, so that the seeds, which are usually carefully stored (in shell) and prepared but seldom protected against soil fungi and insects by dressing, are sown at variable depths and so emerge unevenly. Seed bed losses are usually heavy. Herbicides (like other crop protection chemicals) are rarely used, and weeds are controlled with animal-drawn implements or by hand. The shortages of labour and power at critical times, which are so common in the agriculture of many developing countries, act against timely and efficient sowing and weed control.

The environmental conditions at harvest-time are usually quite different from those in temperate countries: days are becoming hotter and the air is very dry or at least becoming so. Harvesting is done by hand, perhaps with the aid of simple animal-drawn equipment. Methods of drying and curing may include windrowing and/or stacking, but the harvested plants may dry so fast that they become brittle and break when they are handled. The fruits are usually stripped by hand onto mats or even onto the bare soil.

Shelling.   Wherever possible, groundnuts are stored in the shell to protect the seeds from mechanical damage. For marketing, or when they have to be prepared for sowing, the seeds have to be removed from the fruits by shelling. If the fruit walls are too dry or too wet, the fruits are difficult to shell. In tropical seasonally dry climates, the fruit walls may become so brittle that shelling may damage a large proportion of the seeds. It is therefore necessary to add water to the fruits to bring their moisture concentration to 10-12%. At one time, groundnuts were therefore shipped in shell, even to overseas markets, in spite of the cost of transport of the large volume of the produce, and subsequently moistened at the shelling plant. In temperate or wet environments, fruits have to be dried for shelling, storage, or shipment.

Morphology and growth physiology

Effects of branching pattern.   Evidently the skilful use

766

of time is of the greatest importance in groundnut agronomy,
as indeed it is with any other crop [25,60]. As a consequence of the botanically indeterminate nature of the plant,
and of the differences in branching pattern, the sequentially and alternately branched types are characteristically
short- and long-lived, respectively.

Since the first inflorescences are initiated very early
in the life of the plant, reproductive and vegetative growth
are simultaneous, and so compete for nutrients and for the
products of assimilation through the greater part of the
crop cycle. Like other indeterminate crops, the groundnut
is programmed to give priority, in this competition, to the
sinks formed by the increasing number of growing fruits.
This competitive partitioning may be represented by the time
courses of the values of the leaf-to-total-growth ratio
(LTGR) and of the reproductive-to-total-growth ratio (RTGR)
[25,26,48,49,50]. (Conventional growth analysis alone, which
was designed to study growth in the vegetative stage only,
is of little direct use once the crop has entered the reproductive phase, particularly in an indeterminate crop). As
the crop advances, LTGR declines to zero at that time when
all new material is sequestered by the fruits, no new leaves
are expanded, and any further growth of the seeds depends
on translocation of carbon and nitrogen from other parts of
the plant. At this stage RTGR exceeds unity.

The initial values, and the slopes, of the curves of
LTGR and RTGR appear to vary between varieties [48,50]. The
time course of leaf area per plant is affected by branching
pattern. In the sequentially branched forms, the number of
apical meristems which produce new leaves (as well as reproductive branches) is limited, often to no more than five or
six (those of the main axis plus, say, four (n + 1) and one
(n + 2) 'vegetative' branches). This restricts the rate of
expansion and the maximum size of leaf area per plant. As
the number of fruits increases, to perhaps fifteen per plant
in a closely sown crop, growth at these meristems is soon
restricted by internal competition. Leaf area reaches its

maximum early, the life of the crop is short, and the mature dry weight and yield of individual plants are small. These forms are adapted to shorter seasons, and large populations (perhaps 250 000 plants ha^{-1}) are necessary for maximum productivity.

In alternately branched plants, each of which may easily in a normal crop, produce thirty or more apical meristems at the rate of two for every two reproductive branches, more leaves are formed per unit of time (and per fruit) and the maximum leaf area per plant is larger. Though internal competition becomes significant in due course, the plants live longer, become larger, and produce many more fruits and larger yields. These forms are suited to longer seasons, and can attain maximum yield with smaller populations (perhaps 100 000-150 000 plants ha^{-1}).

Moreover, in the sequentially branched forms, densely sown, most of the crop comes from early nodes on the earliest ($n + 1$) branches and so the fruits tend to be of similar age. In the alternately branched forms, in which the fruits are initiated over a longer period of time, many more of them are immature when the crop is harvested. Furthermore, some of the earlier-formed fruits will be morphologically mature while the soil is still wet. Their dormancy prevents the seeds in these fruits from germinating, a risk which is always present in sequentially branched types (particularly Valencias), if the wet season is unduly prolonged, or the crop is sown too early.

Assimilation and crop growth rates [15,134]. The groundnut, like all other grain legumes, is a C3 species, but in bright light its leaves assimilate at rates comparable with those of C4 species [188], presumably because the canopy does not become saturated at relatively small irradiance. Leaf assimilation rates as large as 37 mg $CO_2 dm^{-2} h^{-1}$ have been reported [157]. Nevertheless, the maximum growth rates are generally similar to those of other C3 species [48,95,187, 188], i.e. around 10-20 g dry-matter m^{-2} day^{-1}. These may

be associated with the horizontal orientation of most of
the leaves but, in addition, a proportion of the energy
which is chemically fixed in the leaves may be used to
reduce nitrogen in effectively nodulated plants, so that
the dry-matter in which it is contained cannot be added to
biomass (and see Chapter 4).

Crop yields.   Though the rates, both of CO_2 assimilation
and of crop growth, appear to vary between genotypes, these
variations are not associated with those in economic yield.
Seed yield appears to depend more on the proportion of dry-
matter which is partitioned to the fruits (RTGR) [48]. Of
the conventional components of yield, fruit number is considerably more important than average seed size [96], and
the largest yields in trials were given by cultivars in
which fruit numbers increased more rapidly [96]. Recently
released cultivars, such as Florunner, initiated 10.6 fruits
m^{-2} day^{-1}, while older, smaller-yielding cultivars such as
Dixie Runner initiated only 6.6 fruits m^{-2} day^{-1}. The vegetative growth rates were similar in all the cultivars [48],
but a larger proportion of the accumulated biomass was used
to form fruits in Florunner than in Dixie Runner. Presumably
the larger-yielding forms produced more fruits per inflorescence. The duration of fruit filling is also important
[188], and the multiple correlations of weight of fruits per
plant, shelling percentage, and seed weight with yield are
significant in both more and less fertile soils [42].

It seems possible that the yields of sequentially
branched cultivars, with limited numbers of (n + 1) and few
or no (n + 2) branches, may become source-limited, whereas,
at least until recently, in many alternately branched cultivars, they were more likely to be sink-limited.

SYMBIOTIC FIXATION OF NITROGEN

Ecological aspects

Groundnuts are promiscuously nodulated by strains of *Rhizobium* of the cowpea miscellany group. In the field, plants form root hairs and nodules on the tap root only at the points where lateral roots emerge, or would emerge if the sites were not occupied by nodules. The rhizobia enter the root at the bases of these multicellular hairs, in which no infection threads have been seen [31]. Most soils in which groundnuts have previously been grown appear to contain relatively large populations of effective rhizobia, particularly during the crop season. In the soils of the eastern coastal plain of the USA, for example, peak populations reached 10000 g^{-1} of soil in groundnut fields during July, but were very small from September to May [107]. In such soils, the consequences of inoculation have been variable: yields have been decreased in some studies [172] but increased (along with improvements in seed quality, seed protein and oil concentration) in others [12,32].

Rhizobia isolated from many different legumes can form nodules on groundnuts, although not all of the resulting symbioses are equally effective in fixing nitrogen [192]. Inoculation has led to substantial increases in yield in fields which have not previously carried groundnuts [27,138, 157,180]. On farmers' fields in India, nodulation varies considerably. On half the sites examined, it was poor, and on some of these, the numbers of nodules and the rates of nitrogen fixation (measured by acetylene reduction) were only one-tenth of the values typical on research farms. It may be possible to increase symbiotic nitrogen fixation in such fields by supplying effective and competitive strains of *Rhizobium* [124].

Groundnut (*Arachis hypogaea* L.)

Strain selection, testing, and cultivar/strain interactions

Various methods can be used to select effective strains of
Rhizobium. In North Carolina, isolates are screened on
siratro (*Macroptilium atropurpureum*) in serum bottles in
a growth chamber. Strains which appear promising at twenty-
one days are increased for testing in the glasshouse with
two host genotypes of groundnuts, and later in the field
[192]. Total nitrogen accumulated is regarded as the best
measure of the effectiveness of a particular strain; the
colours and dry weights of the hosts are significantly correlated with their total nitrogen contents [192]. At ICRISAT
in India (see Chapter 18) strains are tested on groundnut cul
tivars in the glasshouse and subsequently in the field,
where native populations of rhizobia are large [124]. Significant yield increases (17-40%) have been obtained in
several seasons and at several sites with the cultivar Robut
33-1 (and in 1981 with ICG 15, a progeny derived from it)
and strain of *Rhizobium* NC92, but not with several other
cultivar/strain combinations [90]. The specific association
between Robut 33-1 and strain NC92 may therefore be heritable [90].

Seasonal and varietal differences in nodulation

Under both rainfed and irrigated conditions in India, the
numbers, growth and mass of nodules have differed between
the long-season Virginia cultivar Kadiri 71-1, and the short
season Spanish cultivar Comet [124]. In the USA, Virginia
types are generally more heavily nodulated than Spanish or
Valencia types [27,46,192], and some Virginia cultivars form
large numbers of nodules on the hypocotyl, which is rare in
Spanish or Valencia cultivars [125]. Nodules form later on
the hypocotyl than on the roots, but they seem to continue
to fix nitrogen during fruit-filling, when root nodules have
senesced. The differences may be heritable [125] and are
perhaps associated with differences in the supply of assimi-

lates, and so with the time-course of leaf area and branching pattern.

Non-nodulating groundnuts

Groundnut genotypes which do not form nodules have recently been reported [71,128]. These genotypes lack the root hairs at the base of which rhizobia enter, and this prevents infection [126]. The 'resistance to nodulation' character is controlled by a pair of duplicate recessive genes [128].

Estimates of nitrogen fixed

Nodules formed by effective strains of *Rhizobium* often fix most of the nitrogen requirements of the crop when environmental conditions are favourable [138,157], but not always [185]. Estimates as large as 240 kg N ha^{-1} fixed by groundnuts have been reported, representing 80% of the plants' total uptake of N [40]. Under optimum soil and climatic conditions in Israel, 222 kg N ha^{-1} were estimated to have been fixed, some 58% of the N accumulated in the plants. The soil contained between 120 and 170 kg N ha^{-1} (as nitrate) in the top 60 cm. Under less favourable conditions, less nitrogen was fixed, but it was still equivalent to 40% of the total nitrogen in the plants [146].

Intercropping with pearl millet, maize or sorghum decreased both the number of nodules and the amount of nitrogen they fixed, presumably because the groundnuts were shaded by the cereals [127].

Although most of the nitrogen in the crop is removed when the seeds are harvested, a few experiments suggest that some N may be left in the soil. In Australia, wheat took up about twice as much nitrogen after intercrops of maize/soyabean and maize/groundnuts, grown without nitrogen fertilisers, than after maize alone (also without N fertiliser) [159]. The N taken up was equivalent to that following maize alone with 100 kg N ha^{-1}. In India, pearl millet yielded 45%

more after groundnuts than after maize [127].

PRINCIPAL LIMITATIONS TO YIELD

Attitudes of farmers and crop priorities

In most tropical, subtropical and winter rainfall areas
where groundnuts are important in traditional farming systems, the climate is seasonally arid and the staple crop is
a cereal. To ensure survival, rural people plant the cereal
as soon as they can when the main rains start [60], in order
to capture as much as possible of the seasonal flush of
nitrate (which is liberated by mineralisation in the uppermost layers of the soil when they have been moistened by
the light early rains), before it is leached beyond reach
by the wetting front which moves down the profile once the
rains are well established. After sowing, the cereal has to
be weeded promptly and thoroughly to minimise the amounts
of nitrate and other nutrients, and of light and water,
which would otherwise be used by the weeds. Only after all
this has been done will the family allocate labour, time and
attention to other crops [80]. A second reason for late
planting is that groundnut seeds are more valuable than
those of the cereal and will not be sown until there is less
risk that the crop will be lost to an early-season dry spell.
Farmers may plant millet dry, or after early showers, but
they will seldom take a chance with their groundnut seeds.

Since the groundnut can provide at least a part of its
own nitrogen, later planting (provided a cultivar of suitable crop duration is used) has less effect on it than on the
cereal. Of course, the farmer and his wife know well enough
that they run two risks - a later-sown crop may run out of
time, and so of water; and it may be more severely affected
by certain diseases, such as Rosette Virus in Africa [18,
80], but these are chances which they deliberately take.
They may be illiterate, but they are not ignorant.

Yields are also limited for lack of the human or animal

labour needed for dense sowing (which is as important as early planting for large yields per unit area, and for some control of Rosette Virus [18,80]), and for weeding. In many developing countries, particularly in Africa, labour at critical times is a scarcer resource than land. Producers' decisions tend to maximise returns to this limiting resource.

Groundnuts are moreover often grown in mixtures. This is sensible from the producer's point of view, but it prevents him from realising the large yields per unit area which the research worker or administrator would like to be able to report.

The first goal of a large-scale mechanised farmer is to maximise profits. This usually means optimising the return to land. Irrigation has consequently been increasing in the USA in recent years. But as the costs of capital, power, equipment and inputs increase, farmers in developed countries may likewise be unable to give as large a priority to the groundnut crop as the groundnut scientist would like [184].

Pests and diseases

Inherited resistances to pests and diseases, and their uses in breeding, are considered in later sections.

Arthropod pests.   Excellent reviews of the pre- and post-harvest pests of groundnuts are available [13,54,150,169]. Some 360 species of arthropod pests attack groundnuts before harvest in different parts of the world.

The incidences of pests often change suddenly. In the USA, the two-spotted mite, *Tetranychus urticae*, has become an important pest since the use of insecticides and fungicides has created suitable conditions for it [28]. In India before 1968, aphids (*Aphis craccivora*), leafminers (*Aproaerema modicella*), hairy caterpillars (*Amsacta* sp.) and species of termites were regarded as the main pests, nationally and regionally. But by 1977-82, five new pests were

774

economically important [6]. White grubs (*Lachnosterna con-sanguinea* and other species) are now serious pests on the sandy soils of north India [104,196] while *Spodoptera litura*, the tobacco caterpillar, has recently become significant, particularly on irrigated groundnuts. The full-grown larvae are very difficult to control with insecticides [6].

On a world scale, the cowpea or groundnut aphid (*Aphis craccivora*) is the most widely distributed of the shoot and foliage feeders. It can damage groundnuts severely by sucking sap from stems and developing pegs [97]. It is also a vector of several harmful viruses [169] (and see below).

Seventeen species of thrips have been reported as pests of groundnuts, but the economics of controlling them with insecticides has long been in doubt [169]. In India, *Frankliniella schultzei* and, to a lesser extent, *Scirtothrips dorsalis*, are vectors of Tomato Spotted Wilt Virus (TSWV) [8,58].

Although soil-inhabiting arthropods are considered to be the most important economic pests of groundnuts, they have not been as well studied as the foliage pests [169]. Their distribution is much affected by soil type and environmental conditions. Leaf-feeding pests are similar in Texas, where the soils are deep and sandy, and in North Carolina, where the soils are heavier, but the soil pests are different [169]. The most important pest in the New World is the lesser cornstalk borer, *Elasmopalpus lignosellus*, particularly in the drier regions such as Texas [169]. Larvae attack all parts of the plant but are particularly damaging on pegs and fruits. *Diabrotica undecimpunctata*, the southern corn rootworm, and other species are important pests in the USA. White grubs are important in many parts of the world - *Lachnosterna* species in India and *Rhopea magnicornis* and *Heteronyx* species in Australia [97]. Millipedes, particularly species of *Peridontopyge*, are common pests in Africa [116].

Nemotodes. Many nematodes are parasitic on groundnuts but

the principal species are in the genera *Meloidogyne*, *Praty-lenchus*, *Belonolaimus* and *Macroposthonia* [140]. Species of *Meloidogyne*, the root-knot nematodes, are important in many parts of the world. *Meloidogyne hapla* is more common in colder regions and *M. arenaria* in warmer ones [140]. *Praty-lenchus brachyurus* is a cosmopolitan pest which attacks roots, pegs and fruits. Damage to the fruit facilitates the entry of fungal saprophytes and pathogens including *Aspergillus flavus* [110]. Control measures include nemati-cides and crop rotations.

Fungal diseases of foliage.   Of these, the most important, worldwide, are the leaf spots and groundnut rust. Two dif-ferent fungal species, *Cercospora arachidicola* (*Mycosphae-rella arachidis*) and *Cercosporidium personatum* (*M. berke-leyii*) produce the early and late leaf spots, respectively. A third species, *Cercospora canescens*, has been recorded in Nigeria and Malawi, but it appears to be of minor importance [93,111]. Although *C. arachidicola* and *C. personatum* can be distinguished by the size and shape of the conidia, and the number of septa [140], and are often said to produce dis-tinctly different types of lesions on the leaves (as well as on stems, petioles and pegs), they may be confused by dif-ferences in the reaction or nutrition of the host, or by environmental effects [140]. On particular cultivars, the symptoms produced by *C. personatum* are almost exactly the same as those generally associated with *C. arachidicola* [174].

The distributions of these two fungi have also changed [168]. Though early leaf spot predominates in Georgia and Virginia [114,189], in other parts of the south-eastern USA *C. arachidicola* appears to have been more important from 1967 to 1976 and *C. personatum* from 1976 to 1979. In many countries the distributions of the two diseases are not precisely known.

The leaf spots are routinely controlled with fungicides in the USA, but it has been estimated that they still

776

decrease yields by about 10%. Losses between 15% and 50% are common where the crop is not sprayed [93]. The fungicide benomyl has hitherto been remarkably effective, but strains of both fungi which are resistant to the active ingredient have now appeared in the USA [33,105]. The pathogens survive in the soil, so rotation can help to lessen the incidence of the diseases. It may not be effective, however, where groundnuts are generally grown in many small plots, at different times, and in mixtures, as they often are in developing countries [110].

Rust, caused by *Puccinia arachidis*, can halve the yields of groundnuts [173]. Before 1970, it was mainly confined to the Caribbean and South America, with occasional outbreaks in the USA [77,178], but during the late sixties and the seventies it spread to both Asia and Africa. It is now a serious threat to production in many parts of the world. The uredinial stage is the most commonly seen. Teliospores have been recorded, but there are no records of spermogonia or aecia [173]. Uredospores are short-lived on infected plant tissues and seldom remain viable for more than a month in hot climates. No physiologic races have yet been reported [173]. When groundnuts are grown under both rainfed and irrigated conditions, as in India, the disease is carried over from crop to crop and so from year to year [173].

Rust can be controlled by several commercial fungicides, but the costs of chemicals and spraying equipment are often beyond the means of small-scale farmers [62].

Fungal diseases of roots, pegs and fruits. The strictly soil-borne pathogens are also economically important, particularly as causes of loss of germinating seeds and of seedlings, but few of them have been adequately studied in tropical countries. *Sclerotium rolfsii* is potentially damaging in the USA but has been largely controlled by cultural practices such as deep ploughing, which limits the supply of substrates for the fungus [20]. *Cylindrocladium* black rot (CBR), associated with *C. crotalariae*, is now serious and widespread in Virginia and North Carolina, and *Sclero-tinia* blight, associated with *S. sclerotiorum*, is serious in Virginia and Oklahoma [140].

Since the outbreak of Turkey X disease in 1960 in the UK, a very large amount of research [43,44,69,70] has been devoted to the toxin-producing soil-borne fungus *Aspergillus flavus* (and related species) on groundnuts (and on other crops also), particularly as at least one of the fungal metabolites, Aflatoxin Bl, is a potent carcinogen in animals and man [44]. The fungus can infect fruits and seeds before harvest, and then develop further during the subsequent handling and in storage. It can be controlled by suitable cultural practices, and by careful drying and storage. Though the toxin can be removed from infected groundnut cake [125], the cake may not be fed to milking cows in the UK.

Viruses.   Many viruses infect groundnuts, but relatively few are sufficiently widespread to cause economic losses every year. The virus disease Groundnut Rosette, first reported in Tanzania in 1907, is thought to be restricted to Africa south of the Sahara [59,202]. The vector is *Aphis craccivora*. The complex may include two viruses, one of which, Groundnut Rosette Virus (GRV), is mechanically transmissible but requires the presence of an aphid-transmitted, symptomless second virus, Groundnut Rosette Assistor Virus (GRAV), to produce symptoms [86]. Crops are infected by virus-bearing alate aphids, but secondary spread by wingless forms may be more important in some countries [59].

Peanut Mottle Virus (PMV) was first reported in the USA in 1965, but it is not thought to be distributed worldwide [101,147]. The symptoms are relatively mild and are often overlooked in the field, although severe strains also exist [133,179]. The virus is transmitted by several aphids, including *A. craccivora* [14,102]. Unlike Rosette, PMV is seed-borne, though not, perhaps, in all cultivars [89,103].

Peanut Stunt Virus (PSV), a cucumo virus, reached epidemic proportions in the USA in 1965 and 1966. *Trifolium repens* is a common alternate host, and so the disease can be controlled by keeping groundnut crops well away from white clover fields [82,181]. This virus is transmitted by three aphids, including *A. craccivora*, in a non-persistent manner [82].

Bud necrosis, caused by Tomato Spotted Wilt Virus (TSWV), is serious in India. It is transmitted by two species of thrips, *Frankliniella schultzei* and *Scirtothrips dorsalis* [8,58]. TSWV was serious on groundnuts in Australia, but control of weeds which were alternate hosts of the virus dramatically decreased the incidence of the disease [153]. Early planting, dense plant populations, and less susceptible cultivars, have lessened its incidence in India [7]. Other diseases of groundnuts, each associated with a different virus, include Peanut Clump, Cowpea Mild Mottle, Peanut Green Mosaic, Crinkle, Eyespot, Yellow Mottle, Marginal Chlorosis, and Rugose Leaf Curl [140].

Bacterial diseases. Though bacterial wilt, caused by *Pseudomonas solanacearum*, has been recorded in most parts of the world where groundnuts are grown, it is serious only in Indonesia, Uganda and China [39,140]. Races of the bacterium vary in pathogenicity.

Competition with weeds

Because leaf area expands relatively slowly, and vertical growth is both slow and inherently limited (particularly in runner types), groundnuts are very sensitive to competition with weeds. The ground is not usually covered until several weeks after sowing, particularly in insufficiently dense sowings of the upright bunch (*vulgaris* and *fastigiata*) forms. Thus, annual weeds are not sufficiently controlled by shading, and weeds of upright habit readily shade the young crop in these critical formative stages. Moreover, groundnuts are more readily damaged by cultivation, because of their underground fruiting habit, than most other annual crops [22]. The risk is particularly great in the Virginia spreading bunch and runner (*hypogaea*) types, because of their more diffuse fruiting habit.

The yields of irrigated groundnuts on light sandy soils in Libya were halved if they were not weeded [132]. Though certain weeds, when not controlled, have decreased yields by as much as 90% in the USA [22], yields were not affected when weeds were removed within three weeks after planting in Oklahoma [85]. The extent of damage by weeds depends on the weed species. In the USA, no more than two weeks of competition with the grass *Panicum dichotomiflorum* decreased yields, but the legumes *Desmodium tortuosum* and *Cassia obtusifolia* did not affect yields until they had been competitive for ten weeks or longer [22]. Effects of habit and shading, and of competition for nitrogen, seem to be important determinants of these differences.

In developing countries, weeds are usually controlled with animal-drawn implements or hand tools. Handweeding with a traditional hoe is effective in Malawi, but the costs in labour and time are considerable [161]. Weeds, and particularly the parasitic witchweed *Alectria vogelii*, are among the most serious causes of loss in Nigeria, where they are estimated, on average, to halve yields [117].

In developed production systems in the USA, Argentina and Venezuela, weeds are controlled by herbicides. Since research on herbicides for groundnuts began in the USA around 1949, many pre- and post-planting formulations have been developed [22]. Herbicides have changed the weed flora of the USA. In the 1970s broad-leaf weeds became serious. When not controlled, they interfere with harvesting operations and also damage combines [22].

Environmental stresses

General adaptation to environment has been considered earlier.

Drought.   Rainfed groundnuts, particularly on sandy soils,
often suffer from drought stress. This has most effect on
yield if it occurs later in the season when the canopy is
dense and fruits are being formed and filled [19]. It may
also hasten maturity, by preventing the development of the
last-initiated fruits [186]. If the fruiting zone is dry,
many fruits will be empty or contain only a single seed.
This is associated with smaller uptakes of nutrients, particularly of calcium [164], and so with typical symptoms of
calcium deficiency [66] (see below), particularly in large-
seeded Virginia cultivars, which need more calcium and water
for satisfactory development of seeds than small-seeded cultivars [19]. Water deficits also decrease fruit number, seed
quality, seed weight and germinability [19]. Intermittent
dry spells before harvest can increase invasion by *Aspergillus flavus* and the consequent accumulation of aflatoxin
[44].

Nutritional stresses.   For maximum yield an adequate
supply of each essential element must be available during
the growing season and toxic conditions must be eliminated
[36]. Shortage of phosphorus is common in tropical soils
and is the most frequent nutrient stress for groundnuts if
fertiliser has not been used. Infection by mycorrhiza can
increase the uptake of phosphorus [120], and added phosphate
and mycorrhiza can have relatively similar beneficial effects on growth, nodulation and nitrogen fixation [38].

Potassium deficiencies have been detected on ferralitic
soils, particularly if they have been over-cultivated. They
may be suspected if many fruits are single-seeded, but this
condition may also indicate a shortage of calcium.

Since calcium is essential for growth of the seeds and
is taken up by the pegs (as well as by the roots), it must
be available in the soil layers in which the pegs form the
fruits. For the aerial parts of the plant, calcium is supplied passively in transpiration, but the pegs and fruits
do not transpire, and so they must rely on diffusion and,

perhaps, active uptake [19]. If calcium is deficient, one or more of the seeds in each fruit may abort or be shrivelled, the plumules become dark in colour, and the seeds germinate poorly. Calcium sulphate applied at the time of flowering can provide an adequate supply for large-seeded Virginia cultivars. The needs of small-seeded cultivars are usually met by the proper use of liming materials [3,4,9]. On strongly acid soils the uptake and utilisation of nutrients are adversely affected.

Nitrogen and phosphate fertilisers which contain sulphur usually meet the needs of the crop for this element, but in many parts of Africa the upland soils are deficient in sulphur and requirements of 12-15 kg S ha^{-1} have been demonstrated [144].

Responses to boron have been recorded in the USA [36], Africa [66] and India [121]. Boron-deficient seeds often have internal damage known as 'hollow heart', but no more than 0.5-1.0 kg B ha^{-1} is needed to correct the deficiency. Iron deficiencies sometimes occur on calcareous soils, particularly when they are wet. A chelated iron source applied to the soil can overcome the shortage.

YIELD AND QUALITY

Yields of fruits generally range from 0.5-4.0 t ha^{-1}. The largest recorded field yields [62,84] approach 10 t ha^{-1}. In the developing countries, which produce 80% of the crop, the mean yields are around 1.0 t ha^{-1} only. In the developed countries, mean yields approach 3.0 t ha^{-1}. About 70% of the weight of the groundnut fruit consists of the seeds, which are good sources of both protein (22-30%) and oil (44-56%) [26].

In general, sequentially branched, and particularly Valencia, cultivars contain more crude protein than Virginia (alternately branched) forms, in which (as a consequence of the branching pattern and the longer life-span) the produce includes a larger proportion of less mature fruits. As in

782

the seeds of other grain legumes, the protein is nutritionally inferior to that of the standard reference protein
(SRP, which approximates the average amino acid profile of
human proteins) because it contains significantly smaller
proportions of lysine, methionine and threonine, and sometimes of isoleucine and valine [109,135,199]. Aspartic and
glutamic acids and arginine constitute about 45% of the
total amino acids, and their proportion is greater than in
the SRP [199]. However, in real diets, proteins from legumes
are complemented by those from cereals and other sources.
Moreover, the so-called 'deficiency' is only relative. It
can be offset by increasing the supply and decreasing the
cost of the product, which can help to increase the total
dietary intake and so ensure a sufficient total supply of
the essential amino acids.

Nearly two-thirds of all groundnuts produced are crushed
for oil [34,35]. The oil is unsaturated and is readily oxidised; the iodine value (82 to 106) and the refractive index
(around 1.47) are relatively large [5]. The unsaturated
acids oleic and linoleic comprise about 80% of the total,
and the fully saturated palmitic acid contributes 10%. Virginia cultivars have more oleic and less linoleic acid than
Spanish and Valencia cultivars (in part, because of differences in average maturity), and this makes their oil more
stable. The proportions of the remaining fatty acids (0.02-
3.55%) change as the seeds mature [21].

One hundred grams of raw groundnut seed provide about
570 kcalories (2.39 mJ) of dietary energy [5]. The mature
seeds contain 20-25% of carbohydrate, of which about 8-10%
is cellulose and hemicellulose, 4% is starch, and 10-12% is
sugars [136]. Sucrose is the principal sugar, varying from
2.86-6.35% depending on the cultivar. When the seeds are
roasted, the sucrose is hydrolysed to fructose and glucose,
which then react with free amino acids to form the numerous
substances which give the characteristic flavour to the
roasted product [108] and the peanut butter which is prepared from it.

783

Groundnuts are useful sources of thiamine, niacin, toco-
pherols (vitamin E) and folic acid in human diets [5]. Like
most other legumes, they contain heat-labile trypsin inhibitors, but harmful effects associated with phyto-haemagglu-
tinins (lectins) are rare. They contain fermentable carbohydrates (flatus factors), but not alkaloids, saponins or
other heat-stable antimetabolites.

GERMPLASM RESOURCES

Extensive collections of germplasm of cultivated groundnuts
have been made from many countries, particularly in South
America. During the past five centuries, tropical Africa
has become an important secondary centre of variation of
both the sequentially and alternately branched types [63].
About twenty-eight institutions in eighteen countries maintain significant collections [145]. The largest collection,
of over 10000 accessions, is maintained by the International
Crops Research Institute for the Semi-Arid Tropics (ICRISAT)
at Patancheru, Hyderabad, India (see Chapter 18). This Institute has been designated by the International Board for
Plant Genetic Resources (IBPGR; see Chapter 18) as a principal
repository for germplasm of *Arachis* [90]. Expeditions in
South America in 1976-81, funded by IBPGR, collected about
480 accessions of cultivated material and 293 of wild relatives, including perhaps 24 previously unknown species [162].
Further collections are needed in northern Bolivia, in
Brazil and in north-east Peru; and in South and South-east
Asia and Central America [87].

Large numbers of accessions of cultivated groundnuts
have been evaluated in respect of potential yield, morphological and chemical characters, tolerance of environmental
stresses, and resistance to pests and diseases, by ICRISAT
(mainly in Asia) [90], by the Institut des Recherches pour
les Huiles et Oléagineux (IRHO) (mainly in Africa) [68], and
by national programmes in several groundnut-growing countries. Genotypes have been found which resist or tolerate

784

several important diseases, including rust [176], Rosette
and other viruses [59,64,67,79,89,103], the two *Cercospora*
leaf spots [1,99,118,160,177], *Cylindrocladium* black rot and
Sclerotinia blight [35,139,190,195], bacterial wilt [39,140]
and *Aspergillus flavus*.

The Virginia cultivars are often thought to be generally
more tolerant of the *Cercospora* leaf spot diseases, but this
may be no more than a consequence of the branching pattern,
which continues to provide young and apparently uninfected
leaves on older plants. In plants resistant to rust, only a
few, sparsely sporulating pustules are formed, and very few
of the uredospores are viable [175].

Resistances or tolerances to several pests have been
found in cultivated forms [30,130,194]. Cultivars which
delay the multiplication of *Aphis craccivora* are known [64,
67,79]. Resistance to tobacco thrips, *Frankliniella fusca*,
is available in cultivars of all three main groups [169].
The Virginia cultivar NC6 is resistant to southern corn
rootworm, *Diabrotica undecimpunctata*, is little affected by
F. fusca, and is resistant to the potato leafhopper, *Empoasca fabae* [191]. Straight leaf trichomes [29] are associated with resistance to *E. kerryi* in India [6]. Modest degrees of resistance have been found to the soil pest *Elasmopalpus*; they may be useful in integrated management systems
[170]. No reliable resistances are yet known to nematodes
[110].

Drought-tolerant cultivars of both early- and late-
maturing types, that outyield other cultivars in dry years
but give average yields in years when rainfall is adequate,
have been developed in Senegal by screening for small values
of leaf water potential under dry conditions [16,56,67,161].
Other means of adaptation to drought include small leaf area
index, escape mechanisms such as early maturity, indeterminate growth, and recovery characteristics [161].

Iron-efficient cultivars which yield reasonably well
without supplemental iron have been found in Israel [81].

A nearly two-fold variation in the proportions of the

785

limiting essential amino acids has been found among groundnut cultivars [199], suggesting that the dietary quality of
the protein could be manipulated by breeding - though, as
in maize, this might carry a yield penalty with it.

Finally, several important sources of disease and pest
resistance have been found in wild species of *Arachis*,
though they are not simple to use in crop improvement programmes. However, resistance, even immunity, to the *Cercospora* leaf spots [1,99,118,160,171,177] and to rust [116]
in wild forms have already been used in breeding. *Arachis
batizocoi* and *A. chacoense* appear to be very resistant to
Aphis craccivora [89], to which little resistance is known
in cultivated forms.

GROUNDNUT BREEDING

Breeding objectives

The main objectives of groundnut breeding are to develop
cultivars which have large potential yield, are adapted to
specific environments (and environmental stresses [62]) and
production systems, and are resistant to diseases and pests.
Breeding is an unceasing endeavour, as the crop is developed
in new environments and production systems and as market
needs change, and the spectrum of pests and diseases continually alters. Where cultivars which meet the primary objectives are already available, breeding for the improved
quality and flavour characteristics desired by processors
and consumers becomes more important.

Although almost two-thirds of the total groundnut harvest in the world is crushed for oil, breeding for larger
oil concentration has so far received little attention.
Inevitably, because of the greater energy density of the oil
[26], this is likely to entail a yield penalty. This may
delay progress unless the price of groundnuts for crushing
is related to oil concentration.

786

Breeding methods

Most groundnut flowers are self-pollinated, but some outcrossing does take place. Moreover, many numbered accessions
and most landrace collections are mixtures of different
forms. Consequently, unimproved cultivars or landraces are
heterogeneous mixtures of mostly homozygous lines.

Breeders first developed improved cultivars by mass
selection from such populations. This may still be satisfactory in the earlier stages of improvement programmes; but
sooner rather than later crossing becomes essential in order
to combine the desirable traits of different cultivars or
genotypes. Crossing is inherently difficult in a cleistogamous, self-fertilised species, and indeed it is far from
easy to feel confident that the seed produced is in fact
hybrid. A crossing and selection programme requires substantial resources, so that exchange of hybrid materials between
breeders is important. The breeding programme at ICRISAT
provides seeds of crosses in early generations so that
breeders in national programmes can evaluate them and select
superior genotypes for their local conditions [61,62].

The breeding and selection methods applied to groundnut
improvement are the same as those used to improve other
self-fertilising species [194], and few novel methods have
been developed in recent years [130]. They include mass
selection, pedigree, bulk population, backcrossing and other
modifications of pure line breeding methods. These methods
restrict recombination among linked genes, as homozygosity is
approached through selfing. Population improvement schemes
such as the convergent cross, recurrent selection and the
diallel selective mating system are presently being explored
[161,194].

To use the wild species of *Arachis* in breeding is a
difficult and extremely skilled task, because of differences
in ploidy, cross-incompatibility and sterility [119,166],
and the need to monitor the crosses cytogenetically. A considerable number of successful crosses have, however, been

made within the section *Arachis*. Moreover, some crosses have been made between diploid species of section *Arachis* and species from sections *Erectoides* and *Rhizomatosae*. So, the genetic resources of more distantly related sections may eventually become accessible using interspecific hybrids as bridges. It is not, however, easy to transfer resistances to diseases and insects from the wild species, not only because of the sterility barriers but because desirable and undesirable traits may prove difficult to separate in the backcrossing sequences.

The possible use of tissue culture and related techniques in the improvement of groundnuts is now being investigated [154]. Both tissue and organ culture have been attempted with some success. These methods may offer means of circumventing incompatibility among interspecific hybrids.

Breeding progress

Advances in groundnut improvement have recently been reviewed [130,144,194]. Cultivars of the large-seeded Virginia types developed for the edible trade in the USA now yield in excess of 30% more than introductions from centres of diversity, and at least 20% more than cultivars developed by selection from introductions [194]. The outstandingly successful cultivars Florigiant and Florunner, which are composites of sister lines selected in F_4 through F_8 generations, give stable yields over seasons and over a wide geographical area [130,158].

Cultivars resistant to Rosette Virus, such as RG1 in Malawi [161], RPM 12 and 91 in Nigeria, and 48-37 and 69-101 in Senegal [57], are derived from crosses with Rosette-resistant materials which have evolved in Africa. Varieties which delay multiplication of aphids and so restrict the spread of viruses in the crop have also been produced [64, 67,79]. The large-seeded Virginia cultivars NC 86 and Va 82B were selected for resistance to *Cylindrocladium crotalariae* and *Sclerotinia* blight, respectively. As mentioned earlier,

Groundnut (*Arachis hypogaea* L.)

NC6, a confectionery type for the Virginia-Carolina area of the USA [30], resists southern corn rootworm and potato leaf-hoppers.

Several programmes have attempted to develop cultivars with shorter growing seasons. In Senegal, line 57-422, a Virginia type derived from a Virginia x Valencia cross, has a growing season of 105 days; its Virginia ancestor would have needed 120 days. Not surprisingly, it has proved diffi-cult to develop alternate-branching types which mature as early as sequential types. A number of types combining early maturity and dormancy have been produced or found in collec-tions in Senegal and Sudan [23]. Earlier-maturing Spanish types such as Pronto have been developed by mass selection from crosses of adapted types with Chico in Oklahoma [98]. Several shorter-duration or drought-tolerant cultivars, such as 47-16, 50-127, 73-33 and 55-437, have been released in Senegal [57,68]. Breeding lines with oil concentrations of 57-58% (7-8% more than the best current Virginia types) have been obtained in Burkina Faso (Upper Volta) in segregates from Virginia x Spanish crosses [144].

The most striking advances in groundnut improvement are now being made at ICRISAT [90], which began a groundnut improvement programme in 1976. Groundnut materials resistant to rust, late leaf spot, bud necrosis, *Aspergillus flavus*, and pests including jassid and thrips have been found, in part among the wild species [90]. Crosses of these materials with adapted or large-yielding breeding lines have been made and late-generation populations or selections are now being evaluated or distributed for use in national programmes. Early-maturing lines and populations with improved quality and potential yields 20-25% greater than those of local cul-tivars have been developed. The resources available, and the continuity of funding suggest that rapid progress will continue.

Nothing the breeder can do has any effect unless seeds of satisfactory quality can reach the producer in the right quantity, and at the right time, place and price. This can

789

be assured only by a reliable, efficient seed industry operated to commercially successful standards. Such industries are taken for granted in developed countries; in many developing countries they do not yet exist.

AVENUES OF COMMUNICATION BETWEEN RESEARCHERS

At present only one journal is devoted exclusively to groundnuts. This is *Peanut Science*, published twice each year by the American Peanut Research and Education Society (APRES) since 1974 (editorial offices: PO Box 31025, Raleigh, North Carolina 27622, United States of America). APRES has more than 500 individual members and a number of sustaining corporate members. Each year APRES holds a conference in one of the groundnut-growing states of the USA and publishes the proceedings. The Society, and its predecessors, have published two significant books - *Peanut Culture and Uses* in 1973 and a very recent (1983) revision, *Peanut Science and Technology* [137]. APRES also publishes a quarterly newsletter, *Peanut Research*, which was originated by the National Peanut Council in 1963.

Oléagineux, published monthly by the Institut des Recherches pour les Huiles et Oléagineux (IRHO), Paris, specialises in annual and perennial oil-bearing crops, and circulates widely. It contains many papers on groundnuts in French, English or Spanish, with summaries in the other two languages. It also includes bibliographies, with summaries of reviewed papers in French. IRHO is a well-staffed research and development organisation sponsored by the government of France. It conducts cooperative projects in developing countries, mostly in Francophone Africa [68].

Abstracts of most of the world literature on groundnuts are brought together in *Tropical Oilseeds Abstracts*, a specialised journal of the Commonwealth Agricultural Bureau, Farnham Royal, Slough, United Kingdom.

The Tropical Development Research Institute, which incorporates the former Tropical Products Institute (TPI) and

790

the Centre for Overseas Pest Research (COPR), is maintained by the United Kingdom government and conducts a wide range of research, development and publishing activities. Much of the work on aflatoxin was done at TPI.

In 1980, an International Groundnut Workshop at ICRISAT was attended by scientists from twenty developed and developing countries. The proceedings have been distributed to most groundnut-growing countries [65].

More recently, in 1983, an international research project, the Peanut Collaborative Research Support Program (CRSP) has been set up by the United States Agency for International Development under Title XII (Famine Prevention and Freedom from Hunger) of the United States International Development and Food Assistance Act 1975. Collaborative projects have been formed with nine host countries in the Caribbean, Africa and Asia, and four universities in the USA [37] (and see Chapter 18).

Other agencies that promote communication among research workers are the African Groundnut Council, with headquarters in Lagos, Nigeria, and the International Development Research Centre, Ottawa, Canada, which has funded work on groundnuts in Africa and Asia.

India maintains communication through a coordinated national programme, with annual planning conferences to disseminate results and information and to articulate programmes for the following season. It is directed by the Indian Council of Agricultural Research, New Delhi [163].

PROSPECTS FOR LARGER YIELDS

It seems likely that groundnut yields will improve in the foreseeable future. In developed countries, they will increase steadily rather than dramatically, as a result of refinements of current techniques rather than new innovations. Yields increased remarkably in the USA between 1950 and 1980, from less than 900 kg ha^{-1} of fruits (the average for the developing countries today) to 3 t ha^{-1}. The improvement was due, not to any one factor, but to a 'package of practices' including new production methods as well as improved cultivars [78]. Further improvement will follow the wider use of irrigation in the drier states.

In the developing countries it is not at present easy to foresee any comparable yield revolution. The crop is grown by small-scale farmers whose resources and access to markets are both limited. Most of them cannot at present afford expensive capital developments or purchased inputs. The physical, financial, technical, commercial, administrative and political supports which farmers in the industrialised countries take for granted are not available to them. However, significant improvements are already technically possible, and larger ones are to be expected, as new cultivars with durable and stable resistances, tolerances and adaptations to biological and environmental constraints are developed. Provided market demand is sustained, and an efficient and reliable seed industry can deliver improved seeds to producers, these cultivars, which will provide a measure of insurance against pests, diseases and drought, should encourage both the small-scale farmers, and the larger-scale entrepreneurs who are entering agriculture, to use more productive methods including timely sowing of dense stands and timely control of weeds. But no one will adopt more productive procedures or continue with them unless they are profitable. Farmers, and particularly 'small' ones, cannot afford to be philanthropists and produce surpluses for the market unless they can rely on an adequate return for their resources and risks.

Much will depend, therefore, on the comparative profitability of this power-intensive crop in production systems which are at present very short of power at critical times, on competition in the market with soyabeans and other vegetable oils, and on the control of *Aspergillus flavus* and of aflatoxin.

LITERATURE CITED

1. Abdou, Y.A., Gregory, W.C. and Cooper, W.C. (1974), *Peanut Science* 1, 6-11.
2. Ablett, G.R. (1978), MSc thesis, Univeristy of Guelph, Ontario, Canada.
3. Adams, F. and Hartzog, D. (1979), *Peanut Science* 6, 73-6.
4. Adams, F. and Hartzog, D. (1980), *Peanut Science* 7, 120-23.
5. Ahmed, E.M. and Young, C.T. (1982), in Pattee, H. and Young, C.T. (eds) *Peanut Science and Technology*. Yoakum, Texas: American Peanut Research and Education Society, 655-88.
6. Amin, P.W. (1983), Occasional Paper 1/83, ICRISAT.
7. Amin, P.W. and Mohammad, A.M. (1980), in *Proceedings of the International Workshop on Groundnuts*. Patancheru, India: ICRISAT, 158-166.
8. Amin, P.W., Reddy, D.V.R. and Ghanekar, A.M. (1981), *Plant Disease Reporter* 65, 663-5.
9. Anderson, G.D. (1970), *Experimental Agriculture* 6, 213-22.
10. Anthony, K.R.M. and Willimott, S.G. (1957), *Empire Journal of Experimental Agriculture* 25, 29-36.
11. Appadurai, R. and Selvaraj, K.V. (1974), *Madras Agricultural Journal* 61, 803-4.
12. Arora, S.K., Saini, J.S., Gandhi, R.C. and Sandu, R.S. (1970), *Oléagineux* 25, 279-80.
13. Bass, M.II. and Arant, F.S. (1973), in *Peanuts: Culture and Uses*. Stillwater, Oklahoma: American Peanut Research and Education Association, 383-428.
14. Behncken, G.M. (1970), *Australian Journal of Agricultural Research* 21, 465-72.
15. Bhagsari, A.S. and Brown, R.H. (1976), *Peanut Science* 3, 1-5.
16. Bockelee-Morvan, A., Gautreau, J., Mortreuil, J.C. and Rouseel, O. (1974), *Oléagineux* 29, 309-14.
17. Bolhuis, G.G. and De Groot, W. (1959), *Netherlands Journal of Agricultural Science* 7, 317-26.
18. Booker, R.H. (1963), *Annals of Applied Biology* 52, 125-31.
19. Boote, K.J., Stansell, J.R., Schubert, A.M. and Stone, J.F. (1982), in Pattee, H.E. and Young, C.T. (eds) *Peanut Science and Technology*. Yoakum, Texas: American Peanut Research and Education Society, 164-205.
20. Boyle, L.W. (1956), *Plant Disease Reporter* 40, 661-5.
21. Brown, D.F., Cater, C.M., Mattil, K.F. and Darroch, J.G. (1979), *Journal of Food Science* 40, 1055-60.
22. Buchanan, G.A., Murray, D.S. and Hauser, E.W. (1982), in Pattee, H.E. and Young, C.T. (eds) *Peanut Science and Technology*. Yoakum, Texas: American Peanut Research and Education Society, 206-49.
23. Bunting, A.H. (1955), *Empire Journal of Experimental Agriculture* 23, 158-70.
24. Bunting, A.H. (1958), *Empire Journal of Experimental Agriculture* 26, 254-8.
25. Bunting, A.H. (1975), *Weather* 30, 312-25.
26. Bunting, A.H. and Elston, J. (1980), in Summerfield, R.J. and Bunting, A.H. (eds) *Advances in Legume Science*, Kew, London: Royal Botanic Gardens, 495-500.
27. Burton, J.C. (1976), in Newton, W.E. and Nyman, W.E. (eds)

Proceedings of the First International Symposium on Nitrogen Fixation, Pullman, Washington: Washington State University Press, 429-46.

28. Campbell, W.V. (1978), *Peanut Science* 5, 83-6.
29. Campbell, W.V., Emery, D.A. and Wynne, J.C. (1976), *Peanut Science* 3, 40-43.
30. Campbell, W.V. and Wynne, J.C. (1980), in *Proceedings of the International Workshop on Groundnuts*. Patancheru, India: ICRISAT, 149-57.
31. Chandler, M.R. (1977), *Journal of Experimental Botany* 29, 749-55.
32. Chesney, H.A.D. (1975), *Agronomy Journal* 67, 7-10.
33. Clark, E.M., Backman, P.A. and Rodriguez-Kabana (1974), *Phytopathology* 64, 1476-7.
34. Cobb, W.Y. and Johnson, B.R. (1973), in *Peanuts: Culture and Uses*. Stillwater, Oklahoma: American Peanut Research and Education Association, 209-63.
35. Coffelt, T.A. and Porter, D.M. (1982), *Plant Disease Reporter* 65, 385-7.
36. Cox, F.R., Adams, F. and Tucker, B.B. (1982), in Pattee, H.E. and Young, C.T. (eds) *Peanut Science and Technology*. Yoakum, Texas: American Peanut Research and Education Society, 139-63.
37. Cummins, D.G. and Jackson, C.R. (eds) (1980), *Special Publication No. 16*. Univeristy of Georgia, Agriculture Experiment Station, 1-165.
38. Daft, M.J. and El Giahmi, A.A. (1976), *Annals of Applied Biology* 83, 273-6.
39. Darong, S. (1981), *Proceedings of the American Peanut Research and Education Society* 13, 21-8.
40. Dart, P.G. and Krantz, B.A. (1977), in Vincent, J.M., Whitney, A.S. and Bose, J. (eds) *Exploiting the Legume-Rhizobium Symbiosis in Tropical Agriculture*. Miscellaneous Publication 145, University of Hawaii, 119-54.
41. De Beer, J.F. (1963), PhD thesis, State Agricultural University, Wageningen, The Netherlands.
42. Dholaria, S.J., Joshi, S.N. and Kabaria, M.M. (1972), *Indian Journal of Agricultural Science* 42; 1084-6.
43. Diener, U.L. (1973), in *Peanuts: Culture and Uses*. Stillwater, Oklahoma: American Peanut Research and Education Association, 523-7.
44. Diener, U.L., Pettit, R.E. and Cole, R.J. (1982) in Pattee, H.E. and Young, C.T. (eds) *Peanut Science and Technology*. Yoakum, Texas: American Peanut Research and Education Society, 486-519.
45. Dryer, J., Duncan, W.G. and McCloud, D.E. (1981), *Crop Science* 21, 686-8.
46. Duggar, J.F. (1935), *Journal of the American Society of Agronomy* 27, 286-8.
47. Duke, J.A. (1981), *Handbook of Legumes of World Economic Importance*. New York: Plenum Press.
48. Duncan, W.G., McCloud, D.E., McGraw, R.L. and Boote, K.J. (1978), *Crop Science* 18, 1015-20.
49. Elston, J. and Bunting, A.H. (1980), in Summerfield, R.J. and Bunting, A.H. (eds) *Advances in Legume Science*. Kew, London: Royal Botanic Gardens, 37-42.
50. Elston, J., Harkness, C. and McDonald, D. (1976), *Annals of Applied Biology* 83, 39-51.

Groundnut (*Arachis hypogaea* L.)

51. Evans, A.C. (1960), *East African Agriculture and Forestry Journal* 26, 1-10.
52. Evans, A.C. and Sreedharan, A. (1962), *East African Agriculture and Forestry Journal* 28, 7-8.
53. FAO (1979 and 1982), *Production Yearbooks*. Vols 33 and 36. Rome: FAO.
54. Feakin, S.D. (1973), *Pest Control in Groundnuts*. Manual No. 2, Third edn. London: PANS.
55. Feldman, S. (1981), *World Farming*, March-April, 52-54.
56. Gautreau, J. (1978), *Oléagineux* 32, 323-32.
57. Gautreau, J. and De Pins, O. (1980), in *Proceedings of the International Workshop on Groundnuts*. Patancheru, India: ICRISAT, 274-81.
58. Ghanekar, A.M., Reddy, D.V.R., Iizuka, N., Amin, P.W. and Gibbons, R.W. (1979), *Annals of Applied Biology* 93, 173-9.
59. Gibbons, R.W. (1977), in Kranz, J., Schmutterer, H. and Koch, W. (eds) *Diseases, Pests, and Weeds in Tropical Crops*. Berlin and Hamburg, Germany: Verlag Paul Parey, 19-21.
60. Gibbons, R.W. (1980), in Summerfield, R.J. and Bunting, A.H., (eds) *Advances in Legume Science*. Kew, London: Royal Botanic Gardens, 483-93.
61. Gibbons, R.W. (1980), in *Proceedings of the International Symposium on Development and Transfer of Technology for Rainfed Agriculture and the SAT Farmer*. Patancheru, India: ICRISAT, 27-37.
62. Gibbons, R.W. (1980), in *Proceedings of the International Workshop on Groundnuts*. Patancheru, India: ICRISAT, 12-16.
63. Gibbons, R.W., Bunting, A.H. and Smartt, J. (1972), *Euphytica* 21, 78-85.
64. Gibbons, R.W. and Mercer, P.M. (1972), *Journal of the American Peanut Research and Education Association* 4, 58-66.
65. Gibbons, R.W. and Mertin, J.V. (eds) (1980), *Proceedings of the International Workshop on Groundnuts*. Patancheru, India: ICRISAT.
66. Gillier, P. (1969), *Oléagineux* 24, 79-81.
67. Gillier, P. (1978), *Oléagineux* 32, 25-8.
68. Gillier, P. (1980), in *Proceedings of the International Workshop on Groundnuts*. Patancheru, India: ICRISAT, 25-30.
69. Goldblatt, L.A. (1969), *Aflatoxin*. New York: Academic Press.
70. Goldblatt, L.A. and Dollear, F.G. (1977), *Pure and Applied Chemistry* 49, 1759-64.
71. Gorbet, D.W. and Burton, J.C. (1979), *Crop Science* 19, 727-8.
72. Gregory, W.C. and Gregory, M.P. (1976), in Simmonds, N.W. (ed.) *Evolution of Crop Plants*. London and New York: Longman, 151-4.
73. Gregory, W.C., Gregory, M.P., Krapovickas, A., Smith, B.W. and Yarbrough, J.A. (1973), in *Peanuts: Culture and Uses*. Stillwater, Oklahoma: American Peanut Research and Education Association, 47-133.
74. Gregory, W.C., Krapovickas, A. and Gregory, M.P. (1980), in Summerfield, R.J. and Bunting, A.H. (eds) *Advances in Legume Science*. Kew, London: Royal Botanic Gardens, 469-81.
75. Gregory, W.C., Smith, B.W. and Yarbrough, J.A. (1951), in *The Peanut—The Unpredictable Legume*. Washington, DC: The National Fertilizer Association, 28-88.
76. Hamat, H.B. and Noor, R.B.M. (1980), in *Proceedings of the International Workshop on Groundnuts*. Patancheru, India: ICRISAT, 233-6.
77. Hammons, R.O. (1977), PANS 23, 300-4.

78. Hammons, R.O. (1982), in Pattee, H.E. and Young, C.T. (*eds*) *Peanut Science and Technology*. Yoakum, Texas: American Peanut Research and Education Society, 1-20.
79. Harkness, C. (1977), Report to African Groundnut Council, Lagos, Nigeria.
80. Harkness, C., Kolawole, K.B. and Yayock, J.H. (1976), *Samaru Conference Paper No.7*. Zaria, Nigeria: Ahmadu Bello University.
81. Hartzook, A., Kartstadt, D., Naveh, M. and Feldman, S. (1974), *Agronomy Journal* 66, 114-15.
82. Hebert, T.T. (1967), *Phytopathology* 57, 461.
83. Henning, R.J., Allison, A.H. and Tripp, L.D. (1982), in Pattee, H.E. and Young, C.T. (*eds*) *Peanut Science and Technology*. Yoakum, Texas: American Peanut Research and Education Society, 123-38.
84. Hildebrand, G.L. (1980), in *Proceedings of the International Workshop on Groundnuts*. Patancheru, India: ICRISAT, 290-96.
85. Hill, L.V. and Santelmann, P.W. (1969), *Weed Science* 17, 1-2.
86. Hull, R. and Adams, A.N. (1968), *Annals of Applied Biology* 62, 139-45.
87. International Board for Plant Genetic Resources (1981), *Revised Priorities Among Crops and Regions*. Rome, Italy: IBPGR.
88. Husted, L. (1936), *Cytologia* 7, 396-423.
89. International Crops Research Institute for the Semi-Arid Tropics (1981), *Annual Report 1979/1980*. Patancheru, India.
90. International Crops Research Institute for the Semi-Arid Tropics (1982), *Annual Report 1981*. Patancheru, India.
91. International Crops Research Institute for the Semi-Arid Tropics (1983), *Annual Report 1982*. Patancheru, India.
92. Ishag, H.M., Ali, M.A. and Ahmadi, A.B. (1980), in *Proceedings of the International Workshop on Groundnuts*. Patancheru, India: ICRISAT, 282-4.
93. Jackson, C.R. and Bell, D.K. (1969), *Research Bulletin 56*. University of Georgia, College of Agriculture.
94. John, C.M., Seshadri, C.R. and Rao, M.B.S. (1943), *Madras Agricultural Journal* 31, 191-200.
95. Kassam, A.H., Kowal, J.M. and Harkness, C. (1975), *Tropical Agriculture, Trinidad* 52, 105-12.
96. Ketring, D.L., Brown, R.H., Sullivan, G.A. and Johnson, B.R. (1982), in Pattee, H.E. and Young, C.T. (*eds*) *Peanut Science and Technology*. Yoakum, Texas: American Peanut Research and Education Society, 411-57.
97. Khan, M.Q. and Husain, M. (1965), *Indian Oilseeds Journal* 9, 67-70.
98. Kirby, J.S. and Banks, D.J. (1980), *Proceedings of the American Peanut Research and Education Society* 12, 49.
99. Kornegay, J.L., Beute, M.K. and Wynne, J.C. (1980), *Peanut Science* 7, 4-9.
100. Krapovickas, A. and Rigoni, V.A. (1960), *Revista de investigaciones agricolas, Buenos Aires* 14, 157.
101. Kuhn, C.W. (1965), *Phytopathology* 55, 880-84.
102. Kuhn, C.W. and Demski, J.W. (1975), *Research Report No. 213*. University of Georgia, College of Agriculture.
103. Kuhn, C.W., Paguio, O.R. and Adams, D.B. (1978), *Plant Disease Reporter* 62, 365-8.
104. Kushwaha, K.S. (1976), *University of Udaipur Research Journal* 14, 75-8.
105. Littrell, R.H. (1974), *Phytopathology* 64, 1377-8.

Groundnut (*Arachis hypogaea* L.)

106. Luo, B.-X., Li, Z.-C., Wen, G.-F., Li, Y.-X., Ye, B.-R. and Chen, Z.-X (1981), *Acta Agronomica Sinica* 7, 1-10.
107. Mahler, R.L. and Wollum, A.G. (1981), *Peanut Science* 8, 1-5.
108. Mason, M.E., Newell, J.A., Johnson, B.R., Koehler, P.E. and Waller, G.R. (1969), *Journal of Agriculture and Food Chemistry* 17, 782-92.
109. Mba, A.V., Njike, M.C. and Dyenuga, V.A. (1974), *Journal de la Science Agronomique* 25, 1547-53.
110. McDonald, D. and Raheja, A.K. (1980), in Summerfield, R.J. and Bunting, A.H. (eds) *Advances in Legume Science*. Kew, London: Royal Botanic Gardens, 501-14.
111. Mercer, P.C. (1977), *Plant Disease Reporter* 61, 55-9.
112. Merrill, E.D. (1954), *Chronica Botanica* 14, 161-384.
113. Middleton, K.J. (1980), in *Proceedings of the International Workshop on Groundnuts*. Patancheru, India: ICRISAT, 223-5.
114. Miller, L.I. (1953), PhD thesis, University of Minnesota, St Paul.
115. Miller, M.L. (ed.) (1983), *Peanut Industry Guide*. Washington, DC: National Peanut Council.
116. Misari, S.M. (1975), *Agriculture Newsletter* 17, 4-9. Zaria, Nigeria: Ahmadu Bello University.
117. Misari, S.M., Harkness, C. and Fowler, A.M. (1980), in *Proceedings of the International Workshop on Groundnuts*. Patancheru, India: ICRISAT, 264-73.
118. Monasterios, T., Jackson, L.F. and Norden, A.J. (1978), *Proceedings of the American Peanut Research and Education Association* 10, 34.
119. Moss, J.P. (1980), in Summerfield, R.J. and Bunting, A.H. (eds) *Advances in Legume Science*. Kew, London: Royal Botanic Gardens, 525-36.
120. Munns, D.N. and Mosse, B. (1980), in Summerfield, R.J. and Bunting, A.H. (eds) *Advances in Legume Science*. Kew, London: Royal Botanic Gardens, 115-25.
121. Muthuswamy, T.D. and Sundararajan, S.R. (1973), *Madras Agricultural Journal* 60, 403.
122. Mutsaers, J.A. (1978), *Netherlands Journal of Agricultural Science* 26, 344-53.
123. Nalampang, A., Charoenwatana, T. and Tiyawalee, D. (1980), in *Proceedings of the International Workshop on Groundnuts*. Patancheru, India: ICRISAT, 237-40.
124. Nambiar, P.T.C. and Dart, P.J. (1980), in *Proceedings of the International Workshop on Groundnuts*. Patancheru, India: ICRISAT, 110-24.
125. Nambiar, P.T.C., Dart, P.J., Srinivasa, B. and Ramanatha Rao, V. (1982), *Experimental Agriculture* 18, 203-7.
126. Nambiar, P.T.C., Nigam, S.N., Dart, P.J. and Gibbons, R.W. (1983), *Journal of Experimental Botany* 34, 484-8.
127. Nambiar, P.T.C., Rao, M.R., Reddy, M.S., Floyd, C., Dart, P.J. and Willey, R.W. (1982), in Graham, P.H. and Harris, S.C. (eds) *Biological Nitrogen Fixation Technology for Tropical Agriculture*. Cali, Colombia: CIAT, 647-52.
128. Nigam, S.N., Arunachalam, V., Gibbons, R.W., Bandyopadhyay, A. and Nambiar, P.T.C. (1980), *Oléagineux* 35, 453-5.
129. Norden, A.J. (1973), in *Peanuts--Culture and Uses*. Stillwater, Oklahoma. American Peanut Research and Education Association, 175-208.

130. Norden, A.J. (1980), in Summerfield, R.J. and Bunting, A.H. (eds)
     Advances in Legume Science. Kew, London: Royal Botanic Gardens,
     515-23.
131. Okigbo, B.N. and Greenland, D.J. (1976), in Multiple Cropping.
     Special Publication No. 27. Madison, Wisconsin: American Society
     of Agronomy, 66-101.
132. Oram, P.A. (1961), Weed Research 1, 211-28.
133. Paguio, O.R. and Kuhn, C.W. (1973), Phytopathology 63, 976-80.
134. Pallas, J.E. Jr and Samish, Y.B. (1974), Crop Science 14, 478-82.
135. Pancholy, S.K., Deshpande, A.S. and Krall, S. (1978), Proceedings
     of the American Peanut Research and Education Association 10,
     30-37.
136. Pattee, H.E., Johns, E.B., Singleton, J.A. and Sanders, T.A.
     (1974), Peanut Science 1, 57-62.
137. Pattee, H.E. and Young, C.T. (eds) (1983), Peanut Science and Tech-
     nology. Yoakum, Texas: American Peanut Research and Education
     Society.
138. Pettit, R.E., Weaver, R.W., Taber, R.A. and Stichler, C.R. (1975),
     in Miller, J.E. (ed.) Peanut Production in Texas. College Station,
     Texas: Texas Agricultural Experiment Station, 26-33.
139. Phipps, P.M. and Beute, M.K. (1977), Plant Disease Reporter 61,
     300-3.
140. Porter, D.M., Smith, D.H. and Rodriguez-Kabana, R. (1982), in
     Pattee, H.E. and Young, C.T. (eds) Peanut Science and Technology.
     Yoakum, Texas: American Peanut Research and Education Society,
     326-410.
141. Potti, V.S.S. and Thomas, A.I. (1978), Proceedings of the National
     Symposium on Intercropping of Pulse Crops. New Delhi, India:
     Indian Agricultural Research Institute.
142. Prevot, P. and Ollagnier, M. (1961), in Proceedings of the 15th
     International Horticultural Congress. Nice, 1958, Vol.2, 217-28.
143. Prevot, P. (1949), Oléagineux Coloniaux, Ser. Sci. 4. Paris, IRHO.
144. Rachie, K.O. and Silvestre, P. (1977), in Leakey, C.L.A. and
     Wills, J.B. (eds) Food Crops of the Lowland Tropics. Oxford:
     Oxford University Press, 40-74.
145. Rao, V.R. (1980), in Proceedings of the International Workshop on
     Groundnuts. Patancheru, India: ICRISAT, 47-57.
146. Ratner, E.I., Lobel, R., Feldbay, H. and Hartzook, A. (1979),
     Plant and Soil 51, 373-86.
147. Reddy, D.V.R., Iizuka, N., Ghanekar, A.M., Murthy, V.K., Kuhn, C.
     W., Gibbons, R.W. and Chohan, J.S. (1978), Plant Disease Reporter
     62, 978-82.
148. Reddy, G.P., Rao, S.C. and Reddy, R. (1965), Indian Oilseeds
     Journal 9, 310-16.
149. Reddy, M.S., Floyd, C.N. and Willey, R.W. (1980), in Proceedings
     of the International Workshop on Groundnuts. Patancheru, India:
     ICRISAT, 133-42.
150. Redlinger, L.M. and Davis, R. (1982), in Pattee, H.E. and Young,
     C.T. (eds) Peanut Science and Technology. Yoakum, Texas: American
     Peanut Research and Education Society, 520-70.
151. Reid, P.H. and Cox, F.R. (1973), in Peanuts: Culture and Uses.
     Stillwater, Oklahoma: American Peanut Research and Education
     Association, 271-97.
152. Richter, C.G. (1899), Inaug.-Diss. Kgl. Bot. Garten. Breslau:
     Scheiber.

Groundnut (*Arachis hypogaea* L.)

153. Saint-Smith, J.H., McCarthy, G.J.P., Rawson, J.E., Langford, S. and Colbran, R.C. (1972), Leaflet No. 1178. Division of Plant Industry, CSIRO, Department of Primary Industries, Australia.
154. Sastri, D.C., Malini, M.S. and Moss, J.P. (1981), in Rao, A.N. (ed.) *Proceedings COSTED Symposium on Tissue Culture of Economically Important Plants.* 42-7.
155. Saxena, N.P., Natarajan, M. and Reddy, M.S. (1983), in *Potential Productivity of Field Crops Under Different Environments.* Los Banos, Philippines: IRRI, 281-305.
156. Schenk, R.U. (1961), *Technical Bulletin N.S.5-53.* Georgia Agricultural Experiment Station.
157. Schiffman, J. and Alpet, Y. (1968), *Experimental Agriculture* 4, 203-8.
158. Schilling, T.T., Mozingo, R.W., Wynne, J.C. and Isleib, T.G. (1983), *Crop Science* 23, 101-5.
159. Searle, P.G.E., Comundom, Y., Shedden, D.C. and Nance, R.A. (1981), *Field Crops Research* 4, 133-45.
160. Sharief, Y., Rawlings, J.O. and Gregory, W.C. (1978), *Euphytica* 27, 741-51.
161. Sibale, P.K. and Kisyombe, C.T. (1980), in *Proceedings of the International Workshop on Groundnuts.* Patancheru, India: ICRISAT, 249-53.
162. Simpsom, C.E. (1980), *Plant Genetic Resources Newsletter* 44, 32-3.
163. Singh, V. (1980), in *Proceedings of the International Workshop on Groundnuts.* Patancheru, India: ICRISAT, 19-24.
164. Skelton, B.J. and Shear, G.M. (1971), *Agronomy Journal* 63, 409-12.
165. Smartt, J., Gregory, W.C. and Gregory, M.P. (1978), *Euphytica* 27, 665-75.
166. Smartt, J. and Stalker, H.T. (1982), in Pattee, H.E. and Young, C.T. (eds) *Peanut Science and Technology.* Yoakum, Texas: American Peanut Research and Education Society, 21-49.
167. Smith, B.W. (1954), *American Journal of Botany* 41, 607-16.
168. Smith, D.H. and Littrell, R.H. (1980), *Plant Disease Reporter* 64, 356-61.
169. Smith, J.W. and Barfield, K.S. (1982), in Pattee, H.E. and Young, C.T. (eds) *Peanut Science and Technology.* Yoakum, Texas: American Peanut Research and Education Society, 250-325.
170. Smith, J.C., Posada, L. and Smith, O.D. (1980), *Peanut Science* 7, 68-71.
171. Sowell, G., Smith, D.H. and Hammons, R.O. (1976), *Plant Disease Reporter* 50, 494-8.
172. Subba Rao, N.S. (1976), in Nutman, P.S. (ed.) *Symbiotic Nitrogen Fixation in Plants.* Cambridge: Cambridge University Press, 255-68.
173. Subrahmanyam, P. and McDonald, D. (1983), *Information Bulletin No. 13,* Patancheru, India: ICRISAT.
174. Subrahmanyam, P., McDonald, D. and Gibbons, R.W. (1982), *Oléagineux* 37, 63-8.
175. Subrahmanyam, P., McDonald, D., Gibbons, R.W. and Subba Rao, P.V. (1983), *Phytopathology* 73, 253-6.
176. Subrahmanyam, P., McDonald, D. and Rao, V.R. (1983), *Plant Disease Reporter* 67, 209-12.
177. Subrahmanyam, P., Mehan, V.K., Nevill, D.J. and McDonald, D. (1980), in *Proceedings of the International Workshop on Groundnuts.* Patancheru, India: ICRISAT, 193-8.

178. Subrahmanyam, P., Reddy, D.V.R., Gibbons, R.W., Rao, V.R. and Garren, K.H. (1979), *PANS* 25, 25-9.
179. Sun, M.K.C. and Hebert, T.T. (1972), *Phytopathology* 62, 832-9.
180. Sundaro Rao, W.V.B. (1971), *Plant and Soil*. Special Volume, 287-91.
181. Tolin, S.A., Isakson, O.W. and Troutman, J.L. (1970), *Plant Disease Reporter* 54, 935-8.
182. Varma, M.P. and Kanke, M.S.S. (1969), *Experimental Agriculture* 5, 223-30.
183. Veeraswamy, R., Rathraswamy, R. and Palaniswamy, G.A. (1974), *Madras Agricultural Journal* 61, 801-2.
184. Watson, S. (1983), *Peanut Farmer* 19(4), 8-12.
185. Weaver, R.W. (1974), *Peanut Science* 1, 23-5.
186. Williams, J.H. (1979), *Rhodesia Journal of Agricultural Research* 17, 57-62.
187. Williams, J.H., Wilson, J.H.H. and Bate, G.C. (1975), *Rhodesia Journal of Agricultural Research* 13, 33-43.
188. Williams, J.H., Wilson, J.H.H. and Bate, G.C. (1975), *Rhodesia Journal of Agricultural Research* 13, 131-44.
189. Woodroof, N.C. (1933), *Phytopathology* 23, 627-40.
190. Wynne, J.C. and Beute, M.K. (1983), *Crop Science* 23, 184.
191. Wynne, J.C., Campbell, W.V., Emery, D.A. and Mozingo, R.W. (1977), *Bulletin No. 458*, North Carolina State University Agricultural Experiment Station.
192. Wynne, J.C., Elkan, G.H. and Schneeweis, T.J. (1980), in *Proceedings of the International Workshop on Groundnuts*. Patancheru, India: ICRISAT, 95-109.
193. Wynne, J.C., Emery, D.A. and Downs, R.J. (1973), *Crop Science* 13, 511-14.
194. Wynne, J.C. and Gregory, W.C. (1981), *Advances in Agronomy* 34, 39-72.
195. Wynne, J.C., Rowe, R.C. and Beute, M.K. (1975), *Peanut Science* 2, 54-6.
196. Yadava, C.P.S., Saxena, R.C., Mishwa, R.K. and Dadheech, L.N. (1977), *Indian Journal of Entomology* 39, 205-10.
197. Yarbrough, J.A. (1949), *American Journal of Botany* 36, 758-72.
198. Yi-Xian, G. (1982), in *Proceedings of the Workshop on Cropping Systems Research in Asia*. Los Banos, Philippines: IRRI, 331-4.
199. Young, C.T., Waller, G.R. and Hammons, R.O. (1973), *Crop Science* 50, 521-3.
200. Young, J.H., Person, N.K., Jr., Donald, J.O. and Mayfield, W.D. (1982), in Pattee, H.E. and Young, C.T. (*eds*) *Peanut Science and Technology*. Yoakum, Texas: American Peanut Research and Education Society, 458-85.
201. Zandstra, H.G. (1982), in *Proceedings of the Workshop on Cropping Systems Research in Asia*. Los Banos, Philippines: IRRI, 43-54.
202. Zimmerman, A. (1907), *Pflanzer* 3, 129-33.

PART III: International Research

18 Recent trends in internationally orientated research on grain legumes

R.J. Summerfield and E.H. Roberts

INTRODUCTION

Many national programmes devoted to the improvement of particular grain legume crops have been referred to throughout this book. These programmes, quite understandably, have usually focussed on specific regions within national boundaries; they have often made laudable progress even though the funding devoted to them is usually small compared with that applied to the improvement of staple cereals and of industrial and export crops. Such programmes, and their links with national extension services and local farmers, are certain to form essential elements in the search for larger and more stable yields from future legume crops. However, few national programmes are strong enough to prosper alone.

A new feature of the last fifteen years has been the rapid development of cooperative, internationally orientated, well-funded research on many of the species of grain legumes of importance in world agriculture. The extent of this expanded effort is not always fully realised. A decade ago, Borlaug [10] lamented the fact that 'neither new high yielding varieties [of grain legumes] nor improved technology have been developed'. But today, as a result of unprecedented efforts, genetically superior genotypes of all the major grain legumes are available - even though few of these genetic advances have yet been translated into agricultural progress on a wide scale, especially in the subsistence farming systems of the tropics and subtropics [2]. Equally clear is the fact that technology developed in Europe or North America cannot easily be transferred to the tropical agricultural systems in which food legumes are so important. Notwithstanding Brazil's experience with soyabeans [39], dramatic and widespread increases in production, similar to those of the so called 'Green Revolution' in cereal crops in certain Asian countries, do not seem likely in most grain legumes or in the subsistence production systems in which they are often cultivated [18,29]. Locale-specific strategies based on detailed understanding of local farming practices and market needs may be required instead [11].

The purpose of this chapter is to summarise the main features of international programmes of research on grain legumes. The materials and methods now emerging from these programmes are already providing valuable resources for research and development in many individual nations.

RESEARCH ON GRAIN LEGUME CROPS AT THE INTERNATIONAL AGRICULTURAL RESEARCH INSTITUTES

A Consultative Group on International Agricultural Research (CGIAR) was established in 1971 as an informal association of governments, international and regional organisations, and private foundations, dedicated to supporting a system of international agricultural research centres (IARCs) and research programmes on food commodities around the world. More than seven thousand staff members, including at least six hundred senior scientists from more than forty countries, work at the thirteen centres supported through CGIAR. Their work on crops, livestock and farming systems is intended to increase the quantity and quality of food produced and to advance the standard of living of poor people in the developing countries. It concentrates on those critical aspects of food production in developing countries that are not adequately covered by other research facilities and *which are of wide actual or potential usefulness, regionally or globally* [12,13,23].

The concept of a Consultative Group (CG) grew from four international agricultural research centres established by the Ford and Rockefeller Foundations in the 1960s to work on rice, wheat, maize, grain legumes, cassava and other root and tuber crops. It became clear from the early successes with rice and wheat that international agricultural research should be expanded and broadened in scope, but that the costs of doing this were beyond the means of the two Foundations. Consequently, the World Bank, the Food and Agriculture Organisation of the United Nations (FAO), and the United Nations Development Programme (UNDP) undertook to sponsor and manage a new kind of international association to support the effort.

The World Bank provides the Chairman and the Executive Secretariat of the CG. The Group is advised on programmes and budgets by a Technical Advisory Committee, whose Secretariat is provided by FAO. The CGIAR has been very successful in mobilising support. In 1972, fifteen donors contributed about US $20 million to support five International Agricultural Research Centres. In 1984, thirty-eight donors provided core budgets totalling about US $175 million to the thirteen International Centres and programmes. In addition, about US $30 million were contributed by donors to support special projects at individual Centres or in particular nations.

Any organisation, public or private, which plans regularly to contribute significant amounts to particular Centres and programmes supported by the CGIAR may become a donor member. Each donor determines for itself the Centres it will support and the level of its funding. Pledges are announced each year at 'International Centres Week' - the annual meeting of the CGIAR held in the autumn.

Operating informally and by consensus, the CGIAR continues to serve as an example of effective, flexible and successful collaboration between the industrialised and

Table 18.1. CGIAR membership (January 1983)

Continuing members

(a) *Countries*

Australia	India	Norway
Belgium	Ireland	Philippines
Brazil	Italy	Saudi Arabia
Canada	Japan	Spain
Denmark	Mexico	Sweden
France	Netherlands	Switzerland
Germany	Nigeria	United Kingdom
		United States

(b) *International Organisations*

African Development Bank
Arab Fund for Economic and Social Development
Asian Development Bank
Commission of the European Communities
Food and Agriculture Organisation of the United Nations
Inter-American Development Bank
International Bank for Reconstruction and Development
International Fund for Agricultural Development
OPEC Fund
United Nations Development Programme
United Nations Environment Programme

(c) *Foundations*

Ford Foundation
International Development Research Centre
Kellogg Foundation
Leverhulme Trust
Rockefeller Foundation

Fixed-term members representing developing countries

Asian region	Indonesia
	Pakistan
Agrican region	Senegal
	Tanzania
Latin American region	Colombia
	Cuba
Southern and eastern European region	Greece
	Romania
Near Eastern region	Iraq
	Libya

---

developing worlds. The membership of the CGIAR, as of
January 1983, is shown in Table 18.1

Research intended to improve the quantity (productivity
and area), quality and reliability of yield of almost all
of the species of grain legumes included in this book, and
of the diverse farming systems in which they are produced,
is included in the programmes of four of the thirteen CGIAR-
supported Centres (Figure 18.1). Almost US $22 million
(about 8% of the Centres' research budgets) is estimated
to have been devoted to the direct costs of these research
programmes and associated training during the four year
period to 1982 (Table 18.2).

Table 18.2 CGIAR budgets (1979-82)

Component of budget	Expenditure (US $million)			
	1979	1980	1981	1982
Total CGIAR Centre budgets	108.0	124.4	139.6	154.7
Core operating programmes	92.1	111.5	125.2	144.0
Research programmes	55.2	65.1	74.5	86.5
Grain legume research programmes	4.4	5.2	6.0	6.1

Each of the programmes summarised below includes
research intended to identify improved sets of agronomic
practices which appear likely to be appropriate as options
for 'resource-poor' farmers in particular geographical
regions or environmental zones. The options include existing
landraces or genotypes as well as improved materials. Technology, culinary and aesthetic attributes of the seeds -
such as colour, size, texture, loss of testas during soaking, protein concentration, cooking time, taste and suitability for particular types of use - are investigated,
taking into account known preferences in different crops
and/or regions. Grain legume germplasm is collected, maintained, evaluated and distributed and up-to-date technical
information is disseminated through workshops, conferences,
information services, research reports and conventional

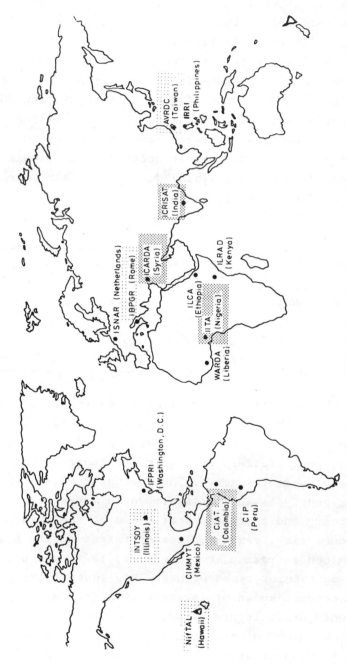

Centres not described are : Centro Internacional de la Papa (CIP); Centro Internacional de Mejoramiento de
Maíz y Trigo (CIMMYT); International Food Policy Research Institute (IFPRI); International Laboratory for Research on Animal
Diseases (ILRAD); International Livestock Centre for Africa (ILCA); International Rice Research Institute (IRRI); International
Service for National Agricultural Research (ISNAR) and West Africa Rice Development Association (WARDA)

Figure 18.1 The CGIAR-supported International Agricultural Research Centres (●)

journal articles and reference text books. Without exception, all of these programmes collaborate wherever possible with legume scientists in national programmes and elsewhere, since this is the only route by which their results can reach the real producers in the nations. Strong emphasis is also placed on technical and academic training, in accordance with the wishes and needs of the cooperating nations.

A fifth institution funded through the CG, The International Board for Plant Genetic Resources (IBPGR), has given substantial support to research on grain legumes as part of the development of a worldwide system to manage the genetic resources of the more important cultivated plants.

Centro Internacional de Agricultura Tropical (CIAT)
Apartado Aéreo 6713, Cali, Colombia, South America

CIAT's primary concern is with the general welfare of poor food consumers, both urban and rural, who live within the tropics of the Western hemisphere: a vast and diverse region. CIAT works on four of the principal food commodities of the region: common beans (*Phaseolus vulgaris*), cassava, rice and beef. The social and economic conditions of the population are at least as diverse as their agro-climatic circumstances, which vary from the cold altiplano, through cool plateaux at moderately high elevations, with productive land and satisfactory climate during a relatively short growing season, to steep Andean slopes and infertile lowland plains subject to alternating drought and floods, and finally to rich river valleys, marshy deltas and dense, multi-storied rain forests [13]. Between one-third [13] and almost one-half [32] of the world's supply of *P. vulgaris* is produced in Latin America, most of which is grown by subsistence farmers, often in association with maize or cassava, on poor soils with limited inputs [14]. Production lags behind domestic need and market demand, and is limited by diseases and pests, physiological 'defects' and periods

of hot, arid weather. In addition to its work for the
American tropics, CIAT is concerned with the improvement
of common beans throughout the world. By 1986, CIAT may be
supporting as many bean scientists in Africa and Asia combined, as in Latin America. Many of these will have been
trained at CIAT or by CIAT scientsts working elsewhere.

Approximately 32000 accessions of *P. vulgaris*, of which
about 10000 have been multiplied, evaluated and catalogued,
are maintained at CIAT. Important research and breeding
objectives of the Bean Improvement Programme [41] include
the synthesis of genotypes or populations with improvement
in the following areas:

(a) virus resistance (especially for Bean Common Mosaic
Virus and Bean Golden Mosaic Virus);

(b) disease resistance (especially for rust (*Uromyces
phaseoli*), anthracnose (*Colletotrichum lindemuthianum*), angular leaf spot (*Isariopsis griseola*) and
common bacterial blight (*Xanthomonas phaseoli*));

(c) insect tolerance/resistance; especially for leafhoppers (*Empoasca kraemeri*), red spider mites
(*Tetranychus desertorum*) and bruchids;

(d) phenological characteristics which use time and
environmental resources more efficiently and with
responses to photo-thermal conditions which are
more completely understood and exploited;

(e) ability to fix nitrogen through the symbiosis with
Rhizobium;

(f) tolerance of moderately acid soils (more rapid uptake and more efficient use of phosphorus, and
tolerance of aluminium and manganese);

(g) adaptation to dry conditions;

(h) plant architecture and yield potential.

In addition to genetic improvement, agronomic research
represents an important activity. The development of
improved agronomic practices and cropping systems,

especially for the superior materials from the breeding programme, is complementary to genetic improvement. Because of the diverse production conditions and seed types required in different regions, a great deal of agronomic and cropping systems research is undertaken away from the Institute's principal site. This work is supported by a collaborative international research network, which links the bean re- search workers of different nations with one another and with CIAT.

International Centre for Agricultural Research in the Dry Areas (ICARDA)
PO Box 5466, Aleppo, Syria

ICARDA is intended to serve the nations in those regions which receive rainfall in the cool months and have dry and often hot summers. Climates in this region (which extends from Turkey in the north to Sudan in the south, and from Morocco in the west to Pakistan in the east) are extremely diverse. They include tropical, subtropical and temperate thermal climates, with warm and cool variations due to lati- tude and altitude, and oceanic and continental differences due to land:sea configurations [30]. Agriculture and animal production are the main food-producing industries of the region. They support the lives of at least 60% of 250 mil- lion people, many of whom live at or near the subsistence level [15]. Such farmers, like small-scale producers every- where, are typically averse to risk [4].

In the rainfed agricultural systems which support most of the rural poor in the region, seasonal changes in tem- perature and photoperiod, in the amount and distribution of rainfall, and in the availability of nitrogen and phosphorus in the soil, coupled, inevitably, with depredations by several potentially devastating pests and diseases, are the major determinants of productivity of the principal annual crops, viz. lentils (*Lens culinaris*), faba beans (*Vicia faba*) [9], chickpeas (*Cicer arietinum*), barley, wheat and triticale. Lentils are grown in the drier regions, faba

beans in wetter areas (and with irrigation in the Nile
Valley of Egypt and Sudan) and chickpeas in the intermediate
rainfall zone.

Approximately 3000 accessions of faba bean populations
and 2500 accessions of pure lines, 6000 accessions of lentils and 5000 accessions of the large-seeded kabuli chickpeas are maintained at ICARDA [44]. Each of these collections is being multiplied, evaluated and catalogued. The
Food Legume Improvement Programme is seeking to exploit
this diversity to develop improved genotypes and populations
of each species, in order to increase total food production
and improve the dietary intake of protein; to increase
rural income; and to make the cropping systems of the region
less dependent on fertiliser nitrogen.

The Programme recognises the following combinations of
traits as those most likely to contribute to more productive
and reliable yields [21, 22];

> (a) increased potential for yields of seeds coupled with
>   more stable yields;
> (b) appropriate adaptation to time (phenology) to make
>   best use of the growing period in different parts
>   of the region;
> (c) resistance to common diseases, pests, the parasitic
>   weed *Orobanche* spp. and the common environmental
>   stresses (particularly temperature and moisture
>   stress); and
> (d) resistance to lodging, pre-harvest fruit shedding
>   and shattering, and a plant morphology that may
>   facilitate harvesting by hand or by machine.

In attempting to achieve these objectives in the shortest
possible time, scientists have concentrated their efforts
on projects seeking to develop:

> (i) improved faba bean genotypes and production methods
>   for areas with assured moisture supply;

810

(ii) improved, drought-tolerant genotypes and production strategies for rainfed faba beans in dry areas (annual rainfall in the region of 350 mm);

(iii) improved drought-tolerant genotypes of lentil either with wide adaptability or well-adapted to different specific agro-ecological situations and better production methods; and

(iv) improved genotypes and production methods for kabuli chickpeas, including the development of genotypes and production 'packages' appropriate for winter planting.

Intensive research and breeding efforts on improving moisture use efficiency, on several major pests and diseases, and on production practices to improve symbiotic nitrogen fixation and weed control form an essential part of this work.

Scientists at ICARDA cooperate closely with their colleagues at ICRISAT (see below) in their work on kabuli chickpeas. Two scientists from ICRISAT are based permanently at ICARDA for this purpose.

International Crops Research Institute for the Semi-Arid Tropics (ICRISAT)
Patancheru PO, Andhra Pradesh 502 324, India

ICRISAT serves the seasonally arid tropical regions which receive rain in the warmer months of the year. Such climates are characteristic of about 20 million km^2 in the tropics, in all or parts of forty-nine countries on five continents. The region is the home of about 700 million people, most of whom live at a subsistence level and depend for their food on production by resource-poor farmers cultivating small farms. The environment is generally harsh; rainfall is limited and erratic and many soils are nutrient-poor.

ICRISAT works on the production and genetic improvement

of five basic crops and seeks to develop farming systems
which will make the best use of the human and animal
resources and the limited rainfall of the region. Research
is concentrated on sorghum and pearl millet, and on three
grain legumes - chickpeas, pigeonpeas (*Cajanus cajan*) and
groundnuts (*Arachis hypogaea*). About 90% of the world's
chickpeas, 96% of the pigeonpeas and 67% of the groundnuts
are produced in the semi-arid tropics. Except for groundnuts (an important cash crop) these legumes are essentially
subsistence food crops consumed by the farm family, but as
economies become more diverse demand increases in both local
and more distant markets.

Germplasm accession collections, exceeding 10000 for
pigeonpeas, 12000 for chickpeas and 10000 for groundnuts,
provide opportunities for hybridisation and selection on a
much more extensive scale than has been possible hitherto.
To complement these activities, researchers are concentrating on the following subject areas as they seek, ultimately,
to release to national programmes (for local selection)
genotypes with potentially larger and more stable yields:

For chickpeas:

(a) development of short-duration, small-seeded desi
    types well adapted to peninsular India, and long-
    duration desi and kabuli types for northern India,
    Pakistan, Nepal and Bangladesh;
(b) development of chickpeas for novel cropping systems
    and environments (e.g. earlier sowings at lower
    latitudes - to avoid heat and water stress
    during the reproductive period, or later sowings
    at northerly latitudes - to produce satisfactory
    yields after rainy season crops such as sorghum or
    pearl millet);
(c) taller plant types better suited to mechanical
    harvest and with satisfactory yield potential;
(d) resistance to major diseases (traditionally referred

to as the 'wilt complex'), now known to be *Fusarium*
wilt (*Fusarium oxysporum* f.sp. *ciceris*), *Ascochyta*
blight (*Ascochyta rabiei*), grey mould (*Botrytis*
cinerea) and stunt (Pea Leaf Roll Virus);

(e) resistance to *Heliothis armigera* pod borers;

(f) more precise adaptation to photo-thermal conditions
at successive stages of development;

(g) more effective and more reliable nodulation (and
appropriate methods for screening for symbiotic
traits).

For pigeonpeas:

(a) development of short-duration types which respond
to inputs such as fertilisers and irrigation and
are suited to sole cropping in rotations with wheat
(and probably relatively insensitive to photo-thermal conditions);

(b) improved medium-duration types well-adapted to the
post-rainy seasons in southern and central India
and West Africa;

(c) long-duration types for intercropping in India and
Africa;

(d) large-seeded 'vegetable' types for the Caribbean;

(e) incorporation of resistance to wilt (*Fusarium udum*),
Sterility Mosaic Virus and blight (*Phytophthora*
drechsleri f.sp. *cajani*) into germplasm within each
maturity range;

(f) tolerance of salinity and to ephemeral waterlogging;

(g) improved symbiotic nitrogen fixation through manipulation of both the plant host and strains of *Rhizo-*
bium capable of survival and reliably infective in
diverse edaphic conditions;

(h) resistance to *Heliothis armigera* pod borers and
Melanagromyza obtusa pod flies;

(i) evaluations of optimum plant populations for genotypes which differ in time, and with more precisely

adapted maturity to the seasonal distribution of
rain.

For groundnuts:

(a) development of genotypes well adapted to cultivation
in the rainy season and/or in the post-rainy season
(with irrigation) and resistant to rust (*Puccinia
arachidis*) and leaf spot (*Cercospora arachidicola*
and *Cercosporidium personatum*) diseases, to *Aspergillus flavus* and the production of aflatoxins, and
to fruit rots caused by a complex of soil-inhabiting
pathogens;

(b) breeding for short-season genotypes which also have
seed dormancy after harvest;

(c) breeding for insect resistance and the formulation
of integrated measures for pest control;

(d) improved ability to withstand drought;

(e) use of wide crossing techniques to transfer resistance to many pests and diseases from wild species
to cultivated forms;

(f) development of suitable strains of *Rhizobium* and
inoculation techniques, and breeding for increased
ability to fix nitrogen symbiotically;

(g) a complete characterisation of groundnut viruses
and identification of genetic resistance or tolerance to them.

The diverse nature of these three grain legume crops
and of the farming systems in which they are cultivated now
or may be cultivated in the future, requires a strong commitment not only to national and international collaboration
but especially to interdisciplinary research on farming systems. ICRISAT scientists and their colleagues elsewhere have
been outstandingly successful in their efforts to develop
appropriate methods for this most challenging type of field
research [8,33,42].

International Institute of Tropical Agriculture (IITA)
PMB 5320, Oyo Road, Ibadan, Nigeria

Construction of IITA began at Ibadan in 1967. This was the first CGIAR Centre in the continent of Africa. CIAT was established at about the same time and these were therefore the first of the International Centres to include signifi-cant work on one or more grain legumes. (ICRISAT and ICARDA came later, in 1972 and 1976, respectively.)

IITA's tasks are directed ecologically to the humid and subhumid tropics in which it is concerned with both farming systems and crop improvement [24,25,26]. It seeks to develop more productive and stable alternatives to the traditional systems of bush fallow and shifting cultivation. In these systems, a relatively brief period of cropping - two or three seasons, with declining returns - was tradi-tionally followed by a much longer period of fallow, during which the regrowth of wild, secondary vegetation restored nutrients and organic matter to the soil. However, as popu-lations have increased, the growing demand for agricultural production decreased and sometimes eliminated the fallow period between cropping seasons. Though producers have evolved more or less stable systems of continuous cropping in some areas, the consequences in others are often severe deterioration of the soil's physical structure, increased erosion, loss of soil fertility, and increasing infestations of weeds, insects, and diseases. The net result is a sharply declining production per unit area of land [13,18].

Although IITA's work was originally concentrated on the lowland tropics of Africa (i.e. within 7° of the equator and below 600 m elevation) it has now expanded considerably to other continents and (largely because IITA also works on crops which are also grown in other regions) into subhumid and even semi-arid environments.

These crops include maize and rice; yams, sweet potatoes and cassava; and several grain legumes. Thus, IITA has begun to redress the bias in many tropical countries in which,

815

until the 1950s, most agricultural research was devoted to export crops, especially fibres, rubber, beverages and oil palm. The Institute is concerned with the improvement of cowpeas (*Vigna unguiculata*) worldwide. Within Africa, it works on soyabeans (*Glycine max*), Lima beans (*Phaseolus lunatus*), winged beans (*Psophocarpus tetragonolobus*) and pigeonpeas. It maintains extensive germplasm resources (about 15000 accessions of cowpeas, for example) to support these programmes. The primary objectives of IITA's work on food legume crops include the following combinations of desirable traits:

For cowpeas:

(a) cultivars with larger and more stable yields of seeds of desirable quality (colour, texture, taste, palatability and cooking time), and less subject than traditional types to deterioration in the field or in store;

(b) resistance to several potentially devastating diseases (e.g. anthracnose (*Colletotrichum lindemuthianum*), *Cercospora cruenta*, *Uromyces appendiculatus* and Cowpea Yellow Mosaic Virus [2]) and pests (e.g. *Megalurothrips sjostedi*, *Maruca testulalis*, several Hemipteran pod borers and the storage pest *Callosobruchus maculatus* [16]) which are especially important in humid and semi-arid tropical regions, respectively;

(c) appropriate maturity durations for different ecological zones.

To serve the principal cowpea growing regions more effectively, IITA has also established regional bases in Brazil, Tanzania and Burkina Faso (Upper Volta). Although many improved genotypes and populations have been selected for use as sole crops, cowpeas are often intercropped with maize, sorghum or millet, and IITA is seeking improved forms for these circumstances also.

For soyabeans:

The spread of soyabeans in tropical regions has been
restricted by lack of adaptation to tropical daylengths and
temperatures, by deterioration of seeds in store (conditioned by maturation environment, timing and technique of
harvest, excessive temperature and humidity during storage,
and hot, arid soils at planting), by incompatibility of
many forms with indigenous strains of *Rhizobium* (other than
R. japonicum), and limited yield potential [20,28]. Work at
IITA has been concerned with each of these facets.

Genotypes have been identified (mainly from South-east
Asia) which are capable of effective symbiosis with indigenous cowpea strains of *Rhizobium* [26]. Current efforts are
concentrating on the expansion of the crop on the diverse
soils of the Guinea savanna zones, and on the effectiveness
of the symbiotic system in hot, dry and, sometimes, acid
soils.

For both legumes, research on methods and techniques
to ensure more effective nodulation, by both indigenous and
exotic strains of *Rhizobium*, is particularly important.

The International Board for Plant Genetic Resources (IBPGR)
Crop Genetic Resources Centre, Food and Agriculture Organ-
ization of the United Nations, Via delle Terme di Caracalla,
00100 Rome, Italy.

The International Board for Plant Genetic Resources was
established by the Consultative Group in 1974. Its task is
to promote and coordinate an international network of centres to further the collection, conservation, evaluation,
documentation and use of plant germplasm. Unlike the other
institutions in the CG system, it is not restricted to food
commodities. Its priorities are defined in terms of the
developmental needs of the nations of the world. Its work
is based on cooperation with a very large number of individuals and with nearly six hundred agricultural institutions,
in one hundred countries, developed as well as developing.

All of the eight International Centres which work on specific crops provide long- or medium-term storage for the species with which they are concerned. The Board is substantially supported by plant breeders and agronomists as well as by geneticists, students of the evolution of cultivated plants, and conservationists.

Its priorities include about one hundred crop species and groups of species. For each species which can be conserved as seed, the Board has either designated or intends to designate at least two centres to hold base stocks under secure conditions of cold temperature and dryness. All material held in such centres, together with information about it, is freely available to all who can make good use of it.

IBPGR is funded by seventeen members of the Consultative Group, who provided nearly US $4 million for its work in 1983. By far the greater part of the costs of the international endeavour is provided by the cooperating governments and institutions. The Board's Secretariat is located in the headquarters of FAO in Rome.

About 15% of the funds of the Board are allocated to food legumes. The principal genera concerned (with the names of the main long-term secure conservation centres in parentheses) are: *Phaseolus* (CIAT); *Cicer* (ICARDA and ICRISAT); *Arachis* (ICRISAT and the Brazilian national agency, CENARGEN, for wild species); *Glycine* (Chinese Academy of Agricultural Sciences, Beijing; National Institute of Agrobiological Resources, Yatabe, Japan; and the National Seed Storage Laboratory, Fort Collins, Colorado, USA); *Vigna* (IITA, the Asian Vegetable Research and Development Centre in Taiwan; and laboratories in Belgium, India, the Philippines, Japan and USA); *Psophocarpus* (Institute of Plant Breeding, University of the Philippines, Los Banos, and Thailand Institute of Scientific and Technological Research, Bangkok); *Cajanus* (ICRISAT and the Indian National Bureau of Plant Genetic Resources); *Lupinus* (Zentralinstitut für Genetik und Kulturpflanzenforschung, Gatersleben, German Democratic Republic); *Vicia* (ICARDA); *Lens* (ICARDA); and *Pisum* (Nordic

Gene Bank, Lund, Sweden), with some attention to *Lathyrus*, *Ervum* and *Trigonella*, for which long-term bases have not yet been designated. Standard lists of descriptors have been prepared for most of these crops.

OTHER CGIAR CENTRES

Two other CGIAR centres are not concerned with specific crops or climatic zones but with issues that affect all agriculture. The International Food Policy Research Institute (IFPRI) was established in 1975 at Washington, DC, USA and focusses on three prime, but sensitive, economic and political issues in national, regional and global food problems: food production, especially as it is affected by technological change; food distribution and consumption, with particular attention to effects on poor people; and international food trade. A fourth IFPRI programme on trends and statistics, compiles and analyses data on past food production and consumption in the developing countries and extrapolates future trends [13].

Five years later, in 1980, the International Service for National Agricultural Research (ISNAR) was established at the Hague in the Netherlands. This service was conceived as a complementary centre to the other Institutes within the CGIAR system; it responds to requests from developing countries for assistance in strengthening their national agricultural research systems.

As stressed repeatedly during the previous discussion of international research on grain legume crops, success depends on strong, competent, and active national research programmes. By identifying and selecting priorities and opportunities for the improvement of crops and livestock and cropping systems in their agro-climatic zones, national programmes contribute significantly to the research programmes of the international centres. In the actual conduct of research and development programmes at the centres, national programmes form a worldwide network for testing,

selecting, and refining new technologies in the agro-
climatic, economic, and social settings for which they are
being developed. Finally, national programmes have the vital
responsibility for adapting new technologies to local needs
and conditions, releasing them to local farmers, and vigor-
ously promoting their adoption.

Thus, ISNAR will seek to offer help to national governments in identifying research needs, setting priorities,
and planning research strategies; designing and installing
research facilities and training scientific and administrative staff; and obtaining technical or financial assistance
from other sources [13,23].

It is clear from these brief summaries that international
research on grain legumes recognises that the first task of
plant breeding is to increase and stabilise seed yields in
order to increase the contribution of these crops to human
diets and the income of producers. New cultivars, appropriately adapted to particular environments and cropping systems, have been and are being produced with increasing
momentum. It is also clear that depredations by pests and/or
diseases are often the primary causes of poor yields from
traditional genotypes. However, as Goldsworthy [16] has
cautioned, selection against these difficulties may increase
the concentrations of protease inhibitors, haemagglutinins
and tannins in seeds or pod walls. Breeders will therefore
need to ensure that the concentrations of these compounds
do not increase to the point where they adversely affect
digestibility and nutritional quality.

A number of international research and development programmes not included within the CGIAR system, but working
closely with the Institutions and programmes summarised
earlier, are now described.

University of Hawaii NifTAL Project and MIRCEN
University of Hawaii, PO Box 0, Paia, Hawaii 96779, USA

NifTAL (an acronym from Nitrogen fixation by Tropical Agricultural Legumes) was established in 1975. A contract from

USAID provided funds for the creation of a facility at the University of Hawaii with objectives to conduct research, and to provide services and training on aspects of the symbioses of strains of *Rhizobium* with legumes. Collectively, these activities are intended to support the development of appropriate methods of management of biological nitrogen fixation (BNF) so that subsistence farmers in developing countries may produce larger amounts of food and feed without increasing their reliance on costly nitrogenous fertilisers.

NifTAL resources in Hawaii are intended to service the needs of national programmes of the developing countries for research support, information and technical services, and for training at different levels. In 1979, the programme undertook to facilitate global, regional, and national networks so that scientists in developing countries might gather multilocation data needed to answer key research questions. These networks have been effective not only in providing the scientists concerned with access to the resources available at NifTAL but also to the expertise of leading BNF researchers throughout the USA, Europe, and in Australia.

Initially, NifTAL dealt only with tropical grain legumes important in human nutrition. Later, the tropical forage legumes were included. Given the many important non-food uses of legumes, especially the tree species [3], NifTAL's concern is now addressed to BNF aspects of all legumes with a role in development.

This expansion is consistent with a role as a flexible and responsive unit, offering development support in a key process (BNF), rather than in a specific discipline or commodity. NifTAL calls upon the expertise of many disciplines to address a process which benefits a range of commodities. The programme balances and complements the efforts of the CGIAR Agricultural Research Centres, Collaborative Research Support Programmes, INTSOY (see below) and others, which are committed to specific commodities and/or defined geographical regions.

Research efforts at NifTAL have shifted gradually from emphasis on the selection of superior strains of *Rhizobium* for use as inoculants for tropical legumes in harsh edaphic environments, towards studies designed to improve understanding of how to manage tropical cropping systems to optimise the use of nitrogen fixed biologically.

Previously, NifTAL aimed its training at 'key' researchers throughout the warmer regions through postgraduate programmes, internships, and short courses. It now works with extension leaders on the design of materials intended to explain the use of BNF technology to farmers, and is also concerned with training on local commercial production of good quality inoculants. Joint training programmes with CGIAR institutes are also organised.

A further change has been the regionalisation of development support. NifTAL's philosophy has been to assemble a resource unit in Hawaii, staffed by BNF specialists with a commitment to international development, and equipped to service the needs of national legume programmes. A recent initiative was to offer research support, technical and material services, and training oriented specifically at crop production constraints within the South and Southeast Asian region. By establishing a BNF Resource Centre at Bangkok, Thailand in 1983, NifTAL expects to capitalise even further on the network approach to accomplish research and facilitate the interchange of technical information. NifTAL is further addressing particular needs in individual countries through assignment of its staff to support positions in national programmes.

In summary, the organisation has ten principal activities (and welcomes enquiries on each one), viz: regionalisation of BNF support; training and communication (e.g. [31]); country and interagency projects; international network trials; legume inoculant development; rhizobial germplasm development; rhizobial anti-sera services; nitrogen management in cropping systems; soil stress tolerance research; and economic analyses of BNF.

International Soybean Program (INTSOY)
University of Illinois at Urbana-Champaign, College of
Agriculture, 113 Mumford Hall, 1301 West Gregory Drive,
Urbana, Illinois 61801, USA

The University of Illinois has a long-established tradition
in soyabean research and educational programmes in the USA
and elsewhere. These programmes were made possible through
cooperation between the University and several agencies,
notably the United States Department of Agriculture (USDA)
and the United States Agency for International Development
(USAID). These international activities involving several
institutions led to the formal establishment of INTSOY in
1973. The International Soybean Program seeks to improve
human nutrition throughout the world by expanding the cultivation and consumption of soyabeans. It cooperates with
national and international organisations in seven principal
activities:

(a) research;
(b) germplasm testing;
(c) feasibility studies;
(d) consultancies;
(e) publications (e.g. [17,28]);
(f) regional and international conferences;
(g) training courses and study programmes.

A major cooperative effort with the University of
Puerto Rico (from 1973 to 1983) helped to focus on the cultivation of food legumes in the tropics and subtropics.
INTSOY provides and disseminates timely information
about variety adaptation, crop management and protection,
seed storage and the processing and utilisation of soya
foods. It also offers direct access to a worldwide collection of soyabean germplasm and breeds and tests genotypes
in diverse ecological regions. Training programmes,

823

conferences and publications combine to ensure that topical and practical information is readily available to soyabean workers throughout the world.

INTSOY has been funded by a series of research contracts from USAID. Special project funds have also been received from the United Nations Development Programme (UNDP), the Food and Agriculture Organisation of the United Nations (FAO), the Rockefeller Foundation, the United Nations International Childrens Emergency Fund (UNICEF), CARE Incorporated (USA), and the International Board for Plant Genetic Resources (IBPGR; see earlier).

The Asian Vegetable Research and Development Centre (AVRDC)
PO Box 42, Shanhua, Tainan 741, Taiwan, Republic of China

The concept of the Asian Vegetable Research and Development Centre (AVRDC) began with a suggestion from USAID in 1963 that a research centre devoted to the improvement of protein- and nutrient-rich vegetable crops should be established in South-east Asia. In 1971 an agreement to establish the Centre was signed by seven countries and The Asian Development Bank, and by 1972 research was under way on a site of 116 hectares in southern Taiwan.

AVRDC's objectives include crop improvement research, training, and the transfer of technology to farmers via the infrastructural networks of national research and extension programmes.

The Centre's work, as designated by its Board of Directors, concentrates on two legumes (soyabeans and mungbeans; *Vigna radiata*, *V. mungo*) and three horticultural crops (Chinese cabbage, tomato and sweet potato). These crops were chosen primarily on the basis of their ability to complement cereal diets, but also on the prospects of significant improvements being realised in a relatively short time.

The original research thrust on soyabeans was confined to Asia but collaborative projects now include many other tropical and subtropical regions. Major constraints to

824

production and so the principal areas of research include:

(a) inappropriate photo-thermal responsiveness and
    adaptation;
(b) depredations by insects and diseases;
(c) poor yield potential;
(d) poor seed quality.

The Centre is seeking to develop cultivars suitable for
the diverse cropping systems within the region (e.g. pre-
and post-cereal cropping and upland monocropping and intercropping).

The primary objectives of the mungbean programme are
to develop large-yielding, stable cultivars with early and
uniform maturity. Such genotypes will need resistance to
Cercospora leaf spot (*Cercospora canescens*), powdery mildew
(*Erysiphe polygoni*) and beanflies (*Ophiomyia* spp. and
Melanagromyza sojae). Research has also emphasised the
development of cultivars relatively insensitive to photoperiod and temperature, not prone to lodge and with good
seed quality.

AVRDC provides segregating populations and germplasm
accessions of both soyabeans and mungbeans to national
breeding programmes for evaluation in different agro-
climates and cropping systems. A network of international
legume trials has also been established.

To complement its research programmes, the Centre conducts various training programmes on legume and horticultural crops, organises symposia and workshops, and publishes
a wide range of literature pertinent to its goals (e.g.
[5,6]).

Core budget funds are provided by the Republic of China,
the USA, the Republic of the Philippines, the Republic of
Korea, Japan, the Federal Republic of Germany and the Kingdom of Thailand. Special project donors include the Federal
Republic of Germany, Canada, Switzerland, Sweden, USDA and
several international organisations and commercial companies.

Formal research agreements have now been signed with organisations in more than twenty countries, and bilateral programmes have been established in the Philippines, Thailand,
the Republic of Korea and Taiwan (and Indonesia and Malaysia
were to be added in 1984).

The Australian Centre for International Agricultural Research (ACIAR)
GPO Box 1571, Canberra City, ACT 2601, Australia

Several internationally-orientated research projects on
grain legumes have been instigated in Australia under the
sponsorship of ACIAR, a statutory authority established in
June 1982 to foster collaborative agricultural research
between Australian institutions and their counterparts in
developing countries. The Centre's concern potentially
covers all developing countries and its priority programme
areas encompass the broad range of agricultural research.
However, many of the projects established to date have
involved South-east Asian countries and the emphasis of
research naturally reflects those areas where special com-
petence exists in Australia.

To date, ACIAR has sponsored several projects on grain
legumes, particularly in its plant improvement, plant protection, plant nutrition, and post-harvest technology programme areas. These projects reflect a diversity of target
species and problems in various countries and are typically
sponsored for an initial period of up to three years, with
further sponsorship subject to review. Examples include the
development of legumes for integration into local farming
systems; improving the yield and reliability of yield of
tropical grain legumes by improving resistance to drought
and heat; investigations of the micronutrient requirements
for biological nitrogen fixation and growth of grain legumes; identification of legume viruses; and ecological studies of root nodule bacteria and the use of inoculants.

Most recently, ACIAR has established a Food Legume

826

Coordination Unit within its plant improvement programme, in order to coordinate projects on pigeonpeas, soyabeans, mung beans, groundnuts and cowpeas. Among the tasks assigned to this Unit are the publication (twice-yearly) of a newsletter outlining latest research developments [1], the conduct of workshops to discuss research progress, and liaison with other international and national research institutes involved in related areas of research on grain legume crop improvement.

Cooperative Research Support Programs (CRSPs) established under Title XII of the United States Foreign Assistance Act

During the last few years substantial funds have been provided under Title XII of the US Foreign Assistance Act to support cooperative programmes of research on food commodities between groups of American agricultural universities and institutions in developing countries. Seven such programmes were in existence in 1983; two of them are concerned with food legumes. In general, the topics studied in the American institutions represent more general and basic facets of common problems, while the more applied aspects, which depend on local natural and social environments, are studied by the cooperators in the developing countries. Training is an important component in every CRSP.

The Bean-Cowpea CRSP, with headquarters at Michigan State University, links nine lead universities (Colorado State University, Cornell University, Michigan State University, the University of California at Davis and Riverside, the University of Georgia, the University of Nebraska, the University of Puerto Rico, the University of Wisconsin and Washington State University) and the Boyce Thompson Institute at Ithaca, New York with research institutions and/or universities in Botswana, Brazil, Cameroon, Dominican Republic, Ecuador, Guatemala, Honduras, Kenya, Malawi, Mexico, Nigeria, Senegal and Tanzania.

The subjects of cooperative research include variety

827

improvement; insect and disease control; farming systems management; nitrogen fixation; utilisation of soil phosphorus; drought and heat tolerance; improved seed and availability of seed; storage and methods of preparation; digestibility and nutrition; and the socio-economic implications of intervention in agronomic topics.

The headquarters of the Peanut CRSP are at the Georgia Experiment Station of the University of Georgia at Experimen American institutions involved are the University of Georgia, Texas A & M University, Alabama A & M University, the University of Florida, and North Carolina State University. The cooperating institutions overseas are in Niger, Burkina Faso, the Caribbean region, Senegal, Nigeria, the Sudan, Thailand, and the Philippines. The principal topics of investigation include nitrogen fixation and mycorrhizal associations; utilisation of groundnuts in food systems; storage; insect and other arthropod pests; viruses; mycotoxins; and disease resistance.

SOME ACHIEVEMENTS OF THE CGIAR GRAIN LEGUME PROGRAMMES

Whilst it is clearly not possible to describe in detail the achievements of a decade, or longer, of research in nine grain legume improvement programmes with a combined annual funding in the region of US $4-6 million (Table 18.2), we have outlined below the principal successes to date of the CGIAR-supported research efforts on these crops.

Common beans at CIAT

Genetic improvement activities have involved two facets: *character development* (which seeks to ensure maximum expression of desired characters in a wide range of genotypes with different seed types); and *character deployment* (which seeks to combine specific characters in cultivars for particular producer and consumer requirements).

Considerable progress has been made in character

828

development. Amongst the large collection of germplasm which is now available, genetic variability which lessens the impact of most of the traditional production constraints in common beans has been identified (e.g. resistances to the major diseases and pests, tolerances of drought and temperature extremes, and improved plant architecture and variations in crop maturity). Collaboration with the CRSP programmes (see above) has led to the development of geno- types almost immune to common bacterial blight. Resistance to all known races of anthracnose and angular leaf spot, and to bruchids, is also available in a range of seed types. These genotypes are now used routinely as parents in the crossing programme.

Advances in character deployment have been equally im- pressive. Genotypes developed for particular production regions are now all resistant to Bean Common Mosaic Virus (BCMV), and those which combine resistance to four or five pests are currently tested widely in on-farm trials. More than forty of these superior genotypes have been named and released as new cultivars by various national programmes. In Costa Rica, economic surveys have revealed that more than 60% of farmers have adopted the new materials. In Guatemala, resistance to BCMV has allowed farmers to re- cultivate lands previously abandoned because of the suscep- tibility of traditional genotypes to this virus. National statistics claim that yields and production have improved sufficiently to eliminate imports of the crop. In Burundi, one of the CIAT releases commands a premium price, and in Argentina black bean production has increased by 30% in those areas planted to the new releases. Then again, local farmers are adopting not only the improved cultivars but also suitable agronomic packages which will better ensure the realisation of their potential. Genotypes tolerant of web blight disease are grown in a no-tillage system and with a fungicide application to produce excellent yields in Costa Rica. These are just a few examples of the many which could be cited to show the rapid adoption of 'new technology' into traditional bean farming systems.

These achievements were made possible only through the combined activities of a research and training network which has been established throughout the region in which CIAT's work is focussed. Some of the advances were developed at CIAT, others by national programmes. The strength of the network is expected to lead to further improvements in common bean production, to the general benefit of subsistence farmers and rural and urban consumers.

Faba beans at ICARDA

As an international centre for the improvement of faba beans [9], ICARDA has assembled the world's largest collection of faba bean populations (the ILB collection), which comprised 2850 accessions by the end of 1983. From these populations, a pure line collection (designated BPL) of about 2600 accessions has been developed by repeated selfing under insect-free conditions. Both the ILB and BPL collections have been freely distributed to national programmes; during 1983, as many as 1365 accessions were despatched.

The BPL accessions have been evaluated for their respective resistance to *Ascochyta* blight (*Ascochyta fabae*), chocolate spot (*Botrytis fabae*), rust (*Uromyces fabae*), the root-rot and wilt complex of diseases, *Orobanche crenata*, and stem nematodes (*Ditylenchus dipsaci*). Sources of resistance to these diseases and parasites have been identified and the durability of resistance has been confirmed through a series of international disease screening nurseries distributed to national programmes in those regions where particular diseases or parasites are endemic. For example, three accessions (BPL 710, 1179, and 1196) were found to be extremely resistant to chocolate spot in Egypt, Sudan and the UK following screening of a Faba Bean International Chocolate Spot Nursery (FBICSN) in these locations. Similarly, through the Faba Bean International *Ascochyta* Blight Nursery (FBIABN) several lines showing resistance both in

Syria and Canada have been identified (BPL 460, 471, 465 and 2485).

Because genetic stocks with multiple disease resistance would be of great importance in developing faba bean cultivars with stable yields, promising BPL selections have been evaluated for resistance to five different diseases (rust, *Stemphylium*, *Alternaria*, chocolate spot and *Ascochyta* blight) and stem nematodes. Some selections have shown multiple resistance. Those genotypes resistant to common diseases have been recombined with locally adapted materials from various regions and the segregating populations sent to appropriate national programmes.

Sources for resistance to the insect pests *Aphis* spp. and *Bruchus dentipes* are also being evaluated. There are indications for the existence of variability in host resistance to these pests.

Genotypes suitable for rainfed cultivation in low rainfall areas have been identified. Some of them have yielded more than 2 t ha^{-1} and so could be an economic alternative for poor farmers who lack adequate options of leguminous crops in such environments.

When soils are fertile and moisture is assured, conventional faba bean genotypes tend to make excessive vegetative growth which results in lodging and reduces seed yield. The determinate habit, where the apex of each shoot terminates in an inflorescence [38], is being evaluated as an alternative plant type for such environments. Since the original source of this character is poorly adapted to the faba bean growing areas within the ICARDA 'mandate' region, extensive crossing has been undertaken and new determinate lines derived from such crosses have been passed on for field evaluation by the national programmes.

Through a special applied research project on faba bean improvement in the Nile Valley of Egypt and Sudan, the new genotypes and improved production technology have been evaluated in a series of on-farm trials. These trials have demonstrated economic increases in yield through the use of

improved genotypes and production techniques compared with
the farmers' traditional practices. Improved genotypes,
weed control and *Orobanche* control have proved to be key
factors in Egypt whereas improved water management, insect
control and weed control have been especially significant
in Sudan. The project has often been referred to as an outstanding example of international collaboration in development research.

Lentils at ICARDA

In keeping with its role as a world centre for lentil crop
improvement, ICARDA has assembled a germplasm collection of
approximately 6000 accessions, most of which have now been
multiplied and evaluated. A catalogue of passport data on
5420 accessions, together with evaluation data on 4550
accessions for nineteen agronomic and stress-related characters, has been prepared. Thus, a prudent selection of accessions can be despatched on request to different national
programmes.

Germplasm evaluation has revealed useful variations for
many of the breeding programme's objectives (as discussed
earlier). For example, tolerance of severe winter cold has
been identified so that autumn planting at high elevations
is now feasible, where traditionally only spring planting
was possible. Then again, genotypes resistant to *Ascochyta*
blight (*Ascochyta lentis*) and rust (*Uromyces fabae*) - two
foliar pathogens of major international importance - have
also been found in the germplasm.

Throughout the Mediterranean region, hand harvesting of
lentils is increasingly uneconomic and so mechanical harvesting techniques need to be developed and genotypes suited to this practice need to be bred. Germplasm evaluation
has been productive: taller, non-lodging plant types with
indehiscent fruits have now been identified. Mowers and
lentil harvest blades have been designed and are being
tested.

Germplasm variability has been exploited indirectly, through the large-scale recombination of characters now underway in the ICARDA breeding programme, and directly by national legume programmes elsewhere. For example, the rust-resistant, productive accession ILL 358 is being multiplied in Ethiopia for release to farmers. Other genotypes from ICARDA are being evaluated in farmers' fields in Syria, Pakistan and the Yemen Arab Republic. National programmes in Argentina, India, Jordan, Morocco, Sudan and Tunisia have each identified material from the International Trials distributed from ICARDA for further testing and evaluation locally. ICARDA participates in an international testing network for improved agronomic packages. Increased economic returns from early planted lentil crops with economic weed control using herbicides have been clearly demonstrated. An International Weed Control Trial has made this technology widely available to national programmes.

Sitona spp. are the most important pests of lentils, causing losses in yield of the order of 20-30%. The economics of controlling *Sitona* with granular insecticides is being evaluated on farmers' fields in west Asia.

Finally, in this brief review of progress with lentils, an early-maturing *macrosperma* type has been discovered which may be particularly useful in the Indian subcontinent, where such material has often been sought but hitherto had not been found.

Kabuli chickpeas at ICARDA

The complementary and cooperative roles of ICRISAT and ICARDA in international crop improvement efforts with chickpeas have been emphasised earlier. ICARDA's efforts are concentrated on the kabuli types of which 5356 accessions have been assembled from thirty-four countries. Of these, more than 3300 have now been evaluated for twenty-seven descriptors and a catalogue has been published. Genotypes tolerant of cold and to leaf miners (*Liriomyza cicerina*) and the parasitic weed *Orobanche* spp. have been identified.

A major advance has been the demonstration of consistently large yield improvements of chickpeas in the Mediterranean basin from sowing in winter (rather than in spring, the traditional practice), provided the crop is protected against the ravages of *Ascochyta* blight [36]. This practice has been adopted by several national programmes for on-farm evaluation.

Field, glasshouse and laboratory screening techniques and disease rating scales for *Ascochyta* blight have been developed. More than 15000 accessions (including both kabuli and desi types) have now been screened; eleven kabulis and six desis are known to be resistant. Six races of *A. rabiei* have been identified and a new gene for resistance to *Ascochyta* has been discovered. Accessions from ICARDA supplied to national programmes in Algeria, Cyprus, Greece, India, Jordan, Lebanon, Morocco, Pakistan, Spain, Syria, Tunisia and Turkey have proved resistant to local races of the pathogen.

Advancing the sowing date means that chickpea cultivation can be extended into drier regions. If and when *Ascochyta* blight remains troublesome, seed dressings (with Calixin M and Thiabendazole) and foliar fungicides (Chlorothalonil) can prove effective.

Now that the major diseases of kabuli chickpeas in west Asia and north Africa have been quantified, efforts to combine resistance to *Ascochyta* blight and wilt (especially *Fusarium* wilt) can be intensified, building on pleasing progress to date.

Groundnuts at ICRISAT

From a modest beginning in 1976, a world groundnut germplasm collection (which includes the wild species of *Arachis*) comprising more than 10000 accessions had been established by the end of 1983. As with the other CGIAR germplasm collections, accessions are readily available to those who request them; in 1983 alone, 7893 accessions were distributed to fifty-two countries. A second major achievement has been to identify among this collection reliable and stable sources of resistances to diseases and pests. Genotypes resistant to rust, leaf spots, seed penetration by *Aspergillus flavus*, thrips and leafhoppers have been found and distributed. Fourteen genotypes resistant to either rust, leaf spots or to both pathogens have been released jointly with the USDA. Advanced disease- and pest-resistant breeding lines have also been widely distributed and several such lines are included in current national yield trials in India. One cultivar (ICGS 11), developed for cultivation with irrigation in the post-rainy season in India, is now in pre-release trials.

Several viruses have been characterised completely and anti-sera have been produced. A major advance was the discovery by ICRISAT researchers that the so-called 'bud necrosis' disease of groundnut crops in India was, in fact, caused by Tomato Spotted Wilt Virus (TSWV); they also identified the thrip vector and have been able to formulate and to recommend integrated control measures to farmers.

For the first time in groundnuts, a specific strain of *Rhizobium* (NC92) has been shown to increase yields of a specific cultivar (Robut 33-1), even when large numbers of indigenous rhizobia are present in the soil. This strain of *Rhizobium* has been released in India and is being evaluated elsewhere. Improved inoculation techniques, using animal-drawn machinery, have also been recommended.

Interspecific hybrids between wild, diploid species of *Arachis*, which have strong resistance to foliar diseases, and susceptible cultivars have been produced; they are stable and tetraploid. Some of these hybrids have given excellent yields of both fruits (about 4000 kg ha^{-1}) and vegetative biomass (about 6000 kg ha^{-1} of 'hay') when grown on research station plots. Groundnut hay is a valuable by-product for feed in the semi-arid tropics.

These achievements within such a short period augur

835

well for the future. Original (1976) expertise in breeding, pathology, virology, entomology and microbiology has been strengthened by the addition of a cytogenetics programme (in 1978) and a team of physiologists (as recently as 1980). In particular, recently initiated work on drought resistance should be of vital importance to subsistence farmers of limited means in the coming years.

Chickpeas and pigeonpeas at ICRISAT

Improvement programmes devoted to these two pulses were instigated in 1973 and most of the posts in an interdisciplinary team of breeders, pathologists, entomologists, physiologists and microbiologists had been filled by the end of the following year. Germplasm botanists have now collected about 13000 accessions of chickpea from fifty-four countries and about 10000 accessions of pigeonpea from thirty-six countries.

Each year, chickpea and pigeonpea breeders from India and elsewhere are invited to tour the breeding plots at ICRISAT. They are able to select materials which interest them and are supplied with seeds of their selections. Recent estimates (1982-83) indicate that of the entries included in the All-India Coordinated Pulse Improvement Project (AICPIP) trials, about 25% of the chickpeas and 50% of the pigeonpeas were either entered directly from ICRISAT or had been derived from ICRISAT germplasm distributed earlier to breeders elsewhere.

Several of the genotypes of both pulses which have been developed at ICRISAT have now been released to farmers; many others are in pre-release, seed-increase trials. With chickpeas, for example, ICCC 4 has been released in Gujarat State, India, ILC 482 in Syria and IC 7357-2-3-IH-BH is being proposed for release in Australia. Numerous other accessions have proved outstandingly and reliably productive in research station, state or national trials, or when tested on farmers' fields, in India, Ethiopia, Bangladesh,

Nepal and Jordan. Formal releases in these countries and
in Lebanon (ILC 482) and Cyprus (ILC 3279) are imminent.
Several kabuli types, generated from the collaboration with
ICARDA (see above) are now being evaluated on farms in
Tunisia, Morocco, Egypt, Pakistan, the USA and in Canada.
With pigeonpeas, two genotypes (ICPL 87 and 92) have been
recommended for release in India, the hybrid ICPH 2 is in
pre-release trials in peninsular India and cultivar Hunt
(from a cross between cvs Prabhat and Baigani made at ICRISAT
in 1974) has been released in Australia, where six other
genotypes are undergoing pre-release seed increases.

Efforts to develop disease- and pest-resistant genotypes of both crops, and to devise screening techniques for
relative susceptibility to various wilt, root-rot, stunt
and blight pathogens have been rewarding. These diverse
sources of resistance are widely utilised in breeding programmes elsewhere and the novel screening techniques have
now become routine in many countries.

Significant, even dramatic, improvements in the yield
of both pulses have been demonstrated by agronomic research.
Timely, albeit limited, irrigation can double chickpea
yields in peninsular India (a finding which has attracted
considerable 'official' attention) and winter-sowing of
Ascochyta-resistant genotypes in the Mediterranean region
can double the yield realised from traditional spring-sown
crops. Yields of pigeonpeas in excess of 2 t ha^{-1} in 130
days can be achieved when these early-maturing types are
cultivated as sole crops in India (where they fit well into
existing cropping patterns) and the potential for post-rainy-
season pigeonpea crops has been clearly established, for
the first time, by ICRISAT research (and has become popular
in the States of Bihar and, with irrigation, Maharashtra).

Strains of *Rhizobium* from the ICRISAT collection have
also been recommended by AICPIP for use in national trials
- IC 76 and IH 195 for chickpeas and pigeonpeas, respectively.

Dialogue and reciprocal exchanges of materials between

ICRISAT scientists and their colleagues in the Indian National Programmes for these two pulses have been a substantial benefit in realising the achievements described briefly here.

Cowpeas at IITA

Traditional forms of cowpeas grown by subsistence farmers are perhaps best thought of as 'horticultural insurance policies'; they represent the results of selections, conscious or otherwise, by successive generations of farmers. Local types, precisely adapted to particular environmental niches, are usually indeterminate, slow to mature, prostrate and leafy, and acutely sensitive to photoperiod and temperature [19] so that they come into flower close to the average date of the end of the rainy season in different locations [40]. They suffer serious depredations by various diseases and pests (especially in humid and semi-arid regions, respectively), receive only nominal crop husbandry and require several harvests of sequentially ripening fruits. Nevertheless, the seeds they produce cook rapidly and have exacting combinations of colour, texture, taste and soaking characteristics in keeping with local preferences (see Chapter 12).

It was against this background that the Cowpea Improvement Programme was instigated at IITA in 1970, with the principal objectives of increasing the productivity and production of the crop (and so boosting farmers' incomes) by developing cultivars with good yield potential, multiple resistance to diseases and pests, and well adapted to various ecological regions and farming systems in the dry Sahel and humid tropics [37]. In working towards these objectives, the programme has benefited from collaboration with several national programmes, particularly in the host country Nigeria since, on a north-south axis, the geographical diversity of that country allows the selection of materials from segregating lines to suit all of the major ecological

838

zones within the regions of particular concern to IITA worldwide.

Original efforts (1970-77) concentrated on germplasm collection and evaluation (the collection has increased from 4200 accessions in 1972 to more than 12000 in 1984), especially in terms of disease and pest resistance, and on crop physiology and agronomic studies designed to evaluate plant characters and husbandry techniques associated with and conducive to larger and more stable yields. A major hybridisation programme was instigated to begin to exploit the diverse germplasm increasingly available, either by releases to national cowpea breeding programmes or for immediate release as registered cultivars (the so-called VITA lines).

The search for disease-resistant genotypes and the subsequent incorporation of multiple disease resistance into lines with good agronomic traits has now largely obviated the need for chemical control of traditionally devastating viruses and diseases [2,43]. These materials are now being used as a basis for further improvements in seed quality, in line with consumer demands, and insect resistance. Indeed, productive varieties with multiple disease resistance and which are also resistant to insect pests such as leafhoppers, aphids, thrips and bruchids have been distributed to national programmes [37].

The synthesis of these resistant materials is a major achievement which fosters considerable optimism during the ongoing search for better sources of resistance to scab (*Sphaceloma* spp.), septoria leaf spot (*Septoria vignae; S. kozo-polzanskii*) and brown pod blotch (*Colletotrichum capsici; C. truncatum*), and to thrips (*Megalurothrips sjostedi*), legume pod borers (*Maruca testulalis*) and bruchids (*Callosobruchus maculatus*).

Cultivars well-adapted to extremely short (sixty-day) and intermediate duration (about eighty-day) cropping seasons have been developed; they are resistant to major diseases, flower synchronously and mature uniformly, and have acceptable seed quality for various regions. These early materials are expected to fit into double and/or relay cropping systems and to improve productivity in those dry savanna regions which have only a very short rainy season. They can also be cultivated in rice fallows, on residual moisture; and several thousand hectares throughout Asia, Africa and Central America, it is anticipated, will be planted in this system in the near future. Furthermore, by escaping the excessive shading and severe competition for water and nutrients imposed on traditional, long-duration cowpeas when intercropped with cereals such as sorghum or maize, the extra-early varieties may have considerable potential to improve traditional expectations of the 'secondary' legume in mixed cropping systems.

International trials and breeding nurseries are distributed to more than 150 cooperators in fifty countries. They are selected within the national programmes and either used as parents or released for general cultivation.

Lines developed by hybridisation at IITA have often proved superior to local material, or those selected from the germplasm collections. They have been released in many countries for cultivation as varieties: for example, in Manaus and other states in Brazil (1981 and 1982), throughout Central America (1979, 1980, 1981 and 1982), in South Yemen (1981), Burkina Faso (1980), Tanzania (1980), Guyana (1982), Botswana (1984), Zimbabwe (1984) and in the different agro-climatic zones of Nigeria. This list of releases, and the scales of cultivation it summarises, are certainly impressive.

By fostering the participation of graduate students from around the world in the Cowpea Improvement Programme, promoting production and training courses, stimulating the creation of joint projects, and basing senior research staff in countries within the 'mandate' region (e.g. in Brazil, Tanzania and, most recently, in the Philippines and in Benin [27]), IITA has not only created but has also ensured the continuity of those channels through which current information and the fruits of its efforts on cowpeas will

need to flow if farmers, the ultimate customers, are to
continue to benefit in the challenging years ahead.

Soyabeans at IITA

Soyabeans can be reliably productive in the tropics but their
undoubted potential is often limited by two major biological
constraints, viz. poor stand establishment and poor nodulation. These were the principal conclusions of exploratory
research initiated on the crop by IITA in 1970. Varieties
in use at that time did not nodulate well and so plants
were often nitrogen-deficient and produced only poor yields.
Those seeds which were produced then lost viability rapidly
when stored in warm, humid conditions - a characteristic
feature not only of soyabeans [35] but also of most other
'orthodox' seeds [34]. So, a research programme involving
plant breeders, physiologists and microbiologists was formed
and began to investigate these and other constraints.
Research revealed that seed deterioration was caused both
by diseases and physiological changes before and after harvest. Genotypes resistant to field-weathering and to loss
of germinability during storage have now been identified
and crossed with high-yielding parents. Screening methods
have also been developed and all advanced breeding lines at
IITA now incorporate improved seed longevity.

The almost complete lack of indigenous strains of *Rhizobium japonicum* was found to be responsible for the poor
nodulation of many soyabeans in tropical soils. Effective
inoculant strains have been identified at IITA and elsewhere but, so far, very few tropical countries have developed an effective inoculant industry. Even in those countries where effective inoculants are produced, these rarely
reach subsistence farmers living away from the commercial
farming areas; such farmers seldom have refrigerators in
which to store the inoculant, and little extension work has
been done to teach them how to use inoculants effectively.
In seeking solutions to these problems, the Soybean Improve-

ment Programme at IITA has taken a novel approach. Lines were identified, mostly of Indonesian origin, which were extremely susceptible to a wide range of *Rhizobium*, including the cowpea cross-inoculum strains indigenous in African soils. This trait is called 'promiscuity' and lines with it are described as 'promiscuous'. Promiscuous soyabeans developed at IITA are able to nodulate effectively with indigenous strains of *Rhizobium*.

Breeding lines are also routinely evaluated for resistance to bacterial pustule (*Xanthomonas phaseoli*); advanced lines incorporating resistance to Soyabean Mosaic Virus are being tested; and pathologists are developing screening methods for frog-eye spot (*Cercospora sojina*). Crosses have been made to incorporate resistance to pod-sucking insects and resistant progeny have been selected at a site where pressures from this pest are severe.

In recent years, there has been a great increase in the number of tropical countries wishing to initiate or increase soyabean production. So IITA researchers meet regularly with representatives from many national research programmes and design the research undertaken at IITA according to these national priorities. One example is the help IITA staff are giving to national programmes concentrated in the highland areas of Rwanda, Ethiopia, Cameroon and Zaire. It was found that lines selected in nurseries in Nigeria were not adapted to elevations greater than 1000 m. Two approaches have been taken to overcome this problem. First, crosses have been made with lines adapted to these high elevation environments to incorporate into them promiscuity and seed longevity from the IITA lines. Populations from these crosses are advanced to near homozygosity by single-pod descent and are then sent to cooperators for selection in their own countries. The second response has been to begin selection on the Jos Plateau in central Nigeria, where the elevation is 1400 m. Selected lines will be screened for promiscuity, seed longevity and disease resistance and sent as yield trials to cooperators in highland environments.

Since the soyabean crop is new to many African house-
holds, and because it could be so important where protein
sources are scarce or prohibitively expensive, IITA is
seeking to develop research on village and home utilisation.
A major objective of this research is to develop appropriate processing technology that will produce food preparations which are acceptable to African consumers.

Soyabean lines well-adapted to the tropics with good
yields, improved seed longevity, promiscuous nodulation,
disease resistance and shattering resistance, and with
maturity ranging from 100-140 days, are now available.
These are disseminated through international trials. In
1984, 193 trials were sent to researchers requesting them
in Africa, Asia and South America. Farmers who grow these
improved materials can certainly expect larger and more
stable yields in the future. To foster rapid progress towards this goal, IITA has recently established regional
coordinators around the world so that the programme based
at Ibadan might serve each of these zones more effectively.

CONCLUDING REMARKS

Grain legume crops harvested as fresh vegetables or, more
often, as pulses (i.e. for their mature, dry seeds) are an
essential component of the diet of an estimated 700 million
people. Clearly, the massive surge in international research
and training devoted to these crops during the past thirty
years, and dependent so heavily on the CGIAR system of
Agricultural Research Centres in the warmer regions, has
potential benefits for a very large number of 'customers'.
Rural consumers living at, or close to, the subsistence
level stand to gain the most - providing that these advances
in research lead to a reliable supply of better products at
prices which they can more easily afford. But, with legumes,
spectacular increases in productivity will come neither as
easily nor rapidly as achieved, for example, with wheat and
rice in the late 1960s. However, as this brief summary of

international research on grain legumes has shown, considerable improvements have been made, and will continue to be forthcoming, that is if funding of these international crop improvement efforts remains secure. Funding, however, has become less secure recently [7] because of financial depression and fluctuating exchange rates. It has also become less flexible because of the tendency of some donors to support specific programmes (which, they judge, will bring local benefits) rather than 'core' research; and occasionally some donors have even failed to honour their pledged financial contributions. The international agricultural research system is indeed a 'Fragile Web' [23] - but not, we suggest, as fragile as the system of agriculture in which grain legume crops are traditionally cultivated in many regions - i.e. by poor farmers managing as best they can on a few hectares of infertile land. Though few such farmers will have heard of the CGIAR and its International Agricultural Research Centres, their future seems increasingly dependent on this network of international research, which includes not only the staff in the international research institutes but also their colleagues in the national programmes and extension services. When farmers decide to plant seeds of a new cultivar, or to adopt an improvement to their traditional farming practices, they are, after all, completing the global network.

ACKNOWLEDGEMENTS

The comments, guidance and contributions of A.H. Bunting, K.E. Dashiell, R.W. Gibbons, P.R. Goldsworthy, J. Halliday, H.E. Kauffman, R.J. Lawn, Y.L. Nene, M.C. Saxena, A.V. Schoonhoven, S.R. Singh and S. Shanmugasundaram during the preparation of this review are gratefully acknowledged.

LITERATURE CITED

1. ACIAR (1984), *ACIAR Food Legume Newsletter* 1, 1-4.

2. Allen, D.J. (1983), *The Pathology of Tropical Food Legumes*. Chichester: Wiley.
3. Allen, O.N. and Allen, E.K. (1981), *The Leguminosae*. London: Macmillan.
4. Anderson, J.R. (1974), *Review of Marketing and Agricultural Economics* 42, 131-84.
5. AVRDC (1984), *Report of the Third External Review of the Asian Vegetable Research and Development Center*. Shanhua, Taiwan: AVRDC (in press).
6. AVRDC (1984), *Progress Report, 1982*. Shanhua, Taiwan: AVRDC.
7. Beardsley, T. (1984), *Nature* 312, 94.
8. Binswanger, H.P., Virmani, S.M. and Kampen, J. (1980), 'Farming systems components for selected areas in India: Evidence from ICRISAT'. *Research Bulletin No. 2*. India: ICRISAT.
9. Bond, D.A. (1979), *Fabis News* 1, 15* (and see reference 32).
10. Borlaug, N.E. (1973), in Milner, M. (*ed.*) *Nutritional Improvement of Food Legumes by Breeding*, New York: Wiley, 7-11.
11. Bunting, A.H. (1981), *Biologist* 28, 161-5.
12. CGIAR (1976), *Consultative Group on International Agricultural Research*. New York: UNDP.
13. CGIAR (1980), *Consultative Group on International Agricultural Research*. Washington, DC: CGIAR.
14. CIAT (1981), *CIAT Report, 1981*. Cali, Colombia.
15. Darling, H.S. (1979), *Span* 22, 55-7.
16. Goldsworthy, P.R. (1982), *Proceedings of the Nutrition Society* 41, 27-39.
17. Goodman, R.M. (*ed.*) (1976), *Expanding the Use of Soybeans*. INTSOY Series No. 10. Urbana, Illinois: INTSOY.
18. Greenland, D.J. (1975), *Science (N.Y.)* 190, 841-4.
19. Hadley, P., Roberts, E.H., Summerfield, R.J. and Minchin, F.R. (1983), *Annals of Botany* 51, 531-43.
20. Hinson, K. and Hartwig, E.E. (1977), *Soybean Production in the Tropics*. FAO Production and Protection Paper No. 4. Rome: FAO.
21. ICARDA (1981), *Food Legume Improvement Program (ICARDA), Research Highlights, 1980-81*. Aleppo, Syria: ICARDA.
22. ICARDA (1981), *Annual Report, 1982*. Aleppo, Syria: ICARDA.
23. IDRC (1983), *The Fragile Web*. Ottawa, Canada: IDRC.
24. IITA (1977), *Decade of Progress and Promise for Tomorrow*. Ibadan, Nigeria: IITA.
25. IITA (1982), *Research Highlights for 1981*. Ibadan, Nigeria: IITA.
26. IITA (1982), *Annual Report 1981*. Ibadan, Nigeria: IITA.
27. IITA (1984), *IITA Research Briefs* 5(2), 1 and 4.
28. INTSOY (1982), *Soybean Seed Quality and Stand Establishment*. INTSOY Series No. 22. Urbana, Illinois: INTSOY.
29. Janzen, D.H. (1973), *Science (N.Y.)* 182, 1212-19.

---

*The English colloquialism 'field bean' (or the German 'feldbohne') is commonly used for *Vicia faba equina* and *minor* in Europe and is of much longer standing than the transatlantic 'field beans' for *Phaseolus vulgaris* in the USA. The editors, plagued elsewhere by inconsistency, ambiguity and confusion among the trivial names given to grain legumes [40], strongly support the adoption internationally of 'faba bean' for *Vicia faba* and 'common bean' for *Phaseolus vulgaris*, with the inclusion of generic terms, at first mention, to confirm unambiguity (and see Chapter 6).

30. Kassam, A.H. (1981), *Plant and Soil* 58, 1-29.
31. King, J. and Ferguson, P. (*eds*) (1983), *BNF Bulletin*. Information Section, NifTAL Project.
32. Laing, D.R., Kretchmer, P.J., Zuluaga, S. and Jones, P.G. (1983), in Smith, W.H. and Banta, S.J. (*eds*) *Potential Productivity of Field Crops Under Different Environments*. The Philippines: IRRI, 227-48.*
33. Mead, R. and Riley, J. (1981), *Journal of the Royal Statistical Society, Series A* 144, 462-509.
34. Roberts, E.H. (1973), *Seed Science and Technology* 1, 499-514.
35. Roberts, E.H. and Ellis, R.H. (1980), in Summerfield, R.J. and Bunting, A.H. (*eds*) *Advances in Legume Science*. London: HMSO, 297-312.
36. Saxena, M.C. and Singh, K.B. (*eds*) (1984), *Ascochyta Blight and Winter Sowing of Chickpeas*. The Hague: Martinus Nijhoff/Dr W. Junk.
37. Singh, S.R., Singh, B.B., Jackai, L.E.N. and Ntare, B.R. (1983), *Cowpea Research at IITA*. Information series No. 14. Ibadan, Nigeria: IITA.
38. Sjodin, J. (1971), *Hereditas* 67, 155-80.
39. Somers, J. (1976), *Foreign Agriculture (USA)* 14, 2-4.
40. Summerfield, R.J., Minchin, F.R., Roberts, E.H. and Hadley, P. (1983), in Smith, W.H. and Banta, S.J. (*eds*) *Potential Productivity of Field Crops Under Different Environments*. The Philippines: IRRI, 249-80.
41. Temple, S.R. and Song, L. (1980), in Summerfield, R.J. and Bunting, A.H. (*eds*) *Advances in Legume Science*, London: HMSO, 365-74.
42. Willey, R.W. (*ed.*) (1981), *Proceedings of an International Workshop on Intercropping*. Hyderabad, India: ICRISAT.
43. Williams, R.J. (1977), *Tropical Agriculture, Trinidad* 54, 53-60.
44. Witcombe, J.R. and Erskine, W. (*eds*) (1984), *Genetic Resources and Their Exploitation - Chickpeas, Faba Beans and Lentils*. The Hague: Martinus Nijhoff/Dr W. Junk.

*See footnote to reference 9.

Index

Index

854

Index

Index

Index